From nonlinearity to coherence:
universal features of nonlinear behaviour in many-body physics

From Nonlinearity to Coherence

Universal Features of Nonlinear Behaviour in Many-Body Physics

J. M. DIXON
*Department of Physics, University of Warwick,
Coventry CV4 7AL, U.K.*

J. A. TUSZYŃSKI
*Department of Physics, University of Alberta,
Edmonton, Alberta, T6G 2J1, Canada*

P. A. CLARKSON
*Institute of Mathematics and Statistics, University of Kent,
Canterbury CT2 7NF, U.K.*

CLARENDON PRESS · OXFORD
1997

Oxford University Press, Great Clarendon Street, Oxford OX2 6DP
Oxford New York
Athens Auckland Bangkok Bogota Bombay Buenos Aires
Calcutta Cape Town Dar es Salaam Delhi Florence Hong Kong
Istanbul Karachi Kuala Lumpur Madras Madrid Melbourne
Mexico City Nairobi Paris Singapore Taipei Tokyo Toronto
and associated companies in
Berlin Ibadan

Oxford is a trade mark of Oxford University Press

Published in the United States
by Oxford University Press, Inc., New York

© J. M. Dixon, J. A. Tuszyński, and P. A. Clarkson, 1997

All rights reserved. No part of this publication may be
reproduced, stored in a retrieval system, or transmitted, in any
form or by any means, without the prior permission in writing of Oxford
University Press. Within the UK, exceptions are allowed in respect of any
fair dealing for the purpose of research or private study, or criticism or
review, as permitted under the Copyright, Designs and Patents Act, 1988, or
in the case of reprographic reproduction in accordance with the terms of
licences issued by the Copyright Licensing Agency. Enquiries concerning
reproduction outside those terms and in other countries should be sent to
the Rights Department, Oxford University Press, at the address above.

This book is sold subject to the condition that it shall not,
by way of trade or otherwise, be lent, re-sold, hired out, or otherwise
circulated without the publisher's prior consent in any form of binding
or cover other than that in which it is published and without a similar
condition including this condition being imposed
on the subsequent purchaser.

A catalogue record for this book is available from the British Library

Library of Congress cataloging in Publication Data
Dixon, J. M., Dr.
From nonlinearity to coherence : universal features of nonlinear
behaviour in many-body physics / J. M. Dixon, J. A. Tuszyński,
P. A. Clarkson.—1st ed.
Includes bibliographical references and index.
1. Condensed matter. 2. Nonlinear theories. 3. Many-body problem.
I. Tuszyński, J. A. II. Clarkson, P. A. III. Title.
QC173.454.D59   1997   530.1'44—dc20   96-36601

ISBN 0 19 853972 X

Typeset by Technical Typesetting Ireland, Belfast, N. Ireland
Printed in Great Britain by Bookcraft Ltd., Midsomer Norton, Avon

'It does not say in the Bible that all laws of nature
 are expressible linearly'

E. Fermi

PREFACE

...in the traditional areas of the physics of gases and crystalline solids, in which a model which treats the behaviour of the whole as essentially the sum of that of its parts (atoms or electrons) has been quite successful; and a few more in which, even if a 'one-particle' picture fails, a description in terms of pairs of particles interacting in a way which is not particularly sensitive to the environment gives good results. But these cases, despite the fact that they totally dominate the presentation of the subject in most elementary textbooks, are actually the exception rather than the rule.

A. J. Leggett (1987), *The problems of physics*
(Oxford University Press, Oxford)

Motivation

During the past quarter of a century there have been numerous dramatic discoveries in very many areas of pure and applied science, including physics, biology, chemistry, medical sciences, and astronomy. Remarkably, at the same time, there have been considerable developments in mathematics, particularly in nonlinear mathematics, and the advent of sophisticated computer technology has made many previously impossible tasks a reality. This unprecedented period in the growth of scientific knowledge has also been characterized by an ever increasing number of interdisciplinary connections which have contributed significantly to a rapid flow of ideas between fields. What is now beginning to emerge, however, is that after a long period of differentiation between disciplines, universal features are becoming more and more pronounced. It appears that, in the not too distant future, integration of intellectual activity may become a dominant trend. Our intention in this book is to make a modest contribution to the drawing together of a number of ideas which originated from previously disparate disciplines. In this way, all of the contributing disciplines can be seen as part of a larger structure which is common to them all. We will apply this approach to several areas, with a particular emphasis on condensed-matter physics through an infusion of ideas from nonlinear mathematics, quantum field theory, many-body physics, and critical phenomena. The four main concepts that we intend to bring together are nonlinearity, quantization, criticality, and co-operativity. The physical substrate on which we wish to demonstrate this convergence is condensed-matter physics in the broadest sense.

Following this line of thought, in recent decades physicists have witnessed the development of an increasing number of parallel concepts in seemingly distinct areas of specialization. Some of the most prominent examples appear in quantum field theory, statistical mechanics and condensed-matter theory. This convergence of ideas is occurring across a diversity of fields and it takes place at two basic levels. First, the mathematical language and models used are either common or very similar (e.g. polynomial field theories) and, second, the new physical concepts employed (localized modes, fractality, chaotic behaviour, pattern formation, etc.) are universal. We feel that there are several reasons to study the question of the interrelation between these approaches and, in particular, the inevitable consequences for the universal physical properties of condensed-matter systems. One of the common features of systems of interest in these fields is their *many-body nature* and the other is their *nonlinearity*. Simultaneous spectacular progress in the area of nonlinear mathematics has made it possible to treat these highly complex systems using very sophisticated analytical methods and to draw extremely useful physical insights from them. The intention of this monograph is to focus on the area of condensed-matter physics and to provide a comprehensive overview of nonlinear phenomena in this rapidly growing field of physics. Furthermore, we shall attempt to provide a self-consistent and systematic methodology which has drawn on ideas from other areas of physics and mathematics. In this respect we have been strongly stimulated by the epoch-making discoveries of three giants of modern many-body physics, namely L. D. Landau (order parameters and symmetry breaking), K. G. Wilson (scaling and universality), and P. W. Anderson (quantum coherence).

Outline

This book is structured in the following way. The first chapter gives a very qualitative overview of the most important concepts of nonlinear science, i.e. bifurcations and multistability, anharmonic motion, separatrix phenomena, chaos, limit cycles, fractals, solitary waves and solitons, etc. Because of severely limited space in a monograph of this size, these topics will be reviewed in a rather qualitative way and no attempt has been made to present a complete review. We begin by motivating a coarse-grained approach to the many-body problem and indicating the possibilities of catastrophes in static, time dependent and spatially homogeneous systems with one or more critical degrees of freedom. Allowing for the dynamics introduces time and small-amplitude oscillations about stable equilibria in these multistable systems and illustrates important differences between harmonic and anharmonic motion. Coupling of two or more degrees of freedom will be shown to have important repercussions for both the development of chaos and the propagation of order. Through a continuum-limit transformation we show how nonlinear wave equations may be derived with the associated new modes of behaviour in the form of solitons and solitary wave solutions. We point out that methods exist that allow for semiclassical quantization of nonlinear modes, which allows one to extend this formalism to a microscopic level.

Having developed a conceptual base in terms of ideas from nonlinear science, we are now in a position to explore their possible manifestations in condensed-matter systems. As has been discussed in Chapter 1, the first important building block in a nonlinear theory consists of multistability and its consequences, i.e. bifurcations. A concrete realization of this phenomenon in many-body physics takes the form of phase transitions and criticality. This, then, is what we undertake to summarize in the following chapter.

Chapter 2 provides a systematic development of the phenomenology of critical systems at a macroscopic level using a fully nonlinear approach. The main concepts involved here are long-range order and broken symmetry, which will be used to clarify the properties of many different critical systems. The physical picture adopted here will be the Landau–Ginzburg model, which will be carefully analysed and extended to cover both first- and second-order phase transitions as well as critical fluctuations and dynamical phenomena. As a necessary prelude to the development of a microscopic theory, a brief overview of lattice models, scaling concepts, and renormalization group theory will be given. What the theory of phase transitions largely ignores, however, is the quantum dynamics of the many-body system studied. It is true to say that this can be safely ignored close enough to the critical point. However, in general the structure of quantum excitations is determined by the equilibrium phase of the system and will play an important role in such effects as responses to external stimuli. The role that quantum effects play becomes increasingly important the further away we are from instability points. The next chapter gives an overview of elementary excitations in solids and, in particular, how nonlinearity may affect them.

In Chapter 3 we offer the reader a brief summary of the main types of collective behaviour manifested in elementary excitations (i.e. phonons, magnons, and excitons) and their mutual interactions in condensed-matter physics. With these physical illustrations we will underline the importance of anharmonicity, with such attendant features as nonlinear waves (spin- and charge-density waves, and elastic soft modes), solitons (optical solitons in fibre optics, and magnetic solitons in spin systems), pattern formation (e.g. vortex lattices), and the formation of a variety of defects. In order properly to account for the situations in which quantum effects are strongly enhanced through mode–mode interactions and anharmonicity, sometimes precipitating an instability in the system, we must venture into the area of quantum fields. The next chapter is intended to give the necessary background from quantum field theory.

Chapter 4 gives a brief overview of a number of relevant developments in quantum field theory that will be of importance to our main topic of interest. Aspects such as field quantization, coherent states, quantum lattice solitons, the Higgs Boson, and the Goldstone theorem will be discussed. This chapter is intended as a forerunner to the systematic development of a nonlinear field-theoretic approach, presented in Chapter 5.

Chapter 5 is the physical centrepiece of this book and it aims to present a microscopic (quantum field theoretical) approach to the nonlinear dynamics of many-body systems which may exhibit broken symmetries. This is intended to be an easily accessible and systematic presentation of the method of coherent structures (MCS), its results, and their physical interpretation. It utilizes two main concepts, is virtually exact, and results in Landau–Ginzburg phenomenology emerging as a special case. Several illustrative examples of the applicability of the MCS are shown in the subsequent chapter (Chapter 6), including the broad features of a theory of superconductivity, the Haldane gap problem, the Hubbard model, metamagnetism, and a nonlinear model of multi-electronic atomic structure. With these last two chapters we have basically completed the development of our physical line of thought. In this development we have made use of the results from investigations of nonlinear differential equations of motion for the order parameter fields. If it were not for the fact that, thanks to recent mathematical discoveries, these problems have become tractable, our methodology would have had only an aesthetic appeal.

Chapter 7 of this monograph is intended to serve as a compendium of mathematical methods (i.e. symmetry reduction, inverse scattering, Painlevé analysis, etc.) and relevant results for physically important equations (e.g. the Sine–Gordon equation, the nonlinear Schrödinger equation, and the nonlinear Klein–Gordon equation, etc.).

The idea behind this order of chapters is to allow an uninterrupted flow of physical developments to precede those in mathematics which are treated in this context as very important tools in understanding physical reality, but nevertheless as tools. Each chapter is complete, with a separate list of references. Several appendices are given at the end of the book in order to facilitate accessibility to multidisciplinary readership. Depending on the particular preference of the reader (more mathematically or more physically oriented), the monograph can be studied using a different path strategy, as follows:

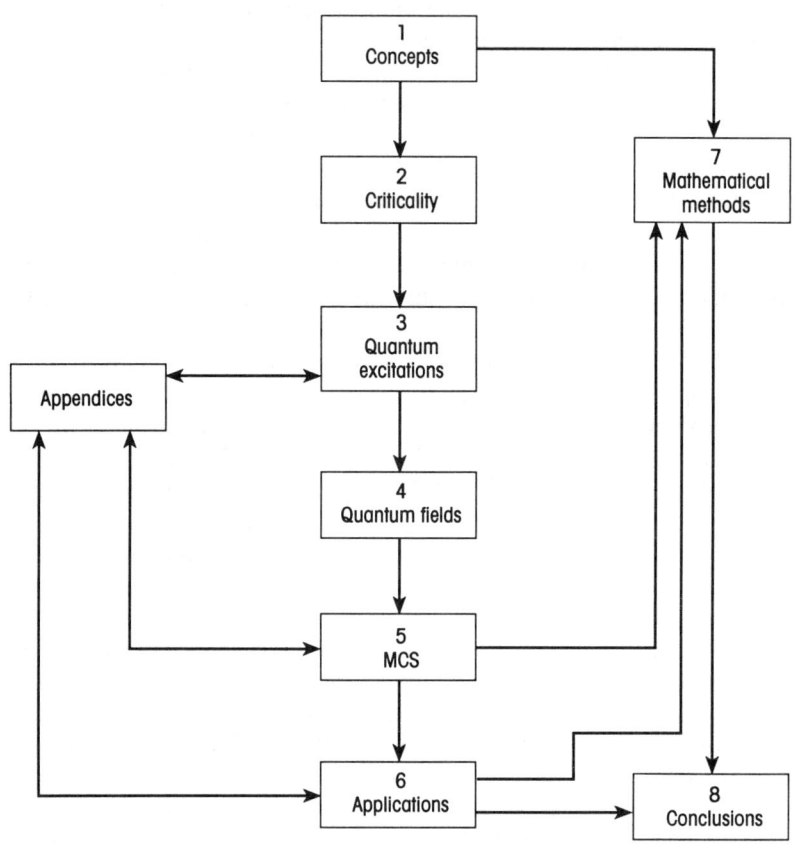

Acknowledgements

Jack Tuszyński and John Dixon wish to express their gratitude to their wives, Ela Tuszyńska and Eileen Dixon, respectively, for their continuous support and endless patience during the preparation of the book, which required long periods of absence from normal family life. We also wish to acknowledge generous financial support of the following granting agencies: NSERC (Canada), SERC (U.K.), the Royal Society (U.K.), NATO and the Nuffield Foundation (U.K.), and the Alexander von Humboldt Stiftung. We are very much indebted to our own institutions, the Departments of Physics at the Universities of Alberta and Warwick, and the Departments of Mathematics at the Universities of Exeter and Kent, for infrastructure support. This book is a result not only of a collaborative effort involving the three authors but also of many discussions with other scientists and students. In this regard we wish to thank Professors M. Grundland and P. Winternitz from the Université de Montréal, J. Stephenson from the University of Alberta and Drs R. F. Pettifer and G. J. Hyland of the University of Warwick, Professor N. H. March of the University of Oxford, M. J. Ablowitz of the University of Colorado, M. D. Kruskal of Rutgers University, and J. B. McLeod of the University of Pittsburgh, and Drs A. P. Bassom of the University of Exeter, M. C. Nucci of the University of Perugia, and E. L. Mansfield of the University of Exeter, as well as Dr M. Otwinowski from the University of Calgary, who was always informing us of new and exciting ideas emerging in the literature. J. A. Tuszyński is indebted to Professor P. Weichman of Caltech for providing him with unpublished lecture notes. The enthusiastic help and hard work of our students, K. Vos, M. Skierski, M. L. A. Nip, D. Sept, A. Hicks, S. Hood, D. K. Ludlow, A. Milne, and M. Kelly, is deeply appreciated. They have helped us to avoid many a headache. The manuscript has been very skilfully typed by Mrs Rosemary Fuller, for which we are enormously grateful. We are very appreciative of the patience and enthusiasm with which Mrs Samantha Stockley tackled the many amendments and additions to the original typescript. We wish to express our deep appreciation to Mrs Jeanette Chattaway, who spent many long hours drawing the diagrams for the book, some of which were extremely complex. Making sense of our scribbles was a major task in itself.

CONTENTS

1 **Introductory concepts: a nonlinear analysis primer**   1
   1.1 Multistability and bifurcations   1
   1.2 Stochastic analogue of a bifurcation   8
   1.3 Harmonic versus anharmonic motion   10
   1.4 External driving and dissipation   13
   1.5 Relaxation dynamics and asymptotic stability   20
   1.6 Coupled systems and limit cycles   22
   1.7 Nonlinear waves and solitons   25
   1.8 Pattern formation   43
   1.9 Chaos and turbulence   54
   1.10 Fractals   63
   1.11 Self-organized criticality   64
   References   68

2 **Phase transitions and criticality**   72
   2.1 Long-range order   72
   2.2 General characteristics of phase transitions   87
   2.3 Broken symmetries and order parameters   89
   2.4 Critical exponents and static scaling   95
   2.5 Two early theoretical models   99
   2.6 Lattice models   103
   2.7 The Landau theory of phase transitions   111
   2.8 Fluctuations and the Landau–Ginzburg (LG) model   119
   2.9 Dynamic scaling and the renormalization group (RG)   122
   2.10 A look beyond   128
   References   136

3 **Elementary excitations in solids**   139
   3.1 Introduction   139
   3.2 Second-quantization techniques   141
   3.3 Lattice vibrations and phonons   151
   3.4 Electronic excitations of the solid   157
   3.5 Other types of elementary excitation   160
   3.6 Solitons and coherent structures   165
   3.7 Defects   170
   References   173

4 **Background to quantum field theory**   175
   4.1 The Bogolyubov transformation for Bosons   176

4.2	Field translation	177
4.3	Bose condensation	179
4.4	The Bogolyubov transformation for Fermions	180
4.5	Boson coherent states	182
4.6	Fermion coherent states	183
4.7	Squeezed states	184
4.8	q-Bosons and quantum algebras	185
4.9	Field quantization	186
4.10	Classical field equations	188
4.11	Quantization of classical solutions	191
4.12	The Goldstone theorem and the Anderson–Higgs mechanism	196
4.13	Tunnelling, instantons, and the false vacuum	198
4.14	Quantum critical phenomena	200
4.15	Quantum lattice solitons	202
4.16	Nonlinear optical phenomena	206
	References	209

5 The method of coherent structures (MCS) — **211**

5.1	Introduction	211
5.2	Motivation for the approach to the method of coherent structures	217
5.3	Derivation of the nonlinear quantum field equations	226
5.4	Conversion to equations of motion for c-number fields	232
5.5	Truncation of terms in the equation of motion using renormalization theory	237
5.6	Classical solutions of the equation of motion	243
5.7	Quantization procedure	263
5.8	The transformation of bases	270
5.9	The role of spin in the development of coherent structures for spin-independent interactions	275
5.10	A generalization of the method of coherent structures to coupled fields	282
5.11	The relativistic extension	290
5.12	Relationship with coherent states	295
	References	296

6 Physical applications of MCS — **300**

6.1	The electron–proton plasma	301
6.2	Superconductivity and MCS	303
6.3	MCS and metamagnetism	321
6.4	The Haldane gap problem for quantum spin chains	333

6.5	Noninteracting electron systems: the hydrogen atom	351
6.6	A nonlinear theory of multi-electron atoms	358
6.7	The Hubbard Hamiltonian	390
References		398

7 Mathematical methods and results — 401
7.1	Symmetry analysis of partial differential equations	401
7.2	The Painlevé tests	456
7.3	The inverse scattering method	484
References		536

8 Conclusions — 554

Appendix A: Solving $\ddot{x} = g(x)$ graphically — 558

Appendix B: The elliptic functions of Jacobi — 561

Appendix C: An extended presentation of semiclassical quantization for classical time-dependent fields — 571

Appendix D: Conversion of annihilators and creators with mixed statistics to Fermions for the spin chain — 580

Appendix E: A derivation of the form of the equation of motion for the classical field from the corresponding equation of motion for the quantum field operator — 583

Appendix F: Intra-shell ordering of energy levels — 591

Index — **597**

1 INTRODUCTORY CONCEPTS: A NONLINEAR ANALYSIS PRIMER

A very small cause which escapes us determines a considerable effect which we cannot fail to see, and then we say that the effect is due to chance. If we could know exactly the laws of nature and the situation of the universe at the initial instant, we could predict exactly the situation of that same universe at a subsequent moment. But even if the natural laws had no longer any secret for us, we would know the initial situation only approximately. If that enables us to predict the subsequent situation with the same approximation, that is all we require, we say that the phenomenon has been predicted that it is governed by laws. But it is not always so; it may happen that small differences in the initial conditions produce very great ones in the final phenomena. A small error in the former will produce an enormous error in the latter. Prediction becomes impossible, and we have the phenomenon of change.

H. Poincaré, *Science et méthode*.
Translated by F. Maitlan, Dover Publ., New York.

1.1 Multistability and bifurcations

Theoretical physics can be viewed as the art and science of the accurate description of physical reality by tractable (i.e. mathematical) means. Unfortunately, the list of exactly solvable models is so short that it would be embarrassing were it not for the great successes that physical models have had when a variety of approximations as well as results from solvable cases are used in calculations. In the area of *classical mechanics*, we can cite the harmonic oscillator, the two-body, and the bilinear many-body problems as cases which are completely solvable. In *quantum mechanics*, the harmonic oscillator again and the central field problem are two of the most noteworthy examples which may be analysed exactly. In statistical mechanics the list includes the one- and two-dimensional Ising models and the mean field model. The great majority of physical problems, however, cannot be directly mapped on to one of the above or a particular one of several other exactly solved problems. Modern physics has developed various types of approximate computational methods in great detail. For example, there are several types of perturbative calculations (Feynman diagrams, Green functions, virial expansions, etc.) as well as formalisms within which they can be conveniently applied. The latter group includes the use of normal modes, second quantization, Fourier and Laplace transforms, linear response approximations, cluster expansions, etc. While these methods have proved to be invaluable in describing a vast array of physical systems and phenomena, a number of shortcomings have recently become quite evident and the development of complementary methods is desirable. In particular, systems of many particles or fields close to instability points, phase changes, or bifurcations are much less amenable to perturbative descriptions than systems close to equilibrium. With the realization that new methods are needed, we observe that nonlinear mathematics has provided us in this century (and especially in its second half) with very many sophisticated calculational techniques which provide powerful tools to overcome problems of calculation for realistic systems and also, at the same time, which supply considerable insight into the conceptual challenges in nonlinear physics. Here, we should mention the emergence of the area of nonlinear waves and solitons, nonlinear stability analysis, dynamical bifurcations, chaotic dynamics, fractals, and symmetry methods for nonlinear differential equations. This monograph is an attempt to give a practical comprehensive 'overview' of the many complex problems arising primarily in modern condensed-matter physics.

The object of our interest in this study is a class of physical systems, in condensed-matter form, composed of a macroscopic number of particles (of the order of Avogadro's number, i.e. $N \sim 10^{23}$), having at least N degrees of freedom. However, the main focus of the study is on the properties close to instability points, which may be manifested in many ways. For example, the creation of new ordered states may highlight *broken symmetries* in the system. In such a situation, *long-range order* may be established. Often one finds the emergence of *nonlinear dynamical modes* of behaviour, which may include electromagnetic *soliton* waves in optical fibres or *magnetic vortices* in type II superconductors. Another phenomenon closely related to the onset of instability is the growth of chaotic regions in the phase space. On the experimental front, one can measure *nonlinear responses* to external stimuli or perturbations, such as those that are exhibited by hysteresis loops. These responses are an indication of the changes to the system's generalized rigidity. This means that average *fluctuations* in physical quantities may be more or less pronounced depending on the structure of the system and on its means of relaxation once stimulated. In particular, the critical state is often associated with an unbounded growth of fluctuations.

Since we have to deal with an otherwise unworkable number of particle co-ordinates, we are forced to take a reductionist point of view and focus on a finite number of most important (driving) degrees of freedom, which are encapsulated in what are referred to, following Landau, as *order parameters* [1]. These describe the modes of behaviour that determine the total properties of the system and, on a particular time scale, may be considered to have a 'slow' time variation. The remaining 'fast' degrees of freedom (also called slaved modes) can either be effectively incorporated through fluctuations, temperature-dependent coefficients, or dissipative terms in the equation of motion, or ignored altogether. This approach has been motivated historically, beginning with Boltzmann's formulation of statistical thermodynamics, which gave microscopic justification for the macroscopic results of classical thermodynamics. In a similar spirit, Landau proposed a first general theory of systems undergoing phase transitions and concentrated on a single thermodynamic quantity which he called the order parameter [1]. The effect of all the other degrees of freedom was incorporated in temperature-dependent coefficients of his famous free energy expansions. The concept of reducing the system's degrees of freedom to only a handful of physically crucial ones was fully justified for critical systems through the development of scaling procedures, known as renormalization group theory, by Kadanoff and Wilson [2]. In addition, a similar ideology permeated the thinking of leading scientists in neighbouring areas of science such as chemical kinetics through the works of Nicolis and Prigogine on the reaction–diffusion equations [3] and biological self-organization expressed through Haken's theory of synergetics [4]. The ideas of *coarse graining*, which is the name given to this reductionist approach, and *scale invariance* are now widely used in such fields as mathematical engineering, fluid dynamics, and astrophysics, to name but a few. M. Feigenbaum's discovery of chaotic maps [5] and B. Mandelbrot's seminal work on fractals [6] are two other illustrations of the power of the notion of scale invariance.

Let us now be more specific. Suppose that the system under consideration can be, in the first instance, adequately described by a macroscopic scalar variable (i.e. an order parameter) η, resulting from some sort of coarse graining procedure. Consequently, under the influence of an external field σ coupled (typically linearly) to η, the state of the system is determined by an algebraic equation which, in thermodynamic terms, is referred to as the equation of state. This could take the generic form

$$P(\sigma, \eta) = 0, \qquad (1.1)$$

where P is usually a polynomial function of η representing the first derivative with respect to η of an associated thermodynamic potential $V(\eta)$. Thus, $P = \partial V/\partial \eta$ can be viewed as an analogue of some sort of conservative but nonlinear force. The question of uniqueness of solutions to eqn (1.1) is clearly linked with that of *multistability* in the system. Assuming the presence of at least one control parameter acting through one of the variable coefficients in $P(\sigma, \eta)$ leads directly to the problem of *catastrophes*, an area thoroughly investigated by René Thom [7] and a generation of mathematicians who followed in his footsteps. In the family of polynomial functions in terms of one or two variables such as η and up to four control parameters, the most important forms of behaviour have been classified as the so-called first seven catastrophe geometries. These are the fold, the cusp, the swallow tail, the butterfly, the elliptic umbilic, the hyperbolic umbilic, and the parabolic umbilic [7]. In terms of their usefulness to physical problems, only those catastrophes which lead to thermodynamic potentials $V(\eta)$ that are bounded from below are acceptable. In the aforementioned group only three would satisfy this requirement, i.e. the cusp, the butterfly, and the parabolic umbilic catastrophes. Interestingly, these three are the ones which were intuitively recognized as essentially important to the physics of phase transitions by L. D. Landau and his co-workers long before catastrophe theory was even conceived. The cusp catastrophe is described by the potential

$$V(\eta) = \frac{\varepsilon}{4}\eta^4 + \tfrac{1}{2}a\eta^2 - \sigma\eta, \qquad (1.2)$$

where $\varepsilon = +1$ or -1 and a is a control parameter. This potential will later

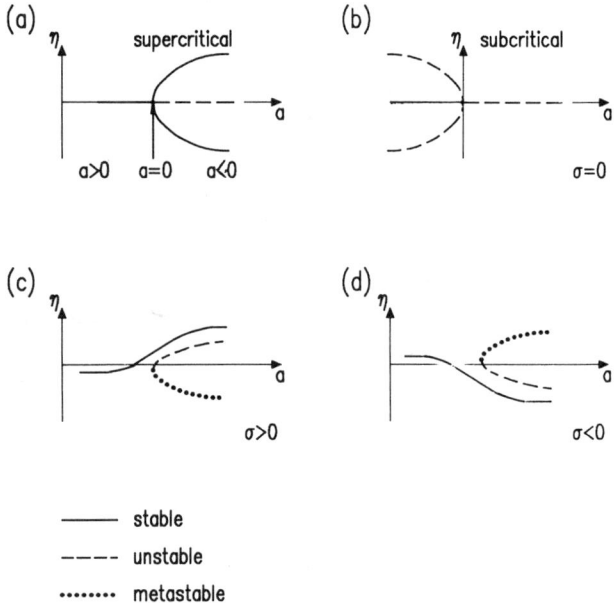

Fig. 1.1 Bifurcation diagrams for the pitchfork bifurcation without an external field σ in the (a) supercritical and (b) subcritical cases, and (c), (d) for two possible signs of the external field.

be shown successfully to describe second order phase transitions both in the absence ($\sigma = 0$) and in the presence ($\sigma \neq 0$) of external fields. On the other hand, the butterfly catastrophe employs

$$V(\eta) = \frac{1}{6}\eta^6 + \frac{a}{4}\eta^4 + \frac{b}{3}\eta^3 + \frac{c}{2}\eta^2 - \sigma\eta, \tag{1.3}$$

and has been used to model first-order phase transitions, both field- and temperature-induced. Once again, the coefficients a, b, and c are variable control parameters of some kind. Finally, the parabolic umbilic catastrophe potential is given by

$$V(\eta_1, \eta_2) = \eta_1^2\eta_2 + \eta_2^4 + a\eta_1^2 + b\eta_2^2 - \sigma_1\eta_1 - \sigma_2\eta_2. \tag{1.4}$$

The applicability of $V(\eta_1, \eta_2)$ in eqn (1.4) has been demonstrated in coupled critical systems such as ferroelastics or ferroelectrics, and entire books [8] have been devoted to the study of the physical consequences of free energies in the form of eqn (1.4), or in similar such forms.

To illustrate the associated phenomenon of *bifurcation*, we take eqn (1.2) as the simplest example, which yields the following equation of state:

$$\varepsilon\eta^3 + a\eta = \sigma, \tag{1.5}$$

where a is the only control parameter present. A plus sign in front of η^3 can be changed to a minus to designate a subcritical bifurcation. In Fig. 1.1

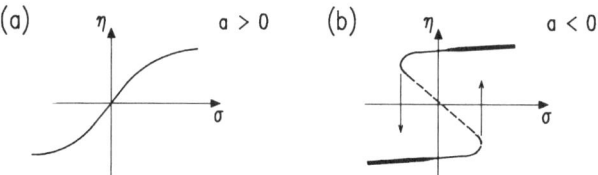

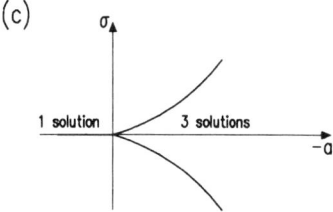

Fig. 1.2 The order parameter's response to an externally applied field in (a) a unistable and (b) a bistable state, and (c) separation of the control space into the solution's multiplicity regions.

are shown bifurcation diagrams for the plots of η as a function of a when $\sigma = 0$ and $\sigma \neq 0$. For $\sigma = 0$, the plot is commonly referred to as a 'pitchfork bifurcation', and it illustrates a transition between a single stable solution for $a > 0$ and a bistable situation when $a < 0$. It will be shown in Chapter 2 that this is a prototype for a second-order phase transition. In the case of $\sigma \neq 0$, the bifurcation still exists and it takes place at $a = 0$. However, a new feature is the phenomenon of external-field induced hysteresis and metastability of the solution, with a higher value of the potential function. Stability corresponds to a solution for which $\partial^2 V / \partial \eta^2 > 0$ and, if more than one solution of the equation of state (1.1) exists that is stable, we call the higher energy solutions metastable. Unstable solutions correspond to $\partial^2 V / \partial \eta^2 < 0$, while the coincidence of $\partial V / \partial \eta = \partial^2 V / \partial \eta^2 = 0$ highlights an instability (or inflection) point. The associated difference in the response of the order parameter to the application of an external field is shown in Fig. 1.2. In unistable situations, η as a function of σ is a smooth single-valued function. On the other hand, multistability results in double-valuedness (at least) in some ranges

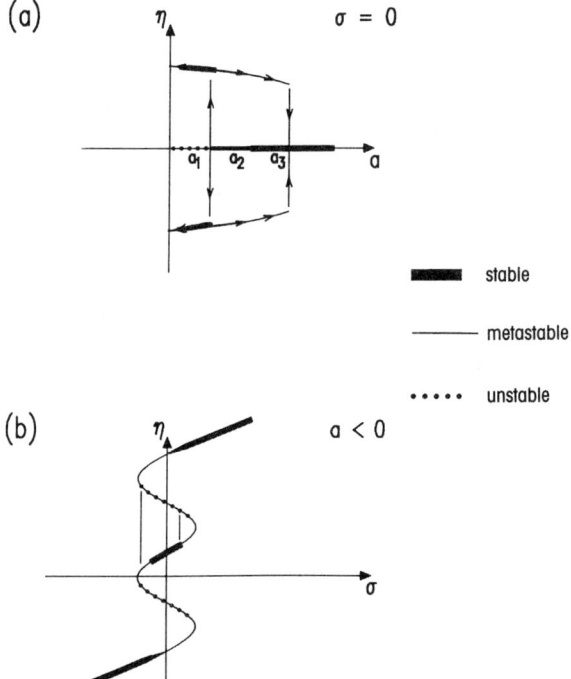

Fig. 1.3 (a) A bifurcation plot and (b) double hysteresis, in the butterfly catastrophe case.

MULTISTABILITY AND BIFURCATIONS

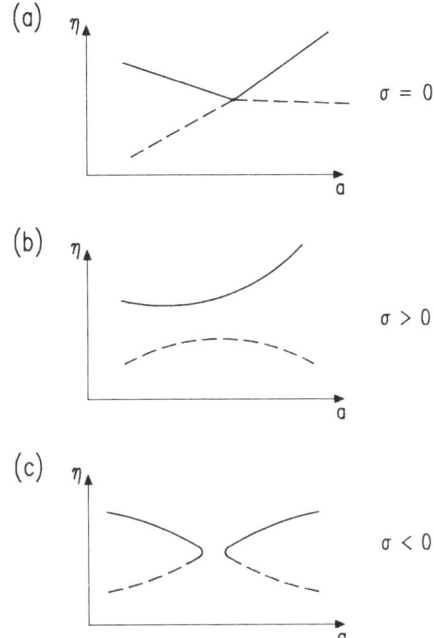

Fig. 1.4 The transcritical bifurcation for (a) $\sigma = 0$, (b) $\sigma > 0$, and (c) $\sigma < 0$.

of the external fields. The regions of multistability in the parameter space are demonstrated in Fig. 1.2(c). The dividing line is given by the equation

$$4a^3 + 27\sigma^2 = 0. \tag{1.6}$$

This indicates the parameter set for which a triple solution of the cubic equation in eqn (1.5) is obtained. Figure 1.3 is an extension of these concepts to the butterfly catastrophe case, which exhibits thermal hysteresis (if the control parameter a involves a temperature dependence) and a double hysteresis under the influence of an external field. In physical terms, we show in Chapter 2 that the butterfly catastrophe is the prototype of a first-order phase transition. Indeed, the simplest catastrophe case (the transcritical catastrophe), given by the quadratic equation of state, is

$$-a\eta + b\eta^2 = \sigma. \tag{1.7}$$

This is illustrated in Fig. 1.4. Obviously, increasing the degree of the polynomial $P(\eta)$ will give rise to a simultaneous presence of more independent real solutions and hence to the possibility of more complicated

8 A NONLINEAR ANALYSIS PRIMER

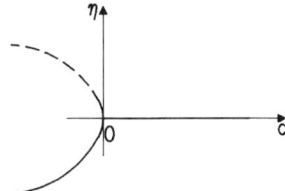

Fig. 1.5 The saddle-node bifurcation.

bifurcation phenomena. Next, we have the saddle-node bifurcation with

$$\eta^2 + c = 0, \tag{1.8}$$

which is shown in Fig. 1.5.

Throughout this section it has been assumed, somewhat artificially, that our nonlinear system can be adequately described with the use of just a single scalar quantity η. This, of course, is a gross oversimplification of real physical systems and we will be gradually developing mathematical means whereby more realistic approaches can be attained. This will involve making the order parameter a time- and later a space-dependent physical quantity. On a different level of theoretical modelling, the order parameter's freedom to fluctuate can be achieved by making it a probabilistic variable. It still follows the general trend charted by the associated bifurcation phenomenon but it does not have to do it slavishly; i.e. deviations from the average are allowed although they may be less probable than the average value itself. This then motivates our next section.

1.2 Stochastic analogue of a bifurcation

If the order parameter η is considered to be a probabilistic variable with the corresponding potential function

$$V(\eta) = a_2\eta^2 + a_4\eta^4 \tag{1.9}$$

then the corresponding probability distribution, according to Boltzmann, is

$$P(\eta) = P_0 e^{-\beta V} = P_0 \exp\left[-\beta(a_2\eta^2 + a_4\eta^4)\right], \tag{1.10}$$

where $\beta = (k_B T)^{-1}$, k_B is the Boltzmann constant, T is the absolute temperature, and P_0 is a normalization function. As the coefficient a_2 changes its sign from positive to negative, the potential function $V(\eta)$ changes form from a single well to a double well and $P(\eta)$ transforms from a single-peaked to a double-peaked function of η (see Fig. 1.6).

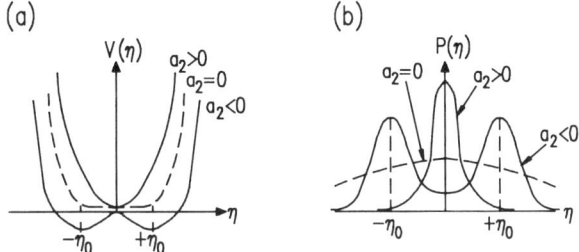

Fig. 1.6 Changes affecting the form of (a) $V(\eta)$ and (b) $P(\eta)$, resulting from a_2 changing its sign.

It is not generally known, but the quartic non-Gaussian distribution function that has appeared in eqn (1.10) can be used in exact calculations of all statistical quantities as follows. Writing $\lambda_2 = \beta a_2$ and $\lambda_4 = \beta a_4$, the partition function Z is found as [9]

$$Z = (2\lambda_4 \beta)^{-1/4} \Gamma(\tfrac{1}{2}) \exp\left(\frac{\lambda_2^2}{8\lambda_4}\right) D_{-1/2}\left[\frac{\lambda_2}{\sqrt{2\lambda_4/\beta}}\right], \qquad (1.11)$$

and Z is analytic *everywhere*. Here, Γ is the gamma function and D_ν is a parabolic cylinder function [10]. In the limit $\lambda_4 \to 0$ or $\lambda_2 \to \infty$ we recover the Gaussian result [11],

$$Z_\infty = \Gamma(\tfrac{1}{2})(\beta\lambda_2)^{-1/2}. \qquad (1.12)$$

Furthermore, the second moment of the quartic non-Gaussian distribution is identified with the static susceptibility function χ and can be obtained as

$$\chi = \langle \eta^2 \rangle = (8\lambda_4 \beta)^{-1/2} D_{-3/2}\left[\frac{\lambda_2}{\sqrt{2\lambda_4/\beta}}\right] / D_{-1/2}\left[\frac{\lambda_2}{\sqrt{2\lambda_4/\beta}}\right]. \qquad (1.13)$$

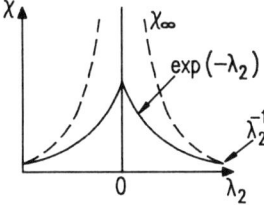

Fig. 1.7 A schematic plot of χ as a function of λ_2, which is proportional to the control parameter a_2.

Again, it is an analytic function everywhere, which tends to its Gaussian limit for $\lambda_4 \to 0$ or $\lambda_2 \to \infty$: that is,

$$\chi_\infty = (2\lambda_2)^{-1}. \tag{1.14}$$

The plot of χ as a function of λ_2 is shown in Fig. 1.7.

The statistical approach sketched here will be elaborated on at the end of Chapter 2, when we discuss its repercussions on the description of phase transitions. For now, however, we wish to return to the simple view of the order parameter as a state variable and we now allow it to be a time-dependent quantity.

1.3 Harmonic versus anharmonic motion

Let us now take another important step in the direction of increasingly realistic model development and assume that the system is not rigidly forced to follow the bifurcation diagram or even be allowed to have stochastic fluctuations from it, as seen in Section 1.2, but rather that it is allowed to execute small oscillations (characterized by inertial mass m) about its *stable* equilibria.

The appropriate equation of motion is then given by Newtonian mechanics as

$$m\ddot{\eta} + a\eta + b\eta^3 = \sigma, \tag{1.15}$$

where $\ddot{\eta} = \partial^2\eta/\partial t^2$. For $a > 0$ (and η small) this can be approximately written as a shifted harmonic oscillator equation,

$$m\ddot{y} \simeq -ay, \tag{1.16}$$

where $y = \eta - \sigma/a$, so that

$$\eta = \sigma/a + A\cos\Omega_0 t, \tag{1.17}$$

with $\Omega_0 = \sqrt{a/m}$ representing small-amplitude (A) periodic oscillations about the stable equilibrium $\eta = \sigma/a$. However, for $a < 0$, the problem is much more complicated (but still solvable), with three types of oscillating motions, as shown in Fig. 1.8.

Figures 1.8(a) and (b) represent the first integral of eqn (1.15), which is

$$\frac{m}{2}\dot{\eta}^2 + \frac{a}{2}\eta^2 + \frac{b}{4}\eta^4 - \sigma\eta + c_0 = 0, \tag{1.18}$$

where c_0 is a constant of integration setting the energy scale. This can also be illustrated in terms of phase-space diagrams and their trajectories, as

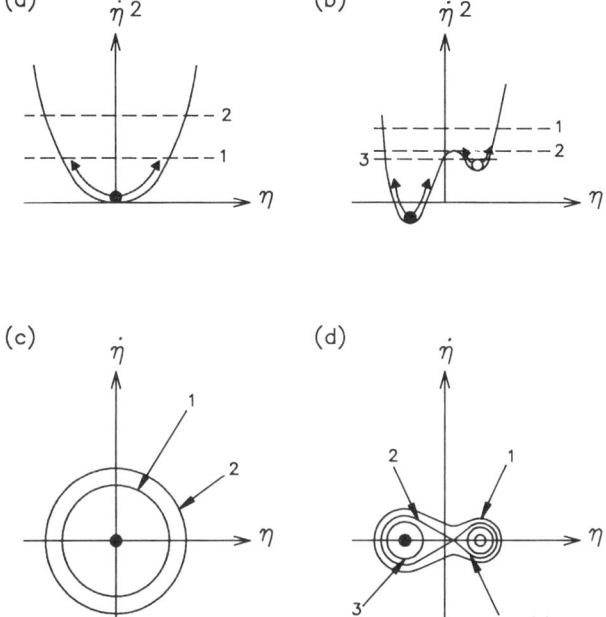

Fig. 1.8 A comparison of (a) a harmonic and (b) an anharmonic potential well, and (c), (d) the corresponding orbits in the $(\dot\eta, \eta)$ phase space.

shown in Figs 1.8(c) and (d). An important special trajectory denoted by number 2 in Fig. 1.8(d) is called a separatrix, and it represents a boundary between oscillations around a single well and those around both wells. All of the orbit numbers in Figs 1.8(a) and (b) and 1.8(c) and (d) correspond to each other.

A new situation arises when we consider periodic nonlinearities. A well-known example is a *simple pendulum*. The equation of motion is the so-called Sine–Gordon (SG) equation [12],

$$\ddot\theta + \frac{g}{l}\sin\theta = 0, \qquad (1.19)$$

which can be integrated once to give

$$\tfrac{1}{2}\dot\theta^2 - \frac{g}{l}\cos\theta + c_0 = 0, \qquad (1.20)$$

where c_0 is an integration constant that sets the energy scale. The existence of periodically located stable elliptic points, around which

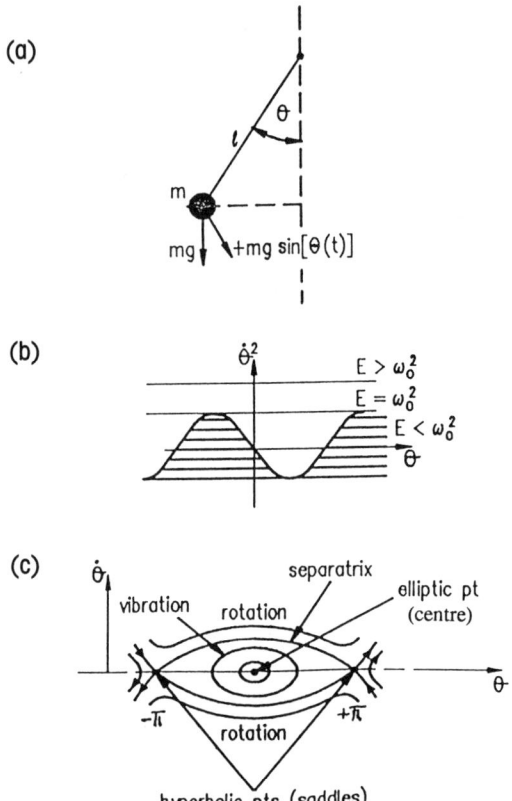

Fig. 1.9 A schematic representation of (a) the simple pendulum of mass m; (b) the first integral; and (c) the phase-space diagram.

closed-orbit vibrations are concentrated, is illustrated in Fig. 1.9. Large-amplitude motion exists in the form of rotations around the point of suspension of the pendulum. These two types of trajectories are once again divided by a separatrix. Note that, for small displacements, $\sin\theta \simeq \theta - \theta^3/6 + \ldots$, thus mapping the problem (partially) on to the previous case. For more details on solutions of the SG equation, the reader is referred to Chapter 7.

A final example in this section is that of anharmonic motion close to an *instability* point. Consider an inverted potential well $V(\eta) = a\eta^2 + b\eta^4$ (see Fig. 1.10(a)), where $a < 0$ and $b < 0$. Then, the situation arises as illustrated in Fig. 1.10. It is clear that no stable, finite amplitude oscillations are possible in this case. This is usually referred to as a *hyperbolic singularity*. More will be said about stability in Sections 1.4 and 1.5.

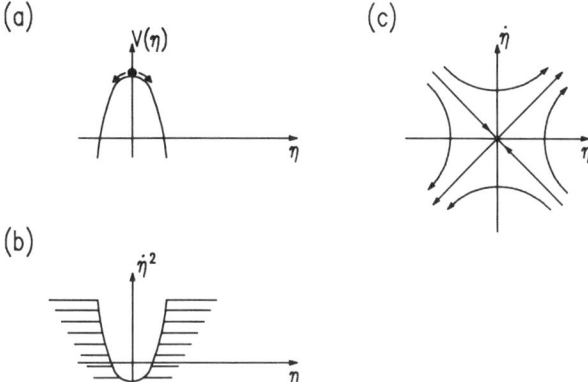

Fig. 1.10 (a) A schematic illustration of the instability point in a quartic anharmonic potential. (b) The first integral. (c) The corresponding phase-space diagram.

1.4 External driving and dissipation

Another important step in the direction of a realistic description of dynamical systems close to instability points is to include effects due to energy *dissipation* and external, time-dependent *driving*. For bistable systems the equation of motion takes the form of the Duffing oscillator,

$$m\ddot{\eta} + a\eta + b\eta^3 + \gamma\dot{\eta} = \sigma(t), \tag{1.21}$$

where $\gamma\dot{\eta}$ is a friction term and the external force $\sigma(t)$ is assumed to be time dependent. Since the system is unstable for $a > 0$, this equation can be linearized about $\eta = 0$ in this regime. Taking $\sigma(t)$ in a harmonic form, we obtain a damped-driven harmonic oscillator equation

$$m\ddot{\eta} + a\eta + \gamma\dot{\eta} = \sigma_0 \cos(\omega_0 t), \tag{1.22}$$

with the well-known periodic solution

$$\eta = \sigma_0 A \cos(\omega_0 t + \theta), \tag{1.23}$$

where

$$A = \left[(m\omega_0^2 - a)^2 + \gamma^2\omega_0^2\right]^{-1/2} \tag{1.24a}$$

and

$$\theta = \tan^{-1}\left[\gamma\omega_0/(m\omega_0^2 - a)\right], \tag{1.24b}$$

signifying the existence of a *resonance* at $\omega_0 = (a/m)^{1/2}$.

The situation changes dramatically if we reincorporate nonlinearity. To see this, let us first take a look at the presence of nonlinearity and forcing,

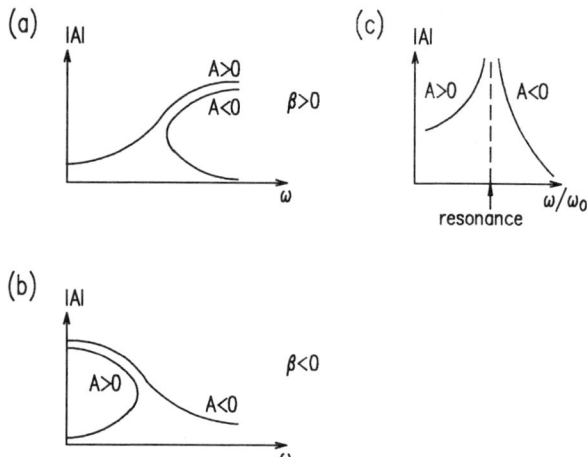

Fig. 1.11 Bifurcation diagrams for the amplitude A of solutions to eqn (1.25) as a function of ω, following eqn (1.26), when (a) $\beta > 0$, (b) $\beta < 0$, and (c) $\beta = 0$.

but *no dissipation*. Scaling the independent variable only in eqn (1.21) according to $s = \omega_0 t$, we obtain

$$\omega^2 \ddot{\eta} + \eta + \beta \eta^3 = F \cos s, \qquad (1.25)$$

where $\beta = b/a$, $F = \sigma_0/a$ and $\omega^2 = \omega_0^2 m/a$. Taking as initial conditions $\eta(0) = A$ and $\dot{\eta}(0) = 0$, we solve the problem for small values of F to find [13]

$$\omega \simeq 1 + \beta\left(\tfrac{3}{8}A^2 - F/2A\right) + \cdots, \qquad (1.26)$$

which results in typical bifurcation diagrams for the amplitude A as a function of the frequency ω (see Fig. 1.11).

Note a stark contrast between the nonlinear cases (a) and (b) and the linear one (c) in Fig. 1.11 where a resonance effect exists. Moreover, taking the forcing term to be large, compared to nonlinearity, gives a different type of solution; namely [13],

$$\omega = 3 + \tfrac{9}{8}\beta(A^2 + AF/8 + F^2/32) + \cdots, \qquad (1.27)$$

which is plotted in Fig. 1.12. P_1 and P_2 are called bifurcation points and they signify the emergence of so-called subharmonics, i.e. oscillations with frequencies which are a fraction of the forcing frequency.

A superposition of Figs 1.11 and 1.12 gives a global range of behaviour, and is shown in Fig. 1.13.

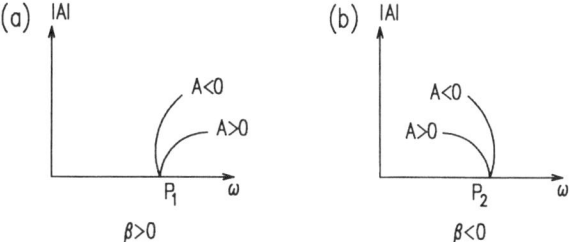

Fig. 1.12 The response of the nondissipative Duffing oscillator to a strong external forcing field for (a) $\beta > 0$ and (b) $\beta < 0$.

Looking at the other limiting case—namely, the presence of nonlinearity and dissipation but *no forcing*—we find that the equation of motion becomes [14]

$$\ddot{\eta} + \gamma \dot{\eta} + \eta + \beta \eta^3 = 0, \tag{1.28}$$

and is integrable only for $\gamma = 3/\sqrt{2}$, in which case we find a so-called kink solution (we have scaled the equation so that $\beta = -1$) in the form

$$\eta = \pm \tfrac{1}{2}\left[1 - \tanh\left(\frac{1}{2\sqrt{2}}(s - s_0)\right)\right], \tag{1.29}$$

and a class of damped oscillatory solutions in terms of elliptic functions (see Fig. 1.14 and Appendix B).

Finally, an analysis of the complete Duffing equation, with nonlinearity, dissipation, *and* forcing, for eqn (1.21) has also been carried out. A description of the resultant behaviour can be found in the book by Cvitanović [5]. What we observe now is the presence of a *limit cycle* for $\gamma = \gamma_0$ (a particular value). Decreasing γ_0 leads to a period doubling cascade and a related bifurcation cascade (see Fig. 1.15), leading eventually to chaos. Much more will be said about this very interesting phenomenon in a later section of this chapter, concerned with chaos. It is

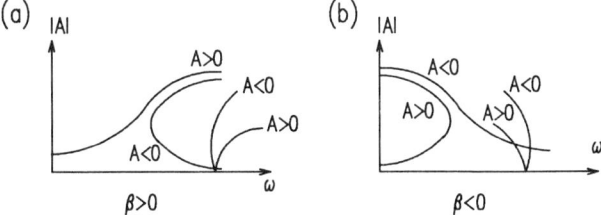

Fig. 1.13 A combination of Figs 1.11 and 1.12 for the Duffing oscillator for (a) $\beta > 0$ and (b) $\beta < 0$.

16   A NONLINEAR ANALYSIS PRIMER

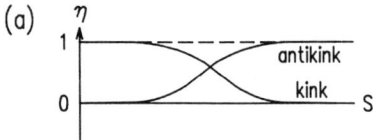

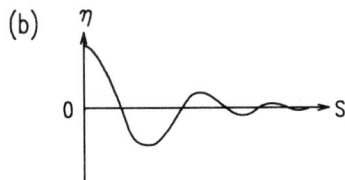

Fig. 1.14 A schematic illustration of (a) the kink and (b) damped oscillatory solutions of eqn (1.28) for $\gamma = 3/\sqrt{2}$ and $\beta = -1$.

worth mentioning here, however, that similar behaviour has been seen in numerical experiments for nonlinear waves, e.g. the nonlinear Schrödinger equation (see Section 1.7).

Another example that we wish to discuss here is that of the *damped-driven pendulum* equation:

$$ml \frac{d^2\theta}{dt^2} + \gamma \frac{d\theta}{dt} + W \sin\theta = A\cos(\omega_0 t), \qquad (1.30)$$

where m is the pendulum's mass, l is the pendulum's length, γ is the damping coefficient, W is the restoring force (gravitation), A is the forcing amplitude, and ω_0 is the forcing frequency. In dimensionless form this equation of motion becomes

$$\frac{d^2\theta}{dt^2} + \frac{1}{q}\frac{d\theta}{dt} + \sin\theta = g\cos(\omega_0 t), \qquad (1.31)$$

where q is the damping parameter and g is the forcing amplitude. First, ignoring both friction and forcing, we find that eqn (1.31) can be integrated once to yield an elliptic-type of differential equation:

$$\tfrac{1}{2}(d\theta/dt)^2 = \cos\theta + c_0, \qquad (1.32)$$

where c_0 is an integration constant. In Fig. 1.16 is shown a phase portrait of eqn (1.32), where we have denoted the angular velocity as $\omega \equiv d\theta/dt$. Note the presence of rotational and vibrational orbits, a separatrix orbit separating the two above rotational and vibrational regimes, an elliptic

EXTERNAL DRIVING AND DISSIPATION 17

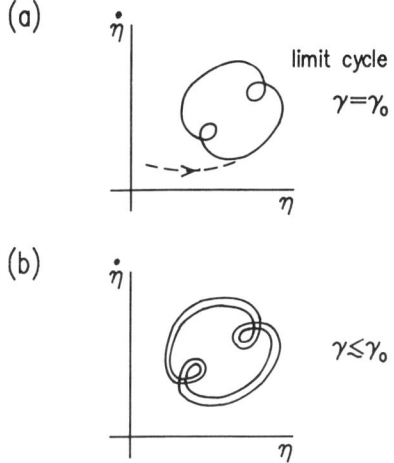

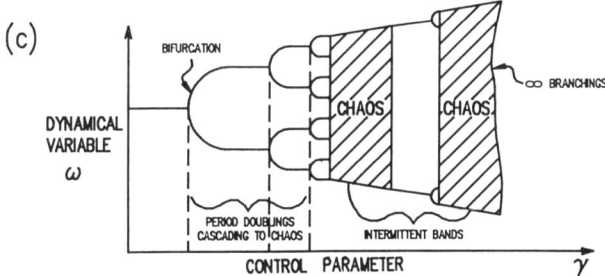

Fig. 1.15 A schematic illustration of the period doubling cascade and a transition to chaos for the frequency ω of the damped driven Duffing oscillator.

focus point, and a hyperbolic instability saddle point. A plot of the angular velocity for the two types of orbit is given in Fig. 1.17.

Also note that the period of motion increases with increasing amplitude (in contrast to the linearized pendulum). This is shown in Fig. 1.17.

The inclusion of friction can be analysed by assuming that $\omega = Be^{\lambda t}$ and linearizing the equation about $\theta = n\pi$ ($\sin\theta \approx \theta - n\pi$ for θ close to $n\pi$, with n even) to obtain a discriminant for the growth rates λ of each mode:

$$\lambda^2 + \lambda \pm 1 = 0, \qquad (1.33)$$

where '+' corresponds to n even and '−' to n odd. The emergence of

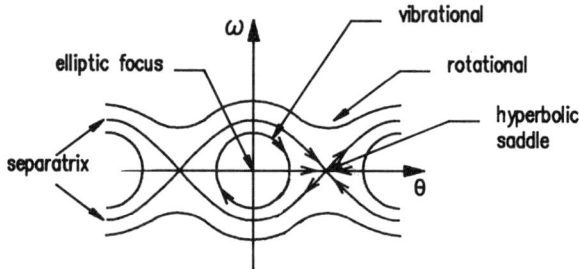

Fig. 1.16 The phase portrait of the free pendulum equation.

focus points and saddle points, as well as the new phase diagram [15] when friction is included, are illustrated in Fig. 1.18. It is important to note the existence of *basins of attraction* associated with each focus point.

The other limiting case, i.e. periodic perturbation without friction, results in a distorted phase portrait with a *stochastic layer* around the separatrix, where motion is virtually unpredictable (see Fig. 1.19). There, separatrices divide oscillatory and rotational trajectories. Periodic perturbation destroys the separatrix, and a stochastic layer forms in its vicinity. The motion is chaotic within the layer. The region occupied by the stochastic layer has holes into which a stochastic trajectory cannot enter. There is an infinite number of stability islands, within which an infinite number of still thinner stochastic layers reside.

We now return to the full damped-driven pendulum equation, i.e. eqn (1.31). By expressing this equation as a system of two coupled first-order

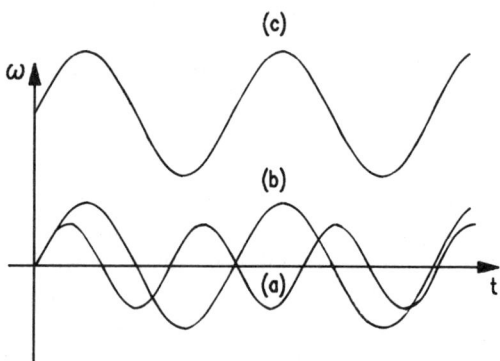

Fig. 1.17 Angular velocity for (a), (b) the pendulum's two particular vibrational orbits, and (c) a rotational orbit.

EXTERNAL DRIVING AND DISSIPATION 19

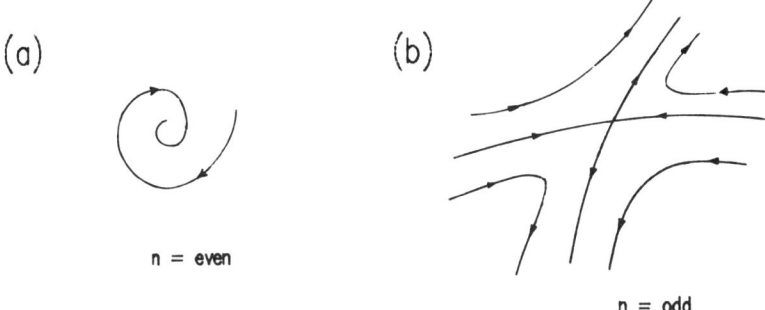

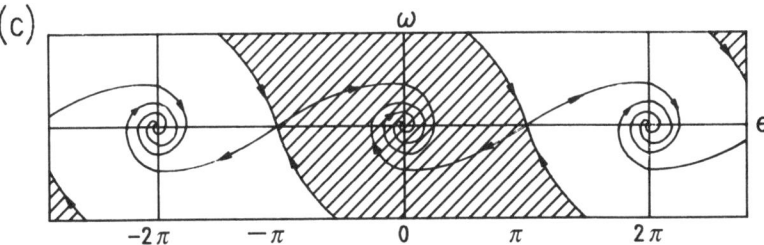

Fig. 1.18 Critical points in phase space: (a) a focal point and (b) a saddle point. In (b) the trajectories going to the saddle point are stable, whereas the trajectories coming from the saddle point are unstable. (c) The phase-space diagram of the damped pendulum. Alternate shaded and unshaded regions are basins of attraction. All points within a particular basin are attracted to the focal point within the basin.

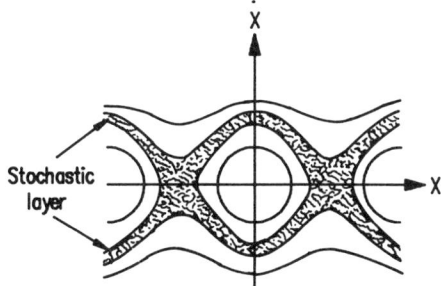

Fig. 1.19 The phase portrait of perturbed pendula.

differential equations, we can clearly see the two independent variables ω and θ and the control parameter ω_0:

$$\frac{d\omega}{dt} = -\frac{1}{q}\omega - \sin\theta + g\cos(\omega_0 t), \qquad \frac{d\theta}{dt} = \omega. \qquad (1.34)$$

The resulting transition to chaos will depend on the values of the system parameters $(\omega, \theta, \omega_0)$ in the phase space. By varying $(\omega, \theta, \omega_0)$ the state of the system can be studied as it undergoes a transition to chaotic behaviour. This is very similar to the Duffing oscillator case, except that the phase space can now be divided into periodic regions. We will return to a discussion on the phenomenon of chaos in Section 1.9.

1.5 Relaxation dynamics and asymptotic stability

In many time-dependent processes in nature, the mechanism of evolution of the concentrations of the molecular species involved is dominated by relaxation dynamics rather than conservative oscillating motion. This means that in the governing equations of motion the kinetic energy term is negligible compared to thermal fluctuations and the process is of a diffusion type. Examples of such phenomena include chemical kinetics of reacting species, exciton migration in molecular crystals, recombination processes in semiconductors, mixing of binary alloys and fluids, etc. This subsection is concerned with the underlying nonlinear dynamics that describe such phenomena.

Considering a single order parameter η relaxing to its equilibrium value with relaxation constant Γ, the evolution equation (in the absence of spatial dispersion) is

$$\eta_t = \frac{1}{\Gamma} P(\eta), \qquad (1.35)$$

where $P(\eta)$ is the nonlinear force function appearing in the equation of state (1.1).

The main point to consider is the number and type of bifurcation points of $P(\eta)$. This has been reviewed in Section 1.1. Depending on whether a given point η_i is a stable or unstable bifurcation point, the system will either tend asymptotically to it, or away from it, and towards the nearest asymptotically stable value of η. In this respect, we now define the asymptotic stability of a solution $\eta(t)$ of a differential equation. For example, an equilibrium, $\eta = 0$, is asymptotically stable if there exists a neighbourhood of initial conditions $0 < |\eta(0)| < \varepsilon$ and if for all $\eta(0)$ in this neighbourhood:

(1) the trajectory $\eta(t)$ satisfies $|\eta(t)| < \varepsilon$ for $t > 0$; and
(2) $|\eta(t)| \to 0$ as $t \to \infty$.

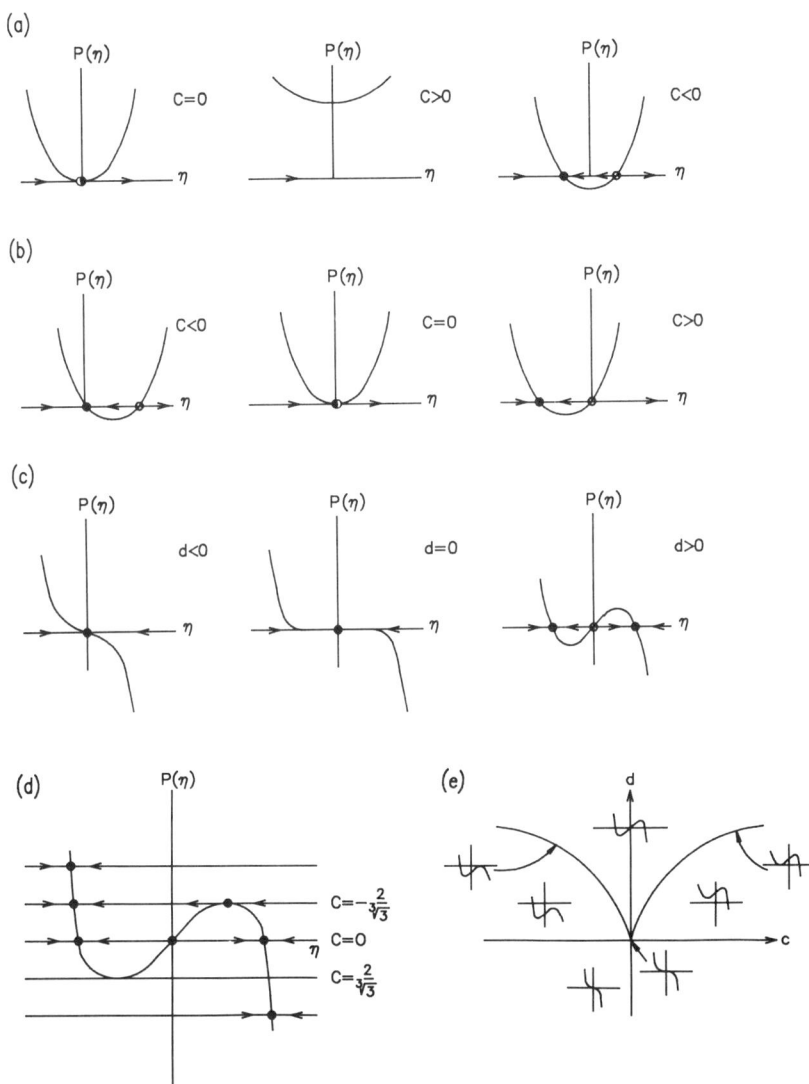

Fig. 1.20 Phase portraits of the main types of bifurcation dynamics: (a) saddle-node bifurcation; (b) transcritical bifurcation; (c) pitchfork bifurcation (supercritical); (d) the single hysteresis bifurcation; (e) the combination diagram for cases (c) and (d).

In Fig. 1.20 we have shown a number of typical forms of behaviour for the most important types of bifurcations of a scalar order parameter η. We show the following types of bifurcations [16]:

(a) the *saddle-node bifurcation* for

$$\dot{\eta} = \eta^2 + c; \tag{1.36}$$

(b) the *transcritical bifurcation* for

$$\dot{\eta} = c\eta + \eta^2; \tag{1.37}$$

(c) the *pitchfork bifurcation* with

$$\dot{\eta} = d\eta \pm \eta^3, \tag{1.38}$$

where '+' corresponds to the subcritical and '−' to the supercritical cases;

(d) the single hysteresis bifurcation

$$\dot{\eta} = c + \eta - \eta^3; \tag{1.39}$$

(e) a combination of (c) and (d) with two control parameters c and d, i.e.

$$\dot{\eta} = c + d\eta - \eta^3. \tag{1.40}$$

In Fig. 1.20, open circles indicate repulsive points (asymptotically unstable equilibria) while full circles denote attractive points (asymptotically stable equilibria). Half-filled circles correspond to attraction from one side and repulsion from the other. The curve in Fig. 1.20(e) is given by the cubic discriminant equation $4d^3 = 27c^2$.

1.6 Coupled systems and limit cycles

We have said earlier that usually a single-component order parameter is an adequate approximation. This is not always true, however. In some situations there may exist competing subsystems, orders, or sets of degrees of freedom (magnetism versus superconductivity, transverse versus longitudinal oscillations, etc.) requiring two or more dependent variables treated on an equal footing. Thus this class of problems is described by the set of coupled equations below:

$$\dot{\eta}_i = f_i(\{\eta_j\}), \quad 1 \leq i, \ j \leq n, \tag{1.41}$$

where n is the number of components of the order parameter (or the

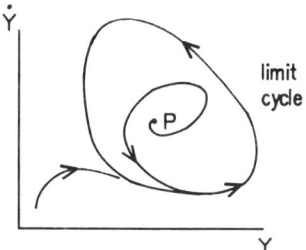

Fig. 1.21 The *Brusselator-type* limit cycle.

number of competing order parameters). Note in this connection that any autonomous nth order differential equation can also be represented in this fashion, since we can write each derivative as a new field (i.e. $\eta^{(i)} \equiv \eta_i$).

An important example of this class of problem is in the area of reaction–diffusion dynamics used to describe chemical reactions where, for obvious reasons, at least two coupled equations are a prerequisite. For example, Prigogine's celebrated Brusselator equations are [3]

$$\dot{X} = X^2 Y - BX + A - X, \qquad (1.42a)$$

and

$$\dot{Y} = -X^2 Y + BX, \qquad (1.42b)$$

which lead to a *limit cycle*, shown in Fig. 1.21. There, point P represents an unstable focus point. It may be shown that the substitution $A \to A + a\cos\omega t$, introduces quasi-periodic and chaotic phases to the Brusselator phase diagram.

Another example is that of a *polar product system*, given by

$$\dot{x}_1 = x_2 + x_1 - x_1^3 - x_1 x_2^2 \qquad (1.43a)$$

and

$$\dot{x}_2 = -x_1 + x_2 - x_1^2 x_2 - x_2^3, \qquad (1.43b)$$

which can be solved exactly by introducing the polar co-ordinates below:

$$x_1 = r\cos\theta, \qquad x_2 = r\sin\theta. \qquad (1.44)$$

Then the solution is

$$r(t) = \frac{1}{\sqrt{1 + r_0^2 e^{-2t}}}, \qquad \theta(t) = \theta_0 - t, \qquad (1.45)$$

and the phase portrait with a stable limit cycle is shown in Fig. 1.22 [16].

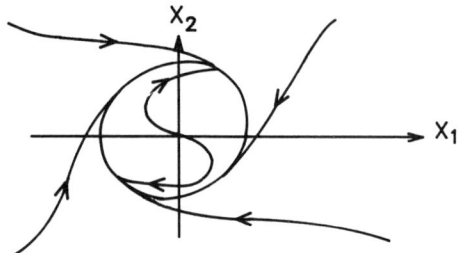

Fig. 1.22 The phase portrait of the polar product's system.

In a *Hopf bifurcation*, the constituent equations (Poincaré–Andronov–Hopf) can be represented as

$$dr/dt = \lambda r - r^3 \quad (1.46a)$$

and

$$d\theta/dt \equiv \omega = -1 \quad (1.46b)$$

in polar co-ordinates and the phase-space picture changes from a stable focus point only to a stable limit cycle as a function of λ, the bifurcation occurring at $\lambda = 0$, as shown in Fig. 1.23. The case in which $\lambda = 1$ corresponds to the previous example, i.e. the polar product system. The radius of the limit cycle has a $\sqrt{\lambda}$ dependence for $\lambda > 0$. A *limit cycle* is a periodic solution of these equations which, in fact, is achieved by choosing arbitrary initial conditions in its basin of attraction. The question of the structural stability of such solutions is very important and can be tackled successfully using Lyapunov stability analysis, at least for two coupled equations. In general, we can find attractive, repulsive, and mixed orbits [17].

A system of two coupled equations which is also of interest is the following:

$$\ddot{x} = -x - 2xy \quad (1.47a)$$

and

$$\ddot{y} = -y \pm y^2 - x^2. \quad (1.47b)$$

The '+' sign option yields the so-called Hénon–Heiles equations with their nonintegrability properties, while the '−' sign case is integrable via a transformation: $\xi = x - y$, $\eta = x + y$.

A summary of the various possibilities for two component systems is given in Fig. 1.24 and shows nodes (sinks and sources), centres, saddles, foci, and limit cycles [16, 17].

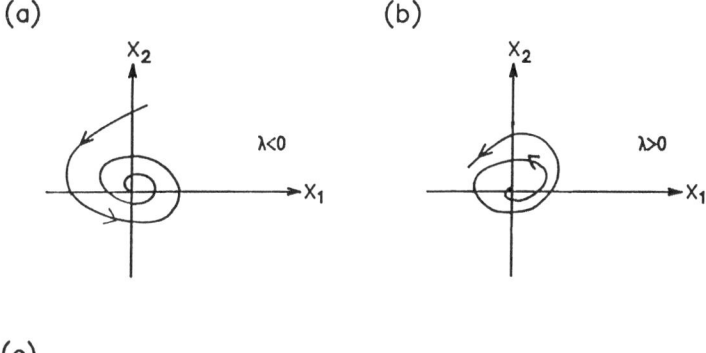

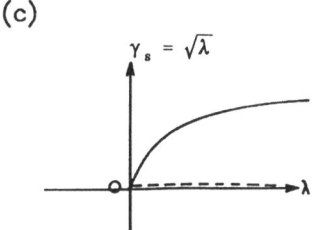

Fig. 1.23 The Hopf bifurcation for (a) $\lambda < 0$ and (b) $\lambda > 0$. (c) The radius of the limit cycle.

Moving on to coupled *second-order equations*, we see a natural generalization of the anharmonic oscillator problem by coupling two such oscillators to find

$$\ddot{\eta}_1 = \alpha_1 \eta_1 + \beta_1 \eta_1^3 + \gamma \eta_1 \eta_2^2 \qquad (1.48a)$$

and

$$\ddot{\eta}_2 = \alpha_2 \eta_2 + \beta_2 \eta_2^3 + \gamma \eta_1^2 \eta_2. \qquad (1.48b)$$

Analysing the solutions obtained numerically, we can see an amazing wealth of behaviour, with chaos developing gradually out of order on increasing the solution's energy and approaching the separatrix (see Fig. 1.25). A return to regular periodic motion is found for high-energy solutions above the separatrix.

1.7 Nonlinear waves and solitons

One can easily imagine a process in which we extend the number of dependent variables to infinity, as can be done for a lattice or a chain. Numerous physical applications of this type of approach can readily be

(a)

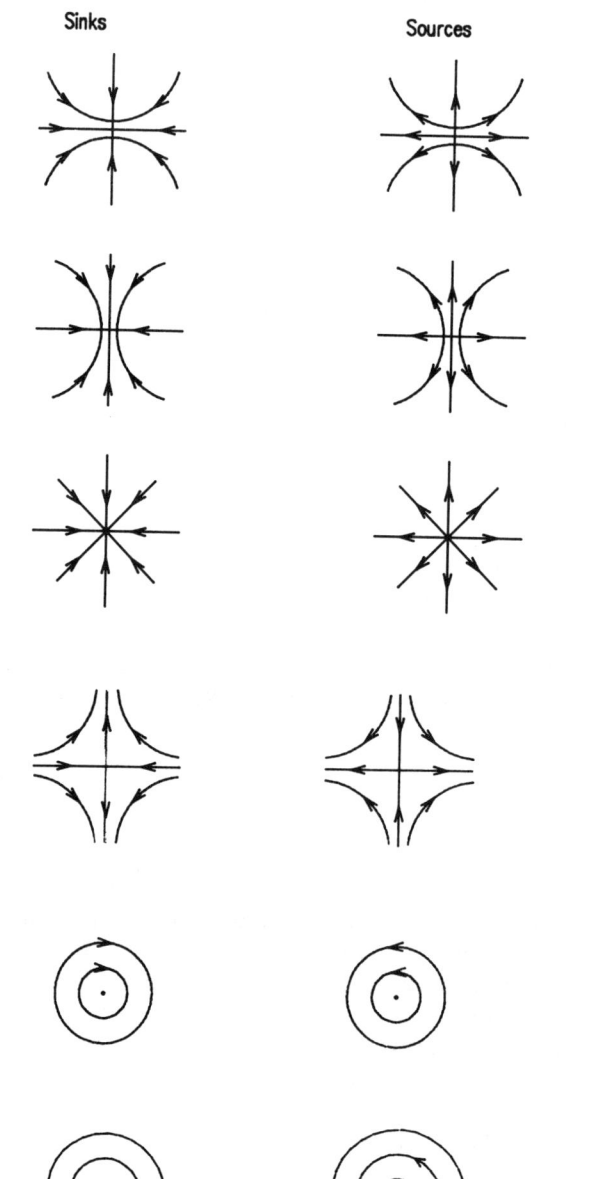

(b)

(c)

(d)

unstable   stable

NONLINEAR WAVES AND SOLITONS

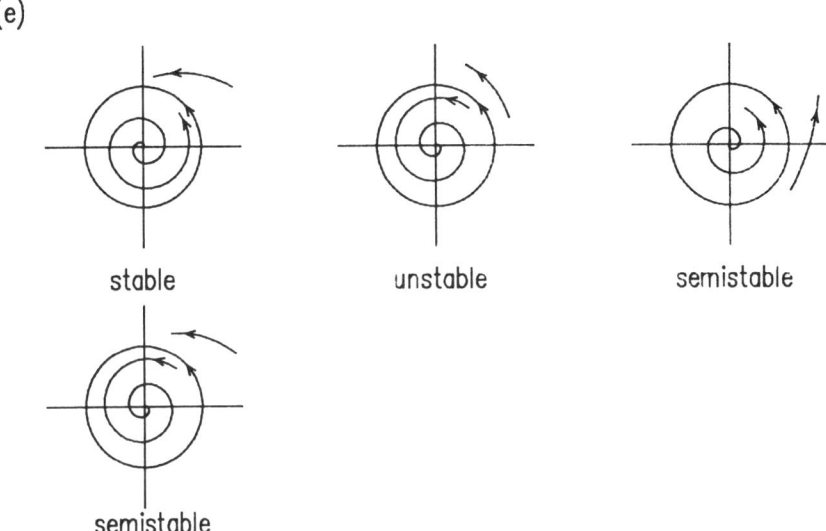

Fig. 1.24 Topologically important points in phase portraits of two-parameter systems: (a) nodes; (b) saddles; (c) centres; (d) foci; (e) limited cycles.

found and embrace phenomena such as (a) electric LC circuits, (b) chains of coupled pendula, (c) masses connected by elastic springs, (d) coupled acoustic resonators, and (e) interacting magnetic moments (see Fig. 1.26), among others.

Historically, the first such construction in terms of a numerical simulation appears to have been attempted by Fermi, Pasta, and Ulam (FPU) [19] in Los Alamos for the set of coupled nonlinear equations

$$\ddot{x}_i = (x_{i+1} + x_{i-1} - 2x_i) + \alpha\left[(x_{i+1} - x_i)^2 - (x_i - x_{i-1})^2\right], \quad (1.49)$$

for $i = 1, 2, \ldots, n-1$ with $x_0 = 0 = x_n$, which represents masses on a lattice that are anharmonically coupled through nonlinear springs. The astonishing result of the computer simulation performed on this system was that the energy initially injected into a chosen mode was not thermalized, as would be expected from the equipartition theorem, but remained concentrated in a few modes. The explanation of this puzzle came much later, following the ground-breaking discovery of what is now called a *soliton* by Zabusky and Kruskal [19–21]. These localized, stable nonlinear modes of behaviour propagate at constant velocities without a change of profile and, remarkably, preserve their shape and identity on collisions, which is schematically illustrated in Fig. 1.27. Unfortunately, discrete nonlinear

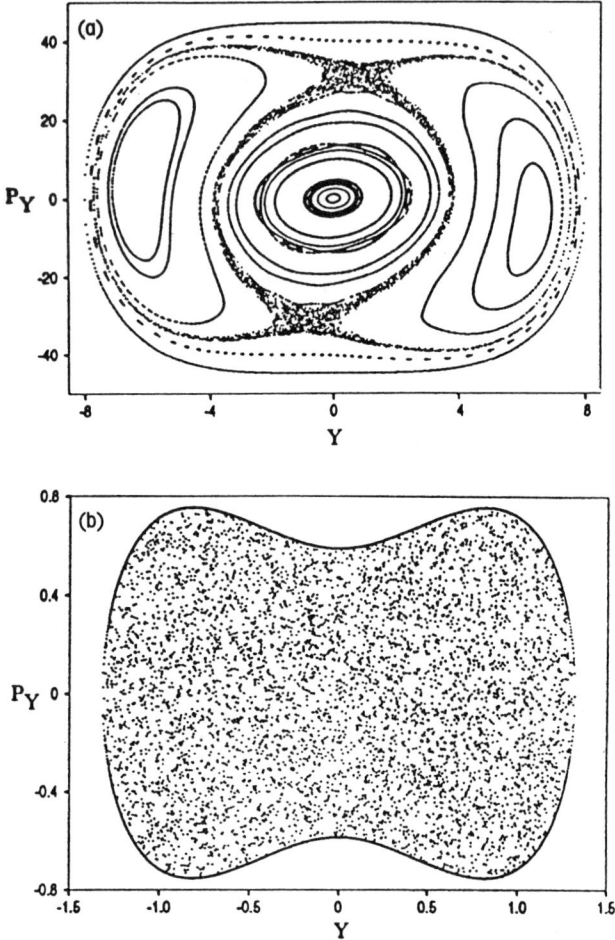

Fig. 1.25 The phase-space behaviour for two coupled anharmonic oscillators given by eqn (1.48): (a) mainly regular; (b) mainly chaotic.

systems of differential equations do not lend themselves to easy analytical treatments. In fact, only in special cases, such as the so-called Toda lattice (see Table 1.1) and the Ablowitz–Ladik equation (see Section 4.15), can exact solutions be found. Thus, a continuum approximation is usually employed in physical modelling. The emphasis, however, is on preserving the nonlinear character of the equations.

The behaviour of the solutions of nonlinear differential equations is in stark contrast to the properties of *linear wave equations* (i.e. those that

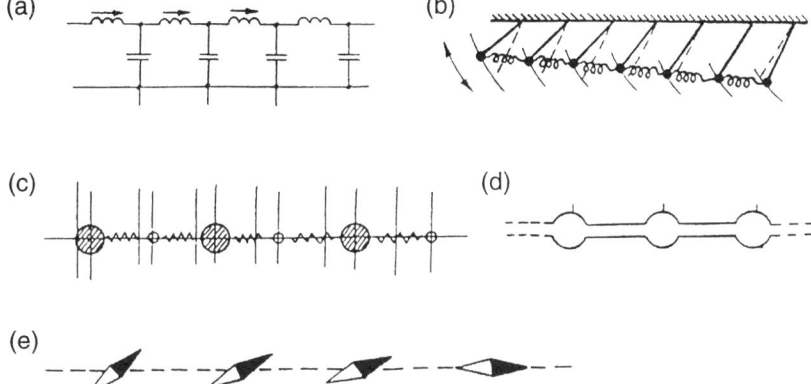

Fig. 1.26 Physical examples of coupled nonlinear oscillators, following reference [18]: (a) *LC* circuits; (b) coupled pendula; (c) masses connected by springs; (d) acoustic resonators; (e) magnetic moments.

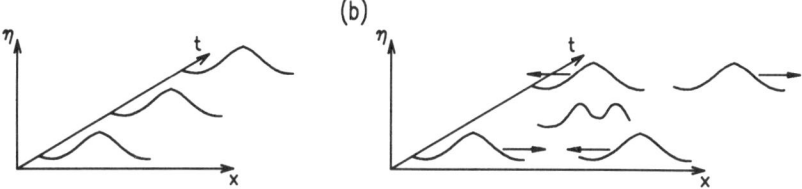

Fig. 1.27 Soliton properties in (a) free propagation and (b) collisions.

have terms proportional to $\psi_t, \psi_{tt}, \ldots, \psi_x, \psi_{xx}, \psi_{xxx}$ taken individually). In the latter case, *dispersion* due to space-dependent derivatives results in the spread of a wave packet, say

$$\psi = \int A_k e^{i(\omega t - kx)} dk, \qquad (1.50)$$

so that with the dispersion relation $\omega = \omega(k)$, the group velocity $v_g = d\omega/dk$ depends on the wave-number k (see Fig. 1.28). It is interesting to note that soliton solutions frequently correspond to singular points of linear perturbation treatments.

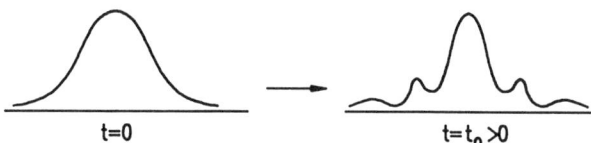

Fig. 1.28 Dispersion for wave packets in linear differential equations.

As an example, we take the linear Klein–Gordon equation,

$$\frac{\partial^2 \eta}{\partial t^2} - c_0^2 \frac{\partial^2 \eta}{\partial x^2} + m^2 \eta = 0, \tag{1.51}$$

the solutions to which are plane waves:

$$\eta(x,t) = A e^{i(\omega t \pm kx)}. \tag{1.52}$$

The associated dispersion relation yields

$$\omega(k) = \pm \sqrt{m^2 + c_0^2 k^2}, \tag{1.53}$$

whereas the group velocity is clearly k-dependent:

$$v_g(k) = \pm \frac{c_0^2 k}{\sqrt{c_0^2 k^2 + m^2}}. \tag{1.54}$$

The wave components (characterized by different values of k) in the wave packet strongly interact. This leads us to the conclusion that, for equations generating soliton-like solutions, some new concept is at work. Solitons play a fundamental role in nonlinear science, just as Fourier modes (normal modes) of propagation were at centre stage in the development of linear analysis.

In many cases, nonlinearity provides a balance to *dispersion*, and a steepening of the wave prevents it from spreading. The diagram below illustrates the behaviour of various types of wave equations [19]:

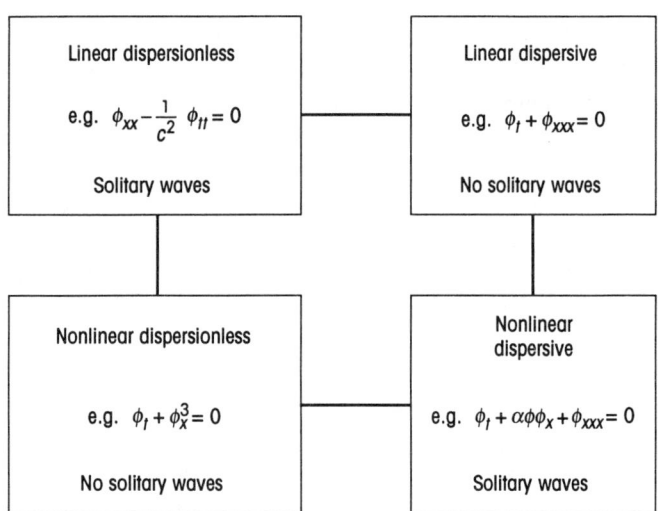

Nonlinear wave equations have four major types of nonsingular solutions: extended travelling waves, nontopological solitary waves, topological solitary waves, and breathers (shown in Fig. 1.29). An in-depth discussion regarding the methods of solving nonlinear differential equations and the properties of their solutions can be found in Chapter 7.

Below, we define those four types of waves:

1. Given an underlying wave equation, a *travelling wave*, $\phi_T(\xi)$, is a solution which depends upon x and t only through $\xi = x - vt$ with v constant. That is, the substitution $\phi = f(\xi)$ with $\xi = x - vt$ and $\partial/\partial x = d/d\xi$, $\partial/\partial t = -v(d/d\xi)$, is called a *similarity transformation*. This is one of the most useful techniques and it reduces a partial differential equation (PDE) to an ordinary differential equation (ODE), both of which are nonlinear. Physically, this means a transition to a moving reference frame (riding on the crest of a wave!).

2. A *solitary wave* $\phi_s(\xi)$ is a localized travelling wave solution of a differential equation which exhibits a transition from one constant asymptotic state at $\xi = -\infty$ to another at $\xi = +\infty$. It is essentially localized in ξ over a region the width of which is Δ.

3. A *topological solitary wave* (a kink or an antikink) connects two asymptotic and different values: $\phi(+\infty) \neq \phi(-\infty)$. They correspond to the emergence of two asymptotic equilibrium plateaux. Their topological charge can be defined as $Q = \phi(+\infty) - \phi(-\infty)$, which can be a conserved quantity.

4. A *breather* is a solution that can be represented as $\phi = f_1(t)f_2(x)$, with $f_1(t)$ being a periodic function of time and $f_2(x)$ a localized function of space. Unlike kinks, breathers require almost no energy to be activated. Hence, they can be seen as bridging the gap between highly nonlinear modes such as topological solitons and the linear (phonon) modes.

A particular type of a solitary wave is called a soliton when it asymptotically preserves its shape and velocity upon collisions with other solitary waves. There are many areas of applicability for the concept of a soliton [22], such as hydrodynamics, in which both deep- and shallow-water waves of soliton type have been observed in canals and on the open sea. In plasma physics, where interactions between electromagnetic fields and charges must be included, a very rich picture of dynamical behaviour is obtained that includes soliton formation. Since the mid-1980s a veritable revolution in fibre optics technology has been taking place, which was triggered by the discovery of a virtually lossless transmission of electromagnetic signals in the form of optical solitons. We discuss this particular

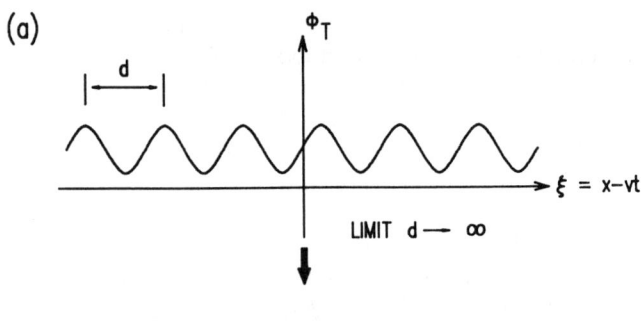

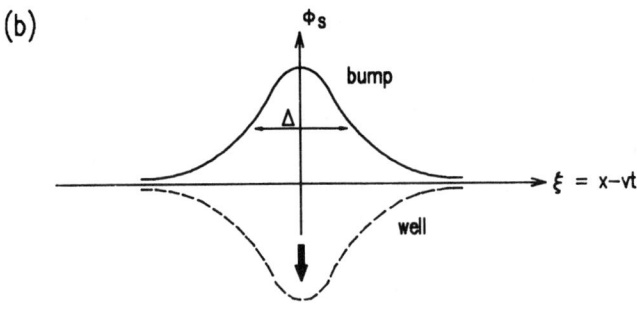

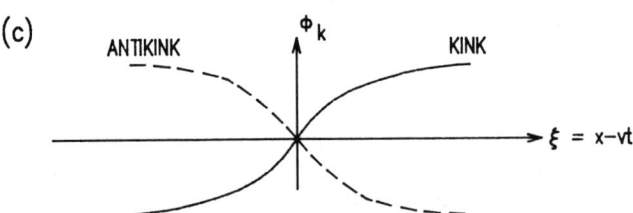

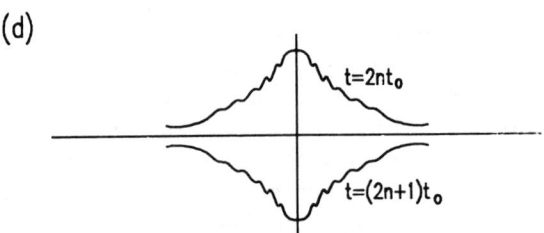

Fig. 1.29 Four types of solution to nonlinear wave equations: (a) travelling waves; (b) solitary waves (nontopological solitons); (c) solitary waves (topological solitons); and (d) a breather.

example in Chapter 4. Similar types of nonlinear modes are also present in the superconducting Josephson junction arrays. In general, the area of condensed-matter physics contains countless examples of nonlinear waves and, in this connection, we mention examples such as spin and charge density waves, nonlinear elastic lattices, domain walls in easy-axis ferromagnets and ferroelectrics, and the various types of vortices in easy-plane ferromagnets, superconductors, and superfluids. It appears that this concept is of such universal applicability that its uses span all physical dimensions, ranging from models of elementary particles (e.g. in the Higgs–Anderson or the Friedberg–Lee models) to descriptions of astrophysical objects and phenomena (e.g. solitons in the solar corona, the Red Spot of Jupiter, or spiral galaxies). In addition, important new applications are being developed that change our perceptions regarding the kinetics of chemical reactions and even life processes (such as solitons in peptides, proteins, or DNA). Obviously, this area itself is in dire need of a conceptual synthesis in the form of a separate multidisciplinary monograph. At the present time, we must content ourselves with only a cursory review of some of the most prominent soliton equations. We will return to a more mathematically rigorous analysis of soliton-type equations in Chapter 7.

1.7.1 The Korteweg–de Vries Equation (KdV)

Kruskal and Zabusky demonstrated that in the continuum limit the FPU lattice problem leads to the KdV equation, which combines weak nonlinearity with weak dispersion,

$$\eta_t + a\eta\eta_x + \eta_{xxx} = 0, \tag{1.55}$$

and the soliton solution of which is

$$\eta = \frac{3u}{a}\text{sech}^2\left[\frac{\sqrt{u}}{2}(x-ut)\right], \tag{1.56}$$

with its amplitude proportional to the velocity of propagation, u, and the width inversely proportional to $\sqrt{u}$. Its generalization in the form of the regularized long wave (RLW) equation,

$$\eta_t + \eta_x + \eta\eta_x - \eta_{xxt} = 0, \tag{1.57}$$

is not integrable; while the generalized KdV equation in the form

$$\eta_t + \alpha\eta^p\eta_x + \frac{\partial^{2r+1}}{\partial x^{2r+1}}\eta = 0, \tag{1.58}$$

with a = constant, $p, r \geq 0$ integers, is only integrable for $p = 2$ and $r = 1$ or $p = 1$ and $r = 1$.

1.7.2 The nonlinear Schrödinger (NLS) equation

The cubic nonlinear version reads

$$\eta_{xx} + i\eta_t + k|\eta|^2\eta = 0, \qquad (1.59)$$

where η is a complex field. It represents self-modulation of an almost monochromatic wave, with linear dispersion and weak nonlinearity. Its soliton solution is given by

$$\eta = \sqrt{2}\, ae^{i\phi}\, \text{sech}[a(x - bt)], \qquad (1.60a)$$

where a and b are arbitrary constants, and ϕ defines the carrier wave, with

$$\phi = \tfrac{1}{2}bx - \left(\tfrac{1}{4}b^2 - a^2\right)t, \qquad (1.60b)$$

and the sech function provides a bell-shaped envelope. It is worth noting that the underlying Lagrangian density has a familiar quartic nonlinear form in the field variable, namely

$$\mathscr{L} = \frac{i}{2}(\eta\eta_t^* - \eta^*\eta_t) + |\eta_x|^2 - \frac{k}{2}|\eta|^4, \qquad (1.61)$$

and is of great potential use in many condensed-matter applications.

A generalized nonlinear Schrödinger equation can be defined as

$$\eta_{xx} + i\eta_t + W(|\eta|)\eta = 0, \qquad (1.62)$$

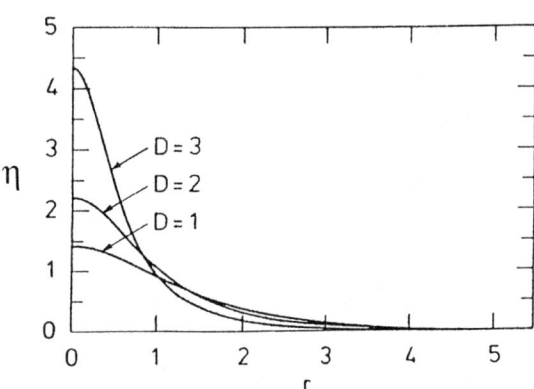

Fig. 1.30 The spatial dependence of the localized solution of a cubic NLS in one, two, and three dimensions [25].

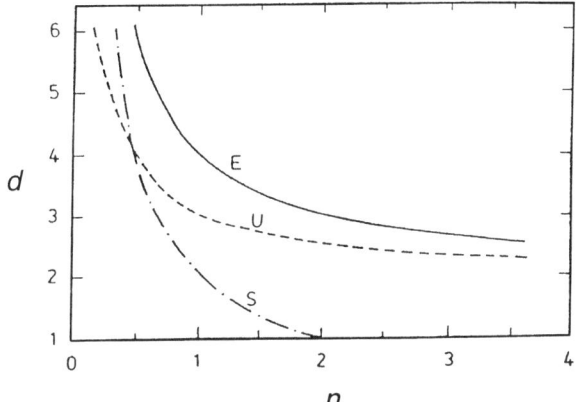

Fig. 1.31 Conditions for the existence, uniqueness, and stability of the ground-state solitary wave solution of the generalized NLS. The solitary wave exists below the curve E: it is unique below and on U for $d \leqslant 4$ and below U for $d < 4$, and is stable below S [25].

where W tends to infinity as $\eta \to \infty$ and is bounded from below everywhere. Although it frequently appears in physical applications, only its cubic version is integrable. A general result has been obtained with $W = |\eta|^{2n}$, which states that stable soliton solutions exist only if $n < 1 + 2/d$, where d is the dimensionality of the physical space [25]. In Fig. 1.30 is shown the spatial extent of a localized soliton-like solution for a cubic nonlinear Schrödinger equation in one, two and three dimensions.

Conditions of the existence, uniqueness, and stability of the ground state solitary wave solution of the generalized NLS of eqn (1.62) have been illustrated in Fig. 1.31.

1.7.3 The Sine–Gordon equation

This equation can be obtained as a continuum limit of a lattice problem with a periodic substrate potential, a linear strain, and a kinetic energy of masses on lattice sites. Its form is

$$\eta_{tt} - c_0^2 \eta_{xx} + m^2 \sin \eta = 0. \tag{1.63}$$

It possesses a wealth of exact solutions, including a kink (antikink)

$$\eta_k = 4 \arctan[\exp(\pm \gamma(\xi - v\tau))], \tag{1.64}$$

where $\gamma = (1 - v^2)^{-1/2}$, $\xi = mx/c_0$, $\tau = mt$, a kink-antikink pair,

$$\eta_{k,\bar{k}} = 4 \arctan\left[\left(\sinh \frac{v\tau}{\sqrt{1-v^2}}\right) \bigg/ \left(v \cosh \frac{\tau}{\sqrt{1-v^2}}\right)\right], \quad (1.65)$$

and a stationary breather solution,

$$\eta_B = 4 \arctan\left[\frac{\varepsilon \sin(t/\sqrt{1+\varepsilon^2})}{\cosh(\varepsilon x/\sqrt{1+\varepsilon^2})}\right], \quad (1.66a)$$

as well as a moving breather,

$$\eta_{MB} = 4 \arctan\left[\frac{\sqrt{1-\lambda^2}}{\lambda} \frac{\sin[\gamma\lambda(t-cx)]}{\cosh[\gamma\sqrt{1-\lambda^2}(x-ct)]}\right]. \quad (1.66b)$$

where ε and λ are parameters defining the breather amplitude. Also, a multikink solution has been obtained [23], and it can be viewed as a nonlinear superposition of individual kinks or antikinks moving in their reference frames at their own velocities. Except for phase shifts, their collisions preserve their shape and velocities. The corresponding Lagrangian density is

$$\mathscr{L} = \tfrac{1}{2}\eta_x^2 - \tfrac{1}{2}\eta_t^2 - \cos \eta, \quad (1.67)$$

and it has found numerous physical applications in problems with periodic potentials (e.g. in adsorption on surfaces).

Its generalization, called the double Sine-Gordon equation, i.e.

$$\eta_{tt} - \eta_{xx} = \sin \eta + \sin 2\eta, \quad (1.68)$$

is not integrable.

1.7.4 The nonlinear Klein-Gordon Equation (NLKG)

In scaled variables, this is defined by

$$\eta_{xx} - \eta_{tt} = F(\eta), \quad (1.69)$$

where η is a real field and $F(\eta) = \pm(\eta - \eta^3)$ is the most frequent choice of the nonlinearity which models bistable properties of the system. Although this equation has no soliton solutions (is not completely integrable), it has two types of solitary waves, kinks (and antikinks) given by

$$\eta = \pm \tanh\left(\frac{1}{\sqrt{2}} \frac{x-vt}{\sqrt{1-v^2}}\right), \quad (1.70)$$

and bumps (and wells) given by

$$\eta = \pm \operatorname{sech}\left(\frac{1}{\sqrt{2}} \frac{x-vt}{\sqrt{1-v^2}}\right). \quad (1.71)$$

These solutions can bounce off each other, lock and annihilate each other on collision or even emit some oscillatory disturbances ('radiation') as a result of collisions [24]. The corresponding Lagrangian density is

$$\mathscr{L} = -\tfrac{1}{2}\eta_x^2 + \tfrac{1}{2}\eta_t^2 \pm \left(\tfrac{1}{2}\eta^2 - \tfrac{1}{4}\eta^4\right), \qquad (1.72)$$

and, as we shall see later, is of great importance in bifurcating dynamical systems, such as those undergoing phase transitions (with a conservative order parameter η).

The total energy of a solution $\eta(x, t)$ is

$$E = \int_{-\infty}^{\infty} dx \left[\tfrac{1}{2}\eta_t^2 + \tfrac{1}{2}\eta_x^2 + U(\eta)\right], \qquad (1.73)$$

where $U(\eta) = \int F(\eta) d\eta$ is the generalized nonlinear potential used. In field theory applications E is referred to as the 'mass' and it is easy to see that only localized solutions have finite mass, since periodic solutions (such as elliptic waves) lead to a total energy divergence. Note that travelling wave solutions with $\xi = x - vt - x_0$ are characterized by relativistic energy invariance,

$$E_v = E_0(1 - v^2/c^2)^{-1/2}, \qquad (1.74)$$

where E_0 is the rest energy and c is the maximum propagation velocity (corresponding to the speed of light). Similarly, the momentum variable $P = \partial \eta/\partial t$ scales according to

$$P = Mv(1 - v^2/c^2)^{-1/2}, \qquad (1.75)$$

where M is the relativistic 'rest' mass, given by

$$M = (\sqrt{2}/L) \int_{\eta_1}^{\eta_2} d\eta \sqrt{U(\eta)}, \qquad (1.76)$$

L is the sample's length, and η_1 and η_2 are the classical turning points for the given solution. In terms of M, the rest energy is, of course,

$$E_0 = Mc^2. \qquad (1.77)$$

A general solution travelling at a velocity v can always be found by integrating formally, to give

$$\frac{\sqrt{2}}{L\sqrt{1-v^2}}(\xi - \xi_0) = \pm \int_{\eta(\xi_0)}^{\eta(\xi)} \frac{d\eta}{\sqrt{\int F(\eta) d\eta + c_0}}, \qquad (1.78)$$

where ξ_0 and c_0 are integration constants and $\xi = x - vt$.

In this connection, it should be said that the question of stability in more than one spatial dimension becomes of paramount importance. This has been somewhat clarified by the following theorem.

DERRICK'S THEOREM. *Let η be a set of scalar fields in $(D+1)$-dimensional space–time. Let the dynamics be described by the Lagrangian*

$$\mathscr{L} = \tfrac{1}{2} \partial_\mu \eta \partial^\mu \eta - U(\eta), \qquad (1.79)$$

with $U \geq 0$ and $U = 0$ for the ground state only. For $D \geq 2$, the only nonsingular time-independent solutions of finite energy are the ground states.

Note also that if $U(\eta)$ has n discrete minima we can have $2(n-1)$ types of (solitary) wave solutions connecting any two neighbouring minima. The kink solution of the ϕ^4-model is singular as the nonlinearity parameter goes to zero. Thus, it cannot be obtained by perturbation-expansions from a linear equation. When $U(\eta)$ has a single minimum, no localized static solutions other that $\eta(x) = 0$ exist.

In general, a classical soliton state $\eta(x)$ corresponds to the degenerate broken-symmetry vacuum state of the system which is macroscopically occupied, i.e. it forms a condensate. The translation mode $d\eta/dx$ describes tunnelling (e.g. instanton-type behaviour) between the various members of the ground state manifold. We shall come back to a field-theoretic discussion of these solutions in Chapter 4.

We have gathered together a number of exactly integrable equations in Table 1.1.

The concept of the soliton has been generalized to higher-dimensional equations. The inverse scattering method has been extended to solve $(2+1)$-dimensional (i.e. two spatial and one temporal dimensions) equations such as the Kadomtsev–Petviashvili [26–28] and Davey–Stewartson (DS) equations [24]. However, whereas the inverse scattering formalism has been developed in $(n+1)$-dimensional space [29], at present there are no known nonlinear PDEs that are solvable by these techniques: see the book by Ablowitz and Clarkson [30] for a detailed discussion of this topic. The KP equation possesses line-solitons, which do not decay as $(x^2 + y^2)^{1/2} \to \infty$ in all directions, and 'lump' solutions, which decay algebraically as $(x^2 + y^2)^{1/2} \to \infty$ (see reference [30]). The DS equations possess several exact solutions with soliton-like behaviour; in particular, 'dromions' which decay exponentially in all directions as $(x^2 + y^2)^{1/2} \to \infty$ [31, 32].

In a different direction, exact solutions of higher-dimensional analogues of soliton equations in $(1+1)$ dimensions have been derived. These are usually derived from the soliton solution of the $(1+1)$-dimensional PDE and are often expressible in terms of special functions [33]. We remark that whereas such solutions of higher-dimensional equations are sometimes called 'solitons', strictly speaking they are not solitons, unless the PDE is completely integrable.

Table 1.1 Some of the important exactly integrable nonlinear equations and systems of equations. The dagger (†) indicates partial integrability.

No.	Name of equation	Form of equation
1	KdV	$\eta_t + a\eta\eta_x + \eta_{xxx} = 0$
	Modified KdV	$\eta_t - 6\eta^2\eta_x + \eta_{xxx} = 0$
2	NLS	$i\eta_t + \eta_{xx} + k|\eta|^2\eta = 0$
	Derivative NLS	$i\eta_t + \eta_{xx} \pm 2i(|\eta|^2\eta)_x = 0$
3	Sine-Gordon	$\eta_{tt} - \eta_{xx} + \sin\eta = 0$
4	NLKG†	$\eta_{xx} - \eta_{tt} = F(\eta)$
5	Burgers equation	$\eta_t + 6\eta\eta_x - b\eta_{xx} = 0 \quad (b > 0)$
6	Hirota equation	$i\eta_t + 3i\alpha|\eta|^2\eta_x + \rho\eta_{xx} + i\sigma\eta_{xxx} + \delta|\eta|^2\eta = 0$
7	Boussinesq equation	$\eta_{xx} - \eta_{tt} + 6(\eta^2)_{xx} + \eta_{xxxx} = 0$
8	Born-Infeld equation	$\eta_{xx}(1 - \eta_t^2) + 2\eta_x\eta_t\eta_{xt} - (1 + \eta_x^2)\eta_{tt} = 0$
9	Landau-Lifshitz equation	$S_t = S \wedge S_{xx} + S \wedge JS,$ where $J = \text{diag}(J_1, J_2, J_3)$ and $S = (S_1, S_2, S_3)$
10	Kadomtsev-Petviashvili (KP)	$\tfrac{3}{4}\beta^2\eta_{yy} + \{\alpha\eta_t + \lambda\eta_x + \tfrac{1}{4}(\eta_{xxx} + 6\eta\eta_x)\}_x = 0$
11	Benjamin-Ono	$\eta_t + (c_0 + A\eta)\eta_x + \dfrac{\gamma c_0}{\pi} P \displaystyle\int_{-\infty}^{\infty} \dfrac{\eta_{\xi\xi}}{\xi - x} d\xi = 0$
12	Toda lattice	$m\dfrac{d^2 r_n}{dt^2} = a[e^{-br_n} - e^{-br_{n+1}}],$ where $r_n = \eta_n - \eta_{n+1}$
13	Reduced Maxwell-Bloch equation	$E_{xx} - E_{tt} = -\alpha P_t$ $P_{tt} + \mu^2 P = (EN)_t$ $N_t = -EP$
14	Three-wave interaction system	$u_{1,x} + c_1 u_{1,t} = iqu_2 u_3^*$ $u_{2,x} + c_2 u_{2,t} = iqu_1 u_3 \qquad q = \text{constant}$ $u_{3,x} + c_3 u_{3,t} = iqu_1^* u_2$
15	Coupled Schrödinger equation	$iu_t + u_{xx} + 2(|u|^2 \pm |v|^2)u = 0$ $iv_t \pm v_{xx} + 2(|v|^2 \pm |u|^2)v = 0$
16	Self-induced transparency equation	$E_t + E_x = \displaystyle\int_{-\infty}^{\infty} P(x, t, \alpha) g(\alpha) d\alpha$ $P_t + 2i\alpha P = E\eta$ $\eta_t = -\tfrac{1}{2}(E_p^* + E_p)$

On the question of how widespread soliton equations are in the description of real physical systems, it is very pertinent to quote from some of the most active practitioners in the field, i.e. Newell and Moloney [39]:

Is it a fluke that many of the equations of mathematical physics like the NLS equation or the Korteweg–de Vries equation or the Sine–Gordon equation, which are derived by standard perturbation analyses as the universal asymptotic description of a wide variety of physical systems, are integrable or close to being integrable? We don't know the answer to this, but the following comment may be relevant. A key observation is that if one starts with an exactly integrable system, then the asymptotic analysis leading to the equation that describes the long-time behaviour of the envelopes of special types of solutions does not destroy this integrable character; rather, the integrability is preserved. Therefore if among the set of all equations that reduce to the same asymptotic description there is one equation that is integrable, then the asymptotic equation is integrable. Therefore whereas integrability is rare in general, the process of reduction to universal, asymptotic equations for wave envelopes increases the probability that the resulting equation has special properties. The reduction process introduces new symmetries and new constraints (conservation laws) and does not destroy existing ones.

A final comment that we wish to make is in regard to the origin of *dissipation* terms (such as η_t, η_t^2, etc.) in nonlinear wave equations. Although a transition from a discrete lattice to a continuum medium has been done repeatedly in the past as a routine procedure by making the following approximation,

$$\eta_{i+1} - 2\eta_i + \eta_{i-1} \simeq a^2 \frac{\partial^2}{\partial x^2} \eta \qquad (1.80)$$

(by Taylor expansion), it is far from trivial and many possible numerical problems may be created through it. An extremely interesting paper [34] recently demonstrated that dissipative terms of the form $\gamma_1 \eta_t$ and $\gamma_2 \eta_t^2$ also arise in the process of taking the continuum limit and their role should be carefully scrutinized. Of course, another source of energy dissipation is the presence of degrees of freedom other than η. As an example, we show the effect of dissipation and forcing on one of the integrable wave equations, i.e. the nonlinear Schrödinger equation, where

$$i\eta_t + \eta_{xx} + 2\eta|\eta|^2 = -i\gamma\eta - iae^{i\omega t}. \qquad (1.81)$$

Note that since η is complex, a term of the form $i\eta$ present above provides an effective dissipation of the envelope. Numerical studies [35–38] have shown that the interplay between nonlinearity, dissipation, and forcing leads to a typical bifurcation diagram with a period doubling cascade,

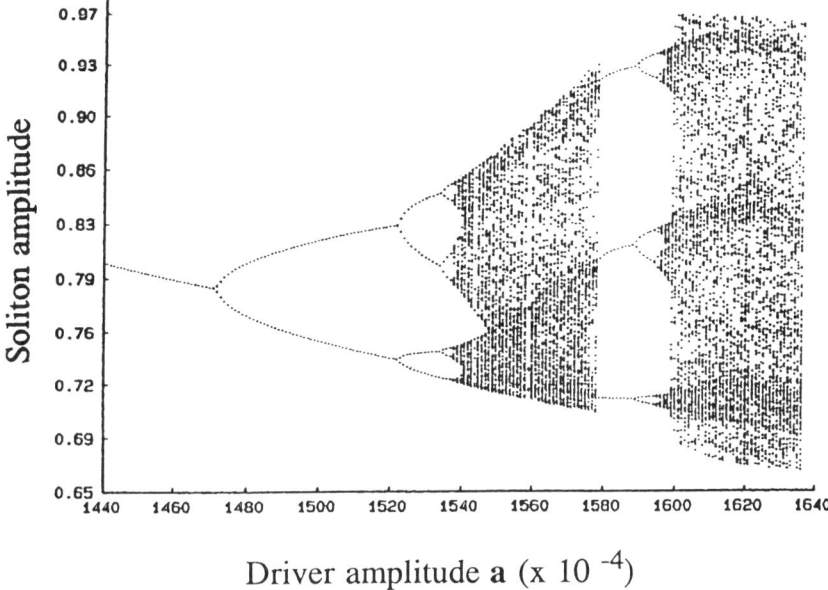

Fig. 1.32 The bifurcation diagram for eqn (1.81) with $\gamma = 0.11$ after the first instability (Hopf bifurcation) (following [35]).

resulting in chaos (see Fig. 1.32), as we have already seen for ordinary differential equations such as the Duffing oscillator. An analogous form of behaviour has been seen for the damped-driven Sine–Gordon equation:

$$\eta_{xx} - \eta_{tt} + \sin \eta = -\gamma \eta_t + \Gamma \sin \Omega t. \tag{1.82}$$

However, for strictly defined multidimensional solitons, questions of stability arise with the associated mathematical complications. We will, instead, be concerned with *coherent structures*, by which we will understand spatially localized, long-lived, stable nonlinear excitations (examples of which include vortices, solitons, and solitary waves). Singular solutions can be thought of as transient coherent structures. Some of these aspects will be discussed from a physical point of view in Chapter 3.

In summary, soliton structures which not only form inherent modes of behaviour of extended systems but also break their symmetry provide a common basis for further perturbation theories, quantization techniques, critical phenomena, and statistical mechanics of numerous nonlinear systems [22].

Obviously, only a small number of PDEs are integrable and thus support solitons. In particular, we do not find soliton solutions in reaction–diffusion-type equations (also referred to in the kinetics of phase transitions as the Landau–Ginzburg equations), i.e. equations of the type

$$\eta_t = \eta_{xx} + F(\eta). \tag{1.83}$$

What is interesting, however, is the existence of travelling wavefronts $\eta = \eta(x - vt)$ with velocity v which connect stable asymptotic states η_0 and η_1 such that $F(\eta_0) = F(\eta_1) = 0$. Three important cases that have been

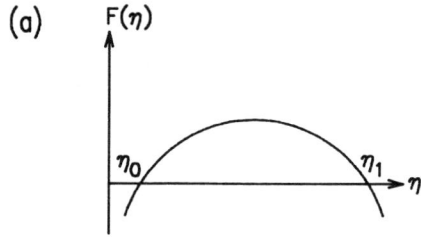

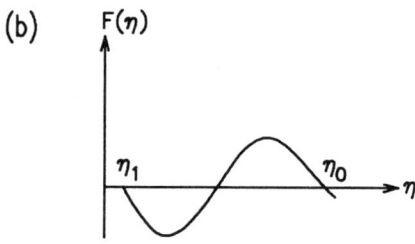

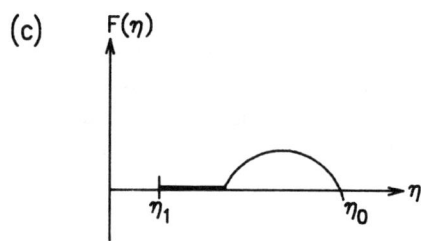

Fig. 1.33 Some generalized nonlinear forces in reaction–diffusion systems: (a) Fisher case; (b) bistable case; (c) ignition case.

thoroughly investigated [40], i.e. the Fisher case, the bistable case, and the ignition case, are illustrated in Fig. 1.33.

It turns out that in the Fisher case (see Fig. 1.33 for explanation), there exists a critical velocity $v^* > 0$ such that for every $v > v^*$ a front exists connecting η_1 and η_0. In the bistable case, however, with $F'(\eta_0) < 0$ and $F'(\eta_1) < 0$, a *unique* wavefront connects η_0 and η_1. Hence, this can be seen as a possible mechanism of velocity selection. In the ignition case a unique wavefront connects the two asymptotically stable equilibria.

In addition to the soliton formation phenomenon discussed above, nonlinear differential equations—and, in particular, systems of equations —have been recently investigated in regard to another fascinating property that originates in nonlinearity. This new type of phenomenon is called pattern formation.

1.8 Pattern formation

In this section we wish briefly to discuss selected physical systems and phenomena that exhibit pattern formation and appear to be describable in terms of the theoretical models—for example, the Landau–Ginzburg model—that will be discussed later in the book.

1.8.1 Fluid dynamics

Numerous types of fluid flow lead to the emergence of pattern formation. For example, convective flows in binary fluid mixtures exhibit interesting transitions of flow patterns, from rolls of linear oscillatory convection that are subject to bending, compression, and pinching to cellular localized structures [41]. Forced, unbounded shear flows possess [42] vortices, elliptic jets, and circular jets, and the flows can be either of periodic or phase-dislocation type. Rayleigh–Bénard convection cells are characterized by transition from linear diffusive convection to convection roll patterns, and they also develop chaotic behaviour and eventually full turbulence [43]. Here, a fluid is placed between horizontal plates and heated from below. When the temperature difference ΔT exceeds a critical value ΔT_c, the heat can no longer be carried up by conduction alone and the fluid is set into motion, with flow in the form of convective rolls the characteristic spacing of which is of order d, the plate separation (see Fig. 1.34).

Another example of pattern formation in fluids is the case of Taylor–Couette flows. Here, a fluid is placed between concentric cylinders and the inner cylinder is rotated. When the angular frequency Ω exceeds a

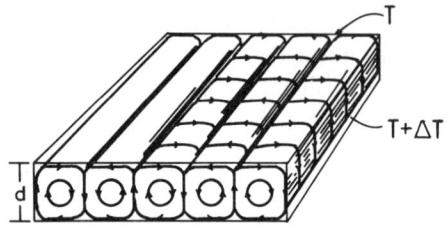

Fig. 1.34 A schematic diagram illustrating Rayleigh–Bénard convection.

critical value Ω_c, the flow is no longer purely azimuthal. An instability occurs and a pattern of Taylor vortices with an axial component of flow arises with a characteristic separation of order d, the distance between cylinders (see Fig. 1.35). In Taylor–Couette flows, it has been recently demonstrated [44] that turbulence may also develop in the form of a spiral band rotating about the axis of symmetry with a constant velocity. It is also noteworthy that it is quite common to describe qualitative features of these phenomena using Landau phenomenology, through so-called Landau–Stuart expansions for the complex amplitude function $A(x,t)$, which satisfies the evolution equation

$$\partial A/\partial t = D\nabla^2 A + \boldsymbol{v}\cdot\nabla A + f(A), \qquad (1.84)$$

where D is the diffusion constant, the second term is the entrainment field, and f is a nonlinear function of A. This then corresponds to the

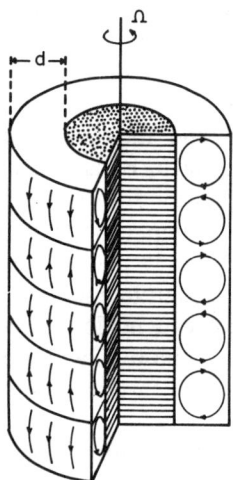

Fig. 1.35 A schematic diagram illustrating Taylor–Couette flow.

PATTERN FORMATION 45

Time-Dependent Landau–Ginzburg (TDLG) equations when a transformation to a moving frame of reference is made. The control parameter is the Reynolds number, and it is believed critically to affect the linear term's coefficient in $f(A)$.

1.8.2 Liquid crystals

It is well known that cholesteric liquid crystals have their symmetry axes shifted by a constant angle on going from plane to plane. This then describes a helical pattern in the system. Perhaps more interestingly, recent experiments on nematic liquid crystals [45] subjected to a variable

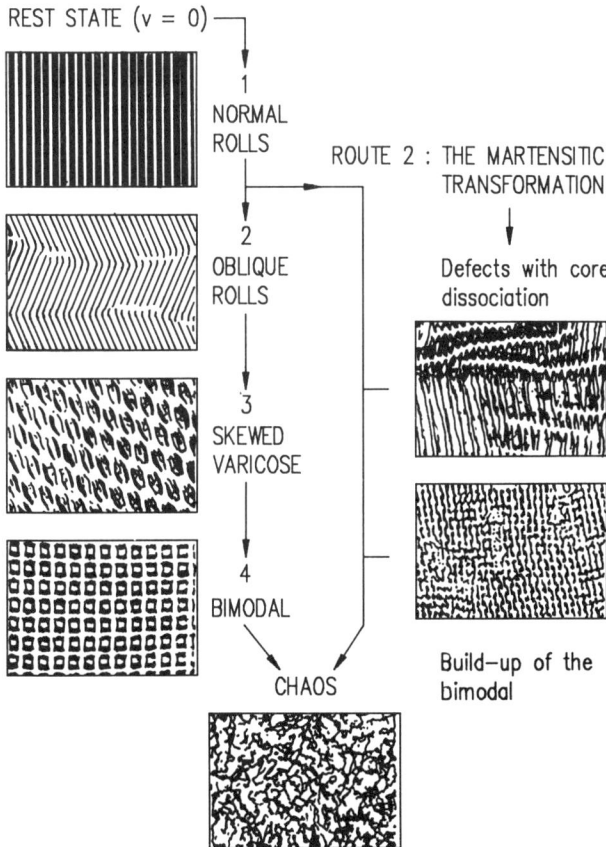

Fig. 1.36 Transitions to the fully chaotic state. Route 1 (left side) occurs when the constraint rate is small. Route 2 below occurs when the constraint is applied suddenly [45].

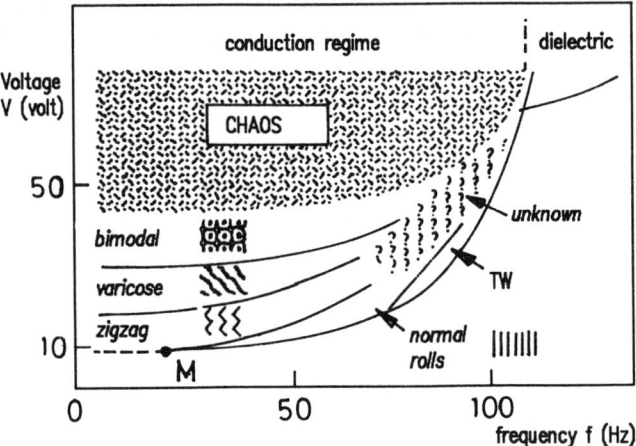

Fig. 1.37 Bifurcation lines in voltage–frequency space. As V increases slowly, space–time chaos is reached through a series of bifurcations ($d = 50$ mm) [45].

potential difference have revealed a series of fascinating pattern transformations with an increasing voltage. The sample's patterns change from the rest state to normal rolls (which could then develop defects in the form of dislocation edges), to undulated rolls, and to oblique rolls, which then produce period pinching ('sheared varicose') and eventually a rectangular cell lattice is shown that disappears at high voltage, leading to chaos. Depending on the way in which the field was applied (suddenly or gradually), these transitions were, respectively, accompanied by defects or were free of them (see Fig. 1.36). The two routes differed in reversibility properties. Joets and Ribotta [45] proposed the so-called Newell–Landau–Ginzburg equations [43] to describe the dynamics of the system (characterized by two components of the order parameter, A and B). These are identical to eqn (1.84) with an added mutual coupling and allowing the coefficients in $f(A)$ and D to be complex numbers. A summary of the relevant experimental results is given in the phase diagram of Fig. 1.37 [45].

1.8.3 Polymers

Among the many possible transformations occurring in polymer systems, two deserve special attention. First, fluctuation-induced first-order phase transitions have been found [46] between a one-dimensional (lamellar) periodic structure and a disordered phase. The transition is accompanied by microphase separation, occurring with a uniform speed.

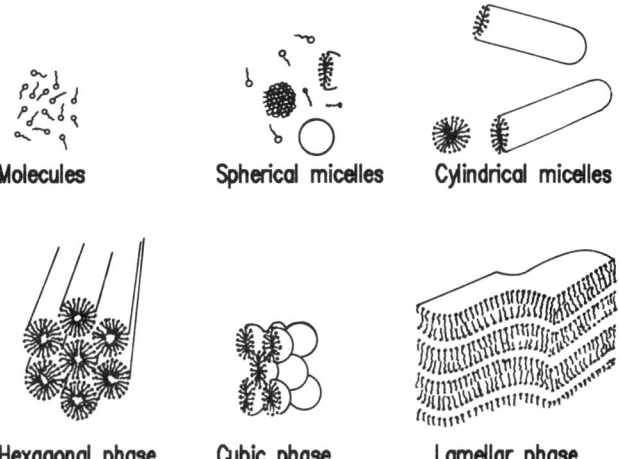

Fig. 1.38 Schematic drawings of the idealized structures of surfactant molecules that can form in solution as the surfactant concentration is increased (after reference [48]).

The second case is a crumpling transition that occurs at a finite temperature and is characterized by a change from a stretched surface to a crumpled tethered one [47]. Here, critical fluctuations lead to a weakly first-order transition. The order parameter is given by the coarse-grained vectors tangent to the surface. The various observed phases that arise in solutions of surfactant molecules [48] are illustrated in Fig. 1.38. In both of these cases a phenomenological Landau–Ginzburg Hamiltonian in terms of the order parameter is postulated.

1.8.4 Crystal growth and structural transitions

The dynamics of crystal growth can be described starting from an Ising-like Hamiltonian (see Chapter 2), given by

$$H = -\sum_{(ij)} J_{ij} s_i s_j - \sum_i \Delta\mu \cdot s_i + V(\{s\}), \quad (1.85)$$

where $s_i = 2c_i - 1 = \pm 1$ is a generalized spin and c_i is the concentration variable for the ith lattice site, such that $c_i = +1$ means that the site is filled (solid) while $c_i = 0$ means that it is empty (vapour). $\Delta\mu$ is the chemical potential, and $V(\{s\})$ is an effective on-site potential. The dynamics are then described using an Onsager relation,

$$\frac{\partial \phi}{\partial t} = -\frac{D}{KT}\frac{\delta F}{\delta \phi}, \quad (1.86)$$

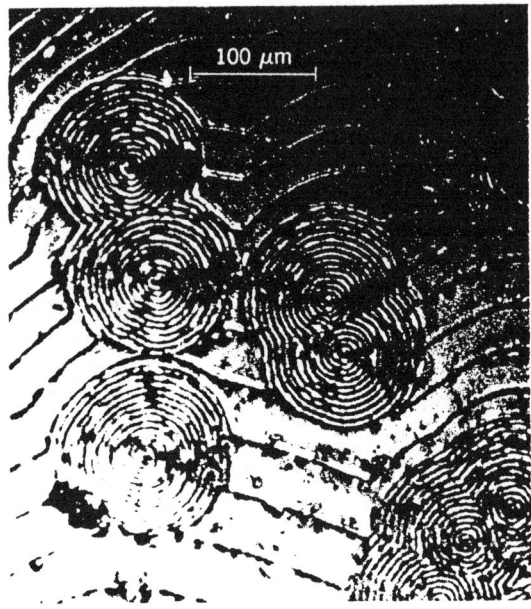

Fig. 1.39 Crystal-growth spirals on the surface of a silicon carbide crystal. Each spiral step originates in a screw dislocation defect (after reference [49]).

where ϕ is the trajectory normal to the surface (interface of an anisotropic medium) playing the role of a coarse-grained order parameter. In the continuum limit the relevant Hamiltonian becomes of the Landau–Ginzburg type and the Onsager equation takes the form of a TDLG equation again. Not surprisingly, typical crystal growth trajectories are spirals (see Fig. 1.39).

Once a crystal lattice is formed, it may undergo various changes and transitions. A typical model Hamiltonian for commensurate-incommensurate transitions is the Frank–van der Merve Hamiltonian [50], which predicts the formation of discommensuration lines and their interactions. The resultant patterns may change from striped to hexagonal, and lattice melting is also possible. The continuum limit of this lattice Hamiltonian leads to the Sine–Gordon equation.

1.8.5 Chemical instabilities

Reaction–diffusion systems of two or more chemically active species are described using coupled TDLG-type equations. The results of computer simulations were confirmed experimentally and often lead to a variety of

PATTERN FORMATION 49

Fig. 1.40 An example of the temporal and spatial development of a chemical reaction (after reference [51]).

patterns, such as those of spiral type in the Belousov–Zhabotinskii system (see Fig. 1.40).

1.8.6 Magnetically ordered systems

The final set of examples of pattern formation in physical systems is concerned with magnetically ordered substances. Various forms of spatial patterns in magnets have been known for a long time, and include such phases as ferromagnetism, antiferromagnetism (collinear, declinational, and canted), and ferrimagnetism, as well as various types of helical order [52, 53] (see Chapter 2 for more details). The basic form of the Hamiltonian used is that of Heisenberg type with nearest- and next-nearest-neighbour interaction terms (and possibly molecular fields due to long-range order as well). This is known to map, in the continuum limit, on the Landau–Ginzberg model and such patterns may once again be traced to various types of exact solution. A recent experimental study [54] has demonstrated the possibility of including metastable spiral domain structures in epitaxially grown single-crystal garnet ferrite films (see Fig. 1.41).

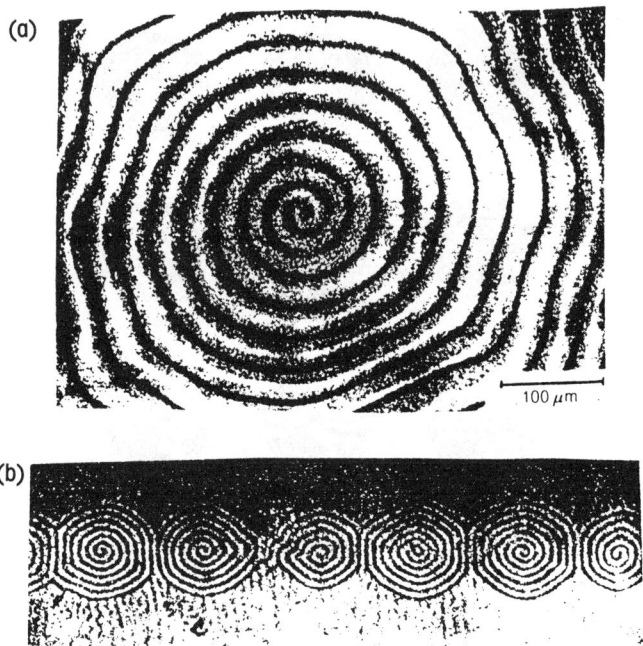

Fig. 1.41 The spiral domain (a) and chain of domains (b) in thin ferromagnetic films (after reference [54]).

1.8.7 Pattern formation in premixed flames

Beautiful experiments [55] on the spatial patterns of premixed flame fronts show a remarkably rich structure of stable nonequilibrium phases, including cells, spirals, pinwheel types, etc. (see Fig. 1.42).

The equation used to model this type of behaviour is the Kuramoto–Shivashinsky equation:

$$\frac{\partial \eta}{\partial t} + \frac{\partial^2 \eta}{\partial x^2} + \frac{\partial^4 \eta}{\partial x^4} + \left(\frac{\partial \eta}{\partial x}\right)^2 = 0. \qquad (1.87)$$

The complexity and diversity of the patterns obtained in the above experiments can only be fully appreciated by viewing a real-time motion picture of the moving flame fronts. This area still awaits a fully satisfactory theoretical account of the phenomena.

1.8.8 Optical instabilities

It has been experimentally demonstrated [56] that counterpropagating

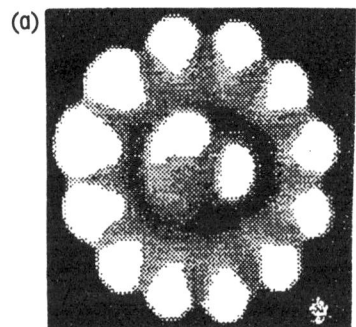 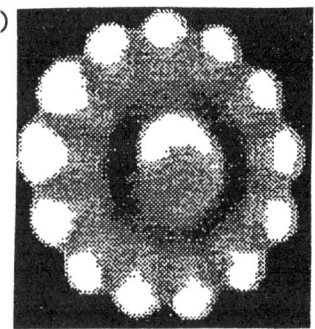

Fig. 1.42 A schematic illustration of some stable phases of the flame front in gas dynamics (after reference [55]) showing (a) a 'triangle dance' and (b) a spiral.

laser beam instabilities in a nonlinear Kerr medium produce both radially symmetric and hexagonal intensity patterns. Simultaneously, analytical efforts [57] have succeeded in modelling such patterns using the higher order nonlinear Schrödinger equation of the form

$$i\psi_t + \nabla^2\psi = a_1\psi|\psi|^2 + a_2\psi|\psi|^4 + ia_3(\psi|\psi|^2)_x + (a_4 + ia_5)\psi(|\psi|^2)_x, \tag{1.88}$$

where a_i ($i = 1,\ldots,5$) are constants and $|\psi|^2$ is the local light intensity (Fig. 1.43). See Section 4.15 for more information on other fascinating nonlinear optical phenomena.

In general, the dynamics of periodic patterns with a particular wavenumber q_0 are analysed by investigating their growth rates λ_0, such that [58]

$$\eta(x,t) = \eta_0 e^{iq_0 x + \lambda_0 t}. \tag{1.89}$$

Then, the real part of λ_0 is the growth rate, and it can be calculated by substituting eqn (1.89) back into the governing equation of motion. The typical behaviour seen is shown in Fig. 1.44. The growth rate Re λ_0 for small disturbances of a reference state with wave vector q is plotted versus q for three different types of linear instabilities. In case (a) the reference state is stable for all q where $\varepsilon = (R - R_c)/R_c < 0$, and R is the control parameter. When the latter reaches its critical value R_c, i.e. when $\varepsilon = 0$, the growth rate is zero at $q = q_0$ and a finite wavelength disturbance becomes marginal. For $\varepsilon > 0$ there is a continuous band of wave vectors for which the reference state is unstable (Re $\lambda > 0$). In case (b) the growth rate vanishes at $q = 0$ for all ε, and the initial instability band has a width of order $\delta q \sim \epsilon^{1/2}$ for $\varepsilon > 0$. In case (c) the maximum growth rate is

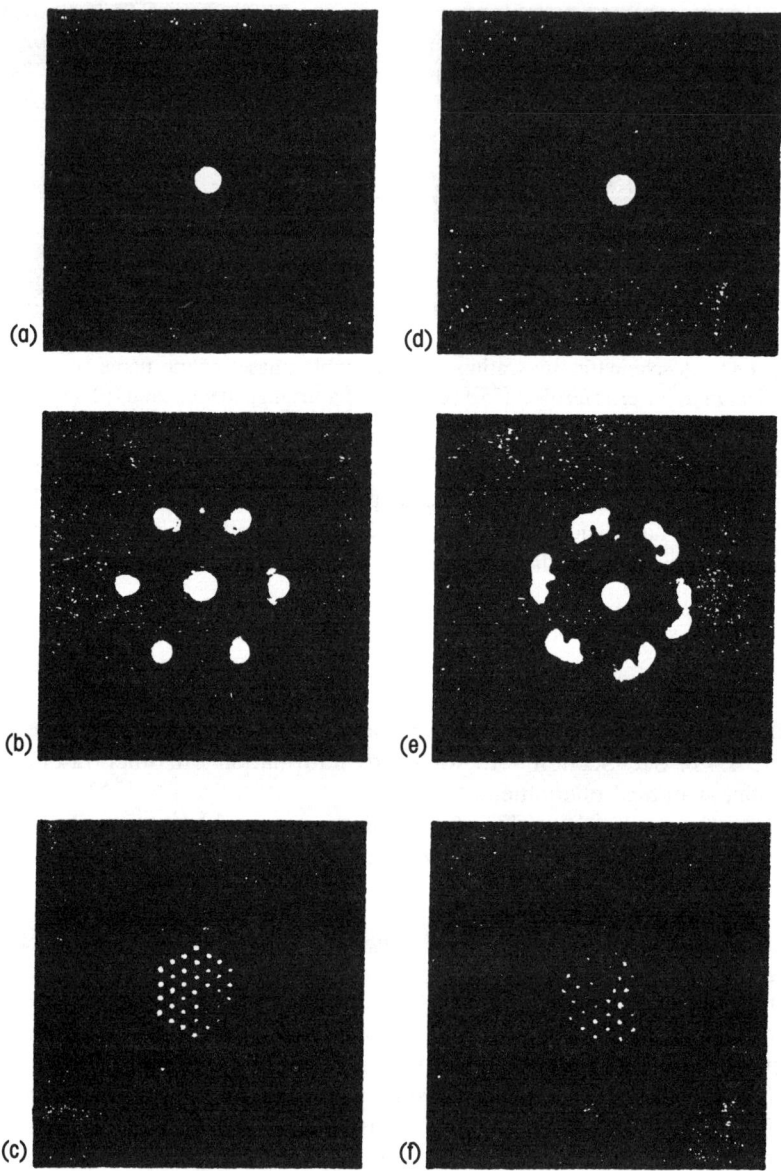

Fig. 1.43 Pattern formation in laser light.

always at $q = 0$ and no intrinsic length scale is singled out by the linear dynamics.

Various types of nonlinear partial differential equations have been

PATTERN FORMATION 53

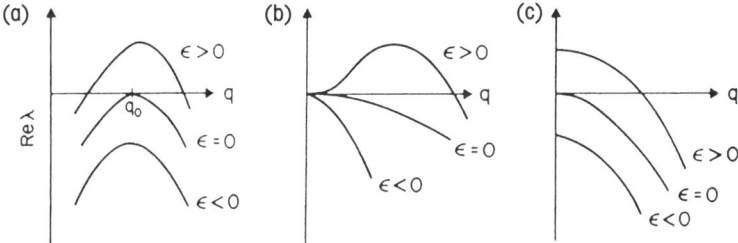

Fig. 1.44 The linear dispersion relation for pattern-forming instabilities.

employed in the studies of pattern forming systems but, due to lack of space, we can only provide a partial list of the most important ones [18]:

1. The generalized Ginsburg–Landau equation:

$$\eta_t = \mu\eta - (1 + i\beta)|\eta|^2\eta + K(1 - ic)\nabla^2\eta, \quad (1.90)$$

which leads to the emergence of simplest spatial patterns near the threshold of instability (parallel rolls, square, or hexagonal cells). In addition, global or local defects may form and grow.

2. The Swift–Hohenberg equation:

$$\eta_t = \left[\mu - (q^2 + \nabla^2)^2\right]\eta + \beta\eta^2 - \eta^3. \quad (1.91)$$

Here, μ is sometimes taken to be a function of position, which leads to the pinning of defect structures (e.g. polygons). It corresponds to the free energy functional of the type; see Fig. 1.45

$$F = \int \left\{-\frac{\mu}{2}\eta^2 + \frac{1}{4}\eta^4 + \frac{1}{2}[(q^2 + \nabla^2)\eta]^2\right\} dx\,dy. \quad (1.92)$$

We have seen here that many of the sophisticated models of condensed-matter dynamics that have been developed, e.g. the Frenkel–Kontorova model, exhibit space–time complexity. There, a nontrivial relationship can be traced between pattern formation, low-dimensional chaos, and remnants of coherent structures. Depending on the strength of the coupling parameter, various intermediate phases arise. Their average separation in terms of control parameter values often exhibits a so-called "devil's staircase', where transition regions between single-phase plateaux occur at incommensurate values. All of the associated phase transitions are of first order [59].

We have seen in perturbed dynamical systems, as well as in pattern-forming systems, that an often-encountered scenario is a period

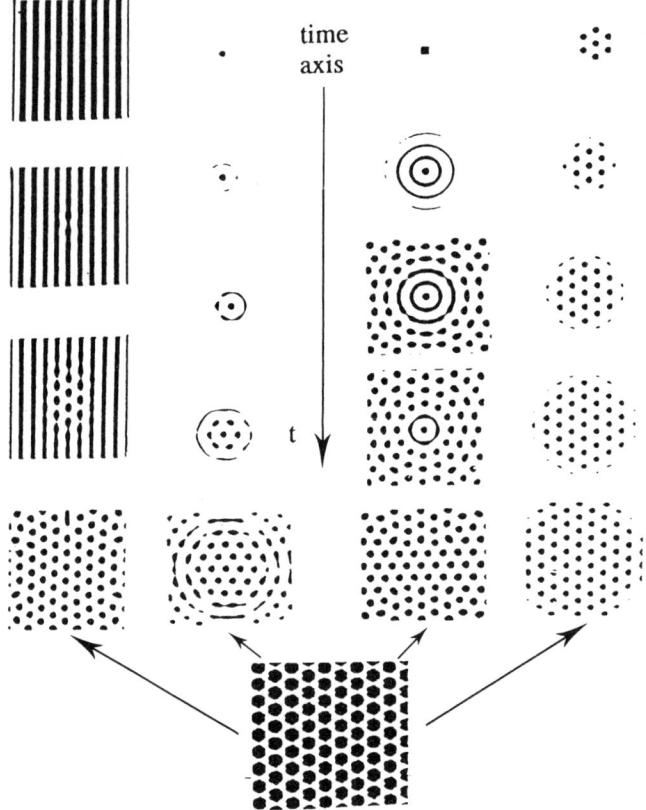

Fig. 1.45 Several distinct routes to the formation of a hexagonal lattice in the Swift–Hohenberg equation (following reference [18]).

doubling cascade, ending with fully developed chaos. This is not only a mathematical concept but has many important physical implications. We now wish to discuss this in more detail.

1.9 Chaos and turbulence

Systems undergoing transitions to chaotic behaviour typically exhibit a variety of universal features. First, plotting the given dynamical variable with respect to changes in the corresponding control parameter results in bifurcations, or splitting into two distinct stable levels. Second, when the dynamical variable is related to the frequency of the system, such bifurcations represent period doubling. Successive period doublings can lead rapidly to an infinite number of branchings, representing one of the possible routes to chaos. As the control parameter is varied, the system may pass through alternating regions of stability and chaos. Such behaviour is called intermittency. Two other features common to chaotic systems are evident in the system's phase-space diagram, and can be treated analytically by discretizing the equations of motion. These features are *sensitivity* to *initial conditions* and *stretching and folding* of phase space regions.

The transition to chaos is clearly shown by the infinite branchings that occur for certain critical values of the control parameters. When the system rapidly branches or period doubles until the number of branches goes to infinity, the system is said to *cascade to chaos*. The bifurcation diagram shows that chaotic behaviour can be intermittent with respect to changes in the system's parameters, as shown by the interspersed bands of chaotic and periodic behaviour. In Fig. 1.15 is shown the actual bifurcation diagram of the damped-driven Duffing oscillator with angular velocity (ω) as the dynamical variable versus the control (friction) parameter γ. The magnified section clearly shows intermittency and the cascade to chaos. For certain values of the control parameter(s), the differential equations of motion may have multivalued solutions. This causes a splitting in the graph of the dynamical variable versus the control parameter and the graphs themselves are called bifurcation diagrams.

Regardless of the physical system that generates it, the bifurcation diagram displays several characteristic features. The successive bifurcations divide the figure into self-similar shapes. Magnification or rescaling of the diagram tends to recover the initial shape. This self-similarity is equivalent to the fact that the geometry of the diagram is *fractal* in nature.

A large class of nonlinear systems exhibit universal features as they undergo transitions to chaos. Therefore, it is logical that a definition of chaos be established before analysing whether a system has undergone a transition into such a state. Chaotic behaviour does not simply occur when the actions of the system become so complex that we are unable to follow a particle, either due to error in measurement or insufficient computing power. Chaos is a state in which the outcome is truly random in relation to defining parameters. Such parameters could be boundary values or initial conditions. The currently accepted definition is based on initial conditions. If the initial conditions are changed by an infinitesimal amount, the behaviour of the system could be drastically affected at later times. We will call a system chaotic if it has this property of very sensitive (exponential) dependence on initial conditions [60].

Another indication of the onset of chaos is through the Fourier transform of any co-ordinate involved in chaotic motion. This is very different from that of quasi-periodic motion, the Fourier transforms of which consist of very sharp peaks. At the turn of the century, Poincaré noted that other bounded systems have continuous broadband spectra. This is very indicative of chaotic motion. In fact, the definition of chaotic motion demands that it have a continuous, broadband component in its spectrum.

During the past two decades, studies of chaotic motion in many systems have found that they typically evolve from quasi-periodic motion into chaos [61]. Two developments in these studies have been most important. The first is that chaos may appear in systems that are governed by very simple rules of behaviour, such as first-order differential equations. Thus, chaos certainly cannot be equated with disorder, as the underlying rules or equations are very well defined and regular. The second development of major concern is the discovery that the transition to chaos can take place only in a few 'universal' ways. These transitions are referred to as the 'routes to chaos' and appear in diverse fields of physics, such as the study of fluids, lasers, and semiconductor devices. We describe them below.

1.9.1. The Landau scenario

The oldest scenario for the transition from laminar to turbulent flow was proposed by Landau in 1944 [62]. In his theory it was assumed that chaos would develop via an infinite sequence of Hopf bifurcations. When a control parameter characterizing the flow (a 'knob') is increased, it causes the number of spikes in the Fourier spectrum to grow. This would constitute a transition to chaos by frequency generation. In general, this is not a realistic approach. The number of spikes, as mentioned above, does not necessarily indicate chaos. The behaviour in the Landau scenario may look more complicated, but is still of deterministic type. There will be no sensitive dependence on initial conditions that is required by the definition of chaos used here. In fact, Landau himself noted that a turbulent state at some time will come close to the initial, periodic state at some later time: 'So a turbulent motion is to a certain extent a quasi-periodic motion' [62].

1.9.2 The Ruelle–Takens–Newhouse scenario

The approach that has been accepted as the most 'generic' was proposed in 1971 by D. Ruelle and F. Takens [5]. The major difference in this perspective is that a system will transform into chaos after only two Hopf bifurcations (each of which has an incommensurate limit cycle)—not infinitely many, as in Landau's scenario. Very small perturbations could result in this sudden shift after only two bifurcations, thus indicating a sensitive dependence on initial conditions. This route to chaos, often called the two-frequency route, has been seen in many experiments.

In order to analyse the transition to chaos, a control parameter must be defined to direct the evolution of the system. In order to be able quantitatively to describe the various regions in the phase space, the quantity called a Lyapunov characteristic exponent (LCE) is used. For the simplest case of the map, $x_{n+1} = f(x_n)$, the associated Lyapunov exponent is defined as

$$\lambda \equiv \lim_{N \to \infty} \sum_{n=0}^{N} \ln|f'(x_n)|, \qquad (1.93)$$

and, in general, stable paths are characterized by $\lambda < 0$, superstable paths by $\lambda = -\infty$, bifurcation points by $\lambda = 0$, and unstable paths by $\lambda > 0$. This exponent can be interpreted in many ways, depending on the system. For example, the LCE associated with a particle's trajectory gives the average rate at which nearby trajectories diverge. The best way of identifying chaotic behaviour for a system is to compute the LCE, or to compute the largest LCE for the system.

1.9.3 The Feigenbaum picture

Feigenbaum's work [63, 64] with discrete mappings of the form

$$x_{n+1} = \Delta f(x_n), \qquad (1.94)$$

has led to the Feigenbaum scenario for the development of chaos. In his scenario, chaos develops as a result of an infinite sequence of period doubling or pitchfork bifurcations. As Δ is varied, there is a transition to chaotic behaviour, and it occurs in a 'universal' period doubling route to chaos. This study of period doubling has been generalized to a class of systems described by equations such as eqn (1.94). The sensitivity to initial conditions that determines chaotic behaviour lies in the location of attractors in the phase space. Points further from the attractors take different paths to approach the stable points. The paths depend very significantly on the location of the initial state in phase space. Attractors in the phase space of the system are determined by the criterion that Feigenbaum discussed in 1978 [64]. Feigenbaum noticed that, in particular, eqn (1.94) has highly bifurcated attractors (stable points). The attractors of the above equation are identified by the relation

$$X^* = \Delta f(X^*). \qquad (1.95)$$

The number, X^*, is the value or position of the attractor. Points near to X^* converge to X^*, and it is considered to be a global attractor when almost all of the points in the system converge towards X^*. In Table 1.2

Table 1.2 Period doublings: experimental and theoretical results, and a summary of the experimental observations of period doublings (after reference [5]).

Experiment	No. of period doublings	δ	α	μ	σ	κ
Hydrodynamic						
Water	4	4.3		4		
Helium	4	3.5		4		
Mercury	4	4.4		5		
Electronic						
Diode	5	4.3	2.4	6		6.3
Transistor	4	4.7				
Josephson junction	3	4.5	2.7		1.5	5
Laser						
Laser feedback	3	4.3				
Laser	3					
Acoustic: helium	3	4.8		6		
Chemical:						
Belousov–Zhabotinskii reaction	3					
Computer						
N–S truncation	5	4.6	2.5			
Brusselator	7	4.6			4.77	1.5
Theory	∞	4.669	2.503	4.58	1.52	6.55

we have summarized both experimental and theoretical facts related to the Feigenbaum map [5].

It should be mentioned in closing that the most important map used in the analysis of chaos is the so-called logistic map, given by the iterative formula

$$X_{n+1} = rX_n(1 - X_n), \tag{1.96}$$

where the control parameter r is contained within $0 < r \leq 4$ and the variable X_n within $0 \leq X_n \leq 1$. The behaviour of its asymmetric solutions is shown in Fig. 1.46.

In the logistic map of eqn (1.96), for values of $r < 3$, the results eventually converge to a steady state called an attractor. However, for values of $r > 3$, the resultant oscillation does not settle down and remains stable; i.e. the behaviour is periodic. The two possible values of $X_n(r)$ never converge and the curve $X_n(r)$ shows what is called a bifurcation.

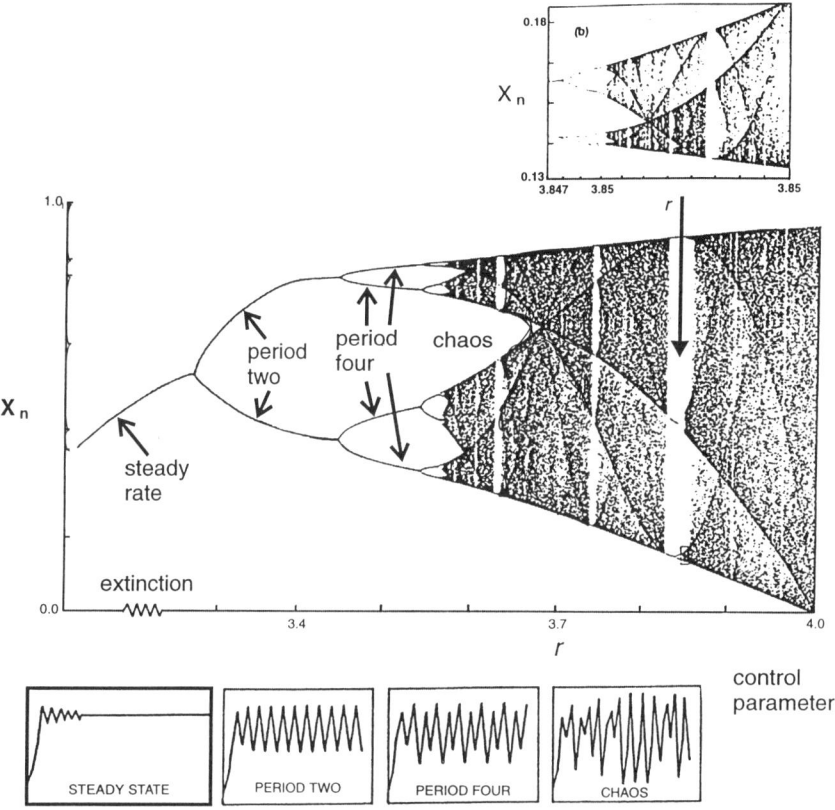

Fig. 1.46 The logistic map and its properties.

Higher values of the control parameter r produce further splitting and doubling of the periodicities involved. Each period doubling is a bifurcation. This is illustrated in Fig. 1.46. Abruptly, at $r_c = 3.58$, the result for X_n no longer oscillates periodically but changes in a chaotic fashion. Although for the values of r above r_c chaos and randomness seem to prevail, a further increase of the control parameter introduces windows of regularity among chaos, which is referred to as intermittency. Computer simulations of the logistic map readily demonstrate (see the inset in Fig. 1.46) that the structure is infinitely deep and self-similar. When magnified, parts of it show identical patterns ad infinitum. These patterns are a very common characteristic of the dynamic behaviour of nonlinear systems leading to chaos and complexity. Bifurcations with successive, infinite period doublings define one of the possible routes to chaos.

The so-called Feigenbaum exponent δ is then found from the location of the period doubling bifurcation points r_k, as

$$\delta = \lim_{k \to \infty} \frac{r_k - r_{k-1}}{r_{k+1} - r_k} = 4.669\ldots . \qquad (1.97)$$

1.9.4 The Pomeau–Manneville scenario

Another scenario related to that due to Feigenbaum is called *the Pomeau–Manneville scenario*, and has a subcritical pitchfork bifurcation as opposed to a supercritical one in the Feigenbaum picture. Here, a periodic equilibrium state is achieved by control parameter changes at unequal intervals, so that quasi-periodic phases occur irregularly and intermittently with chaotic impulses. Increasing the control parameter increases intermittency, eventually removing the quasi-periodic phases altogether and developing chaos. The three described scenarios leading to chaos [5] are summarized in Table 1.3.

The main surprise in the area of nonlinear dynamics was that chaos emerged not as a disordered state with no structure, but, rather, as a *highly hierarchical* form of dynamic ordering. Another important discovery was that of a *strange attractor*. Before we define it, we first make comments about sensitivity to initial conditions. In stationary states (attractors), equilibrium does not depend on initial conditions; i.e. the distance between two nearby points decreases after several iterations. In the chaotic region, two infinitesimally close points depart from each other exponentially with the number of iterations. This signifies strong dependence on initial conditions. Strange attractors, on the other hand, are equilibrium states in a chaotic region where convergence depends strongly on initial conditions. A system of equations exhibiting a strange attractor is, for example, the Lorenz equations

$$\left. \begin{array}{l} \dot{X} = \sigma(X - Y) \\ \dot{Y} = -rX - Y - XZ \\ \dot{Z} = -bZ + XY \end{array} \right\}, \qquad (1.98)$$

which have been used to model weather systems as well as convective instabilities of Rayleigh–Bénard type, among other phenomena. It is interesting to note that they can be transformed to the Maxwell–Bloch equations, which have been used in the modelling of laser instabilities in a polarizable medium (see Section 4.16). The strange attractor for the Lorenz equations, eqns (1.98), is illustrated in Figure 1.47.

Table 1.3 A summary of the three scenarios leading to chaos (after reference [5]).

Scenario	Ruelle–Takens–Newhouse	Feigenbaum	Pomeau–Manneville
Typical bifurcations	Hopf	Pitchfork	(Inverse) saddle-node
Bifurcation diagram (s = stable, u = unstable)			
Eigenvalues of linearization in complex plane as μ is varied			
Main phenomenon	After three bifurcations strange attractor 'probable'	Infinite cascade of period doublings with universal scaling of parameter values $\mu_i - \mu_\infty \sim (4.6692)$	Intermittent transition to chaos; laminar phase lasts $\sim (\mu - \mu_c)^{-1/2}$
Measurement	Power spectrum, correlation	Power spectrum subharmonics ~ 13.5 dB below preceding level	Real-time measurements
Small noise	No influence	High periods disappear (noise level must go down by 6.62 to see one more period doubling)	Time of laminarity scales as $(\mu - \mu_c)^{-1/2} T(\sigma(\mu - \mu_c)^{3/4})$ for noise of standard deviation σ

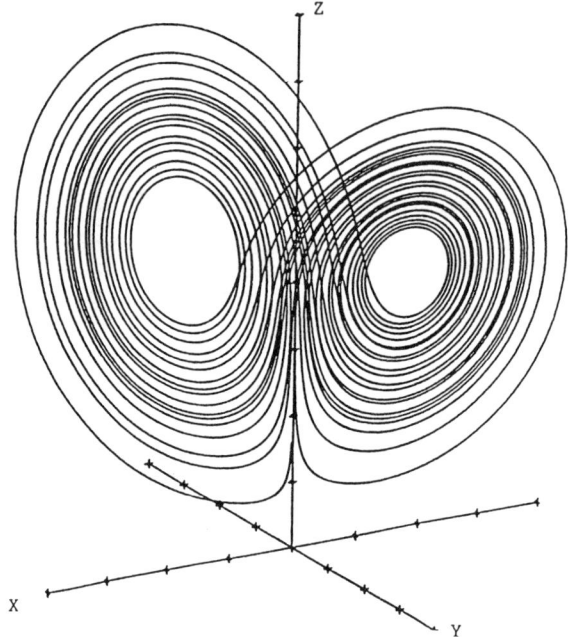

Fig. 1.47 The strange attractor for the Lorenz equation.

To put these theoretical models in perspective, we close this section by showing an actual experimental observation of the development of turbulence in a Rayleigh–Bénard fluid. In Fig. 1.48 it is certainly indicated that there is still a long road towards an understanding of turbulence [12].

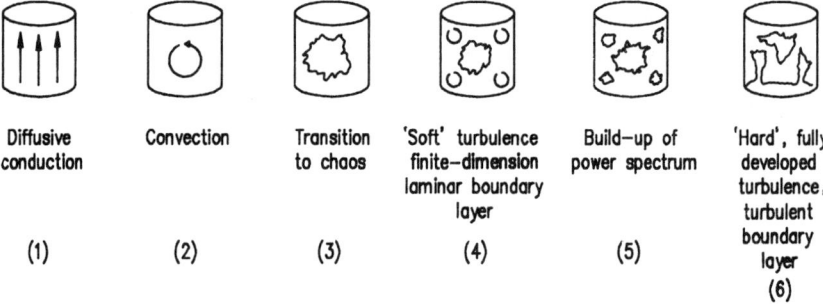

Fig. 1.48 A schematic depiction of the different regimes of flow patterns observed in Rayleigh–Bénard convection cells as the driving force increases from left to right (after reference [12]).

1.10 Fractals

In this section we wish to point to the presence of *fractal objects* in both pattern formation phenomena and the chaotic and turbulent behaviour discussed above. This is especially transparent in the geometries of boundary regions.

Many pattern-forming systems, especially when they are far from thermodynamic equilibrium, exhibit a growth of forms which are of fractal nature. Specific examples include:

- dendritic solidification in an undercooled medium
- viscous fingering phenomena, which occur when two fluids of different viscosities penetrate each other
- aggregation phenomena, such as diffusion-limited aggregation
- electrodeposition patterns of ions on to an electrode.

The basic property of all fractal objects is their self-similarity, i.e. when we cut a part of the object and then magnify it, the resulting object appears the same as (or at least very similar to) the original object. Another property of fractals, which actually earned them their name, is that their dimensionality is not an integer but in general a real number. In the simplest form, the so-called fractal dimension D is given by the relationship

$$V(R) \sim R^D, \tag{1.99}$$

where $V(R)$ is the volume of the region bounded by the interface the radius of which is R. Thus the fractal dimension determines the extent to which an object fills the space. A fractal object fills the space unevenly because its parts are related or correlated. The fractal dimension quantifies the rate at which the object clumps together in space. The notion of a fractal, especially as it pertains to geometrical objects, was principally introduced into science by Mandelbrot [65]. It was subsequently studied by many physicists and generalizations of the definition were proposed, including multifractality. In the physical context it is useful to distinguish two general classes of fractals [66]: (a) *deterministic* (see Fig. 1.49(a)), where a simple iterative rule is present, e.g. involving a procedure to cut a part of the object at each stage and replace it with a fixed element; and (b) *random* (see Fig. 1.49(b)) where a stochastic approach is used so that a given operation, e.g. an aggregation event, is predicted with a preselected probability level.

Fig. 1.49 Examples of two deterministic fractals (following reference [67]: (a) the triangular Sierpinski gasket; (b) the Sierpinski carpet and a random fractal; (c) a diffusion-limited aggregation cluster.

A practical method of determining the fractal dimension of an aggregate is to average the intensity of scattered light $I(q)$, where $q = (4\pi/\lambda)\sin(\theta/2)$ is the wavenumber with λ denoting the wavelength and θ the angle between incident and scattered light. Then [67]

$$I(q) = I_0 q^{-D}, \qquad (1.100)$$

where D is the fractal dimension.

Undoubtedly, fractals have captured the imagination of many physicists and offered a totally new vantage point for classes of complex behaviour. It is too early at this point, however, to be able to provide an integrated picture involving all the previously introduced nonlinear concepts as well as fractality.

1.11 Self-organized criticality

In 1987, Bak *et al.* [68] introduced a new type of critical phenomenon (see

Chapter 2) which they coined self-organized criticality (hereafter SOC). SOC concerns spatially extended dynamical dissipative systems in nature. However, the global features of SOC do not depend on local mechanisms; hence the theory neither predicts nor depends on specific local conditions in the system [69, 70].

The spatially extended dynamical systems that the SOC theory applies to exhibit spatial and temporal self-similar scale-invariance. Spatially this is manifested in the form of fractal properties of the system, while temporally the presence of 'flicker' (also called 'pink') or '$1/f$' noise [71] signals the existence of event lifetimes over all time scales. The authors of SOC theory claim its fractal structure to be a 'snapshot' of a system in the critical state, while the flicker noise results from the combination of the presence of minimally stable structures in the system at criticality and their fractal nature [68].

Physically, such dynamical systems are characterized by power-law spatial and temporal correlations over many orders of magnitude. A likely consequence of this power law behaviour is that systems exhibiting SOC are weakly chaotic (their sensitivity to initial conditions is expressed through a power law and not an exponential formula, as is the case for fully chaotic systems). This hypothesis, if true, suggests that weak chaos may be even more pervasive in nature than previously thought [70].

The concept of self-organization used in the context of SOC refers to the fact that the system evolves to the critical state independently of the initial conditions, i.e. the critical state is an "attractor" for its dynamics. The critical point in SOC is defined as the state for which the system reaches stationary conditions; that is, transport processes occur on all temporal and spatial scales and the system settles into a global state which is virtually invariant under perturbations [72].

It is important to note that while much of the theoretical development of SOC employs terms and concepts adopted from critical phenomena (described in Chapter 2), the underlying physics is quite different. In equilibrium statistical mechanics, criticality is reached by adjusting a control parameter in the system. In SOC, the analogous parameter is self-tuning, as the critical state is an attractor. Additionally, while the critical exponents developed for SOC are derived in analogy to the equilibrium ones, in statistical mechanics, they describe nonequilibrium dynamical behaviour [72].

Cellular automata are mathematical realizations of SOC in which a grid of separate cellular states is defined by a field (such as pressure) and a simple set of rules are employed to set the system in motion (in discrete time steps) and govern evolution. The only influence on the evolution of a square is the value of the field at all its neighbouring sites in the previous

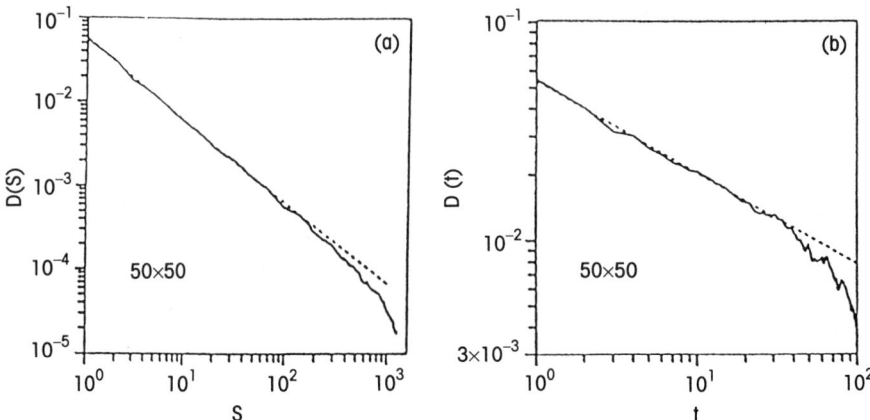

Fig. 1.50 (a) The size distribution $D(s)$ of clusters. (b) The distribution of relaxation times $D(t)$ for the clusters. (Both following reference [69].)

time step. Hence, in the next time step a site may grow or die, depending on the number of live neighbours and other parameters [69]. The physical system to which the cellular automaton approach is most often employed in SOC is the sandpile. The avalanches occurring on a sandpile, where sand is being added or the system is tilted so the slope exceeds some critical value, provide an example of nonlinear diffusion dynamics, and hence its applicability to SOC is well justified [69].

At the critical point not all of the sites have the critical slope. Then, the system evolves in such a way that the minimally stable clusters, defined to be those areas accessible by avalanches starting from a single site perturbation, become intermixed with regions of less-than-minimal stability which impede noise.

When these 'noise filters' become numerous enough to just stop propagation of perturbations over infinite length scales, the system has reached its critical point. At a critical slope of the sandpile, no length or time scale is present and stationary conditions have been reached through self-organization. This is very analogous to second-order phase transitions. Following reference [69], in Fig. 1.50(a) is shown the distribution of cluster sizes at the critical point for a model sandpile. The power-law relationship reflects the self-similarity in the spatial distribution, i.e. its fractal nature. The distribution of relaxation times for a range of perturbations is shown in Fig. 1.50(b).

We conclude that the combination of the scale-invariant sand grain clusters and their minimal stability lead to a power spectrum resembling the $1/f$ type. The fact that these systems scale both in space and time in this manner suggests that the physics behind avalanches and small slides is the same.

In analogy with equilibrium phase transitions (see Chapter 2), the behaviour of crucial parameters of a SOC system near its critical point may be characterized using critical exponents [72, 73]. To this end, we identify the 'slope' θ of the sandpile as the control parameter. The order parameter, on the other hand, corresponds to the current of sand flow, j, or equivalently the number of slides in a given time period. The external field h coupled to the order parameter is identified with the current of sand grains dropped on the pile. Moreover, the correlation length ξ represents here the cut-off in linear cluster size just below criticality. The susceptibility, χ, of the order parameter to a change in the external field is also introduced, and s is the largest cluster volume for a given value of the control parameter. Similarly, with equilibrium critical phenomena we postulate that the above system parameters follow power laws of the form

$$j \propto (\theta - \theta_c)^\beta, \qquad \chi \propto (\theta - \theta_c)^{-\gamma}, \qquad (1.101, 1.102)$$

$$\xi \propto (\theta_c - \theta)^{-\nu}, \qquad s \propto (\theta_c - \theta)^{-1/\sigma} \qquad (1.103, 1.104)$$

and

$$j(\theta = \theta_c) \propto h^{1/\delta}. \qquad (1.105)$$

Several physical observations about the behaviour of the system at criticality can be readily deduced from these relations. First, if $\gamma > 0$, χ is singular at the critical point; hence a small external field h (i.e. a single grain of sand) may cause sand avalanches of all sizes. Also, for $\nu > 0$ the correlation length tends to infinity at the critical point, as in the theory of phase transitions.

Many numerical simulations support the conjecture about power-law scaling. The mean-field theory for SOC [72] leads to the exponents given by $\beta = 1$, $\gamma = 1$, $\delta = 2$, $\sigma = 1/2$, and $\nu = 1/2$. From simulations with cellular automata, values for the exponents in 2-D are found as [72]: $\beta = 0.7$, $\gamma = 1.35$, $\nu = 0.74$, and $\sigma = 0.72$. As the spatial dimension of the simulation is increased, the values converge towards their mean-field predictions, suggesting that the latter may be exact in a sufficiently high dimension.

Further evidence that the systems evolve to a critical point can be found by showing that the temporal and spatial distributions $D(t)$ and $D(s)$, respectively, obey finite-size scaling as in second-order transitions [69].

Hence the corresponding distributions are expected to follow the power laws of the form

$$D(t) = t^{-\alpha} F(L^\sigma/t), \qquad D(s) = s^{-\tau} G(L^d/s), \quad (1.106, 1.107)$$

where α and τ are the slopes of the spatial and temporal distributions, d is the spatial dimension, and L is the linear size of the system. The fact that to within numerical accuracy critical systems do show finite-size scaling at the critical point is direct proof of spatial organization.

The applicability of SOC theory to natural phenomena has received a great deal of attention in recent years [74]. For instance, the empirical Gutenberg–Richter law of earthquake energy distributions is a power law without a dynamical explanation. The proposal that earthquakes may exhibit SOC behaviour [75] suggests that the Earth's crust has evolved to a critical stationary point and that small and large quakes differ only in scale and not in underlying physics. The use of cellular automata to simulate interacting ecosystems has shown that activity settles down to a critical state which may be indicative of SOC [77]. This might explain the long-term stability (and resilience) of life to perturbations, while providing a natural explanation for large biomass extinctions as simply 'avalanches'. The concept that turbulence can be represented as energy dissipation on a fractal has also been suggested in this context [76].

Other phenomena which have been suggested as examples of SOC include financial markets, current flows through a resistor, forest fire growth, sliding vortex lattices in magnetic fields, and charge-density wave systems [74]. Therefore, it is clear that the concept of self-organized criticality is useful in describing the overall characteristics of many dynamical systems. Its impact on science is already extensive, and many challenging problems related to chaos, dissipation, and nonlinearity may become better understood through the use of self-organized criticality.

References

1. L. D. Landau and E. M. Lifshitz, *Statistical Physics*, Pergamon Press, London (1959).
2. K. G. Wilson, *Rev. Mod. Phys.*, **55**, 583 (1983).
3. G. Nicolis and I. Prigogine, *Exploring Complexity*, W. H. Freeman, New York (1989).
4. H. Haken, *Synergetics: an Introduction*, Springer-Verlag, Berlin (1983).
5. P. Cvitanović, *Universality in Chaos*, Adam Hilger, Bristol (1984).
6. R. Jullien and R. Botet, *Aggregation and Fractal Aggregates*, World Scientific, Singapore (1987).

REFERENCES

7. J. M. T. Thompson and H. B. Stewart, *Nonlinear Dynamics and Chaos*, John Wiley and Sons, New York (1986).
8. J. C. Tolédano and P. Tolédano, *The Landau Theory of Phase Transitions*, World Scientific, Singapore (1987).
9. M. Abramowitz and I. A. Stegun (eds), *Handbook of Mathematical Functions*, Dover, New York (1965).
10. J. A. Tuszyński, M. J. Clouter, and H. Kiefte, *Phys. Rev. B*, **33**, 3423 (1986).
11. S.-K. Ma, *Modern Theory of Critical Phenomena*, Benjamin, Reading, MA (1976).
12. D. K. Campbell, *Nucl. Phys.* **B2**, Proc. Suppl. 159 (1987).
13. H. T. Davis, *Introduction to Nonlinear Differential and Integral Equations*, Dover, New York (1962).
14. J. M. Dixon, J. A. Tuszyński, and M. Otwinowski, *Phys. Rev. A*, **44**, 3484 (1991).
15. G. L. Baker and J. P. Gollub, *Chaotic Dynamics*, Cambridge University Press, Cambridge (1990).
16. J. Hale and H. Kocak, *Dynamics and Bifurcations*, Springer-Verlag, Berlin (1991).
17. J. D. Crawford, *Rev. Mod. Phys.*, **63**, 991 (1991).
18. A. V. Gaponov-Grekhov and M. I. Rabinovich, *Nonlinearities in Action: Oscillations, Chaos, Order, Fractals*, Springer-Verlag, Berlin (1988).
19. A. C. Scott, F. Y. F. Chu, and D. W. McLaughlin, *Proc. IEEE*, **61**, 1443 (1973).
20. N. J. Zabusky and M. D. Kruskal, *Phys. Rev. Lett.*, **15**, 240 (1965).
21. P. G. Drazin and R. J. Johnson, *Solitons: an Introduction*, Cambridge University Press, Cambridge (1989).
22. A. R. Bishop, J. R. Krumhansl, and S. E. Trullinger, *Physica*, **D1**, 1 (1980).
23. R. K. Dodd, J. C. Eilbeck, J. D. Gibbon, and H. C. Morris, *Solitons and Nonlinear Wave Equations*, Academic Press, New York (1982).
24. A. S. Fokas and M. J. Ablowitz, *J. Math. Phys.*, **25**, 2494 (1984).
25. J. J. Rasmussen and K. Rypdal, *Phys. Scr.*, **33**, 481 (1986).
26. M. J. Ablowitz, D. BarYaccov, and A. S. Fokas, *Stud. Appl. Math.*, **69**, 35 (1983).
27. A. S. Fokas and M. J. Ablowitz, *Stud. Appl. Math.*, **69**, 211 (1983).
28. S. V. Manakov, *Physica* **3D**, 420 (1981).
29. A. I. Nachman and M. J. Ablowitz, *Stud. Appl. Math.*, **71**, 243 (1984).
30. M. J. Ablowitz and P. A. Clarkson, *Solitons, Nonlinear Evolution Equations and Inverse Scattering*, Cambridge University Press, Cambridge (1991).
31. M. Boiti, J. J.-P. Leon, L. Martina, and F. Pempinelli, *Phys. Lett.*, **123A**, 340 (1988).
32. A. S. Fokas and P. M. Santini, *Phys. Rev. Lett.*, **63**, 1329 (1989); *Physica*, **44D**, 99 (1990).
33. A. Nakamura, *Progr. Theoret. Phys. Suppl.*, **94**, 195 (1988).
34. J. A. Combs and S. Yip, *Phys. Rev. B*, **28**, 6873 (1983).
35. K. H. Spatschek, M. Taki, and T. Eickermann, in *Nonlinear Coherent Structures* (ed. M. Barthes and J. Léon), Springer-Verlag, Berlin (1990).

36. A. R. Bishop, M. G. Forest, D. W. McLaughlin, and E. A. Overmann II, *Physica*, **23D**, 293 (1986).
37. A. R. Bishop, M. G. Forest, D. W. McLaughlin, and E. A. Overman II, *Phys. Lett.*, **127A**, 335 (1988).
38. M. Taki, K. H. Spatschek, J. C. Fernandez, R. Grauer, and G. Reinisch, *Physica*, **40D**, 65 (1989).
39. A. C. Newell and J. V. Moloney, *Nonlinear Optics*, Addison-Wesley, Redwood City, CA (1992).
40. P. C. Fife, Deterministic (continuous and discrete) mathematics of nonlinear problems, in *Nonequilibrium Cooperative Phenomena in Physics and Related Fields* (ed. M. G. Velarde), Plenum Press, New York (1984).
41. P. Kolodner, A. Passner, H. L. Williams, and C. M. Surko, *Nucl. Phys.*, **B2**, Proc. Suppl. 97 (1987).
42. F. K. Browand and C.-M. Ho, *Nucl. Phys.*, **B2**, Proc. Suppl. 97 (1987).
43. A. C. Newell and J. Whitehead, *J. Fluid Mech.*, **38**, 203 (1969).
44. J. J. Hagseth, C. D. Andereck, F. Hayot, and Y. Pomeau, *Phys. Rev. Lett.*, **62**, 1257 (1989).
45. A. Joets and R. Ribotta, *J. Phys. (Paris)*, **47**, 595 (1986).
46. F. S. Bates, J. H. Rosedale, G. H. Fredrickson, and C. J. Glinka, *Phys. Rev. Lett.*, **61**, 2229 (1988).
47. M. Paczulski, M. Kardar, and D. R. Nelson, *Phys. Rev. Lett.*, **60**, 2638 (1988).
48. J. M. Corkhill and J. F. Goodman, *Adv. Colloid Interface Sci.*, **2**, 297 (1969).
49. S. P. Parker (ed.), *Solid-State Physics Resource Book*, McGraw-Hill, New York (1987).
50. R. Eykhott, A. R. Bishop, R. S. Lomdahl, and E. Domany, *Physica D*, **23**, 102 (1986).
51. A. T. Winfree, *Scient. Am.*, **230**, 83 (1974).
52. A. Harpin, P. Meriel, and J. Villain, *J. Phys. (Paris)*, **21**, 67 (1960).
53. G. P. Fletcher, *J. Appl. Phys.*, **37**, 1056 (1966).
54. G. S. Kandaurova and A. E. Sviderskii, *JETP Lett.*, **47**, 490 (1988).
55. M. A. Gorman, M. el-Hamdi, and K. A. Robbins, in *Nonlinear Structures in Physical Systems* (ed. L. Lam and H. C. Morris), Springer-Verlag, Berlin (1989).
56. R. Chang, W. J. Firth, R. Indik, J. V. Moloney, and E. M. Wright, *Opt. Communs*, **88**, 167 (1992).
57. M. Florjanczyk and L. Gagnon, *Phys. Rev. A*, **41**, 4478 (1990).
58. M. C. Cross, in *Far From Equilibrium Phase Transitions* (ed. L. Garrido), Springer-Verlag, Berlin (1988).
59. A. R. Bishop, *Helv. Phys. Acta*, **59**, 811 (1986).
60. P. W. Milloni, M.-L. Shih, and J. R. Ackerhalt, *Chaos in Laser–Matter Interactions*, World Scientific, Singapore (1987).
61. P. Coullet, Stability of the scenarios towards chaos, in *Chaos and Statistical Methods* (ed. Y. Kuramoto), Springer-Verlag, Berlin (1984).
62. D. Ter Haar (ed.), *Collected Papers of L. D. Landau*, Pergamon (1965).
63. M. Feigenbaum, *J. Statist. Phys.*, **21**, 669 (1979).
64. M. Feigenbaum, *J. Statist. Phys.*, **20**, 25 (1978).

REFERENCES

65. B. B. Mandelbrot, *The Fractal Geometry of Nature*, W. H. Freeman, San Francisco (1977).
66. T. Vicsek, *Fractal Growth Phenomena*, World Scientific, Singapore (1989).
67. J. Feder, *Fractals*, Plenum Press, New York (1988).
68. P. Bak, C. Tang, and K. Wiesenfeld, *Phys. Rev. Lett.*, **59**, 381 (1987).
69. P. Bak, C. Tang, and K. Wiesenfeld, *Phys. Rev. A.*, **38**, 364 (1988).
70. P. Bak and K. Chen, *Scient. Am.*, **264**(1), 46 (1991).
71. P. Dutta and P. M. Horn, *Rev. Mod. Phys.*, **53**(3), 497 (1981).
72. C. Tang and P. Bak, *J. Statist. Phys.*, **51**, 797 (1988).
73. C. Tang and P. Bak, *Phys. Rev. Lett.*, **60**, 2347 (1988).
74. P. Bak and K. Chen, *Physica D*, **38**, 5 (1989).
75. P. Bak and C. Tang, *J. Geophys. Rev.*, **94**, 15 635 (1989)
76. B. Mandelbrot, *J. Fluid. Mech.*, **62**, 331 (1974).

2 PHASE TRANSITIONS AND CRITICALITY

...I would claim that the most important advances in this area (condensed matter physics) come about by the emergence of qualitatively new concepts at the intermediate or macroscopic levels—concepts which, one hopes, will be compatible with one's information about the microscopic constituents, but which are in no sense logically dependent on it. These new concepts... have in common... that they provide a new way of classifying a seemingly intractable mass of information of selecting the *important* variables from the innumerable possible variables which one can identify in a macroscopic system.
    A. J. Leggett (1987), *The Problems of Physics* (Oxford University Press, Oxford)

2.1 Long-range order

In this chapter we intend to provide the reader with an overview of the properties of systems undergoing phase transitions, followed by several sections on theoretical developments, with a particular emphasis placed on the Landau and Landau–Ginzburg models. Important elements of lattice models, scaling, and renormalization will also be presented at an accessible level. The motivation here is to prepare the ground for a microscopic approach to the dynamics of strongly interacting many-body systems, which will be the centrepiece of Chapter 5.

One of the main characteristics of most systems which undergo a phase transition is the emergence of *long-range order*. This term refers to a situation in which the value of a physical quantity (e.g. local magnetization) at an arbitrary point in the system is correlated with its value at a point located a long distance away from it (ideally infinitely removed). In many-body systems kept at sufficiently high temperatures, thermal energy is bound to exceed that of the interactions between particles, leading to the predominance of disorder. However, as the temperature is gradually lowered, close to absolute zero the system will eventually settle into its quantum ground state and thermal fluctuations will slowly cease, except for the inevitable zero-point oscillations. If this ground state is nondegenerate, then the entropy will decrease to zero, as predicted by the Third Law of Thermodynamics. As a result, the system will become ordered (perfectly ordered at 0 K); *but* not long-range ordered, since the interactions between its constituents become virtually negligible, as are the correlations between fluctuations at different locations in the sample.

When there are no interactions between the particles (or quasi-particles) of the system, no symmetries are broken and the free energy derivative is a smooth function of any thermodynamic variable (e.g. temperature). The *opposite* situation to the one described above will be the subject of the remainder of this chapter. Thus, if we include an interaction between particles, however small, the system will at some temperature (called the transition temperature) undergo a phase transition to a new thermodynamic ground state. This is usually associated with long-range order.

Long-range order can, in general, arise for two different reasons. The first is the existence of long-range forces (electrostatic, gravitational, etc.) that maintain coupling between particles and result in the 'rigidity' of the system, i.e. its resistance to change induced externally [1]. The second is due to the fact that, even in the absence of long-range forces, when a symmetry present in the disordered phase is broken, uniformity is adopted throughout the sample in the same way on a macroscopic scale. This results from a variational principle that determines the free energy minima. This can be seen in many condensed-matter phenomena, such as superconductivity, ferromagnetism, superfluidity, and liquid crystals [2]. As a consequence, the same state is occupied over a macroscopic domain, giving the appearance of long-range forces; but, in fact, only structural order is of long-range type. However, a local application of a short-range perturbation will be felt throughout the system as a result of communication between its parts and the generalized rigidity that is associated with an ordered phase.

Note that there exists another way through which order *can* be achieved *without* the occurrence of a phase transition. This effect is called a *Schottky anomaly* [3]. In such systems as, for example, semiconductors with vacancies, at low enough temperatures all particles collapse into a single energy level, while at sufficiently high temperatures they possess an equal probability of occupying any of the finite number of available energy levels. The main observable feature of this effect is a sharp peak in the specific heat, which decreases to zero as the temperature approaches 0 K, since eventually all the particles collapse into the nondegenerate ground state. However, this situation does not exhibit manifestations of a phase transition, i.e. thermodynamic singularities, which will be discussed at length later in this chapter.

Finally, we know that liquids do not order; hence it is required that the substance's zero point energy be greater than the energy gained by crystallization in order to prevent spontaneous emergence of an ordered state at low temperatures. This situation is known to be violated only in helium, with the qualification that even helium undergoes a phase transition to a superfluid which is a momentum-ordered phase.

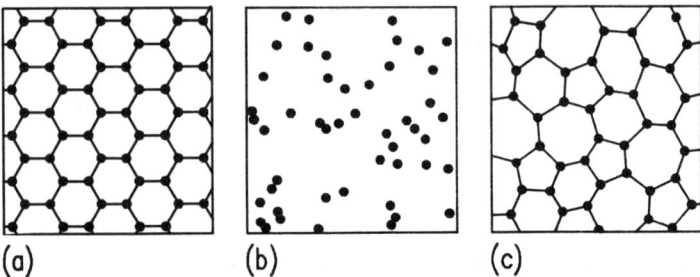

Fig. 2.1 Schematics in two dimensions of atomic arrangements in three phases (after reference [4]). Each dot represents the position of an atom at a given instant of time: (a) crystal; (b) gas; (c) glass.

Below, we list and briefly describe a number of representative examples from condensed matter physics which provide an illustration of long-range order and related concepts.

2.1.1 Crystals

A classical example of long-range order is the perfect crystal, in which the mass density distribution has an infinite correlation length related to the onset of spatial periodicity. Three possible phases that may arise in two dimensions when atomic positions are considered, i.e. a crystal, a gas phase, and a glass phase, are illustrated in Fig 2.1.

Experimental evidence for perfect crystal formation is given by the existence of a regular diffraction pattern associated with the presence of Fourier components in the mass density distribution, with [5]

$$\rho(r) = \sum_G \rho_G e^{iG \cdot r}, \tag{2.1}$$

where the G are vectors in the reciprocal space. This then suggests that the set of numbers ρ_G can be used as infinite-dimensional (in principle) quantities characterizing the low-temperature ordered (crystal) phase. Such quantities are referred to, following Landau, as order parameters [6] and we will say more about them in later sections. However, since only a small number of coefficients in the above expansion in eqn (2.1) are nonzero, they define a multi-component (not an infinite-component) order parameter. This is due to the fact that the effective interaction potential (which could be of van der Waals, screened Coulomb, Lennard–Jones, or another type, depending on the nature of the chemical bond involved) is short-range. As a result, we may obtain one of the four types of crystal lattices in

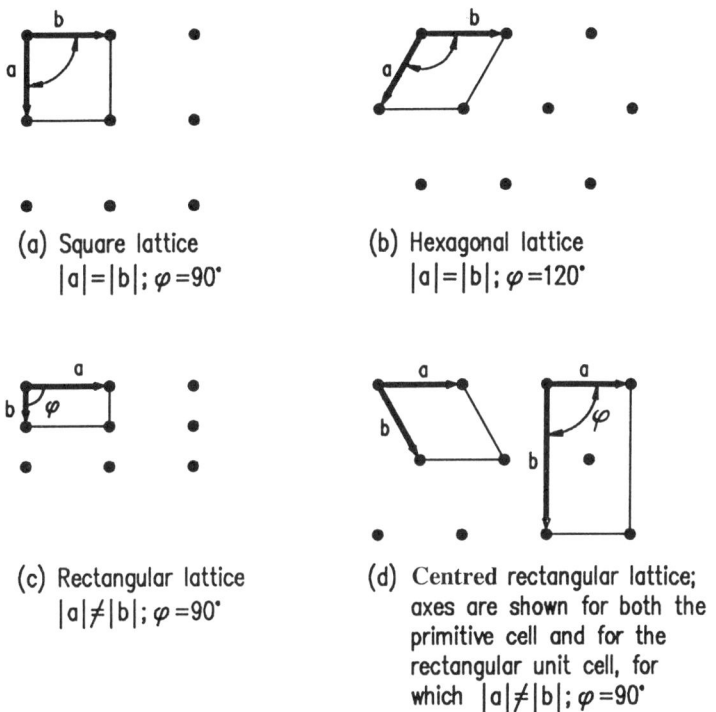

Fig. 2.2 The four types of two-dimensional lattices (after reference [5]).

two-dimensional space (see Fig. 2.2) or one of the 14 Bravais lattice types in three spatial dimensions (see Fig. 2.3).

These 14 Bravais lattice types give rise to the 32 attendant point-group symmetries denoted by

- C_1, C_2, C_3, C_4, C_6
- $C_{2v}, C_{3v}, C_{4v}, C_{6v}$
- $C_{1h}, C_{2h}, C_{3h}, C_{4h}, C_{6h}$
- S_2, S_4, S_6
- D_2, D_3, D_4, D_6
- D_{2d}, D_{3d}
- $D_{2h}, D_{3h}, D_{4h}, D_{6h}$
- $T_1, T_h, T_d, 0, 0_h$.

These are built using rotational axes, reflection planes, and inversion

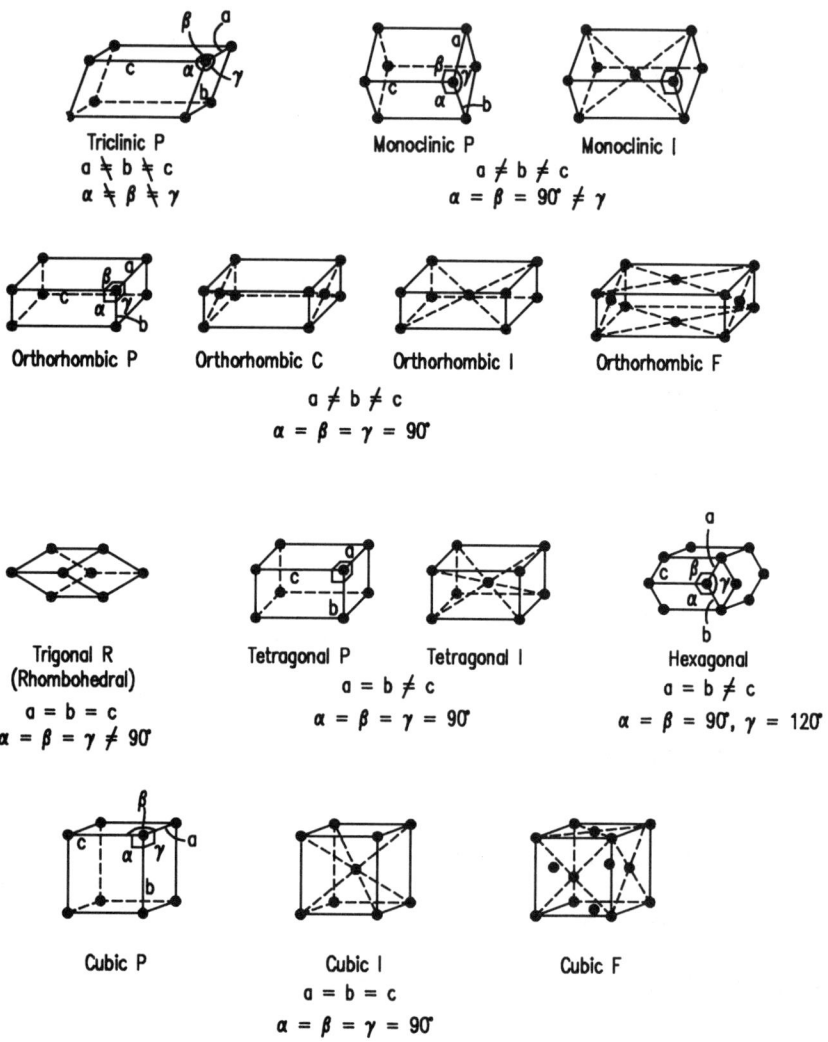

Fig. 2.3 The 14 Bravais lattice types (after reference [5]).

centres. The addition of space translation results in several combination symmetry elements (sliding reflection planes, screw rotational axes, and improper rotational axes) and gives rise to 230 space groups. It is easy to recognize that phase transitions involving crystal symmetries will not only refer to crystallization from a liquid solution or through adsorption on a surface of gas molecules but will also contain the possibility of transformations from one crystal type to another. This latter eventuality is known

2.1.2 Structural phase transitions

Distortions resulting from temperature changes or an application of an external stress may lead to structural phase transitions, as is schematically illustrated in Fig. 2.4. As a consequence of structural phase transitions between various point or space groups of the crystal, new mechanical, electrical, and sometimes magnetic properties may be observed with their associated thermodynamic anomalies at the transition temperature. The

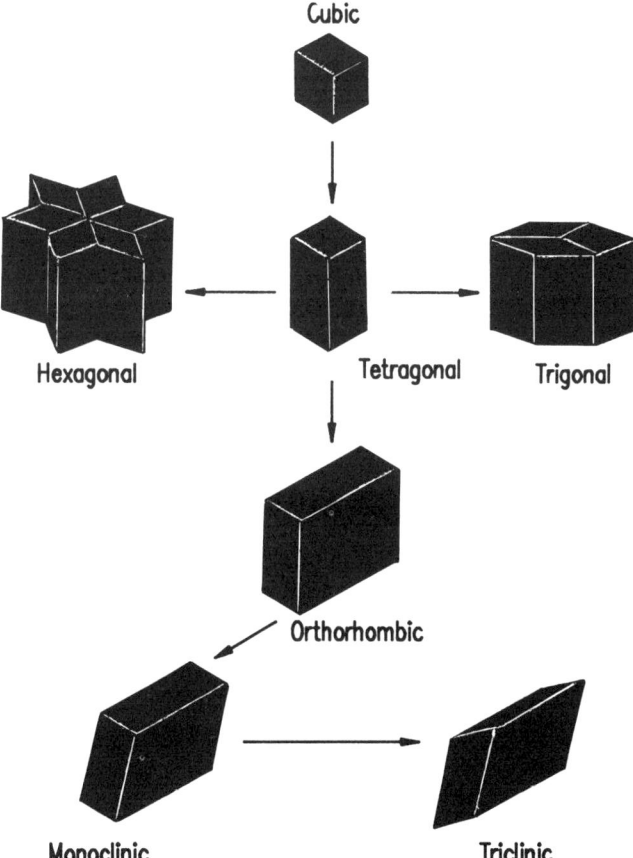

Fig. 2.4 Changes between Bravais lattices resulting from unidirectional distortions.

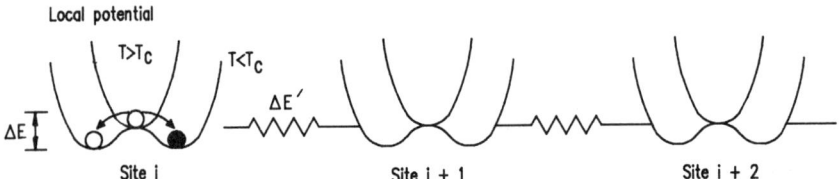

Fig. 2.5 A schematic illustration of the typical situation for structural phase transitions.

two main classes of structural phase transitions are [7] (a) displacive and (b) order–disorder type. The criterion to distinguish between the two types is whether the local energy change ΔE between two equivalent configurations is smaller or greater than the elastic coupling energy $\Delta E'$, respectively (see Fig. 2.5).

In displacive transitions, phonons cause the transition and the order parameter is the amplitude of the related lattice distortion. This yields a change from one lattice ordering to another. Simultaneously, the frequency of the associated phonon mode (called a *soft mode*) tends to zero as $T \rightarrow T_c$, creating a lattice instability. On the other hand, order–disorder transitions feature a transformation between randomly distributed positions of atoms in their local double-well potential bottoms ($T > T_c$) on the one hand and an ordered arrangement ($T < T_c$) on the other. This is usually an abrupt transition associated with a soft diffusive mode characterizing large-amplitude thermal hopping between degenerate potential well bottoms. To describe the associated behaviour, displacive transitions require a continuous model of a Landau–Ginzburg type with ensuing solitary waves, while order–disorder transitions make use of Ising model calculations with an effective (not real) spin variable.

2.1.3 Ferroelectricity

Structural phase transitions are commonly associated with changes in the distribution of electric charges giving rise to a nonzero value of the polarization vector, and result in several types of ferroelectric phenomena. The various possible types of transitions and their definitions are summarized in Fig. 2.6.

▷

Fig. 2.6 A schematic illustration of the most important ferroelastic, ferroelectric, and anti-ferroelectric phase transitions from a centrosymmetric prototype (following reference [8]).

LONG-RANGE ORDER 79

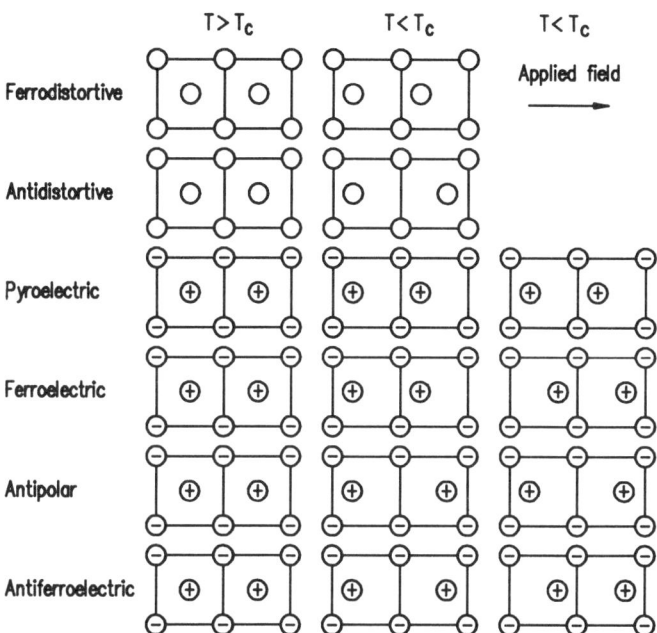

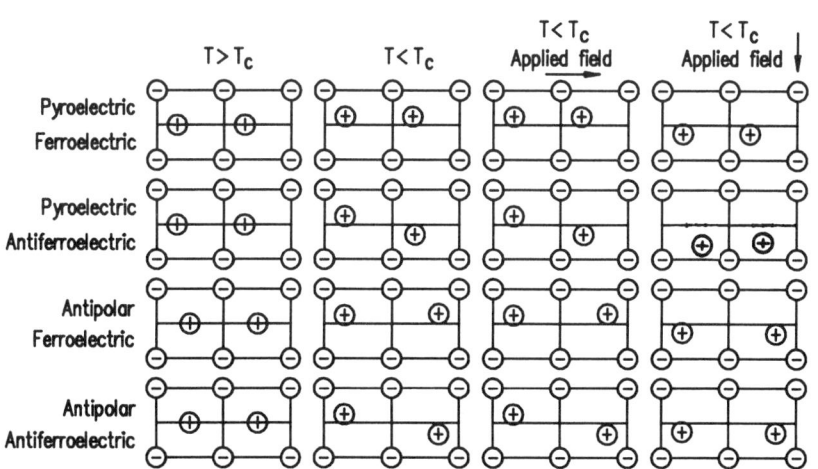

A schematic representation of some more complex ferroelectric and antiferroelectric phase transitions

⊕ ⊖ Charged atoms or groups

○ Uncharged atoms or groups

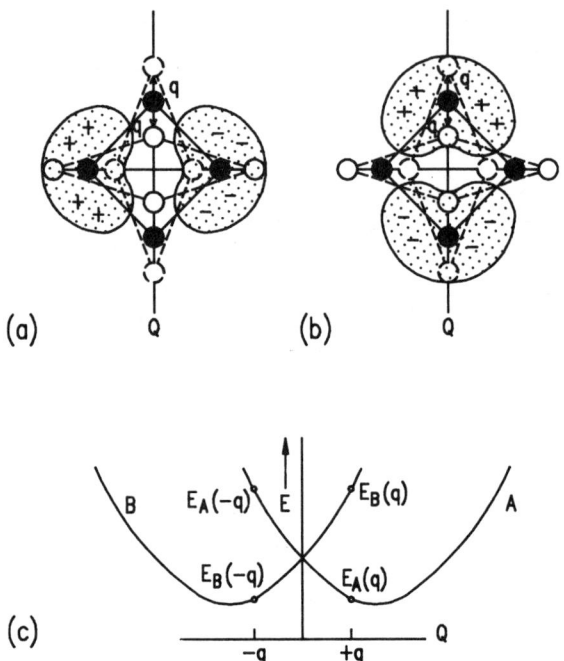

Fig. 2.7 The Jahn–Teller effect for a square planar molecule (following reference [3]). Displacement of the nuclei of the molecule in the electronic field is provided by the wave functions (a) Ψ_A and (b) Ψ_B. (c) The energy E corresponding to Ψ_A and Ψ_B as a function of the displacement co-ordinate Q.

2.1.4 The Jahn–Teller effect

The two main types of degrees of freedom in a solid are electronic and vibrational. The ground state of the electronic part of the Hamiltonian for an insulator can be degenerate. When the electronic energy can be lowered through splitting of the degenerate levels of the ground state manifold resulting from a local elastic distortion at each Jahn–Teller ion site, we deal with the so-called co-operative Jahn–Teller effect. As a result, an electric quadrupole moment is spontaneously generated. The corresponding order parameter can be taken as the energy difference between the two lowest lying electronic states after splitting (see Fig. 2.7 for an illustration).

2.1.5 The Peierls transition

A second type of transition of the above type is the Peierls transition, which differs from the Jahn–Teller effect by the type of broken symmetry present (translational as opposed to point-symmetric). A magnetic analogue of the Peierls transition is called the spin-Peierls transition.

2.1.6 Magnetic orderings

Atomic spins in many compounds and alloys (especially those containing transition group elements) tend spontaneously to form domains within which their individual spins are either parallel or antiparallel. In addition to these two basic magnetic types of order (ferromagnetic and antiferromagnetic, respectively), more complicated (and interesting) regular spatial arrangements of localized magnetic moments are encountered in solids, as illustrated graphically in Fig. 2.8 following reference [9]. Typically, lowering the temperature below the transition temperature T_c (called the Curie temperature for ferromagnets, and the Néel temperature for antiferromagnets) causes the magnetic interactions to overcome thermal fluctuations, and an ordered state is adopted by the system. Thus, in these cases the prototypical (disordered) phase is a paramagnetic one which exists for $T > T_c$ and in which spins are oriented randomly. To measure the degree of order in a complex magnetic phase, as many order-parameter components as there are distinct sublattices may have to be used. For ferromagnets the order parameter is the net magnetization and is a conserved quantity. In antiferromagnets, on the other hand, the order parameter is the so-called staggered magnetization $M_1 - M_2$, where M_1 and M_2 are the magnetization vectors for the two sublattices, and it is not a conserved quantity.

It is also worth mentioning that a group of compounds called metamagnets exhibit order–order type magnetic transitions between several magnetically ordered phases. Moreover, magnetism is not necessarily associated only with localized electrons but also arises due to conduction or delocalized electrons. This latter type is called itinerant electron magnetism and many of the arrangements found for localized magnetic moments also, somewhat surprisingly, exist for itinerant electron magnets.

2.1.7 Ordering transitions in binary alloys

In a brass alloy, for example, which has equal concentrations of copper and zinc, as the temperature is increased, a crystalline structure with ordered atoms undergoes a transition to a disordered structure that differs only in the atomic positions (see Fig. 2.9).

The relevant order parameter may be defined as

$$\eta \equiv \frac{N_1(1) - N_1(2)}{N_1(1) + N_1(2)}, \quad (2.2)$$

DIAMAGNETISM
CRYSTALLINE
AMORPHOUS
Cu, NaCl
SiO2
IDEAL FERROMAGNETISM
|J|≫|D| J>0
CRYSTALLINE
AMORPHOUS
Fe
Fe80P20, FeF2
ANTIFERROMAGNETISM
AF
PARA
J≫D, J<0
CRYSTALLINE
Cr, α-Mn, MnO
INCIPIENT FERROMAGNETISM
PARAMAGNON
Pd, Pt
METAMAGNETISM
FIELD-INDUCED TRANSITIONS
D LARGE
D SMALL
FERRI-MAGNETISM
PARAMAGNETISM
ANTIFERROMAGNETISM
SPIN FLOP
FeCl2·2H2O
FeCl2
MnF2, GdAlO3
SUPERPARAMAGNETISM
77K
200K
H
H/T
P=Ce^-ΔE/kT
ΔE∝V
AMORPHOUS: Fe, Co (~40Å) in Hg
MICTOMAGNETISM
CLUSTER GLASS
A COOLED IN ZERO FIELD
B COOLED IN FIELD
CRYSTALLINE
AMORPHOUS
Fe3Al, AuFe

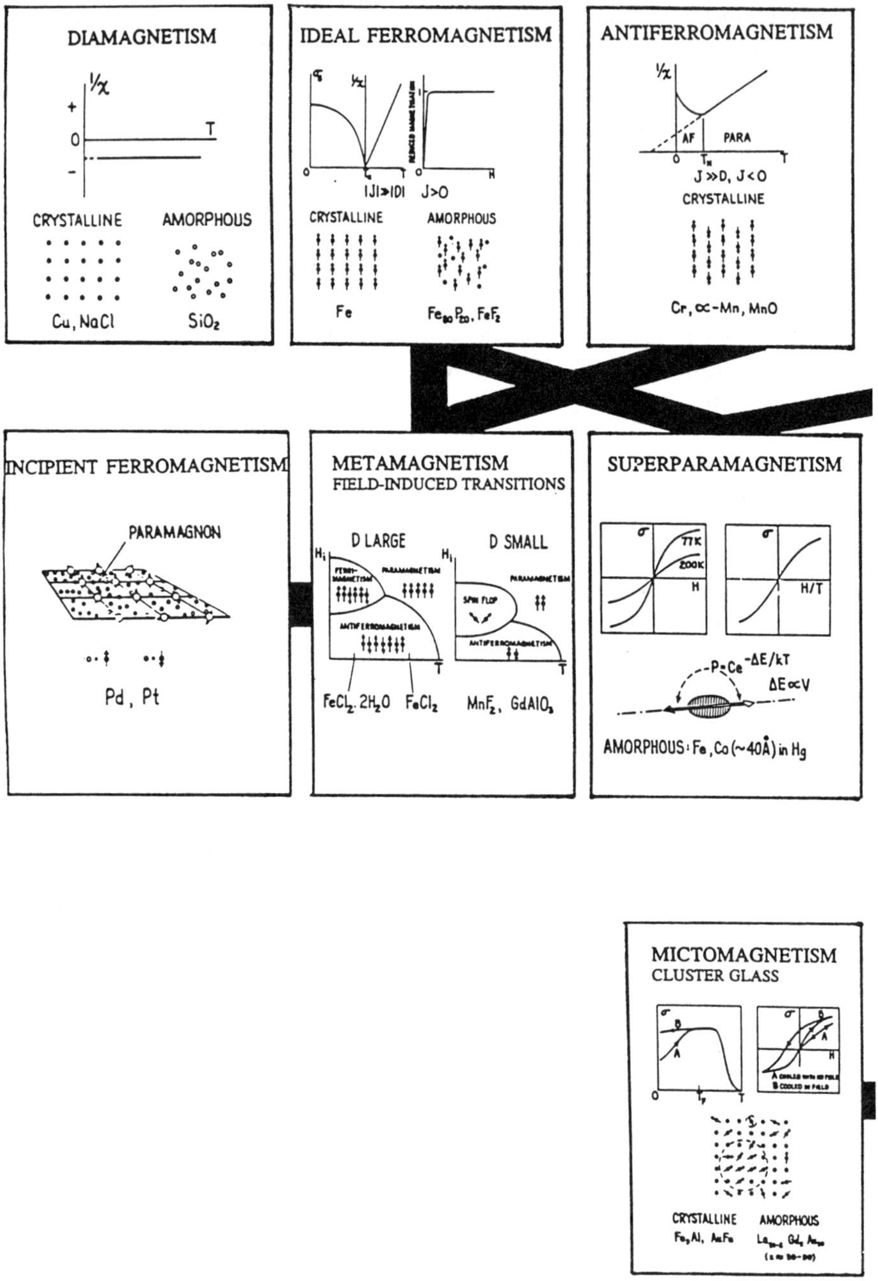

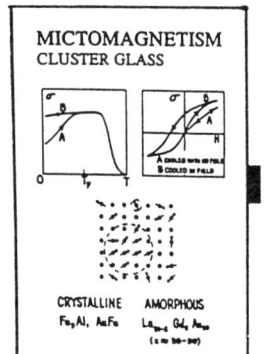

LONG-RANGE ORDER

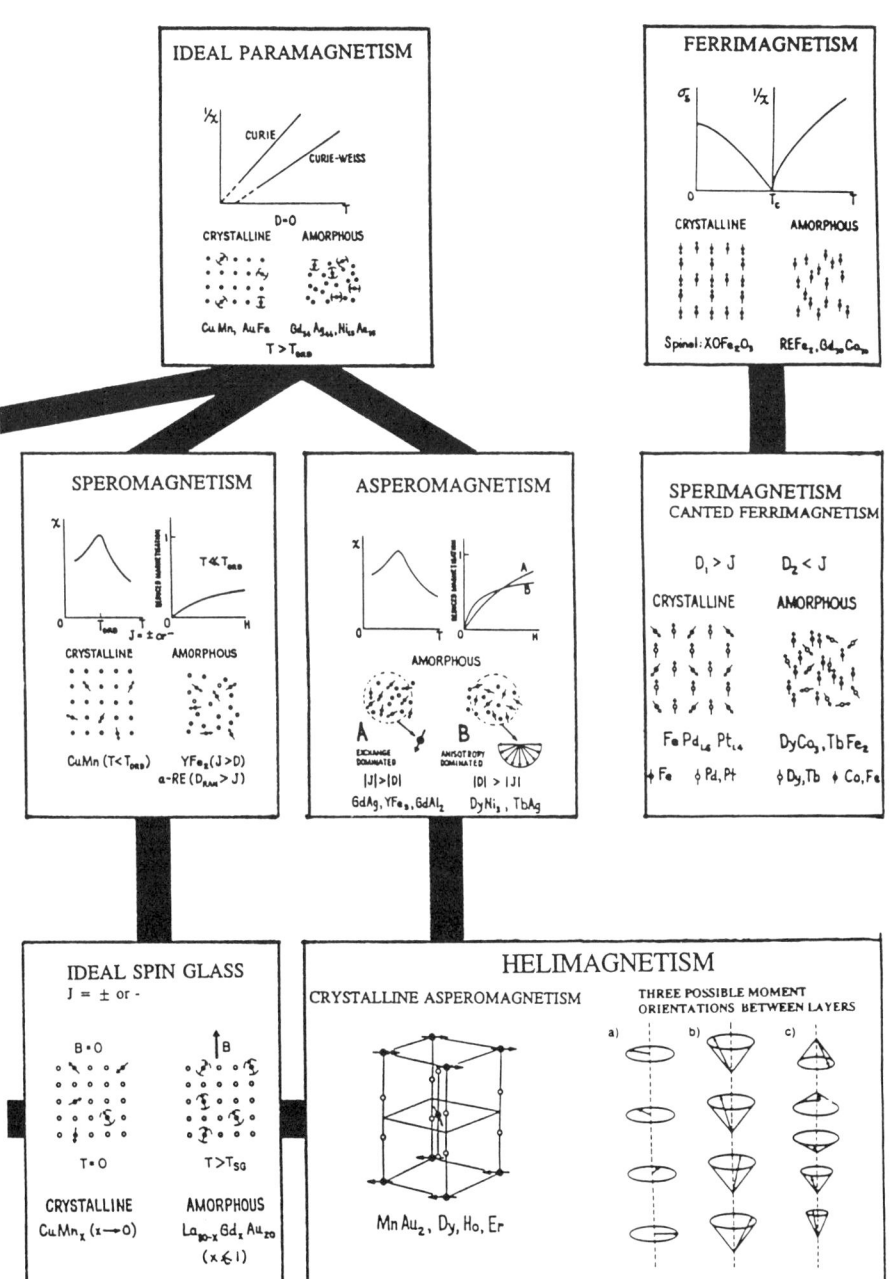

Fig. 2.8 The family tree of magnetic order (following reference [9

84   PHASE TRANSITIONS AND CRITICALITY

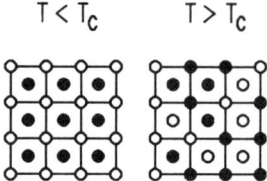

Fig. 2.9 The order–disorder transition in binary alloys.

where $N_i(\alpha)$ is the number of type i atoms in position α ($i = 1, 2$; $\alpha = 1, 2$).

2.1.8 Superconductivity

A number of metals and alloys as well as ceramic materials (the latter group was discovered in 1986 by Bednorz and Müller and referred to as high-T_c superconductors [10]) exhibit a new ordered state in the conduction electron degrees of freedom arising below their characteristic critical

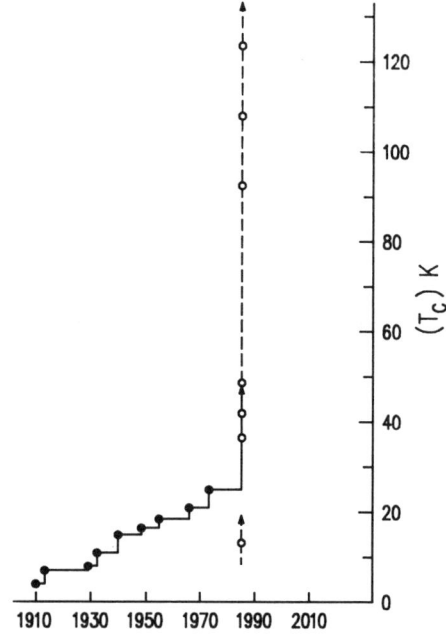

Fig. 2.10 Progress in obtaining ever increasing critical temperatures for superconductors (after reference [10]).

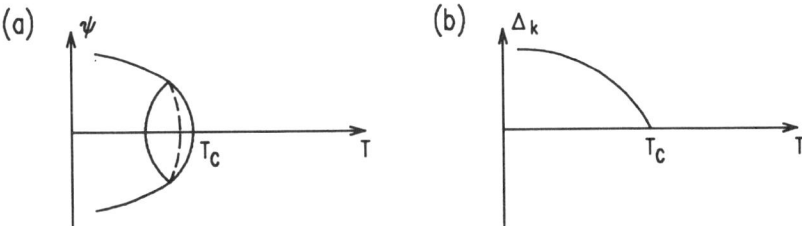

Fig. 2.11 Order parameter (a) and energy gap (b) bifurcation diagrams for superconductors.

temperature T_c. This state has two very important features: ideal conductivity (i.e. zero resistance) and perfect diamagnetism, also called the Meissner effect (i.e. the expulsion of magnetic flux lines). In the case of high-temperature superconductors the Meissner effect is also present but flux expulsion is not complete. The historical progress in obtaining higher and higher critical temperatures is illustrated in Fig. 2.10. Bardeen, Cooper, and Schrieffer demonstrated that the ground state of a (conventional low-temperature) superconductor is formed by bound pairs of electrons (Cooper pairs) with wave vectors k and $-k$ and spins s and $-s$, respectively. The order parameter, therefore, can be chosen as the wave function of the Cooper pair condensate $\psi(r)$ and it exhibits a Hopf bifurcation at $T = T_c$ (see Fig. 2.11(a)). An alternative choice of the order parameter is the energy gap Δ_k, which also bifurcates at $T = T_c$ and scales with $(T - T_c)^{1/2}$ for standard superconductors (see Fig. 2.11(b)). The broken symmetry in this case is the gauge symmetry [2] the generator of which is the Cooper pair number operator, $N = \Sigma_k b_k^+ b_k$. We will return to the topic of superconductivity in Chapter 6, with a much more in-depth discussion.

2.1.9 Superfluidity

The appearance of superfluid properties is manifested by the absence of viscosity, and has been experimentally detected in both ^{4}He and ^{3}He. The atoms of ^{4}He are Bosons and, below a transition temperature T_λ, they undergo the so-called Bose condensation into a $k = 0$ (zero momentum) mode. The associated order parameter is the condensate's quantum wave function. On the other hand, ^{3}He atoms are Fermions and below T_λ ($= 2.7$ mK, which is three orders of magnitude lower than for ^{4}He) are known to form Cooper pairs, very much as superconducting electrons do. Two possible phases can be created, depending on whether the pair's nuclear spin is a triplet with $J = 1$ in which $m_J = 0$ is not present or if all three projections exist for $J = 1$. The order parameter, therefore, is a 2×2 matrix in the spin space.

2.1.10 Liquid crystals

The term *liquid crystal* refers to a large class of organic anisotropic fluids which are composed of strongly elongated molecules. Three basic types of liquid crystals can be distinguished (see Fig. 2.12): nematic, smectic, and cholesteric, with three additional subtypes in the smectic group. The nematic phase is characterized by the existence of a direction to which most of the molecules are parallel, so that the order parameter is a second rank tensor describing correlations along a given direction. In addition to directionality, the smectic phases show a layering pattern. The cholesteric phase is characterized by chirality of the molecules.

Numerous other examples of ordering and phase transition phenomena could be listed here, such as binary fluids, the metal–insulator transition, polymer transitions, and spin- and charge-density waves. What is important, however, is to note the presence of similarities and analogies manifested by: (a) the existence of order parameters; (b) similar features of phase diagrams; and (c) singularities in the responses of each system to external influences at critical temperatures. More detailed explorations of these features will be provided in the subsections that follow.

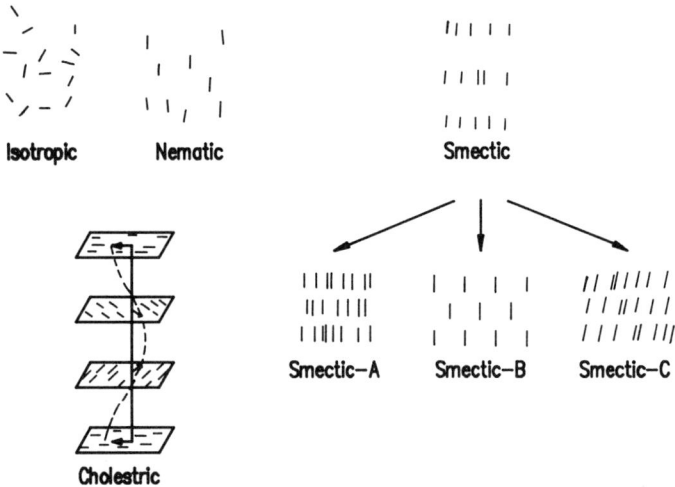

Fig. 2.12 The main types of liquid crystals (after reference [11]).

2.2 General characteristics of phase transitions

One of the common features of systems undergoing phase transitions is the existence of phase diagrams which delineate the regions of stability of equilibrium phases in the space of thermodynamic co-ordinates (pressure p, temperature T, density ρ, etc.). For convenience, usually only a projection on a particular plane is displayed. In Fig. 2.13 are shown several examples of phase diagrams from different condensed-matter phenomena. Note that the following special points have been labelled on these diagrams: (a) a critical (end) point; (b) a triple point; (c) a tricritical point.

The boundaries of the regions of existence of a given phase are drawn as lines of two types: continuous and broken, referring to continuous

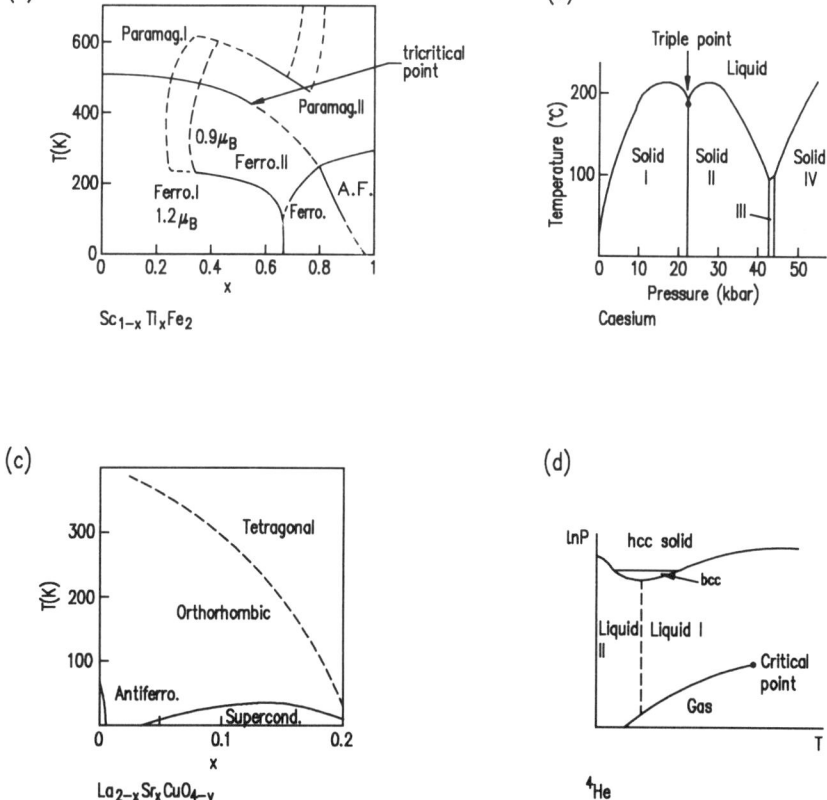

Fig. 2.13 Diverse examples of phase diagrams: (a), (b) following reference [11]; (c) after reference [10]; (d) after reference [2].

(second-order) or discontinuous (first-order) phase transitions, respectively. These are characterized as follows. *First-order phase transitions* are associated with discontinuities of the order parameter, a nonzero latent heat, and hysteresis effects. *Second-order phase transitions* have a continuous order parameter, and no latent heat but a singularity in the specific heat [12]. A more precise definition will be given shortly.

An analytic prescription for finding equilibrium values of thermodynamic quantities is obtained through solving the equation of state of the form

$$f(T, p, \rho) = 0. \tag{2.3}$$

If its solution is single-valued, a unique phase is obtained; on the other hand, if it is multivalued, then provided that all of these solutions are stable, coexistence manifolds (points, lines, planes, etc.) are found for the phase diagram. The general equilibrium conditions for q phases are

$$\mu_1(T, x) = \mu_2(T, x) = \cdots = \mu_q(T, x), \tag{2.4}$$

where μ_i denotes the chemical potential (Gibbs free energy per particle) of the ith phase and x is the generalized thermodynamic force (e.g. pressure). Thus, eqn (2.4) determines a $(3-q)$-dimensional manifold for single-component systems; for example, $q=3$ gives a triple point at which three phase boundaries meet. An extension of this method to multicomponent systems yields the so-called *Gibbs phase rule* [6]. This states that if an s-component system has q phases in equilibrium, then with z generalized thermodynamic forces (including temperature) the dimensionality (dim) of the coexistence manifold is

$$\dim = s - q + z. \tag{2.5}$$

Note that taking $z=2$ and $s=1$ gives the previously obtained results of $3-q$ for a single-component system.

The generally accepted (but still not perfect) way of classifying phase transitions is due to Ehrenfest and is based on the behaviour of the appropriate thermodynamic potential; say, the free energy F. It states that an nth order phase transition occurs when F and its derivatives up to the $(n-1)$th are continuous at T_c while the nth derivative is discontinuous. Differentiation is taken with respect to an arbitrary independent thermodynamic variable (e.g. the order parameter). The difference between $F(T)$ for a first- and a second-order phase transition is schematically shown in Fig. 2.14. Note the existence of a thermal hysteresis (i.e. irreversibility of the phase sequence on changing the temperature in two opposite directions) in the case of first-order transitions and the coexistence of phases a and b between T_a and T_b.

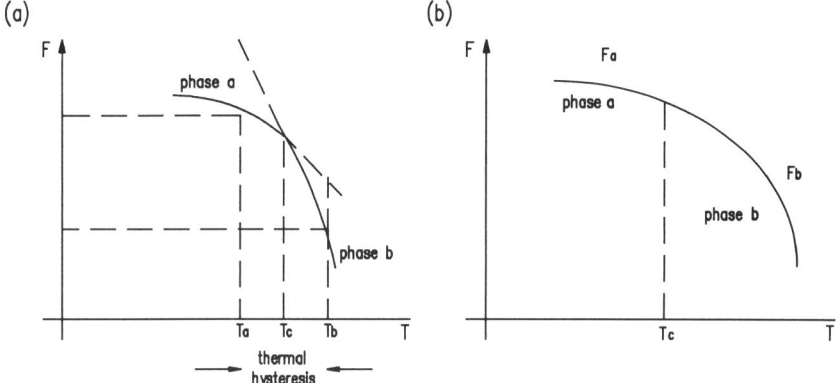

Fig. 2.14 Free energy plots for (a) first- and (b) second-order phase transitions.

The consequences for first-order transitions are the so-called Clausius–Clapeyron equations,

$$dX/dT = \Delta S/\Delta x, \tag{2.6}$$

where S denotes entropy (and hence the latent heat, Q, equals $T\Delta S$) and X is a generalized thermodynamic co-ordinate (with x being its conjugate), e.g. $dP/dT = Q/(V_A - V_B)$ for a fluid and $dH/dT = Q/[T(M_A - M_B)]$ for a magnet, which describes the slope of the transition line. For second-order transitions, *the Ehrenfest equations* are found as

$$dX/dT = \Delta(\partial S/\partial x)/\Delta \chi_T, \tag{2.7}$$

where χ_T is the isothermal susceptibility $\chi_T = \partial X/\partial x$. In the two stated examples we have

$$dP/dT = \Delta C_p/[T\Delta(\partial S/\partial P)] \quad \text{and} \quad dM/dT = -\Delta(\partial M/\partial T)/\Delta \chi_T, \tag{2.8}$$

respectively. In addition to the Ehrenfest classification there exist nonstandard classifications, such as the Münster [13] classification, which focuses on the plots of the entropy and specific heat. It distinguishes a diffusive transition (with a smooth peak of C_v at T_c), an Onsager transition (with a two-sided singularity of C_v at T_c), a lambda transition (with a one-sided singularity of C_v), and an anomalous first-order transition (with a two-sided singularity of C_v and a discontinuity in the enthalpy at T_c).

2.3 Broken symmetries and order parameters

The term *broken symmetry* refers to a situation in which the new (e.g.

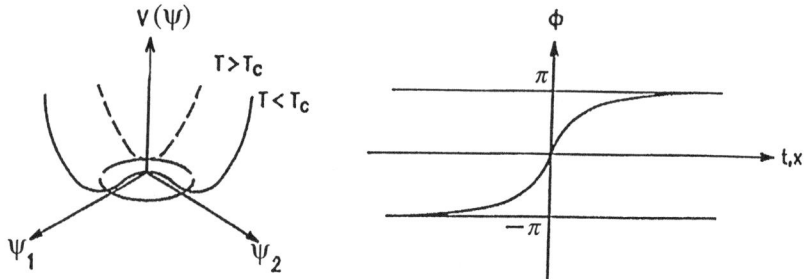

Fig. 2.15 The Goldstone Boson concept.

low-temperature) ground state of the system does not possess the full symmetry group of the Hamiltonian [1]. A broken symmetry can occur either spontaneously (by lowering the temperature, for example) or by the application of an external field or constraint. A classic example is the ferromagnetic-to-paramagnetic phase transition at T_c, in which the full rotational symmetry of the paramagnetic phase is broken by the axial nature of the new ground ferromagnetic state below the Curie temperature. Note that not all phase transitions have to be associated with symmetry breaking (e.g., the liquid–vapour transition in which one can go from the liquid to the vapour phase without singular behaviour), but the vast majority are.

When a symmetry that is broken is continuous (e.g. translational invariance, rotational invariance, gauge symmetry, etc.), then a new excitation may appear, which is called a massless *Goldstone Boson*, the frequency of which goes to zero at long wavelengths [2]. This statement is sometimes referred to as *Goldstone's theorem* [2]. The nature of a Goldstone Boson for a potential $V(\psi)$ that changes form from a single well to a 'Mexican hat' is illustrated in Fig. 2.15. The order parameter ψ is a complex variable, i.e.

$$\psi = \psi_1 + i\psi_2 = |\psi|e^{i\phi}. \tag{2.9}$$

As shown in Fig. 2.15, the associated Goldstone Boson is an excitation that traverses the bottom of the 'Mexican hat' potential in infinite time (or over an infinite range of space, depending on the type of description used). We shall return to this topic from a field-theory viewpoint in Chapter 3.

Examples of Goldstone Bosons include ferromagnetic domain walls and acoustic soft modes in structural phase transitions. However, in superconductivity no Goldstone Bosons are allowed, since the number operator is the generator of the (continuous) gauge transformation, while the Goldstone theorem requires that

$$\int d\mathbf{s} \cdot \langle [j(\mathbf{r},t), \psi(\mathbf{r}')] \rangle = 0, \qquad (2.10)$$

where j denotes the current density [2] and the integral is over a closed loop. The above statement follows from the phase-number quantum uncertainty principle for superconductors. The long-range Coulomb interaction is the reason why the integral does not vanish in superconductors. Also, no Goldstone modes are allowed for plasmons and optical phonons. Note also that Goldstone's theorem is strictly applicable at 0 K only. Approaching a second-order phase transition $T_c \neq 0$, the system will fluctuate into the ordered state from above. The spatial extent of these fluctuations is characterized by the coherence length ξ. The relaxation time for these collective fluctuations is of the order of

$$\tau = \frac{F_{\text{order}} - F_{\text{disorder}}}{\hbar}. \qquad (2.11)$$

Thus, on the scale of $t \ll \tau$, the ordered regions will have collective modes with wavelengths $\lambda < \xi$.

In some cases, such as the onset of ferroelectricity with a transverse optical branch or structural transitions with the softening of a collective excitation, a so-called *soft mode*, rather than a Goldstone mode, is responsible for the transition. Then, the soft mode's frequency ω_k for the wave vector k tends to 0 as $T \to T_c$.

There exist several different types of broken symmetries:

- translational (crystal formation, structural transitions)
- gauge symmetry (superfluidity, superconductivity)
- time reversal (ferromagnets)
- local rotational (liquid crystals)
- rotational (some structural phase transitions)
- space inversion (ferroelectricity).

Gauge symmetry is of special importance, since it is a universal property of Hamiltonians whenever the total number of particles or a generalized charge-like conserved quantity exists. Then, the order parameter ψ is a complex quantity and its local density can be defined as $\rho = \psi^* \psi(r)$, so that a phase shift of ψ according to $\psi \to \psi \, e^{i\varphi}$ leaves the Hamiltonian invariant. Furthermore, *local gauge* symmetry may exist where $\varphi = \varphi(r)$ but is only known to occur in the case of electric charge since it requires a gauge field which couples to the conserved charge (in this case, the electromagnetic field).

Near a second-order phase transition, due to the reduction of rigidity of the system or, conversely, due to its softening, *critical fluctuations* will dominate, no matter what the dimensionality and symmetry of the disordered phase. Their amplitude diverges as $T \to T_c$. This is a classical phenomenon, since it occurs at a finite temperature and the frequencies involved are much less than those of thermal fluctuations. As will be shown later, critical fluctuations become of quasi-macroscopic size and dimension at the critical point. Critical fluctuations cannot be associated with *collective modes* in all cases (and are called diffusive since they have a broad spectrum), but in those which are, they are called soft modes. In addition to critical fluctuations we also have quantum excitations called *collisionless modes* which are required by the symmetries or the conservation laws. In some cases, *hydrodynamic modes* are also present (e.g. sound, second sound, etc.).

We mentioned that an inherent property of systems undergoing phase transitions is the decrease of their rigidity. We may follow Anderson [1] in defining generalized rigidity as the propensity of the system to evolve towards a stable ground state in the low-symmetry phase. Consequently, work has to be expended in order to change this state to another equilibrium. An associated property is the presence of small-amplitude oscillations about the equilibrium state.

As already mentioned several times, an important characteristic quantity that can be identified in all phase transitions is the *order parameter*. Its usefulness lies in reducing the number of variables for a many-body system from the order of Avogadro's number to a small number being equal to the number of the order parameter components and several other independent thermodynamic quantities. Thus, reductionism is at the heart of this idea. The concept of an order parameter was first introduced by

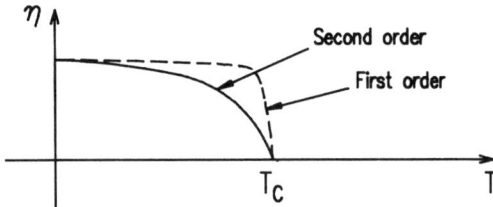

Fig. 2.16 A typical plot of the scalar order parameter as a function of temperature.

Table 2.1 Examples of order parameters.

Phenomenon	Disordered phase	Ordered phase	Order parameter
Equilibrium			
Condensation	Gas	Liquid	Density difference, $\rho_L - \rho_G$
Spontaneous magnetization	Paramagnet	Ferromagnet	Net magnetization, **M**
Antiferromagnetism	Paramagnet	Antiferromagnet	Staggered magnetization, $\mathbf{M}_1 - \mathbf{M}_2$
Superconductivity	Conductor	Superconductor	Cooper pair wave function, ψ
Alloy ordering	Disordered mixture	Sublattice ordered alloy	Sublattice concentration
Ferroelectricity	Paraelectric	Ferroelectric	Polarization
Superfluidity	Fluid	Superfluid	Condensate wave function
Nonequilibrium			
Tunnel diode	Insulator	Conductor	Capacitance charge
Laser action	Lamp (incoherent)	Laser (coherent)	Electric field intensity
Super-radiant source	Noncoherent polarization	Coherent polarization	Atomic polarization
Fluid convection	Turbulent flow	Bénard cells	Amplitude of mode

Landau [6], and to this day it does not have a precise definition. It is a thermodynamic bulk quantity (usually) which is invariant with respect to the symmetry group of the low-temperature (and low-symmetry) phase, and it is zero above the transition temperature while it is nonzero below. It is a quantitative measure of the amount and type of order that is built up in the neighbourhood of the critical point. Below T_c, it can be a degenerate quantity. In order to find the equation of state, a minimization procedure has to be followed for an appropriate thermodynamic potential. From its original application to second-order phase transitions, the idea of an order parameter has been extended to first-order ones (Fig. 2.16). It has been generalized over the years from a scalar to a time- and space-dependent function, and we will show in Chapter 5 that its representation as a quantum field is not only legitimate but quite useful.

Examples of diverse applications of the order parameter concept to both equilibrium and nonequilibrium critical phenomena are listed in Table 2.1.

A summary of the results in this section is provided in Table 2.2, which lists a number of critical phenomena with the type of order parameter, its dimensionality, and the presence of associated excitations. This table has

Table 2.2 An overview of broken symmetry phenomena (after reference [1]).

Phenomenon	Broken symmetry	Order parameter	Dimensionality $T \to 0$	Dimensionality T_c	Goldstone Boson	Fluctuations	Generalised rigidity	Singularities
Ferroelectricity	Translational	P	1	1 or 3	No (optical phonons)	Soft modes	Ferroelectric hysteresis	Domain walls
Ferromagnetism	Time reversal	M	1,3	1,3	Spin waves ($\omega \sim k^2$)	Spin waves	Hysteresis	Domain walls
Antiferromagnetism	Local time reversal	$M_s = M_1 - M_2$	1,3	1,3	Spin waves $\omega \sim \sqrt{k^2 + c}$	Spin waves	AF resonance effects	Domain walls
Superconductivity	Gauge	$\psi = \eta e^{i\phi}$	2	2	No	Diffusive fluctuations of gap	Meissner effect	Flux lines, vortices
Superfluidity	Gauge	$\psi = \eta e^{i\phi}$	2	2	Phonons	Diffusive fluctuations	Superfluidity	Vortex lines
Nematic liquid crystals	Local rotations	Directrix, ($\vec{d}$)	3	3	Yes, in principle	Yes	Orientation elasticity	Disclination points
Cholesteric, smectic liquid crystals	Local rotations	$\rho(Q)$	>3	>3	Yes	Yes	Orientation elasticity	Disclination points and dislocations
Crystal	Translations	ρ_G	3	3	Two transverse and one longitudinal phonons	Phonons	Rigidity	Dislocations, grain boundaries, vacancies, and interstitials
Charge density waves (CDW)	Translations	ρ_G on a triangular lattice	2	2	Phasors	Yes	Sliding of CDWs	Dislocations, discommensurations

2.4 Critical exponents and static scaling

A useful concept in analysing phase transitions both theoretically and experimentally has proved to be that of a critical exponent [12]. In general, if a bulk physical quantity $Q(T)$ either diverges or tends to a constant value (in particular, zero) as the temperature T tends to T_c, then its behaviour in the vicinity of T_c can be characterized by first defining a dimensionless, small quantity ε as

$$\varepsilon \equiv (T - T_c)/T_c, \qquad (2.12)$$

which is called the reduced temperature, and then calculating the associated critical exponent, say a hypothetical exponent called μ, according to

$$\mu = \lim_{\varepsilon \to 0} \frac{\ln Q(\varepsilon)}{\ln \varepsilon}. \qquad (2.13)$$

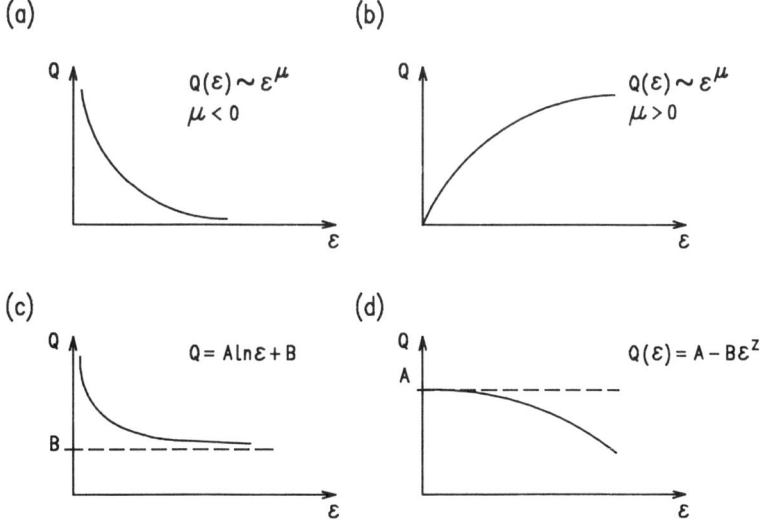

Fig. 2.17 The four generic forms of behaviour near criticality (after reference [12]).

Table 2.3 The definitions of critical exponents for liquid–vapor and magnetic systems (after reference [12]).

Exponent	Definition (liquid–vapour)	Definition (magnetic)
α'	Specific heat at constant volume $C_v \sim (-\varepsilon)^{-\alpha'}$	Specific heat at constant H $C_H \sim (-\varepsilon)^{-\alpha'}$
α	$C_v \sim \varepsilon^{-\alpha}$	$C_H \sim \varepsilon^{-\alpha}$
β	Density difference $\rho_L - \rho_G \sim (-\varepsilon)^{\beta}$	Magnetization $M \sim (-\varepsilon)^{\beta}$
γ'	Isothermal compressibility $\kappa_T \sim (-\varepsilon)^{-\gamma'}$	Isothermal susceptibility $\chi_T \sim (-\varepsilon)^{-\gamma'}$
γ	$\kappa_T \sim \varepsilon^{-\gamma}$	$\chi_T \sim \varepsilon^{-\gamma}$
δ	Pressure–density critical isotherm $P - P_c \sim |\rho_L - \rho_G|^{\delta}$ $(T = T_c)$	Magnetic field–magnetization $H \sim |M|^{\delta}$ $(T = T_c)$
ν'	Correlation length $\xi \sim (-\varepsilon)^{-\nu'}$	Correlation length $\xi \sim (-\varepsilon)^{-\nu'}$
ν	$\xi \sim \varepsilon^{-\nu}$	$\xi \sim \varepsilon^{-\nu}$
η	Density–density pair correlation function $\Gamma(r) \sim |r|^{-(d-2+\eta)}$	Spin–spin pair correlation function $\Gamma(r) \sim |r|^{-(d-2+\eta)}$

In practice, the range of ε that describes critical behaviour is usually between 10^{-7} and 10^{-1}. The four typical forms of behaviour of $Q(\varepsilon)$ are illustrated in Fig. 2.17.

The most important critical exponents are usually denoted (for historical reasons) as α, β, γ, δ, ν, and η, and they describe, respectively, the behaviour of specific heat, order parameter, isothermal susceptibility, response to an external field, correlation length, and pair correlation function. These definitions for liquid–vapour transitions and magnetic systems are given in Table 2.3.

The pair correlation function in Table 2.3 is defined as

$$\Gamma(r) \equiv \int d^3x \langle (\psi(r) - \bar{\psi})(\psi(0) - \bar{\psi}) \rangle \qquad (2.14)$$

and $\langle \ldots \rangle$ denotes ensemble averaging, with $\bar{\psi}$ being the equilibrium value of the order parameter. The correlation length ξ can be defined in many ways, one of which is

$$\xi \equiv \left[-\tfrac{1}{2} \Gamma^{-1}(0)(\partial^2 \Gamma(k)/\partial k^2)_{k=0} \right]^{-1/2}, \qquad (2.15)$$

where $\Gamma(k)$ is the Fourier transform of $\Gamma(r)$.

Through the use of thermodynamic relations for the above quantities, it was determined that critical exponents must satisfy a number of inequalities, some of which are listed below [12]:

$$\alpha' + 2\beta + \gamma' \geq 2, \qquad \alpha' + \beta(\delta + 1) \geq 2, \qquad (2.16a,b)$$

$$\gamma'(\delta + 1) \geq (2 - \alpha')(\delta - 1), \qquad \gamma' \geq \beta(\delta - 1), \qquad (2.16c,d)$$

$$d(\delta - 1)/(\delta + 1) \geq 2 - \eta, \qquad d\gamma'/(2 - \alpha') \geq d\gamma'/(2\beta + \gamma') \geq 2 - \eta, \qquad (2.16e,f)$$

$$(2 - \eta)\nu \geq \gamma, \qquad d\nu' \geq 2 - \alpha', \qquad d\nu \geq 2 - \alpha, \qquad (2.16g,h)$$

where d denotes the spatial dimensionality of the system under consideration.

In Table 2.4 we have gathered a large representative sample of experimental values of critical exponents for diverse physical systems. The results

Table 2.4 Experimental values of critical exponents for selected physical systems (following references [11, 12, 14, 15, 16]).

System	α	β	γ	δ	ν	η
Fluids						
CO_2	0.125	0.347	1.17	4.2	—	—
Xe	0.08	0.345	1.21	4.4	0.57	—
^{3}He	0.11	0.361	1.18	3.5	—	—
^{4}He	0.15	0.354	1.14	—	—	—
Ar	0.13	0.34	1.20	4.2	—	—
Ferromagnets						
Ni	−0.1	0.365	1.35	4.2	—	—
$CrBr_3$	—	0.368	1.215	4.3	—	—
Fe	−0.12	0.33	1.333	4.22	0.64	0.07
$YFeO_3$	—	0.354	1.33	2.8	—	—
Antiferromagnets						
$CoCl_2 \cdot 6H_2O$	≤ 0.11	0.23	—	—	—	—
$RbMnF_3$	−0.139	0.316	1.397	—	—	0.067
Binary mixture						
$Cl_4-C_7F_{14}$	—	0.335	1.2	4.0	—	—
Binary alloy						
Co–Zn	—	0.305	1.25	—	—	—
Spin glass						
AgMn	—	1.0	2.2	3.1	3.0	—
Insulators (typically)	—	0.368	1.215	4.28	—	—
Metals (typically)	—	0.378	1.34	4.58	—	—

for critical exponents calculated in various theoretical models will be discussed in the subsections that follow.

Early on in the development of the theory of phase transitions, it was realized (by Widom, Domb, Hunter, and Kadanoff, among others) that the values of the various critical exponents are not unrelated but satisfy a number of relationships. These relationships can be all derived using scaling concepts. In particular, the so-called *static scaling hypothesis* asserts that the relevant thermodynamic potential—say, the Gibbs potential $G(T, H)$ for magnetic systems—is a generalized homogeneous function, i.e. it satisfies [12]:

$$G(\lambda^{a_\varepsilon}\varepsilon, \lambda^{a_H} H) = \lambda G(\varepsilon, H), \tag{2.17}$$

for an arbitrary value of λ and specific values of a_ε and a_H which are characteristic exponents of a given phase transition. It can be readily demonstrated that eqn (2.17) leads to a number of similar relations for other thermodynamic quantities; namely,

$$\lambda^{a_H} M(\lambda^{a_\varepsilon}\varepsilon, \lambda^{a_H} H) = \lambda M(\varepsilon, H), \tag{2.18}$$

$$\lambda^{2a_H} \chi_T(\lambda^{a_\varepsilon}\varepsilon, \lambda^{a_H} H) = \lambda \chi_T(\varepsilon, H), \tag{2.19}$$

and

$$\lambda^{2a_\varepsilon} C_H(\lambda^{a_\varepsilon}\varepsilon, \lambda^{a_H} H) = \lambda C_H(\varepsilon, H). \tag{2.20}$$

By appropriately choosing the value of λ and making use of the definitions of critical exponents (see Table 2.3), the following equalities are found to be a consequence of the static scaling hypothesis:

$$\beta = (1 - a_H)/a_\varepsilon, \qquad \delta = a_H/(1 - a_H), \qquad (2.21, 2.22)$$

$$\gamma' = (2a_H - 1)/a_\varepsilon, \qquad \gamma = (2a_H - 1)/a_\varepsilon, \qquad (2.23, 2.24)$$

and

$$\alpha' = 2 - 1/a_\varepsilon. \tag{2.25}$$

A number of conclusions can now be drawn on the basis of the above results. First, apart from the distance-sensitive exponents (ν and η), all of the remaining exponents can be expressed in terms of only two independent ones: a_ε and a_H. Second, using eqns (2.21)–(2.25), it can be shown that the following relationships are satisfied:

$$\alpha' + 2\beta + \gamma' = 2, \qquad \alpha + \beta(\delta + 1) = 2, \qquad (2.26a, b)$$

$$\gamma(\delta + 1) = (2 - \alpha)(\delta - 1), \qquad \gamma' = \beta(\delta - 1), \qquad (2.26c, d)$$

and

$$\alpha = \alpha', \qquad \gamma = \gamma', \tag{2.26e, f}$$

TWO EARLY THEORETICAL MODELS

which have their counterparts in the thermodynamics-based inequalities of eqn (2.16)!

It was, therefore, suspected that scaling is indeed an important property of the critical state and that the remaining two exponents ν and η should be involved in a scaling relationship. This gave a motivation for dynamic scaling and the development of renormalization group ideas, which will be discussed in a later subsection. Before we do that, however, we now turn to an overview of specific theoretical models grouped in three separate subsections: early models (van der Waals and Curie–Weiss), lattice models, and the Landau model. This will be followed by an analysis of the role played by fluctuations, leading to the Landau–Ginzburg model and renormalization group methods.

2.5 Two early theoretical models

2.5.1 The van der Waals theory of liquid–vapour transitions

The van der Waals equation of state is given by [17]

$$(P + a/V^2)(V - b) = RT, \qquad (2.27)$$

where a and b are empirical constants related to long-range attraction between molecules and hard-core repulsion, respectively. The critical point (P_c, V_c, T_c) of the system is found as an inflection point of the equation of state; thus it satisfies the following three simultaneous equations:

$$P_c = \frac{RT_c}{V_c - b} - \frac{a}{V_c^2}, \quad \left(\frac{\partial P}{\partial V}\right)_{T=T_c} = 0, \quad \left(\frac{\partial^2 P}{\partial V^2}\right)_{T=T_c} = 0. \qquad (2.28)$$

It is then found that

$$P_c = a/27b^2, \quad V_c = 3b, \quad T_c = 8a/27bR. \qquad (2.29)$$

Note that the critical compressibility ratio is $Z_c = P_c V_c / RT_c = 0.375$ within this model, while in real gases it ranges between 0.230 (for H_2O) and 0.308 (for ^4He).

Introducing reduced parameters according to

$$\tilde{P} = P/P_c, \quad \tilde{V} = V/V_c, \quad \tilde{T} = T/T_c, \qquad (2.30)$$

the van der Waals equation of state becomes dimensionless and substance-independent:

$$(\tilde{P} + 3/\tilde{V}^2)(3\tilde{V} - 1) = 8\tilde{T}. \qquad (2.31)$$

This is shown in Fig. 2.18.

It is important to emphasize that for $T < T_c$ there can be either one or

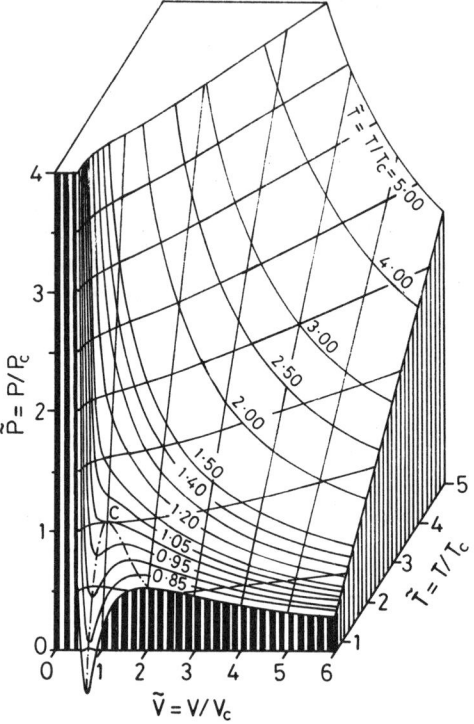

Fig. 2.18 The van der Waals $(\tilde{P}, \tilde{V}, \tilde{T})$ surface (after reference [12]).

three solutions of the equation of state, depending on the values of P and V, unlike in the $T > T_c$, case, in which only a single solution exists. Our interest lies in the vicinity of the critical point and to examine it more closely we introduce a new set of scaled variables; namely,

$$p \equiv (P - P_c)/P_c, \quad v \equiv (V - V_c)/V_c, \quad \varepsilon \equiv (T - T_c)/T_c. \quad (2.32)$$

Then, to the dominant order, the equation of state can be expanded and rearranged to yield

$$2p\left(1 + \tfrac{7}{2}v + 4v^2 + \tfrac{3}{2}v^3\right) \equiv -3v^3 + 8\varepsilon(1 + 2v + v^2). \quad (2.33)$$

It turns out that the following power-law relations can be deduced in the vicinity of the critical point [16]:

$$\rho_L - \rho_G \sim (-\varepsilon)^{1/2}, \quad p \sim v^3, \quad \kappa_T \sim \varepsilon^{-1}, \quad (2.34\text{a–c})$$

and

$$\Delta C_v \simeq \begin{cases} \tfrac{9}{2}Nk + \cdots, & T < T_c, \\ 0, & T > T_c. \end{cases} \quad (2.34\text{d})$$

Thus, the set of corresponding critical exponents is

$$\alpha = 0, \quad \beta = \tfrac{1}{2}, \quad \gamma = 1, \quad \delta = 3, \tag{2.35}$$

which plays a special role in theoretical modelling.

2.5.2 The Curie–Weiss theory of ferromagnetism

An adequate description of a paramagnet consists of a set of N identical noninteracting spin angular momenta $\mathbf{S}_i$ that can only couple to an external magnetic field $\mathbf{H}_0$ through the Zeeman term in the Hamiltonian

$$H = -g\mu_B \sum_{i=1}^{N} \mathbf{S}_i \cdot \mathbf{H}_0, \tag{2.36}$$

where g is the Landé factor and μ_B is the Bohr magneton. Then, the partition function is readily found as

$$Z = \left[\sinh\left(\frac{2S+1}{2S} x\right) \Big/ \sinh(x/2S) \right]^N, \tag{2.37}$$

where $x \equiv g\mu_B H_0 S/kT$. As a consequence, equilibrium magnetization is derived as

$$M = -\left(\frac{\partial F}{\partial H_0}\right) = \frac{\partial}{\partial H_0}(k_B T \ln Z)\big|_T = M_0 B_S\left(\frac{x}{S}\right), \tag{2.38}$$

where F is the free energy, $M_0 = NSg\mu_B$ is the saturation magnetization, and B_S is the Brillouin function, defined by

$$B_S\left(\frac{x}{S}\right) \equiv \frac{2S+1}{2S} \coth\left(\frac{2S+1}{2S} x\right) - \frac{1}{2S} \coth\left(\frac{x}{2S}\right). \tag{2.39}$$

It is apparent from eqn (2.38) that, for all $T > 0$, $M = 0$ if no external field is applied ($H_0 = 0$). This excludes the possibility of ferromagnetism. Weiss proposed a simple generalization which would easily lead to ferromagnetism. Adding interactions amongst the spins in the Heisenberg Hamiltonian form,

$$\Delta H = -J \sum_{\langle i \neq j \rangle} \mathbf{S}_i \cdot \mathbf{S}_j, \tag{2.40}$$

can be shown to result (to the crudest approximation) in an effective molecular field such that

$$\Delta H \simeq -g\mu_B H_e \cdot \sum_i S_i, \qquad (2.41)$$

where

$$H_e = \frac{2J}{g\mu_B} \sum_i S_i = \frac{2zJ}{Ng^2\mu_B^2} M = \gamma M, \qquad (2.42)$$

in which z is the number of nearest neighbours while γ is the so-called Weiss constant. As a consequence, the modified equation of state can be obtained from eqn (2.38) as

$$M = Ng\mu_B S B_S(hS), \qquad (2.43)$$

and $h = g\mu_B H_{\text{eff}} \beta$, with $H_{\text{eff}} = H_0 + H_e$. It can now be seen that at sufficiently high temperatures magnetization is uniquely determined and obeys

$$M = \frac{Ng^2\mu_B^2 S(S+1)}{3k_B T} H_{\text{eff}} = \frac{CH_{\text{eff}}}{T}, \qquad (2.44)$$

yielding a Curie–Weiss law for susceptibility,

$$\chi = \frac{C}{T - T_c}, \qquad (2.45)$$

with the Curie temperature given by

$$T_c = \frac{2zJS(S+1)}{3k_B}. \qquad (2.46)$$

However, below T_c, eqn (2.43) no longer has a unique solution—there are, in fact, three. This is still true for $H = 0$, i.e. in the absence of external fields. One of these solutions is $M = 0$ (paramagnetism) but the remaining two, $\pm M(T)$, signify the onset of ferromagnetism. Analysing eqn (2.43) close to T_c and for small values of H_0, it can be demonstrated that the following relations are satisfied by the ferromagnetic solutions:

$$M \sim (-\varepsilon)^{1/2}, \qquad \chi_T \sim (-\varepsilon)^{-1}, \qquad H_0 \sim M^3, \qquad (2.47\text{a–c})$$

and

$$C_H \sim \begin{cases} \frac{3}{2}NK, & T < T_c, \\ 0, & T > T_c. \end{cases} \qquad (2.47\text{d})$$

Once again, we see that the associated set of critical exponents is the same as in eqn (2.35), i.e. $\alpha = 0$, $\beta = \frac{1}{2}$, $\gamma = 1$, and $\delta = 3$. These values are sometimes referred to as the *classical* (or mean field) *critical exponents*.

In the next section we will explore other important theoretical models based on lattice calculations, but we will return to mean field models in the following section.

2.6 Lattice models

We begin this section by looking at the consequences of the Yang–Lee theorems and related results. This will set the stage for an analysis of particular models.

2.6.1 The Yang–Lee theorems

Yang and Lee [18] considered a system of N classical particles interacting via a two-body potential $u_{ij} = u(|r_i - r_j|)$, where r_i are particle position vectors, such that

$$u(r) = \infty \quad \text{if} \quad r \leqslant a,$$
$$0 > u(r) > -\varepsilon_0 \quad \text{if} \quad a < r < r_0,$$
$$u(r) = 0 \quad \text{if} \quad r \geqslant r_0.$$

Carrying out calculations in the grand canonical ensemble, with fugacity defined by $z = \exp(\beta\mu)$, where μ is the chemical potential, they showed that both pressure P and specific volume $v \equiv V/N$ are analytic functions of z in a region of the complex plane that includes the real positive axis. Moreover, they found that (a) $P \geqslant 0$; (b) $v_0 \leqslant v < \infty$, and (c) $\partial P/\partial v \leqslant 0$. It was then concluded that no singularities occur in the statistical description as long as the volume is finite. Conversely, singularities may appear on the real positive axis in the complex z plane only if we go to the so-called *thermodynamic limit*, i.e. we let $V \to \infty$, $N \to \infty$, and keep N/V constant. This limiting procedure is independent of the shape of the volume and is uniformly convergent, as stated in the two theorems below (where Ξ denotes the grand partition function).

THEOREM 1. $\lim_{V \to \infty} [V^{-1} \log \Xi(z, v)]$ *exists for all $z > 0$, is independent of the shape of the volume V and is a continuous, nondecreasing function of z. It is assumed that as $V \to \infty$, the surface area of V increases no faster than $V^{2/3}$.*

THEOREM 2. *Let R be a region in the complex z plane that contains a segment of the positive real axis and contains no root of the equation $\Xi(z, V) = 0$ for* any V. Then, for all z in R, the quantity $V^{-1} \log \Xi(z, V)$ converges uniformly to a limit as $V \to \infty$. This limit is an analytic function of z for all z in R.

Thus, a single phase is defined by smoothness of the functions $P(z)$ and $v(z)$, and hence $P(v)$. Conversely, a phase transition is always associated with singular behaviour of the above quantities and that necessarily occurs in the thermodynamic limit. Three typical forms of behaviour, a single phase, a second-order phase transition, and a first-order phase transition, are shown in Fig. 2.19.

We note in closing that Jones [20] proved analogous statements for the canonical ensemble in the complex plane of $\beta = (k_B T)^{-1}$. This, of course, is consistent with van Hove's theorem [21], stating that an equation of state calculated in the thermodynamic limit is the same for both canonical and grand canonical ensembles.

2.6.2 The Ising model

The Ising model, despite being very simple, is of crucial importance to the understanding of critical phenomena [22]. It is based on a regular lattice of N points in an n-dimensional space, with spin variables s_i placed at each lattice site and interacting with their nearest neighbours (denoted by $\langle i, j \rangle$) and an external magnetic field H_o parallel to the z-axis. The Ising Hamiltonian is then

$$H = - \sum_{\langle i,j \rangle} J S_i^z S_j^z - H_o \sum_i S_i^z. \tag{2.48}$$

The number of nearest neighbours is labelled z_0 and can be four (for a 2-D square lattice), six (for a 3-D simple cubic or 2-D hexagonal lattice), eight (for a 3-D body-centred cubic), or any other number depending on the dimensionality and geometry of the lattice. The spin–spin interactions in this model are assumed to be isotropic and if $J > 0$ they are of ferromagnetic type, while for $J < 0$ they are antiferromagnetic. The (canonical) partition function

$$Z = \sum_{S_1^z} \sum_{S_2^z} \cdots \sum_{S_N^z} e^{-\beta H(\{S_i\})}, \tag{2.49}$$

has 2^N terms and has been the subject of intensive theoretical investigations over the past several decades. The result in one dimension is relatively simple, and it leads to the free energy [19]

$$F = -N k_B T \ln\left[e^{\beta J} \cosh \beta H_o + \sqrt{e^{2\beta J} \sinh^2 \beta H_o + e^{-2\beta J}}\right], \tag{2.50}$$

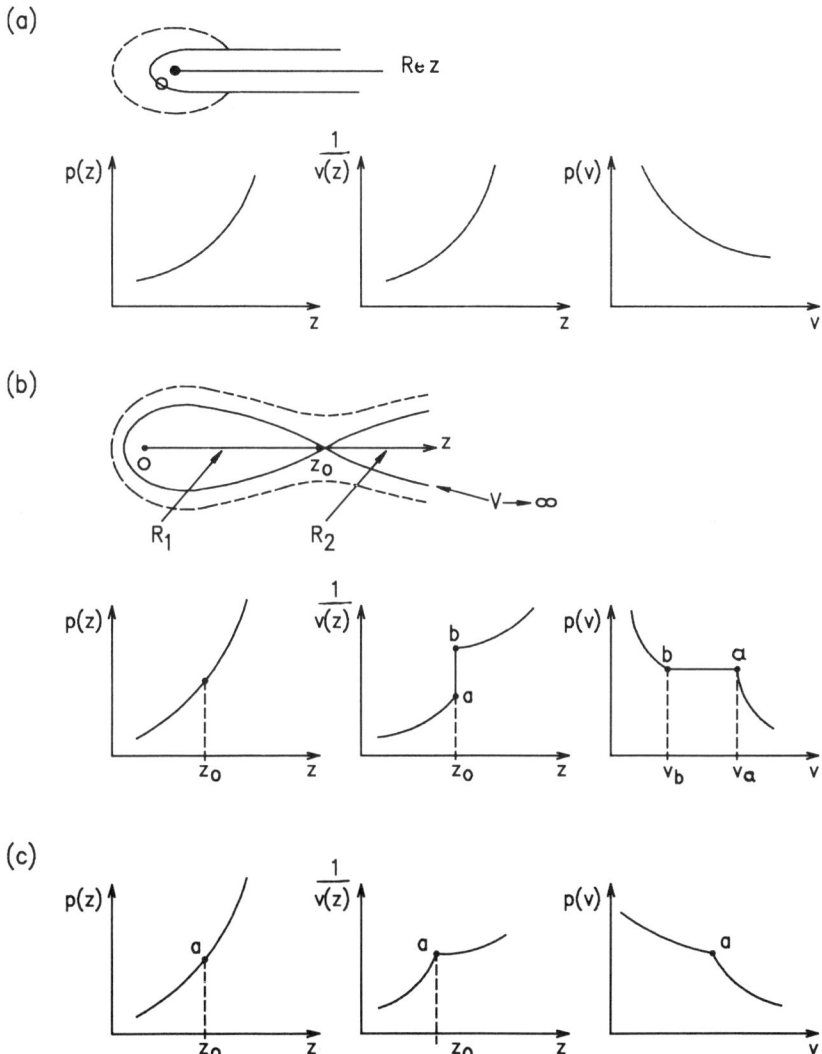

Fig. 2.19 Three typical situations encountered in the Yang–Lee theorem (following reference [19]): (a) a single phase; (b) a first-order phase transition; (c) a second-order phase transition.

and hence the mean magnetization per site is

$$M/N = \sinh(\beta H_o)/[\sinh^2(\beta H_o) + e^{-4\beta J}]^{1/2}. \tag{2.51}$$

However, as can be readily verified, there is no spontaneous magnetization (and hence no phase transition) for $T > 0$.

In two dimensions, Onsager [23] obtained, following rather involved algebra, the free energy

$$F = -Nk_B T \left\{ \ln 2\cosh(2J\beta) + \frac{1}{2\pi} \int_0^\pi d\phi \ln\left[\tfrac{1}{2}\left(1 + \sqrt{1 - k^2 \sin^2 \phi}\right)\right] \right\}, \tag{2.52}$$

where k is the Jacobi elliptic modulus [24], given by

$$k = 2\sinh(2\beta J)/\cosh^2(2\beta J). \tag{2.53a}$$

It can be demonstrated that the associated critical temperature is

$$T_c = 2.269 J/k_B, \tag{2.53b}$$

and the resultant critical exponents were found to be

$$\alpha = 0(\log), \quad \beta = \tfrac{1}{8}, \quad \gamma = \tfrac{7}{4}, \quad \delta = 15, \quad \nu = 1, \quad \eta = \tfrac{1}{4}. \tag{2.54}$$

The first exponent above indicates that there is a logarithmic specific heat divergence. Spontaneous magnetization per spin is 0 above T_c and below T_c (at $H_o = 0$) is given by

$$M/N = (1 + \Delta^2)^{1/4}(1 - 6\Delta^2 + \Delta^4)^{1/8}/(1 - \Delta^2)^{1/2}, \tag{2.55}$$

where $\Delta \equiv \exp(-2\beta J)$.

No exact results have been obtained in three dimensions, but a number of approximations have been worked out and their results are reported later in this chapter.

The Bragg–Williams approximation assumes that there is no short-range order apart from that which follows from long-range order. It is equivalent to the molecular field approximation of Curie–Weiss and it produces an identical equation of state and a mean-field set of critical exponents, i.e. $\alpha = 0$, $\beta = \tfrac{1}{2}$, $\gamma = 1$, and $\delta = 3$. This, again, gives no glimpse into spatial fluctuations. The critical temperature obtained here is $T_c = z_0 J/k_B$.

The Bethe–Peierls approximation is an improvement over the Bragg–Williams result, since it accounts for specific short-range order apart from long-range order. It assumes a certain probability (Boltzmann weighted) of forming correlated spin clusters. As a consequence, the critical temperature is found as $T_c = 2J/k_B \ln[z_0/(z_0 - 2)]$, and it obviously agrees with the exact 1-D result but differs from the 2-D Onsager solution. It also predicts (incorrectly) a finite specific heat discontinuity with an exponential tail for $T > T_c$. The latter statement is qualitatively correct. A comparison of the Onsager, Bragg–Williams, and Bethe–Peierls predictions for

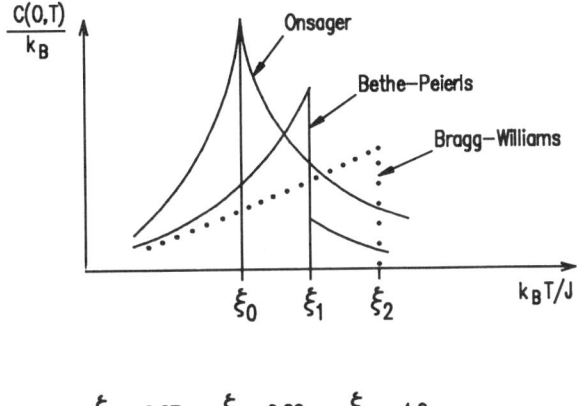

$\xi_0 = 2.27 \quad \xi_1 = 2.88 \quad \xi_2 = 4.0$

Fig. 2.20 A comparison between the specific heats in Ising model calculations [16, 19].

the specific heat is shown in Fig. 2.20. Another interesting feature of the Bethe–Peirels approximation is that it predicts a macroscopic growth of the correlation length at low temperatures according to $\xi \simeq \tfrac{1}{2}e^{2\beta J}$.

Allowing the spins to interact via all three components coupled to form a scalar product leads to the so-called Heisenberg model. However, it exhibits no phase transitions in one- and two-dimensional spaces [25].

A step in the direction of a three-dimensional Ising model is the so-called *spherical model*, which uses the Ising Hamiltonian of eqn (2.48) but also accounts for the constraint [26] that

$$N = \sum_{i=1}^{} S_i^2, \qquad (2.56)$$

which is a conserved quantity. Using the method of steepest descent it was shown that a 3-D lattice (as well as its higher-dimensional extensions; see Table 2.5) possesses ferromagnetic properties in the low-temperature regime, in support of the Curie–Weiss model. Furthermore, Stanley [28]

Table 2.5 Critical exponents in the spherical model (following reference [27]).

Dimensionality	β	δ	γ	ν'	η	ν	ν'	α	α'
$d < 3$				No phase transitions					
$d = 3$	$\tfrac{1}{2}$	5	2	indet	0	1	indet	0	0
$d = 4$	$\tfrac{1}{2}$	3	1	indet	0	0	indet	0	0
$d > 4$	$\tfrac{1}{2}$	3	1	1	0	1	indet	0	0

proved that, generalizing the Ising model to q-component spins, its free energy approaches that of the spherical model when $q \to \infty$, i.e. in the classical limit. This brings us to another important theoretical model, i.e. the X–Y model.

2.6.3 The X–Y model

This is one of the simplest (but highly nontrivial) generalizations of the Ising model. It is assumed that spins have two components and interact through a bilinear scalar product, i.e. the Hamiltonian is taken as

$$H = -J \sum_{\langle i,j \rangle} (S_i^x S_j^x + S_i^y S_j^y) = -JS^2 \sum_{\langle i,j \rangle} \cos(\phi_i - \phi_j), \quad (2.57)$$

where the spin magnitude is S and its angle with respect to the x-axis is ϕ_i for the ith spin. It can be shown that this Hamiltonian, which is invariant with respect to a uniform local spin rotation $\varphi_i \to \varphi_i + \varphi_0$, has a ferromagnetic ground state with energy $E_0 = -z_0 NJS^2/2$. A continuum approximation gives an estimate of eqn (2.57) as

$$H \simeq E_0 + \frac{JS^2}{2a^{d-2}} \int d^d r (\nabla \phi)^2, \quad (2.58)$$

where a is the lattice spacing and d the system's dimensionality. An important calculation in two-dimensional space shows that the correlation function exhibits an algebraic fall-off as

$$\Gamma(r) \sim r^{-\eta(T)}, \quad (2.59)$$

with $\eta(T) = k_b T/2JS^2\pi$, thus demonstrating lack of long-range order. Indeed, Mermin and Wagner [25] and, independently, Hohenberg [29] proved a more general result which states that any 2-D system with short-range interactions, the ordered phase of which has a continuous symmetry, does *not* support long-range order. The reason for this behaviour is that at least one branch of collective excitations has energy that tends to zero continuously as its wavelength goes to zero. This type of asymptotic excitation is indeed the aforementioned *Goldstone Boson*.

However, Kosterlitz and Thouless (KT) [30] also showed that there is a phase transition in the system after all. It takes place at the transition temperature

$$T_{KT} = \pi JS^2/2k_B, \quad (2.60)$$

and it signifies a transformation between bound vortex–antivortex pairs (at low temperatures) and single vortices (at high temperatures). This is illustrated in Fig. 2.21.

LATTICE MODELS 109

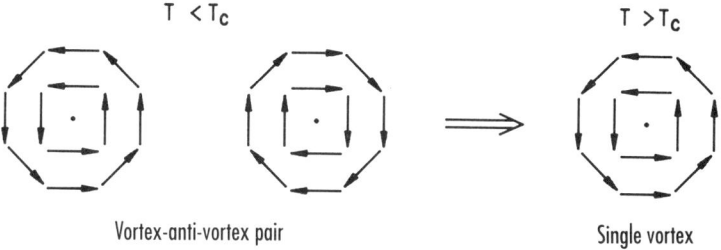

Fig. 2.21 The Kosterlitz–Thouless transition in the X–Y model.

The energy of an isolated vortex is

$$E_{\text{vortex}} = \frac{\pi J S^2}{2} n^2 \ln\left(\frac{L}{a}\right), \qquad (2.61)$$

where L is the linear dimension of the system and n is the vortex strength, which is quantized through

$$\oint \nabla \phi \cdot d\mathbf{l} = 2\pi n. \qquad (2.62)$$

The energy of a vortex–antivortex pair is

$$E_{\text{pair}} = -2\pi J S^2 n_1 n_2 \ln\left|\frac{\mathbf{r}_1 - \mathbf{r}_2}{a}\right|, \qquad (2.63)$$

where n_1 and n_2 are the respective vortex strengths and $\mathbf{r}_1$ and $\mathbf{r}_2$ are the position vectors of their centres. The reason for the KT transition to occur is the form of the entropy of a single vortex,

$$S = k_B \ln(L/a)^2, \qquad (2.64)$$

so that the free energy of the isolated vortex phase is lower than that of paired-up vortices for $T > T_{KT}$.

2.6.4 The Potts model

A generalization of the Ising model to q-components of the spin variable involves the Hamiltonian [31]

$$H = -J \sum_{\langle i,j \rangle} \sum_{\alpha=1}^{q} S_i^\alpha S_j^\alpha. \qquad (2.65)$$

It can be shown that, for a fixed value of q, there is always a dimensionality $d_c(q)$ such that, for $d > d_c(q)$, mean-field behaviour prevails. Conversely, for a fixed dimensionality d there is such a $q_c(d)$ that for $q > q_c(d)$

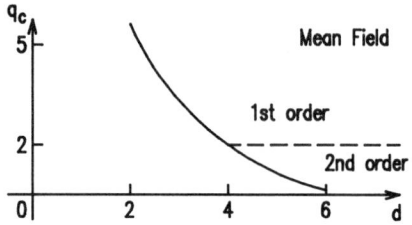

Fig. 2.22 A plot of $q_c(d)$ in the Potts model.

the behaviour is of mean-field type. For example, $d_c(2) = 4$ and $q_c(2) = 4$. In Fig. 2.22 are shown $q_c(d)$ and also a crossover line between second- ($q \leq 2$) and first-order phase transitions ($q > 2$).

A set of associated critical exponents for the Potts model in two spatial dimensions is given in Table 2.6.

2.6.5 Models with competing interactions

The first step in the direction of including distance-dependent interactions in lattice models is to consider both nearest and next-nearest neighbours. A simple one-dimensional Hamiltonian of this type is [32]

$$H = J_1 \sum_i S_i S_{i+1} + J_2 \sum_i S_i S_{i+2}. \qquad (2.66)$$

It is interesting to note that, despite its simplicity, this type of Hamiltonian leads to a new qualitative feature in the form of a helical magnetic phase. It turns out that the zero-temperature ground state energy is

$$E_0 = \begin{cases} -S^2(J_1^2/8J_2 + J_2), & \text{if } J_2 > 0 \text{ and } |J_1| < 4|J_2|, \\ S^2(\pm J_1 + J_2), & \text{otherwise}. \end{cases} \qquad (2.67)$$

Table 2.6 Critical exponents for the Potts model in two spatial dimensions (after reference [31]).

q	$\alpha = \alpha'$	β	$\gamma = \gamma'$	δ	ν	η
0	$-\infty$	$\frac{1}{6}$	∞	∞	∞	0
1	$-\frac{2}{3}$	$\frac{5}{36}$	$2\frac{7}{18}$	$18\frac{1}{5}$	$\frac{4}{3}$	$\frac{5}{24}$
2	0	$\frac{1}{8}$	$\frac{7}{4}$	15	1	$\frac{1}{4}$
3	$\frac{1}{3}$	$\frac{1}{9}$	$\frac{13}{9}$	14	$\frac{5}{6}$	$\frac{4}{15}$
4	$\frac{2}{3}$	$\frac{1}{12}$	$\frac{7}{6}$	15	$\frac{2}{3}$	$\frac{1}{2}$

THE LANDAU THEORY OF PHASE TRANSITIONS

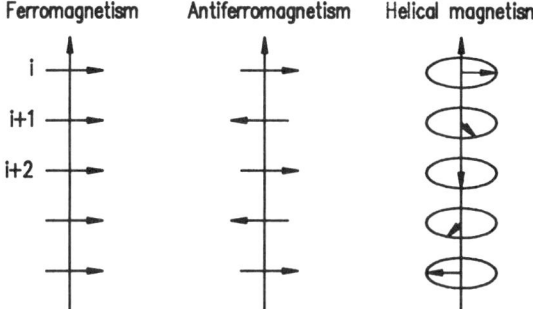

Fig. 2.23 The three possible ordered magnetic phases in the Ising model with competing interactions.

The top expression corresponds to a helical magnetic order with spins tilted by a constant angle with respect to their immediate neighbours (see Fig. 2.23). The bottom expression describes the antiferromagnetic ground state ('−' sign) and the ferromagnetic ground state ('+' sign).

A more intricate situation arises in Dyson's model [33], in which spin–spin interactions decay as a power of spin separation, so that

$$H = - \sum_{i,j} \frac{J_{ij}}{|r_{ij}|^{d+\sigma}} S_i S_j. \qquad (2.68)$$

It has been demonstrated that this model also exhibits a phase transition. When $\sigma > 0$, a second-order phase transition takes place for $d \leq 2$ provided that $0 < \sigma < d$. In the extreme limit of all spins contributing equally to the interaction energy (the so-called *equivalent neighbour model*), it can be shown that the results obtained are identical with those of the mean-field model, e.g. $T_c = 2J/k_B$. This leads us to reconsider the mean-field approach within a more general context provided by the Landau theory [6].

2.7 The Landau theory of phase transitions

The first fundamental theorem of solid state physics states that it is impossible to change the symmetry of a solid (or a condensed-matter system) gradually [6]. Thus, for simple fluids no symmetry breaking takes place, since the vapour pressure curve ends on a critical point that can be circled around on going from a liquid to a gas phase. However, neither the fusion nor the sublimation curve possesses a critical point and so transitions leading to the solid phase are of symmetry-breaking type. On this basis, Landau [6] deduced that second-order phase transitions are associated with symmetry breaking and can be qualitatively described through the use of an order parameter ψ, which is zero in the symmetric (ordered) and nonzero in the dissymmetric (disordered) phase. Assuming that the free energy depends on V, T, and ψ, the conditions on equilibrium are

$$\left.\frac{\partial F(T,V,\psi)}{\partial \psi}\right|_{\psi=\psi_0} = 0 \quad \text{and} \quad \left.\frac{\partial^2 F(T,V,\psi)}{\partial \psi^2}\right|_{\psi=\psi_0} > 0, \quad (2.69)$$

where ψ_0 is the equilibrium value of the order parameter.

The cornerstone of the Landau theory of phase transitions is the assumption that sufficiently close to T_c, on both sides of it, F can be expanded in a Taylor series of ψ as

$$F(T,V,\psi) \simeq F_0(T,V) + F_1(T,V)\psi + F_2(T,V)\psi^2$$
$$+ F_3(T,V)\psi^3 + F_4(T,V)\psi^4 + \cdots. \quad (2.70)$$

Landau then argued that unless there is a generalized force (e.g. a magnetic field) coupled to ψ, $F_1 = 0$, while the coefficient F_3 may also vanish as a result of the existence of invariance conditions: $F \to F$ as $\psi \to -\psi$ (e.g due to time reversal or parity-invariance symmetries). In addition, introducing a control parameter for spontaneous second-order phase transitions in the form of the reduced temperature $\varepsilon \equiv (T - T_c)/T_c$. Landau postulated the simplest possible such expansion as

$$F(T,V,\psi) \equiv F_0 + a\varepsilon\psi^2 + A_4\psi^4, \quad (2.71)$$

where $a > 0$ and $A_4 > 0$ for stability reasons. He thus arrived, through his remarkable physical intuition, at a cusp catastrophe some 30 years before this was recognized as such by mathematicians. As shown in Fig. 2.24, on

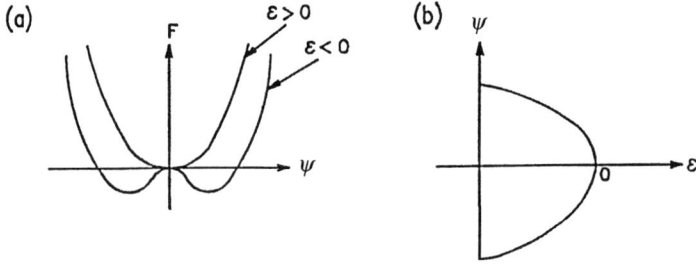

Fig. 2.24 A prototype of second-order phase transitions according to Landau.

THE LANDAU THEORY OF PHASE TRANSITIONS

changing the sign of ε, F transforms from a single- to a double-well shape. Simultaneously, solving the equilibrium conditions for ψ yields

$$\psi = 0, \quad \text{for} \quad \varepsilon > 0, \tag{2.72a}$$
$$\psi = \pm(-a\varepsilon/2A_4)^{1/2}, \quad \text{for} \quad \varepsilon < 0. \tag{2.72b}$$

This then gives the value of the critical exponent β as $\beta = 0.5$. Calculating the entropy gives

$$S = -\frac{\partial F}{\partial T} = \begin{cases} S_0, & \varepsilon > 0, \\ S_0 + \dfrac{a^2}{2A_4}\dfrac{\varepsilon}{T_c}, & \varepsilon \leqslant 0, \end{cases} \tag{2.73}$$

where S_0 is the entropy of the disordered phase. This allows us to evaluate the specific heat as

$$C_v = T\frac{\partial S}{\partial T} = \begin{cases} C_0, & \varepsilon > 0, \\ C_0 + \dfrac{a^2}{2A_4 T_c^2}T, & \varepsilon \leqslant 0, \end{cases} \tag{2.74}$$

where C_0 is the specific heat of the disordered phase. Hence, a discontinuity occurs at T_c, which amounts to

$$\Delta C = \frac{a^2}{2A_4 T_c}. \tag{2.75}$$

The plots of S and C_v are given in Fig. 2.25. Consequently, the second critical exponent α is also found to take on the classical value, i.e. $\alpha = 0$.

In order to find the values of the remaining critical exponents, the free energy must now explicitly include an external field h coupled to ψ, so that

$$F = F_0 + a\varepsilon\psi^2 + A_4\psi^4 - h\psi. \tag{2.76}$$

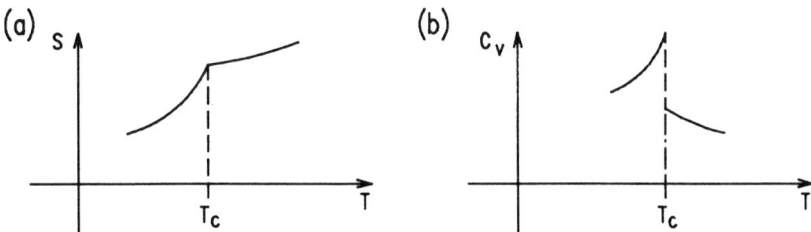

Fig. 2.25 Plots of (a) $S(T)$ and (b) $C_v(T)$ in the Landau model of a second-order phase transition.

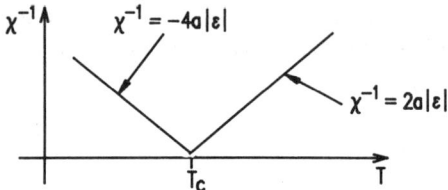

Fig. 2.26 A plot of $\chi^{-1}(T)$ for second-order phase transitions.

Minimizing F with respect to ψ now yields an equation of state in the form

$$h = 2\psi\left[a\varepsilon + 2A_4\psi^2\right]. \tag{2.77}$$

This can be differentiated with respect to h on both sides, and remembering that $\chi \equiv \partial\psi/\partial h$, the following expression is arrived at

$$\chi = \left[2a\varepsilon + 12A_4\psi^2\right]^{-1}. \tag{2.78}$$

Since $\psi \to 0$ as $T \to T_c$, we find that the third exponent is $\gamma = 1$. The plot of χ^{-1} is shown in Fig. 2.26. Note also that at $T = T_c$ the equation of state (2.77) simplifies to

$$h \simeq 4A_4\psi^3, \tag{2.79}$$

and hence $\psi \sim h^{1/3}$, giving $\delta = 3$. We therefore conclude that the quartic expansion in the Landau model invariably leads to classical critical exponents. Thus, both the van der Waals and Curie–Weiss theories can be seen as special applications of a Landau-type mean-field model.

It is very interesting to note that, recently, Ginzburg and Levanyuk [34] pointed to an apparently overlooked property of a sixth power expansion of the free energy for second-order phase transitions

$$F = F_0 + a\varepsilon\psi^2 + A_4\psi^4 + A_6\psi^6, \tag{2.80}$$

where all a, A_4, and A_6 are positive. They found, as a consequence, that depending on the strength of A_4 the critical exponents may take on *any* values from a continuum interval between the critical and tricritical limits. We show an unpublished numerical confirmation of these predictions in Fig. 2.27.

Indeed, the sixth-power expansion has been widely used in the past to model *first*-order phase transitions, but that requires $A_4 < 0$ [35]. In the

---

▷

Fig. 2.27 The influence of the quartic term on the critical exponents in eqn (2.80). CP denotes the critical point limit, while TCP is the tricritical point limit.

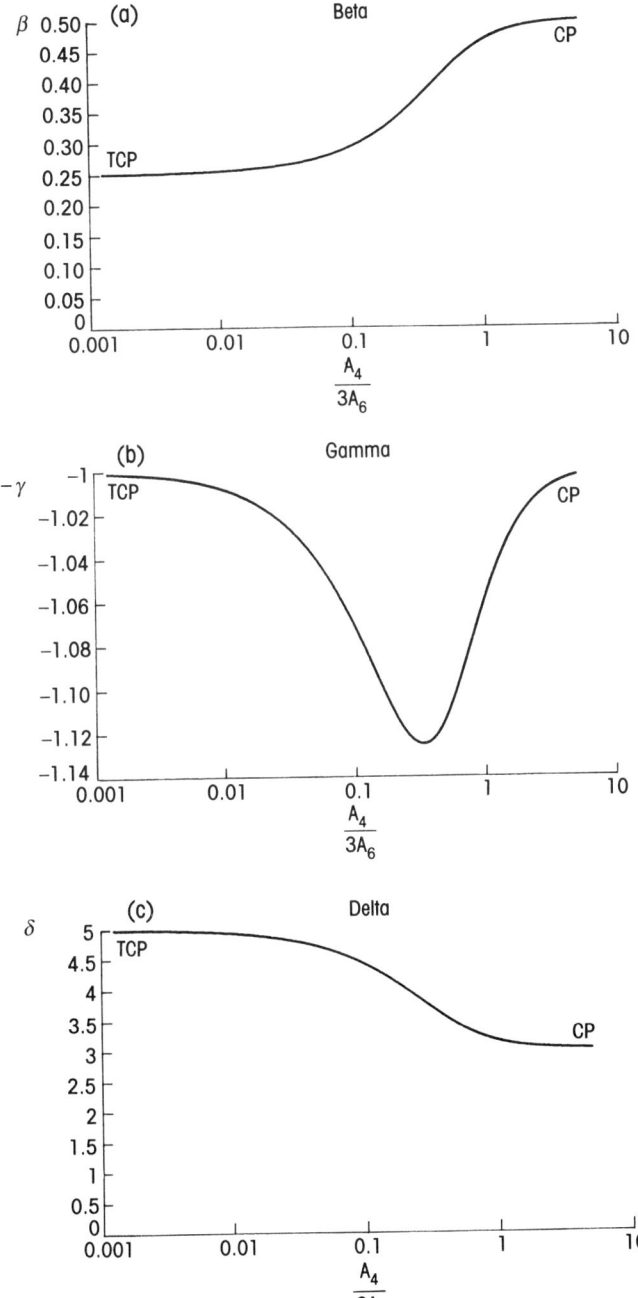

language of catastrophe theory, this latter expansion describes a 'butterfly catastrophe' [36] and a connection between catastrophe theory and the concept of structural stability has been demonstrated in reference [37]. The equation of state obtained from eqn (2.80) can be readily solved to yield

$$\psi = \left\{ \left\{ -A_4 \pm \left[A_4^2 - 3a\varepsilon A_6 \right]^{1/2} \right\} \Big/ 3A_6 \right\}^{1/2}. \quad (2.81)$$

Note that as $A_4 \to 0$, $\psi \sim \varepsilon^{1/4}$ and $\beta = 1/4$ for the *tricritical point* (TCP). The signs '+' or '−' correspond to $A_4 > 0$ or $A_4 < 0$, respectively. The transition temperature is T_c for $A_4 > 0$ (second order) and

$$T_c^* = T_c + A_4^2/4aA_6, \quad (2.82)$$

for $A_4 < 0$ (first order). In the presence of an external field h, we find the corresponding susceptibility becomes

$$\chi = \left[2a\varepsilon + 12A_4\psi^2 + 30A_6\psi^4 \right]^{-1}. \quad (2.83)$$

A schematic plot of $\chi(T)$ in this case is shown below in Fig. 2.30. We know that $\psi \sim \varepsilon^{1/4}$ for $A_4 \to 0$ and $\varepsilon \to 0$. Thus, the exponent γ should also remain at unity close to the TCP. On the other hand, as $T \to T_c$,

$$h \sim 30A_6\psi^5, \quad (2.84)$$

and hence $\psi \sim h^{1/5}$ with $\delta = 5$ for the TCP. Specific heat, likewise, retains its discontinuous behaviour and hence $\alpha = 0$ in both cases. This is consistent with Fig. 2.27.

It is illustrative to point out the existence of a thermal hysteresis phenomenon associated with first-order phase transitions. This is depicted in Fig. 2.28 for the order parameter. Free energy profiles in this case are shown in Fig. 2.29 and the corresponding susceptibility plots are given in Fig. 2.30.

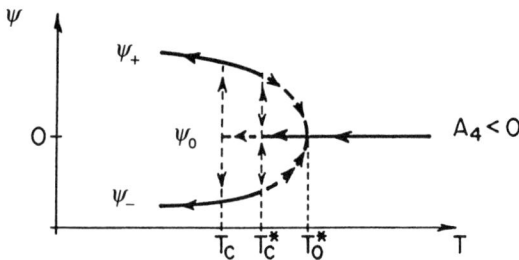

Fig. 2.28 The thermal hysteresis effect for first-order phase transitions.

THE LANDAU THEORY OF PHASE TRANSITIONS 117

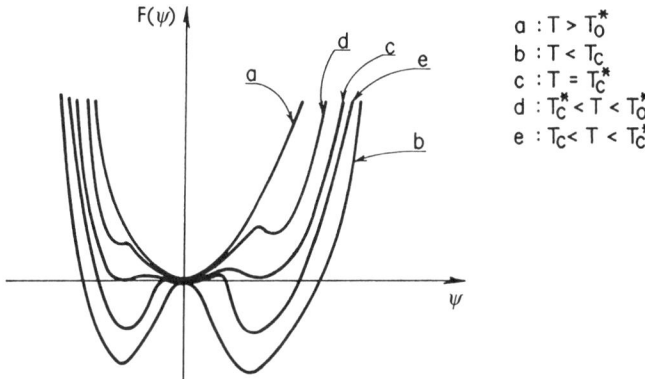

Fig. 2.29 Free energy forms for different temperatures close to a first-order phase transition temperature.

The disordered phase terminates its stability at $T = T_c$, while the ordered one does at $T = T_o^*$, where

$$T_o^* = T_c + A_4^2/3aA_6. \tag{2.85}$$

Field-induced hysteresis effects differ markedly between the two orders of phase transitions. The two possible forms of behaviour for second-order transitions are shown in Fig. 2.31, while the several possibilities that exist for first-order phase transitions are shown in Fig. 2.32.

We have shown in this section that the Landau theory is a very powerful and sophisticated tool in describing a vast array of possible critical forms of behaviour. Contrary to general perceptions, it does not only yield classical critical exponents. When a sixth-power expansion is used, it leads to a range of critical exponents between the classical ones (CP) and the TCP exponents, the values of which change continuously depending on the

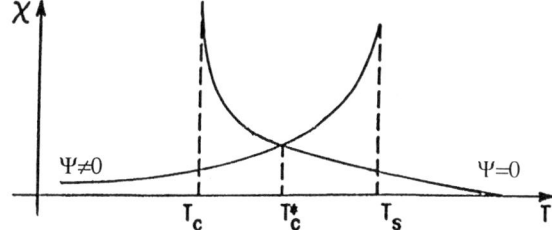

Fig. 2.30 A schematic plot of $\chi(T)$ for first-order phase transitions.

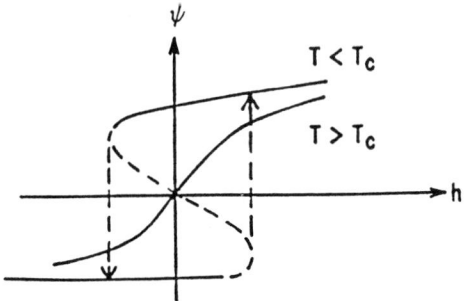

Fig. 2.31 The origin of field-induced hysteresis in second-order phase transitions.

strength of the quartic term. Furthermore, when the quartic coefficient becomes negative, the theory is capable of describing first-order phase transitions. The tricritical point behaviour arises when $A_2 \equiv A_4 = 0$.

Despite these very positive qualities, the Landau theory still suffers from a very serious shortcoming. It is not capable of describing spatial fluctuations, which not only are most certain to exist in all systems at all temperatures, but which dominate the scene close to the transition point.

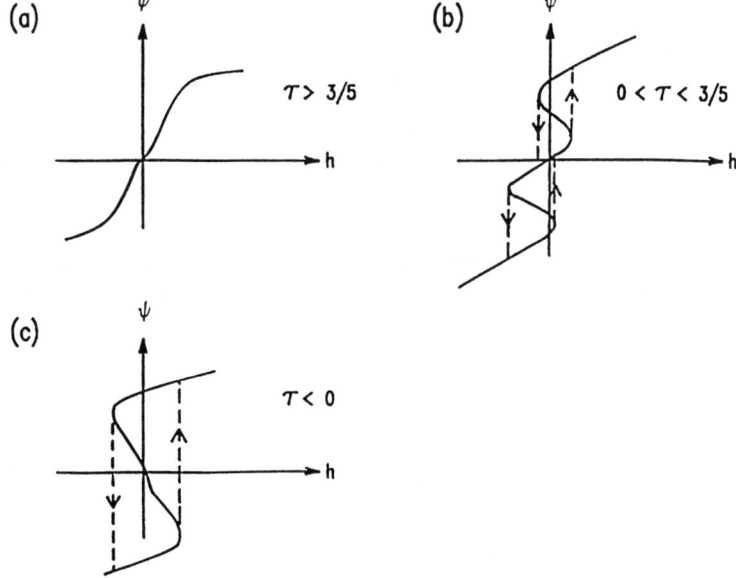

Fig. 2.32 The various hysteresis types for first-order phase transitions. The parameter $\tau \equiv A_2 A_6 / A_4^2$.

THE LANDAU–GINZBURG (LG) MODEL   119

This is why no predictions can be made for the exponents ν and η within the Landau theory. An improvement of this situation is the objective of our discussion in the next section. It will be shown that a relatively simple addition to the free energy improves the model dramatically.

2.8 Fluctuations and the Landau–Ginzburg (LG) model

For the order parameter $\psi(\mathbf{r})$ treated as a function of spatial co-ordinates, the two-point correlation function has been given in eqn (2.14). Through the use of the fluctuation–dissipation theorem [38], it can be shown that the correlation function is proportional to the generalized susceptibility function. A particular example useful in the modelling of the liquid–vapour phase transition describes the number of particles N in volume V. There, we find that

$$\langle (N - \langle N \rangle)^2 \rangle = \langle N \rangle n k_B T \kappa_T, \qquad (2.86)$$

where $n = \langle N \rangle / V$ and $\kappa_T = (-1/V)(\partial V/\partial P)_{T,N}$ is the isothermal compressibility. Ornstein and Zernike [12] demonstrated that in the vicinity of the critical point the particle number correlation function diverges in proportion to $(T - T_c)^{-1}$, so that κ_T must be a divergent function of $T - T_c$ with a critical exponent γ of 1.

To probe the spatial dependence of the correlation function close to criticality, scattering experiments must be performed. If $\mathbf{k}_0$ is the incident wave vector, $\mathbf{k}_s$ is the scattered wave vector and $|\mathbf{q}| = |\mathbf{k}_0 - \mathbf{k}_s| = 2k \sin(\theta/2)$ with $k = |\mathbf{k}_0| \simeq |\mathbf{k}_s|$, then the intensity of scattered radiation taken as a fraction of the intensity of the incident radiation is given by [12]

$$\frac{I(q)}{I_0(q)} = \frac{1}{n} \int d^3 r \exp(-i\mathbf{q} \cdot \mathbf{r}) \Gamma(r) \equiv \frac{1}{n} S(q), \qquad (2.87)$$

where $S(q)$ is called the form factor. Ornstein and Zernike provided a good fit to experimental results using a Lorentzian approximation for $S(q)$, i.e.

$$S(q) \sim (\xi^{-2} + q^2)^{-1}, \qquad (2.88)$$

where ξ is the correlation length. Using eqn (2.87) and assuming a d-dimensional space yields for the correlation function

$$\Gamma(r) = \int S(q) \left[\int_0^\pi e^{iqr \cos \theta} (\sin \theta)^{d-2} d\theta \right] d\Omega_{d-1} q^{d-1} dq, \qquad (2.89a)$$

where $d\Omega_{d-1}$ is a solid angle element in $d-1$ dimensions. The correlation length can be evaluated as

$$\xi(T) = \left[\int r^2 \Gamma(r,T) d^d r / \int \Gamma(r,T) d^d r \right]^{1/2}, \qquad (2.89b)$$

The integral in the square bracket of eqn (2.89a) gives

$$J_{d/2-1}(q^r)/(q^r)^{d/2-1},$$

where J is a Bessel function of the first kind [39]. Using the Lorentzian form of eqn (2.88) yields

$$\Gamma(r) \sim \begin{cases} e^{-r/\xi}, & d = 1, \\ \ln r e^{-r/\xi}, & d = 2, \\ e^{-r/\xi}/r, & d = 3, \end{cases} \qquad (2.90)$$

which fails to describe the situation in two-dimensional space. In order to repair this deficiency, Fisher [12] proposed a correction whereby

$$S(q)|_{T=T_c} \sim q^{-2+\eta} \qquad (q \approx 0), \qquad (2.91)$$

which results in

$$\Gamma(r)|_{T=T_c} \sim r^{-(d-2+\eta)} \qquad (r \to \infty). \qquad (2.92)$$

As was mentioned in the previous section, the Landau theory can be easily extended to account for spatial fluctuations. In their only joint work ever undertaken, Ginzburg and Landau [40] proposed to consider a free energy functional rather than a simple function. The simplest form that would be consistent with Landau's principles for expansions is

$$F(\psi(r),T) = \int d^3 r \left[A_2 \psi^2 + A_4 \psi^4 - h\psi + D(\nabla \psi)^2 \right], \qquad (2.93)$$

where D is a phenomenological parameter describing the contribution due to spatial inhomogeneities. Applying a variational principle to F in order to search for functional free energy minima gives

$$h = 2A_2 \psi + 4A_4 \psi^3 - 2D\nabla^2 \psi, \qquad (2.94)$$

which is a highly nonlinear partial differential equation. Much will be said about this type of equation in Chapter 7, and of its extensions to other possible physical situations in Chapters 5 and 6, but for now we only focus on perturbative solutions obtained as small deviations from the mean-field results (the Landau theory). We thus assume that

$$\psi(r) = \psi_0(T) + \varphi(r), \qquad (2.95)$$

THE LANDAU-GINZBURG (LG) MODEL 121

and linearize eqn (2.94) to obtain

$$\nabla^2 \varphi - \mu \frac{A_2}{D} \psi_0 = -\frac{h_0}{D} \delta(r), \tag{2.96}$$

where we have assumed h to be a localized perturbation of the form $h_0 \delta(r)$ and $\mu = 1$ for $T > T_c$, while $\mu = -2$ for $T < T_c$. A solution of eqn (2.96) in three-dimensional spherical co-ordinates is

$$\varphi = \frac{h_0}{4\pi D} (e^{-r/\xi})/r, \tag{2.97}$$

where $\xi \sim A_2^{-1/2}$ is the correlation length. Note that eqn (2.97) agrees with the result of Ornstein and Zernike given in eqn (2.90).

Another insight into the role of inhomogeneities in the LG model can be gained by Fourier transforming the order parameter according to [41]

$$\psi(r) \equiv L^{-d/2} \sum_{k<k_0} \psi_k e^{ik \cdot r}, \tag{2.98}$$

where k_0 is the cut-off wavelength $k_0 = \Lambda^{-1}$ corresponding to the smallest periodicity in the system (atomic spacing). The free energy then becomes

$$F = \sum_{k<k_0} |\psi_k|^2 (A_2 + Dk^2) + L^{-d} \sum_{k,k',k''<k_0} A_4 \psi_k^* \psi_{k'}^* \psi_{k''} \psi_{k+k'-k''}, \tag{2.99}$$

neglecting the external field h. Ignoring the mode-mode coupling present in the last term above provides the basis for the so-called Gaussian approximation, where

$$F \simeq \sum_{k<k_0} |\psi_k|^2 (A_2 + Dk^2). \tag{2.100}$$

Then, the Fourier transform of the correlation function is found as

$$\Gamma(k) \equiv L^{-d} \int e^{-ik \cdot (r-r')} \langle \psi(r)\psi(r') \rangle \, dr \, dr'$$

$$= \langle |\psi_k|^2 \rangle \simeq \frac{n}{2} (A_2 + Dk^2)^{-1}. \tag{2.101}$$

Therefore, as $T \to T_c$ we find that $\psi = 0$ and consequently $\Gamma(k) \sim k^{\eta-2}$ with $\eta = 0$ in the Gaussian approximation. Hence $\Gamma(r) \sim r^{-(d-2)}$. It can be shown that the Gaussian approximation introduces a new value of the specific heat critical exponent $\alpha = 2 - d/2$. For this reason it is concluded that the Gaussian approximation fails for $d < 4$ but gives a correct result for $d \geqslant 4$ [42].

It should be emphasized that the above results do not invalidate the LG model as such but rather impose limitations on the uses of various approximations applied in conjunction with it. The LG theory itself is limited too, but this arises from the so-called *Ginzburg criterion*, which states that its usefulness is lost when the size of fluctuations exceeds the value of the order parameter, both of which are to be averaged over a volume Ω, i.e.

$$\langle \psi(r)\psi(0)\rangle_\Omega \ll |\langle \psi(0)\rangle|^2. \tag{2.102}$$

In d dimensions we find that $\langle \psi(r)\psi(0)\rangle_\Omega \sim \varepsilon^{(d-2)/2}$ and we have also seen that $\langle \psi(0)\rangle \sim \varepsilon^{1/2}$. Hence the Ginzburg criterion yields

$$\text{constant } \varepsilon^{(d-2)/2} \ll \varepsilon^1. \tag{2.103}$$

Thus, for $d \geqslant d^* = 4$ (the so-called marginal dimensionality) the Landau–Ginzburg theory is exact! This is consistent with our statement regarding the Gaussian approximation.

In order to extend the validity of the Landau–Ginzburg model further, we must now examine dynamic scaling (or real space scaling) and renormalization group theory.

2.9 Dynamic scaling and the renormalization group (RG)

An extension of the concept of scaling at criticality was proposed by Kadanoff to include size scaling understood as course graining. This means that all the essential features of the system should remain unchanged as we increase the length scale of the lattice by a factor λ such that

$$1 \ll \lambda \ll \xi/a,$$

where ξ is the correlation length and a is the lattice spacing (see Fig. 2.33).

If N is the number of lattice sites and d is the dimensionality of the system, then $m = N/\lambda^d$ is the number of blocks obtained as a result of the

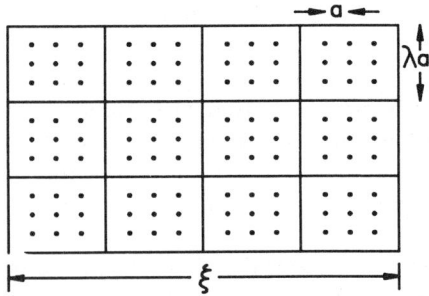

Fig. 2.33 A schematic illustration of the Kadanoff construction.

first rescaling process. Simultaneously, the lattice variables (e.g. spins) and the interaction parameters are redefined. The effective spins now refer to each block and interaction parameters refer to the interacting blocks. Thus, taking the Ising model as an example, we have the cell Hamiltonian

$$H_{\text{cell}} = -J \sum_{\langle i,j \rangle}^{N} S_i S_j - h \sum_i^N S_i. \tag{2.104}$$

Then the block Hamiltonian becomes

$$H_{\text{block}} = -\tilde{J} \sum_{\langle \alpha,\beta \rangle}^{m} \tilde{S}_\alpha \tilde{S}_\beta - \tilde{h} \sum_\alpha^m \tilde{S}_\alpha, \tag{2.105}$$

where the tilde quantities refer to blocks.

A crucial step in this approach is the assumption that thermodynamic potentials scale with block size. Thus, the free energy is assumed to obey the relationship [12, 14]

$$F_{\text{block}}(\tilde{\varepsilon}, \tilde{h}) = \lambda^d F_{\text{cell}}(\varepsilon, h), \tag{2.106}$$

where the tilde field $\tilde{h}$ and reduced temperature $\tilde{\varepsilon}$ satisfy

$$\tilde{h} = \lambda^x h, \qquad \tilde{\varepsilon} = \lambda^y \varepsilon, \tag{2.107}$$

and the exponents x and y are, *a priori*, unknown. It is also conjectured that F is a homogeneous function so that F_{block} and F_{cell} are of the same form, so that we can write

$$F(\lambda^{y/d}\varepsilon, \lambda^{x/d}h) = \lambda F(\varepsilon, h). \tag{2.108}$$

However, earlier on (see eqn (2.18)) when discussing static scaling, a statement was made regarding F that said that

$$F(\lambda^{a_\varepsilon}\varepsilon, \lambda^{a_H}H) = \lambda F(\varepsilon, H). \tag{2.109}$$

This allows us to identify the new exponents x and y with the already known exponents a_ε and a_H. Indeed, we find that

$$y = a_\varepsilon d, \qquad x = a_H d. \tag{2.110}$$

This result is known as the *static scaling hypothesis*. An application of the Kadanoff construction to the pair correlation function

$$\Gamma(r,\varepsilon) \equiv \langle (S_i - \langle S \rangle)(S_j - \langle S \rangle) \rangle, \tag{2.111}$$

gives $y = 1/\nu$ for temperature variable scaling. Recalling that, from static scaling, $y = da_\varepsilon$ and $a_\varepsilon = (2-\alpha')^{-1}$, we find that

$$\alpha = \alpha' \quad \text{and} \quad d\nu = 2 - \alpha. \tag{2.112a}$$

Scaling the field variable yields $x = da_H$ which, with $a_H = \delta/(\delta+1)$, results in

$$(2-\eta)\nu = \delta. \tag{2.112b}$$

We have, therefore, found that the two scaling hypotheses simplify the theory of phase transitions so that we only require the use of *two* independent exponents which were, *a priori*, unknown. The *renormalization group* (RG) theory was developed to calculate the values of these exponents for particular models. Its main objective is to reduce the number of degrees of freedom in a system approaching criticality to a bare minimum. The intention is to extend the Kadanoff construction to its limit, i.e. to carry the scaling procedure up to ξ. However, we also know that as $T \to T_c$, $\xi \to \infty$. Hence, in principle, size scaling should be allowed to continue *ad infinitum* as T approaches T_c.

Denoting the original cell Hamiltonian by H_0 and the Hamiltonian after the nth step of rescaling by H_n, we represent the chain of renormalization transformations R as

$$R(H_0) = H_1, \quad R(H_1) = H_2, \ldots, R(H_n) = H_{n+1}, \ldots, R(H^*) = H^*, \quad (2.113a)$$

where H^* denotes a *fixed point* Hamiltonian which is invariant with respect to scaling transformations. Each step in the RG transformation chain reduces the number of degrees of freedom by λ^d. At the same time the partition function is preserved, so that

$$Z_N(H_n) = Z_m(H_{n+1}), \quad (2.113b)$$

where $m = N/\lambda^d$. It is believed that the values of critical exponents are characteristics of groups of diverse physical models and not individual Hamiltonians. Thus, numerous models may lead to the same fixed point. Indeed, the so-called *universality hypothesis* states that for any two physical systems with the same dimensionality, d, and the same number of order parameter components, n, all such systems belong to the same *universality class*, so that each fixed point corresponds to one universality class. Several important physical examples of the universality class concept are listed in Table 2.7.

In terms of model calculations using RG, a vast literature exists on the

Table 2.7 Examples of universality classes (after reference [15]).

Universality class	System	Order parameter
$d = 2, n = 1$	Absorbed films	Surface density
$d = 2, n = 2$	Superfluid ^{4}He film	Superfluid wave function
$d = 3, n = 1$	Uniaxial ferromagnets	Magnetization
$d = 3, n = 1$	Fluids	Density difference
$d = 3, n = 1$	Mixtures, alloys	Concentration difference
$d = 3, n = 2$	Planar ferromagnets	Magnetization
$d = 3, n = 2$	Superfluids	Superfluid wave function
$d = 3, n = 3$	Isotropic ferromagnets	Magnetization

subject and here we only mention a few most important results. A pertinent discussion on the relevance of RG results to Landau–Ginzburg-type models can be found in Chapter 5.

The so-called *Gaussian model* has the Hamiltonian below, which is a truncated version of the Landau–Ginzburg Hamiltonian,

$$H = \tfrac{1}{2}\int d^d x \{r\phi^2(x) + (\nabla\phi)^2\}, \qquad (2.114)$$

and RG calculations give

$$\alpha = 2 - \frac{d}{2}, \quad \nu = \frac{1}{2}, \quad \delta = \frac{d+2}{d-2}. \qquad (2.115)$$

Note that for $d = 2$, $\alpha = 1$ and $\delta = \infty$!

The S^4-model [42] is a more realistic approximation to the Ising model and it uses the effective Hamiltonian

$$H = -K \sum_{\langle i,j \rangle} S_i S_j + \sum_i \left(\frac{b}{2} S_i^2 + u S_i^4\right). \qquad (2.116)$$

Table 2.8 Summary of the critical exponent values for the most important theoretical models (following references [12], [14], and [42]). Note that $\varepsilon \equiv 4 - d$.

Model	α	β	γ	ν	δ	η
Ornstein–Zernike			2	1	5	0
Classical (MFT)	0 (disc.)	$\tfrac{1}{2}$	1	—	3	—
Classical (TCP)	0 (disc.)	$\tfrac{1}{4}$	1	—	5	—
Spherical, $d = 3$		$\tfrac{1}{2}$	2	1	5	0
Spherical, $\varepsilon > 0$	$-\dfrac{\varepsilon}{2-\varepsilon}$	$\tfrac{1}{2}$	$\dfrac{2}{2-\varepsilon}$	$\dfrac{1}{2-\varepsilon}$	$\dfrac{6-\varepsilon}{2-\varepsilon}$	0
Ising, $d = 2$	0 (log)	$\tfrac{1}{8}$	$\tfrac{7}{4}$	1	15	$\tfrac{1}{4}$
Ising (approx.), $d = 3$	0.12	0.33	1.25	0.64	4.8	0.04
Heisenberg (approx.), $d = 3$	-0.12	0.36	~ 1.39	$+0.71$	4.8	$+0.04$
S^4-model, $d > 4$	$\dfrac{\varepsilon}{2}$	$\tfrac{1}{2} - \dfrac{\varepsilon}{4}$	1	$\tfrac{1}{2}$	$3 + \varepsilon$	0
S^4-model, $d = 4$	0	$\tfrac{1}{2}$	1	$\tfrac{1}{2}$	3	0
S^4-model, $d < 4$	$\dfrac{\varepsilon}{6}$	$\left(\tfrac{1}{2} - \dfrac{\varepsilon}{6}\right)$	$\left(1 + \dfrac{\varepsilon}{6}\right)$	$\left(\tfrac{1}{2} + \dfrac{\varepsilon}{12}\right)$	$3 + \varepsilon$	0
S^4-model, $d = 3$	0.17	0.33	1.17	0.58	4	0
X–Y model, $d = 3$	0.01	0.34	1.30	0.66	4.8	0.04

The partition function is

$$Z = \int_{-\infty}^{\infty} \prod_i dS_i \exp\left\{\beta\left[K \sum_{\langle i,j \rangle} S_i S_j - \sum_i \left(\frac{b}{2} S_i^2 + u S_i^4\right)\right]\right\}, \quad (2.117)$$

and thus can be viewed as that of the Ising model with a double-peaked weighting factor

$$W(S_i) = \exp\left\{-\beta\left[\frac{b}{2} S_i^2 + u S_i^4\right]\right\}. \quad (2.118)$$

The results for the S^4-model, as well as a summary of the results for all other important models, are listed in Table 2.8.

Following the universality hypothesis and RG calculations, a plot can be obtained illustrating the values of all the exponents as a function of d and n (see Fig. 2.34). The relations obtained from RG calculations [43] up to the second order in $\varepsilon \equiv 4 - d$ are

$$\alpha = -\frac{n-4}{2(n+8)}\varepsilon - \frac{(n+2)^2}{4(n+8)^3}(n+28)\varepsilon^2, \quad (2.119)$$

$$\beta = \tfrac{1}{2} - \frac{3}{2(n+8)}\varepsilon + \frac{(n+2)(2n+1)}{2(n+8)^3}\varepsilon^2, \quad (2.120)$$

$$\gamma = 1 + \frac{n+2}{2(n+8)}\varepsilon + \frac{n+2}{4(n+8)^3}(n^2 + 22n + 52)\varepsilon^2, \quad (2.121)$$

$$\delta = 3 + \varepsilon + \frac{1}{2(n+8)^2}(n^2 + 14n + 60)\varepsilon^2, \quad (2.122)$$

$$\eta = \frac{n+2}{2(n+8)^2}\varepsilon^2, \quad (2.123)$$

and

$$\nu = \tfrac{1}{2} + \frac{n+2}{4(n+8)}\varepsilon + \frac{n+2}{8(n+8)^3}(n^2 + 23n + 60)\varepsilon^2, \quad (2.124)$$

▷

Fig. 2.34 Plots of the values of the critical exponents (a) α, (b) β, (c) ν, (d) δ, (e) γ, and (f) η, as functions of d and n (after reference [43]). Note that $\beta = \tfrac{1}{2}$ in the shaded area. The symbols □, △, and ⌂ denote the results for the 3-D Heisenberg, X-Y, and Ising models, respectively. MFA denotes the values obtained in the mean field approximation.

DYNAMIC SCALING

and have been plotted in Fig. 2.34.

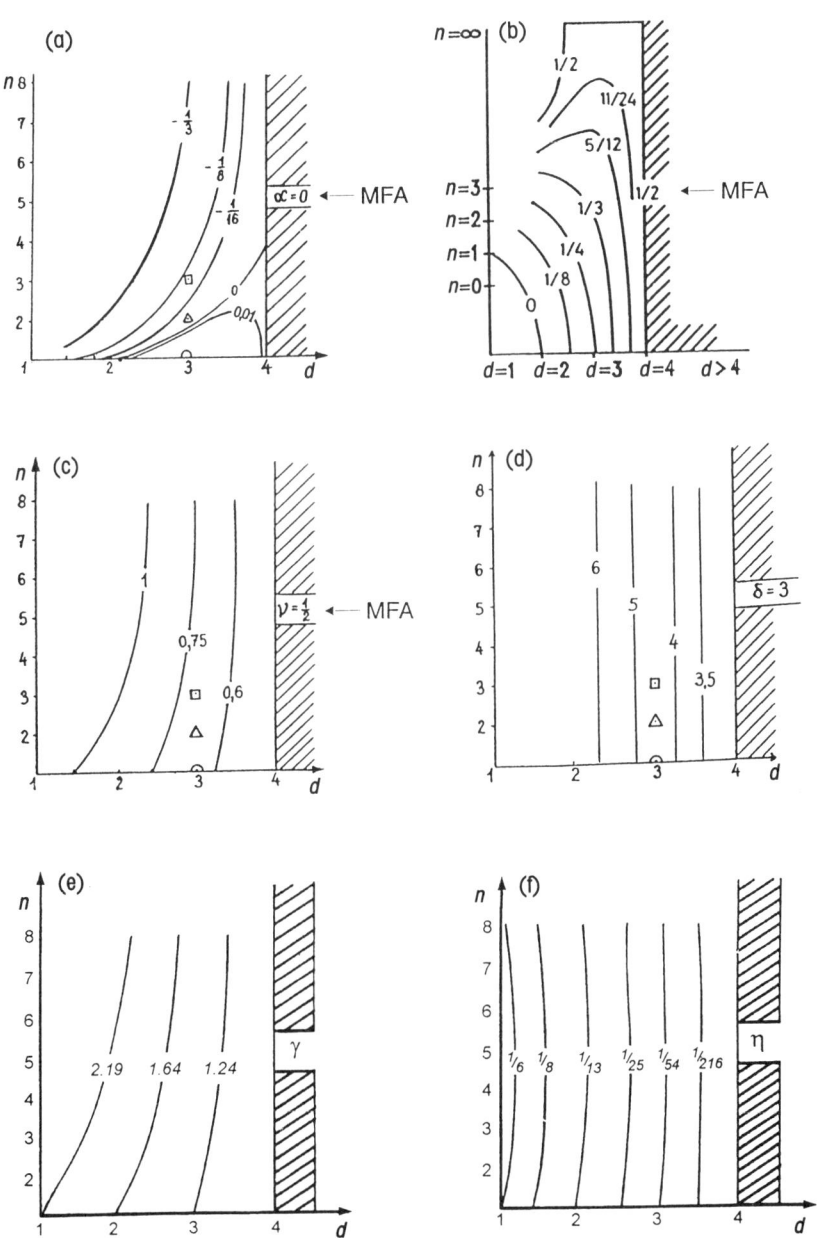

It is noteworthy that for $d \geqslant 4$ the RG approach confirms earlier predictions that the Landau (mean field) model is exact. The founder of the RG theory, K. G. Wilson himself, described his own work as a *second stage* of the Landau theory [43]. He remarked that while the microscopic LG Hamiltonian

$$H = D(\nabla\psi)^2 + A_2\psi^2 + A_4\psi^4, \qquad (2.125)$$

is essentially correct, the macroscopic generalization in the form of the LG free energy,

$$F = \int d^3x \left[\tilde{A}_2\psi^2 + \tilde{A}_4\psi^4 + \tilde{D}(\nabla\psi)^2 \right], \qquad (2.126)$$

is *incorrect*, since it ignores the variations of its parameters (expansion coefficients) with the size of the region sampled.

However, despite the great successes of RG ideas in obtaining critical exponents for virtually arbitrary theoretical models, there is still a number of aspects that require the use of other methods. First of all, the results that one obtains pertain to a small neighbourhood of the critical point and are not easy to extend analytically beyond its immediate vicinity. Second, it cannot be readily applied to first-order phase transitions, where the correlation length does not diverge at the transition temperature [44]. Furthermore, finite-size systems cannot be analysed very successfully using scaling concepts, since there is a limit to the coarse graining operation, and surface energy terms—which are comparable to the bulk energy terms—may contribute significantly. Finally, RG analysis does not tell us much about the nature of classical (and virtually nothing about quantum) excitations that the systems close to an instability point may support. As we have seen in the case of the X–Y model, this type of information is often crucial in understanding the dynamics, the possible pattern formation, and the response to external perturbations. In the closing section of this chapter, we briefly outline several phenomenological attempts aimed at resolving some of these problems.

2.10 A look beyond

As we have tried to demonstrate throughout this chapter, phase transitions are collective phenomena arising from interactions between particles (or quasi-particles) of the system. An associated feature that is transparent at all levels of the theoretical modelling is the existence of *nonlinearities*. This aspect is the main obstacle in coming up with simple solutions to this class of problem.

In terms of incorporating nonlinearities explicitly in the description, several theoretical avenues can be outlined. They can be basically divided into two alternative paths:

1. *The dynamical path* is concerned with solving nonlinear differential equations of motion for the order parameter. This can be done at several levels of description depending on theoretical rigour and the requirements imposed by particular physical applications.

2. *The statistical path* attempts to include contributions due to significant nonlinear modes in particular calculations. For example, an ideal gas of solitons can be included in phase-space integration. Their effects on physical observables can then be evaluated through statistical thermodynamics.

Below, we give a brief outline of substantial building blocks used in the development of the two paths listed above.

2.10.1 The dynamical path

At a microscopic level, Landau–Ginzburg Hamiltonian densities of the type

$$H = \frac{m}{2}\psi_t^2 + \frac{D}{2}(\nabla\psi)^2 + a_2\psi^2 + a_4\psi^4 + \ldots \qquad (2.127)$$

are constructed. Statistically speaking, we know that the most important contribution comes from a mode $\psi(x,t)$ that minimizes the Hamiltonian functional, i.e. from the solution of the two coupled conditions

$$\delta H = 0 \quad \text{and} \quad \delta^2 H > 0, \qquad (2.128)$$

where we have used first and second variations of the Hamiltonian functional. If the order parameter field ψ is real, then we obtain the *nonlinear Klein–Gordon equation*:

$$\Box_\varepsilon \psi = -2\left[a_2\psi + 2a_4\psi^3 + \cdots\right], \qquad (2.129)$$

where the Laplace–Beltrami operator $\Box_\varepsilon$ is defined as

$$\Box_\varepsilon \equiv \frac{\partial^2}{\partial x_0^2} + \varepsilon \sum_{i=1}^{3} \frac{\partial^2}{\partial x_i^2}, \qquad (2.130)$$

with $\varepsilon = -\text{sgn}(D)$, $x_0 = (1/\sqrt{m})t$, and $(x_1, x_2, x_3) = (1/\sqrt{|D|})(x, y, z)$. Stationary solutions or, conversely, massless modes with $m = 0$, lead to the nonlinear Helmholtz equation for a mode $\psi(x)$ that minimizes the Hamiltonian functional, i.e.

$$\nabla^2 \psi = -2[a_2\psi + 2a_4\psi^3 + \cdots]. \tag{2.131}$$

On the other hand, for complex order parameters (as is the case with superconductors and superfluids), the equation of motion is the *complex nonlinear Klein–Gordon equation* (for $m \neq 0$):

$$\Box_\varepsilon \psi = -2[a_2 + 2a_4\psi\psi^* + \cdots]\psi. \tag{2.132}$$

In the case of order parameters which are complex and massless, $m = 0$, the resultant equation is the *nonlinear Schrödinger equation*:

$$-i\psi_t = -D\nabla^2\psi + 2[a_2 + 2a_4\psi\psi^* + \cdots]\psi. \tag{2.133}$$

It is important to note that a large body of mathematical and physical literature exists (see, for example, references [45–47]) dealing with these types of equation and large numbers of exact and approximate solutions have been found, including those in $(3 + 1)$-dimensional space–time. We will give an overview of some of these results in the physical context in Chapter 5 and an exhaustive mathematical discussion will be presented in Chapter 7. For now, it can be said that travelling plane-wave solutions exist of elliptic type with limiting cases in the form of solitary waves (kinks and bumps). Solutions in geometries other than planar also exist, e.g. cylindrical, spherical, or spiral, and their amplitude is usually of damped oscillatory type (i.e. oscillations with a decreasing amplitude).

At a macroscopic level of modelling, the starting point is the phenomenological Landau–Ginzburg expansion for the free energy functional

$$F = \int d^d x \left[a_2\psi^2 + a_4\psi^4 + \cdots + \frac{D}{2}(\nabla\psi)^2 \right]. \tag{2.134}$$

Subsequent derivations of the corresponding equation of state describe the relaxation dynamics of order parameter profiles $\psi(x,t)$ and they obviously crucially depend on the type of order parameter required. The general master equation for the evolution of ψ is

$$\gamma \frac{\partial}{\partial t} P(\psi, t) = \sum_{\delta\psi} [W(\psi - \delta\psi, \psi) P(\psi - \delta\psi, t) - W(\psi, \psi + \delta\psi) P(\psi, t)], \tag{2.135}$$

where P is the probability of a given order parameter value at a particular time instant t, and W is the transition amplitude, assumed to be in the Gaussian form with Boltzmann weighting factor as follows [48]:

$$W(\psi, \psi + \delta\psi) = \alpha \exp\left[-\int \frac{\delta\psi(x)^2}{\Delta} dx\right] \exp\left\{-\frac{\beta}{2}[F(\psi + \delta\psi) - F(\psi)]\right\}, \tag{2.136}$$

where α is a normalization constant, $\beta = 1/k_B T$ as usual, and Δ is related to the energy dissipation rate. It was demonstrated [48] that the equation of motion for ψ takes the integro-differential form of the *Onsager equation*,

$$\frac{\partial \psi}{\partial t} = -\int d^d x \, \Gamma^{-1}(\mathbf{x} - \mathbf{x}') \frac{\delta F}{\delta \psi}, \qquad (2.137)$$

where Γ is a position-dependent relaxation function. If ψ is a global *nonconserved* order parameter (e.g. magnetization or the rate of ordering in a binary mixture), then eqn (2.137) takes the form of the time-dependent Landau–Ginzburg equation (TDLG),

$$\Gamma_0 \frac{\partial \psi}{\partial t} = -\frac{\partial F_0}{\partial \psi} + D \nabla^2 \psi, \qquad (2.138)$$

where the relaxation function Γ has been expanded in reciprocal space as

$$\Gamma(k) \simeq \Gamma_0 + \Gamma_2 k^2 + \cdots \qquad (2.139)$$

and $F_0 = a_2 \psi^2 + a_4 \psi^4 + \cdots$ is the mean-field part of the free energy.

If, on the other hand, the order parameter is globally *conserved* (e.g. diffusionless transitions in alloys) the resultant evolution equation is the *Cahn–Hilliard equation*:

$$\partial \psi / \partial t = \Gamma_2 \nabla^2 (\delta F / \delta \psi). \qquad (2.140)$$

Apart from solving the above equations as they stand, one should consider more sophisticated (and realistic) situations in which:

1. White noise is added to mimic the role of the other (noncritical) degrees of freedom present.
2. A dissipative term $\gamma \psi_t$ may be added for the same reason as in 1.
3. Two or more order parameters are coupled to describe the various cross-coupling phenomena (e.g. piezoelectric, elasto-optical, etc.).
4. External forcing terms $h(x, t)$ are added to the right-hand sides of these dynamical equations to study driving effects (*nonequilibrium phase transitions when h is time dependent*) with constant, or space-, and time-dependent driving forces.

This scenario opens up an extremely large field for investigations. In Table 2.9 we list a number of diverse specific examples of Landau–Ginzburg models in equilibrium and nonequilibrium phase transitions. To do justice to the past investigations in this area, a separate monograph

Table 2.9 Examples of diverse applications of the Landau–Ginzburg model.

System	Order parameter	Free energy functional	Symmetry
Superconductor	Pair wave function	$\int \left\{ a|\psi|^2 + b|\psi|^4 + \left\|\left(i\dfrac{d}{dx} - \dfrac{2e}{c}A\right)\psi\right\|^2 \right\} dx$	$\psi \to e^{i\varphi}\psi$
Liquid crystal	Complex amplitude of density, ψ	$\iint \left\{ a|\psi|^2 + b|\psi|^4 + (\nabla + i\delta\bar{n})\psi^* \dfrac{1}{2M}(\nabla - i\delta\bar{n})\psi \right\} dx^2$	$\psi \to e^{i\varphi}\psi$
Liquid Convection instability	Amplitude of unstable mode, A	$\iint \left\{ a|A|^2 + b|A|^4 + cA^*\left(\dfrac{\partial}{\partial x} - \dfrac{i}{\sqrt{2\pi}}\dfrac{\partial^2}{\partial y^2}\right)^2 A \right\} dx$	$A \to e^{i\varphi}A$, translation, rotation
Laser Single mode	Mode amplitude, u	$a|u|^2 + b|u|^4$	$u \to ue^{i\varphi}$
Multi-mode (no phase relations)	Mode amplitudes, u_λ	$\sum_\lambda a_\lambda |u_\lambda|^2 + \sum_{\lambda\lambda'} b_{\lambda\lambda'} |u_\lambda|^2 |u_{\lambda'}|^2$	$u_\lambda \to u_\lambda e^{i\varphi_\lambda}$ $u \to e^{i\varphi}u$
Mode continuum[†]	$u(x)$	$\int \left\{ a|u|^2 + b|u|^4 + c\left\|\left(iv\dfrac{d}{dx} - v\right)u\right\|^2 \right\} dx$	$u_\lambda \to u_\lambda e^{i\varphi_\lambda}$
Parametric oscillator With coherent pump field		$\alpha(|u_1|^2 + |u_2|^2) + \beta u_1^* u_2^* + cc + \gamma(|u_1|^2 + |u_2|^2)$	$\beta \to \beta e^{i(\varphi_1 + \varphi_2)}$ $u_\lambda \to u_\lambda e^{i\varphi_\lambda}$
With incoherent pump field		$\alpha|u_1|^2 + \beta|u_2|^2 + \gamma(|u_1|^2 - |u_2|^2)^2$	

[†] v is a phase velocity for an individual propagating mode and ν represents the atomic frequency.

should be written on this topic alone. This is clearly impossible within the framework of the present endeavour. Instead, we refer the reader to Chapter 5 for a more fundamental physical theory that underlies these efforts and to Chapter 7 for the mathematical background.

We close this chapter with a brief account of an alternative approach, i.e. the statistical path.

2.10.2 The statistical path (non-Gaussian calculations)

Once again, our starting point is the LG Hamiltonian in eqn (2.125). However, earlier statistical attempts concentrated on somehow factorizing it through Gaussian approximations, in order to obtain rather simple analytical expressions for the partition function and related quantities [41]. The first step is to Fourier transform the order parameter ψ according to eqn (2.98). The result is

$$H = \sum_{k<k_0} (A_2 + Dk^2)|\psi_k|^2 + L^{-d} \sum_{k,k',k''<k_0} A_4 \psi_k^* \psi_{k'}^* \psi_{k''} \psi_{-k-k'-k''}. \tag{2.141}$$

As before, the mode–mode coupling present in the last term causes serious difficulties. Rather than use the Gaussian approximation which treats it in a perturbative manner (and may lead to various problems at criticality where nonlinear couplings dominate), we adopt a different approach. It is based on including only paired-up modes, in the belief that the nonpaired-up terms will mutually cancel out. Then, the approximate Hamiltonian is [49]

$$H \simeq \sum_{k<k_0} H_{(k)} = \sum_{k<k_0} \left[(A_2 + Dk^2)|\psi_k|^2 + 6L^{-d}A_4|\psi_k|^4\right]. \tag{2.142}$$

Hence, the partition function factorizes, but in a highly nontrivial way as

$$Z \simeq \prod_{k<k_0} Z_{(k)} = \prod_{k<k_0} \int_{-\infty}^{+\infty} d\psi_k \exp(-\beta H_{(k)}). \tag{2.143}$$

We then use the very little known (and even less used) *exact* non-Gaussian integral below (see also Section 1.2),

$$\int_0^\infty y^{2mp-1} \exp(-by^{2m} - fy^{4m}) dy$$
$$= (2m)^{-1}(2f)^{-p/2}\Gamma(p)D_{-p}(b/\sqrt{2f})\exp(b^2/2f), \tag{2.144}$$

where D_{-p} is a parabolic cylinder function [39], and evaluate the contributing one-mode partition functions as

$$Z_{(k)} = \sqrt{\pi}(2\lambda_4)^{-1/4}D_{-1/2}(x)\exp(x^2/4). \qquad (2.145)$$

The notation introduced is as follows:

$$x = \lambda_2/\sqrt{2\lambda_4} = \Delta_1 + \Delta_2 k^2, \quad \lambda_2 = \beta(A_2 + Dk^2), \quad \lambda_4 = 6\beta L^{-d}A_4, \qquad (2.146\text{a-c})$$

$$\Delta_1 = a(T - T_c)\left[(L^d/12A_4 k_B T)\right]^{1/2}, \quad \Delta_2 = D\left[(L^d/12k_B T A_4)\right]^{1/2}. \qquad (2.146\text{d,e})$$

Several important observations immediately follow from eqn (2.145). First, for finite or infinite sizes, and for all temperatures including T_c, the obtained expression is *analytic*. Second, the result in eqn (2.145) tends to the Gaussian estimate only when $L \to \infty$ (the thermodynamic limit). Third, all other quantities of interest can be readily derived. For example, the second moment is

$$\Gamma_k \equiv \langle|\psi_k|^2\rangle = (8\lambda_4)^{1/2}D_{-3/2}(x)/D_{-1/2}(x). \qquad (2.147)$$

This is plotted in Fig. 2.35, using several scales to illustrate the fact that this is no longer a homogeneous function and that the latter property is, strictly speaking, of use only at $T \simeq T_c$ and $L = \infty$.

The results in Fig. 2.35 are rather easy to interpret using analytic expansions for the parabolic cylinder function. For $x \simeq 0$, i.e. $L < \infty$ and arbitrary temperatures including T_c,

$$D_{-a-1/2}(\pm x) \sim \left[\sqrt{\pi}\, 2^{-a/2-1/4}/\Gamma\!\left(\tfrac{3}{4} + \tfrac{a}{2}\right)\right]\exp(\mp\sqrt{a}\,x). \qquad (2.148)$$

Consequently, the second moment (proportional to the k-mode susceptibility) gives

$$\Gamma_k \sim \exp\!\left[-(L^d/k_B T A_4)^{-1/2} Dk^2\right], \qquad (2.149)$$

which is of exponential functional form and does not follow a power-law prescription. On the other hand, for $x \to \infty$,

$$D_{-a-1/2}(x) \sim \exp(-x^2/4)x^{-a-1/2}, \qquad (2.150)$$

▷
Fig. 2.35 Plots of the scaled Γ_k as a function of $x = \lambda_2/\sqrt{2\lambda_4}$ following eqn (2.147), using (a) a linear scale, (b) a semi-logarithmic scale, and (c) a log–log scale (after reference [49]).

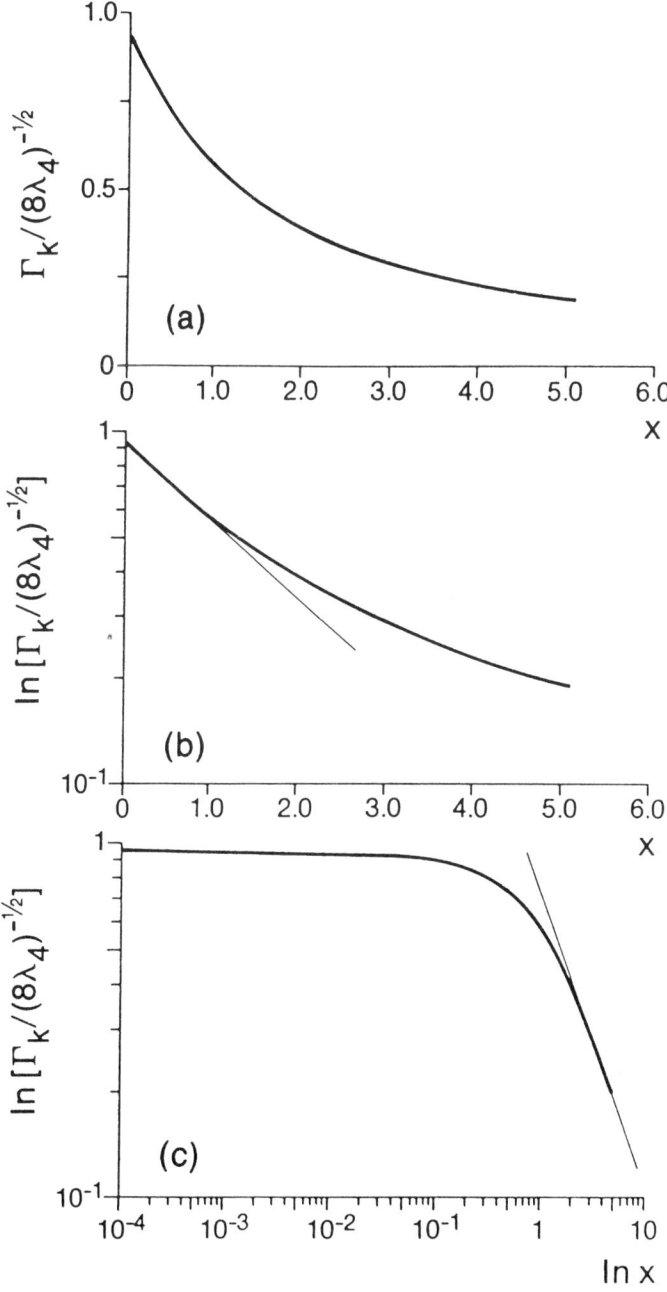

which results in the familiar Gaussian estimate

$$\Gamma_k \sim (2\lambda_2)^{-1}, \tag{2.151}$$

conforming to the scaling hypothesis power-law predictions. Hence, we obtain agreement with the Gaussian theory in the limit of infinite sizes and critical temperatures.

Similar conclusions can be reached for the other quantities of importance, such as the form factor, the correlation length, the specific heat, etc. [49]. What emerges is that in the vicinity of T_c, and for infinite sizes, power-law scaling behaviour becomes exact. On the other hand, if either $L < \infty$ or $T \neq T_c$, these quantities follow highly complicated functional forms of behaviour, expressed through special functions.

In this last section we have tried to convince the reader that the RG group results should and can be complemented by other approaches in the field of critical phenomena, such as nonlinear differential equations and non-Gaussian statistics. We believe that critical phenomena can be viewed as a special subfield of dynamical instabilities in the many-body problem, which will be emphasized in Chapters 3 and 4. Since this is *per se* a quantum-mechanical problem with its many manifestations (discrete energy levels, collective excitations, etc.), we feel that an excursion into the realm of quantum many-body physics is required.

References

1. P. W. Anderson, *Basic Notions of Condensed Matter Physics*, Benjamin/Cummings, Menlo Park, CA (1984).
2. R. H. White and T. Geballe, *Long Range Order in Solids*, Academic Press, New York (1979).
3. S. P. Parker (ed.), *Solid-State Physics Resource Book*, McGraw-Hill, New York (1987).
4. E. W. Montroll and J. L. Leibowitz, *Fluctuation Phenomena*, North-Holland, Amsterdam (1979).
5. C. Kittel, *Introduction to Solid State Physics*, John Wiley, New York (1956).
6. L. D. Landau and E. M. Lifshitz, *Statistical Physics*, Pergamon Press, London (1959).
7. A. D. Bruce and R. A. Cowley, *Structural Phase Transitions*, Taylor and Francis, London (1981).
8. M. E. Lines and A. M. Glass, *Principles and Applications of Ferroelectrics and Related Materials*, Clarendon Press, Oxford (1972).
9. D. C. Mattis, *The Theory of Magnetism II*, Springer-Verlag, Berlin (1985).
10. J. C. Phillips, *Physics of High-temperature Superconductors*, Academic Press, New York (1989).

REFERENCES

11. A. Isihara, *Condensed Matter Physics*, Oxford University Press, Oxford (1991).
12. H. E. Stanley, *Introduction to Phase Transitions and Critical Phenomena*, Oxford University Press, Oxford (1972).
13. A. Münster, *Statistical Thermodynamics*, Springer-Verlag, Berlin (1969).
14. J. M. Yeomans, *Statistical Mechanics of Phase Transitions*, Oxford University Press, Oxford (1992).
15. G. Carreri, *Order and Disorder in Matter*, Benjamin/Cummings, Menlo Park, CA (1984).
16. M. Plischke and B. Bergersen, *Equilibrium Statistical Mechanics*, Prentice Hall, Englewood Cliffs, NJ (1989).
17. F. Reif, *Fundamentals of Statistical and Thermal Physics*, McGraw-Hill, New York (1965).
18. C. N. Yang and T. D. Lee, *Phys. Rev.*, **87**, 404 (1952).
19. K. Huang, *Statistical Mechanics*, John Wiley, New York (1987).
20. G. L. Jones, *J. Math. Phys.* **7**, 200 (1966).
21. L. van Hove, *Physica*, **15**, 951 (1949).
22. R. J. Baxter, *Exactly Solved Models in Statistical Mechanics*, Academic Press, New York (1982).
23. L. Onsager, *Phys. Rev.*, **65**, 117 (1944).
24. P. E. Byrd and M. D. Friedman, *Handbook of Elliptic Integrals for Engineers and Scientists*, Springer-Verlag, Berlin (1971).
25. N. D. Mermin and H. Wagner, *Phys. Rev. Lett. A*, **17**, 1133 (1966).
26. T. H. Berlin and M. Kac, *Phys. Rev.*, **86**, 821 (1952).
27. G. S. Joyce, in *Phase Transitions and Critical Phenomena*, Vol. 2 (ed. C. Domb and M. S. Green), Academic Press, New York (1972).
28. H. E. Stanley, *Phys. Rev.*, **176**, 718 (1968).
29. P. C. Hohenberg, *Phys. Rev.* **158**, 383 (1967).
30. J. M. Kosterlitz and D. J. Thouless, *J. Phys. C*, **6**, 1181 (1973).
31. F. Y. Wu, *Rev. Mod. Phys.*, **54**, 235 (1982).
32. T. Niemeijer, *J. Math. Phys.*, **12**, 1487 (1971).
33. F. J. Dyson, *Communs. Math. Phys.*, **12**, 91; 212 (1969); **21**, 269 (1979).
34. V. L. Ginzburg, A. P. Levanynk, and A. A. Sobyanin, *Ferroelectrics*, **73**, 171 (1987).
35. K. Binder, *Rep. Progr. Phys.* **50**, 783 (1987).
36. J. M. T. Thompson and H. B. Stewart, *Nonlinear Dynamics and Chaos*, John Wiley, Sons, New York (1986).
37. V. de Alfaro and M. Rasetti, *Fort. Phys.* **26**, 143 (1978).
38. R. K. Pathria, *Statistical Mechanics*, Pergamon Press, Oxford (1972).
39. M. Abramowitz and I. A. Stegun (eds.), *Handbook of Mathematical Functions*, Dover, New York (1965).
40. V. L. Ginzburg and L. D. Landau, *Zh. Eksp. Teor. Fiz.*, **20**, 1064 (1950).
41. S.-K. Ma, *Modern Theory of Critical Phenomena*, Benjamin, New York (1976).
42. L. E. Reichl, *A Modern Course in Statistical Physics*, University of Texas Press, Austin, Texas (1979).

43. K. G. Wilson, *Rev. Mod. Phys.*, **55**, 583 (1983); *Scient. Am.*, **241** (1979); *Phys. Rev. Lett.*, **28**, 248 (1972).
44. A. J. Leggett, *The Problems of Physics*, Oxford University Press, Oxford (1987).
45. R. K. Dodd, J. C. Eilbeck, J. D. Gibbon, and H. C. Morris, *Solitons and Nonlinear Wave Equations*, Academic Press, New York (1982).
46. A. Hasegawa, *Optical Solitons in Fibres*, Springer-Verlag, Berlin (1990).
47. M. J. Ablowitz and P. A. Clarkson, *Solitons, Nonlinear Evolution Equations and Inverse Scattering*, Cambridge University Press, Cambridge (1991).
48. H. Metiu, K. Kitahara, and J. Ross, *J. Chem. Phys.*, **64**, 292 (1976).
49. J. A. Tuszyński and A. Wierzbicki, *Phys. Rev. B.*, **43**, 8472 (1991).

3 ELEMENTARY EXCITATIONS IN SOLIDS

The most fundamental question that one might be expected to answer is 'why is a solid?'...

We are ill-equipped to answer these questions in any other than a qualitative way, for they demand the solution of the many-body problem in one of its most difficult forms...

But, of course, such a task is impossible. Methods have not yet been developed that can find even the lowest-lying energy level of such a complex system. The best that we can do at present is to guess at the form that states will take and then to try and calculate their energy.

...the philosophy of the quantum theory of the solid state is often to accept the crystal structure as one of the given quantities of any problem. We then consider the wave functions of electrons in this structure, and the dynamics of the atoms as they undergo small displacements from it.

> P. L. Taylor (1970), *A Quantum Approach to the Solid State*
> (Prentice Hall, Englewood Cliffs)

3.1 Introduction

In this section we review what quantum excitations actually are and how we describe them in solids. This will enable us to introduce the idea of second quantization in a simple way, and how it may be used to great advantage to study a variety of different types of excitation in condensed matter. We shall later use second quantization when we develop the Method of Coherent Structures (MCS), so it is important that the reader thoroughly understands what is being done. The technique will be described at two different levels. In this part of the book we concentrate on physical ideas and how they may be employed to provide simple rules for the setting up of operators and quantum states in second quantization. We may consider this to be the simplest level. At a more advanced level, ideas such as completeness, what a representative is, how the occupation formalism follows from a configuration or momentum picture, and so on, can readily be described in more detail. The interested reader may also find in the literature how second quantized operators, within a given configuration, may be written in terms of orbital angular momentum and spin operators, which are perhaps more familiar to spectroscopists and those interested in magnetic resonance or magnetism generally.

A solid consists of an array of atoms or ions in close proximity, sometimes interfused with a sea of almost free electrons. The array may or may not be periodic and each ion is made up of a positively charged massive nucleus surrounded by groups of lighter electrons which are bound to the particular nucleus. Furthermore, the nuclei are not stationary at any temperature and vibrate about some mean position. A solid can exhibit a myriad of fascinating physical features, e.g. being a metal, a superconductor, a semiconductor, a semi-metal, an ionic or covalent material, and so on. These properties can, in principle, all be extracted from the following Hamiltonian:

$$H_1 = \sum_n \left(\frac{-\hbar^2}{2M_n}\right)\nabla_n^2 + \frac{1}{2}\sum_{n,m}{}' \frac{Z_n Z_m e^2}{|\mathbf{R}_n - \mathbf{R}_m|}$$
$$+ \sum_i \left(\frac{-\hbar^2}{2m}\right)\nabla_i^2 + \frac{1}{2}\sum_{i,j}{}' \frac{e^2}{|\mathbf{r}_i - \mathbf{r}_j|}$$
$$- \sum_{n,i} \frac{Z_n e^2}{|\mathbf{r}_i - \mathbf{R}_n|}, \qquad (3.1)$$

(provided that smaller terms such as spin–orbit coupling are also included). The positions of the nuclei are denoted by position vectors $\mathbf{R}_n, \mathbf{R}_m, \ldots$ etc. relative to some fixed origin and, similarly, the electron positions are denoted by $\mathbf{r}_i, \mathbf{r}_j$, etc. with respect to the same origin. The magnitude of the positive charge at the nth site is written as $Z_n|e|$. Each electron has a mass m and the mass of the nucleus at the nth site is M_n. The dashes on two of the summations indicate that we must exclude terms with either $i=j$ or $m=n$ and thus avoid self-interactions. Subscripts n or i on Laplacians indicate that they are defined using the co-ordinates of that particular nucleus or electron respectively. The first term in eqn (3.1) represents the kinetic energy of the nuclei and the second describes the mutual repulsion between them. Similarly, the third contribution in eqn (3.1) may be interpreted as the kinetic energy of the electrons and the next as the repulsion between them. The last term is the potential energy of attraction between electrons and the various nuclei.

The Hamiltonian appears to be rather awkward as it is, and is not in an appropriate form to easily extract physical effects. In fact, it probably contains too much information but in any case it is too unwieldy to work with. Often it is better to consider simpler models when we wish to describe elementary excitations in a simple way. As will become clear from Chapters 5 and 6, the MCS does *not* do this but makes an exact transformation on either eqn (3.1) or H_1 with fixed nuclei, although it is not necessary to even make this approximation. The properties of a solid will depend chiefly on only a fraction of the electrons in the solid; namely, the outer valence electrons in the outermost shells. They will also depend on those electrons, if present in any significant numbers, which are able to move more freely between nuclear sites. If each ion has a large number of bound electrons in inner orbitals, then when a nucleus moves, these virtually follow the nucleus instantaneously. This is not entirely true, however, and when the behaviour of the outer electrons is modified by the presence of other nearby ions, the inner electrons must relax to some degree to accommodate.

Experimentalists are usually not particularly interested in the nature of the ground state of a physical system itself, but they investigate it by applying a variety of stimuli to raise its energy from a low-lying to an excited state of higher energy. In the simplest cases the processes involved are said to lead to the creation of *elementary excitations*. These excitations may be classified into two main groups depending on whether the physical system involves noninteracting particles or strongly (or appreciably) interacting particles. When a group of particles are noninteracting or weakly coupled together, one can raise the energy of one particle without seriously affecting the others. As the coupling is increased between particles, the particle excitations will acquire some sort of lifetime due to the interaction: that is, any momentum possessed by one particle will be quickly transmitted to the others. Provided that the lifetime is sufficiently long, excitations are particle-like and we call them *quasi-particles*. However, for sufficiently strong interactions it becomes quite meaningless to talk about the motion of individual particles. In such a case we speak of a *collective excitation*, a typical example of which would be the excitation of the electrons as a whole in a metal, which are collectively oscillating at the plasma frequency [1]. Before we provide an overview of the types of elementary excitation present in solids, we first give an elementary introduction to the method of second quantization, which has proved itself to be a very powerful tool in condensed matter theory.

3.2 Second-quantization techniques

In this section we shall attempt an introduction to the methods of second quantization so that the reader may easily follow the development of the Method of Coherent Structures (MCS) described later in the book. The technique of second quantization itself is a very elegant way of accounting for the symmetry of states and operators when describing N-particle systems. The formalism developed retains the appropriate statistics of all

142   ELEMENTARY EXCITATIONS IN SOLIDS

the particles throughout. Furthermore, it is simple to interpret physically and much less cumbersome than employing wave functions and operators in configuration space. Redundant information is extracted by the method, i.e. if we describe a system of N identical particles we do *not* need to know in which state *each* particle is, but only how many are in each state. Furthermore, the number of particles that we deal with may be variable, a factor which is essential in field theory but, as in our case, in the nonrelativistic domain, the physical content is the same.

3.2.1 The harmonic oscillator

The simplest way to begin is to consider a simple one-dimensional quantum harmonic oscillator, the quantization of which is very well known. The Hamiltonian of such an oscillator may be written as

$$H = \frac{1}{2m}(p^2 + m^2\omega^2 q^2), \tag{3.2}$$

where p is the momentum and q is the displacement of a mass m which is undergoing the oscillatory motion. The angular frequency of the oscillator is ω. The momentum and displacement satisfy the commutation relations

$$[q, p]_- = i\hbar, \quad [q, q]_- = [p, p]_- = 0. \tag{3.3}$$

As a consequence of eqns (3.2) and (3.3), we have

$$[H, p]_- = i\hbar m \omega^2 q \quad \text{and} \quad [H, q]_- = -i\hbar p/m. \tag{3.4}$$

Now suppose that $|n\rangle$ is an eigenstate of H, so that

$$H|n\rangle = E_n|n\rangle. \tag{3.5}$$

By using the relations in eqn (3.4), we now investigate what happens if we operate on $|n\rangle$ with $H(p + im\omega q)$. We find that

$$H(p + im\omega q)|n\rangle = \{Hp - pH + im\omega(Hq - qH)$$
$$+ (p + im\omega q)H\}|n\rangle, \tag{3.6}$$

where the right-hand side of eqn (3.6) has been written in such a way that the use of the commutators in eqn (3.4) becomes obvious. The right-hand side of eqn (3.6) becomes

$$\left\{i\hbar m\omega^2 q + im\omega \frac{(-i\hbar p)}{m} + (p + im\omega q)E_n\right\}|n\rangle,$$

where we have used eqn (3.5) as well as eqn (3.4), and rearranging we obtain

$$(E_n + \hbar\omega)(p + im\omega q)|n\rangle. \tag{3.7}$$

In a similar way,

$$H(p - im\omega q)|n\rangle = (E_n - \hbar\omega)(p - im\omega q)|n\rangle. \tag{3.8}$$

Equations (3.7) and (3.8) show that both $(p + im\omega q)|n\rangle$ and $(p - im\omega q)|n\rangle$ are also eigenfunctions of H but have energies $E_n + \hbar\omega$ and $E_n - \hbar\omega$, respectively. We call $p + im\omega q$ a *raising operator* or a *creator* since, when it acts on a given eigenfunction of H, it produces a new eigenfunction the energy of which is raised by $\hbar\omega$; or, equivalently, we say that a *vibrational quantum of energy* $\hbar\omega$ *has been created*. Similarly, $p - im\omega q$ is a *lowering operator* or an *annihilator*, since its effect on a given eigenfunction is to lower its energy by $\hbar\omega$, or we say that it annihilates a quantum of energy $\hbar\omega$.

We will now use the raising and lowering operators to find the eigenvalue spectrum by considering the lowering operator applied to $|n\rangle$ repeatedly. Eventually, we will come to the lowest state (or the ground state), beyond which it will be impossible to go. Thus, for the ground state $|G\rangle$ we must have

$$(p - im\omega q)|G\rangle = 0. \tag{3.9}$$

Now we act on eqn (3.9) with a raising operator and use the fact that $|G\rangle$ is an eigenstate of H with energy E_G. Thus

$$(p + im\omega q)(p - im\omega q)|G\rangle = (2mE_G - m\hbar\omega)|G\rangle. \tag{3.10}$$

Hence,

$$E_G = \hbar\omega/2, \tag{3.11}$$

and continuing to operate on $|G\rangle$ with raising operators will clearly give the relation

$$E_n = \hbar\omega(n + \tfrac{1}{2}), \tag{3.12}$$

for the energy of the oscillator, where n is zero or a positive integer.

Returning now to our formulation of annihilators and creators, it is convenient to 'normalize' them with a factor N, so that a raising operator is written as

$$\beta^+ = N(-ip + m\omega q), \tag{3.13}$$

and a lowering operator as

$$\beta = N(+ip + m\omega q). \tag{3.14}$$

144   ELEMENTARY EXCITATIONS IN SOLIDS

We choose N so that

$$\beta^+ |n\rangle = \sqrt{n+1}\, |n+1\rangle, \qquad (3.15)$$

where $|n+1\rangle$ is an eigenstate of H with energy $E_n + \hbar\omega$, and

$$\beta |n+1\rangle = \sqrt{n+1}\, |n\rangle. \qquad (3.16)$$

From eqns (3.15) and (3.16), we have

$$\beta\beta^+ |n\rangle = (n+1)|n\rangle. \qquad (3.17)$$

On the other hand, from eqns (3.14), (3.15), (3.12), and (3.3) we obtain

$$\beta\beta^+ |n\rangle = mN^2\{(2n+1)\hbar\omega + \hbar\omega\}|n\rangle. \qquad (3.18)$$

Moreover, from eqns (3.18) and (3.17), $N = (2m\hbar\omega)^{-1/2}$ and β and β^+ become

$$\beta = \frac{1}{\sqrt{2m\hbar\omega}}(ip + m\omega q) \qquad (3.19)$$

and

$$\beta^+ = \frac{1}{\sqrt{2m\hbar\omega}}(-ip + m\omega q). \qquad (3.20)$$

Using eqns (3.19), (3.20), and (3.3), the fundamental commutation rules become

$$[\beta, \beta^+]_- = 1, \quad [\beta, \beta]_- = [\beta^+, \beta^+]_- = 0. \qquad (3.21)$$

In terms of annihilators and creators, the Hamiltonian itself may be written as

$$H = \hbar\omega(\beta^+\beta + \tfrac{1}{2}). \qquad (3.22)$$

Equation (3.22) is said to be the Hamiltonian of the oscillator in *second-quantized form*, β^+ and β being called a *second-quantized creator and annihilator*, respectively. The eigenstates $|n\rangle$ are said to form an *occupation number space*.

3.2.2 A collection of harmonic oscillators

Generalizing eqn (3.2), the Hamiltonian for a collection of noninteracting harmonic oscillators now becomes

$$H = \sum_k \frac{1}{2m}(p_k^2 + m^2\omega_k^2 q_k^2). \qquad (3.23)$$

SECOND-QUANTIZATION TECHNIQUES 145

In view of the fact that the canonical momenta and displacements satisfy

$$[q_{k'}, p_k]_- = i\hbar \delta_{k'k} \tag{3.24}$$

and

$$[p_{k'}, p_k]_- = [q_{k'}, q_k]_- = 0, \tag{3.25}$$

where $\delta_{kk'}$ is a Kronecker delta function, we can define new creators and annihilators by

$$\beta_k^+ = \frac{1}{\sqrt{2m\hbar\omega_k}}(-ip_k + m\omega_k q_k) \tag{3.26}$$

and

$$\beta_k = \frac{1}{\sqrt{2m\hbar\omega_k}}(+ip_k + m\omega_k q_k), \tag{3.27}$$

for each oscillator separately. We find that their commutation relations are

$$[\beta_k, \beta_{k'}^+]_- = \delta_{k,k'} \tag{3.28}$$

and

$$[\beta_k^+, \beta_{k'}^+]_- = [\beta_k, \beta_{k'}]_- = 0. \tag{3.29}$$

The eigenfunctions of the Hamiltonian in eqn (3.23) are simply products of single harmonic oscillator eigenfunctions, one for each oscillator. They may be labelled by occupation numbers n_1, n_2, n_3, etc. and written as

$$|n_1, n_2, n_3, \ldots, n_k, \ldots\rangle. \tag{3.30}$$

The reason for this is that eqn (3.23) does *not* contain an interaction between the separate oscillators. An operator such as H, in eqn (3.23), which is only the *sum* of operators for each oscillator, each term mutually commuting with every other, is said to be a *one-body operator*. A creation operator β_k^+ acts only on the occupation number in eqn (3.30) which refers to the kth oscillator, and no other. Thus

$$\beta_k^+ |n_1, n_2, n_3, \ldots, n_k, \ldots\rangle = \sqrt{n_k + 1}\, |n_1, n_2, \ldots, n_k + 1, \ldots\rangle. \tag{3.31}$$

Similarly

$$\beta_k |n_1, n_2, \ldots, n_k + 1, \ldots\rangle = \sqrt{n_k + 1}\, |n_1, n_2, \ldots, n_k, \ldots\rangle. \tag{3.32}$$

Hence

$$H|n_1, n_2, \ldots, n_k, \ldots\rangle$$
$$= \sum_k \hbar\omega_k (\beta_k^+ \beta_k + \tfrac{1}{2})|n_1, n_2, \ldots, n_k, \ldots\rangle. \tag{3.33}$$

We can represent the ground state of the collection of oscillators by

$$|G\rangle = |0,0,0,0\ldots\rangle, \qquad (3.34)$$

where each oscillator has no quanta excited above its own vacuum state $|0\rangle$. With this definition we can show that the state in eqn (3.30) may be obtained using creators β_k^+, n_k times for the kth oscillator. Thus

$$|n_1, n_2, \ldots, n_k, \ldots\rangle = \prod_k \frac{(\beta_k^+)^{n_k}}{\sqrt{n_k!}} |G\rangle. \qquad (3.35)$$

What is perhaps of greater interest is that the operator H in eqn (3.23) may be written as

$$H = \sum_{k,k'} \langle n_k | \frac{1}{2m}(p^2 + m^2\omega^2 q^2) | n_{k'} \rangle \beta_k^+ \beta_{k'}, \qquad (3.36)$$

provided that each oscillator has the same angular frequency ω.

However, in the summand of eqn (3.36), n_k and $n_{k'}$ denote the occupancy quantum numbers of one (the kth) oscillator, and both the annihilator $\beta_{k'}$ and the creator β_k^+ refer to the *same* oscillator. The operator in eqn (3.36) then acts on a ket the labels of which are the occupancies for each oscillator separately. Equation (3.36) is said to represent the Hamiltonian in *second-quantized form*. Using the result in eqn (3.22) we have that the Hamiltonian in eqn (3.36) is

$$H = \sum_{k,k'} \hbar \omega_k (\beta_k^+ \beta_k + \tfrac{1}{2}) \delta_{k,k'}. \qquad (3.37)$$

This operator is diagonal in the new occupancy kets

$$|n_1, n_2, \ldots, n_k, \ldots\rangle, \qquad (3.38)$$

i.e. those which denote occupancies for each oscillator separately, giving the total energy as

$$E = \sum_k \hbar \omega_k (n_k + \tfrac{1}{2}). \qquad (3.39)$$

However, eqn (3.37) can be viewed in another way, namely that k labels not a different oscillator but the kth eigenstate of *one* oscillator and the n_k now denote the occupancy of these states for *one* oscillator.

So far, the technique that we have described above has assumed that the particles or quasi-particles involved satisfy Bose–Einstein commutation rules; i.e. those in eqn (3.21), where any eigenstate can be occupied by an arbitrary number of particles. The method itself is much more general than this and below we show how it may be extended to Fermions.

3.2.3 Second-quantization in the Fermion particle case

Consider the Hamiltonian of a single Fermion in the presence of a space-dependent potential. We may write this as

$$H = p^2/2m + V(\mathbf{r}). \tag{3.40}$$

Here, we follow the notation of reference [2] and utilize a complete set of orthonormal one-particle states which we label $|\alpha\rangle$, so that the property of closure is expressed by

$$\sum_{\alpha} |\alpha\rangle\langle\alpha| = I, \tag{3.41}$$

where I is the identity operator. We introduce also a vacuum state for Fermions, denoted by $|0\rangle$, which we assume is normalized so that

$$\langle 0|0\rangle = 1. \tag{3.42}$$

The Hamiltonian may now be written, using eqn (3.41), as

$$H = \sum_{\alpha',\alpha''} |\alpha'\rangle\langle\alpha'|(p^2/2m + V(r))|\alpha''\rangle\langle\alpha''|. \tag{3.43}$$

Inserting eqn (3.42) between the ket $|\alpha''\rangle$ and the bra $\langle\alpha''|$, eqn (3.43) becomes

$$H = \sum_{\alpha',\alpha''} \langle\alpha'|H|\alpha''\rangle|\alpha'\rangle\langle 0|0\rangle\langle\alpha''|. \tag{3.44}$$

It is now convenient to define operators $C_{\alpha''}$ and $C_{\alpha'}^+$ by the relations

$$C_{\alpha''} = |0\rangle\langle\alpha''|, \quad C_{\alpha'}^+ = |\alpha'\rangle\langle 0|. \tag{3.45}$$

It is easily verified that the operator $C_{\alpha''}$ is an annihilator which destroys the one-particle state $|\alpha''\rangle$, since

$$C_{\alpha''}|\alpha''\rangle = |0\rangle\langle\alpha''|\alpha''\rangle = |0\rangle, \tag{3.46}$$

and $C_{\alpha'}^+$ creates the one-particle state $|\alpha'\rangle$ from the vacuum state, i.e.

$$C_{\alpha'}^+|0\rangle = |\alpha'\rangle\langle 0|0\rangle = |\alpha'\rangle. \tag{3.47}$$

The Hamiltonian in eqn (3.44) can be written in second-quantized form in terms of annihilators and creators, as

$$H = \sum_{\alpha',\alpha''} \langle\alpha|H|\alpha''\rangle C_{\alpha'}^+ C_{\alpha''}. \tag{3.48}$$

3.2.4 The many-Fermion case

We now consider N noninteracting Fermion particles, e.g. electrons, so that a general N-particle state may be written as

$$\psi = \sum_{\beta_1 \ldots \beta_N} C(\beta_1 \ldots \beta_N) \Phi(\beta_1, \beta_2, \ldots, \beta_N), \tag{3.49}$$

where the labels β_i ($i = 1, \ldots, N$) identify the one-particle states which appear in each state determinant Φ, where Φ is given by

$$\Phi = \frac{1}{\sqrt{N!}} \begin{vmatrix} \phi_{\beta_1}(1) & \phi_{\beta_1}(2) & \cdots & \phi_{\beta_1}(N) \\ \phi_{\beta_2}(1) & \phi_{\beta_2}(2) & \cdots & \phi_{\beta_2}(N) \\ \vdots & \vdots & & \vdots \\ \phi_{\beta_N}(1) & \cdots & \cdots & \phi_{\beta_N}(N) \end{vmatrix} \quad (3.50)$$

Equation (3.49) represents a linear combination of such determinants, the coefficients C being appropriate for a particular set of quantum numbers which identify the single-particle states ϕ_{β_i} in eqn (3.50). By construction, the interchange of two particles, i.e. the permutation of arguments in the one-electron functions of eqn (3.49), simply transposes two columns of the Slater determinant, thereby changing its sign. We define an occupation number n_β, which is equal to one if the corresponding state ϕ_β appears in the Slater determinant and is zero otherwise. Thus, we can abbreviate our notation and write

$$\Phi = |\{n_\beta\}\rangle, \quad (3.51)$$

where the brackets $\{\ \}$ denote the set of occupation numbers in Φ which can only be zero or one. Taking four particles as an example, this might be written as

$$\Phi = |1,0,0,1,1,1\rangle. \quad (3.52)$$

Each slot in eqn (3.52) represents a particular one-particle state: only the first one and the last three states are occupied and the rest are empty.

In a similar way to that for single particles, the states $\Phi = |\{n_\nu\}\rangle$ satisfy an orthogonality relationship, namely

$$\langle \{n_\nu\} | \{n_{\nu'}\} \rangle = \delta_{\nu,\nu'}, \quad (3.53)$$

which is interpreted to mean that unless the *same* states are occupied in the string $\{n_\nu\}$ and $\{n_{\nu'}\}$, the overlap of the two states is zero and otherwise the set $\{n_\nu\} \equiv \{n_{\nu'}\}$ and it is normalized. Since the occupation-number representation spans a complete set, we can express this by the closure property

$$\sum_{\{n_\nu\}} |\{n_\nu\}\rangle \langle \{n_\nu\}| = I. \quad (3.54)$$

In a virtually identical way to that presented in the previous subsection, we now rewrite the Hamiltonian using projection operators as

$$H = \sum_{\{n_\nu\}\{n_{\nu'}\}} |\{n_\nu\}\rangle \langle \{n_\nu\}| H |\{n_{\nu'}\}\rangle \langle \{n_{\nu'}\}|, \quad (3.55)$$

SECOND-QUANTIZATION TECHNIQUES   149

where H is a sum of one-particle Hamiltonians of the type in eqn (3.40). For many Fermions, we define a creator operator by

$$C_p^+ = \sum_{\{n_i\} \text{ for } i \neq p} (-1)^{N_p} |n_i, \ldots, n_p = 1, \ldots\rangle\langle n_i, \ldots, n_p = 0, \ldots|, \quad (3.56)$$

where the integer N_p is defined by

$$N_p = \sum_{j=1}^{p-1} n_j. \quad (3.57)$$

An annihilator is similarly defined by

$$C_p = \sum_{\{n_i\} \text{ for } i \neq p} (-1)^{N_p} |\ldots, n_i, \ldots, n_p = 0, \ldots\rangle\langle \ldots, n_i, \ldots, n_p = 1, \ldots|. \quad (3.58)$$

Clearly, from eqns (3.56) and (3.58), $C_p^2 = 0 = (C_p^+)^2$ and the effect of these operators on a multi-electron state is given by

$$C_p^+ |\ldots, n_i, \ldots, n_p = 0, \ldots\rangle = (-1)^{N_p} |\ldots, n_i, \ldots, n_p = 1, \ldots\rangle \quad (3.59a)$$

and

$$C_p^+ |\ldots, n_i, \ldots, n_p = 1, \ldots\rangle = 0, \quad (3.59b)$$

which follow directly from the definitions in eqn (3.56) and (3.58). Similarly,

$$C_p |\ldots, n_i, \ldots, n_p = 1, \ldots\rangle = (-1)^{N_p} |\ldots, n_i, \ldots, n_p = 0, \ldots\rangle \quad (3.60a)$$

and

$$C_p |\ldots, n_i, \ldots, n_p = 0, \ldots\rangle = 0. \quad (3.60b)$$

It may be readily shown that these operators satisfy the required Fermion anticommutation rules

$$[C_p^+, C_q^+]_+ = [C_p, C_q]_+ = 0 \quad \text{and} \quad [C_p, C_q^+]_+ = \delta_{pq}. \quad (3.61, 3.62)$$

Since the Hamiltonian takes the form

$$H = \sum_i p_i^2/2m + V(\mathbf{r}_i), \quad (3.63)$$

it acts equally on all particles and we find that the only matrix elements of the form

$$\langle\{n_\nu\}| H |\{n_{\nu'}\}\rangle$$

$$= \int \Phi^*(\beta_1, \beta_2, \ldots, \beta_N) H \Phi(\beta_1', \beta_2', \ldots, \beta_N') d^3r_1 \cdots d^3r_N, \quad (3.64)$$

which are nonzero are those when just *one* of the β'_i's is different from the β_i or the matrix element is diagonal in the ith position. The integral in eqn (3.64) then reduces to

$$\langle \phi_{\beta_i} | p^2/2m + V(r) | \phi_{\beta_i} \rangle, \tag{3.65}$$

for a particular β_i. Thus, from eqn (3.55), the Hamiltonian becomes

$$H = \sum_{\substack{\beta_i, \beta'_i \\ \{n_j\}(j \neq \beta_i, \beta'_i)}} H_{ii'} |\ldots n_{\beta_i} = 1, n_{\beta'_i} = 0 \ldots\rangle\langle\ldots n_{\beta_i} = 0, n_{\beta'_i} = 1 \ldots |, \tag{3.66}$$

where

$$H_{ii'} = \langle n_1, \ldots, n_{\beta_i} = 1, n_{\beta'_i} = 0 \ldots | p^2/2m$$
$$+ V(\mathbf{r}) | n_1, \ldots, n_{\beta_i} = 0, n_{\beta'_i} = 1 \ldots \rangle. \tag{3.67}$$

Hence, using the definitions in eqns (3.56) and (3.58),

$$H = \sum_{\beta_i, \beta'_i} H_{ii'} C^+_{\beta_i} C_{\beta'_i}. \tag{3.68}$$

In other words, the second-quantized form of H may be written as

$$H = \sum_{\nu, \mu} \langle \nu | p^2/2m + V(r) | \mu \rangle C^+_\nu C_\mu, \tag{3.69}$$

where ν and μ denote the quantum numbers of one-particle states which form a complete set.

When interactions are introduced between particles, the formalism of second quantization becomes the only practicable formalism to use which contains essential physics and is easy to interpret. It is straightforward to demonstrate [2] that, in this technique, an interaction of the form

$$V = \tfrac{1}{2} \sum_{\substack{i,j \\ i \neq j}} V(r_i - r_j), \tag{3.70}$$

which is called a *two-body operator*, becomes

$$V = \tfrac{1}{2} \sum_{\lambda, \mu, \nu, \delta} \langle \lambda, \mu | V(r_1 - r_2) | \nu, \delta \rangle C^+_\lambda C^+_\mu C_\delta C_\nu, \tag{3.71}$$

where $\langle \lambda, \mu | V(r_1 - r_2) | \nu, \delta \rangle$ is to be interpreted as

$$\iint \lambda(r_1) \mu(r_2) V(r_1 - r_2) \nu(r_1) \delta(r_2) \, d^3r_1 \, d^3r_2.$$

This completes our brief excursion into the formalism of second quantization. What we intend to do next is to present an overview of particular types of elementary excitations in solids, in which we will make full use of the technique of second quantization. Two of these examples merit special exposure due to their significance. These are lattice vibrations, the quanta of which are termed phonons, and the case involving an electron gas.

3.3 Lattice vibrations and phonons

3.3.1 Introducing vibrations in solids

Let us now go back to the Hamiltonian H_1 in eqn (3.1) and consider the masses of the particles involved. It is important to realize that the nuclei are much more massive than the electrons and hence their kinetic energy is considerably lower than the electronic kinetic energy. The electrons react very rapidly when the nuclei move, so a good starting point is to neglect the kinetic energy of the nuclei entirely and solve for the electronic eigenvalues and eigenfunctions, treating the positions of the nuclei as fixed parameters. This is the well-known Born–Oppenheimer (BO) approximation. A very clear exposition of this approach was given by Born and Huang many years ago [3]. The electronic eigenfunctions ϕ_s and energies E_s are therefore functions of the positions of the nuclei, R_n, and the electron position vectors r_i. We may express this by writing

$$\phi_s = \phi_s(r_1, r_2, \ldots, r_{N'}, R_1, R_2, \ldots, R_N) \tag{3.72}$$

and

$$E_s = E_s(r_1, r_2, \ldots, r_{N'}, R_1, R_2, \ldots, R_N), \tag{3.73}$$

where N' denotes the number of electrons and N is the number of nuclei in the system. In the BO approximation the nuclear motions are then treated by perturbation theory using an effective expansion parameter which turns out to be $\varepsilon = (m/M)^{1/4}$, where m is the electronic mass and M is the mean nuclear mass. It transpires that, in second order, a good approximation to the nuclear Hamiltonian is the first term of H_1; namely, the kinetic energy of the nuclei, plus the second order contribution from the second term when the nuclear potential energy is expanded about equilibrium positions R_{n0} and it is assumed that

$$R_n - R_{n0} = \varepsilon \mathbf{u}_n, \tag{3.74}$$

in which ε indicates that the displacement from equilibrium is small relative to internuclear separations.

To see this more clearly, we express the first two terms of H_1 as

$$H_N = \sum_n \frac{|P_n|^2}{2M} + \tfrac{1}{2} {\sum_{n,m}}' U(R_n - R_m), \tag{3.75}$$

where we have written the nuclear momenta as P_n, the internuclear interaction as U, and assumed, for convenience, that each nucleus has the same mass. To simplify matters, we assume that the solid described by eqn (3.1) is periodic and that there is only one nucleus per unit cell.

3.3.2 Harmonic vibrations

It is a relatively simple matter to generalize to the case with more than one atom per unit cell. However, we will not describe this in detail here, but draw the reader's attention to a fuller account in the book by Jones and March [4]. To obtain the second-order part of the potential energy, we expand about the equilibrium positions and write

$$U(R_n - R_m) = U(R_{n0} - R_{m0})$$
$$+ \tfrac{1}{2} \sum_{\alpha,\beta} \left[\frac{\partial^2 U(R_n - R_m)}{\partial R_{n\alpha} \partial R_{m\beta}} \right]_0 u_{n\alpha} u_{m\beta}, \quad \text{for } n \neq m, \quad (3.76)$$

where the labels α and β denote directions so that, for example, $u_{n\alpha}$ denotes the displacement of the nth nucleus in the α-direction (so α could be x, y, or z if the system were cubic). The first-order term disappears because it is assumed to be zero at equilibrium.

Normal co-ordinates q_k are now defined by

$$u_{n\alpha} = \frac{1}{\sqrt{N}} \sum_k n_{k\alpha} q_k \exp(i k \cdot R_{n0}), \quad (3.77)$$

where the wave vectors k are those in the first Brillouin zone. The unit vectors n_k denote the polarization direction for a mode with wave vector k. In a similar way, we may define a component of momentum for the nth ion by

$$P_{n\alpha} = \frac{1}{\sqrt{N}} \sum_k n_{k\alpha} P_k \exp(i k \cdot R_{n0}). \quad (3.78)$$

If we now insert eqns (3.77) and (3.78) into the nuclear Hamiltonian in eqn (3.75), using eqn (3.76) and the fact that

$$\sum_{R_{n0}} e^{i k \cdot R_{n0}} = N\delta(k, K), \quad (3.79)$$

where K is any reciprocal lattice vector, we find that

$$H_N = \sum_k \frac{P_k^+ P_k}{2M} + \frac{M}{2} \sum_k D_{\alpha\beta}(k) n_{k\beta} n_{k\alpha} q_k^+ q_k. \quad (3.80)$$

In the second term, $D_{\alpha\beta}$ is defined by

$$D_{\alpha\beta}(k) = M^{-1} \sum_{R_{n'_0}} \phi_{\alpha\beta}(R_{n'_0}) \exp(-i k \cdot R_{n'_0}) \qquad (3.81)$$

and is called the dynamical matrix. We have used the shorthand

$$\phi_{\alpha\beta}(R_{n'_0}) = \left(\frac{\partial^2 U}{\partial R_{n_0} \partial R_{m_0}}\right)_0 = \phi_{\alpha\beta}(R_{n_0} - R_{m_0}), \qquad (3.82)$$

and the translational invariance of $\phi_{\alpha\beta}$. The eigenfrequencies for the normal modes satisfy

$$\sum_\beta D_{\alpha\beta}(k) n_{k\beta} = \omega^2 n_{k\alpha}. \qquad (3.83)$$

Using the fact that the polarization vectors are unit vectors and exploiting eqn (3.83), the nuclear Hamiltonian becomes

$$H_N = \sum_k \frac{P_k^+ P_k}{2M} + \sum_k \tfrac{1}{2} M \omega_k^2 q_k^+ q_k. \qquad (3.84)$$

This corresponds to a large set of three-dimensional independent harmonic oscillators, the kth having an angular frequency of ω_k. In an elastic continuum model (the case in which wavelengths are large, so that they do not 'see' the lattice), the three polarization vectors $n_{k\sigma}$ correspond to the two transverse modes and one longitudinal mode. For the case of an isotropic continuum the eigenvectors of the dynamical matrix are either exactly parallel or perpendicular to the wave-vector k but, in general, in a discrete lattice this is not always true but happens only along symmetry directions in the first Brillouin zone. When $k = 0$ these three modes are degenerate and the situation corresponds to identical displacements of all the atoms or ions in the crystal. These latter modes cannot cause any energy change when $k = 0$, so $\omega(k = 0, \sigma) = 0$, where σ labels the modes. The gradient of the dispersion curves ($E(k)$ versus k) for low k is of the order of the velocity of sound and hence these modes are called acoustic modes.

When there is more than one ion per unit cell, it becomes possible for these ions to vibrate out of phase with one another, so that the relation $\omega(k, \sigma) = 0$ need not hold at $k = 0$. However, there will still be three acoustic modes, but other higher energy modes at $k = 0$ will arise which have frequencies in the infrared and are called optical modes. In fact, if there are r ions per unit cell, there will be $3r - 3$ optical modes, the dynamical matrix now being of dimension $3r \times 3r$. To obtain the result in

154 ELEMENTARY EXCITATIONS IN SOLIDS

eqn (3.84) we have made what is called the *harmonic approximation*, neglecting terms that are cubic and higher in the nuclear displacements. Real solids are not harmonic in general but, nevertheless, many of their properties can be analysed within this approximation, co-joined with that of Born and Oppenheimer. Anharmonic terms, $O(U^3)$, become important when displacements are large, or when the electron-ion coupling is so strong that to talk of electronic and nuclear motions separately would be quite wrong. This situation may arise in the study of the Jahn–Teller effect, and one then speaks not of vibrational or electronic states but of vibronic states [5].

Returning now to the second-quantized picture, we may proceed in a very similar way to that described in the previous section, and apply it to the Hamiltonian in eqn (3.84), which has the same form as eqn (3.23). Then, using eqns (3.24)–(3.29), we reduce eqn (3.84) to diagonal occupation form, namely

$$H = \sum_k \hbar \omega_k (\beta_k^+ \beta_k + \tfrac{1}{2}), \qquad (3.85)$$

and consequently the eigenenergies are given by

$$E = \sum_k \hbar \omega_k (n_k + \tfrac{1}{2}) \qquad (3.86)$$

where $n_k = 0, 1, 2, \ldots$. A single quantum of vibrational energy is called a *phonon*.

Another example of this type of phenomenon occurs in insulators which are made up of positive and negative ions. If we imagine the ions *not* vibrating (this is, of course, impossible physically) then the net dipole moment per unit volume averages to zero. However, if we now 'turn on'

(a)

(b)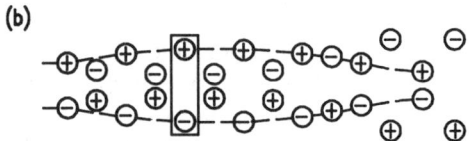

Fig. 3.1 Lattice polarization waves in a polar crystal: (a) a crystal at rest; (b) a transverse polarization wave.

LATTICE VIBRATIONS AND PHONONS

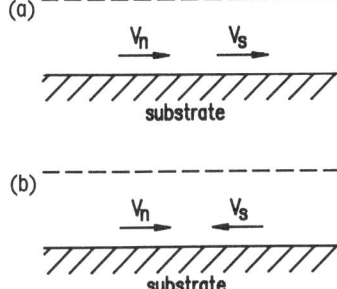

Fig. 3.2 (a) First and (b) second sound modes in ^{4}He where V_n and V_s denote the normal and superfluid velocities, respectively.

the vibrations of positive and negative ions in a unit cell, their corresponding dipole moment per unit volume will begin to fluctuate locally, leading to a net polarization, the polarization vibrations appearing as waves in the crystal. These polarization waves may be quantized in a similar way to sound (see Fig. 3.1).

More complicated types of vibration can be found in superfluid ^{4}He, and are conventionally referred to as the first, second, third, etc. sound. Like polarization modes, they arise as a result of different phase relations between the distinct vibrating components of the material. In this latter case the medium considered has both normal and superfluid components which execute independent motion (see Fig. 3.2).

3.3.3 Effects due to anharmonicity

If we consider phonons as an example, there will, of course, be interactions between them resulting from anharmonicity, since we have truncated the expansion of the internuclear potential in eqn (3.76). In fact, were it not for these interactions a solid would not expand when heated! Another way of putting this is to say that thermal resistance is generated by the interaction of sound waves with one another. Below, we discuss the role that anharmonicity plays in solids.

In the lowest order of approximation, anharmonic effects lead to three-phonon scattering processes, which can be of two types. 'Normal' processes conserve momentum when phonons are scattered by the anharmonic potential and lead to zero thermal resistance. On the other hand, so-called Umklapp processes transfer some of the incoming phonon momentum to the crystal lattice, effectively resulting in a loss of momentum, so that this process generates thermal resistance [6]. The thermal conductivity coefficient, K, is defined through

$$Q = -K \nabla T, \qquad (3.87)$$

where Q is the heat flux and ∇T represents a temperature gradient. In the simplest case of a cubic lattice it can be demonstrated that

$$K = \tfrac{1}{3} \frac{1}{(2\pi)^3} \sum_\sigma \int d^3k\, C_\sigma(k)\tau_\sigma(k) v_\sigma^2(k), \qquad (3.88)$$

where σ labels a particular phonon branch, C_σ is the corresponding specific heat, and v_σ is an associated group velocity. Thus

$$v_\sigma(k) \equiv \nabla_k \omega_\sigma(k). \qquad (3.89)$$

In eqn (3.88), $\tau_\sigma(k)$ is the relaxation time, which obviously becomes finite when collision processes, due to anharmonicity, are included. In general, τ should include contributions due to the presence of defects, due to boundary effects, as well as phonon–phonon scattering. Focusing on the latter effect only, it is found that

$$\tau_{\text{ph}}^{-1} = BT^3\omega^2, \qquad (3.90)$$

where τ_{ph} is solely due to phonon scattering. This process appears to be dominant in the low-temperature regime.

In addition to nonzero thermal resistance, the presence of anharmonicity leads to several other measurable effects which sometimes can be calculated successfully in a perturbative manner. First of all, the presence of anharmonicity in the lattice Hamiltonian introduces, in second-quantization formalism, terms of the type

$$(\beta^+_{-k\alpha} + \beta_{k\alpha})(\beta^+_{-k'\alpha'} + \beta_{k'\alpha'})(\beta^+_{-k''\alpha''} + \beta_{k''\alpha''}), \qquad (3.91)$$

from third-order interaction terms in the displacements. These and higher-order corrections will have an effect on the positions of energy levels and the density of states as a function of the phonon frequencies. That in turn will introduce corrections to the specific heat over its harmonic value. There exists a considerable body of literature dealing with this topic alone and we refer the reader to references [7–9] for more information. What is perhaps more important than striving for a quantitatively accurate picture is the fact that new *qualitative physical features* appear when anharmonicity is present. One of these features is thermal expansion of the crystal lattice, which can only be present if the lattice is anharmonic. To illustrate this in simple terms we can consider a one-dimensional example, with a potential energy given by [6]

$$V(x) = c_2 x^2 + c_3 x^3 + c_4 x^4, \qquad (3.92)$$

where x represents the deviation from an equilibrium position and the c_i's ($i = 1, 2, 4$) are force constants. Using the Bolzmann distribution to calculate the average displacement $\langle x \rangle$ at a given temperature T, we obtain

$$\langle x \rangle \simeq -\frac{3c_3}{4c_2^2} k_B T + \cdots, \tag{3.93}$$

indicating that c_3 must be negative to obtain a correct result, i.e. a linear thermal expansion of the solid.

As will be pointed out later in this chapter, anharmonicity in solids may eventually result in much more dramatic effects, such as structural phase transitions, in which the equilibrium positions of atomic nuclei are destabilized and new symmetry elements are introduced to the structure (or some old elements removed, depending on the direction in which the changes are occurring). This effect is not only static but can be dynamical in nature, so that long-wavelength modes can be seen as precursors leading to a structural change. In addition, there exist situations in which lattice vibrations and electronic degrees of freedom are so strongly coupled together, e.g. in Jahn–Teller systems, that it is not meaningful to talk about vibrational and electronic states separately.

In the next section we examine how electronic degrees of freedom may be conveniently discussed in terms of collective excitations such as plasmons.

3.4 Electronic excitations of the solid

We now examine the opposite extreme in which nuclei in our solid Hamiltonian of eqn (3.1) are fixed, providing a positively charged rigid background. We admit, however, a spatial variation of the density of, more or less, freely moving electrons. Consequently, the spatial charge distribution is locally no longer neutral. This induces the presence of repulsive fields that tend to draw the electrons back to their original positions. Due to their inertia they overshoot these positions and thereby induce oscillatory motion, termed plasma oscillations, and their quanta are called plasmons [10] (see Fig. 3.3). Their distinct feature is a very high frequency compared with optic phonon frequencies.

The physical picture can be realized by again utilizing the Hamiltonian H_1 in eqn (3.1) and introducing a density of particles of charge $-|e|$, given by

$$\rho(\mathbf{r}) = \sum_i \delta(\mathbf{r} - \mathbf{r}_i) - \sum_n Z_n \delta(\mathbf{r} - \mathbf{R}_{n0}). \tag{3.94}$$

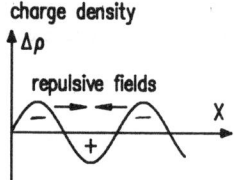

Fig. 3.3 The mechanism of plasma oscillations.

The potential energy terms in H_1 may now be written as

$$\sum_k \frac{2\pi|e|^2}{k^2}(\rho_k^+ \rho_k - \bar{n}), \tag{3.95}$$

where ρ_k is the kth Fourier component of the particle density $\rho(r)$, so that

$$\rho(r) = \sum_k \rho_k \exp(i\mathbf{k}\cdot\mathbf{r}), \tag{3.96}$$

and $\bar{n}$ is given by

$$\bar{n} = n(1+Z). \tag{3.97}$$

We have, for convenience, put all nuclear charges equal $Z_n = Z$ and n is the average electron density [11–13]. By adding an extra term to the Hamiltonian which contains additional degrees of freedom and applying an appropriate unitary transformation, with subsidiary conditions to limit the number of degrees of freedom, H_1 becomes

$$\begin{aligned}\tilde{H}_1 = &\sum_i \frac{p_i^2}{2m} \\ &+ \tfrac{1}{2} \sum_{\substack{i,j \\ i\neq j}} |e|^2 \tilde{U}(r_i - r_j) - |e|^2 \sum_{i,n} Z_n \tilde{U}(r_i - R_n) \\ &+ \sum_{\substack{k \\ k<k_c}} \left[\frac{\pi_k^+ \pi_k}{2} + \omega_p^2 \frac{Q_k^+ Q_k}{2}\right] + \ldots,\end{aligned} \tag{3.98}$$

where $\tilde{U}(\mathbf{r}_i - \mathbf{r}_j)$ is an effective screened Coulomb interaction [11] and k_c is the maximum wave vector allowed for plasmons. Thus the electron–nucleus attraction is *also* screened as a result of this transformation. The fourth term describes plasma oscillations, where π_k is the plasmon momentum, Q_k is the corresponding canonical 'displacement', and ω_p is the plasma frequency, given by

$$\omega_p^2 = 4\pi e^2 n/m. \tag{3.99}$$

ELECTRONIC EXCITATIONS OF THE SOLID 159

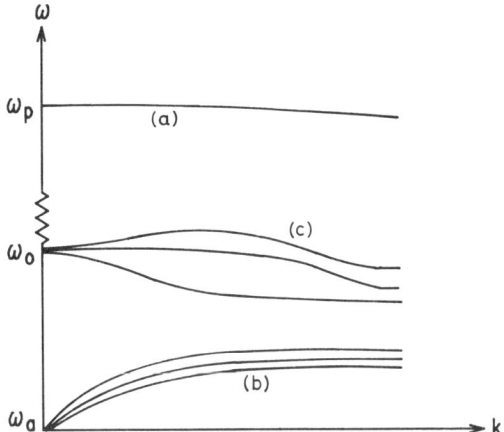

Fig. 3.4 The schematic dependence of the frequency of oscillation versus wave vector k for (a) plasmons, (b) acoustic phonons, and (c) optical phonons.

This frequency is typically orders of magnitude greater than the infrared optical frequencies and is usually independent of the wave vector k, although higher-order corrections introduce a quadratic and a quartic dependence on k, which are rather weak [13]. We have depicted the dispersion relations for plasmons and phonons in Fig. 3.4. When the nuclei are allowed to move, this superimposes a much lower frequency component on the electronic plasma frequency. Despite what might appear to be drastic simplifications above, a model employing plasma oscillations approximates reality quite well in simple metals such as alkalis. However, very much analogous to the need for interaction between phonons, the above picture is incomplete if we want to account for several observed phenomena.

For example, the need to account for finite electrical conductivity implies that the picture of freely moving electrons is inadequate. First, due to the distance dependence of the Coulomb force, plasmon–phonon scattering processes will frequently occur provided that energy and momentum considerations are satisfied. Plasmons are not the only type of excitation for the electronic degrees of freedom. If the electron–electron interaction is treated as a scattering mechanism, then only when the momentum transfer is small do plasmon oscillations take place [2] but, if the momentum transfer is larger, particle–hole pair excitations are formed. This can be viewed, provided that the electron–electron interaction is weak enough, as simply lifting an electron out of the Fermi sphere of states, leaving a hole behind. In Fig. 3.5 we show a typical excitation spectrum of a uniform

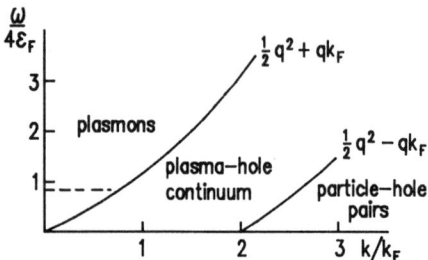

Fig. 3.5 A typical phase diagram of excitations in the interacting electron gas (following reference [14]).

interacting electron gas, indicating the regions in which plasmons are the appropriate excitation, others in which the particle-hole picture is sensible, and a plasmon-hole region [14].

In addition to these two most important types of excitation in a solid, i.e. phonons and electron excitations, a large number of other types have been discovered over the years [15]. Due to limitations of space, we shall only provide a brief overview of some of these other types.

3.5 Other types of elementary excitation

3.5.1 Spin waves

So far, we have not included spin degrees of freedom in the description. In materials such as ferromagnets this is, of course, an unacceptable omission. We have seen in Chapter 2 that below the Curie temperature the unquenched magnetic moments order themselves spontaneously even in the absence of an external magnetic field. The picture developed in Chapter 2, however, did not incorporate any dynamics whatever. Just as phonons have been obtained as small oscillations about equilibrium positions in an elastic lattice, spin waves can be derived as small-amplitude fluctuations around the average value of the magnetization vector (see Fig. 3.6).

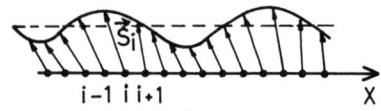

Fig. 3.6 A schematic version of a spin wave in a ferromagnet.

OTHER TYPES OF ELEMENTARY EXCITATION

A mathematically convenient way to describe spin waves is first to consider a Heisenberg-type Hamiltonian

$$H = -\tfrac{1}{2} \sum_{\substack{i,j \\ i \neq j}} J_{ij} S_i \cdot S_j, \tag{3.100}$$

introduce a spin wave with a wave vector k as a Fourier transform

$$S(k) = N^{-1/2} \sum_{j=1}^{N} S_j \exp(i k \cdot R_j), \tag{3.101}$$

and simultaneously define the Fourier transform of the exchange constants:

$$J(k) = \sum_{j=1}^{N} J_{j0} \exp\bigl(i k \cdot (R_j - R_0)\bigr). \tag{3.102}$$

We then introduce 'spin-up' and 'spin-down' ladder operators for a spin wave, as

$$S_\pm(k) = S_x(k) \pm i S_y(k), \tag{3.103}$$

so that the spin–spin interaction can be expressed as

$$S_j \cdot S_l = S_{jz} S_{lz} + \tfrac{1}{2}(S_{j+} S_{l-} + S_{j-} S_{l+}). \tag{3.104}$$

Defining the vacuum state $|0\rangle$ as that in which all spins are pointing along the z-axis, we obtain

$$S_{jz}|0\rangle = S|0\rangle, \qquad S_{j+}|0\rangle = 0. \tag{3.105}$$

These properties of the S_{jz} and S_{j+} operators resemble those of the second-quantized operators appearing in the last few sections. In order to make this correspondence more direct, the so-called Holstein–Primakoff transformation is applied, in which Bose–Einstein creators and annihilators, B_j^+ and B_j, respectively, enter into the definitions of the spin operators:

$$S_{jz} = S - B_j^+ B_j, \qquad S_{j+} = \sqrt{(2S - B_j^+ B_j)}\, B_j,$$
$$S_{j-} = B_j^+ \sqrt{2S - B_j^+ B_j}. \tag{3.106}$$

The final step in diagonalizing the Heisenberg Hamiltonian is to expand the square roots in eqn (3.106), which is followed by Fourier transforming the B operators, so that

$$B_j = N^{-1/2} \sum_k b_k \exp(-i k \cdot R_j). \tag{3.107}$$

The new operators b_k and their Hermitian conjugates b_k^+ are called *magnon* annihilators and creators, respectively, and the Hamiltonian may now be represented (approximately), in a diagonal form as

$$H = E_0 + \sum_k \varepsilon(k) b_k^+ b_k, \qquad (3.108)$$

where

$$\varepsilon(k) = S[J(0) - J(k)], \qquad (3.109)$$

and the corresponding magnon state that b_k^+ might create from the vacuum is

$$|k\rangle = S_-(k)|0\rangle. \qquad (3.110)$$

3.5.2 Excitons

In crystal lattices, localized excitations caused by a displacement of electronic charges from the vicinity of their original nuclei are also known to exist. These types of excitation are termed *excitons* and they often exhibit mobility. Two distinct classes of exciton can be defined, and these are called either *Frenkel excitons* or *Wannier excitons* [15, 16]. Frenkel excitons occur when the atoms forming the lattice are weakly interacting (e.g. in organic crystals) and they support the presence of excited polarized states of isolated molecules. Their mode of propagation through the crystal is best characterized as hopping. Wannier excitons, on the other hand, involve the creation of weakly bound electron–hole pairs separated by distances that are large compared to the lattice constant. For this reason, Wannier excitons may form periodic **k** waves of polarization, somewhat similar to the idea of a magnon. The two types of exciton are compared graphically in Fig. 3.7.

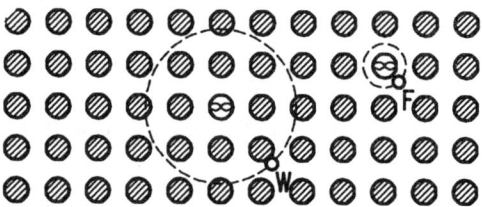

Fig. 3.7 The large-radius Wannier exciton (W), as compared to the small-radius Frenkel exciton (F).

OTHER TYPES OF ELEMENTARY EXCITATION 163

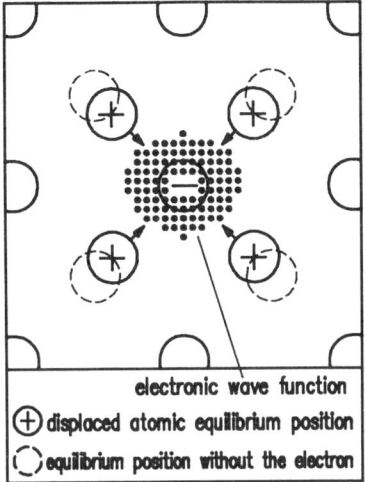

Fig. 3.8 A self-trapped state of the electron in a deformable lattice (a small polaron).

3.5.3 Polarons

In a deformable solid, an excess charge carrier may either form a small polaron state or remain relatively free. The polaron state occurs when an electron displaces surrounding atoms or ions from their equilibrium positions and thereby lowers the total energy of the system. This effect is frequently called *self-trapping*. Simultaneously, the electron's effective mass is increased, since polaron formation is associated with a lattice deformation in the form of a 'polarization cloud' (see Fig. 3.8 for an illustration).

As the atoms of the solid are progressively moved from their equilibrium positions, the energy of the system at first increases but then falls upon reaching the peak of the energy barrier for self-trapping. Hence the excess charge carrier will initially remain untrapped until the surrounding atoms experience sufficiently large displacements from their equilibrium positions. Therefore, the effective mass of the electron depends on the velocity with which it propagates being nearly equal to the bare mass at high velocities. When the nuclei cannot respond and displace quickly enough, and at low velocities, it is largely due to the effect of the electron dragging the polarization cloud along behind it.

3.5.4 Polaritons

Coupled modes may be formed not only between different types of elementary excitation of a solid but also between one of them and an external field, e.g. an electromagnetic wave. Polaritons are a new type of excitation which couple polarized lattice vibrations and photons travelling through the crystal. Since both of these subsystem excitations can be treated harmonically, their interaction can be likened to the coupling of two pendulums. As a result, a new mode of vibration is produced which involves the joint propagation of light and polarization waves. In order to derive the dispersion relation for the polariton we consider an isotropic solid with dielctric constant $\varepsilon(\omega)$ and the equation for the transverse part of the electromagnetic field in matter, namely

$$c^2 \nabla^2 E = \partial^2 D / \partial t^2, \tag{3.111}$$

where E is the electric field, and the displacement field is given by

$$D = \varepsilon' E + 4\pi P. \tag{3.112}$$

In eqn (3.112), P denotes the polarization and ε' the real part of the dielectric constant. Inserting a plane-wave solution, $\exp(i(\mathbf{k} \cdot \mathbf{r} - \omega t))$, into eqn (3.111), we find that

$$c^2 k^2 / \omega^2 = \varepsilon(\omega). \tag{3.113}$$

So far, we have not described coupling between the two oscillations. To this end, we select a particular polarization mode with its characteristic frequency ω_t and couple it to the electric field, assuming that the static polarizability is α_0. Then, the coupled system of fields may be described by

$$\ddot{P} + \omega_t^2 P = \omega_t^2 \alpha_0 E \tag{3.114}$$

and

$$\varepsilon' \ddot{E} - c^2 \nabla^2 E = -4\pi \ddot{P}. \tag{3.115}$$

Equation (3.115) is obtained by differentiating eqn (3.112) twice with respect to time and inserting eqn (3.111). These two equations can be simultaneously solved by a plane wave in both P and E, since P and E are parallel. The resultant determinant, from which the angular frequency may be obtained, is simply

$$\begin{vmatrix} -\omega^2 + \omega_t^2 & -\alpha_0 \omega_t^2 \\ -4\pi\omega^2 & -\varepsilon'\omega^2 + c^2 k^2 \end{vmatrix} = 0. \tag{3.116}$$

This can be solved for k as a function of ω, so that

$$\frac{c^2 k^2}{\omega^2} = \varepsilon' + \frac{4\pi \alpha_0 \omega_t^2}{\omega_t^2 - \omega^2}. \tag{3.117}$$

Inverting this relationship produces the required dispersion relation for the polariton, as shown graphically in Fig. 3.9.

SOLITONS AND COHERENT STRUCTURES

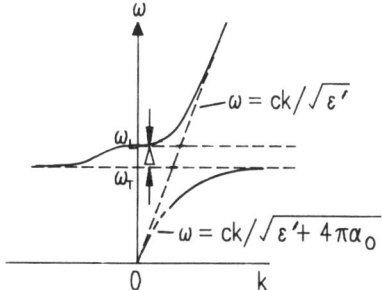

Fig. 3.9 The dispersion relation for the polariton.

It should by now be apparent that elementary excitations are indeed relative concepts, the definition of which frequently depends on the choice of the nonperturbed Hamiltonian selected for reference, i.e. on mathematical expediency. For reasonable choices of unperturbed Hamiltonians, that is when perturbations *are* small enough to be absorbed in a few steps as corrections, this approach is indeed valid and very useful. However, this is not always the case, for at least three reasons. First, it may happen that when the system is close to a phase transition the concept of a unique equilibrium reference point becomes suspect and subsequent perturbation calculations are very slowly, if at all, convergent. Second, it may not be possible to perturb the reference Hamiltonian using a power series expansion, e.g. as in superconductivity where the electron–phonon interaction is very strong and causes dynamic symmetry breaking (more on this in Chapter 4). Third, in addition to the presence of nearly harmonic excitations other types of behaviour, such as solitons and defects, may be present in the system. The coexistence of all of these types of excitation may result in numerous collisions, strong scattering, and hence finite lifetimes and finite mean free paths, thereby complicating the original picture a great deal. In the next two sections of this chapter we intend to describe these other, highly nonlinear modes and how they arise.

3.6 Solitons and coherent structures

A frequently employed model of dynamical effects associated with structural phase transitions is based on the Hamiltonian [17]

$$H = \int dx \left[\frac{m}{2} \dot{u}^2 - \frac{|A|}{2} u^2 + \frac{B}{4} u^4 + \frac{mc_0^2}{2} \left(\frac{du}{dx} \right)^2 \right], \qquad (3.118)$$

where u is a time- and position-dependent displacement and c_0 is the velocity of sound. This Hamiltonian can be easily obtained in the continuum limit for a system of harmonically coupled masses with bistable on-site potentials. The variational principle, when applied to H, gives the following equation of motion for $u(x,t)$:

$$m\ddot{u} - |A|u + Bu^3 - mc_0^2 \frac{\partial^2 u}{\partial x^2} = 0, \qquad (3.119)$$

which is of a nonlinear Klein–Gordon (NLKG) type. We know from Chapter 1 that, in addition to constant solutions signalling a perfectly ordered broken-symmetry state, elliptic wave solutions of eqn (3.119) also exist and can be interpreted as classical, nonlinear analogues of phonons. More interestingly, a nonperturbative solution is also found in the form of a tanh-kink, which can be interpreted as a domain wall between two equally important sign choices for the amplitude of the lattice distortion. In terms of physical meaning this can, for example, represent so-called twinning. A mathematical analysis of the NLKG equation in $(3+1)$-dimensional space–time is provided in Chapter 7.

Analogous phenomena are indeed quite widespread throughout solid-state physics, and for domain-wall examples we need only look to ferromagnetism, ferroelectricity, or virtually any situation in which sharp interfaces exist [17]. In Figs 3.10 and 3.11 examples are shown of domain walls from ferromagnetism and liquid–solid transitions, respectively.

Another example of domain-wall phenomenon can be found in one-dimensional conductors such as polyacetylene $(CH)_n$ which, when doped with donors or acceptors, generate sequences of two different configurations A and B, as shown in Fig. 3.12, where the two configurations differ by the position of double and single bonds but have the same energy per site. The presence of unbalanced charges creates the so-called topological defects which are analogous to domain walls in ferromagnets.

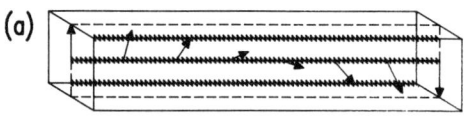

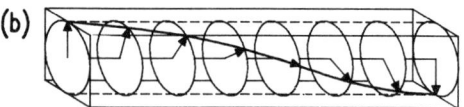

Fig. 3.10 A Néel wall (a) and a Bloch wall (b) in ferromagnets.

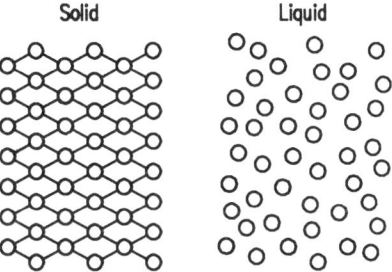

Fig. 3.11 A sharp solid–liquid interface.

Somewhat similar phenomena can also be seen in long Josephson-junction transmission lines, where the superconducting phase ϕ tunnels through periodic regions of an insulator interspersed with regions of superconductivity [18]. The difference is that, in the continuum limit, the equation of motion for ϕ is a Sine–Gordon equation

$$\phi_{xx} - \phi_{tt} = \sin \phi. \tag{3.120}$$

Here, unlike in the previous cases, the kink-type solutions are true solitons and they survive collisions with other solitons and nonlinear excitations. Furthermore, analytical means exist to obtain multi-soliton solutions which do not exist in the NLKG equation.

In optical fibres, not only have solitons been detected but they are now at the centre of a technological revolution taking place in the telecommunications industry. The basic mechanism for soliton formation in optical fibres is a combination of dispersive properties of the material and the presence of nonlinearity through the Kerr effect [19], where the index of refraction, n, is a nonlinear function of the electric field $\boldsymbol{E}$:

$$n(\omega, |\boldsymbol{E}|^2) = n_0(\omega) + n_2 |\boldsymbol{E}|^2 + \cdots . \tag{3.121}$$

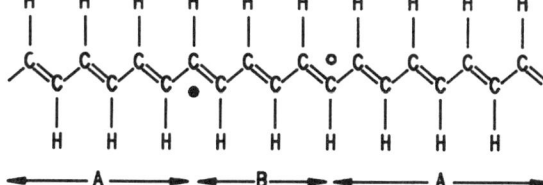

Fig. 3.12 Soliton (●) and antisoliton (○) in polyacetylene.

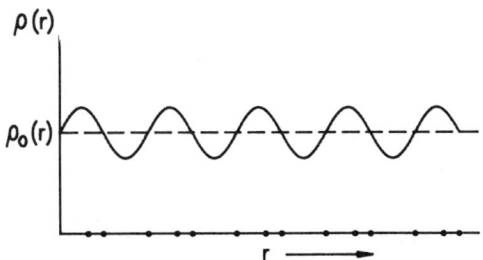

Fig. 3.13 The charge density wave $\rho(r)$ and the associated lattice distortion, represented by the dots.

Defining D_L as the linear part of the electric displacement field, we obtain, from one of the Maxwell equations,

$$\partial_{xx}^2 E - \frac{1}{c^2}\partial_{tt}^2 D_L - \frac{2n_0 n_2}{c^2}\partial_{tt}^2 |E|^2 E = 0. \tag{3.122}$$

It turns out that the amplitude of the electric field then satisfies the nonlinear Schrödinger equation (NLS) with the attendant soliton solutions.

In low-dimensional conductors such as $NbSe_3$ or $TaSe_3$, below a characteristic temperature T_t, an electronic *charge density wave* (CDW) is spontaneously generated, in which the position-dependent charge distribution $\rho(r)$ can be expressed as (see Fig. 3.13 for an illustration)

$$\rho(r) \simeq \rho_0(r)[1 + \alpha \cos(q_0 \cdot r + \phi)], \tag{3.123}$$

where ρ_0 is the high-temperature phase charge distribution, α is the amplitude of the CDW (which can, in fact, be used as an order parameter), q_0 is its wave vector, and ϕ is its phase. This effect is stabilized by an accompanying lattice distortion of the ions [20]. The periodicity of the wave, determined by q_0, depends on the topology of the Fermi surface. The formula (3.123) expresses a linear approximation of the effect which, in fact, arises from a nonlinear elliptic wave. The lattice distortion opens a gap in the electronic energy spectrum, thereby reducing the number of conduction electrons, which in some cases may trigger a metal–insulator transition.

A related phenomenon is called a *spin density wave* (SDW), which occurs in a CDW state in which the electronic density modulation is 180° out of phase between 'spin-up' (+) and 'spin-down' (−) charges. This leads to an effective removal of CDW modulation but a SDW distribution becomes space-dependent, so that

$$\rho_\pm(r) = \tfrac{1}{2}\rho_0[1 \pm \alpha \cos(q_0 \cdot r + \phi)], \tag{3.124}$$

SOLITONS AND COHERENT STRUCTURES 169

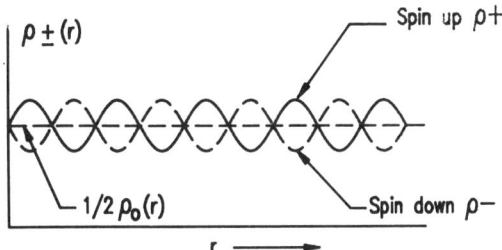

Fig. 3.14 A spin density wave.

where the notation is analogous to that in eqn (3.123) (see Fig. 3.14 for an illustration).

Generalization of solitons and nonlinear waves to higher dimensions are not easy, but there exist structures with properties strongly resembling the former two types of excitation. In particular, various types of vortices in fluids, superfluids, superconductors, and easy-plane magnets merit the designation 'two-dimensional solitons'. In type II superconductors [17] individual vortices arise at a lower critical field, H_{c1}, as a magnetic flux line penetrates into the superconducting state (see Fig. 3.15).

In the range $H_{c2} > H > H_{c1}$, where H_{c2} is the upper critical field, a so-called Abrikosov lattice of magnetic vortices is formed and has a hexagonal symmetry (see Fig. 3.16). This can be thought of as a highly nonlinear standing wave with peaks at vortex centres and valleys between neighbouring vortices. The amount of magnetic flux contained within a single vortex must necessarily be an integer multiple of an elementary flux quantum.

Vortices have been found to be a characteristic feature of two-dimensional anisotropic Heisenberg and X–Y models (see Chapter 2).

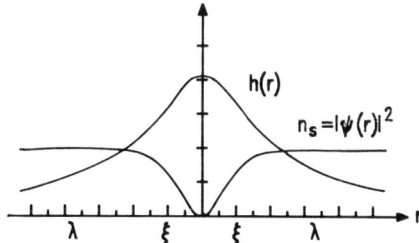

Fig. 3.15 An isolated vortex in a type II superconductor. Here $h(r)$ denotes the magnetic field intensity, n_s is the superconducting charge density, and ξ and λ are the coherence length and penetration depth, respectively.

170   ELEMENTARY EXCITATIONS IN SOLIDS

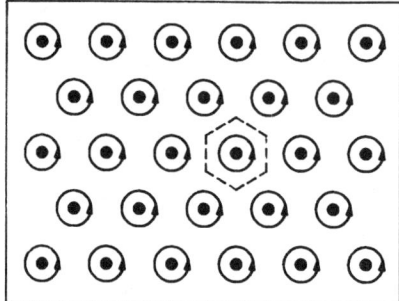

Fig. 3.16 An Abrikosov vortex lattice.

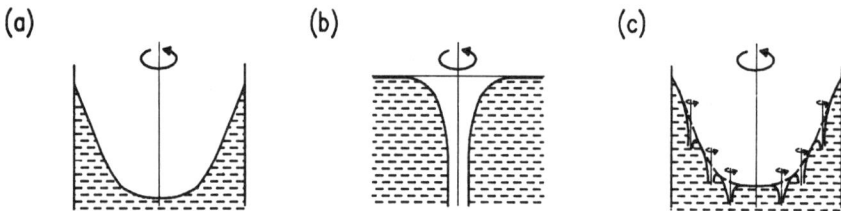

Fig. 3.17 (a) The paraboloid shape of a classical fluid surface under rotation. (b) A single quantum of superfluid vorticity. (c) An array of quantized superfluid vortices.

Two general types have been observed: (a) in-plane and (b) out-of-plane vortices [21].

Quantized vortices are also known to exist in superfluid helium. A comparison between the behaviour of a rotating classical fluid, a single superfluid vortex, and an array of superfluid vortices is shown in Fig. 3.17.

In summary, we wish to emphasize that soliton solutions and their generalizations can *only* be obtained by incorporating fully nonlinear contributions, since these solutions appear as singular points of linear perturbation theories and thus cannot be reached in a finite number of steps. Another type of highly nonlinear structure, which is an integral part of any solid, is a defect. The last section of this chapter is devoted to this topic.

3.7 Defects

If we take as a starting point a homogeneous order parameter in a broken-symmetry phase, the previously discussed domain-wall soliton can also be viewed as a defect or imperfection in an otherwise perfectly

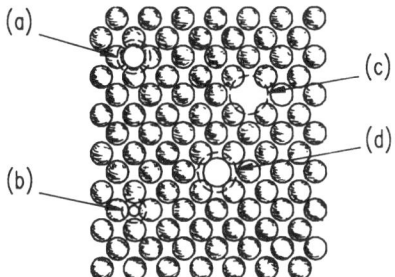

Fig. 3.18 Four types of point defect: (a) proper interstitial; (b) interstitial impurity; (c) lattice vacancy; (d) substitutional impurity.

ordered state. Indeed, the same argument applies to vortices which exist in systems with complex order parameters and can be generated by a complete rotation of a domain wall around the axis going through its centre [22].

However, what is usually thought of as a defect can be defined as an imperfection from a completely symmetric state, especially that of a translationally invariant crystal lattice. Therefore, we can distinguish crystal-lattice defects of several types depending on their dimensionality. A zero-dimensional or point defect can occur as one of four distinct types, i.e. a proper interstitial, an interstitial impurity, a substitutional impurity, or a lattice vacancy (see Fig. 3.18).

One-dimensional lattice defects come in two types, i.e. a screw dislocation and an edge dislocation. An edge dislocation may be thought of as the result of adding or subtracting a half-plane of atoms. A screw dislocation, on the other hand, results from cutting part way through the crystal and displacing it parallel to the edge of the cut. Usually, combinations of the two types of defect occur (see Fig. 3.19) [20].

Dislocations are mechanically deleterious to the extent determined by

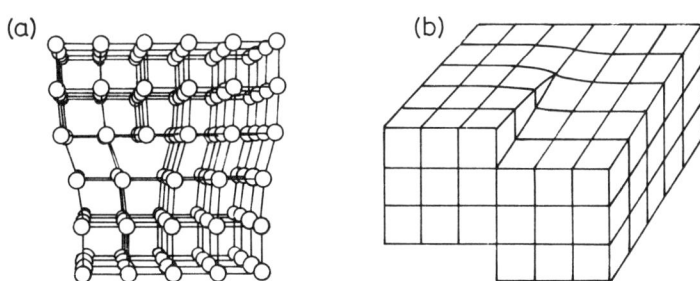

Fig. 3.19 (a) An edge dislocation. (b) a screw dislocation.

172   ELEMENTARY EXCITATIONS IN SOLIDS

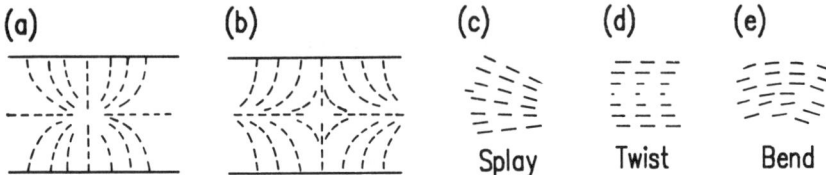

Fig. 3.20 Defects in nematic liquid crystals. (a) A positive point defect. (b) A negative point defect. (c)–(e) Three types of line deformations; splay, twist, and bend respectively.

their mobility. Thus pinning dislocations and tangling improve mechanical strength. Frequently, crystals grow from a defect, forming 'whiskers' or 'spirals'. In addition, dislocations are spontaneously generated during solidification of crystals from the melt, which is a nonequilibrium process [23]. The result of defect formation in crystals is not only a randomly distorted state of translational order but, more interestingly, it is frequently the case that defects order themselves, forming various types of defect lattices.

Defects are not limited to crystal structures but are an integral part of other types of ordering in nature. For example, in liquid crystals we can distinguish both point defects and linear deformations of new and interesting kinds, as shown in Fig. 3.20.

In ferromagnets, domain structures exhibit a remarkable wealth of complexity and usually are manifested by a so-called labyrinthine structure [17]. When an external field is applied, the latter may be deformed to show a predominance of nearly linear and point-like defects (see Fig. 3.21). As we have already shown in Chapter 1, an application of alternating magnetic fields may even result in the generation of spiral domains, and in special cases an impressive emergence of regular lattices, the building blocks of which are magnetic spiral domains, has been observed [23].

In recent years it has been possible to demonstrate that defects such as dislocations in crystal lattices can be modelled using nonlinear analysis. For example, the one-dimensional Frenkel–Kontorova model [25] has recently been revived and employed to describe defects. It is based on the Hamiltonian, H, given by

$$H = \sum_k A[1 - \cos(2\pi x_k/a)] + \sum_k B(x_{k+1} - x_k)^2 + \tfrac{1}{2}\sum_k m\dot{x}_k^2. \quad (3.125)$$

The first term describes a periodic substrate potential, the second a harmonic restoring force, and the third a kinetic energy. In the continuum limit, the equation of motion based on H possesses multi-kink-type solutions, which can be interpreted as edge dislocations. Extensions of this type of model (e.g. to higher dimensions or more components) have been

 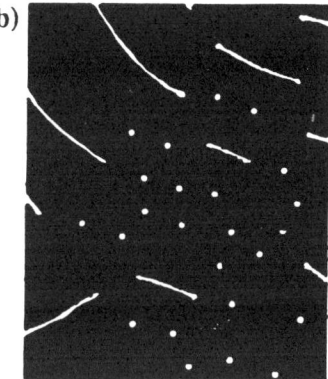

Fig. 3.21 A magnetic domain structure: (a) when no external field is applied; (b) when a strong magnetic field is applied.

used to explain phenomena such as defect mobility, commensurate–incommensurate phase transitions, and even the glassy behaviour of vortices.

Defect states are often mechanisms for transport or flow, including charge, mass, and flux transfer (e.g. dislocations and Josephson fluxons). Unstable nonlinear modes such as defects may be long-lived and thus physically significant (e.g. clusters, heterophase, or antiphase fluctuations, droplet states, nucleation centres, etc.). Thus, a nonlinear theory of condensed many-body systems requires a formalism that accounts not only for elementary excitations but also for localized and indeed 'singular' modes.

Throughout this chapter, we have introduced the reader to the methods of second quantization and tried to emphasize their importance and usefulness in condensed-matter physics. However, we have also indicated that there exist a wealth of forms of behaviour which escape rigorous treatments within conventional (i.e. perturbative) quantum approaches. In Chapters 5 and 6 we will attempt to indicate that a possible 'compromise' method is available which, on the one hand, originates from well-defined quantum Hamiltonians and, on the other, properly accounts for the presence of nonlinearities due to anharmonicity, two-particle interactions, etc. Before we give an exposition of this method, it is appropriate to review some earlier theoretical developments, especially in quantum field theory, which gave the necessary impetus and inspiration to the MCS approach.

References

1. D. Pines, *The Many-body Problem*, Benjamin/Cummings, Reading, MA (1962).

2. P. L. Taylor, *A Quantum Approach to the Solid State*, Prentice-Hall, Englewood Cliffs, NJ (1970).
3. M. Born and K. Huang, *Dynamical Theory of Crystal Lattices*, Oxford University Press, Oxford (1954).
4. W. Jones and N. H. March, *Theoretical Solid State Physics*, Wiley-Interscience, London (1973).
5. I. B. Bersuker and V. Z. Polinger, *Vibronic Interactions in Molecules and Crystals*, Springer-Verlag, Berlin (1989).
6. C. Kittel, *Introduction to Solid State Physics*, John Wiley, New York (1968).
7. J. Callaway, *Quantum Theory of the Solid State*, Academic Press, New York (1974).
8. R. F. Wallis, *Lattice Dynamics*, Pergamon Press, London (1965).
9. A. Maradudin, E. W. Montroll, G. H. Weiss, and I. P. Ipatova, *Theory of Lattice Dynamics in the Harmonic Approximation*, Academic Press, New York (1971).
10. H. Haken, *Quantum Field Theory of Solids, an Introduction*, North-Holland, New York (1976).
11. J. M. Dixon and R. S. Wardlaw, *Physica*, **135A**, 105 (1986).
12. S. Raimes, *Many-electron Theory*, North-Holland, Amsterdam (1972).
13. D. Bohm, K. Huang, and D. Pines, *Phys. Rev.*, **107**, 71 (1957).
14. N. H. March and M. Parinello, *Collective Effects in Solids and Liquids*, Adam Hilger, Bristol (1982).
15. S. Nakajima, Y. Toyozawa, and R. Abe, *The Physics of Elementary Excitations*, Springer-Verlag, Berlin (1980).
16. A. Isihara, *Condensed Matter Physics*, Oxford University Press, Oxford (1991).
17. R. H. White and T. Geballe, *Long Range Order in Solids*, Academic Press, New York (1979).
18. M. Barthes and J. Léon (eds), *Nonlinear Coherent Structures*, Springer-Verlag, Berlin (1990).
19. A. C. Newell and J. V. Moloney, *Nonlinear Optics*, Addison-Wesley, Redwood City, CA (1992).
20. S. P. Parker (ed.), *Solid-state Physics Resource Book*, McGraw-Hill, New York (1987).
21. M. E. Gouvea, G. M. Wysin, A. R. Bishop, and F. G. Mertens, *Phys. Rev. B*, **39**, 11 840 (1989).
22. P. W. Anderson, *Basic Notions of Condensed Matter Physics*, Benjamin/Cummings, Menlo Park, CA (1984).
23. C. Brice, *The Growth of Crystals from Liquids*, North-Holland, Amsterdam (1973).
24. G. S. Kandaurova and A. E. Sviderskii, *JETP Letts.*, **47**, 490 (1988).
25. F. Frenkel and T. Kontorova, *Phys. Z. Sovjet.*, **13**, 1 (1938).

4 BACKGROUND TO QUANTUM FIELD THEORY

> ...in the various series expansions considered, one always encounters a lack of analyticity in the dependence on the coupling constant... It could seem that nature does not believe in power-series expansions in the many-body problem; one suspects the same may hold true in other areas of physics (quantum electrodynamics, for example).
>
> ...one cannot be sure, even when one is in a region in which well-defined series expansion appears to be converging, that there does not exist a ground state of the system with quite different character (more order, for example), from that assumed in the perturbation-theoretic calculation.
>
> What one can hope for from an incorrect perturbation-theoretic approach is that it will contain built-in danger signals. These danger signals take the form of inconsistencies; a quantity that should, but does not, satisfy causality requirements, or an instability of a given excitation mode of the system.
>
> ...I have proposed caution in the acceptance of perturbation-theoretic results because we are at a rather primitive stage in our ability to recognise danger signals. Thus, the stability of rather simple kinds of collective modes has been investigated using rather simple approximations. The use of more-sophisticated modes or more sophisticated approximations may well show up new kinds of instabilities.
>
> D. Pines (1962), *The Many-body Problem*
> (Benjamin/Cummings)

In this chapter we present several developments in quantum field theory that will prove useful in our approach to the physics of quantum many-body systems, which follows in Chapter 5. As has been indicated in Chapter 3, the main difficulties in dealing with quantum many-body systems are directly related to two-body interactions between particles or quasi-particles. In cases in which the quantum ground state is very stable over the range of parameters considered, a perturbative approach may go some distance towards extracting the main features of the system. However, over the past half-century or so, more and more physical examples have come to the fore that involve a dramatic change in the ground state brought about either by externally applied fields or simply internal couplings. These situations are generally referred to as symmetry-breaking phenomena. Much of our discussion in the present chapter will be directed at a quantum-level description of broken symmetries. We present some introductory concepts in this area by first describing a number of well-known physical phenomena. Although some of the examples that we bring up may

4.1 The Bogolyubov transformation for Bosons

We begin by outlining a method to deal with Hamiltonians that do not conserve the number of quasi-particles in a given mode. This situation has been briefly mentioned earlier in connection with anharmonic phonons. It also exists in optical mode pumping, which occurs in lasers. In this section we are particularly concerned with Hamiltonians possessing terms of the type b^+b^+ or bb, which clearly do not conserve particle numbers. Using the method developed by Bogolyubov and outlined below, we can deal with this type of Hamiltonian successfully. For example, the effective Hamiltonian for superfluids can take the form [1]

$$H_{\text{eff}} = \tfrac{1}{2}\sum_k [\varepsilon_k(b_k^+ b_k + b_{-k}^+ b_{-k}) + \chi(b_k^+ b_{-k}^+ + b_{-k} b_k)]. \qquad (4.1)$$

The Bogolyubov transformation is a means of diagonalizing this Hamiltonian, introducing a new set of second-quantized operators (η_k, η_k^+), such that the commutation relations remain of the Bose–Einstein type, i.e.

$$[\eta_k, \eta_k^+]_- = [\eta_{-k}, \eta_{-k}^+]_- = 1. \qquad (4.2)$$

This new set (η_k, η_k^+) is related to the old one (b_k, b_k^+) through the relationships

$$b_k = \eta_k \cosh\theta_k - \eta_{-k}^+ \sinh\theta_k \quad \text{and} \quad b_{-k} = \eta_{-k}\cosh\theta_k - \eta_k^+ \sinh\theta_k, \qquad (4.3, 4.4)$$

and similarly for their Hermitian conjugates. This transformation is canonical, since it preserves the commutation relations of the original set for any real values of the adjustable coefficients θ_k. By this, we mean that

$$[b_k, b_k^+] = \cosh^2\theta_k - \sinh^2\theta_k \equiv 1. \qquad (4.5)$$

Substituting eqns (4.3) and (4.4) into the Hamiltonian (4.1) gives, for its off-diagonal part,

$$H_{\text{eff}}^{\text{off-diag}} = -\sum_k (\varepsilon_k + \chi)\cosh\theta_k \sinh\theta_k (\eta_k^+ \eta_{-k}^+ + \eta_{-k}\eta_k)$$
$$+ \chi \sum_k (\cosh^2\theta_k + \sinh^2\theta_k)(\eta_k^+ \eta_{-k}^+ + \eta_{-k}\eta_k), \qquad (4.6)$$

which can be put identically to zero upon a suitable choice of θ_k. In fact, we can show that $H_{\text{eff}}^{\text{off-diag}} = 0$ if

$$\tanh(2\theta_k) = \frac{\chi}{\varepsilon_k + \chi}. \tag{4.7}$$

Then, the diagonal part of H_{eff} becomes

$$H_{\text{eff}} = H_0 + \sum_k \sqrt{\varepsilon_k^2 + 2\chi\varepsilon_k}\ \eta_k^+ \eta_k, \tag{4.8}$$

where H_0 is a constant energy shift. Thus, the awkward terms in the original form of the effective Hamiltonian have been removed and the Hamiltonian has been diagonalized in its new operator set. The method can be viewed as a generalization of a unitary transformation of the 'co-ordinate system' and is applicable whenever products of two ladder operators lead to off-diagonal terms. Our next example is concerned with another form of difficulty frequently encountered in many-body physics.

4.2 Field translation

Whenever the effective Hamiltonian considered includes the external fields f_k and f_k^*, coupled to creators and annihilators, respectively, the Hamiltonian has to be made diagonal through some type of unitary transformation. This situation corresponds to interactions of quasi-particles with electromagnetic fields or pumping modes (such as those existing in lasers). The effective Hamiltonian is then given by [2]

$$H_{\text{eff}} = \sum_k [\varepsilon_k a_k^+ a_k + f_k a_k^+ + f_k^* a_k]. \tag{4.9}$$

Here, f_k and f_k^* are possibly time-dependent c-functions, the moduli of which describe the amplitudes of the externally applied field. In order to cast H_{eff} in a diagonal form, we define a new set of operators, α_k and α_k^+, such that

$$a_k = \alpha_k + C_k \quad \text{and} \quad a_k^+ = \alpha_k^+ + C_k^*, \tag{4.10, 4.11}$$

where C_k and C_k^* are c-numbers to be specified. This transformation corresponds to a uniform but time-dependent shift operation. Substituting eqns (4.10) and (4.11) into eqn (4.9) yields an effective Hamiltonian in terms of the new operators

$$H_{\text{eff}} = \sum_k [\varepsilon_k \alpha_k^+ \alpha_k + \alpha_k^+ (\varepsilon_k C_k + f_k) + \alpha_k (\varepsilon_k C_k^* + f_k^*)] + H_0, \tag{4.12}$$

where H_0 is a constant energy shift. We can now conveniently choose f_k as

$$f_k = -\varepsilon_k C_k, \qquad (4.13)$$

leading to

$$H_{\text{eff}} = H_0 + \sum_k \varepsilon_k \alpha_k^+ \alpha_k. \qquad (4.14)$$

Thus, H_{eff} is now diagonal in α_k operators. Note that the Bose–Einstein commutation relations of the original creators and annihilators are preserved for the new operators, i.e.

$$[a_k, a_{k'}^+]_- = [\alpha_k, \alpha_{k'}^+]_- = \delta_{k,k'}. \qquad (4.15)$$

Another property is that eqns (4.10) and (4.11) can also be viewed as a unitary transformation U such that

$$a_k = U^{-1} \alpha_k U, \qquad (4.16)$$

with U defined by [2]

$$U \equiv \exp\left\{-\int dk [C_k^* \alpha_k - C_k \alpha_k^+]\right\}. \qquad (4.17)$$

Finally, another interesting feature arises from the fact that the expectation value of α_k, the new annihilator, in the ground state $|0\rangle$ is nonzero. This may seem somewhat paradoxical because in standard, i.e. harmonic and diagonal, cases the expectation value of a single annihilator or creator vanishes. However, that occurs only in perfectly symmetrical situations or, in other words, when the symmetry of the Hamiltonian is unbroken. With external pumping or, as we see later, with anharmonic terms from two-body interactions, the above feature can no longer be taken for granted (for the ground state!). The deciding factor is the presence or absence of symmetry breaking, spontaneous or field induced. This means that although

$$a_k |0\rangle = 0, \qquad (4.18)$$

and consequently $\langle 0| a_k |0\rangle = 0$, the new expectation value yields

$$\langle 0| \alpha_k |0\rangle = -C_k \neq 0. \qquad (4.19)$$

This is because the true ground state $|0\rangle\!\rangle$ of H_{eff} is

$$|0\rangle\!\rangle = U^{-1} |0\rangle = \exp\left(-\tfrac{1}{2}\int dk\, |C_k|^2\right) \exp\left(-\int dk C_k \alpha_k^+\right)|0\rangle, \qquad (4.20)$$

and it corresponds to a superposition of arbitrarily many a-Bosons ('old' Bosons), since

$$\langle 0| \alpha_k^+ \alpha_k |0\rangle = |C_k|^2, \qquad (4.21)$$

is the number of α-Bosons with momentum k. The c-number introduced describes a *classical* extended object which is formed from quantum excitations as a result of condensation. This brings us to the related topic of Bose condensation.

4.3 Bose condensation

A system of noninteracting Bosons, such as that described by the Hamiltonian in eqn (4.1) with $\chi = 0$, will undergo a phase transition, if contained in a finite volume V at the transition temperature T_c, given by [1]

$$k_B T_c = \frac{2\pi \hbar^2}{m} \left[\frac{\langle N \rangle}{\zeta(\frac{3}{2}) V} \right]^{2/3}, \qquad (4.22)$$

where m is the mass of a single molecule, $\langle N \rangle$ is the mean number of particles, and ζ is the Riemann zeta function. The transition is from a high-temperature phase, in which the quasi-particles are distributed according to the Bose–Einstein distribution. Here, $\langle n_k \rangle$ is the mean number of quasi-particles in the kth state, given by

$$\langle n_k \rangle = \{\exp[\beta(\varepsilon_k - \mu)] - 1\}^{-1}, \qquad (4.23)$$

and μ denotes the chemical potential. The low-temperature phase is a mixture of the normal phase and the condensed phase. In the condensed phase all of the quasi-particles are in the $k = 0$ state and their number density n_0 depends on temperature via

$$n_0(T) = \frac{\langle N \rangle}{V} \left[1 - \left(\frac{T}{T_c} \right)^{3/2} \right]. \qquad (4.24)$$

The density of particles occupying the $k \neq 0$ states is

$$n_n(T) = \left(\frac{m k_B T}{2\pi \hbar^2} \right)^{3/2} \zeta(\tfrac{3}{2}). \qquad (4.25)$$

Thus, a macroscopic number of quasi-particles condense into the ground state, which is responsible for a number of special, and indeed exotic (see Chapter 2), properties, e.g. superfluidity. One of the more interesting properties of a Bose condensate is the emergence of *off-diagonal long-range order*. Choosing the order parameter for the transition to be the condensate's wave function,

$$\psi = |\psi| e^{i\phi}, \qquad (4.26)$$

so that n_0 is proportional to $|\psi|^2$, we find that the density matrix is given by [3]

$$\rho(r,r') = \frac{1}{V} \sum_k \langle n_k \rangle \exp(i k \cdot (r - r')). \quad (4.27)$$

It vanishes in the normal phase, above T_c, which is equivalent to saying that

$$\lim \rho(r,r') \to 0 \quad \text{as } |r - r'| \to \infty; \quad (4.28)$$

while it is nonzero in the superfluid phase below T_c:

$$\lim \rho(r,r') \to n_0 \quad \text{as } |r - r'| \to \infty. \quad (4.29)$$

The off-diagonal long-range order which arises as a result of Bose condensation also implies that quantum correlations acquire macroscopic meaning, so that

$$\langle \psi^* \psi \rangle \cong \langle \psi^* \rangle \langle \psi \rangle + \cdots, \quad (4.30)$$

where

$$\langle \psi \rangle = \sqrt{n_0}\, e^{i\phi}. \quad (4.31)$$

This is what is meant by long-range order. Hence, in a condensed state, quantum systems behave predominantly in a classical way, with corrections due to quantum noncommutation properties disappearing in the thermodynamic limit. The phenomena of Bose condensation and superfluidity represent important examples of quantum behaviour which have macroscopic manifestations. Both quantum and classical features are interrelated. We shall now consider an analogous example involving Fermions.

4.4 The Bogolyubov transformation for Fermions

Superconductivity is another phenomenon which exhibits macroscopic quantum behaviour. The BCS theory has proved very successful over the years in describing the mechanism of superconductivity. It provides an *effective* Hamiltonian for conduction electrons interacting with lattice phonons [4]

$$H_{\text{BCS}} = \sum_{k,\sigma} \varepsilon_k a^+_{k,\sigma} a_{k,\sigma} - \tfrac{1}{2} \sum_{\substack{k,k',q \\ \sigma,\sigma'}} W_{k,k',q} a^+_{k+q,\sigma} a^+_{k',\sigma'} a_{k'+q,\sigma'} a_{k,\sigma}, \quad (4.32)$$

where the Fermion ladder operators $a^+_{k,\sigma}$ and $a_{k,\sigma}$ create and annihilate a conduction electron state with momentum k and spin projection σ, respectively. The interaction parameter W is made up of a repulsive term between bare electrons and an attractive term due to the electron–phonon coupling. The ground state of this Hamiltonian is unstable with respect to Cooper pair formation at $T \leqslant T_c$. The Cooper pair wave function is taken as

$$|\phi\rangle \equiv N \exp\left\{\sum_k c_k a^+_{k\uparrow} a^+_{-k\downarrow}\right\}|0\rangle, \qquad (4.33)$$

and is made up of pairs of electrons with equal and opposite momenta and spin projections. Lengthy algebra then shows that the system's energy can be divided into the ground state energy and an additional energy due to excitations. The Hamiltonian treating the latter contribution can be diagonalized using a form of Bogolyubov transformation adapted to electron ladder operators, so that

$$\tilde{a}^+_{k\uparrow} = a^+_{k\uparrow} u_k - a_{-k\downarrow} v_k, \qquad \tilde{a}_{k\uparrow} = a_{k\uparrow} u_k - a^+_{-k\downarrow} v_k, \quad (4.34, 4.35)$$

$$\tilde{a}^+_{-k\downarrow} = a^+_{-k\downarrow} u_k + a_{k\uparrow} v_k, \qquad \tilde{a}_{-k\downarrow} = a_{-k\downarrow} u_k + a^+_{k\uparrow} v_k. \quad (4.36, 4.37)$$

A variation of the expectation value of the Hamiltonian with respect to the postulated wave function determines the coefficients u_k and v_k. This procedure results in optimum parameters which lead to an effective Hamiltonian, H_{BCS}, given by

$$H_{BCS} \simeq \sum_k \tilde{\varepsilon}_k \left(\tilde{a}^+_{k\uparrow} \tilde{a}_{k\uparrow} + \tilde{a}^+_{-k\downarrow} \tilde{a}_{-k\downarrow} \right) + \cdots \text{ (smaller terms)}, \quad (4.38)$$

where the excitation energy $\tilde{\varepsilon}_k$ is found to be

$$\tilde{\varepsilon}_k = \sqrt{E'^2_k + \Delta^2_k}, \qquad (4.39)$$

with $E'_k = \varepsilon_k - \mu$, μ being the chemical potential (i.e. the Fermi energy), and Δ_k is the so-called energy gap function for each mode k. The average value of the energy gap satisfies the equation

$$\Delta = \begin{cases} 0, & \text{if } |E'_k| > \hbar\omega_D, \\ 2\hbar\omega_D \exp(-(1/N(0)V_0)), & \text{if } |E'_k| < \hbar\omega_D. \end{cases} \qquad (4.40)$$

Here, $N(0)$ is the density of states at the Fermi level, ω_D is the Debye frequency of phonons, and V_0 is the approximate potential strength ($W_{kk'}$). This means that the only electrons that are involved in the Cooper pair formation lie within a narrow energy shell $2\hbar\omega_D$ wide about the Fermi level.

Once again, faced with a highly nonlinear phenomenon arising from the

4.5 Boson coherent states

We have seen in quantum systems that there exist conditions under which ground states can be approximated by nearly classical functions, i.e. that the Heisenberg uncertainty principle may sometimes play a negligible role. In order to find the best state representation for such systems, the concept of a coherent state has been developed and used to great advantage in areas such as laser physics, for example. A Boson coherent state, $|\alpha\rangle$, is defined as a superposition of harmonic oscillator eigenstates $|n\rangle$ by [5]

$$|\alpha\rangle = \sum_{n=0}^{\infty} \frac{\alpha^n}{\sqrt{n!}} \exp\left(-\tfrac{1}{2}|\alpha|^2\right)|n\rangle. \quad (4.41)$$

This is frequently referred to as the Glauber state representation. It can be shown that $|\alpha\rangle$ is an eigenstate of the annihilation operator a, i.e.

$$a|\alpha\rangle = \alpha|\alpha\rangle. \quad (4.42)$$

It can also be demonstrated that $|\alpha\rangle$ is akin to a generalized rotation applied to the vacuum state:

$$|\alpha\rangle = \exp(\alpha a^+ - \alpha^* a)|0\rangle \equiv D(\alpha)|0\rangle. \quad (4.43)$$

However, unlike the states $|n\rangle$ above, coherent states, $|\alpha\rangle$, are not orthogonal since

$$\langle \beta|\alpha\rangle = \exp\left[-\tfrac{1}{2}(|\alpha|^2 + |\beta|^2) + \beta^*\alpha\right], \quad (4.44)$$

and hence

$$|\langle \beta|\alpha\rangle|^2 = \exp(-|\beta - \alpha|^2). \quad (4.45)$$

Coherent states form, on the other hand, a complete set since

$$\frac{1}{\pi} \int d^2\alpha \, |\alpha\rangle\langle\alpha| = 1. \quad (4.46)$$

The probability of finding n quanta in a coherent state is

$$P_n(\alpha) = |\langle n|\alpha\rangle|^2 = \frac{|\alpha|^{2n}}{n!} \exp(-|\alpha|^2), \quad (4.47)$$

which happens to be a Poisson distribution with the average number $\langle n \rangle = |\alpha|^2$.

A simple extension of this concept to multimode coherent states is possible with the product definition

$$|\{\alpha_k\}\rangle \equiv \prod_k |\alpha_k\rangle. \qquad (4.48)$$

It is also noteworthy that coherent states are good approximations to the ground states of both Bose-condensed and field-translated systems, since from eqn (4.42) it follows that

$$\langle \alpha | a | \alpha \rangle = \alpha \neq 0 \quad \text{and} \quad D(\alpha)^{-1} a D(\alpha) = a + \alpha. \qquad (4.49a)$$

The first of the above relationships is identical in form to eqn (4.19). It can be proved that coherent states are quantum states with the least possible uncertainty, i.e. they most closely approximate classical states. In fact, the mean square fluctuation in the generalized position operator in state $|\alpha\rangle$,

$$x(\psi) = \tfrac{1}{2}[ae^{-i\psi} + a^+ e^{i\psi}], \qquad (4.49b)$$

is

$$\Delta x^2(\psi) \equiv \langle x^2(\psi) \rangle - \langle x(\psi) \rangle^2 = \tfrac{1}{4}. \qquad (4.49c)$$

The latter two features give us a motivation to investigate these and similar classical objects as playing important roles in the boundary region between classical and quantum physics.

4.6 Fermion coherent states

Having recognized the importance of coherent states, we are keenly interested in extending the validity of this concept to Fermion systems. This is the objective of the present section.

In order to construct coherent states for Fermions, a Grassmann algebra of anticommuting generator numbers ξ_a is defined through [5]

$$\xi_\alpha \xi_\beta + \xi_\beta \xi_\alpha = 0 \qquad (4.50)$$

and

$$(\lambda \xi_{\alpha_1} \cdots \xi_{\alpha_n})^* = \lambda^* \xi^*_{\alpha_n} \cdots \xi^*_{\alpha_1}, \qquad (4.51)$$

where λ is a complex number.

A Fermion coherent state $|\xi\rangle$ is built as

$$|\xi\rangle = \exp\left[-\sum_\alpha \xi_\alpha a^+_\alpha\right]|0\rangle. \qquad (4.52)$$

It can be verified that $|\xi\rangle$ is an eigenstate of an annihilation operator, so that
$$a_\alpha |\xi\rangle = \xi_\alpha |\xi\rangle. \tag{4.53}$$
As in the case of Boson coherent states, orthogonality is *not* satisfied by two such states $|\xi\rangle$ and $|\xi'\rangle$, since
$$\langle \xi | \xi' \rangle = \exp\left[-\sum_\alpha \xi_\alpha^* \xi_\alpha\right]. \tag{4.54}$$
However, the closure relation is preserved:
$$\int \prod_\alpha d\xi_\alpha^* \, d\xi_\alpha \exp\left[-\sum_\alpha \xi_\alpha^* \xi_\alpha\right] |\xi\rangle\langle\xi| = \mathbf{1}, \tag{4.55}$$
where **1** denotes the unit operator in the Fermion Fock space.

4.7 Squeezed states

An extension of the concept of coherent states has been accomplished by defining so-called squeezed states. In the single-mode case, a squeezing operator is first introduced as [6]
$$S(r,\varphi) \equiv \exp\left[\frac{r}{2}(a^{+2}e^{2i\varphi} - a^2 e^{-2i\varphi})\right], \tag{4.56}$$
in analogy to the operator $D(\alpha)$ for coherent states. Then, a squeezing operator is applied to annihilators through
$$c \equiv S(r,\varphi)^{-1} a S(r,\varphi)$$
$$= a \cosh r + a^+ \sinh r e^{2i\varphi}, \tag{4.57}$$
so that a squeezed state is obtained by applying $S(r,\varphi)$ to the vacuum
$$|c\rangle = S(r,\varphi)|0\rangle. \tag{4.58}$$
It can be demonstrated that the generalized position operator
$$x(\psi) = \tfrac{1}{2}[c e^{-i\psi} + c^+ e^{i\psi}] \tag{4.59}$$
has a distance- and angle-dependent mean square deviation value, given by
$$\Delta x^2(\psi) = \tfrac{1}{4}|\cosh r + e^{2i(\psi-\varphi)} \sinh r|^2 \tag{4.60}$$
which, for particular choices of angles, yields
$$\Delta x^2(\psi) = \tfrac{1}{4}\begin{cases} \exp(-2r) & \text{for } \psi - \varphi = \dfrac{\pi}{2}, \dfrac{3\pi}{2}, \\ \cosh(2r) & \text{for } \psi - \varphi = \dfrac{\pi}{4}, \dfrac{3\pi}{4}, \\ \exp(2r) & \text{for } \psi - \varphi = 0, \pi. \end{cases} \tag{4.61}$$

Indeed, when $\psi - \varphi = \pi/2$ or $3\pi/2$, the uncertainty in position is reduced below the $\frac{1}{4}$ value typical of coherent states.

We have seen in these several examples the development of coherent and squeezed states or other nearly classical objects of quantum origin. What is somewhat unsatisfactory, however, is the rather *ad hoc* manner in which the reality of these concepts has been established: that is to say that a given formalism was expressly developed to meet the 'theoretical demand' created by experimental discoveries. It would be more satisfying to find a common unifying scheme for all these coherence phenomena under one umbrella. One such pathway involves q-deformed algebras.

4.8 q-Bosons and quantum algebras

Quantum algebras and quantum groups were first introduced to mathematical physics in the 1970s and 1980s [7–11]. Some of the main motivations can be traced to the inverse scattering techniques (see Chapter 7) and to the exactly solvable models in statistical mechanics (see Chapter 2). The underlying notion is that of a deformation; hence these objects are commonly referred to as q-deformed (algebras, groups, and operators). Indeed, the concept of deformation is common in physical theory since, for example, quantum mechanics may be considered to be a deformation of classical mechanics (with $\hbar$ representing the deformation parameter), and the theory of relativity is a deformation of classical mechanics (with $1/c$ being the deformation parameter here).

A q-deformed algebra of linear operators a^+, a, and N acting in the occupation (Fock) space $F = \{|n\rangle;\ n \in N\}$ is defined through the following relations:

$$a^+|n\rangle = \sqrt{[n+1]}\,|n+1\rangle, \qquad a|n\rangle = \sqrt{[n]}\,|n-1\rangle, \qquad N|n\rangle = n|n\rangle,$$
$$(4.62\text{–}4.64)$$

where we denote

$$[c] \equiv [c]_q = \frac{q^c - q^{-c}}{q - q^{-1}} = \frac{\sinh(c \ln q)}{\sinh(\ln q)} \qquad (4.65)$$

and c is a complex number.

Note that the limit $q \to 1$ brings us back to their ordinary counterparts. The operators that we have introduced above satisfy the standard relations

$$(a)^+ = a^+, \qquad (N)^+ = N, \qquad [N, a^+] = a^+, \qquad [N, a] = -a, \qquad (4.66)$$

where the bracket denotes a commutator, i.e. $[X, Y] \equiv XY - YX$.

A new set of properties can be readily derived which significantly alters the q-deformed algebra:

$$aa^+ = [N+1], \qquad a^+a = [N], \tag{4.67}$$

$$aa^+ - q^{-1}a^+a = q^N, \qquad aa^+ - qa^+a = q^{-N}, \tag{4.68}$$

Thus, the operators a and a^+ reduce to ordinary Boson operators as $q \to 1$.

Particular representations of q-deformed algebras have been applied to a number of important models, including a q-deformed harmonic oscillator, in which

$$[x, p_x] = i\hbar([N+1] - [N]), \tag{4.69}$$

and q-deformed angular momenta, with

$$[J_3, J_-] = -J_-, \qquad [J_3, J_+] = +J_+, \qquad [J_+, J_-] = [2J_3] \tag{4.70}$$

The field of quantum algebras is developing rapidly and applications to many-body physics are gradually being introduced [12].

4.9 Field quantization

In the remainder of this chapter we shall concern ourselves with quantum fields and their classical analogues or equivalents, rather than with second-quantized operators, as has been the case so far. The use of quantum fields rather than occupation number representation provides an extra advantage of being able to investigate both localized and extended states and densities. Furthermore, as will be shown, it is possible to treat quantum fields as composed of two parts, one of which is truly quantum-like, with associated noncommutation properties, while the other is virtually a classical field amplitude. The relative strength of these two components depends on the type of system and the regime studied. Rather paradoxically, however, in many instances the *classical* component of the quantum field dominates. Before we outline how this can be described in mathematical language, we first equip the reader with some necessary background information on field quantization. We shall consider two cases, which differ because the associated particles obey Bose–Einstein statistics in one case and Fermi–Dirac in the other.

4.9.1 The Bose–Einstein case

Our starting point here is the time-dependent Schrödinger equation for a

FIELD QUANTIZATION 187

Bose–Einstein particle moving in a potential V, which may be written as

$$\left[-\frac{\hbar^2}{2m}\nabla^2 + V\right]\Phi_n = i\hbar \frac{\partial}{\partial t}\Phi_n. \tag{4.71}$$

The eigenfunctions $\Phi_n(x, t)$ are represented as

$$\Phi_n = b_n(0)e^{-iE_n t/\hbar}\phi_n(x) = b_n(t)\phi_n(x), \tag{4.72}$$

so that an arbitrary wave function $\psi(x)$ can be expanded as

$$\psi(x, t) = \sum_n b_n \phi_n(x). \tag{4.73}$$

Assuming ϕ_n to be a complete and orthonormal set of functions, we can identify the coefficients b_n with annihilation operators of energy quanta, so that the Hamiltonian H can formally be recast as

$$H = \sum_n E_n b_n^+ b_n, \tag{4.74}$$

and $\psi(x, t)$ of eqn (4.73) becomes a quantum field with its Hermitian conjugate

$$\psi^+ = \sum_n b_n^+ \varphi_n^*(x). \tag{4.75}$$

Then, ψ and ψ^+ satisfy the canonical commutation properties for Bosons,

$$[\psi(x'), \psi^+(x)]_- = \delta(x - x') \tag{4.76}$$

and

$$[\psi(x'), \psi(x)]_- = [\psi^+(x'), \psi^+(x)]_- = 0, \tag{4.77}$$

provided that b_n^+ and b_m satisfy

$$[b_n, b_m^+]_- = \delta_{m,n}, \quad [b_m, b_n]_- = [b_m^+, b_n^+]_- = 0. \tag{4.78}$$

Thus, the commutation properties of the fields are closely related to the quantum statistics of the particles that they describe.

4.9.2 The Fermi–Dirac case

If the particle in question is a Fermion, then the same Schrödinger equation as in eqn (4.71) can be used to describe its dynamics. However, when it is second quantized, this necessitates the introduction of operators a_n and a_m^+, which satisfy Fermi–Dirac anticommutation rules. That is,

$$[a_m, a_n^+]_+ = \delta_{m,n}, \quad [a_m^+, a_n^+]_+ = [a_m, a_n]_+ = 0, \tag{4.79}$$

where $[A, B]_+ \equiv AB + BA$ is an anticommutation bracket. Quantum fields may now be defined by

$$\psi(x) = \sum_n a_n \varphi_n(x) \tag{4.80}$$

and

$$\psi^+(x) = \sum_n a_n^+ \varphi_n^*(x), \tag{4.81}$$

where φ_n are the eigenfunctions of the time-independent Schrödinger equation, which are orthonormal and form a complete set. It is easy to show that these fields satisfy the following anticommutation rules:

$$[\psi(x), \psi^+(x')]_+ = \delta(x - x') \tag{4.82}$$

and

$$[\psi^+(x), \psi^+(x)]_+ = [\psi(x), \psi(x')]_+ = 0. \tag{4.83}$$

Our intention in the remainder of this chapter is now to provide an overview of the important properties of these quantum fields for strongly interacting quantum systems in the vicinity of symmetry breaking, where classical critical fluctuations are of paramount importance. Based on what has been said thus far about the vicinity of broken-symmetry situations, we shall first examine the behaviour of the classical components of these fields.

4.10 Classical field equations

Makhankov and Fedyanin [13] considered a very broad class of quasi-one-dimensional condensed matter systems described by the generic Hamiltonian

$$H = E_0 + p \sum_i N_i + \mu \sum_i (a_i^+ a_{i+1} + a_{i+1}^+ a_i)$$
$$+ \mu' \sum_i (a_i^+ a_{i+1}^+ + a_i a_{i+1}) + q \sum_i N_i N_{i+1}, \tag{4.84}$$

where $N_i = a_i^+ a_i$ is the number operator, p is the magnitude of a single energy level for each quasi-particle situated at lattice point i, the symbols μ and μ' denote overlap integrals, and q is the interaction constant between subsystems. The ladder operators a_k and a_k^+ may be assumed to satisfy either the Fermi–Dirac or Bose–Einstein commutation relations. The approach to the problem of dynamics of such systems adopted by these authors is based on the Heisenberg equation of motion for the

CLASSICAL FIELD EQUATIONS 189

ladder operators. This is followed by a choice of a trial one-particle wave function $|\psi(t)\rangle$ and a subsequent calculation of the expectation values for all terms in the Heisenberg equation. The final steps are to take the continuum limit and a truncation of the expansion to include only terms of order three or less in the field variable, defined as

$$\langle 0|a_f(t)|\psi(0)\rangle = \varphi_f(t) \to \varphi(x,t). \tag{4.85}$$

This last step is very reminiscent of coherent states, which we discussed earlier. In the lowest nontrivial order of approximation, the resultant equation of motion for $\varphi(x,t)$ is a nonlinear Schrödinger equation of the form

$$i\hbar \frac{\partial \varphi}{\partial t} = (p + 2\mu + 2q)\varphi + \mu a_0^2 \varphi_{xx} - 2q|\varphi|^2 \varphi, \tag{4.86}$$

where a_0 is the lattice spacing. What these authors have shown thereby is that interactions between the constituents of an interacting many-body system become *nonlinear* in the field and hence condensation effects may take place. It will be shown in Chapter 5 that a similar result can be obtained using the method of coherent structures within a more general framework.

In the context of field theory and elementary particles, the usual approach to obtaining underlying classical field equations has been either to impose a particular ansatz (e.g. the φ^4-ansatz in the Yang–Mills theory [14]) or simply to postulate a symmetry-based Lagrangian form. For relativistically invariant models, the Lagrangian density is usually of the type

$$\mathscr{L} = \tfrac{1}{2} \partial_\mu \phi \partial^\mu \phi - U(\phi), \tag{4.87}$$

and the corresponding energy functional becomes

$$E = \int \mathrm{d}x \left[\tfrac{1}{2}(\partial_0 \phi)^2 + \tfrac{1}{2}(\partial_\mu \phi)^2 + U(\phi) \right], \tag{4.88}$$

where the time variable denoted by the zero subscript has been separated out from space derivatives. The associated Euler–Lagrange equation takes the form of the nonlinear Klein–Gordon equation:

$$\Box \phi + U'(\phi) = 0. \tag{4.89}$$

Special cases of massless and tachyonic particles can readily be incorporated in this general approach by choosing particular forms of the symmetry variable $\xi = k_\mu x^\mu$ such that $\phi = \phi(\xi)$ (see Chapters 5 and 7 for more details). For elementary-particle modelling purposes one is particularly interested in localized (ideally solitonic) solutions of eqn (4.89). However, according to *Derrick's theorem* (see Chapter 1) no nonsingular, time-independent, finite-energy stable solutions exist beyond *one* spatial dimension for scalar fields φ and positive-definite potentials $U(\varphi)$. However, this may be too restrictive a statement to be guided too literally by in physics, since quite often we are interested in *long-lived* states which evolve slowly with time.

In the familiar φ^4-case, for example, we take

$$U(\phi) = \frac{\lambda}{2} \phi^4 - \mu^2 \phi^2 + \frac{\mu^4}{2\lambda}, \tag{4.90}$$

and obtain a static kink (antikink) of the form

$$\phi = \pm a \tanh(\mu x), \tag{4.91}$$

the energy of which is found to be

$$E = 4\mu^3/3\lambda. \tag{4.92}$$

For the Sine–Gordon potential,

$$U(\phi) = \frac{m^4}{g^2}\left[1 - \cos\left(\frac{g\phi}{m}\right)\right], \tag{4.93}$$

a static single-kink solution can be derived and takes the form

$$\phi_0 = \pm 4(m/g)\tan^{-1}\exp[\pm m(x - x_0)]. \tag{4.94}$$

Since eqn (4.89) is relativistically invariant, propagating solutions can be generated from static ones as [15]

$$\phi = \phi(\xi), \tag{4.95}$$

where $\xi = \pm(x - x_0 - vt)/\sqrt{1 - v^2}$ is a symmetry variable. The rest energy may be defined by

$$E_0 = \int dx \left[\tfrac{1}{2}(\partial\phi_0/\partial x)^2 + U(\phi_0)\right], \tag{4.96}$$

where ϕ_0 is a static solution. Given this definition we then find the energy–velocity relationship as

$$E = (1 - v^2)^{-1/2} E_0, \tag{4.97}$$

and similarly the momentum–energy connection as

$$p = v(1 - v^2)^{-1/2} E_0. \tag{4.98}$$

Models of this type, sometimes with a more complicated nonlinear potential $U(\phi)$, have also been used to investigate phase transitions in the Early Universe [16]. In general, the classical solutions discussed here can be grouped into a number of categories depending on their physical interpretation:

1. Constant solutions (i.e. space- and time-independent). These correspond to mean fields (homogeneous and static), such as those, for example, which are encountered in classical theories of critical phenomena.

2. Static solutions (time-independent in their moving reference frame but space-dependent). These are nonlinear waves, some of which (if localized) are solitary waves. They have been the object of intense investigations as primary candidates for models of elementary particles.

3. Time- and space-dependent solutions, such as breathers or relaxation modes, with finite lifetimes. Their physical interpretation and significance are still largely unclear.

4. Instanton solutions, which are nonsingular solutions of the modified field equations, where a transformation is made of the form $t \to ix_4$. These solutions are interpreted as quantum tunnelling phenomena where the system, over an infinite time domain, switches from one state to another within the degenerate ground manifold.

The existence of nonzero, nonsingular, classical field solutions signals the presence of field condensation leading, for example, to particle formation. This can also be understood as a first approximation to the vacuum expectation value of the quantum field which does not vanish when spontaneous symmetry breaking has occurred and is zero otherwise [17].

The semiclassical approaches to quantum field theory described in this section fulfil their roles as providers of phenomenological models of condensation phenomena. What they have so far delivered lacks the quantum nature of the original field. This can indeed be recovered also at an heuristic level, and one such method is briefly outlined below.

4.11 Quantization of classical solutions

Jackiw [18] and others (see, for example, reference [19]) have developed a method which, to a first approximation, ignores quantum effects and begins by solving classical nonlinear field equations. Heisenberg operator field equations are solved for c-number fields, which are subsequently quantized semiclassically. Quantization of static classical solutions can be achieved by perturbation, i.e. through linearization:

$$\phi(x,t) = \phi_0(x) + \psi_k(x)e^{i\omega_k t}. \quad (4.99a)$$

A more formal approach has been developed which is referred to as a canonical transform or a collective co-ordinate method [18–21]. In this approach, the quantum field $\phi(x,t)$ is decomposed into classical and quantum parts ϕ_c and χ, respectively:

$$\phi(x,t) = \phi_c(x - \hat{x}) + \chi(x - \hat{x}, t), \quad (4.99b)$$

where $\hat{x}$ is a quantum position operator and χ is a phonon (meson) field in the presence of a classical soliton ϕ_c. Solitons survive the quantization procedure. Returning to eqn (4.99a), substituting it into eqn (4.89), and linearizing the latter with respect to ψ_k gives a stationary Schrödinger equation for the quantum corrections ψ_k:

$$\left[-\left(\frac{d^2}{dx^2}\right) + U''(\phi_0(x)) \right] \psi_k(x) = \omega_k^2 \psi_k(x). \quad (4.100)$$

Stability of ϕ_0 is indicated by $\omega_k^2 > 0$, which means that perturbations do not grow with time but oscillate. The corresponding quantum state is stable. In this connection, note that the Schrödinger equation (4.100) always has a zero-frequency ($\omega_k = 0$) solution, also referred to as the 'translation mode' (the lowest energy mode). This is found to be

$$\psi_0 = d\phi_0/dx. \quad (4.101)$$

If ϕ_0 is a solitary-wave solution, then ψ_0 is normalizable, since $\int dx (d\phi_0/dx)^2 < \infty$. It follows from translational invariance that

$$\phi(x,t) = \phi_0(x) + \alpha \frac{d\phi_0}{dx}(x), \quad (4.102)$$

where α is a normalization constant related to the system's periodicity. This is apparent since if $\phi_0(x)$ is a solution, so is $\phi_0(x + x_0)$, where $x_0 = na$ and a is the lattice period. As for the other quantum bound states, their form depends on the particular model considered. For example, in the ϕ^4-model with the potential

$$U(\phi) = \frac{m^4}{2q^2}\left(1 - \frac{q^2\phi^2}{m^2}\right)^2 \quad (4.103)$$

and the classical static kink solution

$$\phi_0 = \pm(m/q)\tanh[m(x - x_0)], \quad (4.104)$$

QUANTIZATION OF CLASSICAL SOLUTIONS

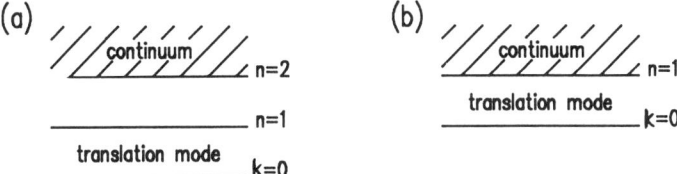

Fig. 4.1 A schematic depiction of the quantum energy levels in the two models discussed: (a) the 'φ^4'-model; (b) the SG model.

the corresponding stationary Schrödinger equation is

$$\left[-\frac{d^2}{dx^2} + 4m^2 - \frac{6m^2}{\cosh^2 mx}\right]\psi_k = \omega_k^2 \psi_k. \quad (4.105)$$

Similarly, in the SG model described by eqns (4.93) and (4.94), the stationary Schrödinger equation is

$$\left[-\frac{d^2}{dx^2} + m^2 - \frac{2m^2}{\cosh^2 mx}\right]\psi_k = \omega_k^2 \psi_k. \quad (4.106)$$

Equations (4.105) and (4.106) are the $L = 2$ and $L = 1$ cases, respectively, of the eigenequation

$$\left[-\frac{d^2}{dx^2} + L^2 - \frac{L(L+1)}{\cosh^2 x}\right]\psi_k = \omega_k^2 \psi_k. \quad (4.107)$$

It is known that the above equation has discrete eigenvalues when $L^2 - n^2$ is nonnegative, i.e. when $n = L, L-1, \ldots, 1$, and a continuous spectrum with $\omega_k^2 = k^2 + L^2$ ($k^2 > 0$). The two energy spectra of interest are shown in Fig. 4.1.

The quantum perturbations ψ_k are the internal degrees of freedom (if they are bound) that may coexist with a soliton (envelope) state. This is why they are referred to as 'meson' states in quantum field theory [18, 19]. They can be analyzed using standard perturbative methods of quantum field theory.

Thus, we can summarize by saying that a new (overcomplete) 'Hilbert' space of states has been afforded by nonlinearity, which consists of the following sectors:

1. Soliton states (including N-soliton states) which represent classical localized envelope solutions of the field equation.

Table 4.1 A comparison between a diatomic molecule and a quantum lump (following reference [22]).

Order of approximation	Diatomic molecule	Quantum lump
0	Ground state	Classical lump energy
1	Vibrational states	Normal modes
2	Rotational states	Kinetic energy (momentum eigenstates)

2. Quantum bound states for each soliton.

3. Scattering states that scatter off classical solutions and correspond to the continuum of Fig. 4.1.

4. Other classical states such as extended (elliptic) waves which, as will be seen in Chapter 5, correspond to collective modes. Although these periodically modulated fields are of little significance to particle physicists due to their infinite energies, they can be especially important in condensed matter physics.

5. Singular solutions which can be interpreted as various types of defect (see Chapter 3).

This picture resembles the atomic structure with a classical envelope giving the electron charge distribution (localized in space) and the various possible internal quantum states of the atomic electrons corresponding to the 'meson states'. Coleman [22] envisaged 'quantum lumps' as composite systems similar to diatomic molecules. This is highlighted in Table 4.1.

In general, the classical solution ϕ_0 may take a number of functional forms, not necessarily corresponding to a localized state. If ϕ_0 is a constant, then quantum excitations based on it form a continuum of free states. If ϕ_0 is a soliton, we have seen the emergence of a finite number of bound states, above which a continuum was formed. If ϕ_0 is periodic, then we know, from the Bloch theory of metals, that bands of states are formed which are separated by energy gaps. Finally, for singular solutions we do not expect bound states but only scattering to occur.

The situation in higher-dimensional spaces is much more complicated. A general theory regarding the density of eigenvalues for Hamiltonians of the type

$$H(p,q) = \sum_{n=1}^{d} \frac{p_n^2}{2} + V(q), \qquad (4.108)$$

where $V(q) \geq 0$ and d is the physical space dimensionality, can be found in reference [23]. If $V(q) \to \infty$ as $|q| \to \infty$, the spectrum of H is discrete and, for $d > 1$, the eigenvalues may be degenerate when $V(q)$ is symmetric. Defining the number of eigenvalues of H less than E as

$$N(E) = (2\pi\hbar)^{-d} \int d^d p \, d^d q \, \theta[E - H(p,q)], \qquad (4.109)$$

where θ is the Heaviside step function, we can also find the density of eigenvalues as [23]

$$\rho(E) = \frac{d}{dE} N(E) = \frac{(2\pi)^{-d/2} \hbar^{-d}}{\Gamma(d/2)} \int_{V(q) \leqslant E} d^d q [E - V(q)]^{d/2 - 1}. \qquad (4.110)$$

As an example, consider the static, spherically symmetric ϕ^4-field theory with the equation of motion for the classical field ϕ:

$$-\nabla^2 \phi = -\phi + \phi^3. \qquad (4.111)$$

Then, taking only a unique, positive, spherically symmetric solution $\phi(r)$, such that $\phi(r) \to 0$ as $r \to \infty$, we find the spectrum of $H = T + gV(r)$, where $V(r) = -3\phi^2(r)$, g is an adjustable coupling constant, and T is the kinetic energy. The results for $N(E)$ in this case [24] are shown in Fig. 4.2.

The approximation used to obtain the solid curve in Fig. 4.2 estimates $N(E)$ as

$$N(E) \sim g^{d/2} (2\pi\hbar)^{-d} \int d^d q \, \theta(-V(q)). \qquad (4.112)$$

Based on these ideas and a more detailed look at criticality, in Chapter 5

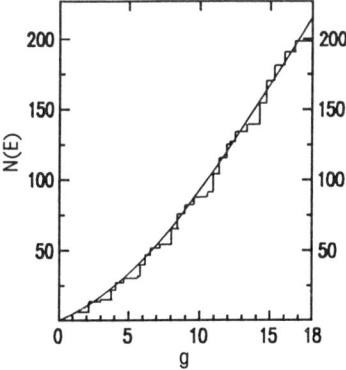

Fig. 4.2 The plot of $N(E)$ as a function of the strength g in the case of eqn (4.111) (following reference [24]).

we will develop a parallel nonlinear formalism within which a quantum many-body theory can emerge. One of the examples that will be shown in Chapter 6 is that of the atomic structure. The analogy traced here will then become very clear.

4.12 The Goldstone theorem and the Anderson–Higgs mechanism

The Goldstone theorem states [3, 15] that for Lorentz-invariant field theories with a positive-definite inner product, if there is a local conserved current (or, equivalently, if the Lagrangian is invariant under a continuous transformation, e.g. a local rotation or a gauge shift), then necessarily there exists a branch of massless, spinless Bosons. They are called Goldstone Bosons. This simply means that for scalar, complex field theories one cannot expect an energy gap above the ground state (represented by the classical field). We have given concrete physical examples of Goldstone Bosons in Chapter 1.

However, the following qualification must be made [23]. Massless Goldstone Bosons occur only for finite-range interactions. For infinite-range interactions, on the other hand, a positive excitation energy will be produced at any distance, so that the energy in the limit of the wavenumber $k \to 0$ will be finite and the mode will acquire a mass. Thus, systems with electromagnetic interactions and broken symmetries do not have massless Goldstone modes.

An even more remarkable phenomenon occurs for a field theory with a broken symmetry and a local gauge invariance, including electromagnetic interactions. This is known as *the Anderson–Higgs* mechanism. The Lagrangian used is given by

$$\mathscr{L} = [(\partial^\mu + \mathrm{i} e A^\mu)\phi]^*[(\partial_\mu + \mathrm{i} e A_\mu)\phi] \\ - \alpha |\phi|^2 - \beta |\phi|^4 - \tfrac{1}{4} F_{\mu\nu} F^{\mu\nu}, \quad (4.113)$$

where $F^{\mu\nu} = \partial^\mu A^\nu - \partial^\nu A^\mu$ and A^μ is the vector potential. When treated independently, the (charged) order parameter field ϕ and the electromagnetic field $F^{\mu\nu}$ possess massless modes ($\eta_\|$, A^μ) and a massive mode $\eta_\perp$. The Lagrangian in eqn (4.113) is invariant under local gauge transformations:

$$\phi \to \mathrm{e}^{-\mathrm{i}\theta(x)}\phi(x), \quad \phi^* \to \mathrm{e}^{\mathrm{i}\theta(x)}\phi^*(x) \quad (4.114)$$

and

$$A^\mu(x) \to A^\mu(x) + \frac{1}{e} \partial^\mu \theta(x), \quad (4.115)$$

as well as under global gauge transformations where $\theta(x)$ is constant. For $\alpha > 0$ (disordered phase), there exist two massive modes for $\phi(x)$ and two massless modes for the electromagnetic field. On the other hand, for $\alpha < 0$ (broken symmetry phase), the following results are found:

1. The mode $\eta_\parallel$, which in the noninteracting case was massless (a Goldstone Boson corresponding to excitations along the potential's minimum), no longer appears independently. The massive mode $\eta_\perp$ remains and it is unaffected by A^μ. These modes are found by linearizing the coupled equations of motion about their constant equilibrium solutions.

2. The electromagnetic field is transformed into

$$\tilde{A}^\mu = A^\mu + \frac{1}{e} \partial^\mu \eta_\parallel, \qquad (4.116)$$

its mass is $m^2 = -2e^2\alpha/\beta$, and it satisfies the Lorentz gauge

$$\partial_\mu \tilde{A}^\mu = 0. \qquad (4.117)$$

Thus, we have three independent massive modes. The situation is summarized in Table 4.2.

An application of the Anderson–Higgs mechanism to superconductivity gives the Meissner effect as a manifestation of finite mass creation, since an externally applied magnetic field only penetrates a finite distance (inverse mass) into the superconductor.

Table 4.2 A summary of the results in the Anderson–Higgs phenomenon (following reference [5]).

Case	Field	Modes	Number of components	m^2
Noninteracting fields	Scalar: ϕ, ϕ^* e – m: A^μ	$\eta_\parallel$ $\eta_\perp$ A^μ	1 1 2 (transverse)	-2α 0 0
Interacting fields ($\alpha > 0$)	Scalar: e – m	ϕ, ϕ^* A^μ	2 2 (transverse)	α 0
Interacting fields ($\alpha < 0$), broken symmetry	Scalar: e – m	$\eta_\perp$ $\tilde{A}^\mu$	1 3	-2α $-2e^2\alpha/\beta$

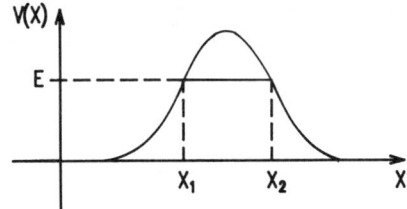

Fig. 4.3 The quantum tunnelling phenomenon.

4.13 Tunnelling, instantons, and the false vacuum

Having degenerate minima of the classical energy functional, as in the case with broken-symmetry potentials, implies a finite probability of quantum tunnelling between them. For an effective Hamiltonian of the type

$$H_{\text{eff}} = \frac{p^2}{2m} + V(x), \qquad (4.118)$$

where $V(x)$ has a potential barrier between points x_1 and x_2, as shown in Fig. 4.3, the WKB approximation predicts the transmission amplitude on going from one side of the barrier to the other:

$$T(E) \simeq \exp\left\{-\frac{1}{\hbar}\int_{x_1}^{x_2} dx \sqrt{2m(V-E)}\right\}. \qquad (4.119)$$

As a result, the energy levels E_n with eigenfunctions $\psi_n(\pm x)$ undergo splitting, so that the new wave functions are

$$\psi_n^{\pm}(x) = \psi_n(x) \pm \psi_n(-x), \qquad (4.120)$$

and the new eigenenergy separation is

$$E_n^+ - E_n^- = \Delta E_n = \exp\left[-\frac{1}{\hbar}\int_{x_1}^{x_2} \sqrt{2mV(x)}\,dx\right]. \qquad (4.121)$$

A particular situation of this type arises when the classical field potential $V(\phi)$ is doubly degenerate. Although there may be no classical solution to the field equation (4.89), a formal transformation into the imaginary time axis,

$$t \to -ix_0, \qquad (4.122)$$

results in a nonsingular solution highlighting quantum tunnelling. This type of solution is called an instanton, and it represents an asymptotic

TUNNELLING, INSTANTONS, FALSE VACUUM 199

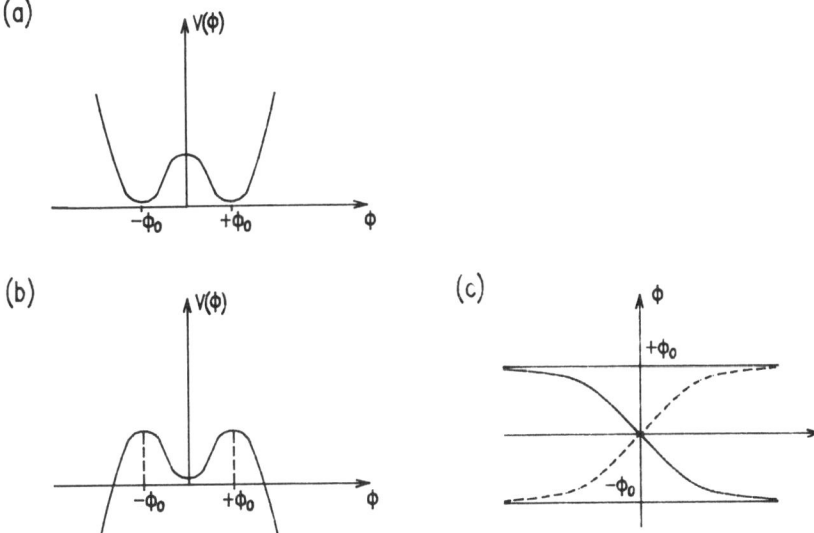

Fig. 4.4 A schematic illustration of the instanton concept: (a) the actual potential; (b) the effective potential after the transformation (4.122); (c) the form of the solution.

(quantum) transition between the two potential minima as $t \to \pm\infty$. This is shown schematically in Fig. 4.4.

For large t, the instanton solution can be approximated by

$$\phi \simeq \pm\phi_0 + Ae^{-\omega t}, \tag{4.123}$$

where $|(d^2V/d\phi^2)|_{\phi=\pm\phi_0} = \omega^2$ and A is an amplitude which depends on the nonlinear potential used. For periodic potentials such as the one in the SG equation, we obtain a continuum of eigenstates forming bands, and this is a direct result of barrier penetration.

A final example is that of metastable states, or so-called 'false vacua'. An example of their emergence is in the ϕ^6-model. There, in addition to an absolutely stable state, say ϕ_0, we have a classical locally stable minimum of $V(\phi)$, say ϕ_1 (see Chapter 2). If we now allow quantum effects to take place, quantum decay will lead to tunnelling from ϕ_1 to ϕ_0 at a rate [22]

$$\Gamma/V = Ke^{-S_0}, \tag{4.124}$$

where V is the volume, Γ is the decay probability per unit time, and K is given by

$$K = \sqrt{\hbar S_0/2\pi}, \tag{4.125}$$

while S_0 is the classical action of a quasi-particle, calculated as

$$S_0 = \int_{\phi_0}^{\phi_1} dx \sqrt{2mV(\phi)} = \int_{-\infty}^{\infty} dt \left[\tfrac{1}{2}(dx/dt)^2 + V(\phi) \right]. \quad (4.126)$$

In summary, classical solutions of field equations can be useful concepts in broken-symmetry phenomena, where they can be used as lowest-order approximations to quantum ground states. Subsequent semi-classical quantization methods have been developed which can account for a number of properties. The quantum nature of the original field can be largely restored with such attendant properties as discrete internal energy levels and quantum tunnelling. A quantum field theory of solitons, and especially their stability against perturbations, has been thoroughly discussed in references [25, 26].

In the next section of this chapter we present a short overview of a related topic which has recently received considerable attention in the physical literature. The question addressed in this connection is whether quantum fluctuations are at all important close to criticality in any many-body systems. This is especially relevant for superfluids and (but not only) superconductors.

4.14 Quantum critical phenomena

All of the spin lattice models described in Chapter 2 in connection with phase transitions are rooted in the quantum-mechanical origin of the spin. However, to bring out the quantum nature of these models one needs to account for noncommutation properties of Pauli spin matrices. Experimentally, this can be realized by applying a transverse magnetic field to a spin lattice. The simplest model that can be studied in this connection is the Ising model in a transverse field, with the Hamiltonian

$$H = -J \sum_{\langle i,j \rangle} s_i^z s_j^z - H^x \sum_i s_i^x. \quad (4.127)$$

The eigenstates of the s_i^z operator, $|s_i\rangle$, no longer diagonalize the above Hamiltonian. Instead, a single spin precession takes place, so that the wave function is

$$|\psi(t)\rangle = \cos(tH^x)|\uparrow\rangle - i\sin(tH^x)|\downarrow\rangle. \quad (4.128)$$

Therefore, the field H^x induces the time-periodic fluctuations in the z-component of each spin, even at $T = 0$ K. The relevant phase diagram for the spin system is shown in Fig. 4.5.

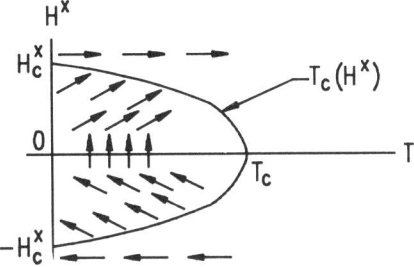

Fig. 4.5 A schematic phase diagram for the spin system in a transverse magnetic field.

The question that can be asked is about the nature of the transition for $H_c^x \neq 0$. It turns out that the critical exponents remain the same as for $H_c^x = 0$ (i.e. are classical) except at $T = 0$. Thus, except for renormalizing the critical amplitudes, the nature of the phase transition is unchanged for all nonzero temperatures, which marginalizes quantum effects from the point of view of criticality.

Another important result that links this discussion with the remainder of this chapter is the form of the effective Hamiltonian H_{eff} for the quantum Ising model. Curiously, through the use of the quantum Kac–Hubbard–Stratanovich transformation, this leads us back to the Landau–Ginzburg form discussed at length in Chapter 2! Namely, it takes the form given below:

$$H_{\text{eff}} = \int \frac{d^d x}{a_0^d} \int_0^\beta dt \left[\tfrac{1}{2} e |\nabla \psi|^2 + \tfrac{1}{2} e_t (\partial_t \psi)^2 + \tfrac{1}{2} r \psi^2 + u \psi^4 \right], \quad (4.129)$$

where the coupling constants are

$$e = \frac{a_0^2}{J_0}, \quad e_t = \frac{\tanh(\beta H^x)}{16(H^x)^3}, \quad r = \frac{1}{J_0} - \frac{\tanh(\beta H^x)}{H^x}, \quad (4.130\text{–}4.132)$$

and u is a complicated integral involving two-body interactions, but $u > 0$. Above, a_0 denotes the lattice spacing, J_0 is the nearest-neighbour exchange constant and $\beta = (k_B T)^{-1}$. The parameter u is a relevant variable describing the strength of long-range interactions. The order parameter field $\psi(x, t)$ represents a time- and space-dependent continuum spin variable defined on a d-dimensional infinite real space and over a finite-sized time interval $(0, \beta)$. Thus, the d-dimensional quantum Ising model appears to be formally equivalent to the classical Landau–Ginzburg model in $(d + 1)$-dimensional space–time! The time dimension is, of course, finite unless $T = 0$, in which case $\beta \to \infty$. This explains why the associated critical exponents are not affected unless $T = 0\,\text{K}$. The corresponding partition function could then be calculated according to

$$Z = \int d\psi(x,t) e^{-\beta H_{\text{eff}}(\psi)}. \tag{4.133}$$

However, a similar procedure applied to the X–Y model in a transverse magnetic field yields an extra term in H_{eff} which is proportional to $\varphi^* \partial_t \varphi$, where φ replaces $\psi(x,t)$ as an order parameter and is a complex variable. This indicates that the quantum aspect of the model adds an additional two dimensions over its classical analogue!

To summarize, perfectly ordered quantum systems do not add any new critical behaviour at $T > 0\,\text{K}$, while at $0\,\text{K}$ their behaviour is either trivial or corresponds to a classical system of higher dimensionality [27].

4.15 Quantum lattice solitons

A question that is still looming very large in the context of nonlinear modelling of physical systems is the effect that quantum corrections may have on classical solutions. A recent study [18] of this problem for one-dimensional lattices of several types provides an important insight and will be briefly discussed in this section. Three specific models will be considered, their eigenfunctions and eigenenergies calculated, and comparisons with classical solitons made. What follows is based entirely on reference [28]. The three examples reviewed below represent three general types of quasi-particle: (a) Bosons, (b) q-deformed Bosons, and (c) Fermions. Assuming periodic boundary conditions, each of the systems considered below commutes with the translation operator and, for a fixed number of quanta, n, the energy eigenstates for a particular value of momentum k are separated into bands. The lowest energy band, called the soliton band, is denoted by $E = E_n(k)$, and it will be characterized by its binding energy, its effective mass, and the maximum group velocity of the corresponding quantum soliton. In all three systems, $E_1(k) = -2\cos k$.

4.15.1 The quantum discrete nonlinear Schrödinger equation (QDNLS)

The Hamiltonian postulated here is

$$H_1 = -\sum_{i=1}^{N} \left[b_i^+ b_{i+1} + b_i^+ b_{i-1} + \frac{\gamma}{2} b_i^+ b_i^+ b_i b_i \right], \tag{4.134}$$

where b_i^+ and b_i are standard Boson creators and annihilators, respectively, located at site i. In the limit $N \to \infty$, the soliton band has the energy

$$E_2(k) = -\sqrt{\gamma^2 + 16\cos^2(k/2)}. \tag{4.135}$$

The binding energy is calculated as the difference between $E_2(0)$ and the bottom of the continuum band, and it is found as

$$E_b = \sqrt{\gamma^2 + 16} - 4. \tag{4.136}$$

Hence, there is always an energy gap in this system. The effective mass of the quantum soliton is found from

$$E_2(k) = E_2(0) + \frac{k^2}{2m^*} + O(k^4), \tag{4.137}$$

and we obtain $m^* = (\sqrt{\gamma^2 + 16})/4$, where we adopt units in which $\hbar = 1$. The maximum group velocity is

$$V_m = (dE_2/dk)_{k=\pi/2} = 4/\sqrt{\gamma^2 + 8}. \tag{4.138}$$

If γ is small (i.e. $\gamma \ll 1$), then the QDNLS is well approximated by the continuum NLS, the n-quantum soliton binding energy of which is

$$E_b = \frac{\gamma^2}{48} n(n^2 - 1). \tag{4.139}$$

The corresponding effective mass is $m^* = n/2$ and the maximum group velocity is $V_m = 2\sin(\pi/2n)$. On the other hand, for strong anharmonicities ($\gamma \gg 1$), one finds that

$$E_n(k) = -\tfrac{1}{2}n(n-1)\gamma - \left[\frac{2n}{(n-1)!\gamma^{n-1}}\right]\cos k. \tag{4.140}$$

The binding energy is then found as

$$E_b = \tfrac{1}{2}n(n-1)\gamma, \tag{4.141}$$

the effective mass as

$$m^* = \frac{(n-1)!}{2n}\gamma^{n-1}, \tag{4.142}$$

and the maximum group velocity as

$$V_m = \frac{2n}{(n-1)!\gamma^{n-1}}. \tag{4.143}$$

Thus, the soliton solution becomes pinned both as the number of quanta $n \to \infty$ and as the nonlinearity coefficient γ increases.

4.15.2 The quantum Ablowitz–Ladik equation (QAL)

The reduced Hamiltonian operator in this case is

$$H_2 = -\sum_{i=1}^{N} [b_i^+(b_{i+1} + b_{i-1})], \qquad (4.144)$$

where the operators b_i^+ and b_j satisfy the q-deformed commutation relations

$$[b_j^+, b_k^+] = [b_j, b_k] = 0, \qquad [b_j, b_k^+] = \left(1 + \frac{\gamma}{2} b_j^+ b_k\right)\delta_{jk}. \qquad (4.145)$$

It can be shown that two soliton bands exist here, and that the top one corresponds to the classical soliton solutions. In the limit as $N \to \infty$, both bands are characterized by

$$E_2(k) = \pm \frac{2\cos(k/2)(\gamma + 4)}{\sqrt{2\gamma + 4}} \qquad (4.146)$$

and the binding energy is found as

$$E_b = \frac{2(\gamma + 4)}{\sqrt{2\gamma + 4}} - 4, \qquad (4.147)$$

while the effective masses are

$$m^* = \pm \frac{2\sqrt{2\gamma + 4}}{\gamma + 4}. \qquad (4.148)$$

Note that the positive mass corresponds to the bottom band, while the negative mass corresponds to the upper band. In both cases the maximum group velocity equals

$$V_m = \frac{\gamma + 4}{\sqrt{2\gamma + 4}}. \qquad (4.149)$$

The limit of $\gamma \ll 1$ yields the continuum nonlinear Schrödinger equation as a good approximation and hence the results obtained in the limit are identical to those in the QDNLS case. Consequently, eqn (4.139) should be adopted for the binding energy E_b with $m^* = n/2$ and $V_m = 2\sin(\pi/2n)$ for the effective mass and the maximum group velocity, respectively.

In the limit of γ large, however, it has been found that the energy eigenvalues are

$$E_n(k) = -2\cos(k/n)(\gamma/2)^{(n-1)/2} \qquad (4.150)$$

and the binding energy is

$$E_b = 2(\gamma/2)^{(n-1)/2}. \qquad (4.151)$$

The effective mass is

$$m^* = (n^2/2)(2/\gamma)^{(n-1)/2} \qquad (4.152)$$

and the maximum group velocity is

$$V_m = (2/n)\sin(\pi/n)(\gamma/2)^{(n-1)/2}. \qquad (4.153)$$

Note that while for small γ QAL solitons closely approximate QDNLS ones, for large γ the situation is quite different and QAL solitons increase their velocity with the increase of both n and γ.

4.15.3 A fermionic polaron model (FP)

The reduced effective Hamiltonian is chosen as

$$H_3 = -\sum_{i=1}^{N} [a_i^+ a_{i+1} + a_i^+ a_{i-1} + \gamma a_i^+ a_i a_{i+1}^+ a_{i+1}], \qquad (4.154)$$

where a_i^+ and a_i are standard fermionic creators and annihilators, respectively. It can be shown that in the limit as $N \to \infty$, the soliton band energy satisfies the formula

$$E(k) = -\left[\gamma + \frac{4}{\gamma}\cos^2\left(\frac{k}{2}\right)\right], \qquad (4.155)$$

provided that $\gamma > 2\cos(k/2)$. The corresponding binding energy is

$$E_b = \gamma + 4/\gamma, \qquad (4.156)$$

the effective mass is given by

$$m^* = \gamma/2, \qquad (4.157)$$

and the maximum group velocity is given by

$$V_m = 2/\gamma. \qquad (4.158)$$

Since in the small γ limit the soliton band merges into the energy continuum, both E_b and m^* are not defined in this limit. In the large γ limit, however, we find that

$$E_n(k) = -(n-1)\gamma - \frac{2}{\gamma^{n-1}}\cos k, \qquad (4.159)$$

and, consequently,

$$E_b = (n-1)\gamma, \qquad (4.160)$$

the effective mass is

$$m^* = \gamma^{n-1}/2, \qquad (4.161)$$

and the group velocity is given by

$$V_m = 2/\gamma^{n-1}, \qquad (4.162)$$

and it decreases both with an increase in γ and n. For the FP model, there is no proper correspondence limit (from quantum to classical objects) because the number of Fermions can at most equal the number of degrees of freedom N.

We have demonstrated in this section that solitons exist in quantum lattices with bosonic, fermionic and q-Boson character. It has been emphasized that their properties strongly depend on: (a) the strength of anharmonicity; (b) the type of commutation relationship satisfied by quantum operators; and (c) the number of degrees of freedom. Many results obtained in classical theory have been confirmed by quantum calculations [28].

In the last section of this chapter, we wish to discuss a particular physical example in which many of the nonlinear ideas come together in a very powerful way. Moreover, several key aspects of this conceptually new nonlinear behaviour are being utilized technologically at present. The area of physics that we have chosen, as so far the most promising testing ground for the emerging paradigms of new physics, is nonlinear optics.

4.16 Nonlinear optical phenomena

Lasers and, in particular, the interactions between coherent light and dielectric media, provide an almost ideal substrate for the investigations of nonlinear effects. Laser action is relatively easy to control and analyse, is characterized by a high level of substance purity, and operates at a borderline between quantum and classical physics. Most importantly, lasers and nonlinear media have given visual illustrations of soliton propagation, deterministic instabilities, pattern formation, quantum coherence, and chaos [29].

The theory of laser emission strives to understand how order evolves from randomness. Lasing atoms or molecules can be viewed as undergoing small internal motions, producing an oscillating dipole moment. Spontaneous emissions and absorptions of energy quanta at this stage result in thermal equilibrium and can be described using a linear theory. However, above the lasing threshold of pumping, co-operative effects dominate, with the emission field exceeding a critical field producing a coherent signal. We can decompose the electric field E as

$$E = [b(t)e^{-i\omega t} + b^*(t)e^{i\omega t}]g(x), \qquad (4.163)$$

where ω is the frequency of the spontaneously emitted wave, $b(t)$ is its amplitude, and $g(x)$ describes the spatial behaviour. The field amplitude at the semi-classical level satisfies the evolution equation

$$db/dt = -\alpha b + \gamma b(\sigma - \sigma_0)b^*b + F(t), \qquad (4.164)$$

where σ describes the degree of occupation inversion for atomic energy levels, which is related to the pumping rate, σ_0 is the critical value of σ leading to laser action, and $F(t)$ is a fluctuating force field. The above equation has an underlying effective potential that changes from a single well for $\sigma < \sigma_0$ to a double well for $\sigma \geq \sigma_0$, which signifies the onset of bistability at $\sigma = \sigma_0$. A macroscopic coherent polarization field above the threshold arises from individual quantum phenomena.

In the next stage of theoretical modelling, account must be taken of the spatial inhomogenity of the nonlinear dielectric medium. The decomposition in eqn (4.163) is replaced by the more accurate one below [30]:

$$E(x,t) = \hat{e}\{A(x,t)\exp[i(k \cdot x - \omega t)] + A^*(x,t)\exp[-i(k \cdot x - \omega t)]\}. \qquad (4.165)$$

It is then assumed that the so-called Kerr medium has a refractive index which depends on the field intensity, i.e. that to the lowest order

$$n = n_0 + n_2|A|^2 + \cdots. \qquad (4.166)$$

Furthermore, in the long-wavelength limit of one-dimensional propagation (e.g. along an optical fibre), Maxwell's equations for E result in the following envelope equation:

$$0 = \frac{\partial A}{\partial z} + \left(\frac{\partial k}{\partial \omega}\right)\frac{\partial A}{\partial t} + \frac{i}{2}\left(\frac{\partial^2 k}{\partial \omega^2}\right)\frac{\partial^2 A}{\partial t^2} - ik\left(\frac{\delta n}{n} + \frac{n_2}{n}|A|^2\right)A, \qquad (4.167)$$

where $k = k(\omega)$ is obtained from the medium's dispersion relation $\omega = \omega(k)$ and the derivatives evaluated at the lasing frequency to give constant coefficients above. The symbol δn represents fluctuations in the space-dependent refractive index. It is of utmost importance to realize that eqn (4.167) takes the form of a nonlinear Schrödinger equation, with the attendant self-focusing property leading to the propagation of solitons [31]. An experimental detection of soliton transmission in optical fibres indicates that they can travel distances in excess of 10^7 m [32]!

Beyond the weak limit producing the NLS of eqn (4.167), one can introduce higher-order Kerr effects leading, for example, to quintic nonlinearities and odd nonlinear dispersion effects of terms due to delayed response of the medium (the Raman effect). In many instances (see Chapter 1), even that may result in exactly solvable equations which exhibit spatial localization of coherent light in reduced geometries (e.g. axial, cylindrical, spherical, etc.) as well as specific types of pattern formation [33, 34]. Experimental evidence for these types of behaviour is accruing [35, 36].

It is important to note that what we have discussed so far is virtually concerned with single-mode lasers. Once our discussion embraces multi-mode behaviour a number of new phenomena emerge. At the level of quantum-mechanical description, the corresponding Hamiltonians must account for distinct modes (with the attendant creation and annihilation operators) and interactions between them. Hence, in addition to the field-translation effects bringing about coherent modes for single-mode lasers, multi-mode lasers call for the use of four-legged interaction terms and this yields squeezed states [37] (see Section 4.7). However, apart from the potentially beneficial effects, hidden pitfalls also exist. Since multi-mode lasers require at least three independent macroscopic variables for their description (e.g. the electric field E, the polarization P, and the population inversion D), this may under certain circumstances create conditions in which deterministic chaos can arise. Indeed, the first-order evolution equations (the Maxwell–Bloch equations) [29],

$$\dot{E} = -(i\omega_c + \kappa)E + gP, \qquad \dot{P} = -(i\omega_0 + \gamma_\perp)P + gED, \quad (4.168, 4.169)$$

and

$$\dot{D} = -\gamma_\parallel (D - D_0) - 2g(E^*P + EP^*) \qquad (4.170)$$

are mathematically equivalent to the famous Lorenz equations (see Chapter 1) which yield strange attractors and chaos. Here, κ is the cavity decay rate determined by the cavity reflectivity, and $\gamma_\perp$ and $\gamma_\parallel$ are the decay rates of atomic polarization and population inversion of the gain

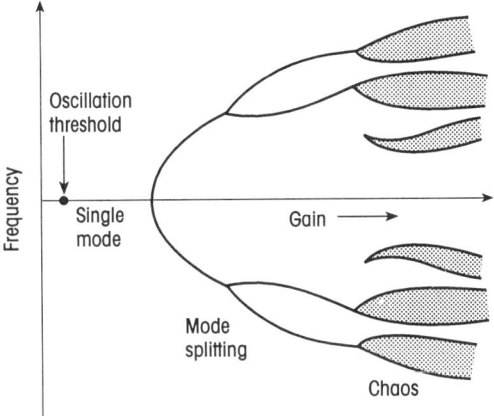

Fig. 4.6 A schematic bifurcation diagram for a multi-mode laser (following reference [29]).

medium. g is the unsaturated gain of the medium at gain line centre ω_0, and ω_c is the laser cavity frequency which, together with the initial population difference D_0, applies to the conditions before the laser medium is pumped. Fortunately, as it turns out, the conditions required for the onset of chaos in this type of laser call for a cavity of high transmission and a gain of at least nine times the threshold value for lasing, which is very difficult to implement. In Fig. 4.6 we show a schematic bifurcation diagram for a multimode laser, where an increase in the gain ratio opens a path to chaos.

References

1. M. Plischke and B. Bergersen, *Equilibrium Statistical Physics*, Prentice Hall, Englewood Cliffs, NJ (1989).
2. H. Umezawa, M. Matsumoto, and M. Tachiki, *Thermo Field Dynamics and Condensed States*, North-Holland, Amsterdam (1982).
3. A. Isihara, *Condensed Matter Physics*, Oxford University Press, Oxford (1991).
4. H. Haken, *Quantum Field Theory of Solids: an Introduction*, North-Holland, Amsterdam (1976).
5. J. W. Negele and H. Orland, *Quantum Many-particle Systems*, Addison-Wesley, Redwood City, CA (1988).
6. D. Stoler, *Phys. Rev.*, **D1**, 3217 (1970).
7. V. G. Drinfeld, *Sov. Math. Dokl.* **32**, 254 (1985).
8. M. Jimbo, *Lett. Math. Phys.* **10**, 63 (1985).

9. S. L. Woronowicz, *Communs. Math. Phys.*, **111**, 613 (1987).
10. A. J. Macfarlane, *J. Phys. A*, **22**, 4581 (1989).
11. L. C. Biedenharn, *J. Phys. A*, **22**, L873 (1989).
12. J. A. Tuszyński and M. Kibler, *J. Phys. A*, **25**, 2425 (1992).
13. V. G. Makhankov and V. K. Fedyanin, *Phys. Rep.*, **104**, 1 (1984).
14. A. Actor, *Rev. Mod. Phys.*, **51**, 461 (1979).
15. B. Sakita, *Quantum Theory of Many-variable Systems and Fields*, World Scientific, Singapore (1985).
16. A. D. Linde, *Rep. Progr. Phys.*, **42**, 389 (1979).
17. R. N. Cahn, *Rep. Progr. Phys.*, **52**, 389 (1989).
18. R. Jackiw, *Rev. Mod. Phys.*, **49**, 681 (1977).
19. R. Rajaraman, *Solitons and Instantons*, North-Holland, Amsterdam (1982).
20. E. Tomboulis, *Phys. Rev. D*, **12**, 1678 (1975).
21. J. L. Gervais and B. Sakita, *Phys. Rev. D*, **11**, 2943 (1975).
22. S. Coleman, *Aspects of Symmetry*, Cambridge University Press, Cambridge (1985).
23. G. Parisi, *Statistical Field Theory*, Addison-Wesley, Redwood City, CA (1988).
24. E. Brézin and G. Parisi, *J. Statist. Phys.*, **25**, 273 (1978).
25. V. G. Makhankov, *Phys. Rep.*, **35**, 1 (1978).
26. V. E. Zakharov, E. A. Kuznetsov, and A. M. Rubenchik, in *Solitons* (ed. S. E. Trullinger, V. E. Zakharov, and V. L. Pokrousky), Elsevier, Amsterdam (1986).
27. J. A. Hertz, *Phys. Rev.*, **B14**, 1165 (1976).
28. A. C. Scott, J. C. Eilbeck, and H. Gilhoj, *Physica* **78D**, 194 (1994).
29. R. G. Harrison, *Contemp. Phys.* **29**, 341 (1988).
30. A. C. Newell and J. V. Moloney, *Nonlinear Optics*, Addison-Wesley, Redwood City, CA (1992).
31. V. E. Zakharov and A. B. Shabat, *Sov. Phys. JETP*, **34**, 62 (1972).
32. L. F. Mollenauer, M. J. Neubelt, S. G. Evangelides, J. P. Gordon, J. R. Simpson, and L. G. Cohen, *Opt. Lett.*, **15**, 1203 (1990).
33. M. Florjanczyk and L. Gagnon, *Phys. Rev. A*, **41**, 4478 (1990).
34. L. Gagnon, *J. Opt. Soc. Am.*, **B7**, 1098 (1990).
35. K. J. Blow and N. J. Doran, Solitons in optical fibres, in *Nonlinear Waves in Solid State Physics* (ed. A. D. Boardman *et al.*), Plenum Press, New York (1990).
36. S. John, *Physics Today*, May 1991, p. 32.
37. P. D. Drummond and S. J. Carter, *J. Opt. Soc. Am.*, **B4**, 1565 (1987).

5 THE METHOD OF COHERENT STRUCTURES (MCS)

...I disagree with most physicists at the present time just on this point [of removing divergent diagrams]. I cannot tolerate departing from the standard rules of mathematics. Of course, the proper inference from this work is that the basic equations are not right. There must be some drastic change introduced into them so that no infinities occur in the theory at all and so that we can carry out the solution of the equations sensibly, according to ordinary rules and without being bothered by difficulties. This requirement will necessitate some really drastic changes: simple changes will not do, just because the Heisenberg equations of motion in the present theory are all so satisfactory. I feel that the change required will be just about as drastic as the passage from the Bohr orbit theory to the quantum mechanics.

P. A. M. Dirac (1978), *Directions in Physics* (J. Wiley, New York)

5.1 Introduction

The description of a quantum many-body system, containing particles or quasi-particles of only one type, e.g. electrons only, is made notoriously difficult by the operators appearing in the Hamiltonian which describe the interaction between different particles. In the vast majority of cases such interactions are of a two-body type, and interactions which are three-, four- or higher-body rarely appear. In the absence of any coupling between particles, the study reduces, as is well known, to a number of noninteracting one-body systems, the simplest of which is the case of hydrogen when the proton and electron are treated as if they were point charges. Historically, Bohr's model of the atom provided a valuable transition to the more abstract theory of quantum mechanics. It is no accident, as the reader will see later, that the Bohr model, despite its inadequacies, used *classical* ideas for the most part and, to prevent the electron radiating energy and spiralling in on to the nucleus, postulated that its orbits contained an integral number of de Broglie wavelengths or, equivalently, that the angular momentum was an integral multiple of $\hbar$. Even this may be viewed classically by considering the orbits as circular and noting that a fractional number of wavelengths around the orbit cannot persist because destructive interference will occur [1].

Another requirement, which Bohr called the Correspondence Principle, was that quantum physics, and his model in particular, should give the *same result as classical physics* in the limit of large principal quantum numbers. This type of idea will be made full use of later, i.e. that there are regimes, points in particular, in quantum-number space, in which the behaviour of the system is classical.

There have been considerable developments in the path-integral technique of quantum simulation, which incorporates quantum effects in a rigorous manner [2–6]. What is of interest here is that the method rests on a correspondence which can be shown to exist between a given quantum system and an equivalent classical system [3, 7]. The classical system is constructed by replacing each quantum particle by a cyclic harmonic chain of P classical particles, sometimes referred to as 'beads' [8], with a spring constant between them given by $mP/\beta^2\hbar^2$, where m is the mass of the quantum particle and β is given by $\beta = 1/k_B T$, with k_B being Boltzmann's constant and T the absolute temperature. 'Beads' on different chains interact via the potential between particles in the original quantum problem, but scaled down by a factor of P. In the limit when P tends to infinity, the above correspondence becomes exact and is an isomorphism.

Returning now to the case of more than one particle, if we consider two or three particles, we may consider the problem to be numerically soluble whether or not the particles mutually interact; but once the number of particles is large, sheer computational difficulty prevents an analytical or even numerical solution. Alternatively, a many-electron atom, for example, is often studied using a central-field approximation, on which zero order eigenfunctions are based and which works most satisfactorily [9]. That is, there appears to be an effective potential in which each electron moves or —and this will become clearer in our discussion—the many-electron problem reduces to an *effective* one-body problem with an associated potential energy. There is, in fact, a rigorous method [10–12] of demonstrating that an effective one-body potential energy exists using density functional formalism, but it is not clear how to find the exchange potential energy unambiguously.

The most straightforward method of tackling many-body systems is to employ perturbation theory in one guise or another. Although this method has been used very successfully in countless instances in solid state physics and physics as a whole, it does have many subtle pitfalls for the unwary. For one thing, there may well be no small parameter to expand in or, as more often appears to be the case, the phenomenon being investigated may not be described by formulas which can be expanded in some parameter. In fact, such expansions may be singular [13]. The well-known example in superconductivity readily springs to mind, in which the electron–phonon interaction I, as $I \to 0$, exhibits physical characteristics which do not present themselves functionally as a zero but as an essential

INTRODUCTION

singularity, thereby prohibiting the use of an expansion in I, at least to finite order. Another problem is that an unperturbed Hamiltonian, H_0, must be chosen, and it has often been the case in the literature [14] that the perturbation has a higher degree of symmetry that H_0. Hence, use of a perturbation expansion in low order is most inappropriate. To retrieve all such symmetries in the original Hamiltonian, H, one is forced to go to infinite order. The latter problem may largely be overcome by *postulating* a form of H_0 so that it contains as much of the underlying physics as possible.

There are situations in which the unperturbed Hamiltonian, H_0, it is felt, should contain, at least in part, a continuum of levels. This can arise, for example, when dealing with a coupled system of electrons and phonons. If the phonons are put into the perturbation to avoid this problem so that H_0 has well separated levels, another difficulty apparently arises. Writing the Hamiltonian as $H = H_0 + \lambda V$, where λ is varied between 0 and 1, then when $\lambda \to 0$ it is often assumed that the resulting eigenvalues of H, emanating from two different manifolds of H_0, must not overlap [15]. However, if V contains harmonic oscillators, this latter condition cannot be met. Fortunately, there is an iterative form of perturbation theory [16] which overcomes this type of problem, but it very soon becomes too laborious to be practical much beyond second order.

One of the very first approaches to many-electron systems, which provides a simple physical picture, was the Hartree method, in which each electron is associated with its own wave function. For example, the ith particle is related to $\psi_i(r)$ in such a way that $|\psi_i(r)|^2$ denotes the probability per unit volume of finding the ith particle at position r. The effective potential of the ith electron may then be written as

$$V_i(r_i) = \sum_s{}' \int \frac{e^2 |\psi_s(r_j)|^2}{|r_i - r_j|} d^3 r_j, \tag{5.1}$$

where the prime on the summation indicates that the particular s for the ith particle is to be excluded and that s ranges only over those 'states' $\psi_s(r_j)$ which are occupied. In this picture the wave function for the whole many-electron system is then simply a product $\Pi_s \psi_s(r_i)$ over the number of occupied states s. As is well known, this is physically inadequate, since this total wave function is not antisymmetrical with respect to interchange of electrons between different states [18]. What is also of great significance is that *this approach attempts to allocate wave functions for individual particles in the system*. Knowing how to solve for energies and eigenfunctions of noninteracting systems, this idea is physically attractive. However, the method of coherent structures (MCS) *deliberately avoids this idea and effectively associates a 'wave function' or field to the many-electron system as a whole.*

A more sophisticated method of attack is perhaps to begin with a known antisymmetric wave function for the system as a whole. In the Hartree–Fock technique, this is taken to be a normalized Slater determinant containing N orbitals (both spin and orbital quantum numbers label these) occupied in different ways by N electrons. This is done simply to make the particles indistinguishable while still allowing nonuniform charge distributions. *Once again, the idea of individual orbitals is introduced* and the expectation value of the Hamiltonian is then minimized to obtain complicated nonlinear coupled equations for the individual orbitals. These are solved self-consistently by inserting suitable wave functions into the potential energies, solving the complicated equations for the orbitals, and iterating until the orbitals obtained and those inserted are the same to within some degree of accuracy. Although this method incorporates antisymmetry requirements and the Pauli Principle, whereby electrons with parallel spins are kept apart, it does not incorporate correlations due to the Coulomb repulsion properly. There are a number of other comments about this approach which make it unsatisfactory. At the simplest level the coupled equations have a very complicated nature and are of an integro-differential type, but they do have one significant feature, which is that 'exchange' terms (those additional to those in the Hartree method which appear again) lead immediately to a *nonlocal potential energy*. These exchange contributions alter the properties of the Hartree model drastically and, in fact, the Hartree–Fock (HF) technique produces results which bear no resemblance to physical reality for solids, while for atoms the method works reasonably well. This latter method (HF) has a most spectacular failure in that it predicts a vanishing density of states at the Fermi level, thereby ensuring that it will be completely useless to describe physical properties which are sensitive to the number of states at the Fermi level, e.g. thermal and electrical conductivities [18]. That HF is far from exact, principally since the total wave function cannot be written as a single Slater determinant, is seen in a most dramatic way when one calculates the overlap between the exact wave function and the HF wave function, and it is found that it tends exponentially to zero with the number of electrons [19]! Thus three features which we particularly pick out and note are: (a) that again it uses the idea of individual wave functions (albeit occupied in different ways); (b) that it is *variational* in nature; and (c) that it leads to nonlocal potential energies. Furthermore, (d) the HF equations are highly nonlinear.

Another method which should be mentioned is the Thomas–Fermi theory. This way of looking at, in particular, a many-electron atom, is

INTRODUCTION

semi-classical, an important feature in itself. The idea is to relate a self-consistent potential energy $V(r)$ to the electron density $\rho(r)$, and a very readable account of this approach can be found in reference [20]. As pointed out there, it cannot be exact, since when $V(r)$ is greater than the 'fastest electron's' energy E_f, $\rho(r)$ becomes imaginary, so that the Thomas–Fermi relation between $\rho(r)$ and $V(r)$ cannot be used in this regime. In atoms and ions the difference $V(r) - E_f$ is expressed as

$$V(r) - E_f = -\frac{Ze^2}{r}\phi. \tag{5.2}$$

What is so significant is that the *equation for $\phi(x)$ is nonlinear* and takes the form

$$d^2\phi/dx^2 = \phi^{3/2}/x^{1/2}, \tag{5.3}$$

where $r = bx$, x is dimensionless, and b is an appropriately chosen constant. The qualitative features of, for example, atomic binding energies do emerge from this method, but are definitely not in such good agreement with experiment as from a Hartree method calculated by Foldy [20, 21].

Furthermore, quite good comparisons with experiment of $\sqrt{-(\text{energy})}$ versus atomic number plots (obtained from the Thomas–Fermi approach) can be made, particularly for atomic states with low principal quantum numbers and high atomic numbers. This method is again self-consistent, but does not lend itself readily to extensions. However, it does have many similarities with MCS. It may be modified to include correlation effects [22]. However, the relationship between $\rho(\mathbf{r})$ and $V(\mathbf{r})$ may be derived variationally [20]. As March [20] has pointed out, the work of Hohenberg and Kohn [12] has shown that the philosophy behind the Thomas–Fermi method is completely justified even in a many-electron problem. These latter theorists have shown that the ground energy is a unique functional of the particle density. More recently, a rigorous density functional theory for inhomogeneous Bose-condensed fluids has been presented [111].

According to March [20], 'the major problem, of course, is to find the functional and this is equivalent to solving the many-body problem'. The MCS goes a long way in the right direction to do this, the functional, as will become apparent shortly, being one in $|\psi|^2$ and its derivatives, where ψ is the field that we define later. The function $\langle|\psi|^2\rangle$ also plays the role of a density of particles, where $\langle\ \rangle$ denotes a suitable average! If the field is written in modulus–argument form as $\psi = \eta \exp(i\chi)$, the density is essentially the square of its modulus. In our study, we shall see later that the equation of motion may be cast into a form containing ψ and its gradients only, and when the above form is inserted into our equation of motion and real and imaginary parts separated, the equation for the imaginary part, dropping a very small residual term, contains only η, gradients of η and a term in $(\nabla \cdot (\eta^2 \nabla \chi))/\eta$. Thus, $\nabla \chi$ may be written in terms of η and its derivatives only in the equation for the real parts, apart from the residual term, and becomes an equation for χ_t in η and its gradients only. Thus, the argument of the field is also a function of η and its gradients and hence, so is the Hamiltonian density, which to first order, as it turns out, is of *Landau–Ginzburg form*!

To go beyond HF and incorporate further correlations, other than a generalized version of the Thomas–Fermi method, one immediately becomes involved with the whole panoply of Green functions or propagators, Feynman diagrams, self-energies, vertex functions, and dielectric or screening functions [18, 22, 23]. There is a great deal of attraction in this approach in that the various diagrams can be readily interpreted physically and the different types are attractively described by bubbles, oysters, etc. Summing over particular sets of diagrams appears to correspond to well-known approximations, met with elsewhere, such as the Random Phase Approximation (over ring diagrams) and the Hartree or Hartree–Fock schemes. Nevertheless, there are subtleties to be reckoned with. For example, it is impossible to sum over all series of diagrams and often a given set, designated 'the most divergent', is chosen and resummed as an approximation. Admittedly, some diagrammatic summations are preselected by screening or dielectric functions and are chosen well using physical principles. However, to select a given set of diagrams might be viewed as a type of rearrangement of a series. The series clearly must be convergent on physical grounds, but it may not be absolutely but only conditionally convergent. If the series is conditionally convergent, then a well-known theorem in analysis [24] shows that it is always possible to rearrange such a series to converge to any sum that we please! That is, the result of our selected summation may simply be governed by what we expect from physical intuition and does not follow via well-argued approximations, the degree of validity of which is easy to evaluate. Even when equation-of-motion methods for Green functions are used, it is not always clear exactly what approximations are being made to decouple equations [25]. Moreover, this approach soon becomes very complex indeed.

There are a whole host of other techniques which are utilized in many-body problems, including collective field techniques [26], the methods of constrained classical solutions [27–29], and an approach based on the Dyson–Schwinger equations [30]. One such very successful method is referred to as a large-N expansion and is used in quantum mechanics and atomic physics [31]. In this context N is usually a number of internal degrees of freedom which parametrize the symmetry group of the problem considered. The solution to a particular problem is obtained exactly in the large N limit and then finite N corrections are obtained by expansion in $1/N$. This method has received considerable attention in recent years and has proved remarkably successful in tackling single particle-potential problems; for example, critical phenomena described by multicomponent field theories, quantum spin models, and investigations in nuclear physics.

Perturbation expansions certainly lose much of their credibility if the coupling constant can be scaled out of the theory, and this situation is indeed encountered with power-law potentials in quantum mechanics [17]. Furthermore, such techniques are particularly inappropriate when dealing with critical phenomena.

What is important from our point of view is that Yaffe [17] has demonstrated that for a quantum theory satisfying certain assumptions it is possible to determine a set of generalized coherent states which can be used to construct a *classical phase space* and obtain a *classical Hamiltonian* such that the resulting dynamics agrees with the large-N quantum dynamics. Similarly, Berezin [32] has also used a coherent state method to show that when $N \to \infty$, the resulting theory is a *classical* theory. The classical nature of the large-N limit is confirmed by a number of authors. The method of expansion in $1/N$ is often referred to as nonperturbative, since the interactions appearing in the model need not be small [31]. In many instances what is very surprising and most interesting is that quite accurate results are obtained by retaining only the first few terms in the expansion. Furthermore, the method may be generalized to the relativistic situation.

5.2 Motivation for the approach to the method of coherent structures

The goal of the MCS is to set up a complementary procedure for analysing many-body systems which is sufficiently accurate, simple in essence, and avoids some of the problems which the methods discussed in the last section may encounter. We wish to incorporate the following features in this approach, and we list them below so that we can refer back to them easily:

(a) There exists a *classical picture* associated with a quantum-mechanical many-body system, close to which physical characteristics can be determined accurately. This we have seen in thc last section in Bohr theory, the path-integral method of quantum simulation, the Hartree Method, the Thomas–Fermi picture and large-N expansions.

(b) Interacting many-body systems inevitably involve *nonlinearity*, which must be properly accounted for. This feature is present in the physically intuitively appealing Thomas–Fermi approach, the density functional studies, and in both Hartree and Hartree–Fock models.

(c) There is a *nonperturbative* element and many fascinating physical phenomena would disappear from our theoretical investigations if we ignore it. We have described some phenomena in Chapter 1, where this aspect is apparent, and we comment later on its significance.

(d) While traditionally wave functions are associated with *individual* particles, in this approach a 'field' is assigned to the system as a whole. The quantum component of this field satisfies a Schrödinger equation with what appears to be an effective potential energy determined by its classical envelope.

(e) Sets of *coherent states* appear naturally in MCS. As stated in Section 5.1 for the large-N expansion method, sets of generalized coherent states can be used to construct a classical phase space and a *classical Hamiltonian* such that the associated dynamics agree with that for large N. Coherence phenomena in MCS will be closely associated with *localized structures* in the phase space ($(\nabla \eta)^2$ versus η) of the classical envelope η of the field ψ.

(f) Many laws of physics, expressed as differential relationships, may be derived as an Euler–Lagrange equation from an appropriate Hamiltonian, Lagrangian or free energy functional. The equation of motion for the 'field' that we introduce will be seen to derive from a *Landau–Ginzburg*-like functional, sometimes requiring additional terms. Hence, there is a strong connection with the *variational*, functional, or self-consistent approaches discussed in Section 5.1.

(g) The form of the equation of motion is exactly the same whatever base of states we use to define the 'field' ψ. This is true despite the fact that these equations are highly nonlinear. Therefore, we can say that the form of these equations is *invariant* under a change of base functions—as it should be.

(h) Even when *spin* is incorporated, the form of the equations of motion is invariant for Hamiltonians, which are themselves independent of spin, a very large number of which are found in physics.

(i) Our resulting equation of motion for the *classical* component of the field is of the same form whether associated particles satisfy Bose–Einstein or Fermi–Dirac statistics. The statistics of the particles are restored using the physical boundary conditions and/or quantization procedures.

(j) A relativistic extension of the method is obtained easily and it follows directly from the single particle dispersion relation. As a result, instead of the nonlinear Schrödinger type of equation of motion for nonrelativistic particles, a nonlinear Klein–Gordon form of equation arises in the relativistic case.

Our starting point is really part of our motivation and is an appropriate second-quantized Hamiltonian. The reader may quite properly ask why we have adopted this approach. The reason is that there are numerous examples in condensed-matter physics which possess an *effective Hamiltonian* structure in second-quantized notation which is of essentially the *same form*. This, of course, does not mean that all these systems possess identical physical states, but that only a single mathematical procedure may be used to investigate these problems. All of these Hamiltonians contain one-body parts which describe components such as kinetic or potential energies, but no interaction terms. The latter are provided by additional two-body anharmonic interactions. The formalism itself can be used for a variable number of particles, statistics are built in, and extraneous information about individual particles is excluded. Elementary excitations appear naturally [33] and to a first approximation the approach diagonalizes Hamiltonians of a large number of many-particle systems when interactions are weak. The method is well-known and well-established, and provides a very convenient and succinct approach to many-body physics [34, 35]. However, if interactions are sufficiently strong a new phenomenon often occurs, which cannot be treated perturbatively with elementary excitations used as a zero-order approximation (see (c) above). The new phenomenon is the so-called *symmetry-breaking* effect [36, 37] when a degenerate (broken symmetry) ground state of the total Hamiltonian is not invariant with respect to an application of one or more symmetry operations of the total Hamiltonian, such as the application of parity or gauge invariance. When this occurs such physical systems exhibit extraordinary properties such as Bose condensation, long-range order, and the existence of soft (Goldstone) modes and coherence, leading to such fascinating phenomena as superconductivity, superfluidity, and structural phase transitions (see Chapters 1, 2, and 3).

To make our point clearer let us consider some specific examples. One of the simplest models that one can use to examine the behaviour of conduction electrons in a metal, in an elementary way, is a system of N spinless electrons in unit volume, mutually interacting via a Coulomb repulsion. To take account of the overall charge neutrality of the metal, it is assumed that the electrons move in a uniform positive background. Using second quantization with a complete set of plane waves (we shall henceforth call this a *base*) the corresponding Hamiltonian takes the form [38]

$$H_1 = \sum_k \varepsilon_k c_k^+ c_k + \sum_{k,k',q \neq 0} \frac{2\pi e^2}{q^2} c_{k-q}^+ c_{k'+q}^+ c_{k'} c_k, \quad (5.4)$$

where ε_k represents the expectation value of the kinetic energy for the plane wave $|k\rangle$, or

$$\varepsilon_k = \langle k| \frac{p^2}{2m} |k\rangle = \frac{\hbar^2 k^2}{2m}, \quad (5.5)$$

and q is proportional to the momentum transfer. Notice that, for this simple case, the one-body component of H_1 is *diagonal* and, making the substitutions

$$k - q = K, \quad k' + q = L, \quad k' = M,$$

the two-body operator may be written as

$$c_K^+ c_L^+ c_M c_{K+L-M}: \quad (5.6)$$

that is, the interaction conserved linear momentum.

Another deceptively simple example [38] is that due to Fröhlich to describe the *interaction between electrons and phonons*. In this model ions interact with each other and with electrons via a short-range screened potential. The associated Hamiltonian may be written as

$$H_2 = \sum_k \varepsilon_k c_k^+ c_k + \sum_q \hbar \omega_q b_q^+ b_q + \sum_{k,k'} M_{k,k'} (b_{-q}^+ + b_q) c_k^+ c_{k'}. \quad (5.7)$$

In H_2, the first term on the right-hand side describes the electrons, the second the phonons, and the third the interaction between them (linear in the displacement of the ions indicated by the $(b_{-q}^+ + b_q)$ factor). When a unitary transformation is applied to H_2 of the form $\exp(iS) H_2 \exp(-iS)$, and the series in S is truncated by dropping terms in S^3 and higher, an effective Hamiltonian, $H_{\text{eff}}^{(2)}$, is found, given by

$$H_{\text{eff}}^{(2)} = \sum_k \varepsilon_k c_k^+ c_k + \sum_q \hbar \omega_q b_q^+ b_q$$

$$+ \sum_{k,k',q} W_{k,k',q} c_{k'+q}^+ c_{k-q}^+ c_k c_{k'}$$

$$+ \text{(smaller two-body electron–electron contributions)}. \quad (5.8)$$

Again, notice that the one-body terms are diagonal and that the main electron–electron term may be cast into the form of eqn (5.6). Apart from serving as a motivation for the BCS theory of superconductivity, which introduces an effective Hamiltonian instead (see Chapter 6), the Fröhlich Hamiltonian of eqn (5.7) has recently been utilized to describe coherent phonon oscillations in semiconductors [39]. In this latter example, a two-band semiconductor is considered, in which electrons interact with optical phonons [40].

In the theory of *liquid ^4He*, the atoms are treated as a system of interacting Bosons with a Hamiltonian of the form

$$H_3 = \sum_k \varepsilon_k b_k^+ b_k + \tfrac{1}{2} \sum_{k,k',q} V_q b_{k-q}^+ b_{k'+q}^+ b_k \cdot b_k, \quad (5.9)$$

where $\varepsilon_k = \hbar^2 k^2 / 2m$, V_q is the qth Fourier component of a short-range potential, and b_k^+ creates a Boson from the vacuum. As in the two previous examples, the interaction may easily be transformed into the form of eqn (5.6).

Another example is provided by the *Heisenberg model of a ferromagnet* [41], in which the interaction between spins of a scalar type, in the presence of an externally applied magnetic field proportional to ω_0, for which the Hamiltonian is

$$H_4 = - \sum_l s(l) \cdot \left(\omega_0 + \sum_{l'} J_{ll'} s(l') \right), \quad (5.10)$$

where the vectors l denote the position vectors of the spin sites and $J_{ll'}$ is the exchange interaction constant. There are usually assumed to be a finite number of spins at fixed sites on a chain, so that kinetic energies can be neglected and periodic boundary conditions are assumed. In a standard way, Boson operators b_l and b_l^+ are introduced using a Holstein–Primakoff transformation, so that the z-component of a particular spin is written as

$$s_z(l) = \hbar(\tfrac{1}{2} - n_l), \quad (5.11)$$

with raising and lowering operators as

$$s_+(l) = \hbar(1-n_l)^{1/2} b_l, \qquad s_-(l) = \hbar b_l^+ (1-n_l)^{1/2}, \quad (5.12)$$

where $n_l = b_l^+ b_l$ is the corresponding number operator. At very low temperatures the magnetization in the z-direction will be close to its saturation value and the z-component of spin close to $\tfrac{1}{2} N \hbar$, where N is the number of spins. When the deviations from saturation n_l are small, we may linearize and find that, in this approximation, the Hamiltonian becomes

$$H_4^0 = \sum_q \hbar \omega_q b_q^+ b_q, \quad (5.13)$$

where b_q is the qth component of a Fourier series for b_l (further details may be found in reference [38]). However, when the magnons or spin-wave excitations created by b_q^+ begin to interact significantly (or, alternatively, the deviation from saturation becomes larger), an effective interaction between the magnons arises, of the form

$$\Delta H_4 = \frac{\hbar^2}{2N} \sum_{q,q',p} (J_q + J_{q'+p} - 2J_{q'+p-q}) b_{q-p}^+ b_{q'+p}^+ b_{q'} b_q, \quad (5.14)$$

where J_q is defined by

$$J_q = \sum_{l'} J_{l,l'} \exp(iq \cdot (l' - l)). \quad (5.15)$$

Yet again, the interaction in ΔH_4 is of the form of eqn (5.6), which is easily shown by a simple transformation.

There are even biophysical examples in which such effective Hamiltonians occur. One of them, proposed by Fröhlich [42], was intended to describe the transition to a *metabolic state occurring in living cells*. The effective second-quantized Hamiltonian put forward by Wu and Austin [43] introduces a membrane's dipolar modes described by operators a_i^+ and a_i, a heat bath's mode by b_i^+ and b_i, and energy pumping operators P_i^+ and P_i. This Hamiltonian may be written as

$$\begin{aligned}H_5 = &\sum_i \omega_i a_i^+ a_i + \sum_i \Omega_i b_i^+ b_i + \sum_i \theta_i P_i^+ P_i \\&+ \tfrac{1}{2} \sum_{i,j,k} (\chi a_i^+ a_j b_k^+ + \chi^* a_j^+ a_i b_k) \\&+ \sum_{i,j} (\lambda b_i a_j^+ + \lambda^* b_i^+ a_j) \\&+ \sum_{i,j} (\xi P_i a_j^+ + \xi^* P_i^+ a_j). \quad (5.16)\end{aligned}$$

Remarkably [44], the Hamiltonian H_5 may be transformed, by a series of unitary transformations, to an effective one which can be written in the form

$$H_{\text{eff}}^{(5)} = \sum_k W_k a_k^+ a_k + \sum_{k,k',q} \Delta_{k,k',q} a_{k'+q}^+ a_{k-q}^+ a_k a_{k'}. \quad (5.17)$$

Like all the other examples, the interaction in $H_{\text{eff}}^{(5)}$ can be written like the form in eqn (5.6).

Another biophysical example is called the Davydov model and was used relatively recently to investigate the problem of *energy transfer in one-dimensional molecular chains* [45]. The Hamiltonian is given by

$$H_6 = \sum_m \hbar\Omega_m a_m^+ a_m + \sum_q \hbar\omega_q\left(b_q^+ b_q + \tfrac{1}{2}\right)$$
$$+ \sum_{q,m} \chi_{q,m}(b_q^+ + b_q) a_m^+ a_m, \qquad (5.18)$$

where the operators b_q^+ and b_q are phonon ladder operators and the a_m^+ (a_m) are exciton creators (annihilators). Performing a unitary transformation [46] on H_6 reduces it to the form

$$H_{\text{eff}}^{(6)} = \sum_{m,n} (E_{m,n}\delta_{m,n} + J_{m,n}) a_m^+ a_n + \sum_q \hbar\omega_q\left(b_q^+ b_q + \tfrac{1}{2}\right)$$
$$+ \sum_{m,n,q} \Delta_{m,n,q} a_{m+q}^+ a_{n-q}^+ a_m a_n. \qquad (5.19)$$

Again the exciton interactions take the form of eqn (5.6), but the corresponding one-body exciton component is no longer necessarily diagonal. However, there are bases in which such a one-body part can be made diagonal and the interaction remains momentum conserving, as we see later when we consider changes of base.

There are innumerable other examples employing the form of a Hamiltonian which has diagonal one-body contributions and momentum conserving two-body interactions. Sometimes this form is obscured by the type of base normally used. For example, in an *atom* a typical Hamiltonian could be written as

$$H_7 = \sum_\alpha \varepsilon_\alpha a_\alpha^+ a_\alpha + \tfrac{1}{2} \sum_{\alpha\beta\gamma\delta} V_{\alpha\beta\gamma\delta} a_\alpha^+ a_\beta^+ a_\delta a_\gamma, \qquad (5.20)$$

where each of the subscripts α, β, γ, and δ is shorthand for one particular set of the four quantum numbers n, l, m_l, and m_s [35]. Although the one-body part is diagonal and its coefficient is explicitly given by

$$\varepsilon_\alpha = \left\langle \alpha \left| \frac{p^2}{2m} - \frac{Ze^2}{r} \right| \alpha \right\rangle, \qquad (5.21)$$

the two-body interaction does not apparently conserve momentum in this form, and the coupling matrix element has the more complicated form

$$V_{\alpha\beta\gamma\delta} = \left\langle \alpha, \beta \left| \frac{e^2}{r_{12}} \right| \gamma, \delta \right\rangle. \qquad (5.22)$$

Nevertheless, as we will see later, the equation of motion for the appropriate 'field' that we define, using a Hamiltonian of this apparently nonconserving type, still takes the same form! Furthermore, the one-body part can always be chosen to be diagonal, since contributions that are nondiagonal merely modify the phase of the 'field' ψ and disappear, as we later explain, in the quantum equation.

Another slightly anomalous case is the effective *BCS Hamiltonian for superconductors*, which because of spin components ($\uparrow$ and $\downarrow$) takes the form

$$H_8 = \sum_k \varepsilon_k (c^+_{k\uparrow} c_{k\uparrow} + c^+_{-k\downarrow} c_{-k\downarrow})$$
$$- \sum_{k,k'} V_{k,k'} c^+_{k'\uparrow} c^+_{-k'\downarrow} c_{-k\downarrow} c_{k\uparrow}. \tag{5.23}$$

If the terminology is slightly modified [38], so that we can write

$$c^+_{k'} \equiv c^+_{k'\uparrow}, \qquad c^+_{-k'} \equiv c^+_{-k'\downarrow},$$

the form in eqn (5.6) is retrieved by putting

$$k' \to K, \qquad -k' \to L, \qquad -k \to M.$$

This may seem somewhat artificial, but we show later that when spin is introduced, a spin-dependent field may be defined, and the equations of motion for this are of the *same form* as that derived without considering the spin components. We note in passing that similar considerations may be applied to the Hamiltonian describing anharmonic lattice vibrations [47].

All of the above examples, representing a diversity of physical phenomena, can be reduced to the same second-quantized form of the underlying Hamiltonian. In particular, the two-body interaction can always be written in a form that conserves linear momentum. Hence, our starting point is the effective generic Hamiltonian

$$H = \sum_k \omega_k q^+_k q_k + \sum_{k,l,m} \Delta_{k,l,m} q^+_k q^+_l q_m q_{k+l-m}, \tag{5.24}$$

where the complete set of 'one-particle' functions (or a base) that we have used are plane waves normalized over a volume Ω. For reasons given earlier, we choose the one-body part of H in eqn (5.24) to be diagonal and the interaction to conserve linear momentum. At this stage it is not obvious that the interaction can be made momentum conserving and at the same time the one-body component diagonal. However, when we consider an arbitrary base later, the reason why we can do this will become clear. We simply choose a plane-wave basis for simplicity, as later on we show that our result, our equation of motion, is independent of base selection.

Our method of attack relies on standard techniques that have now become well established in different disciplines but perhaps are very rarely used in conjunction: that is, these techniques are well known in particular areas of physics but practitioners of theoretical condensed matter physics are usually, but not always, unaware of their power and applicability. There are four areas that we shall utilize to analyse the problem encapsulated by eqn (5.24). The first of these concerns *critical phenomena* and the basic ideas of renormalization group theory, although for our purposes the latter will only impinge on our study in a peripheral way for purposes of truncation. The reader will be able to follow what we do closely, since this particular area, involving the notions of a critical point, an order parameter, and critical exponents, has been already reviewed in this book (see Chapter 2). To make matters more concrete we will draw the reader's attention to particular concepts by referring to the very readable account given by Ma [48], giving specific pages on which the appropriate idea used in the text is discussed. We will also refer to particular sections of the book by Amit [49] and make full use of the ideas of *quantum field theory*. Here, we would draw the reader's attention to the excellent and 'easy to read' accounts by Jackiw [13] and Rajaraman [50] in his book. Our approach will employ the many recent mathematical developments concerning the solutions of highly *nonlinear partial differential equations* (PDEs). With high-speed computers, and especially symbolic computational techniques, great advances have been made using what is called the Symmetry Reduction Method, which is later fully described in this book (in Chapter 7). These methods have been used to study continuous symmetries of many important PDEs in both Euclidean and Minkowski multidimensional space-times; for example, in references [51–54]. Using symmetry methods they either reduce a PDE to one of lower order or to an ordinary differential equation (ODE) by judicious choice of new dependent and independent variables. Often, the resulting ODEs may be solved exactly, but even when this cannot be done the method reduces the PDE to an irreducible equation, for which it may be necessary to use numerical procedures. Recently, there has been considerable interest in the development of generalized symmetry methods, yielding many new reductions and exact solutions of physically significant PDEs [57–61]. We provide the reader with an accessible overview of these in Chapter 7. The methods of *second quantization* discussed elsewhere in this book (for example, in Chapters 3 and 4) form the last component that we utilize in our approach. We also include in this category knowledge of the different representations which may be used in an attack on a many-body problem, such as the Schrödinger, Heisenberg, and interaction pictures [18, 55]. We shall utilize the Heisenberg representation in which time dependences are transformed from wave functions to operators (whereas in the Schrödinger picture it is the wave functions that are time-dependent and satisfy the time-dependent Schrödinger equation).

5.3 Derivation of the nonlinear quantum field equations

As we explained above, for simplicity, we use a base of normalized plane waves knowing that we can demonstrate that the same form of field equations will appear in any 'base'. We introduce annihilators q_k and creators q_l^+, the subscripted vectors denoting a particular plane wave. These are second-quantized operators, so we need their commutation relations in both a Fermi–Dirac and a Bose–Einstein context. As is well known they are, for Bose–Einstein statistics,

$$[q_k^+, q_l^+]_- = [q_k, q_l]_- = 0 \quad \text{and} \quad [q_k, q_l^+]_- = \delta_{k,l}, \qquad (5.25)$$

and, for the Fermi–Dirac case,

$$[q_k^+, q_l^+]_+ = [q_k, q_l]_+ = 0 \quad \text{and} \quad [q_k, q_l^+]_+ = \delta_{k,l}, \qquad (5.26)$$

where we use the symbol $[A, B]_-$ to denote a commutator $AB - BA$ and $[A, B]_+$ for the anticommutator $AB + BA$. These relations will be frequently referred to in both Chapters 5 and 6. All of the operators in eqns (5.25) and (5.26) are time-independent, as is the Hamiltonian H in eqn (5.24).

The next step is to use Heisenberg's equation of motion for a space- and time-dependent operator $A_H(r, t)$. In a Heisenberg representation this takes the form

$$i\hbar \partial_t A_H(r, t) = -[H, A_H(r, t)]_-. \qquad (5.27)$$

In the Heisenberg picture [18], when the Hamiltonian is specifically time-independent, the time-dependent operator $A_H(r, t)$ is related to the corresponding Schrödinger time-independent operator $A_S(r)$ by

$$A_H(r, t) = \exp(iHt/\hbar) A_S(r) \exp(-iHt/\hbar). \qquad (5.28)$$

In particular, for a specific annihilator, we can use the above to give

$$q_\eta(t) = \exp(iHt/\hbar) q_\eta \exp(-iHt/\hbar), \qquad (5.29)$$

where we have written $q_\eta(t)$ to represent the Heisenberg time-dependent operator and q_η is the corresponding Schrödinger operator. Inserting eqn (5.29) into (5.27) yields

$$\begin{aligned} i\hbar \partial_t q_\eta(t) &= -[H, q_\eta(t)]_- \\ &= -\exp(iHt/\hbar)[H, q_\eta]_- \exp(-iHt/\hbar), \end{aligned} \qquad (5.30)$$

since H is time-independent. From the time-independent nature of H it is clear that we can evaluate the right-hand side of eqn (5.30) by simply calculating the commutator of q_η with H, and then replace particular resulting q_ξ's by $q_\xi(t)$.

Thus, from now on the explicit time dependence of the annihilators and creators will be omitted, but implicitly assumed in our study. It is easy to show from eqns (5.25) and (5.30), for the Bose–Einstein case, that

$$i\hbar \partial_t q_\eta = \omega_\eta q_\eta + \sum_{k,m} (\Delta_{\eta,k,m} + \Delta_{k,\eta,m}) q_k^+ q_m q_{k+\eta-m}. \tag{5.31}$$

Similarly, we find for Fermions that

$$i\hbar \partial_t q_\eta = \omega_\eta q_\eta - \sum_{k,m} (\Delta_{k,\eta,m} - \Delta_{\eta,k,m}) q_k^+ q_m q_{k+\eta-m}. \tag{5.32}$$

We have only presented the equations for the annihilators, as these are all that we need to use later, but clearly corresponding equations may be readily found for creators.

By interchanging q_k^+ and q_l^+ in eqn (5.24), supposing that they describe Bosons, we see that we may put

$$\Delta_{klm} = \Delta_{lkm}. \tag{5.33}$$

Similarly, in eqn (5.32), for Fermions we may put

$$\Delta_{klm} = -\Delta_{lkm}. \tag{5.34}$$

Actually, $\Delta_{k,l,m} = f(|l - m|)$ for some function f related to the nature of the two-body interaction, so another way of justifying the use of eqns (5.33) and (5.34) is to write the two-body interaction in eqn (5.32) as

$$\sum_{k,m} f(|k - m|) q_k^+ q_m q_{k+\eta-m} - \sum_{k,m} f(|\eta - m|) q_k^+ q_m q_{k+\eta-m}.$$

In the second term of the above operator, we change the summations to those over k' and m', where $k = k'$ and $m = k' + \eta - m'$, and the second operator above becomes

$$- \sum_{k',m'} f(|m' - k'|) q_{k'}^+ q_{k'+\eta-m'} q_{m'}.$$

The anticommuting properties of the two annihilators then demonstrate that this second term is the same as the first. Similar arguments may be advanced for the Boson case.

Hence, both eqns (5.31) and (5.32) may be written as

$$i\hbar \partial_t q_\eta = \omega_\eta q_\eta + 2 \sum_{k,m} \Delta_{\eta k m} q_k^+ q_m q_{k+\eta-m}. \tag{5.35}$$

We see from the above that this equation *is of the same form whether the associated particles satisfy Bose–Einstein or Fermi–Dirac statistics. Thus, feature (i) mentioned in Section 5.2 is satisfied automatically.*

The next part of the study proceeds in a standard way [18, 38, 55, 56], and a quantum field operator is defined by

$$\psi(r,t) = \Omega^{-1/2} \sum_k \exp(-i k \cdot r) q_k(t), \tag{5.36}$$

where Ω is the volume over which the plane waves are normalized and we explicitly include the time dependence of $q_k(t)$. Although we shall use eqn (5.36) because it is easy to work with, we could have defined $\psi(r,t)$ in a more general base as [18]

$$\psi(r,t) = \sum_m \phi_m(r) q_m(t), \tag{5.37}$$

where the base is now formed with the complete set of functions ϕ_m and the corresponding creators q_m^+ create base states in occupation space by

$$q_m^+ |0\rangle = |\phi_m\rangle. \tag{5.38}$$

We shall return to this more general form (i.e. that in eqn (5.38)) later, when we consider equations of motion in any base. Going back to eqn (5.36), it is important to note that the field operator acts in occupation space on the occupation number wave function and that the plane-wave part is *not* an operator in this space but merely acts as a premultiplicative phase factor [18] which weights the different annihilation operators. In other words, in eqn (5.36) the components of r are not considered to be operators but parameters, the eigenvalues of the position operator's components, $\hat{x}$, $\hat{y}$, and $\hat{z}$. Thus, the plane waves and annihilators act as c-numbers relative to each other, i.e. they mutually commute. One should also recognize the fact that $\psi(r,t)$ in eqn (5.36) is *not* a wave function but an operator and, from the commutators in eqns (5.25) and (5.26), the fields (i.e. taken with different space arguments) satisfy the standard commutation (anticommutation) relations

$$[\psi(r,t), \psi^+(r',t)]_\pm = \delta(r - r') \tag{5.39}$$

and

$$[\psi(r,t), \psi(r',t)]_\pm = [\psi^+(r,t), \psi^+(r',t)]_\pm = 0, \tag{5.40}$$

where '+' refers to Fermions and '−' to Bosons, and the right-hand side of eqn (5.39) is a Dirac delta function.

Both sides of eqn (5.35) are now multiplied by $\exp(-i\eta \cdot r)$, divided by

$\Omega^{1/2}$, and summed over η. Using the definition of the field in eqn (5.36), we find that the result may be expressed as

$$i\hbar \partial_t \psi(r,t) = \frac{1}{\sqrt{\Omega}} \Bigg\{ \sum_\eta \omega_\eta \exp(-i\eta \cdot r) q_\eta$$
$$+ 2 \sum_{k,m,\eta} \Delta_{\eta,k,m} \exp(+ik \cdot r) q_k^+ \exp(-im \cdot r) q_m$$
$$\times \exp(-i(k+\eta-m) \cdot r) q_{k+\eta-m} \Bigg\}, \tag{5.41}$$

where the second term on the right follows from the fact that exponentials and annihilators (or creators) mutually commute. This may be written in a simpler, perhaps more identifiable form (relative to the definition of $\psi(r,t)$ by changing the variables that we sum over to k', m', and η', where

$$k \to k', \quad m \to m', \quad \eta + k - m \to \eta', \tag{5.42}$$

and then changing back $k' \to k$, $m' \to m$, and $\eta' \to \eta$, to give

$$i\hbar \partial_t \psi(r,t) = \frac{1}{\sqrt{\Omega}} \Bigg\{ \sum_\eta \omega_\eta \exp(-i\eta \cdot r) q_\eta$$
$$+ 2 \sum_{k,m,\eta} \Delta_{\eta+m-k,k,m} \exp(+ik \cdot r) q_k^+$$
$$\times \exp(-im \cdot r) q_m \exp(-i\eta \cdot r) q_\eta \Bigg\}. \tag{5.43}$$

Our objective is to recast eqn (5.43) using only *field operators and their derivatives* with respect to positional eigenvalues. There is a problem, however, in that ω_η depends on η and the interaction coefficient $\Delta_{\eta+m-k,k,m}$ is a function of all the components of η, m, and k, in general. The simplest way to proceed is to expand each of these parameters i.e. ω_η and Δ—the values of which depend on the particular physical system under consideration—as a Taylor series about some 'point' in the space spanned by η, k, and m. Of course, one can only do this if the associated series are valid everywhere in the space of η, k, and m. This is in fact so, because the wave-vector dependence of *the matrix elements* arises from *the basis* we used to define the field and *not* the interaction operator itself. In a plane-wave basis these are exponentials with an infinite radius of convergence and hence the result. Our analysis will continue, for the moment, as if this point is arbitrary, and we denote it by (η_0, k_0, m_0). For convenience, we also define

$$f(\eta, k, m) = 2\Delta_{\eta+m-k,k,m}. \tag{5.44}$$

To all orders, for ω_η and Δ, the Taylor expansions, are straightforward but quite lengthy, but we give them below for completeness:

$$\omega_\eta = \omega_{\eta_0} + \sum_{s=1}^{\infty} \frac{\left[(\eta - \eta_0) \cdot \nabla_\eta\right]^s \omega_0}{s!} \qquad (5.45)$$

and

$$f(\eta, k, m) = f(\eta_0, k_0, m_0) + (\eta - \eta_0) \cdot (\nabla_\eta f)_0$$
$$+ (k - k_0) \cdot (\nabla_k f)_0 + (m - m_0) \cdot (\nabla_m f)_0$$
$$+ \sum_{s=2}^{\infty} \sum_{r=0}^{s} \sum_{t=0}^{s-r} \frac{{}^sC_r {}^{s-r}C_t}{s!} \left[(\eta - \eta_0) \cdot \nabla_\eta\right]^r$$
$$\times \left[(k - k_0) \cdot \nabla_k\right]^t \left[((m - m_0) \cdot \nabla_m)^{s-r-t} f\right]_0. \qquad (5.46)$$

Some explanation of the notation in the above formulas is now required. Each of the gradients, written as ∇_k with a different subscript vector, is defined by

$$\nabla_k \equiv i \partial_{k_x} + j \partial_{k_y} + k \partial_{k_z}, \qquad (5.47)$$

where i, j, and k are unit vectors in the k_x, k_y, and k_z directions, respectively. In eqn (5.45), ∇_η operates *only* on ω_η to the right and *not* on $(\eta - \eta_0)$ in a combination such as

$$\left((\eta - \eta_0) \cdot \nabla_\eta\right)\left((\eta - \eta_0) \cdot \nabla_\eta\right) \omega_\eta$$

and the subscript zero on ω means that these operations are performed and the result evaluated at η_0. Similarly, in eqn (5.46) the partial gradient operators ∇_η, ∇_k, and ∇_m operate only on f and the result evaluated at (η_0, k_0, m_0). In eqn (5.46), the ${}^sC_r$ are binomial coefficients, given by

$${}^sC_r = s!/(r!(s-r)!). \qquad (5.48)$$

The point of making these expansions is that, for example, in the one-body part of eqn (5.43), involving ω_η, powers of η_x, η_y, and η_z may then be replaced by derivatives of the field $\psi(r, t)$. Thus,

$$\frac{1}{\sqrt{\Omega}} \sum_\eta \eta_x \exp(-i\eta \cdot r) q_\eta \equiv i \partial_x \psi(r, t), \qquad (5.49)$$

or

$$\frac{1}{\sqrt{\Omega}} \sum_\eta \eta_x \eta_y \exp(-i\eta \cdot r) q_\eta \equiv -\partial_{xy} \psi(r, t). \qquad (5.50)$$

Consequently, the equation of motion in eqn (5.43) may be converted entirely into one for the field $\psi(r,t)$ and its derivatives. At this stage the Taylor expansion could be considered to be infinite in the sense that we could include all orders but, as we shall see, this is unnecessary. The expansion above may be viewed as well as one about the classical–quantum divide using the Correspondence Principle. It is also worth pointing out that the result of this procedure is an equation of motion for a fully quantum operator, and hence the *ordering* of terms such as those in eqns (5.49) and (5.50), when inserted back into eqn (5.43), is important. To illustrate the above procedure, consider the result of using the zeroth, $s = 1$ and $s = 2$ terms from ω_η and suppose initially that the interaction terms in Δ are zero. This corresponds to the case of noninteracting particles. The reader will be able to verify (with a little algebra!) that the result will be

$$i\hbar \partial_t \psi = \lambda_0 \psi + i\boldsymbol{\lambda}_1 \cdot \nabla\psi - \tfrac{1}{2} \sum_{i,j} (\lambda_2)_{ij} \partial_{x_i x_j} \psi, \qquad (5.51)$$

where

$$\lambda_0 = \omega_{\eta_0} - \boldsymbol{\eta}_0 \cdot (\nabla_\eta \omega)_0 + \tfrac{1}{2} \sum_{i,j} \eta_{0i}\eta_{0j}\left(\partial^2_{\eta_i \eta_j} \omega\right)_0, \qquad (5.52a)$$

$$(\boldsymbol{\lambda}_1)_i = -\sum_j \eta_{0j}\left(\partial^2_{\eta_i \eta_j} \omega\right)_0 + (\partial_{\eta_i} \omega)_0, \qquad (5.52b)$$

and

$$(\lambda_2)_{ij} = \left(\partial^2_{\eta_i \eta_j} \omega\right)_0. \qquad (5.52c)$$

The notation $(x_1, x_2, x_3) \equiv (x, y, z)$ has been used and the summations run over $i, j \equiv 1, 2, 3$. A judicious transformation on the matrix λ_2 may now be performed to make it diagonal [64], with the result that eqn (5.51) may readily be rewritten as

$$i\hbar \partial_t \psi = \lambda_0 \psi + i\boldsymbol{\gamma}_1 \cdot \nabla_\varepsilon \psi - \tfrac{1}{2} \nabla_\varepsilon^2 \psi, \qquad (5.53)$$

where a scaling of the independent variables can be chosen to make the coefficient of the third term on the right in eqn (5.53), $-\tfrac{1}{2}$ and to modify the vector $\boldsymbol{\lambda}_1$ to $\boldsymbol{\gamma}_1$. The ε subscripts on ∇ and ∇^2 are to denote signatures, and are defined by

$$\nabla_\varepsilon = \boldsymbol{i}\varepsilon_1 \partial_x + \boldsymbol{j}\varepsilon_2 \partial_y + \boldsymbol{k}\varepsilon_3 \partial_z \qquad (5.54)$$

and

$$\nabla_\varepsilon^2 = \varepsilon_1 \partial^2_{xx} + \varepsilon_2 \partial^2_{yy} + \varepsilon_3 \partial^2_{zz}, \qquad (5.55)$$

where ε_i ($i = 1, 2, 3$) $= \mp 1$. The operator ∇_ε^2 is called a Laplace–Beltrami operator. The form of eqn (5.53) is a time-dependent (linear) Schrödinger-like equation for the field ψ, with a dispersive term in $\nabla_\varepsilon \psi$. Note that ψ

here is $\psi(r,t)$ and is *not* a wave function for an individual particle but an operator for the system as a whole (see property (d) in section 5.2).

We only have to go to zeroth order in the Taylor expansion of f to obtain a nonlinear term in the equation of motion. In fact, it is fairly obvious that all other terms from the interaction will be nonlinear (see property (b) in Section 5.2). In zeroth order (by 'order' we mean the power of wave vectors appearing in the Taylor expansion of the interaction term, the one-body part being truncated at k^2) we find that

$$i\hbar \partial_t \psi = \lambda_0 \psi + i\boldsymbol{\gamma}_1 \cdot (\nabla_\varepsilon \psi) - \tfrac{1}{2}\nabla_\varepsilon^2 \psi + \Omega f(\boldsymbol{\eta}_0, k_0, m_0)\psi^+ \psi\psi. \quad (5.56)$$

For an analysis of this type of PDE, the reader is referred to references [57–61] and an extensive discussion in Chapter 7.

5.4 Conversion to equations of motion for c-number fields

The quantum field equations that we derived in the previous section would be extremely difficult to solve as they stand, since the field and its derivatives with respect to the eigenvalues of the components of the position vector are operators, so we need a method of converting them to c-number fields. Our results so far have been exact but for an arbitrary point ($\boldsymbol{\eta}_0, k_0, m_0$). We now confront the question of choosing this point in the realization that there may be more than one such point at our disposal, as will become clear shortly. In reciprocal space this happens to be a *critical point* of the system, or a *fixed point*. To see how such a point is found, consider initially what might appear to be a completely different system, namely a system of spins located on a regular square array of sites and coupled together by exchange interactions. In its Hamiltonian there will be a series of coupling constants, e.g. J_1 for nearest neighbours, J_2 for next-nearest neighbours, and so on (see Chapter 2). We now follow the ideas of the pioneering work of Kadanoff [62] and perform a series of transformations. A very accessible account of this procedure has been given by Reichl [63] in her book. Basically, what one does is set up a vector, K say, the components of which are the parameters in the Hamiltonian for the spins. Transformations are then defined by constructing larger and larger blocks of spins so that, at each stage, the Hamiltonian becomes modified *and is of the same form* between blocks of spins as between the original spins. After each transformation the parameters in the corresponding Hamiltonian are changed—or the components of K are different. This process is repeated until components of the vector K at the nth transformation R_n are the same as those at the $(n-1)$th. These latter components are then said to denote a vector K^* which represents a 'fixed' point of the system. This can be written as

$$\lim_{n \to \infty} R_n K = K^*. \tag{5.57}$$

The subspace in parameter space the points of which have the above property is called the *critical surface* [M, 136] of the fixed point K^*, where references here and hereafter, denoted by M, are references to Ma's book [48], the numbers following being the page numbers. Thus, all points on the critical surface are eventually driven to K^*. The critical surface is expected to cover a large proportion of parameter space and the behaviour of different materials at their critical points can be represented by appropriate points on the same critical surface. This property is often referred to as universality of critical behaviour. Critical exponents are the properties of the R_n in the neighbourhood of the 'fixed' point and will thus have the *same* values [M, 161]. What is of great significance for us is that the *critical region* control parameter space is *defined* as the range over which $O(\xi^{+y_2})$ is negligible, where ξ is the correlation length and y_2 is a negative parameter near the fixed point [M, 139, 157]. That is, it is the region in which the correlation length is long or corresponding reciprocal vector components are very small. Hence, we identify (η_0, k_0, m_0) as the critical point or fixed point of the system, since we expanded in small deviations from this point in reciprocal space, and the fields in our field equation are those for the situation close to the critical point.

As we said earlier, a classical field $\phi(r,t)$ may be found for which a Hamiltonian exists so that the partition function of the classical Hamiltonian is the same as the corresponding quantum mechanical Hamiltonian [M, 531] and thus physically equivalent. Therefore, the classical field $\phi_c(r,t)$ should be an excellent approximation to the quantum field $\Psi(r,t)$, since one can obtain from $\phi_c(r,t)$ all of the pertinent thermodynamic properties. Thus, initially at least, in our quantum field equations (5.53) or (5.56) we can treat the field as classical (i.e. the quantum field is made up of a relatively large classical component compared to the quantum part), being a function both of space and time. *The importance of an associated classical picture (see item (a) in Section 5.2) now becomes very apparent for a wide range of many-body systems.* The inherent nonlinearity (item (b) of Section 5.2) is always present due to the crucial role played by the interaction terms.

We now intend to incorporate quantum field theory in our description, having already taken advantage of ideas from the critical phenomena area and second-quantization techniques. The reader's attention is drawn to the very readable review article by Jackiw [13] in this respect. As Amit [49] (hereafter, given as A, plus the page number) has pointed out, in quantum field theory [A, 105] full use is made of an expansion in powers of $\hbar$ around the solutions of a classical theory. A linear expansion about the classical field $\phi_c(r, t)$ is now appropriate and extremely accurate. That is, we write

$$\psi = \phi_c(r,t)\hat{I} + \hat{\phi}(r,t). \tag{5.58}$$

In eqn (5.58) above, $\hat{\phi}$ is a quantum-mechanical operator, the amplitude of which is of order $\hbar$ relative to the magnitude of the classical field ϕ_c, and $\hat{I}$ is the unit operator in Fock space. In fact, Jackiw has argued that it is most definitely possible to gain information about the physical content of nontrivial quantum field theories by semi-classical methods. Furthermore, there are phenomena which arise that cannot be viewed via a perturbative expansion and have brought to light a very rich and unexpected structure in the quantal Hilbert space, e.g. the emergence of Fermions from Bosons and spontaneous symmetry breaking without Goldstone Bosons. In a first approximation quantal effects are ignored [13] and, as we have seen above, from the way in which our field equation has been constructed via the choice of the point $(\boldsymbol{\eta}_0, \boldsymbol{k}_0, \boldsymbol{m}_0)$, this should be an excellent starting point, corrections being of order $\hbar$, relative to the 'zero order' classical field ϕ_c. Quantum effects are subsequently regained by linearization in $\hbar$. This type of procedure is standard in quantum field theory. Our field equations, therefore, become equations of motion for c-number fields which can then be analysed by the methods of mathematical physics, so we can now use all the recent developments in the area of nonlinear differential equations [51–54]. Thus, for example, in eqn (5.56), ψ is merely replaced by its classical counterpart $\phi_c(r, t)$, treated as a c-number (see Appendix E for a justification of this procedure in terms of coherent states).

For completeness, we give the second order equations of motion for ψ. This was obtained in exactly the same way as the derivation of eqn (5.56), but a number of other unitary transformations are required to get it into an elegant form [64]. We find that

$$i\hbar \partial_t \psi = \mu_0 \psi + i\boldsymbol{\mu}_1 \cdot (\nabla_\varepsilon \psi) + \mu_2 \nabla_\varepsilon^2 \psi + \sum_{i,j} R_{ij} \partial_{x_i x_j}^2 \psi$$
$$- 2(\nabla_\varepsilon \psi^+)\psi(\nabla_\varepsilon \psi) + \mu_3 \psi^+ \psi\psi$$
$$\times i[\psi^+ \psi(\boldsymbol{\mu}_4 \cdot \nabla_\varepsilon)\psi + \psi^+((\boldsymbol{\mu}_4 \cdot \nabla_\varepsilon)\psi)\psi]$$
$$+ ((\nabla_\varepsilon^2 \psi^+)\psi\psi + \psi^+ \psi \nabla_\varepsilon^2 \psi), \tag{5.59}$$

where μ_0, μ_2, and μ_3 are constant parameters, determined in terms of matrix elements in the original second-quantized Hamiltonian, and $\boldsymbol{\mu}_1$ and

CONVERSION TO EQUATIONS OF MOTION

μ_4 are constant vectors derived from the particular system in a similar way. The matrix of constants, R_{ij}, is a small residual term that vanishes in cubic or spherical symmetry. The interested reader should look up the literature [64] for more details of the appropriate transformations. The reason why we need not go to higher order in the interaction is provided by renormalization group theory and will become clear in the next section.

It should be stressed at this point that a given physical system may have a number of critical points, each of which corresponds to a phase transition associated with particular long-range behaviour. The position of critical points, in reciprocal space, will, in general, be different, i.e. η_0, k_0, m_0 will have different values, and hence the parameters in the equation of motion (see eqn (5.59)) will differ, *but* the *form* of this equation will be the same whichever critical point is chosen to expand about. The appropriate parameter values to use in given circumstances will be dictated by the physical phenomenon being studied. Alternatively, and equivalently, parameters may be found from the boundary conditions on the field determined from the equation of motion.

It is also worth pointing out that the Hamiltonian itself, i.e. eqn (5.24), may be written in terms of the field and its derivatives by using exactly the same procedure as in Section 5.3. Let us first consider the one-body part of the Hamiltonian density. From this, four different types of energy density term occur:

1. When we consider the zero order expansion of the associated matrix element (its value at the critical point), it is obvious that a component of the form $\psi^+\psi$ will arise.

2. Terms from the Taylor expansion of the matrix element which are linear (or of the order one) in the reciprocal vector components, for example, k_x, will generate contributions of the type

$$(\partial_x\psi^+)\psi \quad \text{and} \quad \psi^+(\partial_x\psi).$$

   For example, see the term in λ_1 in eqn (5.52b). This, in fact, may be eliminated by using a Galilean transformation [64].

3. In second order, in say k_x^2, we expect to find terms such as

$$(\partial_{xx}^2\psi^+)\psi, \quad (\partial_x\psi^+)(\partial_x\psi), \quad \psi^+(\partial_{xx}^2\psi),$$

   and similar other components by changing from k_x^2 to k_y^2 or k_z^2.

4. In second order there will also be cross-terms, such as

$$\left(\partial_{xy}^2\psi^+\right)\psi, \quad (\partial_y\psi^+)(\partial_x\psi), \quad (\partial_x\psi^+)(\partial_y\psi), \quad \psi^+\left(\partial_{xy}^2\psi\right),$$

   and whether such terms do arise in a specific instance will depend on the symmetries of the Hamiltonian.

It is obvious from (3) and (1) above that terms such as

$$H_{OB} = a\psi^+ \psi + b \nabla\psi^+ \cdot \nabla\psi, \qquad (5.60)$$

can arise in the Hamiltonian density, where the coefficients a and b are constants (OB denoting one-body).

Selecting k_x and k_y as examples, we can easily see that from the two-body or interaction term the following can appear in a similar way:

(a) $\qquad \psi^+ \psi^+ \psi\psi;$

(b) $(\partial_x\psi^+)\psi^+\psi\psi, \quad \psi^+(\partial_x\psi^+)\psi\psi, \quad \psi^+\psi^+(\partial_x\psi)\psi, \quad \psi^+\psi^+\psi(\partial_x\psi).$

From the second order terms in k_x and k_y, very complicated components arise as follows:

(c) $(\partial_{xx}^2\psi^+)\psi^+\psi\psi, \quad (\partial_x\psi^+)(\partial_x\psi^+)\psi\psi, \quad (\partial_x\psi^+)\psi^+(\partial_x\psi)\psi,$

$(\partial_x\psi^+)\psi^+\psi(\partial_x\psi), \quad \psi^+(\partial_{xx}^2\psi^+)\psi\psi,$

$\psi^+(\partial_x\psi^+)(\partial_x\psi)\psi, \quad \psi^+(\partial_x\psi^+)\psi(\partial_x\psi), \quad \psi^+\psi^+\psi(\partial_{xx}^2\psi),$

$\psi^+\psi^+(\partial_{xx}^2\psi)\psi, \quad \psi^+\psi^+(\partial_x\psi)(\partial_x\psi);$

and

(d) $\left(\partial_{xy}^2\psi^+\right)\psi^+\psi\psi, \quad (\partial_x\psi^+)\psi^+(\partial_y\psi)\psi,$

$(\partial_y\psi^+)(\partial_x\psi^+)\psi\psi, \quad \psi^+\left(\partial_{xy}^2\psi^+\right)\psi\psi, \quad \psi^+(\partial_x\psi^+)(\partial_y\psi)\psi,$

$(\partial_y\psi^+)\psi^+(\partial_x\psi)\psi, \quad \psi^+(\partial_y\psi^+)(\partial_x\psi)\psi, \quad \psi^+\psi^+\left(\partial_{xy}^2\psi\right)\psi,$

$\psi^+\psi^+(\partial_x\psi)(\partial_y\psi), \quad \psi^+(\partial_y\psi^+)\psi(\partial_x\psi), \quad \psi^+\psi^+\psi\left(\partial_{xy}^2\psi\right),$

$(\partial_y\psi^+)\psi^+\psi(\partial_x\psi).$

We observe immediately that one particularly important component of the two-body Hamiltonian density, from (a), is

$$H_{TB} = c\psi^+ \psi^+ \psi\psi. \qquad (5.61)$$

Combining eqns (5.60) and (5.61), we find that

$$H_{LG} = H_{OB} + H_{TB} = a\psi^+\psi + b(\nabla\psi^+)\cdot(\nabla\psi) + c\psi^+\psi^+\psi\psi. \qquad (5.62)$$

That is, a *Landau–Ginzburg* form of Hamiltonian density, H_{LG}, has appeared as a subcase above and has played such a very prominent role in the development of field-theoretical approaches to critical phenomena [49, 50]. This has emerged in our study in an *exact* way, although ψ which enters in eqn (5.62) is a field *operator* and *not* an order parameter. Admittedly, when replaced by the spatial part of the classical field, $\phi_c(r,t)$

5.5 Truncation of terms in the equation of motion using renormalization theory

We proceed in this section for the case in which the critical point is associated with a second-order phase transition. If we look again at the expressions in (a), (b), (c), and (d) of the last section for both one-body and interaction components in the Hamiltonian density, we see that, from the one-body terms, we always obtain a pair of operators one, or both, of which could be derivatives of the field. In fact, for higher powers of k the derivatives increase in order. Similarly, because of the second quantized two-body nature of the interaction, we always obtain a product of four operators. One, two, three, or indeed four of the terms of this product could involve derivatives of the field. We also observe that terms in each group, from one- and two-body components of the Hamiltonian density, are invariant with respect to the transformation $\psi \to -\psi$. Below we show that, in principle, all the terms including derivatives, may be written as a power series in ϕ_c and hence renormalization arguments can be used to truncate the expansion in ϕ_c and its derivatives. In fact, what we try to do is to represent the Hamiltonian density as $|\nabla \phi_c|^2$ plus a polynomial in ϕ_c. To do this, the main aim is to be able to write $\nabla \phi_c$ as a function of ϕ_c, with very general assumptions made about ϕ_c.

We now revert to the classical field and suppose initially that it depends on only one spatial variable but not on time:

$$\phi_c = F(x), \tag{5.63}$$

The gradient of ϕ_c, namely $\nabla \phi_c$, has the magnitude $d\phi_c/dx$ which is some other function of x, let us say $G(x)$:

$$\nabla \phi_c = i\, dF/dx = iG(x), \tag{5.64}$$

where i is a Cartesian unit vector along the x-axis. If ϕ_c is continuous and monotonic in some closed interval of x, say $a \leq x \leq b$, and $F(a) = \alpha$ and $F(b) = \beta$, then it possesses a single-valued inverse function $x = F^{-1}(\phi_c)$ for $\alpha \leq \phi_c \leq \beta$ which is also continuous and monotonic [24]. Thus, with these very general requirements on ϕ_c and using eqns (5.63) and (5.64), $\nabla \phi_c$ may be written as a function of ϕ_c alone, i.e.

$$\nabla \phi_c = iG(F^{-1}(\phi_c)), \tag{5.65}$$

and under quite general circumstances this may be expressed as a Taylor series in ϕ_c. Hence, the classical Hamiltonian density, H_{CD}, may be written as

$$H_{CD} = \lambda_1 |\nabla \phi_c|^2 + \left(\text{a polynomial in } |\phi_c|^2\right). \tag{5.66}$$

This is all very well known, but suppose now that ϕ_c is a function of *three* independent spatial variables, so that

$$\phi_c(\mathbf{r}) = \tilde{F}(x, y, z). \tag{5.67}$$

We again ask ourselves under what circumstances $\nabla \phi_c(r)$ will be a constant vector k (not necessarily a unit vector) multiplied by a function of $\phi_c(r)$:

$$\nabla \phi_c(\mathbf{r}) = k\tilde{G}(\phi_c). \tag{5.68}$$

This is now clearly equivalent to

$$k_1 \partial_x \phi_c + k_2 \partial_y \phi_c + k_3 \partial_z \phi_c = \tilde{G}(\phi_c), \tag{5.69}$$

for nonzero constants k_1, k_2, and k_3. Such an equation may be solved using the method of Lagrange [65] by considering the subsidiary ordinary equations

$$dx/k_1 = dy/k_2 = dz/k_3 = d\phi_c/\tilde{G}(\phi_c). \tag{5.70}$$

One form of solution is

$$\chi\left(\frac{x}{k_1} - \int \frac{1}{\tilde{G}(\phi_c)} d\phi_c, \frac{x}{k_1} - \frac{y}{k_2}, \frac{x}{k_1} - \frac{z}{k_3}\right) = 0, \tag{5.71}$$

where χ *is an arbitrary function* of the three variables u, v, and w as defined in the above equation. The reader may easily verify that this is a solution of eqn (5.69) by differentiating eqn (5.71) successively with respect to x, y, and z. The very arbitrariness in χ of eqn (5.71) is an indication of the broad range of functions $\phi_c(r)$ for which eqn (5.68) may be solved. Thus, a more general form for eqn (5.68) may be postulated, as

$$\nabla \phi(r) = F(\phi) = i f_1(\phi) + j f_2(\phi) + k f_3(\phi), \tag{5.72}$$

where i, j, and k are the standard Cartesian unit vectors. Henceforth, we drop the label c on ϕ_c for convenience. Equation (5.72) implies that

$$\partial_x \phi = f_1(\phi), \qquad \partial_y \phi = f_2(\phi), \qquad \partial_z \phi = f_3(\phi). \tag{5.73}$$

If ϕ has 'well-behaved' differentiability properties we must have

$$\partial^2_{yx} \phi = \partial^2_{xy} \phi,$$

so that: $f_1' f_2 - f_2' f_1 = 0$, the primes indicating differentiation with respect to ϕ, or

$$f_2^2 \frac{d}{d\phi}\left(\frac{f_1}{f_2}\right) = 0. \tag{5.74}$$

If we assume that all of f_1, f_2, and f_3 do not vanish, then clearly eqn (5.74) implies that f_1 is proportional to f_2. In a similar way, it is easy to show that f_2 is also proportional to f_3. Thus, what appears to be a more general case, i.e. in eqn (5.72), degenerates, with very general assumptions, into the case that we studied earlier in eqn (5.68).

The important point to emerge here is that for a large class of functions $\nabla \phi_c(r)$ may be written as a vector function of ϕ_c and hence may be represented in the form of a Taylor series about an appropriate point in space. It therefore follows from eqns (5.68) and (5.72) that $\nabla^2 \phi_c$ may *also* be put into this same form, so that even when there is more than one independent variable eqn (5.71) still holds. To preserve the even parity (under the reversal $\phi_c \to -\phi_c$ of the Hamiltonian density), $\tilde{G}(\phi_c)$ must be odd in ϕ_c. This clearly makes all terms in (a), (b), (c), and (d), for both one- and two-body contributions, *even in ϕ_c*.

Having demonstrated that the idea of a Landau–Ginzburg type expansion for the classical Hamiltonian density follows from very general considerations, we now turn to the results from renormalization group analysis [A, 12] which concludes that, in the generic case, we only need to consider $|\nabla \phi_c|^2$, $|\phi_c|^2$, and $|\phi_c|^4$ in H_{CD}, since higher-order monomials may be considered to be *irrelevant* in the description of the associated critical behaviour. The main tool in this approach is an expansion in powers of coupling constants about some quadratic form, which is exactly soluble, and here we take $|\nabla \phi|^2 + |\phi|^2$. It is here that Feynman diagrams enable one to keep track of the various terms generated by perturbation theory. Traditionally, an infinite set of diagrams are chosen in some way which can be resummed in some regime of the parameter space. The result is then analytically continued to other regions. All of the terms in the expansion of a given physical property can be mapped directly on to an identifiable set of graphs.

What we attempt to find is the number of space dimensions (or in our case space–time dimensions) in which a momentum cut-off can be absorbed into a *finite* number of parameters. A well-defined limiting procedure, called *regularization*, enables one to obtain finite expressions corresponding to infinite Feynman integrals. After a theory is regularized, it is said to be *renormalized*. An easy method of looking at whether or not a particular theory is renormalizable is to consider the dimension of the various terms appearing in the Lagrangian density. For example, if the fundamental physical constants h, c, and k_B are set equal to unity [A, 148], the field ϕ then has the dimensions $[\phi] = L^{1-(d/2)}$ in real space or $\Lambda^{(d/2)-1}$ in momentum space, where L is a particular length and d is the number of space–time dimensions. As the Lagrangian density $\mathscr{L}$ must have the dimensions of inverse volume in this scheme, it is clear that $[\mathscr{L}] = L^{-d}$ or Λ^d. Thus, if λ_r is the coupling coefficient for a particular monomial ϕ^r, since the term $\lambda_r \phi^r$ is in the Lagrangian $\mathscr{L}$, it follows that the dimensions of λ_r are given by $[\lambda_r] = \Lambda^{r+d-(1/2)rd}$. Hence, for each coupling coefficient, there corresponds a space–time dimension, d, which makes it dimensionless. In fact, for a particular monomial, a dimensionless coupling constant is the necessary and sufficient condition for what are called 'primitive divergences' of all vertex parts to be independent of the order in the perturbation expansion. The reason why this is important is that if the primitive divergences of the graphs of a vertex function increase with the order of perturbation theory, there is no way in which a strong Λ-dependence can be absorbed in a finite number of constants. It can be deduced that the critical dimension $d = d_c$ is obtained as $d_c = 2r/(r-2)$. When $d > d_c$, for a theory with a weak dependence on cut-off, it turns out that an *infinite* number of arbitrary constants must be introduced. In our case we have four space–time independent variables, so $d_c = 4$ and therefore $r = 4$. That is, when all the gradients and Laplacians have been reduced to monomials we can truncate at $r = 4$ or ϕ^4 and the whole theory is renormalizable. Thus, we have shown that the Hamiltonian density $H_{\rm CD}$ may be renormalized so that it only contains $(\nabla \phi_c)^2$, ϕ_c^2, and ϕ_c^4 terms, and all higher order monomials of the type ϕ_c^r $(r > 4)$ merely 'redress' those retained.

Let us now see what consequence this entails in our equation of motion. In what follows, we assume that ϕ is real. If the space-dependent part of the classical field is complex, it is easy to see that arguments analogous to those below can be used. The term linear in $\nabla \phi$ in the equation of motion (see eqn (5.59)) may be removed by a Galilean transformation, so we omit it here. The equation of motion takes the form

$$0 = \lambda_0 \phi \| + \nabla^2 \phi \| + \mu_0 \phi^3 \| + \mu_1 \phi^2 \nabla \phi \|$$
$$+ \mu_{2a} \phi^2 \nabla^2 \phi + \mu_{2b} \phi (\nabla \phi)^2 \| \ldots, \qquad (5.75)$$

to second order, where λ_0, μ_0, μ_1, μ_{2a}, and μ_{2b} are constants. That is, the first term on the right of eqn (5.75) is a zero-order term in an expansion about the critical point, the linear term is transformed away, and the next term is of second order. Both of these first two terms arise from the one-body interaction. Above, we have inserted vertical parallel lines to distinguish order and the subscripted numbers on parameters indicate the particular order from the interaction. The remaining terms with prefactors μ_i ($i = 0, 1, 2$) are from the two-body interaction.

We now go back to the very general assumption that $\nabla \phi$ is some odd function of ϕ itself multiplied by a constant vector k_1, i.e.

$$\nabla \phi = f(\phi) k_1 = \{a_1 \phi + a_3 \phi^3 + a_5 \phi^5 + a_7 \phi^7 \ldots\} k_1. \quad (5.76)$$

This, of course, must be consistent with the equation of motion itself. From eqn (5.76) it is easily verified that

$$\nabla^2 \phi = ff' k_1^2, \qquad \nabla(\nabla^2 \phi) = \left[(f')^2 f + f^2 f''\right] k_1^2 k_1, \quad (5.77, 5.78)$$

$$\nabla^2(\nabla^2 \phi) = \left[4 f^2 f' f'' + f(f')^3 + f^3(f''')\right] k_1^4, \quad (5.79)$$

etc. for higher gradients, where a single prime denotes differentiation with respect to ϕ, two primes a second differentiation with respect to ϕ, and so on. Since eqn (5.76) converges, the coefficients a_i, as the index i increases, become smaller and smaller, or $a_i \to 0$ as $i \to \infty$. The predominant or leading terms in eqns (5.76)–(5.79), respectively, are

$$\nabla \phi \simeq a_1 \phi k_1, \qquad \nabla^2 \phi \simeq a_1^2 \phi k_1^2, \quad (5.80, 5.81)$$

$$\nabla(\nabla^2 \phi) \simeq \left[a_1^3 \phi + 6 a_1^2 a_3 \phi^3\right] k_1^2 k_1, \quad (5.82)$$

and

$$\nabla^2(\nabla^2 \phi) \simeq \left[30 a_1^3 a_3 \phi^3 + a_1^4 \phi\right] k_1^4. \quad (5.83)$$

For convenience of exposition, it is easy to see that the independent spatial variables may be scaled in such a way that, without loss of generality, we can take $a_1 k_1 \simeq 1$. Using eqns (5.80) and (5.81), we can write the leading terms of the second order components of eqn (5.75), from the two-body interaction, as approximately $\mu_{2a} \phi^3 + \mu_{2b} \phi^3$. This contribution must be clearly retained in the equation of motion since it corresponds to a ϕ^4 term in the Hamiltonian density via the appropriate Euler–Lagrange equation, i.e. it represents a relevant variable. Now let us consider the types of term which would arise in third and fourth order, i.e.

third, $\qquad \mu_{3a} \cdot \phi \nabla \phi \nabla^2 \phi + \mu_{3b} \cdot \phi^2 \nabla(\nabla^2 \phi) + \mu_{3c} \cdot (\nabla \phi)^2 \nabla \phi \quad (5.84)$

and

fourth, $\qquad \mu_{4a} (\nabla \phi)^2 \nabla^2 \phi + \mu_{4b} \phi (\nabla^2 \phi)^2$

$\qquad\qquad + \mu_{4c} \phi \nabla \phi \cdot \nabla(\nabla^2 \phi) + \mu_{4d} \phi^2 \nabla^2(\nabla^2 \phi). \quad (5.85)$

In eqn (5.84), $\mathbf{\mu}_{3a}$, $\mathbf{\mu}_{3b}$, and $\mathbf{\mu}_{3c}$ are constant vectors, whereas μ_{4a}, μ_{4b},

μ_{4c}, and μ_{4d} are constant parameters in eqn (5.85). We may now use eqns (5.80)–(5.83) to find the leading terms in eqs (5.84) and (5.85). First, consider third-order terms and obtain

$$\mu_{3a} \cdot k_1 a_1 \phi^3 + \mu_{3b} \cdot k_1 a_1 \left[\phi^3 + \frac{6a_3}{a_1} \phi^5 \right]$$
$$+ \mu_{3c} \cdot k_1 a_1 \phi^3, \tag{5.86}$$

assuming $a_1 \neq 0$. Here, nearly all the terms are of the form ϕ^3 (which already appeared in second order) but an additional term now occurs, namely a contribution proportional to ϕ^5. This latter term can only arise from a component ϕ^6 in H_{CD} and, therefore, may be renormalized away into $(\nabla\phi)^2$, ϕ^2, or ϕ^4 terms. Hence, it is *not* necessary to retain third order, since these terms merely redress those of lower order because either they are of monomial type ϕ^r with $r > 3$, and the renormalization argument may be used, or they are of the same type as in lower order. By construction, they are of order k^3 in a wave vector, with magnitude k, away from the critical point. This argument can also be used for all higher orders in the expansion.

We should point out that should the associated critical point be related either to a different dimensionality d_c of the physical space or to a different order of the transition, and not as has been assumed above, then it becomes necessary to retain terms up to ϕ^6 in H_{CD}. This eventuality can and does arise when $d_c = 3$. The important point is that the Taylor series in the wave vector space away from the critical point could be truncated and the coefficients of terms below the truncation point are modified. A similar situation may also arise if the Hamiltonian density contains odd terms in ϕ, but the appropriate series may still be truncated by the above device. This type of situation may arise when the previously assumed invariance of the Hamiltonian with respect to the $\phi \to -\phi$ replacement does not hold, either because the system itself does not possess this symmetry or because an external field has been applied that breaks it.

The reader should be aware that nowhere have fluctuations in the particular physical system been mentioned. The MCS is essentially a zero temperature concept although fluctuations have been partially incorporated through the gradients which appear when matrix elements are expanded about the critical point before renormalizing. To incorporate thermal fluctuations in a more accurate way it would be necessary to incorporate entropy in the energy density, thereby introducing a term linear in temperature and a function of the field, ψ. Appropriate contributions in the equation of motion would then appear when the energy functional is minimized.

5.6 Classical solutions of the equation of motion

In this section we guide the reader through a sequence of steps in solving the field equations of motion derived in previous sections, with an increasing level of complexity from step to step. The first attempt we make is to analyse them in the absence of two-body interactions and consequently in the absence of nonlinearity. Second, we introduce nonlinearity in what we call the 'zeroth order' in the simplest nontrivial way, i.e. through interactions of constant strength. As will be shown in detail later, a massive amount of information about the system, even in a three-dimensional space, can be retrieved through nonlinear analysis. We then move on to the more complicated first and second order cases where exact solutions can also be found, but in a very restrictive number of circumstances. We believe, however, that even though we cannot solve some of these equations completely at present, the day will come when solutions will be found.

5.6.1 The noninteracting case

Writing the classical part of the field as ϕ, dropping the 'c' subscripts, we have seen that eqn (5.53) will reduce to

$$i\hbar \partial_t \phi = \lambda_0 \phi + i\boldsymbol{\gamma}_1 \cdot \nabla_\varepsilon \phi - \tfrac{1}{2}\nabla_\varepsilon^2 \phi. \tag{5.87}$$

To simplify this equation, we make the Galilean transformation

$$t' = t, \qquad x'_i = x_i - v_i t, \quad i = 1, 2, 3,$$

and choose the propagation velocity so that $-\hbar v_i = (\boldsymbol{\gamma}_1)_i \varepsilon_i$, to eliminate the gradient term. It is straightforward to show that eqn (5.87) reduces to

$$i\hbar \partial_t \phi = \lambda_0 \phi - \tfrac{1}{2}\nabla_\varepsilon^2 \phi, \tag{5.88}$$

where the prime on t has been dropped for convenience and the Laplacian is really with respect to the primed independent co-ordinates. The spatial variables can now be separated from the time by writing

$$\phi = u(t) f(\boldsymbol{r}). \tag{5.89}$$

Substituting this into eqn (5.88) we obtain

$$u = u_0 \exp(-it\lambda/\hbar), \tag{5.90}$$

where λ is an arbitrary separation constant and u_0 is an arbitrary amplitude. The spatial component $f(\boldsymbol{r})$ now satisfies

$$\nabla^2 f + \Omega f = 0, \tag{5.91}$$

where $\Omega = 2(\lambda - \lambda_0)$ and we have put each signature to unity. Equation (5.91) is a Helmholtz equation which, of course, has been studied extensively, and may be solved readily since it separates in Cartesian, spherical polar and cylindrical polar co-ordinates [65]. In *Cartesian co-ordinates* we find that

$$\phi = u_0 \prod_{i=1}^{3} \left\{ A_i \exp\left[\mu_i^{1/2}(x_i - v_i t) \right] + B_i \exp\left[-\mu_i^{1/2}(x_i - v_i t) \right] \right\} \exp(-it\lambda/\hbar), \quad (5.92)$$

where A_i and B_i are arbitrary constants and the μ_i ($i = 1, 2, 3$) are separation constants. If the μ_i's are positive, the terms in braces either exponentially increase or decrease spatially. On the other hand, if the $\mu_i < 0$, oscillatory functions in space and time result.

For *spherical polar co-ordinates* we find solutions of the form

$$f = r^{-1/2} Z_{l+1/2}(\sqrt{\Omega}\, r) Y_l^m(\theta, \Phi), \quad (5.93)$$

where $Z_{l+1/2}$ is a Bessel function [66]. If $\Omega > 0$, $Z_{l+1/2}$ decays with increasing r. However, for $\Omega < 0$ Bessel functions with imaginary arguments result (in the usual notation, I_ν or K_ν) and either diverge or decay for large r. More specifically, when r is small and $\Omega < 0$, K_ν diverges and I_ν tends to zero as r^ν. In eqn (5.93), $Y_l^m(\theta, \Phi)$ is a spherical harmonic. In fact, because of our initial Galilean transformation, the independent variables in eqn (5.93) are time dependent and

$$r(t) = \left[\sum_i (x_i - v_i t)^2 \right]^{1/2}, \quad \theta(t) = \cos^{-1}\left(\frac{x_3 - v_3 t}{r(t)} \right),$$

$$\Phi(t) = \tan^{-1}\left(\frac{x_2 - v_2 t}{x_1 - v_1 t} \right).$$

Thus, for large times and a fixed co-ordinate, r, the angles θ and Φ approach the constant values θ_0 and Φ_0. We note that, as the time dependence in $r(t)$, $\theta(t)$, and $\Phi(t)$ originates from the gradient term in the PDE that we began with, we can interpret it as a directional source which makes the solution tend asymptotically into a particular direction (θ_0, Φ_0). A similar phenomenon arises in *cylindrical polar co-ordinates*. What is important in this particular case (i.e. of noninteracting particles), however, is that the fields ϕ do *not* exhibit time-independent spatial localization or dispersion-free behaviours which characterize coherent structures. As we see in the next section, the two-body interaction term brings about *nonlinear terms in the equation of motion* and does indeed lead to a wealth of different types of coherent structure.

5.6.2 Incorporating the two-body interaction to zeroth order

In this section, as in all succeeding ones, we retain all the contributions from the one-body term up to second order, but take the two-body component only to zeroth order. This situation is epitomized by eqn (5.56). Defining $F_0 = \Omega f(\eta_0, k_0, m_0)$ and making a similar Galilean transformation to that in Subsection 5.6.1 above, we find that the equation of motion becomes

$$i\hbar\,\partial_t \phi = \lambda_0 \phi - \tfrac{1}{2}\nabla_\varepsilon^2 \phi + F_0 \phi^+ \phi\phi. \qquad (5.94)$$

The linear term in eqn (5.94) can be eliminated by writing $\phi \to \phi \exp(i\Lambda t)$ and choosing $-\hbar\Lambda = \lambda_0$. By judicious scaling of dependent and independent variables after this latter transformation, eqn (5.94) may be modified to produce the form

$$i\partial_t \phi + \nabla_\varepsilon^2 \phi + q|\phi|^2 \phi = 0, \qquad (5.95)$$

where q is a parameter to be chosen. In $(1 + 1)$-dimensional space–time this is a nonlinear Schrödinger equation (NLS), which is well known to be an integrable system [68] and which admits both single and multiple soliton-like solutions. It has been studied extensively in the past [67–71] and, if $q > 0$, N-envelope soliton solutions exist [72], with a single soliton in the form of a ball-shaped sech function. If the parameter $q < 0$, on the other hand, multi-soliton solutions have a kink shape with a single soliton envelope proportional to the tanh function. The reader is referred to Chapter 7 for an extensive overview of the properties of the NLS.

We now turn to the case in which eqn (5.95) is considered in more than one spatial dimension. One method of approach to this type of problem is to use the method of symmetry reduction, which is discussed in detail in Chapter 7. In this method solutions to associated PDEs are written in the form

$$\phi(\mathbf{r},t) = \alpha(\mathbf{r},t) f(\xi) \quad \text{with } f(\xi) = M(\xi)\exp(i\chi(\xi)). \qquad (5.96)$$

In eqn (5.96), $f(\xi)$ is referred to as the 'envelope', with $\chi(\xi)$ being the 'carrier' wave. The function $\xi = \xi(\mathbf{r},t)$ is called the *symmetry variable* and the prefactor, $\alpha(\mathbf{r},t)$, is found for each symmetry variable separately. A very interesting physical case has been studied [73] in which there are two 'symmetry variables', one for the envelope depending on $\xi_1 = x - v_1 t$ and the other for the carrier depending on $\xi_2 = x - v_2 t$. It is found that the envelope velocities for solitary wave-like solutions form two bands separated by a gap which represents forbidden velocities. Curiously, this is very reminiscent of electronic bands in periodic solids. Below, we discuss in substantial detail the form of multidimensional solutions found for the equation of motion in zeroth order. We do this in two parts, distinguished by the two signatures in the Laplacian.

Euclidean signatures ($\varepsilon_1 = \varepsilon_2 = \varepsilon_3 + 1$)

Unfortunately, the NLS equation is not integrable in more than (1 + 1) dimensions. However, there are techniques that enable one to find special types of exact solution. We consider these cases to be 'partially integrable', meaning by this that they are integrable when extra symmetry or boundary conditions are imposed. For the special case of Euclidean signatures, the NLS has been studied in great detail by a number of authors, particularly Gagnon and Winternitz [51–53]. The PDE to be solved is reduced to an ODE for the envelope in the case of a single symmetry variable ξ (i.e. in mathematical language this is described by orbits with co-dimension one). The reduced equations, which are ODEs, are investigated using various techniques such as the so-called Painlevé analysis [74] or the inverse scattering method [75–78], and in many cases explicit solutions can be found in terms of elementary functions, elliptic functions, and Painlevé transcendents. In some instances one may have to resort to numerical integration techniques if the ODE fails the Painlevé test (see Chapter 7). In either case, because of the exact nature of the equations of motion, the symmetry reduction method will reveal inherent symmetries and topologies and hence may yield important physical insight into the system under investigation.

To show the astonishing wealth of solutions that the symmetry reduction method generates, we illustrate some results in Tables 5.1 and 5.2. In particular, in Table 5.1 we have listed solutions having generic orbits of co-dimension one in the space of independent variables, indicating a single symmetry variable, the form of the carrier function and the differential equation (an ODE) that the amplitude of the envelope, $M(\xi)$, must satisfy. Referring to Table 5.1, it is interesting to comment on some of the physical characteristics of the solutions. For example, number 1 is a cylindrical solution having a constant amplitude but a time- and angular-dependent phase. On the other hand, number 2, imagining planes for which ξ is constant, is accelerating along the z-axis, again with a constant amplitude, but the phase varies in a more complicated way with time. Notice that in numbers 1–4 there is no damping in the prefactor α, either as a function of time or space. All other solutions have a decaying factor. Thus, numbers 8 and 9 are axial solutions but accelerate along the z-axis as $\sqrt{t}$ and their amplitude is damped with time as $t^{-1/2}$. Cylindrical solutions in numbers 7 and 5 are expanding since, for constant ξ, ρ is proportional to $\sqrt{t}$, but amplitudes in these two cases are reduced with

Table 5.1 Solutions having generic orbits of co-dimension one in the space of independent variables (x, y, z, t) having the form $\chi(\mathbf{x}, t) = \alpha(\mathbf{x}, t) f(\xi)$, $f(\xi) = M(\xi)\exp(i\chi(\xi))$ following references [51–53].

Solution number	$\alpha(\mathbf{x}, t)$	ξ	χ	ODE
1	$\exp\{i(a\theta - bt)\}$	$(x^2 + y^2)^{1/2} = \rho$	$S_0 \int \dfrac{d\rho}{\rho M^2} + \chi_0$	$\ddot{M} - \dfrac{S_0^2}{\rho^2 M^2} + \dfrac{\dot{M}}{\rho} - \dfrac{a^2}{\rho^2} M$ $= (a_0 - b)M + a_1 M^3$, $a \geq 0,\ b \in R$
2	$\exp[\tfrac{1}{6} iat(3z - at^2)]$	$z - \tfrac{1}{2} a t^2$	$S_0 \int \dfrac{d\xi}{M^2} + \chi_0$	$\ddot{M} - \dfrac{S_0^2}{M^3} - \tfrac{1}{2} a \xi M$ $= a_0 M + a_1 M^3$, $a > 0$
3	$\exp(-iat)$	z	$S_0 \int \dfrac{dz}{M^2} + \chi_0$	$\ddot{M} - \dfrac{S_0^2}{M^3}$ $= (a_0 - a)M + a_1 M^3$, $a \in R$
4	$\exp(-ibt)$	$(x^2 + y^2 + z^2)^{1/2} = r$	$S_0 \int \dfrac{dr}{r^2 M^2} + \chi_0$	$\ddot{M} - \dfrac{S_0^2}{r^4 M^3} + \dfrac{2}{r}\dot{M} + bM$ $= a_0 M + a_1 M^3$, $b \in R$
5	$t^{-1/2} \exp\left[i\left(-a_0 t + \dfrac{z^2}{4t} + a\theta - \tfrac{1}{2} b \ln t \right) \right]$	t/ρ^2		Third order, $a \geq 0,\ b \geq 0$

Table 5.1 (Continued)

Solution number	$a(\mathbf{x}, t)$	ξ	χ	ODE
6	$z^{-1}\exp[-i(a_0 t + a\theta + b\ln z)]$	z/ρ	$b > 0$ $b = 0$ $S_0 \int \dfrac{\xi^2}{M^2(1+\xi^2)^{3/2}}\,d\xi + \chi_0$	Third order $\xi^2(1+\xi^2)\ddot{M} + \dfrac{-\xi^6}{(1+\xi^2)^2}$ $\times \dfrac{S_0^2}{M^3} + \xi(\xi^2-2)\dot{M}$ $+(2-a^2\xi^2)M = a_1 M^3$ $a \geq 0, b \geq 0$
7	$t^{-1/2}\exp[i(-a_0 t + a\theta - \tfrac{1}{2}b\ln t)]$	t/ρ^2	$\chi = S_0 \int \dfrac{d\xi}{\xi M^2} + \dfrac{1}{8\xi} + \chi_0$	$4\xi^3\ddot{M} + 4\xi^2\dot{M} - 4\xi\left(\dfrac{1}{16\xi}\right)\dfrac{S_0^2}{M^3}$ $+\left(\tfrac{1}{2}b - a^2\xi + \dfrac{1}{16\xi}\right)M$ $= a_1 M^3,$ $a \geq 0, b \geq 0$
8	$t^{-1/2}\exp\left(\left(\dfrac{i\rho^2}{4t} - a_0 t - \tfrac{1}{2}b\ln t\right)\right)$	t/z^2		Third order, $b \geq 0$
9	$t^{-1/2}\exp[-i(a_0 t + \tfrac{1}{2}b\ln t)]$	t/z^2		Third order, $b \geq 0$
10	$\rho^{-1}\exp[-i(a_0 t + b\ln\rho)]$	$a\ln\rho + \theta$	$b = 0$ $\chi = S_0 \int M^{-2}$ $\times \exp[2a\xi/(a^2+1)]\,d\xi$ $+\chi_0$	$(a^2+1)\ddot{M} - (a^2+1)M^{-3}$ $\times \exp[4a\xi/(a^2+1)]S_0^2$ $-2a\dot{M} + M = a_1 M^3,$ $a \geq 0, b \geq 0$
11	$t^{-1/2}\exp\left(i\left(\dfrac{z^2}{4t} - \dfrac{b}{2}\ln t - a_0 t\right)\right)$	t/y^2		Third order, $b \geq 0$
12	$t^{-1/2}\exp[-i(a_0 t + \tfrac{1}{2}b\ln t)]$	t/r^2		Third order, $b \geq 0$

Table 5.2 Solutions having generic orbits of co-dimension two in the space of independent variables (x, y, z, t), reducing to a PDE in two variables, having the form $\chi(\mathbf{x}, t) = (\xi_1, \xi_2)$ $\alpha(\mathbf{x}, t)$ (following references [51—53]).

Solution number	$\alpha(\mathbf{x}, t)$	ξ_1	ξ_2	PDE
1	1	$r = (x^2 + y^2 + z^2)^{1/2}$	t	$i\partial_t f + \partial_{rr} f + \dfrac{2}{r}\partial_r f$ $= a_0 f + a_1 f \|f\|^2$
2	$\exp[i(b\theta - at)]$ $\theta = \tan^{-1}(y/x)$	$\rho = (x^2 + y^2)^{1/2}$	$z + c\theta = \xi$	$\partial_{\rho\rho}f + \dfrac{1}{\rho}\partial_\rho f + \left(1 + \dfrac{c^2}{\rho^2}\right)\partial_{\xi\xi}f$ $+ \dfrac{b}{\rho^2}(2ic\partial_\xi f - bf)$ $= (a_0 - a)f + a_1 f\|f\|^2,$ $a \in R, b \geq 0, c \geq 0$
3	$\exp(ai\theta)$	ρ	$t + b\theta = \xi$	$i\partial_\xi f + \dfrac{b^2}{\rho^2}\partial_{\xi\xi}f + \dfrac{a}{\rho^2}$ $\times (2ib\partial_\xi f - af)$ $+ \partial_{\rho\rho}f + \dfrac{1}{\rho}\partial_\rho f$ $= a_0 f + a_1 f\|f\|^2,$ $a = 0, b \geq 0 \text{ or } a > 0, b \in R$
4	$\exp[i y^2/4t]$	t	$x - \dfrac{a}{t}y = \xi$	$i\partial_t f + \left(1 + \dfrac{a^2}{t^2}\right)\partial_{\xi\xi}f + \dfrac{i}{2t}f$ $= a_0 f + a_1 f\|f\|^2,$ $a \geq 0$
5	$\exp[\tfrac{1}{6}iat(3z - at^2)]$	y	$\xi = z - \tfrac{1}{2}at^2$	$\partial_{yy}f + \partial_{\xi\xi}f - \tfrac{1}{2}a\xi f$ $= a_0 f + a_1 f\|f\|^2,$ $a > 0$

Table 5.2 (Continued)

Solution number	$\alpha(x,t)$	ξ_1	ξ_2	PDE
6	$\exp\left[\dfrac{i}{4t}\left(x^2 + \dfrac{y^2 t^2 - 2bxyt + b^2 x^2}{t^2 - b^2 + ct}\right)\right]$	t	$\xi = \dfrac{a(bx - yt)}{t^2 - b^2 + ct} + z$	$i\partial_t f + \left[1 + \dfrac{a^2(b^2 + t^2)}{(t^2 - b^2 + ct)}\right]\partial_{\xi\xi} f$ $+ \dfrac{i}{2t}\left[1 + \dfrac{(b^2 + t^2)}{(t^2 - b^2 + ct)}\right] f$ $= a_0 f + a_1 f|f|^2$ $a > 0, b \geq 0, c \geq 0$ or $a = b = 0, c \geq 0$
7	$\exp[i b\theta + \tfrac{1}{6} at(3z - at^2)]$	ρ	$\xi = z - \tfrac{1}{2} at^2$	$\partial_{\rho\rho} f + \dfrac{1}{\rho}\partial_\rho f - \dfrac{b^2}{\rho^2} f$ $+ \partial_{\xi\xi} f - \tfrac{1}{2} a\xi f$ $= a_0 f + a_1 f|f|^2,$ $a > 0, b \in R$
8	$\exp[i(b\theta + z^2/4t)]$	t	ρ	$i\partial_t f + \dfrac{1}{\rho}\partial_\rho f + \partial_{\rho\rho} f$ $- \dfrac{b^2}{\rho^2} f + \dfrac{i}{2t} f$ $= a_0 f + a_1 f|f|^2,$ $b \geq 0$
9	$\exp(-iat)$	x	y	$\partial_{xx} f + \partial_{yy} f = (a_0 - a) f$ $+ a_1 f|f|^2,$ $a \in R$
10	1	t	z	$i\partial_t f + \partial_{zz} f = a_0 f + a_1 f|f|^2$

Table 5.2 (*Continued*)

Solution number	$\alpha(x,t)$	ξ_1	ξ_2	PDE
11	$t^{-1/2}\exp[\mathrm{i}(a\theta - a_0 t - \tfrac{1}{2}b\ln t)]$	t/ρ^2	z/ρ	$\xi_1(\mathrm{i}+4\xi_1)\partial_{\xi_1}f + \xi_1\xi_2\partial_{\xi_2}f$ $+4\xi_1^3\partial_{\xi_1\xi_1}f + \xi_1(1+\xi_2^2)$ $\times \partial_{\xi_2\xi_2}f + 4\xi_1^2\xi_2\partial_{\xi_1\xi_2}f$ $-(-\tfrac{1}{2}b + a^2\xi_1 + \tfrac{1}{2}\mathrm{i})f$ $= a_1 f|f|^2,$ $a \geqslant 0, b \geqslant 0$
12	$t^{-1/2}\exp[\mathrm{i}(z^2/4t - \tfrac{1}{2}b\ln t - a_0 t)]$	t/ρ^2	$a\ln\rho + \theta$	$\xi_1(\mathrm{i}+4\xi_1)\partial_{\xi_1}f + 4\xi_1^3\partial_{\xi_1\xi_1}f$ $+\xi_1(1+a^2)\partial_{\xi_2\xi_2}f$ $-4a\xi_1^2\partial_{\xi_1\xi_2}f$ $+\tfrac{1}{2}bf = a_1 f|f|^2,$ $a \geqslant 0, b \geqslant 0$
13	$z^{-1}\exp[-\mathrm{i}(a_0 t + b\ln z)]$	z/ρ	$a\ln\rho + \theta$	$\xi_1[\xi_1^2 - 2(1+b\mathrm{i})]\partial_{\xi_1}f$ $+\tfrac{\xi_1}{5}(1+a^2)\partial_{\xi_2\xi_2}f$ $-2a\xi_1^3\partial_{\xi_1\xi_2}f$ $+\xi_1^2(1+\xi_1^2)\partial_{\xi_1\xi_1}f$ $+(1+b\mathrm{i})(b\mathrm{i}+2)f = a_1 f|f|^2,$ $a \geqslant 0, b \geqslant 0$
14	$t^{-1/2}\exp[\mathrm{i}(-\tfrac{1}{2}b\ln t - a_0 t)]$	t/ρ^2	$a\ln\rho + \theta$	$\xi_1(\mathrm{i}+4\xi_1)\partial_{\xi_1}f + 4\xi_1^3\partial_{\xi_1\xi_1}f$ $+\xi_1(1+a^2)\partial_{\xi_2\xi_2}f$ $-4a\xi_1^2\partial_{\xi_1\xi_2}f$ $+\tfrac{1}{2}(b-\mathrm{i})f = a_1 f|f|^2,$ $a \geqslant 0, b \geqslant 0$

time as $t^{-1/2}$. Number 6 is a fascinating case in that it represents a stationary cone, the amplitude of which decreases as z^{-1}. Similarly, number 10 describes a spiral surface parallel to the z-axis, its amplitude being reduced as ρ^{-1} as it unwinds. There are also two spherical solutions, namely numbers 4 and 12, the first stationary and undamped and the second expanding as $t^{1/2}$ and damped as $t^{-1/2}$, respectively.

In Table 5.2 we have represented solutions in a similar format, but with two symmetry variables, as in Table 5.1, but here they are associated with generic orbits of co-dimension two. These solutions are significantly more complicated and we discuss below some of the more physically meaningful ones in the order of increasing complexity. Solution number 10 is reduced to the solution of a NLS equation for f, one symmetry variable being time and the other z, i.e. in $(1+1)$ dimensions. In number 4 the situation is similar except that putting ξ_2 to be a constant produces a surface which undergoes a rotation as time increases—very like a moving domain wall. Number 6 exhibits three-dimensional tilting plane waves. The case in number 9 illustrates time-independent symmetry variables, the corresponding PDE for f being the complex nonlinear Klein–Gordon equation in the x–y plane. Numbers 8 and 1 denote cylindrical and spherical time-dependent solutions. Cases numbered 11–14 are much more complex, since putting each symmetry variable to a constant describes two different surfaces. In number 11, ξ_2 describes a conical surface, whereas ξ_1 can be interpreted as a cylinder expanding with time. Numbers 12 and 14 denote expanding cylinders and stationary spiral surfaces, and in 13 there are cones and spiral surfaces.

Having indicated briefly what types of ODE or PDE result for orbits with co-dimension one or co-dimension two, it is of interest to see what sort of explicit analytical solutions can be found. They are numerous, so we merely divide them up into physical types. The interested reader should consult the original papers [51–54], in which explicit solutions have been listed in terms of constants given by Gagnon and Winternitz [52, 53], who have also specified the régimes of applicability for each solution's constants. There appear to be five main types:

1. First, there are *spatially homogeneous* solutions of the form

$$\phi(r,t) = \phi_0 \exp(-ibt).$$

2. The classical solutions may also be of the *unidirectional quasi-linear* type and have the form

$$\phi(\mathbf{r},t) = \phi_0 f(c_2\eta + d_2)\exp(\tfrac{1}{6}iat\eta),$$

SOLUTIONS OF THE EQUATION OF MOTION 253

where the symmetry variable is $\eta = 3z - at^2$ and $f(\eta)$ satisfies the second *Painlevé transcendent* equation [74]. In both of these types, ϕ_0, b, and a are constants but are *not* arbitrary.

3. Third, the classical solution may have a *cylindrically symmetric* envelope, so that

$$\phi(r,t) = \phi_0 \rho^{-1/3} F(\xi) \exp[i(c_3\theta + d_3 t + g_3(\xi))],$$

where $\xi = a_3 \rho^{2/3} + b_3$, *the function F can have several different forms* [64], and $g_3(\xi)$ has been given in each case.

4. As a fourth type we list *stationary z-dependent* solutions

$$\phi(r,t) = \phi_0 H(\omega) \exp[i(c_4 t + g_4(\omega))].$$

In this case $\omega = a_4 z + b_4$ is the symmetry variable and $H(\omega)$ can have many different functional forms [64].

5. Lastly, solutions exist of the *angle-dependent multivalued* type, which may be written as

$$\phi(r,t) = \phi_0 \rho^{-1} B(\Omega) \exp[i(c_5 t + g_5(\Omega))],$$

where $\Omega = a_5\theta + b_5$ is the symmetry variable and the whole host of dependences for $B(\Omega)$ have been given by other authors [52, 53, 64].

The parameters a_i, b_i, c_i, d_i, e_i, f_i, and h_i ($i = 1, 2, \ldots, 5$) are again *not* arbitrary. The very fact that such a wealth of solutions appear exactly for a Euclidean signature (possibly the particular one most used in physical applications) when we include the zeroth order nonlinearity (the first nontrivial introduction of nonlinearity) strongly supports the correctness and virtually exact nature of our development. However, we now briefly consider another signature in order to see what may emerge. Again the reader will be struck by the enormous wealth of form in the obtained solutions.

Minkowski signatures ($\varepsilon_1 = -1$, $\varepsilon_2 = +1$, $\varepsilon_3 = +1$)

Clearly, there will be in general eight different signatures since ε_1, ε_2, and ε_3 can each have two different values. However, by scaling, interchanging variables, etc., it is clear that solutions in these cases will be related and here we are considering the simplest nontrivial non-Euclidean case. Unfortunately, there is no in-depth analysis of solutions for the (3 + 1) dimensional nonlinear Schrödinger equation for a Minkowski signature. Obviously, one could try, in a simplistic way, to replace these independent variables which correspond to a negative signature, let us say x_k, by ix_k and substitute this into solutions already obtained above. This will not, however, generate all the solutions since the PDE is invariant with respect to different symmetry groups for different signatures. As it turns out, from preliminary studies, only the subgroups of the symmetry group (see Chapter 7) which involve dilation operations (scale invariant transformations) will be different in the two cases of differing signatures.

There is another way we can show the existence of a multitude of special exact solutions and that is to use recent investigations of the nonlinear Klein–Gordon (NLKG) equation [54, 79, 80]. To do this we represent the classical field as

$$\phi = \eta \exp(i\chi), \qquad (5.97)$$

where η and χ are assumed to be real. Substituting eqn (5.97) and separating real and imaginary parts produces the following two equations:

real part, $\qquad \nabla_\varepsilon^2 \eta = 2\left[\lambda_0 + \hbar \chi_t + \tfrac{1}{2}(\nabla_\varepsilon \chi)^2\right]\eta + 2F_0 \eta^3 \qquad (5.98)$

and

imaginary part, $\qquad \hbar \eta_t = -\nabla_\varepsilon \chi \cdot \nabla \eta - \tfrac{1}{2}(\nabla_\varepsilon^2 \chi)\eta. \qquad (5.99)$

Equation (5.98) becomes a cubic NLKG equation provided that we require

$$\hbar \chi_t + \tfrac{1}{2}(\nabla_\varepsilon \chi)^2 = \text{constant}. \qquad (5.100)$$

We can easily satisfy the requirement in eqn (5.100) and, for example, we can do this in Cartesian co-ordinates by putting

$$\chi = \mathbf{k} \cdot \mathbf{r} - wt + \chi_0. \qquad (5.101)$$

In a similar way, in cylindrical polar co-ordinates,

$$\chi = k_\rho \rho + k_z z + wt + \chi_0 \qquad (5.102)$$

or, in spherical polar co-ordinates,

$$\chi = k_r r - wt, \qquad (5.103)$$

will suffice to ensure that eqn (5.100) is satisfied. Given that eqn (5.100) is satisfied, we must now ensure, with the particular choice of χ, that eqn (5.99) is satisfied. Clearly, the last term of eqn (5.99) vanishes from this particular choice so we only require that the two gradients are orthogonal if the amplitude, η, does not vary with time ($\eta_t = 0$). One of the simplest ways of doing this is to choose η to be a function of a particular set of independent variables on which χ does not depend. Thus, eqn (5.98) becomes a one-, two-, or three-dimensional NLKG equation when the carrier, χ, depends on two, one, or no spatial independent variables, respectively. These cases have been analysed in detail [54, 79, 80] and discussed at length when related to the MCS picture [64].

When η depends on only one independent spatial variable, solutions involve elliptic waves, both singular and nonsingular, and their limiting cases (when the elliptic modulus is zero or unity, i.e. trigonometric solutions and localized hyperbolic functions such as sech x and tanh x). In this case, of course, signature plays no role. For η depending on two independent spatial variables we have two subcases, namely when the independent variables have the same or different signatures. In the former case, the NLKG equation has to be solved in the two-dimensional Euclidean (E(2)) space, whereas in the latter it is the two-dimensional Minkowski space (M(1.1)). For E(2) and assuming

$$2\left[\lambda_0 + \hbar\chi_t + \tfrac{1}{2}(\nabla_\varepsilon \chi)^2\right] = \text{constant} = E_0 \neq 0, \qquad (5.104)$$

one finds the the symmetry variables may only be

$$\xi_0 = \left(|x_i|^2\right)^{1/2} \quad \text{or} \quad \xi_1 = (x_1^2 + x_2^2)^{1/2}, \qquad (5.105)$$

and the reduced ODE is

$$\eta_{\xi\xi} + (k/\xi)\eta_\xi = \lambda(E_0\eta + F_0\eta^3), \qquad (5.106)$$

where $\lambda = 1$ and $k = 1$ or 0 for the two variables ξ_k, respectively. In M(1, 1) the allowed symmetry variables are

$$\xi_0 = \left(|x_1|^2\right)^{1/2}, \quad \xi_1 = (x_1^2 - x_2^2)^{1/2}, \quad \xi_2 = x_1 + x_2. \quad (5.107)$$

In all these cases the reduced ODE is eqn (5.106) with $k = 1$ for ξ_1 and $k = 0$ otherwise. It is found that $\lambda = 1$ for ξ_1, and that $\lambda = -1$ in other cases. When $E_0 = 0$, the symmetry group of eqn (5.104) acquires scaling transformations in addition to the various rotations and translations when $E_0 \neq 0$. This results in a large number of new solutions. We present the extra solutions in Table 5.3. When η depends on three spatial variables, and $E_0 \neq 0$, all of the reductions lead to an equation such as eqn (5.106) and in Table 5.4 we have given the values of k, λ, and also the associated symmetry variables. An important new feature is the presence of what are called degenerate symmetry variables. These variables depend on arbitrary functions. In Fig. 5.1, some of the more important symmetry variables obtained in our analysis are graphically illustrated.

Table 5.3 The results of symmetry reduction for eqn (5.106) with $E_0 = 0$ (see eqn (5.114)) in two- and three-dimensional Euclidean space, where $\eta = \sigma(x) f(\xi)$ and h is a constant.

$\sigma(x)$	ξ	Reduced ODE
$2\left(-\dfrac{1}{F_0}\right)^{1/2} \dfrac{1}{x_3}$	$(x_1^2 + x_2^2)/x_3^2$	$\xi(\xi+1)\ddot{f} + (1 + \tfrac{5}{2}\xi)\dot{f} + \tfrac{1}{2}f + f^3 = 0$
$\left(-\dfrac{1}{F_0}\right)^{1/2} \dfrac{1}{x_1}$	x_2/x_1	$(1+\xi^2)\ddot{f} + 4\xi\dot{f} + 2f + f^3 = 0$
$2h[-(1/F_0)(h^2 + 4)(x_1^2 + x_2^2)]^{-1/2}$	$\dfrac{4h}{h^2+4}[-\tfrac{1}{2}h\ln(x_1^2 + x_2^2)$ $+ \tan^{-1}(x_2/x_1)]$	$\ddot{f} + \dot{f} + \dfrac{h^2+4}{4h^2}f + f^3 = 0$

Table 5.4 Reduction of eqn (5.98) with $E_0 \neq 0$ for E(3) and M(2, 1) (following reference [54]) (p arbitrary).

ξ	k	λ
E(3)		
$(x_1^2)^{1/2}$	0	1
$(x_1^2 + x_2^2)^{1/2}$	1	1
$(x_1^2 + x_2^2 + x_3^2)^{1/2}$	2	1
M(2, 1)		
$(x_1^2)^{1/2}$	0	1
$(x_1^2 - x_2^2)^{1/2}$	1	1
$(x_1^2 - x_2^2 - x_3^2)^{1/2}$	2	1
$(x_2^2)^{1/2}$	0	-1
$(x_2^2 + x_3^2)^{1/2}$	1	-1
Degenerate variables $x_2 + f(x_1 + x_3)$	0	-1
$x_2 + p \ln(x_1 + x_3)$	0	-1
$x_2 + \frac{1}{4}(x_1 + x_3)^2$	0	-1

5.6.3 The interaction included to first order

In this section we incorporate contributions which originate from two-body interactions that are linearly dependent on the wave vector. This means, in terms of spatial dependence, that long-range interactions between particles are now included to a certain extent. Therefore, in most cases this constitutes an improvement over the zeroth-order approximation.

The equation of motion for the classical field in this case takes the form of a generalized nonlinear Schrödinger equation (GNLS), given by

$$i\hbar \, \partial_t \phi = \gamma_0 \phi + i\gamma_1 \cdot (\nabla_\varepsilon \phi) - \tfrac{1}{2} \nabla_\varepsilon^2 \phi$$
$$+ \gamma_2 \phi^+ \phi \phi + 2i\gamma_3 \cdot [\phi^+ \phi \nabla_\varepsilon \phi], \quad (5.108)$$

where γ_0 and γ_2 are constant parameters while γ_1 and γ_3 are fixed vectors. We *may* proceed by writing the classical field in terms of its envelope η and carrier χ, as in eqn (5.97). Having substituted eqn (5.97) into eqn (5.108), and separating real and imaginary parts, we find the following:

real part,

$$\tfrac{1}{2} \nabla_\varepsilon^2 \eta = \left[\hbar \chi_t + \gamma_0 - \gamma_1 \cdot \nabla_\varepsilon \chi + \tfrac{1}{2}(\nabla_\varepsilon \chi)^2 \right] \eta$$
$$+ [\gamma_2 - 2\gamma_3 \cdot \nabla_\varepsilon \chi] \eta^3 \quad (5.109)$$

258 THE METHOD OF COHERENT STRUCTURES (MCS)

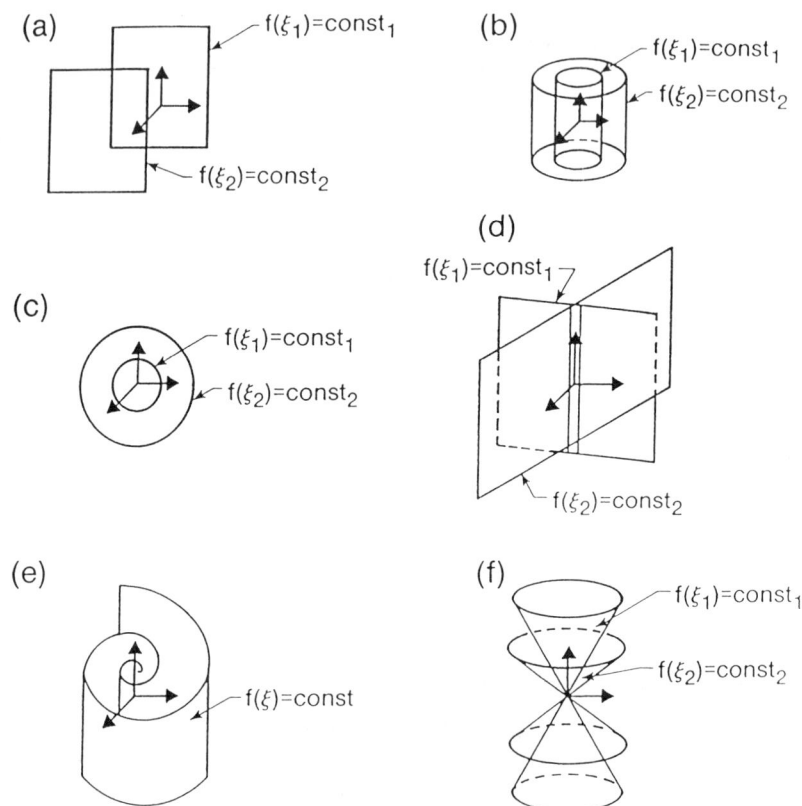

Fig. 5.1 Surfaces of constant symmetry variables for selected cases analysed in this chapter: (a) parallel planes; (b) concentric cylinders; (c) concentric spheres; (d) intersecting planes; (e) spiral surfaces; (f) a family of cones.

and

imaginary part,

$$\hbar \eta_t = (\boldsymbol{\gamma}_1 + 2\eta^2 \boldsymbol{\gamma}_3) \cdot \nabla_\varepsilon \eta - (\nabla_\varepsilon \eta) \cdot (\nabla_\varepsilon \xi) - \tfrac{1}{2} \eta \nabla_\varepsilon^2 \chi. \quad (5.110)$$

As in Subsection 5.6.2, this particular approach now imposes extra constraints, so that eqn (5.109) again becomes a NLKG equation. One of the simplest ways to do this is to require that χ_t is a constant and $\nabla_\varepsilon \chi$ a constant vector, so that

$$\chi = \sum_{i=0}^{3} K_i x_i + \chi_0, \quad (5.111)$$

SOLUTIONS OF THE EQUATION OF MOTION 259

where the K_i ($i = 0, 1, 2, 3$) are constants, as is χ_0. With the assumption in eqn (5.111), eqn (5.109) becomes a cubic time-independent three-dimensional NLKG equation. That is,

$$\nabla_\varepsilon^2 \eta = A\eta + B\eta^3, \tag{5.112}$$

where

$$A = 2\left(\hbar K_0 + \gamma_0 - \sum_{i=1}^{3} \gamma_{1i} \varepsilon_i K_i + \tfrac{1}{2} \sum_{i=1}^{3} K_i^2\right)$$

and

$$B = 2\left(\gamma_2 - 2 \sum_{i=1}^{3} \gamma_{3i} \varepsilon_i K_i\right).$$

Equation (5.112) must still be consistent with eqn (5.110). That is, η must be chosen in such a way that eqn (5.110) is satisfied but without making too many restrictive constraints on the functional form of η, i.e. so that eqn (5.112) can, in principle, be satisfied by η. One way in which this may be achieved is to require that

$$\boldsymbol{\gamma}_3 \cdot \nabla_\varepsilon \eta = 0, \tag{5.113}$$

when one spatial variable is removed parallel to $\boldsymbol{\gamma}_3$. If this is done, then eqn (5.110) can be satisfied by choosing η to have the functional dependence:

$$\eta = \eta(x_1 - v_1 t, x_2 - v_2 t), \tag{5.114}$$

where x_1 and x_2 are two linearly independent co-ordinates in the plane perpendicular to $\boldsymbol{\gamma}_3$, and v_1 and v_2 are constant velocities so that when eqn (5.114) is inserted into eqn (5.110), it is satisfied identically. Clearly, eqn (5.114) is not in conflict with eqn (5.112) and we merely seek two-dimensional solutions of the NLKG equation. This has been done [79, 80], so a similar procedure to that in the previous section may be followed. In fact, in addition to solutions of the form in eqn (5.114) one can also find symmetry reductions to an ODE by using the single variables:

$$\xi_1 = k_1 x_1 + k_2 x_2 + \omega t \quad \text{and} \quad \xi_2 = k_3 \rho - \omega t,$$

where $\eta(\xi_1)$ will then represent plane waves in two dimensions and $\eta(\xi_2)$ radial cylindrical waves.

It is found that the profiles of the envelope solutions for eqn (5.112) represent Jacobi elliptic functions and have periods depending on the integration constants—the latter are fixed in particular applications by appropriate boundary conditions. Special cases are the well-studied sech and tanh solitary waves. Once again, it is obvious that an enormous number of analytical solutions are possible from the two-body interaction in first order.

5.6.4 The two-body interaction to second order

Finally, we consider the most general case, in which we expand the two-body interaction matrix element to quadratic terms in the wave vectors. We have already seen earlier that an attempt to go beyond this level will not introduce any new qualitative features, due to the fact that all the higher order terms merely redress those in second order and below.

The procedure here is to follow what we did in Subsection 5.6.3 above for first order, χ being a solution of the equation like eqn (5.111). In second order—see eqn (5.59)—η is chosen so that its gradient and μ_4 are mutually orthogonal. One now finds that the real equation (after separating into real and imaginary parts as before) takes the form

$$\nabla_\varepsilon^2 \eta = \left[\bar{A}\eta + \bar{B}\eta^3 + 2\eta(\nabla_\varepsilon \eta)^2\right]/(\mu_2 + 2\eta^2), \qquad (5.115)$$

where $\bar{A}$ and $\bar{B}$ are constant coefficients. Equation (5.115) is clearly of the form

$$\nabla_\varepsilon^2 \eta = Q(\eta, (\nabla_\varepsilon \eta)^2), \qquad (5.116)$$

for some function Q. Remarkably, such a nonlinear PDE has been studied earlier [81–83]. Symmetry reduction analysis on this type of equation has been performed, assuming only that the symmetry variable ξ has the property

$$(\nabla \xi)^2 = f(\xi), \qquad (5.117)$$

i.e. $(\nabla \xi)^2$ is constant on every level of the function ξ. Note this is the same assumption that we used in our renormalization, so $\nabla^2 \xi$ also has this property, and really all we require is that f shall be invertible. With these very general assumptions, all the symmetry variables of co-dimension one in Euclidean and Minkowski metrics were found. In our case, in two-dimensional space, the only known symmetry variables are listed in Table 5.5. In all cases, the reduced ODE takes the form

$$\varepsilon\eta'' + K\eta'/\xi = (\bar{A}\eta + \bar{B}\eta^3 + 2\varepsilon\eta(\eta')^2)/(\mu_2 + 2\eta^2). \qquad (5.118)$$

From Table 5.5, two-thirds of the cases have $K = 0$, so we tackle them first. Putting $\eta' = p$ and $\eta'' = p\,\mathrm{d}p/\mathrm{d}\eta$, eqn (5.118) becomes

$$\varepsilon p(\mathrm{d}p/\mathrm{d}\eta) = g(\eta) + f(\eta)p^2, \qquad (5.119)$$

SOLUTIONS OF THE EQUATION OF MOTION   261

Table 5.5 Symmetry variables ξ, $(\nabla\xi)^2$, $\nabla^2\xi$, K and ε for eqn (5.116) for the Euclidean and Minkowski cases in two-dimensional space.

	ξ	$(\nabla\xi)^2$	$\nabla^2\xi$	K	ε
Euclidean (+ +)					
(a)	$ax_1 + bx_2$ $a^2 + b^2 = 1$	1	0	0	1
(b)	$x_1^2 + x_2^2$	1	$1/\xi$	1	1
Minkowski (+ −)					
(a)	x_1	1	0	0	1
(b)	x_2	−1	0	0	−1
(c)	$x_1 + x_2$	0	0	0	0
(d)	$x_1^2 - x_2^2$	1	$1/\xi$	1	1

where $g(\eta) = (\bar{A}\eta + \bar{B}\eta^3)/(\mu_2 + 2\eta^2)$ and $f(\eta) = 2\varepsilon\eta/(\mu_2 + 2\eta^2)$, in which, of course, we assume that, $\mu_2 + 2\eta^2 \neq 0$. By changing the dependent variable to W, where $p = W^{-1}$, we obtain the Bernoulli equation

$$\varepsilon(\mathrm{d}W/\mathrm{d}\eta) = -g(\eta)W^3 - f(\eta)W. \tag{5.120}$$

It is straightforward to find a solution of this equation, which is

$$W^{-2} = (\mu_2 + 2\eta^2)\left[\frac{1}{4\varepsilon}\bar{B}\ln(\mu_2 + 2\eta^2) - C\right] - \frac{1}{4\varepsilon}\bar{B}(2\bar{A}/\bar{B} - \mu_2)$$

$$= \left(\frac{\mathrm{d}\eta}{\mathrm{d}\xi}\right)^2 = F(\eta), \tag{5.121}$$

where C is an integration constant. One further integration gives an implicit solution:

$$\xi - \xi_0 = \int \mathrm{d}\eta\, F(\eta)^{-1/2}. \tag{5.122}$$

We can obtain the qualitative features of the various types of solution by examining the roots η_0 of the equation $F(\eta) \equiv 0$ (see Appendix A for a description of the method). We consider four cases only and bear in mind that if all roots are real, they will appear in symmetrical pairs at $\mp\eta_0$. The following possibilities can be found. There could be no real roots of $F(\eta)$. There could be two symmetrically located distinct single real roots. On the other hand, there could be a pair of double roots or four distinct single real roots. The equation $F(\eta) = 0$ can be cast into the form

$$z \ln z = \frac{4\varepsilon}{\bar{B}} Cz + \frac{4D}{\bar{B}}, \qquad (5.123)$$

where $\mu_2 + 2\eta^2 = z$ and

$$D = \tfrac{1}{4}\bar{B}\left(\frac{2\bar{A}}{\bar{B}} - \mu_2\right).$$

By considering the form of the solution near a particular root (see Appendix A) one can easily conclude, by analogy with elliptic integrals, that localized solutions $\eta(\xi)$ can only exist when the corresponding root is *double*, whereas periodic solutions occur between two single distinct roots. Thus, if there are no real roots, singular solutions only will ensue. If there are either four distinct single real roots or two distinct single real roots, one obtains periodic solutions η between two real roots for which $F(\eta) \geq 0$. A pair of double roots will indeed occur whenever $4D/\bar{B} < 0$ and the double root is located at $z = -4D/\bar{B}$, the integration constant C being $C = (1/4\varepsilon)\bar{B}[\ln|4D/\bar{B}| + 1]$. In this case we obtain a solution which interpolates between the two double roots when ξ varies from $-\infty$ to $+\infty$. This can be interpreted as a kink-type solitary wave.

When $K = 1$ and $\varepsilon = 1$ (see Table 5.5), it is more difficult to find exact solutions but a simple transformation makes the situation easier to handle, namely by putting $\eta = y(\lambda \ln \xi)$. The equation may now be cast in the form

$$y'' = [\exp(2u/\lambda)/(\lambda^2)]g(y) + f(y)(y')^2, \qquad (5.124)$$

where the primes in eqn (5.124) now indicate differentiation with respect to $u \equiv \lambda \ln \xi$, which is the new independent variable, while λ is a parameter to be chosen and

$$g(y) = \frac{\bar{A}y + \bar{B}y^3}{\mu_2 + 2y^2}, \qquad f(y) = \frac{2y}{\mu_2 + 2y^2}. \qquad (5.125)$$

Provided that $y' \neq 0$, eqn (5.124) may be divided through by y' and directly integrated with respect to u in the special limit when $\lambda \to \infty$, since in this particular case, the first term on the right tends to zero. The result is

$$y' = y_0(\mu_2 + 2y^2)^{1/2}, \qquad (5.126)$$

where y_0 is a constant of integration. Equation (5.126) is separable and its solution is

$$y = \sqrt{\frac{|\mu_2|}{2}} \sinh\left[\sqrt{2}\, y_0 (u - u_0)\right] \quad \text{for } \mu_2 > 0,$$

while

$$y = \sqrt{\frac{|\mu_2|}{2}} \cosh\left[\sqrt{2}\, y_0 (u - u_0)\right] \quad \text{for } \mu_2 < 0,$$

where u_0 is a second integration constant.

To sum up our deliberations in Subsections 5.6.1–5.6.4, we note that in the noninteracting (linear) regime (i.e. in Subsection 5.6.1), although a number of solutions were found in the form of oscillatory, decaying, and exponential functions, the important feature was that *no localization* phenomena were found (there was always a tendency to disperse with no compensating mechanism to maintain the overall shape). In Subsection 5.6.2, in 'zeroth order', an enormous wealth of solutions was found, based largely on the complete symmetry reduction analysis for the Euclidean signature. This latter is the first, nontrivial order in which localization phenomena are found. This is a direct result of the presence of nonlinearity, which has the effect of balancing out dispersion-like characteristics. The types of solution found range from the spatially homogeneous, to quasi-linear, to cylindrical, and even to multivalued solutions, with a wide variety of functional forms often expressed in terms of Jacobi elliptic functions (see Appendix B for their definitions and properties). In the Minkowski signature, solutions were found by a particular ansatz that enabled results for the NLKG equation to be used. In first order (Subsection 5.6.3), we have presented solutions the envelopes of which are generally elliptic functions. Their carrier waves are of plane-wave type and the envelopes themselves have arguments which are plane wave-like or cylindrical wave-like and propagate along a direction normal to the carrier wave. Both first- and second-order approximations in the MCS (Subsections 5.6.3 and 5.6.4) utilize the idea of an ansatz to reduce the solution to one for a NLKG in two dimensions and indicate a clear possibility of localization. More will be said about such solutions in Chapter 7.

5.7 Quantization procedure

We showed in an earlier section that close to a critical point classical effects are by far the most dominant. This is very well known [49] and first-order quantum corrections then have a very small relative magnitude compared to the classical field and can be treated very accurately in a perturbative manner. Thus, what we have done so far is to adopt the standard approach in quantum field theory [13, 50]; we have initially ignored quantum effects and solved the equations of motion for the classical c-fields, ϕ_c. The classical solutions that we have found fall into four main categories. First, there are constant solutions for ϕ_c and, obviously, because of their magnitude relative to the quantum component, they correspond to expectation values of the full quantum field. When these values are nonzero, they signal spontaneous symmetry-breaking in the system [A, 17]. The solutions can also be static in the sense that their shape remains constant or that they are static within their own (moving) frame of reference. Although the shape remains static, they are still space-dependent and represent elementary excitations when they are periodic or possess a localized envelope within which quantum states may be bound. Some solutions, ϕ_c, can be both time- and space-dependent and correspond to metastable excitations with finite lifetimes or unstable decay patterns. A fourth category of classical solution are kink-like solutions which may describe quantum tunnelling phenomena (instantons) or domain wall formation.

5.7.1 The simplest example

We will illustrate the quantization procedure in one spatial dimension using the NLS equation arising in the zeroth order of the MCS, since it is much easier to follow than its higher-order extensions. However, we bear in mind that the present method can be utilized in any number of dimensions. In fact, in Chapter 6, a three-dimensional example is considered (the multi-electron atom).

We begin with the equation of motion for the field, ψ, namely the NLS:

$$i\hbar \partial_t \psi = -\frac{\hbar^2}{2m} \nabla^2 \psi + \omega \psi + \omega_0 \psi^+ \psi \psi, \qquad (5.127)$$

where we have retained terms from the two-body interaction to zero order only and used Euclidean signatures for simplicity. The operator, ψ, is now expressed as

$$\psi = \hat{M} \exp(i\hat{N}), \qquad (5.128)$$

where we assume that $\hat{M}$ and $\hat{N}$ are real commuting operators. Inserting eqn (5.128) into (5.127) and equating real and imaginary parts yields the following:

real part,

$$-\hat{M}\hat{N}_t\hbar = -\frac{\hbar^2}{2m}\left(\nabla^2\hat{M} - \hat{M}(\nabla\hat{N})^2\right) + \omega\hat{M} + \omega_0\hat{M}^3 \quad (5.129)$$

and
imaginary part,

$$\hbar\hat{M}_t = \frac{-\hbar^2}{2m}\left(2\nabla\hat{M}\cdot\nabla\hat{N} + \hat{M}\,\nabla^2\hat{N}\right). \quad (5.130)$$

To make our assumptions even more specific, we assume that

$$\hat{M} = \phi_c(\mathbf{r},t)\hat{I} + \varepsilon U(\mathbf{r},t)\hat{0}, \qquad \hat{N} = -\frac{E_0 t}{\hbar}\hat{I} + \varepsilon\Omega(\mathbf{r},t)\hat{0}, \quad (5.131\text{a,b})$$

where $\hat{0}$ is a real operator, not depending on the time or spatial independent variables, i.e. commuting with them, and *common* to both $\hat{M}$ and $\hat{N}$, and $\hat{I}$ is the unit operator in Fock space. The space- and time-dependence of the two quantum components is described by $U(\mathbf{r},t)$ for the amplitude of ψ and $\Omega(\mathbf{r},t)$ for its phase. The symbol ε is merely inserted to indicate the smallness of quantum components. Equations (5.131), when inserted into eqns (5.129) and (5.130), after linearizing and equating terms separately of order unity and ε, produce three equations (see Appendix C for details). The first of these, which arises from terms of order ε^0, is the classical field equation as required and the other two are coupled space–time equations. Since we have chosen $\hat{0}$ to be common to both $\hat{M}$ and $\hat{N}$, and real, it may be effectively cancelled through these latter two equations and what remains are coupled partial differential equations containing no operators. The solution of these equations is discussed in Appendix C, where $\hat{0}$ is dropped for convenience, and it is shown, subject to ω, E_0, and ω_0 being constant parameters and $\omega - E_0 \gg \omega_0\phi_c^2$, that an effective (linear) Schrödinger equation for the quantum corrections results of the form

$$-\tfrac{1}{2}\nabla_\varepsilon^2\Lambda + 3F_0\phi_c^2\Lambda = E\Lambda. \quad (5.132)$$

Looking at eqn (5.132) we see that $3F_0\phi_c^2$ is playing the role of an effective potential and E is the eigenvalue formed from ω, E_0, and separation constants which arise in the solution of the coupled PDEs. From a practical point of view, it is important to establish the functional dependence in the effective potential which is directly affected by the type of classical solution ϕ_c forming it. In general, in one space dimension, the equation for the classical field ϕ_c will have three basic kinds of nonsingular elliptic wave solutions. These are expressed in terms of the Jacobi elliptic functions sn, cn, and dn [79]. They may be straightforwardly obtained by integrating the classical field equation directly (in one space dimension $\nabla^2 \phi_c \to \partial_{xx} \phi_c$, so if it is multiplied through by $\partial_x \phi_c$, it may be integrated immediately). This integrated equation may then be compared with standard forms given by Byrd and Friedman [84] and a number of distinct cases arise, which are analysed below.

5.7.2 The elliptic functions in the effective potential

dn-waves

When all the signatures are $\varepsilon_i = -1$, the polynomial which appears in the once integrated classical equation, say $P(\phi_c)$, where $(d\phi_c/dx)^2 = P(\phi_c)$, is

$$P(\phi_c) = 2E_0 \phi_c^2 - F_0 \phi_c^4 + C, \tag{5.133}$$

where C is an integration constant. When $P(\phi_c)$ has four distinct real roots $\pm\eta_1$ and $\pm\eta_2$ ($|\eta_2| > |\eta_1|$), then the solution can be written as

$$\phi_c = \mp \eta_2 \, \mathrm{dn}\left[(-\varepsilon_i F_0)^{1/2} \eta_2 (x_i - x_i^0), k\right], \tag{5.134}$$

where the elliptic modulus is $k = (1 - (\eta_1^2/\eta_2^2))^{1/2}$, x_i is one of the independent variables and the x_i^0 are integration constants.

cn-waves

On the other hand, if all the signatures are $\varepsilon_i = -1$ but $P(\phi_c)$ has two real roots $\mp\eta_1$ and two purely imaginary roots $\mp i\eta_2$, the solution is

$$\phi_c = \mp \eta_1 \, \mathrm{cn}\left\{\left[-\varepsilon_i F_0 (\eta_1^2 + \eta_2^2)\right]^{1/2} (x_i - x_i^0), k\right\}, \tag{5.135}$$

where the elliptic modulus is given by $k^2 = \eta_1^2/(\eta_1^2 + \eta_2^2)$.

sn-waves

The last of the three elliptic cases is when all the signatures are $\varepsilon_i = +1$ and the quartic polynomial has four real roots $\mp\eta_1$ and $\mp\eta_2$ ($|\eta_2| > |\eta_1|$). Then the solution is

$$\phi_c = \pm \eta_1 \, \mathrm{sn}\left[(\varepsilon_i F_0)^{1/2} \eta_2 (x_i - x_i^0), k\right], \tag{5.136}$$

with the elliptic modulus $k = \eta_1/\eta_2$.

The effective potential in the quantum equation is proportional to ϕ_c^2,

QUANTIZATION PROCEDURE   267

where ϕ_c is one of the three elliptic functions above. Fortunately, there are two identities for elliptic functions (see Appendix B) given by

$$\text{sn}^2(x,k) + \text{cn}^2(x,k) = 1, \quad k^2 \text{sn}^2(x,t) + \text{dn}^2(x,k) = 1, \quad (5.137)$$

which enables us to write the potential in each case in terms of sn^2 only.

5.7.3 The Lamé equation and Lamé functions

We have seen that the quantum equation for the correction Λ may be written in the same form for *all* three cases ((i), (ii), and (iii)) above. This may be expressed as

$$\frac{d^2\Lambda}{d\alpha^2} = (a + b\,\text{sn}^2(\alpha,k))\Lambda, \qquad (5.138)$$

and is the well-studied Jacobian form of the Lamé equation [85] when $b = n(n+1)k^2$. The independent variable α and the constants a and b are listed in Table 5.6 for each of the above three cases. This equation and its solutions have been discussed at length in the book by Whittaker and Watson [86]. The Lamé equation has two other forms; namely, an algebraic one,

$$\frac{d^2\Lambda}{d\xi^2} + \left(\frac{\frac{1}{2}}{\xi-e_1} + \frac{\frac{1}{2}}{\xi-e_2} + \frac{\frac{1}{2}}{\xi-e_3}\right)\frac{d\Lambda}{d\xi}$$
$$= \frac{[n(n+1)\xi + B]\Lambda}{4(\xi-e_1)(\xi-e_2)(\xi-e_3)}, \qquad (5.139)$$

and the Weierstrassian form,

$$d^2\Lambda/du^2 = [n(n+1)P(u) + B]\Lambda, \qquad (5.140)$$

Table 5.6 Independent variables and constants from each of three classical elliptic function solutions ϕ_c for the quantum equation (5.138).

Case	Type	α	a	b
(i)	dn	$\eta_2(-\varepsilon_i F_0)^{1/2}(x_i - x_i^0)$	$\dfrac{+2\varepsilon_i(E_0 - 3F_0\eta_2^2)}{\eta_2^2(-\varepsilon_i F_0)}$	$\dfrac{+6\varepsilon_i F_0 \eta_2^2 k^2}{\eta_2^2(-\varepsilon_i F_0)}$
(ii)	cn	$[-\varepsilon_i F_0(\eta_1^2 + \eta_2^2)]^{1/2}(x_i - x_i^0)$	$\dfrac{+2\varepsilon_i(E_0 - 3F_0\eta_1^2)}{[-\varepsilon_i F_0(\eta_1^2 + \eta_2^2)]}$	$\dfrac{+6\varepsilon_i F_0 \eta_1^2}{[-\varepsilon_i F_0(\eta_1^2 + \eta_2^2)]}$
(iii)	sn	$\eta_2(\varepsilon_i F_0)^{1/2}(x_i - x_i^0)$	$\dfrac{-2\varepsilon_i E_0}{\eta_2^2(\varepsilon_i F_0)}$	$\dfrac{6\varepsilon_i F_0 \eta_1^2}{\eta_2^2(\varepsilon_i F_0)}$

where B is related to a in eqn (5.138) by

$$B + e_3 n(n+1) = a(e_1 - e_2),$$

and $P(u)$ is the Weierstrass P function.

When n is a positive integer, it is found that there are $2n + 1$ values of B for which eqn (5.140) has a solution which can belong to any one of four species. If one of these solutions is expanded in descending powers of ξ, the coefficient of the term $\xi^{n/2}$ being taken as unity, the function is called a Lamé function of degree n (of one or other of the four species). Lamé functions are often denoted by $E_n^m(\xi)$, where $m = 1, 2, \ldots, 2n + 1$. If a series solution of the form

$$\Lambda = \sum_{r=0}^{\infty} b_r (\xi - e_2)^{1/2n - r}, \tag{5.141}$$

is inserted into eqn (5.139), a recurrence relation may be obtained among the coefficients b_r, each of which is a polynomial in B of degree r. In reference [86] the coefficients of b^r and also their signs have been provided. If such a series terminates, e.g. if n is even and $b_{1+n/2} = 0$, then all succeeding terms vanish and B is a root of an algebraic equation of degree $\tfrac{1}{2}n + 1$. There are $\tfrac{1}{2}n + 1$ real and distinct values of B when n is even and $\tfrac{1}{2}(n+1)$ when n is odd. It is a curious fact that for each value of n there are $2n + 1$ values of m which provide solutions E_n^m. This is very reminiscent of the spherical harmonics arising from quantum eigenfunctions in a spherical potential.

When eqn (5.138) has periodic solutions, they may be expressed by

$$\left\{ 1, \begin{array}{lll} \text{sn } \alpha, & \text{cn } \alpha \text{ dn } \alpha, & \\ \text{cn } \alpha, & \text{dn } \alpha \text{ sn } \alpha, & \text{sn } \alpha \text{ cn } \alpha \text{ dn } \alpha \\ \text{dn } \alpha, & \text{sn } \alpha \text{ cn } \alpha, & \end{array} \right\} \prod_p \left(\text{sn}^2 \alpha - \text{sn}^2 \alpha_p \right). \tag{5.142}$$

Each of the columns inside { } denotes a species, by analogy with the four species of ellipsoidal harmonics [86]. Each of the eight prefactors in { } may multiply the product to the right and the α_p are constants.

Complete solutions for an arbitrary value of B (which may be fixed by appropriate boundary conditions) may be written in the form

$$\Lambda_\mp = \prod_{r=1}^{n} \left(\frac{\sigma(a_r \mp u)}{\sigma(u)\sigma(a_r)} \right) \exp\left(\mp u \sum_{r=1}^{n} \zeta(a_r) \right). \tag{5.143}$$

In eqn (5.143) above, u is the independent variable, and σ and ζ are the sigma and zeta functions of Weierstrass, where ζ is defined by

$$d\zeta(z)/dz = -P(z).$$

The reader will notice immediately that for arbitrary values of B the quantum solutions in eqn (5.143) contain decaying factors and, indeed, in some cases they could be singular.

5.7.4 Hyperbolic and trigonometric potentials

For particular values of the integration constant C and the parameters in eqn (5.133), one can obtain *hyperbolic solutions* when the elliptic modulus $k \to 1$. This corresponds to $C = 0$. One then obtains two different types of solution which corresponds to solitary waves, namely (a) the kink for $\varepsilon_i = +1$,

$$\phi_c = \mp(-\lambda_0/F_0)^{1/2} \tanh\left[(-\lambda_0)^{1/2}(x_i - x_i^0)\right], \quad (5.144)$$

and (b) the bump for $\varepsilon_i = -1$,

$$\phi_c = \mp(-2\lambda_0/F_0)^{1/2} \mathrm{sech}\left[(-2\lambda_0)^{1/2}(x_i - x_i^0)\right]. \quad (5.145)$$

The x_i^0 are integration constants. When either eqn (5.144) or eqn (5.145) is substituted into the equation above eqn (5.133), the following quantum equation is derived:

$$-d^2\Lambda/d\alpha^2 + \left[(L^2 - \omega_n^2) - L(L+1)\mathrm{sech}^2\alpha\right]\Lambda = 0, \quad (5.146)$$

where in both cases (a) and (b) $L = 2$ and λ_0 is negative. However, in case (a),

$$\alpha = (-\lambda_0)^{1/2}(x_i - x_i^0) \quad \text{with } \omega_n^2 = 2E_n/|\lambda_0|, \quad (5.147)$$

but in case (b),

$$\alpha = (-2\lambda_0)^{1/2}(x_i - x_i^0) \quad \text{with } \omega_n^2 = 3 - E_n/|\lambda_0|. \quad (5.148)$$

The type of Schrödinger equation in eqn (5.146) has been studied by many authors [87]. An explicit solution is obtained as

$$\Lambda = \cosh^{-K}(\alpha) F(K+3, K+\tfrac{1}{2} - \tfrac{5}{2}|K+1|e^\alpha/(e^\alpha + e^{-\alpha})), \quad (5.149)$$

where F is a hypergeometric function and $K = 4 - \omega_n^2$. The solution is well behaved, i.e. finite when $\alpha \to -\infty$ as long as $K - 2$ is a negative integer. This condition leads to quantized energies which are $E_n = ((4 - n^2)/2)|\lambda_0|$ for case (a) and $E_n = (n^2 - 1)|\lambda_0|$ for case (b) with $n = 1, 2$ in both cases. This fact will be utilized in the Haldane gap problem application discussed in Chapter 6. Solutions for higher eigenvalues form a continuous spectrum, are unbounded, and take the form $\exp(in\alpha)$ multiplied by a Jacobi polynomial in $\tanh \alpha$ [13]. It is interesting that there are only two bound levels for the sech potential and that the levels do not bunch and approach the continuum as in the hydrogen limit. This, in fact, follows from the Bargmann criterion [88], since $\int_0^\infty x \operatorname{sech}^2 x \, dx$ is finite [89] and $\operatorname{sech}^2 x + \tanh^2 x = 1$ (i.e. the $\tanh^2$ potential problem may be written as a sech^2 potential with a modified eigenvalue).

It is also possible to obtain *trigonometric potentials* in the quantum equation by considering the elliptic functions when their elliptic modulus $k \to 0$. From Appendix B, $\operatorname{cn}(u, k) \to \cos u$, $\operatorname{sn}(u, k) \to \sin u$, and $\operatorname{dn}(u, k) \to 1 - (k^2/2)\sin^2 u$ in this limit. It is clear that in all these cases the quantum equation reduces to

$$\frac{d^2 \Lambda}{d\alpha^2} + [-(a + \tfrac{1}{2}b) + \tfrac{1}{2}b \cos 2\alpha]\Lambda = 0. \tag{5.150}$$

This is a Mathieu equation, on which there is a vast literature. Because of its periodic potential, it leads to Bloch-like eigenfunctions and the presence of both allowed and forbidden bands [85, 112]. We notice, incidentally, that in all three cases, i.e. eqns (5.104), (5.105), and (5.106), when k is actually zero, the amplitude of $\phi_c \to 0$. When $k = 0$ the quantum equation becomes

$$d^2 \Lambda / d\alpha^2 = a\Lambda, \tag{5.151}$$

which will produce exponentially increasing or decreasing functions (for $a > 0$) and oscillatory forms when $a < 0$.

To summarize this part of this chapter, we have found that, at all three orders of approximation to the classical equation in the MCS, a wealth of analytic solutions can be obtained and such properties as localization can be routinely studied. Furthermore, the quantization procedure used subsequently for several classes of solution works surprisingly well in view of the fact that the effective potential procedure emerges quite naturally as a special case.

What we have done so far in this chapter completes the conceptual development of the method as such. The inquisitive reader, however, may have posed questions regarding the generality of the approach, e.g. the naive plane-wave base may have been put in question and nowhere is spin mentioned. Also, the possibility of deriving relativistically invariant equations will be discussed. We shall address these questions in full detail and close this chapter with an important generalization to the case of two different types of elementary particle interacting together.

5.8 The transformation of bases

In Section 5.3.2 describing the MCS we have used a plane-wave basis to

THE TRANSFORMATION OF BASES

obtain results. If we were dealing with a linear theory, there would be no problem about changing from one basis to another. However, the MCS is intentionally and importantly nonlinear, so it is not obvious what will happen to our equations of motion in a basis other than a plane-wave basis. We shall show here that any complete set of one-particle states may be utilized and results in equations of motion of the same form.

To proceed, we use the standard definition of a field $\psi(r)$ in an arbitrary basis [90], namely

$$\psi(r) = \sum_p U_p(r) a_p. \tag{5.152}$$

Here, the function $U_p(r)$ replaces the factor $\Omega^{-1/2} \exp(-i\boldsymbol{k} \cdot \boldsymbol{r})$ which we used in our main development (see eqn (5.36)) in a plane-wave basis. The subscript, p, is a label to identify one member of a complete set and is shorthand for a string of an appropriate set of quantum numbers. We suppose that, in the U-basis (using the functions $U_p(r)$), the one-body part of the Hamiltonian is diagonal. In a plane-wave basis (PB) labels corresponding to p in the nonplane-wave basis (NPB) become wave vectors which we denote by $\boldsymbol{k}_1, \boldsymbol{k}_2, \boldsymbol{k}_3, \ldots$, etc. In a NPB the basis functions will be denoted by $|p\rangle, |q\rangle, |r\rangle, |s\rangle, \ldots$; and the Hamiltonian, in second quantized form, becomes

$$H_1 = \sum_p \omega_p a_p^+ a_p + \sum_{p,q,r,s} \langle p, q | V | r, s \rangle a_p^+ a_q^+ a_s a_r, \tag{5.153}$$

where, in general, the two-body interaction is now no longer explicitly of the linear momentum conserving form. If the one-body part of H_1, i.e. in ω_p, only contains the kinetic energy, this *will* be diagonal in PB, but in general this will not be so. The form of H_1 is a special one in that we have taken the one-body part to be diagonal, but even if we began with a Hamiltonian where this was not so, it could always be reduced to this form by a unitary transformation. This is well known, so we take H_1 as our starting point. We assume, in the standard way, that the a_p^+ operators create Fermion basis functions $|U_p(r)\rangle \equiv |p\rangle$ from the vacuum $|0\rangle$, so that $a_p^+ |0\rangle = |p\rangle$. A similar treatment may also be given for Boson operators.

There is clearly a unitary transformation which links the basis functions in the NPB to those for a PB. We express this by

$$|p\rangle = \sum_{k_1} C_{k_1,p} |k_1\rangle, \tag{5.154}$$

or

$$a_p^+ = \sum_{k_1} C_{k_1,p} a_{k_1}^+. \tag{5.155}$$

The two bases, i.e. PB and NPB, are complete and can always be chosen to be orthonormal. Thus, from eqn (5.154), we have

$$\langle q|p\rangle = \sum_{k_1} C^*_{k_1,q} C_{k_1,p} = \delta_{q,p}. \tag{5.156}$$

The transformation in eqn (5.154) may be inverted to give

$$a^+_{k_1} = \sum_p C^*_{k_1,p} a^+_p, \tag{5.157}$$

which is clearly consistent with eqn (5.154), since multiplying both sides of eqn (5.157) by $C_{k_1,p'}$, summing over k_1, and using eqn (5.156) reduces to eqn (5.154).

To obtain the equation of motion for the field defined in eqn (5.152) we use Heisenberg's equation of motion for a particular annihilator a_η in the NPB, just as we did in our development of the MCS, using plane-wave annihilators. We obtain

$$i\hbar \partial_t a_\eta = [a_\eta, H_1]_- = \omega_\eta a_\eta + \sum_{q,r,s} \langle \eta, q|V|r,s\rangle a^+_q a_s a_r$$

$$- \sum_{p,r,s} \langle p, \eta|V|r,s\rangle a^+_p a_s a_r. \tag{5.158}$$

We now multiply eqn (5.158) on both sides by $U_\eta(\mathbf{r})$ and sum over η. To see how the calculation proceeds, we shall consider only one of the terms in eqn (5.158) from the two-body interaction—the others may be tackled in a similar way. Thus, taking the second term on the right-hand side of eqn (5.158), we obtain

$$\sum_{q,r,s,\eta} U_\eta(\mathbf{r}) \langle \eta, q|V|r,s\rangle a^+_q a_s a_r$$

$$= \sum_{\eta, k_1, k_2, k_3, k_4} U_\eta(\mathbf{r}) C^*_{k_4,\eta} \langle k_4, k_1|V|k_2, k_3\rangle a^+_{k_1} a_{k_3} a_{k_2}, \tag{5.159}$$

where $\langle \eta|$ has been converted to plane waves using the dual of eqn (5.154) and each of q, r, and s has been similarly replaced utilizing eqn (5.156). The sum over η may be now performed using a real-space version of eqn (5.157), namely

$$\frac{1}{\sqrt{\Omega}} \exp(-i k_4 \cdot \mathbf{r}) = \sum_\eta C^*_{k_4,\eta} U_\eta(\mathbf{r}). \tag{5.160}$$

and the orthonormality of the $|k_i\rangle$'s.

The reader will notice that a basis function in the PB is defined deliberately with a minus in the exponent, so that these transformations are consistent with our earlier development of the MCS. With the aid of eqn (5.160), the right-hand side of eqn (5.159) now becomes

$$\sum_{k_1,k_2,k_3,k_4} \frac{\exp(-i\mathbf{k}_4 \cdot \mathbf{r})}{\sqrt{\Omega}} \langle k_4, k_1| V |k_2, k_3\rangle a^+_{k_1} a_{k_3} a_{k_2}. \qquad (5.161)$$

All of the wave-vector labels in eqn (5.161) now refer explicitly to plane waves and, for nonzero two-body matrix elements, the conservation of linear momentum will hold; namely,

$$\mathbf{k}_4 = \mathbf{k}_2 + \mathbf{k}_3 - \mathbf{k}_1. \qquad (5.162)$$

Clearly, using eqn (5.162), eqn (5.161) may now be written in terms of fields in a PB in the usual way. For example, when the matrix element in eqn (5.161) is expanded, i.e. when the matrix element is a constant, the leading term clearly gives a term proportional to $\psi^+_{PB}\psi_{PB}\psi_{PB}$ and derivatives for higher order terms may be obtained in the usual way. All we have to do now is to ensure that the fields so obtained are related directly to those in a NPB. To do this we write the field creator in the NPB as

$$\psi^+_{NPB} = \sum_p U^*_p(r) a^+_p, \qquad (5.163)$$

and then replace on the right-hand side of eqn (5.163) using eqn (5.154). Note, before we do this, that when we multiplied both sides of eqn (5.158) by $U_\eta(r)$ and then summed over η, the left-hand side became $i\hbar(\partial_t \psi_{NPB})$ and the term from the one-body contribution is $\sum_\eta \omega_\eta U_\eta a_\eta$. We continue by returning to eqn (5.163), and we use eqn (5.154) to obtain

$$\psi^+_{NPB} = \sum_p U^*_p(r) a^+_p = \sum_{p,k_1} U^*_p(r) C_{k_1,p} a^+_{k_1}$$

$$= \sum_{k_1} \frac{\exp(+i\mathbf{k}_1 \cdot \mathbf{r})}{\sqrt{\Omega}} a^+_{k_1} = \psi^+_{PB}, \qquad (5.164)$$

where in the last step we use eqn (5.160).

We see from the above that we may again expand matrix elements in eqn (5.161), about the critical point, with respect to the **k**-vectors, as previously for the PB case, and then convert the resulting expression to one involving the field in the NPB and its derivatives. From eqn (5.164) this may be transformed directly into one involving ψ_{NPB} and its gradients. Thus, due to eqn (5.164), terms in the equation of motion originating from the two-body interactions will now be identical in form in terms of ψ_{NPB} and its gradients to those for ψ_{PB} and its gradients.

There is, however, what appears to be a difficulty with the one body contribution to the equation of motion which now, in the NPB, takes the form

$$\sum_{\eta} \omega_{\eta} U_{\eta}(r) a_{\eta}. \tag{5.165}$$

The problem is to express eqn (5.165) as a function of ψ_{NBP} and its gradients. To move forward we note that when the one-body part of the Hamiltonian is second quantized using either a PB or a NPB, we must have the equivalence

$$\sum_{\eta} \omega_{\eta} a_{\eta}^{+} a_{\eta} = \sum_{\eta} \langle \eta | \omega | \eta \rangle a_{\eta}^{+} a_{\eta}$$

$$= \sum_{k_1, k_2} \langle k_1 | \omega | k_2 \rangle a_{k_1}^{+} a_{k_2}, \tag{5.166}$$

and, from eqn (5.164),

$$\sum_{\eta} U_{\eta} a_{\eta} = \sum_{k_3} \frac{1}{\sqrt{\Omega}} \exp(-ik_3 \cdot r) a_{k_3}. \tag{5.167}$$

We now take the commutator of the left-hand side of eqn (5.167) with the left-hand side of eqn (5.166), and similarly for the right-hand sides, to obtain

$$\sum_{\eta} U_{\eta} \omega_{\eta} a_{\eta} = \sum_{k_1, k_2} \langle k_1 | \omega | k_2 \rangle \frac{1}{\sqrt{\Omega}} \exp(-ik_1 \cdot r) a_{k_2}. \tag{5.168}$$

In the simplest situation it is obvious that if the operator, ω, involves kinetic energy or an operator proportional to ∇, perhaps due to an external field, then k_1 will be equal to k_2, so that eqn (5.168) will give rise to terms such as $\nabla^2 \psi$ or $\mu \cdot \nabla \psi$, with μ a constant vector in the equation of motion. More often, however, this will not be the case and we have some more general operator. As kinetic energy has already been discussed, let us suppose that ω is a pure potential energy $V(r)$. The right-hand side of eqn (5.168) becomes

$$\frac{1}{\Omega \sqrt{\Omega}} \sum_{k_1, k_2} \int_{\Omega} e^{+ik_1 \cdot (-r + r')} V(r') e^{-ik_2 \cdot r'} d^3 r' a_{k_2}. \tag{5.169}$$

If we use the relation

$$\sum_{k_1} \exp(-k_1 \cdot (-r + r')) = \Omega \delta(r - r'), \tag{5.170}$$

in eqn (5.169), it reduces to

$$\frac{V(r)}{\sqrt{\Omega}} \sum_{k_2} \exp(-ik_2 \cdot r) a_{k_2} = V(r) \psi_{\text{PB}}. \tag{5.171}$$

The result in eqn (5.171) is exact. However, we would like to point out that there is another way in which to re-express eqn (5.168) in terms of the field in a PB. Going back to eqn (5.168) and putting $\omega = V(r)$, we see that it may be written in the form

$$\sum_{k_2} \left[\sum_{k_1} e^{i(k_2 - k_1) \cdot r} \langle k_1 | V(r) | k_2 \rangle \right]$$
$$\times \frac{1}{\sqrt{\Omega}} \exp(-i k_2 \cdot r) a_{k_2}. \quad (5.172)$$

The terms in the summation inside the brackets [ ] are a function of $k_1 - k_2$ and, therefore, this may be expanded as a series in the components of this vector. Having done this, the summation over k_1 may then be performed *leaving an expansion in powers of the components of* $\mathbf{k}_2$ *alone*. Thus, eqn (5.172) may again be written in terms of the PB field and its derivative only. This means that the one-body party of the Hamiltonian can not only be written as in eqn (5.171) but in an autonomous way involving ψ and its derivative, i.e. with no multiplicative factor involving functions of r. The latter form has been used throughout our discussion in the main text for the MCS. Unfortunately, this latter method results in a possible significant redressing of the term in $\nabla^2 \psi$ and others, whereas if $V(r)$ is retained, this is not so.

In summary, we have shown in this section that the field equation of motion retains the same form in any orthonormal base of states (because of the relationship in eqn (5.164)) and not only for the plane waves which were used in the main discussion of the method. The equations obtained in the plane wave basis were very convenient to analyse, but this may not always be the case, as will become clear when we study the problem of multi-electron atoms in Chapter 6.

In the next section we investigate another important facet of the many-body problem, i.e. the role of spin variables in the equations of motion.

5.9 The role of spin in the development of coherent structures for spin-independent interactions

In our discussions so far we have developed a virtually exact description of many-body systems using the MCS, but we have not discussed the possibility of the fields being described by discrete quantum numbers such as spin components. The prescription in the MCS was to begin with the generic

Hamiltonian in eqn (5.24). What emerges from this latter study is that the two-body part of H is chiefly responsible for new qualitative behaviour of the physical system. This behaviour is closely associated with broken symmetries and can lead to the emergence of a new type of degenerate ground state with a coherent structure. Obviously, this phenomenon is very closely connected to the fact that various parts of the physical system being investigated are becoming correlated in some way. In all of this, it may appear to be somewhat surprising that spin has not so far been mentioned at all. Even if it had been included in the quantum labels, it is difficult to see what meaning could be attached to the expansion about the special point above. Furthermore, it is well known that spin is certainly an important correlation component in a number of phenomena, especially in superconductivity and magnetism. Thus, in this section we endeavour to include spin components *but only for two-body interactions which do not depend on spin explicitly* (for details, see reference [108]). To include spin operators in the interaction makes this problem much more complicated, so we do not discuss this here. In this section we will find the surprising result that the equation of motion for the quantum field ψ, whether the associated particles have spin $S = \tfrac{1}{2}$ or higher, is unchanged in its *form*.

We consider first the case of Fermions with spin $S = \tfrac{1}{2}$. There are only two spin components here, which in general we label as σ or σ', and explicitly they are $m_s = +\tfrac{1}{2}$ and $m_s = -\tfrac{1}{2}$, which we abbreviate to '+' and '−', respectively. The spin independence of the interactions is expressed by the following equations:

$$\omega_{k,l} = \left\langle \overset{+}{k} \middle| \omega \middle| \overset{+}{l} \right\rangle = \left\langle \overset{-}{k} \middle| \omega \middle| \overset{-}{l} \right\rangle = \langle k | \omega | l \rangle \qquad (5.173)$$

and

$$\Delta_{k,l,m} = \left\langle \overset{\sigma}{k}, \overset{\sigma'}{l} \middle| \Delta_{1,2} \middle| \overset{\sigma}{k+l-m}, \overset{\sigma'}{m} \right\rangle$$
$$= \langle k, l | \Delta_{1,2} | k+l-m, m \rangle. \qquad (5.174)$$

In equations above, for each ket, we have written the spin components above the relevant spatial labels k, l, and m, the operator Δ has been given subscript labels 1 and 2 to denote the fact that in second quantization it will contain operators referring to particles 1 and 2, and in the bra and ket the first label refers to particle 1 and the second to particle 2. Our generic starting Hamiltonian becomes

$$H_2 = \sum_{\substack{k,l \\ \sigma}} \omega_{k,l} q^+_{k\sigma} q_{l\sigma}$$
$$+ \sum_{\substack{k,l,m \\ \sigma,\sigma'}} \Delta_{k,l,m} q^+_{k\sigma} q^+_{l\sigma'} q_{m,\sigma'} q_{(k+l-m)\sigma}. \qquad (5.175)$$

THE ROLE OF SPIN 277

The annihilators and creators refer to Fermions. We now use Heisenberg's equation of motion for the spin-up (+) and spin-down (−) annihilators. When the commutators are calculated, it may be shown, in a relatively straightforward way, that the above equations take the explicit form

$$i\hbar \partial_t q_{\eta+} = \sum_k \omega_{\eta,k} q_{k+}$$
$$+ \sum_{k,m} [\Delta_{\eta,k,m} q^+_{k+} q_{m+} q_{(\eta+k-m)+}$$
$$- \Delta_{k,\eta,m} q^+_{k+} q_{m+} q_{(k+\eta-m)+}$$
$$+ \Delta_{\eta,k,m} q^+_{k-} q_{m-} q_{(\eta+k-m)+}$$
$$- \Delta_{k,\eta,m} q^+_{k-} q_{m+} q_{(k+\eta-m)-}], \qquad (5.176)$$

and similarly for the $q_{\eta-}$ operator.

At this point we define spin-dependent fields by analogy with the spin-independent case, by

$$\psi_+(r) = \Omega^{-1/2} \sum_k q_{k+} \exp(-i k \cdot r) \qquad (5.177)$$

and

$$\psi_-(r) = \Omega^{-1/2} \sum_k q_{k-} \exp(-i k \cdot r), \qquad (5.178)$$

and similarly for their Hermitian conjugates ψ^+_+ and ψ^+_-. Here, Ω is a normalizing constant volume, and it is to be understood that the components of the position vector r are c-numbers relative to both the spin-up and spin-down annihilators. The one-body term in $\omega_{\eta,k}$ can always be chosen to be diagonal by using a unitary transformation, so what is *normally* done now in the MCS follows five main steps:

1. $\Delta_{\eta,k,m}$, $\Delta_{k,\eta,m}$, and $\omega_{\eta,k}$ (now assumed diagonal) are expanded as a Taylor series about the particular point (k_0, l_0, m_0) in reciprocal space, or quantum-number space, in powers of the components of η, k, and m.

2. Then both sides of eqn (5.176) are multiplied by $\exp[-i\eta \cdot r]$.

3. Both sides of eqn (5.176) are divided by $\Omega^{1/2}$ and summed over η.

4. A prefactor of $\exp(i k \cdot r)$ is inserted to the left of each $q^+_{k_\mp}$ and a factor $\exp(-i m \cdot r)$ to the left of $q_{m_\mp}$, so that what remains is a prefactor of $\exp[-i(\eta + k - m) \cdot r]$ for $q_{(\eta+k-m)_\mp}$.

5. Then simply using the definitions of the fields makes eqn (5.176) an equation in fields entirely.

Returning to the part of eqn (5.176) involving $\omega_{k,l}$ and the five steps enunciated above we see that we must obtain, in the spin-dependent case,

$$i\hbar \partial_t \psi_+ = \lambda_0 \psi_+ + i\boldsymbol{\lambda}_1 \cdot \nabla \psi_+ - \tfrac{1}{2} \sum_{i,j} (\lambda_2)_{ij} \partial_{x_i x_j} \psi_+ + \text{interaction terms.}$$

(5.179)

Going back to eqn (5.24), if we consider the interaction term, interchange the labels k and l, and remember the spin-independent form of $\Delta_{k,l,m}$ in eqn (5.174), from the Fermion commutation rules we may put

$$\Delta_{k,l,m} = -\Delta_{l,k,m}.$$

Using the above, the interaction terms in eqn (5.186), therefore, take the form

$$\sum_{k,m} \Delta_{\eta,k,m} \{ 2q_{k+}^+ q_{m+} q_{(\eta+k-m)+} + q_{k-}^+ q_{m-} q_{(\eta+k-m)+}$$
$$+ q_{k-}^+ q_{m+} q_{(k+\eta-m)-} \}.$$

If the spin subscripts are omitted in the above term, the first term would simply correspond to what we obtained in the spinless case, as we discussed earlier. Furthermore, using the five main steps above, it is clear that in the spinless case a component q_k with an associated exponential factor transforms into the field ψ, i.e. $q_k \to \psi$, where '$\to$' means transforms into and q_k itself means not just q_k but q_k premultiplied by $\exp(-i\mathbf{k}\cdot\mathbf{r})$. In a very similar way, in the spinless case, $k_x q_k \to \partial_{x_1} \psi$. In the spin-dependent case, remembering that the matrix elements are spin-independent and are expanded in terms of, for example, powers of k_x, k_y, and k_z, and powers of the other quantum label components, we must have $q_{k\sigma} \to \psi_\sigma$ and $k_x q_{k\sigma} \to \partial_{x_1} \psi_\sigma$, and $\psi^+ \psi \nabla \psi$ in the spinless case must be replaced by $\psi_+^+ \psi_+ \nabla \psi_+ + \tfrac{1}{2} \psi_-^+ \psi_- \nabla \psi_+ + \tfrac{1}{2} \psi_-^+ \psi_+ \nabla \psi_-$ in the spin-dependent case. Similarly, $\psi^+ \psi \psi$ must be replaced by

$$\psi_+^+ \psi_+ \psi_+ + \tfrac{1}{2} \psi_-^+ \psi_- \psi_+ + \tfrac{1}{2} \psi_-^+ \psi_+ \psi_-$$

and $\psi^+ \partial_{x_i x_j} \psi$ by

$$\psi_+^+ (\partial_{x_i x_j} \psi_+) \psi_+ + \tfrac{1}{2} \psi_-^+ (\partial_{x_i x_j} \psi_-) \psi_+ + \tfrac{1}{2} \psi_-^+ (\partial_{x_i x_j} \psi_+) \psi_-. \quad (5.180)$$

The rule in the equation of motion for ψ_+ is to replace each term in the spinless case by a first term in which all the ψ's become ψ_+, one half of a second term in which the first ψ becomes ψ_-, the second ψ_-, and the third is replaced by ψ_+, the order in which the ψ's appear being preserved. Lastly, a third term arises which is premultiplied by one half, the first ψ becomes ψ_-, the second ψ_+, and the third ψ_-, in that order. A corresponding rule clearly exists for ψ_-. The net result of all these rules, for the

THE ROLE OF SPIN

equation of motion for ψ_+, is

$$i\hbar \partial_t \psi_+ = \lambda_0 \psi_+ + \boldsymbol{\lambda}_1 \cdot \nabla \psi_+ - \tfrac{1}{2} \sum_{ij} (\lambda_2)_{ij} \partial_{x_i x_j} \psi_+$$

$$+ \nu_2 \{ \psi_+^\dagger \psi_+ \psi_+ + \tfrac{1}{2} \psi_-^\dagger [\psi_- \psi_+ + \psi_+ \psi_-] \}$$

$$+ \Omega i (\nabla_\eta f)_0 \{ \psi_+^\dagger \psi_+ \nabla \psi_+ + \psi_+^\dagger (\nabla \psi_+) \psi_+ + \tfrac{1}{2} \psi_-^\dagger \nabla [\psi_- \psi_+ + \psi_+ \psi_-] \}$$

$$+ \frac{\Omega}{2!} \sum_{ij} (\partial_{m_i m_j} f)_0 \left\{ \psi_+^\dagger \left(-\partial_{x_i x_j} \psi_+ - 2i(\partial_{x_i} \psi_+) m_j^0 + \psi_+ m_i^0 m_j^0 \right) \psi_+ \right.$$

$$+ \psi_+^\dagger \psi_+ \left(-\partial_{x_i x_j} \psi_+ - 2i(\partial_{x_i} \psi_+) \eta_j^0 + \eta_i^0 \eta_j^0 \psi_+ \right)$$

$$\left. + 2 \psi_+^\dagger \left(i \partial_{x_i} \psi_+ - m_i^0 \psi_+ \right) \left(i \partial_{x_j} \psi_+ - \eta_j^0 \psi_+ \right) \right\}$$

$$+ \frac{\Omega}{2!} \sum_{ij} (\partial_{m_i m_j} f)_0 \left\{ -i m_j^0 \psi_-^\dagger \partial_{x_i} [\psi_- \psi_+ + \psi_+ \psi_-] \right.$$

$$+ \tfrac{1}{2} m_i^0 m_j^0 \psi_-^\dagger [\psi_- \psi_+ + \psi_+ \psi_-]$$

$$- \tfrac{1}{2} \psi_-^\dagger \partial_{x_i x_j} [\psi_- \psi_+ + \psi_+ \psi_-] + \tfrac{1}{2} \eta_i^0 \eta_j^0 \psi_-^\dagger [\psi_- \psi_+ + \psi_+ \psi_-]$$

$$\left. - i \eta_j^0 \psi_-^\dagger \partial_{x_i} [\psi_- \psi_+ + \psi_+ \psi_-] + m_i^0 \eta_j^0 \psi_-^\dagger [\psi_- \psi_+ + \psi_+ \psi_-] \right\}. \quad (5.181)$$

It is easy to see from the definition of the spin-dependent fields in eqn (5.177) and the commutation rules for the annihilators q_k that

$$\psi_\sigma \psi_{\sigma'} + \psi_{\sigma'} \psi_\sigma = 0. \quad (5.182)$$

Thus, we see that all of the square-bracketed operators vanish in eqn (5.181) and that the equation of motion for ψ_+ is identical to that for ψ in the spinless field case. We can also show by the same means that the equation of motion for ψ_- is also of the same form as for ψ. As we pointed out earlier, the equations of motion for Boson annihilators and creators were identical to those for Fermions, since for Bosons we may put $\Delta_{\eta,k,m} = \Delta_{k,\eta,m}$. Hence, *for Bosons with spin $S = 0$ and Fermions with total spin $S = 1/2$ the equations of motion are of the same form*, namely identical to the corresponding equations in the spinless field case which we discussed earlier.

What we must now ask is what will happen to the equations of motion when we consider Fermions with higher half-integral spin or Bosons with integral spin larger than $S = 0$. Clearly, the Hamiltonian in eqn (5.175) will still be the appropriate one to use, but the interpretation of σ and σ' must be modified so that the summations for Fermions over these discrete variables are not just over '+' and '−' but over the $2S + 1$ components of the total spin S. For Bosons these latter summations will also be over the associated components. For the Fermion case, the equation of motion for $q_{\eta\sigma''}$ becomes, using Heisenberg's equations of motion and eqn (5.175) (but necessarily with $\mathbf{k} \neq \mathbf{l}$ in $\omega_{\mathbf{k},\mathbf{l}}$)

$$i\hbar \partial_t q_{\eta\sigma''} = \sum_k \omega_{\eta,k} q_{k,\sigma''}$$

$$+ \sum_{k,m,\sigma} \Delta_{\eta,k,m} \{q_{k,\sigma}^+ q_{m,\sigma} q_{(\eta+k-m),\sigma''} + q_{k,\sigma}^+ q_{m,\sigma} q_{(k+\eta-m),\sigma''}\}.$$

(5.183)

The above equation holds for Fermion and Boson annihilators (for Boson operators $\Delta_{k\eta m} = \Delta_{\eta k m}$). Once again, this may be rewritten in terms of fields ψ_σ and ψ_σ^+. One can verify with some very tedious algebra that eqn (5.183) can be reduced to a form like that in eqn (5.181), where '+' is replaced by σ'', '$-$' by σ, and the right-hand side should be summed over σ. This amounts to leaving those terms involving σ'' alone (since they may be divided by the non-zero constant $\Sigma_\sigma 1$) and summing over σ the square bracket-like terms, where now $\sigma \neq \sigma''$. Because of the last restriction each of the sums over the bracketed terms become zero from the commutation rules for the fields, i.e. they vanish in the same way as for the $S = 1/2$ case. Thus, the equations of motion, for *any Fermionic spin S*, and component σ'' are the same as in the $S = 1/2$ case which, in turn, is identical to the spinless field situation.

We are thus left with the case of Bosons with spin $S \neq 0$. This is not such a simple system to analyse as the Fermion case, since the field annihilators commute whether or not they possess the same spin components. The one-body terms give rise to $\omega_{\eta,k}$, so when expanded and converted to fields will yield eqn (5.179), without the interaction terms and '+' replaced by σ'', as before. Thus, we only need to consider the interaction terms. The summations over σ arise in only six different types, which are as follows:

(a) $\quad \sum_\sigma \psi_\sigma^+ (\partial_{x_i x_j} \psi_{\sigma''}) \psi_\sigma + \sum_\sigma \psi_\sigma^+ (\partial_{x_i x_j} \psi_\sigma) \psi_{\sigma''}$;

(b) $\quad \sum_\sigma \psi_\sigma^+ (\partial_{x_i} \psi_{\sigma''}) \psi_\sigma + \sum_\sigma \psi_\sigma^+ (\partial_{x_i} \psi_\sigma) \psi_{\sigma''}$;

(c) $\quad \sum_\sigma \psi_\sigma^+ \psi_{\sigma''} \psi_\sigma + \sum_\sigma \psi_\sigma^+ \psi_\sigma \psi_{\sigma''}$;

(d) $\quad \sum_\sigma \psi_\sigma^+ \psi_{\sigma''} \partial_{x_i x_j} \psi_\sigma + \sum_\sigma \psi_\sigma^+ \psi_\sigma \partial_{x_i x_j} \psi_{\sigma''}$;

(e) $\quad \sum_\sigma \psi_\sigma^+ \psi_{\sigma''} \partial_{x_i} \psi_\sigma + \sum_\sigma \psi_\sigma^+ \psi_\sigma \partial_{x_i} \psi_{\sigma''}$;

(f) $\quad \sum_\sigma \psi_\sigma^+ (\partial_{x_i} \psi_{\sigma''}) \partial_{x_j} \psi_\sigma + \sum_\sigma \psi_\sigma^+ (\partial_{x_i} \psi_\sigma) \partial_{x_j} \psi_{\sigma''}$.

Consider each of these cases and fix σ''. The second summation in each may be clearly written as a particular operator summed over σ times a function involving $\psi_{\sigma''}$ or its derivative. For example, for case (a) we have $(\sum_\sigma \psi_\sigma^+ (\partial_{x_i x_j} \psi_\sigma)) \psi_{\sigma''}$, or in case (e) $(\sum_\sigma \psi_\sigma^+ \psi_\sigma) \partial_{x_i} \psi_{\sigma''}$. For each case, the sum over σ in the rounded brackets *will be identical for each fixed σ'' separately*. Now let us examine the first term in each case and begin with case (c). Clearly, ψ_σ may be commuted through $\psi_{\sigma''}$ and the sum over σ performed. The term will then again take the form of the examples (a)–(f) and the sum over σ will be identical for each fixed σ''. In case (e) as $\partial_{x_i} \psi_\sigma(r)$ may be defined as a limit,

$$\lim_{\delta x_i \to 0} \frac{\psi_\sigma(r + \delta x_i) - \psi_\sigma(r)}{\delta x_i},$$

$\partial_{x_i} \psi_\sigma(r)$ may be commuted through $\psi_{\sigma''}$ and the sum over σ again performed. A similar argument may be put forward for the cases containing second derivatives. Thus, for the first term in each case, *the summation over σ will give the same result for each component of spin σ''*. Hence, the *equation of motion for each spin component of the Bosons will be identical*. If the boundary conditions for each field $\psi_{\sigma''}$ are assumed to be the same, then we take each of these fields to be identical. The summations over σ may, therefore, be performed. For example

$$\sum_\sigma \psi_\sigma^+ \psi_{\sigma''} \psi_\sigma = \left(\sum_\sigma \psi_\sigma^+ \psi_\sigma \right) \psi_{\sigma''} = (2S+1) \psi_{\sigma''}^+ \psi_{\sigma''} \psi_{\sigma''}. \quad (5.184)$$

The reason for the factor $(2S+1)$ is that there are $(2S+1)$ components σ of the spin S. This same factor will appear in all other cases too, e.g. case (a) yields

$$\sum_\sigma \psi_\sigma^+ \left(\partial_{x_i x_j}^+ \psi_{\sigma''} \right) \psi_\sigma = (2S+1) \psi_{\sigma''}^+ \left(\partial_{x_i x_j}^2 \psi_{\sigma''} \right) \psi_{\sigma''}. \quad (5.185)$$

We, therefore, conclude that the equation of motion for each spin component field $\psi_{\sigma''}$ will be the same and only differs from the spinless field case by a factor $(2S + 1)$ which scales the two-body interactive terms. Thus, for Bosons with $S = 1$ (as, for example, is the case for Cooper pairs in superconductors) the nonlinearity in the equation of motion is strengthened by a factor of three over the spin-zero case or, indeed, for any of the situations for Fermions. This effect might lead to higher critical temperatures if quasi-particles with larger values of integral spin could be formed. Interestingly, we note that for superfluid ^{3}He the transition temperature is much lower than for ^{4}He, in agreement with our predictions.

The arguments put forward in this subsection will be found useful and justify our approach to superconductivity in which, in Chapter 6, we used the Landau–Ginzburg model but spin is notably absent. An explicit incorporation of the spin could merely redress the coefficients.

5.10 A generalization of the method of coherent structures to coupled fields

In this section we apply the methodology of MCS to physical problems that involve two distinct degrees of freedom. By this we refer to the large variety of condensed matter systems the physics of which may be analysed by studying the interactions between *two* different order parameters or *two* distinct fields. Examples might include superconductors (the interaction between electrons described by one field and phonons described by a second field) [90], ferroelectric–ferromagnetic systems [91], ferroelectric–piezoelectric crystals [92], crystalline–superfluid systems [93], and also orientation–position ordering phenomena in molecular liquid crystals [94], to name but a few. There are many different physical effects which result from the interplay between two different fields, such as critical temperature shifts and crossover phenomena. The coupling of two fields in the first instance can be readily analysed approximately by some sort of mean field approach to reduce the problem to one involving one effective field. However, MCS does emphatically *not* do this and great store is placed on attempts to solve *nonlinear* equations as fully as possible, using recent mathematical developments in such areas as symmetry reduction. Here, we adopt the same approach but for two fields, beginning with an appropriate second-quantized Hamiltonian containing annihilators and creators for two different degrees of freedom. The general situation described here is indeed ubiquitous in condensed matter physics with, for example, various phonon modes interacting among themselves and leading to structural instabilities [94]. In a practical sense, any combination of two types of excitation present in a solid may be described in this way, e.g. electron–plasmon, magnon–phonon, electron–polaron and photo-electron, cross-coupled excitations, or subsystems.

5.10.1 Generalization of the method

We describe the entire physical system to be investigated by two parts, A and B, with strong two-body and hence nonlinear interactions within each of the systems (i.e. within A or within B) *and also* between the two subsystems. In the second-quantized formalism, system A will be described by annihilators a and creators a^+, whereas system B will be associated with operators b and b^+. We shall again use a plane-wave basis for both systems, so that each annihilator will be labelled with a wave vector. In the original MCS, the forms of the corresponding equations of motion were independent of basis, so we adopt here the simplest. Furthermore, at least at this stage, all spin labels are dropped. Again, in the one-field MCS, the form of equations of motion was also independent of spin quantum numbers. This may or may not be so, but here we drop spin for convenience.

Our model system is to be described by the generic Hamiltonian

$$H = H_a + H_b = H_{ab}. \tag{5.186}$$

In the above equation, H_a describes system A in isolation and H_b similarly for system B. Obviously, H_{ab} provides an interaction between the two subsystems. We shall include three-legged operators in both H_a and H_b, and the coupling, H_{ab}, will incorporate one- and two-quantum exchanges between systems A and B. Particles, quasi-particles, or excitations described by annihilators and creators, in either system A or system B, will be allowed to satisfy either Fermi–Dirac or Bose–Einstein commutation rules, and the whole Hamiltonian will, of course, be Hermitian. Thus, H_a and H_b are in general defined by

$$H_a = \sum_k \omega_k a_k^+ a_k + \sum_{k,l,m} \delta_{k,l,m} a_k^+ a_l^+ a_m a_{k+l-m}$$
$$+ \sum_{k,l,m} \{\varepsilon_{k,l,m} a_k^+ a_l a_{k-l} + \varepsilon_{k,l,m}^* a_{k-l}^+ a_l^+ a_k\}. \tag{5.187}$$

and

$$H_b = \sum_k \Omega_k b_k^+ b_k + \sum_{k,l,m} \Delta_{k,l,m} b_k^+ b_l^+ b_m b_{k+l-m}$$
$$+ \sum_{k,l,m} \{\Lambda_{k,l,m} b_k^+ b_l b_{k-l} + \Lambda_{k,l,m}^* b_{k-l}^+ b_l^+ b_k\}. \tag{5.188}$$

The reader will notice that each term conserves momentum and that three-legged terms are included to incorporate, for example, phonon–phonon interactions. In all cases we investigate, the reader should not imagine that H_a and H_b or indeed H_{ab} will always take the same form; e.g. if H_a referred to electrons the three-legged terms would disappear. However, we can use the above equations as a basis and drop terms as appropriate. Thus, we will not require a separate Hamiltonian for each case and simply refer back to H_a and H_b, which will always contain the terms we need. On this basis we take the interaction to be in its most general form

$$
\begin{aligned}
H_{ab} = &\sum_{k,l} \{\xi_{k,l} a_k^+ b_l + \xi_{k,l}^* b_l^+ a_k\} \\
&+ \sum_{k,l} \{\eta_{k,l} a_k^+ a_l b_{k-l} + \eta_{k,l}^* b_{k-l}^+ a_l^+ a_k\} \\
&+ \sum_{k,l} \{\lambda_{k,l} b_k^+ b_l a_{k-l} + \lambda_{k,l}^* a_{k-l}^+ b_l^+ b_k\} \\
&+ \sum_{k,l,m} \{\mu_{k,l,m} b_k^+ a_l^+ a_m a_{k+l-m} + \mu_{k,l,m}^* a_{k+l-m}^+ a_m^+ a_l b_k\} \\
&+ \sum_{k,l,m} \{\gamma_{k,l,m} a_k^+ b_l^+ b_m b_{k+l-m} + \gamma_{k,l,m}^* b_{k+l-m}^+ b_m^+ b_l a_k\} \\
&+ \sum_{k,l,m} \{\alpha_{k,l,m} a_k^+ b_l^+ b_m a_{k+l-m} + \alpha_{k,l,m}^* a_{k+l-m}^+ b_m^+ b_l a_k\}. \quad (5.189)
\end{aligned}
$$

As systems A and B can either describe Fermions or Bosons, there are four cases to consider. We may write them as follows:

(1) Boson–Boson (meaning system A–system B);
(2) indistinguishable Fermion–Fermion (this could be electron–electron);
(3) distinguishable Fermion–Fermion (e.g. electron–proton);
(4) Boson–Fermion.

Cases (2) and (3) are distinguished because of the obvious reason that cross-commutation rules, i.e. between the two systems A and B, will be different. Let us now consider the most complicated situation, namely Boson–Boson.

5.10.2 The Boson–Boson system

Both sets of operators, i.e. a_k, a_k^+, b_l, and b_l^+, satisfy Bose–Einstein commutation relations within each set and commute between sets. Notice that in H_{ab} we have specifically excluded all terms within a subsystem, which alone create one particle from the vacuum. However, we have allowed transformation of one 'particle' to another in H_{ab} and within H_a. For example, two of one type scatter off one to another. In the one-body part of H_{ab} we have included not only interactions which preserve different particles separately but also which incorporate an interaction when, for example, two B-particles are destroyed, a B-particle is created but an

GENERALIZATION OF MCS

A-excitation or quasi-particle also appears to conserve momentum. We have excluded particularly processes of the type in which a B-particle is destroyed at the expense of creating two or three A-particles.

As in MCS, we first employ Heisenberg's equation of motion but for each of the annihilators a_η and b_ξ separately. The reader may readily verify (with some algebra!) that

$$i\hbar \partial_t a_\eta = \omega_\eta a_\eta + \sum_{l,m} (\delta_{\eta,l,m} + \delta_{l,\eta,m}) a_l^+ a_m a_{\eta+l-m} + \xi_\eta b_\eta$$

$$+ \sum_l \{\eta_{\eta,l} a_l b_{\eta-l} + \eta^*_{l,\eta} b^+_{l-\eta} a_l\} + \sum_l \lambda^*_{\eta+l,l} b^+_l b_{\eta+l}$$

$$+ \sum_{l,m} (\varepsilon_{\eta,l,m} a_l a_{\eta-l} + \varepsilon^*_{\eta+l,l,m} a_l^+ a_{\eta+l} + \varepsilon^*_{l,\eta,m} a^+_{l-\eta} a_l)$$

$$+ \sum_{l,m} \{\mu_{l,\eta,m} b_l^+ a_m a_{l+\eta-m} + \mu^*_{\eta+m-l,l,m} a_m^+ a_l b_{\eta+m-l}$$

$$+ \mu^*_{l,m,\eta} a^+_{l+m-\eta} a_m b_l\}$$

$$+ \sum_{l,m} \gamma_{\eta,l,m} b_l^+ b_m b_{\eta+l-m}$$

$$+ \sum_{l,m} \{\alpha_{\eta,l,m} b_l^+ b_m a_{\eta+l-m} + \alpha^*_{\eta+m-l,l,m} b_m^+ b_l a_{\eta+m-l}\} \quad (5.190)$$

and, similarly,

$$i\hbar \partial_t b_\xi = \Omega_\xi b_\xi + \sum_{l,m} \{\Delta_{\xi,l,m} b_l^+ b_m b_{\xi+l-m} + \Delta_{l,\xi,m} b_l^+ b_m b_{l+\xi-m}\}$$

$$+ \xi^*_\xi a_\xi + \sum_l \eta^*_{\xi+l,l} a_l^+ a_{\xi+l} + \sum_l \{\lambda_{\xi,l} b_l a_{\xi-l} + \lambda^*_{l,\xi} a^+_{l-\xi} b_l\}$$

$$+ \sum_{l,m} \{\Lambda_{\xi,l,m} b_l b_{\xi-l} + \Lambda^*_{\xi+l,l,m} b_l^+ b_{\xi+l} + \Lambda^*_{l,\xi,m} b^+_{l-\xi} b_l\}$$

$$+ \sum_{l,m} \{\mu_{\xi,l,m} a_l^+ a_m a_{\xi+l-m}\} + \sum_{l,m} \{\gamma_{l,\xi,m} a_l^+ b_m b_{l+\xi-m}$$

$$+ \gamma^*_{\xi+m-l,l,m} b_m^+ b_l a_{\xi+m-l} + \gamma^*_{l,m,\xi} b^+_{l+m-\xi} b_m a_l\}$$

$$+ \sum_{l,m} (\alpha_{l,\xi,m} a_l^+ b_m a_{l+\xi-m} + \alpha^*_{l,m,\xi} a^+_{l+m-\xi} b_m a_l). \quad (5.191)$$

In the two above equations, the two subsystems, A and B, are assumed to be *isolated* from the environment, so that, from momentum conservation, ξ, ω, and Ω must be diagonal.

To proceed further we now define *two* fields in a way much akin to MCS, as

$$\phi(r) = \frac{1}{\sqrt{V}} \sum_k b_k \exp(-i\mathbf{k} \cdot \mathbf{r}), \qquad (5.192)$$

and

$$\psi(r) = \frac{1}{\sqrt{V}} \sum_k a_k \exp(-i\mathbf{k} \cdot \mathbf{r}), \qquad (5.193)$$

where V is the volume over which the plane waves are normalized. Each coefficient in the two equations of motion is now Taylor expanded about some 'point' in reciprocal space. It is, of course, assumed that there is at least one such 'point' which is common to both subsystems. This 'point' may be chosen to be a critical point of the joint system, as in MCS, or it may be a point close to which the system behaves classically. That such a latter point exists follows from the Correspondence Principle. Both sides of eqn (5.190) are multiplied by $(1/\sqrt{V})\exp(-i\boldsymbol{\eta} \cdot \mathbf{r})$ and summed over $\boldsymbol{\eta}$ and, similarly, both sides of eqn (5.191) are multiplied by

$$(1/\sqrt{V})\exp(-i\boldsymbol{\xi} \cdot \mathbf{r})$$

and summed over $\boldsymbol{\xi}$. If the first nontrivial nonlinear terms which arise from interactions within each subsystem and between them are retained, then we find, from eqn (5.190), that

$$\begin{aligned} i\hbar \partial_t \psi = {} & \omega_0 \psi + i\boldsymbol{\omega}_1 \cdot \nabla\psi + \omega_2 \nabla^2 \psi + 2\delta_0 \psi^+ \psi\psi + \xi_0 \phi \\ & + \eta_0 \psi\phi + \eta_0^* \phi^+ \psi + \lambda_0^* \phi^+ \phi \\ & + \varepsilon_0 \psi^2 + 2\varepsilon_0^* \psi^+ \psi + \mu_0 \phi^+ \psi^2 + 2\mu_0^* \psi^+ \psi\phi \\ & + \gamma_0 \phi^+ \phi\phi + (\alpha_0 + \alpha_0^*)\phi^+ \phi\psi, \end{aligned} \qquad (5.194)$$

where the new parameters are related in an obvious fashion to those in eqn (5.190) (obtained as in MCS, discussed earlier). We could have gone further and retained two-body terms within the subsystem up to first and second order. Here, we retain only the simplest terms for ease of exposition. Similarly, from eqn (5.191), the equation of motion for ϕ may be obtained as

$$\begin{aligned} i\hbar \partial_t \phi = {} & \Omega_0 \phi + i\boldsymbol{\Omega}_1 \cdot \nabla\phi + \Omega_2 \nabla^2 \phi + 2\Delta_0 \phi^+ \phi\phi + \xi_0^* \psi \\ & + \Lambda_0 \phi^2 + \eta_0^* \psi^+ \psi + \lambda_0 \phi\psi + \lambda_0^* \psi^+ \phi + 2\Lambda_0^* \phi^+ \phi + \mu_0 \psi^+ \psi\psi \\ & + \gamma_0 \psi^+ \phi\phi + 2\gamma_0^* \phi^+ \phi\psi + (\alpha_0 + \alpha_0^*)\psi^+ \phi\psi. \end{aligned} \qquad (5.195)$$

Thus, we have arrived at two highly nonlinear coupled equations of motion which must be solved for the individual fields. They are, in fact, quite complicated in this case but, as we see later, for each succeeding case they become simpler and more tractable due to changes principally in commutation relations, but also because certain terms must be dropped on physical grounds.

5.10.3 The Fermion–Fermion system for indistinguishable particles

In this case the two subsystems might be the electrons on two different atoms in a solid. As these particles are indistinguishable, the associated annihilators and creators now satisfy Fermi-Dirac commutation relationships within each set and anti-commute between sets. We can now apply the same procedure as in the previous case except that the ε, Λ, η, and λ coefficients vanish, since charge must be preserved. We readily find the corresponding Heisenberg equations of motion for the annihilators in both the A and B sets. By the same procedure as in the last section, the two resultant equations may again be transformed into field equations of motion. For the field ψ, we obtain

$$i\hbar \partial_t \psi = \omega_0 \psi + i\boldsymbol{\omega}_1 \cdot \nabla \psi + \omega_2 \nabla^2 \psi + \xi_0 \phi + 2\delta_0 \psi^+ \psi\psi$$
$$+ (\alpha_0 + \alpha_0^*)\phi^+ \phi\psi + \gamma_0 \phi^+ \phi\phi - \mu_0 \phi^+ \psi\psi. \quad (5.196)$$

For Fermions we have used the property that we may put $\delta_{\eta,l,m} = -\delta_{l,\eta,m}$. Furthermore, the two terms in μ^* do cancel out, since otherwise they either create or destroy a particle. For the other field, ϕ, the equation of motion becomes

$$i\hbar \partial_t \phi = \Omega_0 \phi + i\boldsymbol{\Omega}_1 \cdot \nabla \phi + \Omega_2 \nabla^2 \phi + 2\Delta_0 \phi^+ \phi\phi$$
$$+ \xi_0^* \psi - (\alpha_0 + \alpha_0^*)\psi^+ \phi\psi - \gamma_0 \psi^+ \phi\phi + \mu_0 \psi^+ \psi\psi. \quad (5.197)$$

Already the equations are becoming less complex—at any rate they have fewer terms! Let us now turn to the other Fermion–Fermion case.

5.10.4 The Fermion–Fermion system for distinguishable particles

Here, as an example, we might be considering a plasma made up of electrons and protons as distinguishable types of particle. Unlike the particles in the last section, the numbers of each particle type must now be separately conserved. Thus, the second-quantized operators satisfy the Fermi-Dirac commutation rules within each set, and *between* subsystems they commute (not anti-commute as in Section 5.10.3). For this particular case the ε-, Λ-, ξ-, η-, λ-, μ-, and γ-type interactions are absent, so equations of motion are greatly simplified. We first find the Heisenberg equations of motion for the annihilators and then, in the usual way, the field equations. The field equations here are particularly simple and become

$$i\hbar \partial_t \psi = \omega_0 \psi + i\boldsymbol{\omega}_1 \cdot \nabla\psi + \omega_2 \nabla^2 \psi + 2\delta_0 \psi^+ \psi\psi + (\alpha_0 + \alpha_0^*)\phi^+ \phi\psi \quad (5.198)$$

and

$$i\hbar \partial_t \phi = \Omega_0 \phi + i\boldsymbol{\Omega}_1 \cdot \nabla\phi + \Omega_2 \nabla^2 \phi + 2\Delta_0 \phi^+ \phi\phi + (\alpha_0 + \alpha_0^*)\psi^+ \psi\phi. \quad (5.199)$$

5.10.5 The Fermion–Boson case

This is one of the most important types, as it epitomizes a system of interacting electrons and phonons. It may be a little more complicated than that in the last section, but it has enormous potential. If 'a' operators refer to the Fermions and 'b' operators to Bosons, we have the respective commutation rules to abide by. For this particular case, interactions of ε, ξ, λ, γ, and μ type have to be eliminated from the Hamiltonian. This case is straightforward, as are the others (if not a little tedious!), and the corresponding equations of motion for the annihilators are easy to obtain. The field equations are derived in the same way as in earlier sections, and give

$$i\hbar \partial_t \psi = \omega_0 \psi + i\boldsymbol{\omega}_1 \cdot \nabla\psi + \omega_2 \nabla^2 \psi + 2\delta_0 \psi^+ \psi\psi + \eta_0 \psi\phi$$
$$+ \eta_0^* \phi^+ \psi + (\alpha_0 + \alpha_0^*)\phi^+ \phi\psi, \quad (5.200)$$

with

$$i\hbar \partial_t \phi = \Omega_0 \phi + i\boldsymbol{\Omega}_1 \cdot \nabla\phi + \Omega_2 \nabla^2 \phi + 2\Delta_0 \phi^+ \phi\phi + \eta_0^* \psi^+ \psi$$
$$+ (\alpha_0 + \alpha_0^*)\psi^+ \phi\psi + \Lambda_0 \phi^2 + 2\Lambda_0^* \phi^+ \phi. \quad (5.201)$$

5.10.6 Methods of solution

We shall not go into great detail here about particular solutions, but merely indicate that solutions *can* indeed be found. Some examples of the procedure used can be found in Chapter 6. To solve these equations we must treat the fields as classical, as we did in the MCS earlier, and then somehow decouple them—assuming a particular analytical orbit in the phase space, given by

$$G(\phi, \psi) = 0. \quad (5.202)$$

This can be done, and some commonly used examples involve straight-line, parabolic, elliptic, or hyperbolic functions in the space of dependent variables ϕ and ψ. It is, of course, not at all guaranteed that any of these approaches will produce results, but in a number of cases this type of procedure works quite well. When a particular orbit is investigated, compatibility conditions have to be satisfied for stationary orbits: that is,

$$|\partial G/\partial \phi|^2 |\nabla \phi|^2 = |\partial G/\partial \psi|^2 |\nabla \psi|^2 \qquad (5.203)$$

may be used effectively to decouple the two field equations. Several applications of this general technique have been given by Rajaraman in his monograph [50], where several cases have been worked out in detail. It should be pointed out, however, that in addition to the analytical solutions one can obtain in this way, large segments of the phase space for coupled systems can be filled with chaotic behaviour [96]. As an example, it may be established that for the distinguishable Fermion–Fermion case a linear function corresponding to G may be used to show that the two fields may be effectively decoupled, their equations of motion both being of the form of a stationary nonlinear Schrödinger equation. The Fermion–Boson situation may also be attacked using a series of ansätze. Further, conditions on the electron–phonon coupling and the strength of currents present for Cooper pair formation (or localization phenomena) may be established. This promises to be a very fruitful topic for future investigations, but due to lack of space we draw this section to a close and, instead, make a connection with the Landau–Ginzburg theory.

5.10.7 Connection with the Landau–Ginzburg formalism

It was shown earlier in this book on the MCS for a single field (for subsystem A only in this case) that the equations of motion can be derived, for a subcase in which the two-body interaction is treated in zero order, via the Euler–Lagrange equations, from a Hamiltonian density of the Landau–Ginzburg form

$$H^a_{\text{LG}} = \alpha_a \psi^+ \psi + \beta_a \psi^+ \psi^+ \psi \psi + \gamma_a (\nabla \psi^+) \cdot (\nabla \psi). \qquad (5.204)$$

It can be shown that, for *two* coupled fields ψ and ϕ, as we have considered here, the corresponding Hamiltonian density (to the same order) can be written in the same spirit, as

$$H_D = H^a_{\text{LG}}(\psi, \nabla \psi) + H^b_{\text{LG}}(\phi, \nabla \phi) + H^{ab}_{\text{LG}}(\phi, \psi, \nabla \phi, \nabla \psi), \qquad (5.205)$$

where $H^a_{\text{LG}}(\psi, \nabla \psi)$ is H^a_{LG} in eqn (5.204) and H^b_{LG} is obtained from eqn

(5.204) by replacing ψ by ϕ and a by b. The last term in eqn (5.205) obviously arises from the interaction and has the general form

$$H_{LG}^{ab} = \mu_1 |\phi|^2 |\psi|^2 + \mu_2 (\phi + \phi^*) |\psi|^2 + \mu_3 (\psi + \psi^*) |\phi|^2$$
$$+ \mu_4 (\psi\phi^* + \phi\psi^*) + \mu_5 (\phi^* |\psi|^2 \psi + \phi |\psi|^2 \psi^*)$$
$$+ \mu_6 (\psi^* |\phi|^2 \phi + \psi |\phi|^2 \phi^*). \tag{5.206}$$

where the μ_i ($i = 1, \ldots, 6$) are parameters. In some systems the terms such as $(\phi + \phi^*)|\psi|^2$ can be excluded on symmetry grounds. However, this term can be physically highly significant in electron–phonon systems, since then $\phi + \phi^*$ might represent a lattice displacement and $|\psi|^2$ an electron density distribution.

Chapter 5, therefore, has been devoted entirely to the development of the MCS formalism, starting from the simplest second-quantized Hamiltonian with one type of quasi-particle only and no spin degrees of freedom. In these last sections we have demonstrated that the applicability of the MCS is equally valid in multi-component systems and extensions to spin-dependent fields are rather tedious but nevertheless routine. Finally, the preliminary choice of the plane-wave base has been shown to produce generic equations, despite its being a special choice. Specifically, we have shown that in any other orthonormal base the form of equations is identical.

5.11 The relativistic extension

It is important to emphasize the connection between the dispersion relation for the single particle energy ω_k and the form of the field equation of motion obtained in the MCS. In particular, a Schrödinger-type equation is Galilean invariant and hence its validity would be limited to massive quasi-particles moving at velocities much less than the speed of light, c. However, in view of recent experimental observations [97] and their theoretical interpretation [98], pattern formation in electromagnetic beams travelling through nonlinear media have indicated that quasi-particle interactions play a very significant part even in photon ensembles. This then suggests that our Hamiltonian in eqn (5.24) should, in principle, be applicable to both massive and massless quasi-particles. Moreover, we shall assume that the Hamiltonian is made up of a Dirac-type one-particle contribution plus two-body interactions [109]. Our main objective will be to investigate the form of the one-body term in the field equation as compared to the nonrelativistic case. We fully realize that, due to the transition from the Coulomb to the Breit-like interaction, the two-body part of the Hamiltonian, and hence in the equation of motion, nonlinear terms, will also be affected [113].

Our objective in this section is to demonstrate that a relativistic field equation can be obtained using the generic Hamiltonian in eqn (5.24) in conjunction with an appropriate dispersion relation for ω_k. The starting point in this derivation is to assume a dispersion relation for the relativistic case based on the energy–momentum relationship in relativity theory [99]:

$$E^2 = p^2 c^2 + m_0^2 c^4, \qquad (5.207)$$

where p is the quasi-particle's momentum, and m_0 its rest mass. The next step is to use the de Broglie relations for energy and momentum, $E = \hbar \omega$ and $p = \hbar k$. Substituting into eqn (5.207), taking the square root on both sides, and assuming a negligible mass ($m_0 \simeq 0$) yields $\omega_k = \pm ck$. Consequently, the field equation has two branches corresponding to the two signs in the equation above. Our field, ψ, will be defined using, as a basis, the free-particle eigenstates of the Dirac equation [110]. Only the autonomous term in ψ from the one-body potential will be retained for simplicity and we keep only the zero-order part of the two-body interaction. We shall also assume that projectors are utilized so that only positive energy eigenstates are incorporated [109].

It should be stressed here that the method outlined in this section is quite general and does not exclusively apply to light relativistic particles. Although a detailed exposition of the relevant derivation seems to exceed the format of this monograph, we nonetheless wish to outline the main steps in such a procedure. The starting point is to consider a model Hamiltonian for relativistic particles (massive) interacting with radiation (massless). The particle terms in the Hamiltonian should be of Dirac form and their interactions involve two-body Breit–Coulomb terms as well as one-body terms, due to their interactions with electromagnetic fields. Photons are treated as free particles. The next step would be to express the above Hamiltonian in second-quantized form. This would be followed by a unitary transformation effectively to decouple photons from the particle variables, very much analogous to the Fröhlich Hamiltonian treatment of the electron–phonon problem for superconductors (see reference [38]). Once an effective Hamiltonian is constructed for relativistic particles, Heisenberg equations of motion are easy to derive using their commutation relations. The last step involves defining quantum fields ψ for the particle in a similar way to the procedure presented in the rest of this chapter for nonrelativistic fields. Consequently, essential nonlinearities will emerge as always, due to two-particle (here Breit–Coulomb) interactions between electrons. However, as we show in this section, there will also be higher-order nonlinear terms in the equation of motion, which arise as a result of the relativistic dispersion relation which couples the two types of field variable. We obtain, for the field equation,

$$i\nabla_\pm \psi = \alpha\psi + \beta|\psi|^2\psi, \tag{5.208}$$

where

$$\nabla_\pm \equiv \frac{\partial}{\partial t} \pm c\frac{\partial}{\partial x}, \tag{5.209}$$

and the real coefficients α and β describe interactions of the quasi-particles with the background (e.g. a nonlinear dielectric medium) and inter-particle couplings, respectively [100]. The Hermitian conjugate of eqn (5.208) is given by

$$-i\nabla_\pm \psi^* = \alpha\psi^* + \beta|\psi|^2\psi^*. \tag{5.210}$$

We then act with the operator $i\nabla_\mp$ on both sides of eqn (5.208), to yield

$$-\Box\psi = i\alpha\nabla_\mp\psi + i\beta\nabla_\mp(|\psi|^2\psi), \tag{5.211}$$

where

$$\Box \equiv \frac{\partial^2}{\partial t^2} - c^2\frac{\partial^2}{\partial x^2}. \tag{5.212}$$

is the d'Alembertian. One can then use both eqn (5.208) and eqn (5.209) to reduce the right-hand side of eqn (5.211) to polynomial form in ψ. This produces the final result:

$$\Box\psi = -\psi(\alpha + \beta|\psi|^2)^2 = \delta U/\delta\psi^*. \tag{5.213}$$

We have therefore obtained a relativistically-invariant field equation of motion for ψ, in which the field propagation is determined solely from the dispersion relation and the dominant *quintic* nonlinearity results from self-interactions in the quasi-particle (photon) field. Lower-order nonlinearities arise from a combination of background and self-interaction processes. Integrating the right-hand side of eqn (5.213) with respect to the field variable gives an effective potential $U(\psi)$, given by

$$U(\psi) = -\alpha^2|\psi|^2 - \alpha\beta|\psi|^4 - \frac{\beta^2}{3}|\psi|^6. \tag{5.214}$$

In Fig. 5.2 we have plotted the form of $U(\psi)$ for the two distinct cases out of the four possible sign combinations of α and β corresponding to attractive or repulsive background and quasi-particle interactions. Note that only the first case ($\alpha = -1$, $\beta = +1$) corresponds to a broken symmetry and hence would lead to self-localized solutions. The limit of $\alpha \to 0$

THE RELATIVISTIC EXTENSION   293

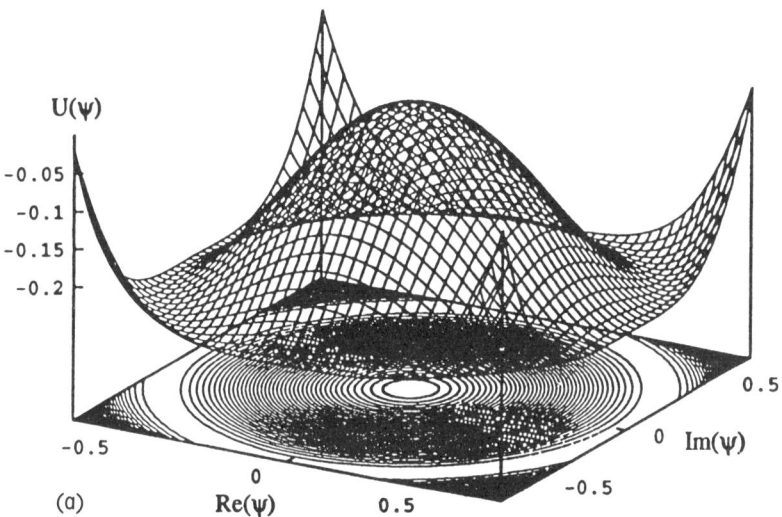

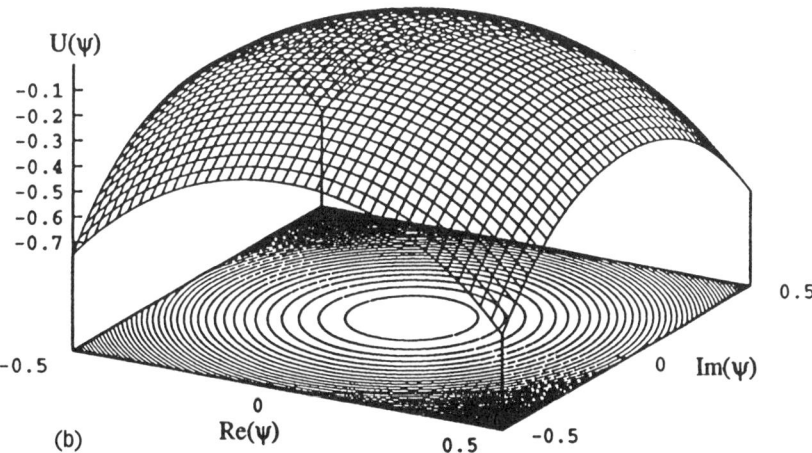

Fig. 5.2 Plots of the potential $U(\Psi)$ in eqn (5.214) for the two distinct cases: (a) $\alpha = -1$ or $\alpha = +1$, $\beta = -1$; (b) $\alpha = +1$, $\beta = +1$ or $\alpha = -1$, $\beta = -1$.

describes a tricritical point in the language of phase transitions and is known to be associated with a wealth of exact solutions resulting from the presence of dilational symmetry [80]. This is manifested by a change of symmetry of the equation from Poincaré to similitude, and the new geometries the field variable ψ may now reflect include spiral and conical ones [80]. For an exhaustive analysis of exact solutions to eqn (5.224) in $(3 + 1)$-dimensional space–time, the reader is referred to reference [80].

In the case of massive relativistic particles, we once again use eqn (5.207) and for relatively large rest energies expand the dispersion relation as

$$\hbar \omega_k = \pm m_0 c^2 \left[1 + \frac{\hbar^2 k^2}{2 m_0^2 c^2} + \cdots \right]. \tag{5.215}$$

Proceeding in the same way as before would lead to a relativistically invariant field equation, but terms with space derivatives would have to go to fourth order. The opposite case of nearly massless relativistic particles poses a more serious difficulty, since negative powers of k appear in the expansion. However, in this case the initial equation will be of integrodifferential type due to the negative powers in k. Subsequent space differentiation will therefore eliminate the integrals and increase the overall rank of the resulting equations.

Interactions between quasi-particles, either directly or mediated by the background, appear to be responsible for nonlinearities in the underlying equations of motion for quantum fields. We have shown earlier how quadratic dispersion relations lead to Galilean-invariant nonlinear Schrödinger-type equations of motion. In this section we have demonstrated how a similar procedure applied to quasi-particles with a linear dispersion relation results in a Lorentz-invariant nonlinear Klein–Gordon form of the field equation of motion. The motivation for our work is both experimental and theoretical. Ever since Chiao *et al.* [101] discovered self-localization effects of electromagnetic beams in nonlinear media, much attention has been given to the nonlinear behaviour of coherent photons under controlled experimental conditions inducing self-focusing [103]. A variety of interesting effects have been seen, including dynamical instabilities, chaos [102], and pattern formation [104]. On the theoretical front, both soliton propagation and pattern formation have been linked to various nonlinear differential equations, including the generalized nonlinear Schrödinger [95] and the nonlinear Klein–Gordon equations [105, 106]. The latter equation has been shown to support the formation of four-fold radially symmetric as well as various spiral-symmetry localized patterns. It is therefore quite interesting that, as argued here, a direct connection seems to exist between the form of a second-quantized Hamiltonian and the resultant spatially localized pattern for the field distribution defining the quasi-particle number density function.

5.12 Relationship with coherent states

As has been pointed out by Landau and Lifshitz [107], there is a close analogy between the passage from quantum mechanics to classical mechanics and the relationship between electrodynamics and geometrical optics. When the de Broglie wavelengths of particles are small by comparison with the typical dimensions that determine the physical conditions of a given system, then we say the properties of a system are close to being classical. Electromagnetic waves, on the other hand, are described by electric and magnetic induction field vectors that satisfy Maxwell's equations. The limiting case of geometrical optics corresponds to small wavelengths or to the fact that the phase, ϕ, of a particular field component, written as an $\exp i\phi$, varies by a large amount over small distances. Furthermore, the limiting procedures are closely related to a variational procedure whereby a functional is minimized. Thus, in mechanics the functional is the action the least possible value of which determines the path of a particle. In the case of geometrical optics it is the optical path length, and the principle of minimizing it is Fermat's principle. In the limiting procedure from quantum mechanics to classical physics, if we wish to obtain a definite path for a particle (there being no such concept in quantum mechanics), we must begin with a wave function of a particular form. This is the wave packet which has a nonzero amplitude over a very small region of space. When $\hbar \to 0$ this region must also tend to zero and the wave packet will then move along a classical path. To give an interpretation to the classical field, ψ_c, as we see later, we must use a particular type of ground state. This is similar to the situation for the wave packet, but the wave packet is a wavefunction, whereas the field is an operator. Quantum-mechanical operators, in fact, must reduce in the classical limit simply to multiplying by the corresponding physical quantity.

The classical nature of quantum field operators is tied, in our study, to a critical point associated with a phase transition, but *not* necessarily with long-range order since there are critical points at which long-range order is not established. We have seen in Chapter 4 that in the case of off-diagonal long-range order, the ground state expectation value of $|\psi|^2$ may be approximately expressed as

$$\langle |\psi|^2 \rangle \simeq \langle \psi^* \rangle \langle \psi \rangle + \cdots .$$

In particular, the ground state expectation value of the field itself does *not* vanish. This is a seemingly paradoxical situation, since if the ground state is simply a linear superposition of determinantal states, each containing the *same* number of particles, then clearly, from the original definition of the field in eqn (5.36), the expectation value of the quantum field, $\psi(r,t)$, *is* zero. However, if we consider the ground state as a superposition of states with a *different* number of particles, this is not so in general and $\langle \psi \rangle \neq 0$. One way in which this can be achieved is to choose the ground state as a coherent state (see Appendix E and Chapter 4 for further details).

References

1. A. Beiser, *Concepts of Modern Physics*, McGraw-Hill, New York (1967).
2. J. A. Barker, *J. Chem. Phys.*, **70**, 2914 (1979).
3. D. Chandler and P. G. Wolynes, *J. Chem. Phys.*, **74**, 4078 (1981).
4. D. M. Ceperley and M. H. Kalos, *Monte-Carlo Methods in Statistical Physics* (ed. K. Binder), Springer-Verlag, Berlin (1981) p. 145.
5. B. J. Berne and D. Thirumalai, *Ann. Rev. Phys. Chem.*, **37**, 401 (1986).
6. M. J. Gillan, in *Computer Modelling of Fluids, Polymers and Solids* (ed. C. R. A. Catlow, S. C. Parker, and M. P. Allen), *NATO ASI Series C: Mathematical and Physical Sciences*, **293** (Kluwer, Dordrecht, 1990), p. 155.
7. R. P. Feynman, *Rev. Mod. Phys.*, **20**, 376, 1948; *Statistical Mechanics*, Benjamin, New York (1972).
8. F. Christodoulos and M. J. Gillan, *Phil. Mag. B*, **63**, 641 (1991).
9. B. R. Judd, *Second Quantisation and Atomic Spectroscopy*, G. H. Dieke Memorial Lectures, The Johns Hopkins Press, Baltimore (1967).
10. W. Kohn, *Phys. Rev. Lett.*, **56**, 2219 (1986).
11. W. Kohn and L. J. Sham, *Phys. Rev.*, **140**, A1133 (1965).
12. P. Hohenberg and W. Kohn, *Phys. Rev.*, **136**, B864 (1964).
13. R. Jackiw, *Rev. Mod. Phys.*, **49**, 681 (1977).
14. K. W. H. Stevens, *Phys. Rep.*, **24**, 1 (1976).
15. A. Messiah, *Quantum Mechanics*, North-Holland, Amsterdam (1962).
16. C. Bloch, *Nucl. Phys.*, **6**, 329 (1958).
17. L. G. Yaffe, *Rev. Mod. Phys.*, **54**, 407 (1982).
18. J. C. Inkson, *Many-body Theory of Solids: an Introduction*, Plenum Press, New York (1984).
19. L. Hedin and S. Lundqvist, in *Solid State Physics, Advances in Research and Application* (ed. F. Seitz, D. Turnbull, and H. Ehrenreich), Vol. 23 (1969), Academic Press, New York and London, pp. 1–181.
20. N. H. March, *Self-consistent Fields in Atoms, Hartree and Thomas–Fermi Atoms*, Pergamon Press, New York (1975).
21. L. L. Foldy, *Phys. Rev.*, **83**, 397 (1951).

22. N. H. March, W. H. Young, and S. Sampanthar, *The Many-body Problem in Quantum Mechanics*, Cambridge University Press, Cambridge (1967).
23. R. D. Mattuck, *A Guide to Feynman Diagrams in the Many-body Problem*, McGraw-Hill, New York (1976).
24. R. Courant, *Differential and Integral Calculus*, Vol. 1, Blackie, Glasgow (1966), p. 373.
25. S. Doniach and E. H. Sondheimer, *Green's Functions for Solid State Physicists*, *Frontiers in Physics*, Benjamin, New York (1974).
26. A. Jevicki and B. Sakita, *Nucl. Phys. B*, **165**, 511 (1980).
27. A. Jevicki and N. Papanicolaou, *Nucl. Phys. B*, **171**, 362 (1980).
28. A. Jevicki and H. Levine, *Ann. Phys.*, **136**, 113 (1981).
29. K. Bardakci, *Nucl. Phys. B*, **178**, 263 (1981).
30. S. R. Wadia, Enrico Fermi Institute preprint 80/47, Chicago (1980).
31. A. Chatterjee, *Phys. Rep.*, **186**, 249 (1990).
32. F. A. Berezin, *Communs. Math. Phys.*, **63**, 131 (1978).
33. S. Nakajima, Y. Toyozawa, and R. Abe, *The Physics of Elementary Excitations*, Springer-Verlag, Berlin (1980).
34. F. A. Berezin, *The Method of Second Quantisation*, Academic Press, New York (1966).
35. B. G. Wybourne, *Classical Groups for Physicists*, John Wiley, London (1974).
36. P. W. Anderson, *Basic Notions of Condensed Matter Physics*, *Frontiers in Physics Series*, Benjamin/Cummings, Menlo Park, CA (1984).
37. V. G. Makhankov and V. K. Fedyanin, *Phys. Rep.*, **104**, 1 (1984).
38. P. L. Taylor, *A Quantum Approach to the Solid State*, Prentice-Hall, New York (1970).
39. G. C. Cho, W. Kutt, and H. Kurz, *Phys. Rev. Lett.*, **65**, 764 (1990).
40. A. V. Kuznetsov and C. J. Stanton, *Phys. Rev. Lett.*, **73**, 3243 (1994).
41. R. M. White, *Quantum Theory of Magnetism*, McGraw-Hill, New York (1983).
42. H. Fröhlich, *Int. J. Quant. Chem.*, **2**, 641 (1968).
43. T. M. Wu and S. Austin, *Phys. Lett.*, **64A**, 151 (1977).
44. J. A. Tuszyński, R. Paul, R. Chatterjee, and S. R. Sreenivasan, *Phys. Rev. A*, **30**, 2666 (1984).
45. A. S. Davydov, *Biology and Quantum Mechanics*, Pergamon Press, Oxford (1982).
46. J. A. Tuszyński, *Int. J. Quant. Chem.*, **29**, 379 (1986).
47. A. D. Bruce and R. A. Cowley, *Structural Phase Transitions*, Taylor and Francis, London (1981).
48. S.-K. Ma. *Modern Theory of Critical Phenomena*, Benjamin, Reading, MA (1976).
49. D. J. Amit, *Field Theory, the Renormalisation Group and Critical Phenomena*, McGraw-Hill, New York (1978).
50. R. Rajaraman, *Solitons and Instantons: an Introduction to Solitons and Instantons in Quantum Field Theory*, North Holland, Amsterdam (1987).
51. L. Gagnon and P. Winternitz, *J. Phys. A: Math. Gen.*, **21**, 1493 (1988).
52. L. Gagnon and P. Winternitz, *Phys. Lett.*, **134A**, 276 (1989).

53. L. Gagnon and P. Winternitz, *Phys. Rev.*, **A39**, 296 (1989).
54. P. Winternitz, A. M. Grundland, and J. A. Tuszyński, *J. Math. Phys.*, **28**, 2194 (1987).
55. S. Raimes, *Many Electron Theory*, North-Holland, Amsterdam (1972).
56. H. Haken, *Quantum Field Theory of Solids*, North-Holland, Amsterdam (1976).
57. G. W. Bluman and J. D. Cole, *J. Math. Mech.*, **18**, 1025 (1969).
58. P. A. Clarkson and M. D. Kruskal, *J. Math. Phys.*, **30**, 2201 (1989).
59. P. A. Clarkson, *J. Phys. A: Math. Gen.*, **22**, 2355 (1989); **22**, 3821 (1989).
60. P. A. Clarkson and P. Winternitz, *Physica*, **49D**, 257 (1991).
61. D. Levi and P. Winternitz, *J. Phys. A: Math. Gen.*, **22**, 2915 (1989).
62. L. P. Kadanoff, *Ann. Phys. (New York)*, **2**, 263 (1966).
63. L. E. Reichl, *A Modern Course of Statistical Physics*, University of Texas Press, Austin (1979).
64. J. M. Dixon and J. A. Tuszyński, *J. Phys. A: Math. Gen.*, **22**, 4895 (1989); J. A. Tuszyński and J. M. Dixon, *J. Phys. A: Math. Gen.*, **22**, 4877 (1989).
65. F. Ayres, Jr., *Theory and Problems of Differential Equations*, Schaum's Outline Series, McGraw-Hill, New York (1952).
66. W. Magnus, F. Oberhettinger, and R. P. Soni, *Formulas and Theorems for Special Functions of Mathematical Physics*, Springer-Verlag, Berlin (1966).
67. V. E. Zakharov and A. B. Shabat, *Sov. Phys. JETP*, **34**, 62 (1972).
68. A. C. Scott, F. Y. F. Chu, and D. W. McLaughlin, *Proc. IEEE*, **61**, 1443 (1973).
69. R. K. Bullough and P. J. Caudrey (eds), *Solitons, Topics in Current Physics*, Springer-Verlag, Berlin (1980).
70. G. L. Lamb, *Elements of Soliton Theory*, John Wiley, New York (1980).
71. M. J. Ablowitz and H. Segur, *Solitons and the Inverse Scattering Transform*, SIAM, Philadelphia, PA (1981).
72. R. Hirota, in *Backlund Transformations* (ed. L. A. Dodd and B. Eckman), Springer-Verlag, Berlin (1976), p. 48.
73. J. A. Tuszyński, M. Otwinowski, R. Paul, and A. P. Smith, *Phys. Rev. B*, **36**, 2190 (1987).
74. E. L. Ince, *Ordinary Differential Equations*, Dover, New York (1956).
75. L. M. Alonso, *Lett. Math. Phys.*, **8**, 111 (1984); *J. Math. Phys.*, **25**, 1735 (1987).
76. T. Konishi and M. Wadati, *J. Phys. Soc. Japan*, **55**, 1075 (1986).
77. M. Wadati and A. Kuniba, *J. Phys. Soc. Japan*, **55**, 76 (1986).
78. M. Wadati, A. Kuniba, and T. Konishi, *J. Phys. Soc. Japan*, **57**, 1710 (1985).
79. P. Winternitz, A. M. Grundland, and J. A. Tuszyński, *J. Phys. C: Solid State Phys.*, **21**, 1931 (1988).
80. A. M. Grundland and J. A. Tuszyński, *J. Phys. A: Math. Gen.*, **20**, 6243 (1987).
81. A. M. Grundland, J. Harnad, and P. Winternitz, *KINAM Rev. Fis.*, **4**, 333; in *Symmetries in Science II* (ed. B. Gruber and R. Lenczewski), Plenum Press, New York (1982), p. 197.
82. G. Cieciura and A. M. Grundland, *J. Math. Phys.*, **25**, 3460 (1984).
83. A. M. Grundland, *J. Math. Phys.*, **25**, 791 (1984).
84. P. F. Byrd and M. E. Friedman, *Handbook of Elliptic Integrals for Engineers and Scientists*, Springer-Verlag, Berlin (1971).

85. F. M. Arscott, *Periodic Differential Equations*, Pergamon Press, London (1964).
86. E. T. Whittaker and G. N. Watson, *A Course of Modern Analysis* Cambridge University Press, Cambridge (1963).
87. P. M. Morse and H. Feshbach, *Methods of Theoretical Physics, Part I*, McGraw-Hill, New York (1953).
88. V. De Alfaro and T. Regge, *Potential Scattering*, North-Holland, Amsterdam (1965).
89. I. S. Gradshteyn and I. M. Ryzhik, *Table of Integrals, Series and Products*, Academic Press, New York (1980).
90. J. Callaway, *Quantum Theory of the Solid State*, Academic Press, New York (1976).
91. C. J. Gorter and T. van Peski-Tinbergen, *Physica*, **22**, 273 (1956).
92. G. A. Smolenski, *Sov. Phys. Solid State*, **4**, 807 (1962).
93. K. S. Liu and M. E. Fisher, *J. Low Temp. Phys.*, **10**, 655 (1973).
94. D. S. Webster and M. J. R. Hoch, *J. Phys. Chem. Solids*, **32**, 2663 (1971).
95. J. C. Toledano and P. Toledano, *The Landau Theory of Phase Transitions*, World Scientific, Singapore (1987).
96. J. Hale and H. Kocak, *Dynamics and Bifurcations*, Springer-Verlag, Berlin (1991).
97. R. Chang, W. J. Firth, R. Indik, J. V. Moloney, and E. M. Wright, *Opt. Communs*, **88**, 167 (1992).
98. M. Florjanczyk and L. Gagnon, *Phys. Rev. A*, **41**, 4478 (1990).
99. A. L. Fetter and J. D. Walecka, *Quantum Theory of Many-particle Systems*, McGraw-Hill, New York (1971).
100. V. G. Makhankov and V. K. Fedyanin, *Phys. Rep.*, **104**, 1 (1984).
101. R. Y. Chiao, E. Garmire, and C. Townes, *Phys. Rev. Lett.*, **13**, 479 (1964).
102. K. J. Blow and N. J. Doran, Solitons in optical fibres, in *Nonlinear Waves in Solid State Physics* (ed. A. D. Boardman *et al.*), Plenum Press, New York (1990).
103. R. G. Harrison, *Contemp. Phys.*, **29**, 341 (1988).
104. A. Yariv and D. M. Pepper, *Optics Lett.*, **1**, 16 (1977).
105. I. L. Bogolyubskii and V. G. Makhankov, *JETP Lett.*, **24**, 12 (1976).
106. G. L. Alfimov, V. M. Eleonsky, N. E. Kulagin, L. M. Lehrmann, and V. P. Silin, *Phys. Lett.*, **A138**, 443 (1989).
107. L. D. Landau and E. M. Lifshitz, *Quantum Mechanics*, Pergamon Press, London (1958).
108. J. M. Dixon and J. A. Tuszyński, *Phys. Lett.*, **A155**, 107 (1991).
109. M. H. Mittleman, *Phys. Rev. A*, **24**, 1167 (1981).
110. W. Greiner, *Relativistic Quantum Mechanics; Wave Equations*, Vol. 3 in *Theoretical Physics Series*, Springer-Verlag, Berlin (1990).
111. A. Griffin, *Can. J. Phys.*, **73**, 755 (1995).
112. L. Ruby, *Am. J. Phys.*, **64**, 39 (1996).
113. J. A. Tuszyński and J. M. Dixon, to be published, *Int. J. Mod. Phys. B* (1997).

6 PHYSICAL APPLICATIONS OF MCS

In the education of a physicist, we recount the bold voyages of great explorers—Newton and Einstein, Faraday and Bohr—in search of new laws of nature. They found and charted the continents on which we have built our cities of the mind and of art. Does anyone really suppose that similar vast and fertile territories are still waiting to be discovered and colonized? The unaccustomed rules that govern black holes and quasars in the cosmic deeps affect our lives no more than the icy crags of the Himalayas or the conjunctions of the planets...

Think of physics simply as the 'fundamental' science and it is oversubscribed almost to bankruptcy. But define it as the science whose aim is to describe natural phenomena in the most mathematical or numerical language, and you will understand its past and have confidence in its future... The task of the modern physicist is to determine the mathematically comprehensible characteristics of the natural world and of human artifacts...

Ziman, J. M., Phys. Bull., 25, 280 (1974)

In this chapter we wish to analyse in some detail a number of important examples of physical systems using the method of coherent structures (MCS), which has been developed in Chapter 5. Our intention is to illustrate how the method is actually used in concrete examples of many-particle systems. We have selected seven particular cases to study and ordered them in sections arranged as follows. We begin by providing a direct link with the last section of Chapter 5 and show how calculations can be carried out for a system of two distinguishable Fermion particles. The next example addresses the problem of superconductivity and here we demonstrate how the MCS may be used to connect the BCS theory with the Landau–Ginzburg (LG) approach. The next two sections are concerned with spin chains but differ with respect to the phenomena of interest. In these sections the Hamiltonians themselves *are* spin-dependent and we show how the MCS may still be used to our advantage despite our description involving *spin-independent* Hamiltonians in Chapter 5. In the first of these two sections, the phenomenon of metamagnetism is treated and the level of modelling is classical, since the main objective is to construct a phase diagram. The second of the two sections aims at explaining the so-called Haldane gap problem for antiferromagnetic spin chains. Here, the very nature of the question posed requires an inclusion of quantum interactions. The next two sections of the chapter deal with the problem of quantum energy levels of multi-electron atoms. They differ from the previous applications in several significant ways, the most important of which is that the system under study is spherically symmetric and, second, that it is of finite size. Our model of the atom is based on a nonlinear spatially coherent structure in which quantum excitations take place. In the last section the Hubbard Hamiltonian, which is very important in the theory of strongly correlated electron systems, is discussed.

6.1 The electron–proton plasma

In Section 5.10 we have shown how two sets of distinct quasi-particles can be described using the MCS method. A simple illustration of this approach can be provided in the case of distinguishable Fermion–Fermion sets. We can take as an example the interacting electron–proton plasma. Assuming that the electron subsystem is described by the quantum field ψ and proton field by ϕ, and that the interactions between them are of two-body type, the field Hamiltonian becomes biquadratic in form and the equations of motion are

$$i\hbar \partial_t \psi = \omega_0 \psi + i\boldsymbol{\omega}_1 \cdot \nabla \psi + \omega_2 \nabla^2 \psi$$
$$+ 2\delta_0 \psi^+ \psi\psi + (\alpha_0 + \alpha_0^*)\phi^+ \phi\psi \qquad (6.1)$$

and

$$i\hbar \partial_t \phi = \Omega_0 \phi + i\boldsymbol{\Omega}_1 \cdot \nabla \phi + \Omega_2 \nabla^2 \phi$$
$$+ 2\Delta_0 \phi^+ \phi\phi + (\alpha_0 + \alpha_0^*)\psi^+ \psi\phi. \qquad (6.2)$$

We refer the reader to Section 5.10 for a derivation of these two equations.

In this case it is relatively easy to obtain exact solutions to the systems of equations of motion for ψ and ϕ, namely eqns (6.1) and (6.2). It is assumed for simplicity that the co-ordinate system is chosen in such a way that $\boldsymbol{\omega}_1 \cdot \nabla \psi = \boldsymbol{\Omega}_1 \cdot \nabla \phi = 0$. This can be achieved by either a convenient rotation of the co-ordinates or the choice of a moving frame of reference. The time dependence of the two fields is taken in the form

$$\phi \to \phi e^{iE_2 t/\hbar} \quad \text{and} \quad \psi \to \psi e^{iE_1 t/\hbar}, \qquad (6.3)$$

where the phase factors above are the only time-dependences associated with the fields ϕ and ψ. Then, the two fields are assumed to be linearly dependent, which yields

$$\phi = \lambda \psi \qquad (6.4)$$

transforming eqns (6.1) and (6.2) into two equations of identical form.

Since they now refer to the same field, compatibility is required. The equations take the forms:

$$0 = (\omega_0 + E_1)\psi + \omega_2 \nabla^2 \psi + \left[2\delta_0 + \lambda^2(\alpha_0 + \alpha_0^*)\right] \psi^+ \psi\psi \qquad (6.5)$$

and

$$0 = (\Omega_0 + E_2)\psi + \Omega_2 \nabla^2 \psi + \left[2\Delta_0 \lambda^2 + (\alpha_0 + \alpha_0^*)\right] \psi^+ \psi\psi. \qquad (6.6)$$

Multiplying eqn (6.5) on both sides by an arbitrary constant β and comparing corresponding terms in the two equations gives

$$\beta\omega_2 = \Omega_2, \qquad \beta(\omega_0 + E_1) = \Omega_0 + E_2, \qquad (6.7, 6.8)$$

and

$$\beta\left[2\delta_0 + \lambda^2(\alpha_0 + \alpha_0^*)\right] = 2\Delta_0 \lambda^2 + (\alpha_0 + \alpha_0^*). \qquad (6.9)$$

From eqn (6.7), β is fixed as $\beta = \Omega_2/\omega_2$. Substituting this into eqn (6.8), we may solve for E_2, to give

$$E_2 = \frac{\Omega_2}{\omega_2}(\omega_0 + E_1) - \Omega_0, \qquad (6.10)$$

in which, of course, E_1 is still arbitrary. Then, in eqn (6.9), substituting for β from eqn (6.7) we obtain

$$\lambda^2 = \left[\frac{\omega_2}{\Omega_2}(\alpha_0 + \alpha_0^*) - 2\delta_0\right] \bigg/ \left[(\alpha_0 + \alpha_0^*) - 2\frac{\omega_2}{\Omega_2}\Delta_0\right]. \qquad (6.11)$$

The net result is that the two expressions in eqns (6.5) and (6.6) are made compatible, and both become stationary *nonlinear Schrödinger equations*. There are a number of solutions of these equations which have been thoroughly investigated in the past, and we refer the reader to Chapter 7 for more information. Suffice it to say that among the spatially inhomogeneous solutions one finds elliptic waves of several kinds (i.e. as sn, cn, and dn functions) as well as hyperbolic localized solutions (sech and tanh functions), and there are obviously constant solutions for both ψ and ϕ independent of the compatibility relations obtained above.

An important question still remains regarding the validity of such analytical solutions. Some insight into this problem may be gained by assuming that both ψ and ϕ are real or, equivalently, that only their moduli have nontrivial space dependence while the phases of ψ and ϕ are identical. This situation is illustrated in a series of diagrams, where in Fig. 6.1 we show the effective nonlinear potential as a function of these two fields. Moreover, in Fig. 6.2, we illustrate the Poincaré sections for the coupled equations as we gradually increase the total energy of the solution

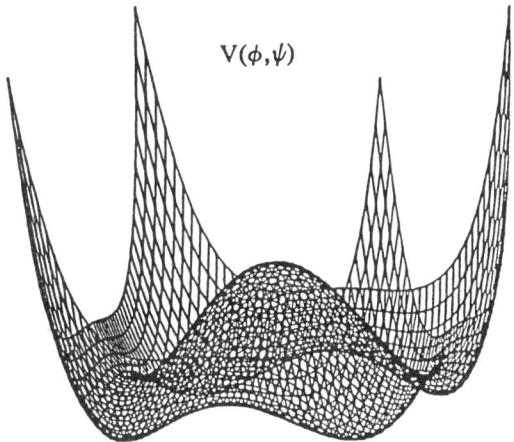

Fig. 6.1 A plot of the nonlinear potential in terms of the fields ϕ and ψ associated with eqns (6.1) and (6.2).

[1]. What appears to be taking place is that the low lying orbits are regular and fill the phase space almost completely. As we increase the energy and approach the potential peak in the centre (i.e. for $\psi = \phi = 0$), the dynamics of the system becomes more and more chaotic and eventually, at what is called the separatrix level, the entire phase space is filled with stochastic motion. Increasing the energy even further, we somewhat surprisingly recover regularity of motion and the chaotic region gradually disappears.

What we have shown here is that even the simplest coupled systems exhibit a fascinating range of nonlinear behaviour, the details of which depend on the model parameters and the energies that these conservative systems may have.

6.2 Superconductivity and MCS

It is very well known that in certain materials there is an effective attractive interaction between electrons which is mediated by lattice vibrations. This latter attractive coupling overcomes the natural repulsion of the electrons below a certain transition temperature $T = T_c$ when a second-order phase transition takes place. One of the remarkable consequences of this transition is that it leads to a phase in which there is almost a complete disappearance of the resistance to weak steady currents. Such materials are usually referred to as low-temperature *superconductors* and,

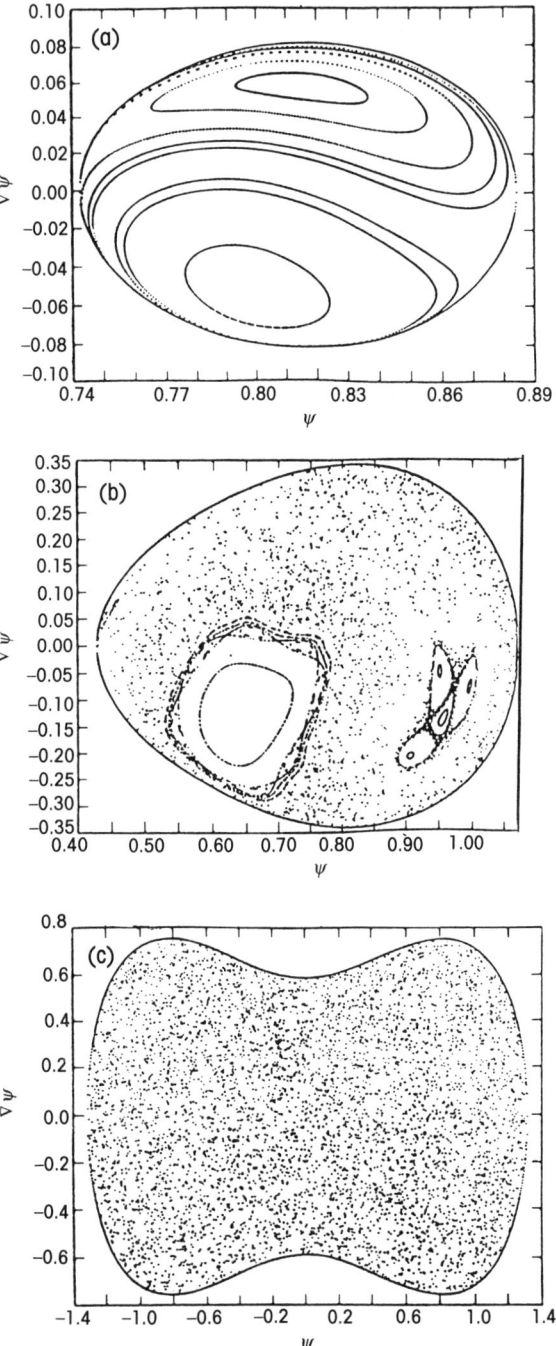

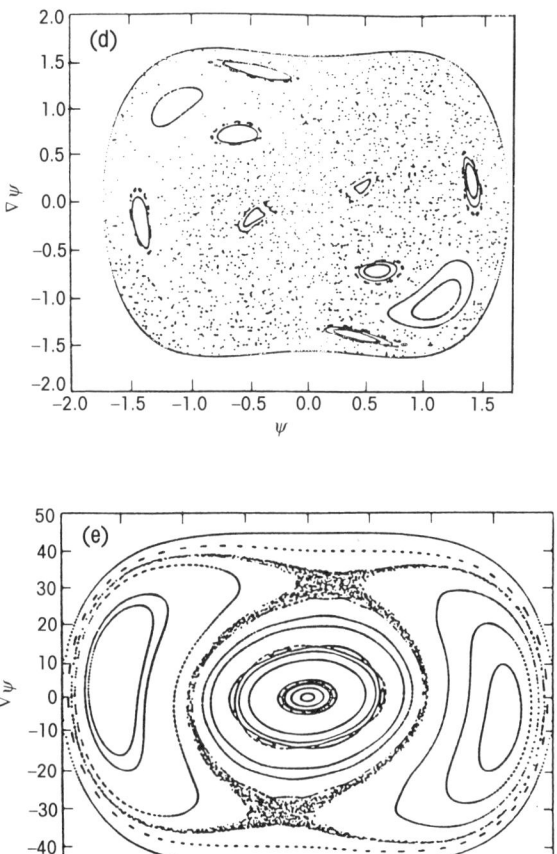

Fig. 6.2 Poincaré sections for eqns (6.5) and (6.6) in one-dimensional space with an increasing energy of the solution. The energy E corresponds to the Landau–Ginzburg Hamiltonian in the fields ϕ and ψ of eqns (6.5) and (6.6). (a) Poincaré section at $E = -0.33$; (b) Poincaré section at $E = -0.275$; (c) complete chaos at $E = -0.05$; (d) Poincaré section at $E = 1.00$; (e) Poincaré section at $E = 1000$.

above $T = T_c$, the specific heat contribution from electrons, for example, is linear with temperature, but below $T = T_c$ there is an exponential dependence. A magnetic field can only penetrate a superconductor to within a short distance of its surface, and this effect is known as the Meissner effect. If the applied field is increased sufficiently, the superconductivity is destroyed. However, superconductors may be categorized into two main groups: in one case, or type I superconductors, the whole superconductor is restored to its normal state at some sufficiently high field H_c; whereas for type II superconductors the internal induction field B is reduced gradually from its initial value. In type II materials the magnetic field partially penetrates between two fields H_{c1} and H_{c2}, giving rise to what is called the mixed state. Interactions between electrons and phonons in a metal were first described in a quantum-mechanical formalism [2] using the Fröhlich Hamiltonian

$$H = \sum_k \frac{\hbar^2}{2m^*} k^2 a_k^+ a_k + \sum_q \hbar \omega_q b_q^+ b_q \\ + \sum_{k,l} M_{k,l} [b_q + b_{-q}^+] a_k^+ a_l, \qquad (6.12)$$

where $q = k - l$. Here, the operators a_k^+ and a_k refer to the electrons, while b_q^+ and b_q refer to the phonons. The effective coupling constant $M_{k,l}$ can be written [2] as

$$M_{k,l} = i\sqrt{\frac{N\hbar}{2M\omega_q}} |l - k| V_{k-l}, \qquad (6.13)$$

where V_{k-l} is the Fourier transform of a screened Coulomb potential due to a particular ion at the origin and N is the number of distinct collective co-ordinates associated with lattice displacements. Typically, V_{k-l} takes the form

$$V_{k-l} = \frac{2\pi e^2}{|k - l|^2 + q_c^2}, \qquad (6.14)$$

where q_c^{-1} is of the order of the inter-particle distance and plasma waves only exist when their wavelength is greater than this. In this simple model the ions of the metal are assumed to interact with one another and with conduction electrons via a short-range screened potential. The conduction electrons, on the other hand, are considered to be essentially independent Fermions. Note that the 'bare' Coulomb interaction between the ions and the conduction electrons is *not* used and, to incorporate screening, repulsive terms have been built in to some extent via the short-range nature of the effective interaction remaining. The Fröhlich Hamiltonian, despite its approximate nature, has played an important role in the development of the theory of superconductivity by leading directly to the BCS model. We could, in principle, use the Hamiltonian in eqn (6.12) as our starting point and adopt a coupled field approach in the MCS [3]. Here, however, we prefer to take a simpler and perhaps easier-to-interpret route and follow the historical path leading next to the BCS Hamiltonian.

The BCS Hamiltonian can be obtained directly from the Fröhlich Hamiltonian through a sequence of unitary transformations effectively to decouple the electrons from phonons. This results in the effective Hamiltonian

$$H_{\text{eff}} = \sum_{k,\alpha} \omega_k a^+_{k,\alpha} a_{k,\alpha}$$
$$- \sum_{\substack{k,l,m \\ \alpha,\beta}} V_{k,l,m} a^+_{k,\alpha} a^+_{l,\beta} a_{m,\beta} a_{k+l-m,\alpha}, \qquad (6.15)$$

where k, l, and m refer to wave vectors and α and β are spin labels. The BCS theory of superconductivity arose from an original variational calculation in which the variational state was taken to be one in which electrons with equal but opposite momenta and opposite spins were paired together. An effective Hamiltonian to incorporate this phenomenon is called the BCS 'reduced' Hamiltonian, which may be written as

$$H_{\text{BCS}} = \sum_k \varepsilon_k (a^+_k a_k + a^+_{-k} a_{-k})$$
$$- \sum_{k,k'} V_{k,k'} a^+_{k'} a^+_{-k'} a_{-k} a_k, \qquad (6.16)$$

where the convention has been adopted that an operator with an explicit minus sign in the subscript refers to a spin-down state but an operator with a plus sign refers to a spin-up state. That is, for example,

$$a^+_{k'} \equiv a^+_{k'\uparrow} \quad \text{and} \quad a^+_{-k'} \equiv a^+_{-k'\downarrow}.$$

After this, the standard analytical approach is to use a Bogolyubov–Valatin transformation to bring the Hamiltonian into diagonal form. A number of other approximations are also made [2] finally to describe a transition to a superconducting state at $T = T_c$, where

$$k_B T_c = 1.14 \hbar \omega_D \exp[-1/\rho(E_F) V_0], \qquad (6.17)$$

where $V_{k,k'}$ is put equal to a constant V_0 provided that wave vectors have magnitudes such that their kinetic energies are in a thin shell about the Fermi energy E_F of width $2\hbar \omega_D$, ω_D being the Debye frequency. In eqn (6.17), $\rho(E_F)$ represents the density of states at the Fermi level.

It is clear from eqn (6.16) that the BCS Hamiltonian falls into the category of theoretical models (a very large class indeed!) which has been analysed in Chapter 5 using the MCS. The approximation that the BCS theory corresponds to is. to take the coupling coefficient $V_{k,k'}$ as a constant. In MCS this represents the 'zeroth order' and consequently the field equation for the field operator ψ, where ψ is defined in the usual way by

$$\psi(r) = \Omega^{-1/2} \sum_k e^{-ik \cdot r} a_k, \qquad (6.18)$$

is the nonlinear Schrödinger equation

$$-i\hbar \partial_t \psi = \varepsilon \psi + \frac{\hbar^2}{2m^*} \nabla^2 \psi - 2\Omega V_0 \psi^+ \psi \psi, \qquad (6.19)$$

where V_0 is the constant value of the coupling coefficient and ε is the energy associated with the one-body terms in the original Hamiltonian. As we pointed out in Chapter 5, if the starting Hamiltonian itself is not explicitly spin-dependent, then the field equations for a spin-dependent field take the same form as without spin. Thus, in the present context we are fully justified in ignoring the spin label.

The first interesting consequence of eqn (6.19) is that it directly links the BCS theory with another fundamental approach to superconductivity, namely the LG theory. This comes about since the Hamiltonian density leading to eqn (6.19), through a variational principle, [69] is

$$H_{LG} = \frac{\hbar^2}{2m^*} |\nabla \psi|^2 - \varepsilon |\psi|^2 + \Omega V_0 |\psi|^4. \qquad (6.20)$$

It should be noted in this connection that Gor'kov [4] came to the same conclusion, that BCS theory implies the LG model, more than 30 years ago, using completely different methods. In fact, strictly speaking, the LG model considers the field ψ to be a classical space-dependent order parameter and the associated energy functional is a *free* energy and *not* a Hamiltonian density: that is,

$$G_S(r) = G_N(r) + A(T)|\psi(r)|^2 + \tfrac{1}{2}C |\psi(r)|^4 + \frac{\hbar^2}{2m^*} |\nabla \psi(r)|^2. \qquad (6.21)$$

In eqn (6.21), $G_N(r)$ is the free energy of the normal state, m^* is an effective mass, and the parameter $C > 0$. The coefficient A, we assume, is a linear function of temperature T, and we write $A(T) = \bar{a}(T - T_c)$, where T_c is the superconducting transition temperature. The expression

$$n_s = |\psi(r)|^2 \qquad (6.22)$$

is usually interpreted as the density of superconducting electrons, m^*

would correspond to the mass of two electrons forming a *Cooper pair*, and the superconducting current density is given by

$$j = -i\frac{e^*\hbar}{2m^*}(\psi^*\nabla\psi - \psi\nabla\psi^*), \tag{6.23}$$

in the absence of the magnetic vector potential. The two characteristic length scales in the LG model are the coherence length ξ_{LG} and the penetration depth λ_{LG}. The coherence length ξ_{LG} is

$$\xi_{LG} = \sqrt{\frac{\hbar^2}{2m^*\bar{a}}}(T - T_c)^{-1/2}, \tag{6.24}$$

and it defines the spatial extent of correlated behaviour within the superconducting condensate. The penetration depth equals

$$\lambda_{LG}(T) = \left(\frac{m^*c^2C}{8\pi(e^*)^2\bar{a}}\right)^{1/2}(T - T_c)^{-1/2}, \tag{6.25}$$

and it defines the mean distance over which an externally applied magnetic field will penetrate into the superconductor. The ratio

$$K = \lambda_{LG}/\xi_{LG} \tag{6.26}$$

delineates the boundary between type I and type II superconductors in parameter space, with $K < 1/\sqrt{2}$ resulting in type I and $K \geq 1/\sqrt{2}$ in type II superconductivity.

Having derived a field equation of motion, we now proceed to analyse its solutions in the classical approximation. We now adopt the conventional interpretation that

$$m^* = 2m, \quad \varepsilon = -A(T), \quad \text{and} \quad \Omega V_0 = C/2, \tag{6.27}$$

where m is the electronic mass and the model parameters ε and V_0 are now treated in a statistical sense as 'dressed' parameters to introduce temperature dependence into the model. We now assume that $\psi(r)$ is complex, so that we can write it in modulus–argument form as

$$\psi(r) = \eta(r)\exp(i\phi(r)). \tag{6.28}$$

The function $\phi = \phi(r)$ is usually referred to as the carrier and $\eta = \eta(r)$ as the envelope. Equation (6.28) may be substituted into eqn (6.19) to yield the following two equations for the real (Re) and imaginary (Im) parts:

$$\text{Re,} \quad A(T)\eta + C\eta^3 - \frac{\hbar^2}{2m^*}\nabla^2\eta + \frac{\hbar^2}{2m^*}\eta(\nabla\phi)^2 = 0; \tag{6.29}$$

$$\text{Im,} \quad 2\nabla\eta\cdot\nabla\phi + \eta\nabla^2\phi = 0. \tag{6.30}$$

Assuming $\eta \neq 0$, it is obvious that eqn (6.30) may be written as a type of continuity equation, i.e.

$$\frac{1}{\eta} \nabla \cdot [\eta^2 \nabla \phi] = 0, \qquad (6.31)$$

and hence it may be integrated directly to give

$$\eta^2 \nabla \phi = C_0 + \nabla \times F, \qquad (6.32)$$

where C_0 is an arbitrary constant vector and F is an arbitrary vector function, $\nabla \times F$ denoting its curl. That a curl should be present is obvious, since $\nabla \cdot \nabla \times F = 0$ whatever the value of the vector F. We may now substitute $\nabla \phi$ from eqn (6.32) into eqn (6.29) and obtain

$$A(T)\eta + C\eta^3 - \frac{\hbar^2}{2m^*} \nabla^2 \eta + \frac{\hbar^2}{2m^*\eta^3}(C_0 + \nabla \times F)^2 = 0. \qquad (6.33)$$

To distinguish two physical cases, we recall that a current j is defined through eqn (6.23). Alternatively, we may write it as

$$j = \frac{e^*\hbar}{m^*} \eta^2 \nabla \phi. \qquad (6.34)$$

With this, eqn (6.33) can be solved exactly in one dimension, whether or not superconducting currents are present. These two cases will now be briefly discussed in turn.

1. Assuming *no superconducting currents*, eqn (6.33) simplifies to

$$\frac{\hbar^2}{2m^*} \nabla^2 \eta = A(T)\eta + C\eta^3, \qquad (6.35)$$

which is a stationary NLKG equation. In one-dimensional space it can be immediately integrated to yield

$$(d\eta/dx)^2 = \Delta(\eta^2 - \eta_1^2)(\eta^2 - \eta_2^2), \qquad (6.36)$$

where $\Delta = m^*C/\hbar^2$ and $\mp\eta_1$ and $\mp\eta_2$ are the roots of the quartic polynomial

$$P(\eta) = \eta^4 + (2A(T)/C)\eta^2 + k_2. \qquad (6.37)$$

Here, k_2 is an integration constant. The real, nonsingular solutions of eqn (6.36) may be summarized as follows:

(a) When $T > T_c$ and $m^* > 0$, the only such solution is $\eta = \eta_0 = 0$, representing the disordered or normal phase.

SUPERCONDUCTIVITY AND MCS

(b) When $T < T_c$ and $m^* > 0$, we have a constant solution $\eta = \eta_0 = \mp\sqrt{-A/C}$, describing the ordered or superconducting phase in the mean field approximation. This is also the lowest energy solution of eqn (6.35). In addition, the remaining space-dependent solutions can all be written as

$$\eta = \left|\eta_1 \operatorname{sn}\left[\mp \Delta^{1/2}\eta_2(x - x_0), k\right]\right|, \qquad (6.38)$$

where x_0 is another integration constant, the Jacobi modulus $k^2 = \eta_1/\eta_2$, and the real period of this solution is $\lambda = 4K(k)/(\Delta^{1/2}\eta_2)$, where $K(k)$ is a complete elliptic integral (see Appendix B). In the limit when $k \to 1$, eqn (6.38) takes the form of a cusp, given by

$$\eta = \left|\eta_1 \tanh\left[\Delta^{1/2}\eta_1(x - x_0)\right]\right|. \qquad (6.39)$$

(c) When $T > T_c$ and $m^* < 0$ then, apart from the constant solution $\eta_0 = 0$, which happens to have the highest energy, the remaining solutions are given by

$$\eta = \left|\eta_1 \operatorname{cn}\left\{\mp\left[-\Delta(\eta_1^2 + |\eta_2|^2)\right]^{1/2}(x - x_0), k\right\}\right|, \qquad (6.40)$$

where $\mp\eta_1$ are the real roots of $P(\eta)$, $\mp\eta_2$ are the imaginary ones, and the Jacobi modulus, k, is given by $k^2 = \eta_1^2/(\eta_1^2 + |\eta_2|^2)$.

(d) When $T < T_c$ and $m^* < 0$, one obtains four general types of solution. First, the constant solution $\eta = \eta_0$ as in case (b) corresponds to the superconducting mean field. Second, the cnoidal waves of eqn (6.40) represent oscillations of the order parameter between $+\eta_1$ and 0. As demonstrated in Fig. 6.3, for physical reasons one has to introduce a cut-off value in the space of cnoidal oscillations for both cases (c) and (d). This means that solutions the period of which is less than twice the lattice spacing, d, are not admissible, since otherwise they would correspond to an unphysical situation in which the nodes of a wave would fall between lattice sites.

Third, in the limit when $k \to 1$ the solution in eqn (6.40) becomes a solitary wave of the form

$$\eta = [-2A/C]^{1/2} \operatorname{sech}\left[(2A\Delta/C)^{1/2}(x - x_0)\right]. \qquad (6.41)$$

Note that as $T \to T_c$ these latter solutions disappear and thus one may interpret them as manifesting the onset of criticality. The fourth type of solution is a set of doubly periodic dnoidal waves of the form

$$\eta = \left|\eta_2 \operatorname{dn}\left[(-\Delta)^{1/2}\eta_2(x - x_0), k\right]\right|, \qquad (6.42)$$

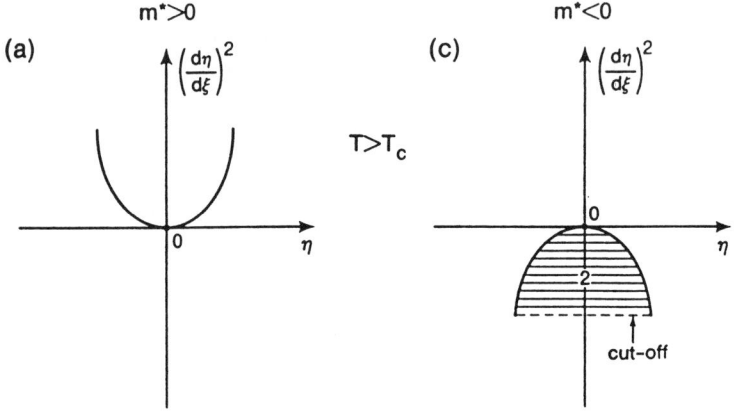

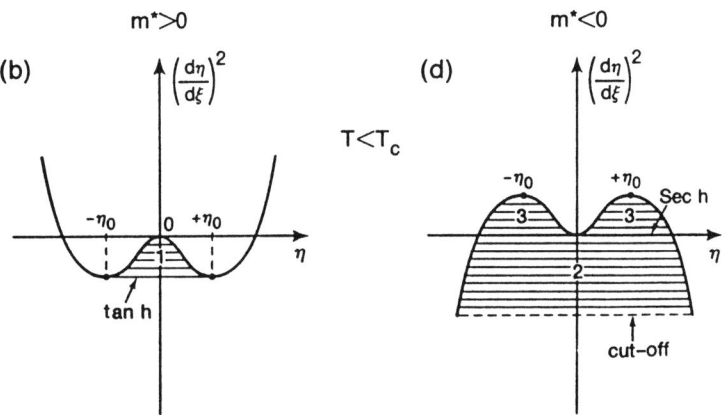

Fig. 6.3 A graphical illustration of the types of solutions to eqn (6.35). (a) $m^* > 0$, $T > T_c$: here the only solution is $\eta = 0$. (b) $m^* > 0$, $T < T_c$: here the constant solutions are $\eta = \mp \eta_0$, (1) denotes $\eta \sim \text{sn}(\ldots)$, and $\tanh(\ldots)$ is the limit of (1) when $k \to 1$. (c) $m^* < 0$, $T > T_c$: the constant solution is $\eta = 0$ and (2) denotes $\eta \sim \text{cn}(\ldots)$. (d) $m^* < 0$, $T < T_c$: the constant solutions here are $\eta = \mp \eta_0$, (2) is the same as in (c), (3) denotes $\eta \sim \text{dn}(\ldots)$, and $\text{sech}(\ldots)$ is the limit of (2) when $k \to 1$.

where both η_1 and η_2 are real roots of $P(\eta)$ and the Jacobi modulus is $k = (1 - \eta_1^2/\eta_2^2)^{1/2}$. The solution in eqn (6.42) represents oscillations with an amplitude in the régime $\eta_1 \leq \eta \leq \eta_2$. For graphical illustrations of all the types of solution of eqn (6.35), we draw the attention of the reader to Figs. 6.3 and 6.4 for physical interpretation. The common characteristic of cases (c) and (d), in contrast to the two previous cases, is the existence of

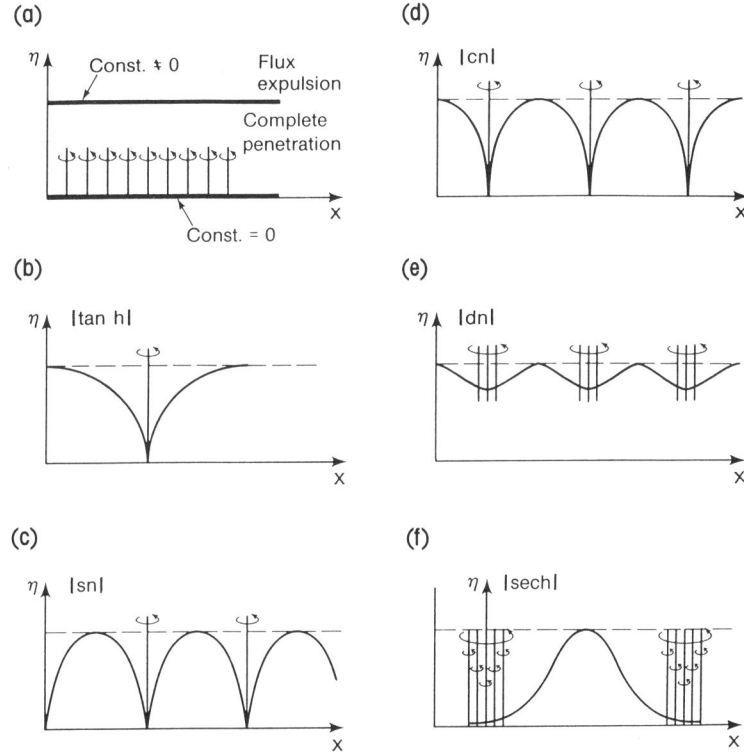

Fig. 6.4 The physical meaning of the nonsingular solutions to the LG equations.

what we interpret as modulational instability in the space of solutions. This can be arrived at when calculating the energies of the latter. It transpires that in case (c) the cnoidal waves always have an energy lower than the mean field solution $\eta = 0$ and, in fact, their energies decrease unboundedly with the decrease of their period which is associated with a corresponding increase in their amplitude. Similarly, in case (d) the cnoidal waves have the same property. On the other hand, the energies of dnoidal waves fill in the region between the energies of the ordered phase and that of the solitary wave. The energy of the solitary wave G_{SW}, given with respect to that of the disordered phase $\eta_0 = 0$, is

$$G_{SW} - G_0 = \tfrac{4}{3}(-\hbar^2/2m^*)^{1/2}(-A)^{3/2}/C. \qquad (6.43)$$

The energy of an N-soliton condensate is lower than that of the normal phase by the approximate amount

$$\Delta E = -\tfrac{4}{3}(-A)^{3/2}(-\hbar^2/2m^*)^{1/2}/C, \qquad (6.44)$$

where we have assumed the picture of a nearly free soliton gas. In this approximation the mean number of soliton quasi-particles is proportional to $T - T_c$, close to T_c, as can be found from the appropriate expression for the average number N_c of Bosons in a Bose condensate. Therefore, assuming, in the usual way, that $A = a(T - T_c)$, we obtain the correct scaling of the energy gap equation with temperature, i.e.

$$\Delta(T) = \Delta E/N_c \sim (T - T_c)^{1/2}. \tag{6.45}$$

Between the energy of the superconducting mean field and the normal phase one also finds a continuum of cnoidal waves. The latter ones could be envisaged as forming from a periodic arrangement of soliton–anti-soliton pairs in the limit of high soliton density. The dynamical picture that emerges involves destabilization of the normal phase on approaching T_c from above. At T_c and below, soliton energy levels become available to the system in the form of fluctuations. Their number and energy (negative with respect to the normal phase) increase as the temperature is lowered, forming initially well-separated regions, which later become characteristically cnoidal with small oscillations about the ordered mean field solution. Eventually, with the temperature approaching 0 K, the classical oscillations of the amplitude vanish completely and the only remaining oscillations are of a quantum nature.

It should be emphasized that the energy difference between the cusp and the mean-field solution η_0 is given by

$$G_{\text{cusp}} - G_0 = \tfrac{4}{3}(-A)^{3/2} \sqrt{\frac{\hbar^2}{2m^*}} /C, \tag{6.46}$$

and all the remaining space-dependent solutions form a continuum, starting just above the cusp-energy. In conclusion, we see that the case in which $m^* > 0$ seems strongly to favour the creation of long-range order through the constant solution η_0 and the plateau regions of the cusp. This suggests applicability to those systems which exhibit very long coherence lengths, as for a standard superconductor. Obviously, the existence of modulational instability for $m^* < 0$, when taken at face value, may be interpreted as a serious deficiency of the proposed model. However, it may not be so if the continuum model employed here is subject to certain characteristic boundary conditions, e.g. translational periodicity of the crystal lattice, where lattice periodicity imposes a natural cut-off of the lower end of the energy spectrum in each case.

2. *Assuming a constant superconducting current density*, using eqn (6.34), our field equation (6.33) reduces to

$$A(T)\eta + C\eta^3 - \frac{\hbar^2}{2m^*}\nabla^2\eta + \frac{m^*j_s^2}{2(e^*)^2\eta^3} = 0, \tag{6.47}$$

which, in one-dimensional space, can be integrated once more to yield

$$(d\eta/dx)^2 = (m^*C/\hbar^2)\{\eta^4 + \alpha\eta^2 + \beta + \gamma/\eta^2\} \equiv R(\eta), \quad (6.48)$$

where $\alpha = 2A/C$, β is an integration constant, and $\gamma = -m^*j_s^2/C(e^*)^2$. The polynomial $R(\eta)$ on the right-hand side of eqn (6.48) is displayed in Fig. 6.5 for various cases of m^*, T, and j_s. This provides a basis for the analysis of the solutions of eqn (6.48). The first step of this analysis is to find the number and location of the extrema of this function which satisfy the cubic equation

$$2y^3 + \alpha y^2 - \gamma = 0,$$

where $y = \eta^2$ and the roots of the equation are denoted as a, b, and c. The discriminant of this cubic is

$$D = (\gamma^2/16)[-1 + \alpha^3/27\gamma],$$

and the product of the roots a, b, and c is $\gamma/2 = abc$. We, therefore, infer that when $m^* > 0$ and $\alpha > 0$ (corresponding to (a) in Fig. 6.5), the only real root of the cubic equation is negative and hence corresponds to no real value of η for an extremum. On the other hand, when $m^* > 0$ and $\alpha < 0$, the situation may vary depending on the value of j_s through the value of γ. It is easy to show that the critical value of γ (which corresponds to a point of inflexion on the curve) is $\gamma_c = \alpha^3/27$. Working this out in terms of j_s gives the scaling of the critical current j_s as

$$j_s = \frac{2e^*\bar{a}}{3C}\sqrt{\frac{2\bar{a}}{3m^*}}(T - T_c)^{3/2}, \quad (6.49)$$

which is of the classic form as a function of $T - T_c$! In Fig. 6.5, when $m^* < 0$ for both cases (c) and (d) there is only one real positive root of the cubic equation and hence in the figures only one pair of equally spaced maxima. However, as j_s increases, the solutions, which will be discussed subsequently, are affected by the value of j_s both in their form and periodicity. The important effect is that for a certain critical value of j_s, no spatially dependent solutions are allowed to exist with a period exceeding twice the lattice spacing. This will determine what we call the cut-off in Fig. 6.5.

In Table 6.1 we have listed only the *real* solutions to eqn (6.47). The solutions 1 and 2 are nonsingular and oscillate periodically between two real roots of the polynomial. Solutions 3–6 are periodically singular. Solution 7 is an algebraic singular solution corresponding to a triple root of the cubic polynomial of eqn (6.47) or the inflexion point of $R(\eta)$ in eqn (6.48). Solution 8 corresponds to one double and one single root of $R(\sqrt{y})$

and is also periodically singular. Finally, solution 9 also corresponds to such a root condition. However, the order of the roots is different. As a result, it describes a well-behaved solitary wave ('bump'). All of these solutions have been illustrated graphically in Fig. 6.4 through the use of various hatchings (see Appendix A). It is obvious that real solutions are given by horizontal line intervals located under the curve $(d\eta/dx)^2$ versus η. Those intervals that join two points on this latter curve lead to finite solutions with singularities. Among the latter class one can distinguish several distinct functional forms, as we have indicated both in Table 6.1 and, correspondingly, through the type of hatching in Fig. 6.4. The 'bump' solitary wave corresponds to one double and one single root, and is also indicated in a separate way.

We shall now address the question of the critical current j_c when $m^* < 0$. The starting point is to develop an equation for the cut-off. From Table 6.1 the period of solution number 2 which describes the order parameter modulation for $m^* < 0$ is

$$\lambda = 4K\left(\sqrt{\frac{a-b}{a-c}}\right)\bigg/\sqrt{(a-c)}. \tag{6.50}$$

The position of the cut-off is determined by the requirement that $\lambda_{\min} = 2d$, where d is the lattice spacing. This, of course, is an implicit equation on the integration constant β provided that the remaining constants α and γ are kept fixed.

It is a natural question to ask if the effective mass, m^*, is allowed to be negative on physical grounds. We feel that this is a real possibility which could be realized through a coupling to certain types of elementary excitations, e.g. optical but *not* acoustic phonons. We recall that the effective mass is related to the dispersion relation by

$$m^* = \hbar^2/(d^2E_k/dk^2). \tag{6.51}$$

▷

Fig. 6.5 A graphical illustration of the types of solutions to eqn (6.48) and their existence for various values of current density j. (a) $m^* > 0$, $T > T_c$: for any value of j_s, only constant (unstable) solutions exist. (b) $m^* > 0$, $T < T_c$: for $j_s = j_1 < j_c$, both 'bump' and elliptical wave solutions exist—for $j_s = j_c$, all the allowed solutions disappear except for the constant solutions (unstable). (c) $m^* < 0$, $T > T_c$: for $j_s = j_1 < j_c$, the allowed elliptical solutions exist as long as their period is larger than the cut-off value—the cut-off moves upwards until $j_s = j_c$, when it then dispels all the space-dependent solutions. (d) $m^* < 0$, $T < T_c$: for $j_s = j_1 < j_c$, the allowed elliptical solutions exist as long as their period is larger than the cut-off value—the cut-off moves upwards until $j_s = j_c$, when it then dispels all the space-dependent solutions. Numbers correspond to those of Table 6.1.

SUPERCONDUCTIVITY AND MCS   317

The effective approximate energy [2] is

$$E_k = \left(\hat{\varepsilon}_k^2 + \hat{\Delta}^2\right)^{1/2}, \qquad (6.52)$$

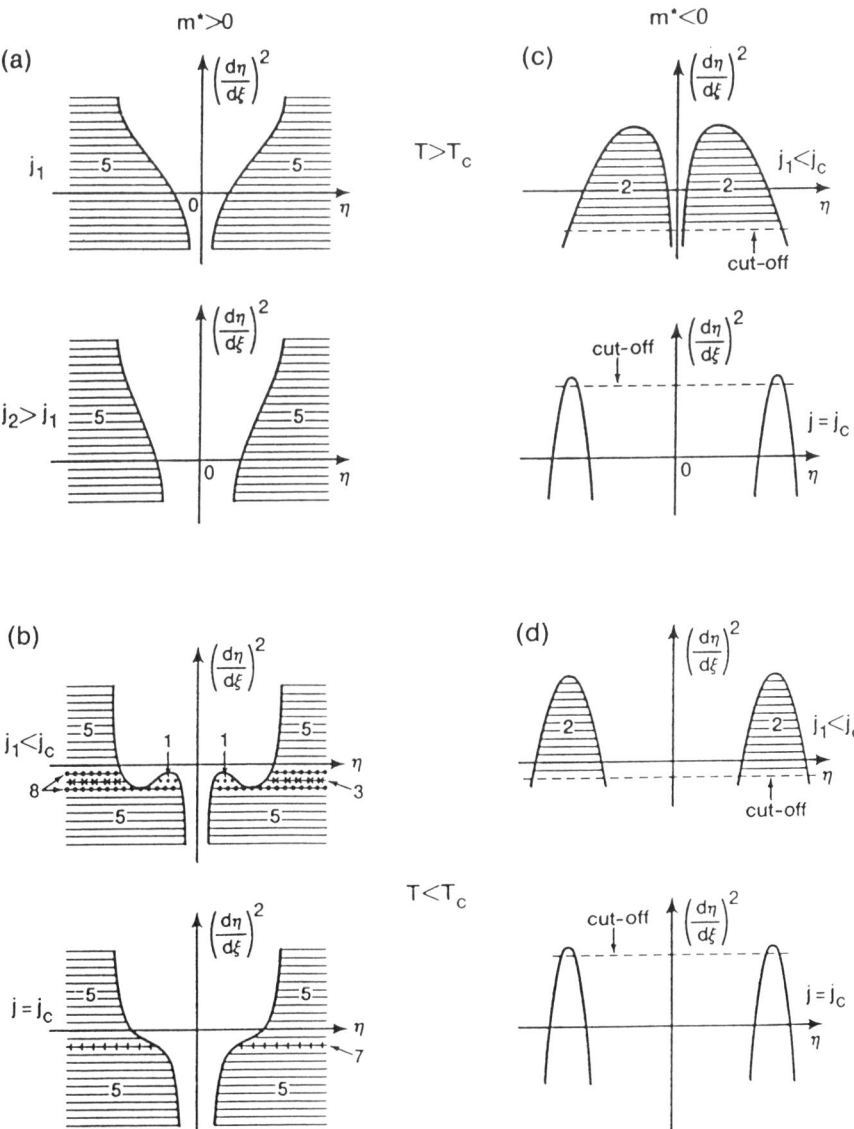

Table 6.1 The general form of solutions to eqn (6.48). Here x is the independent variable defined as $x \equiv \sqrt{\Delta_0}(\xi - \xi_0)$, with $\Delta_0 = 4|m^*|C/\hbar^2$.

No.	ε	Root condition	Solution	g	k^2
1a	$+1$	$a > b \geq y > c$	$\eta = \pm[c + (b-c)\mathrm{sn}^2(x/g, k)]^{1/2}$	$2/\sqrt{a-c}$	$(b-c)/(a-c)$
1b	$+1$	$a > b > y \geq c$	$\eta = \pm\left[\dfrac{a(b-c)\mathrm{sn}^2(x/g, k) - b(a-c)}{(b-c)\mathrm{sn}^2(x/g, k) - (a-c)}\right]^{1/2}$	$2/\sqrt{a-c}$	$(b-c)/(a-c)$
2a	-1	$a \geq y > b > c$	$\eta = \pm\left[\dfrac{b(a-c) - c(a-b)\mathrm{sn}^2(x/g, k)}{(a-c) - (a-b)\mathrm{sn}^2(x/g, k)}\right]^{1/2}$	$2/\sqrt{a-c}$	$(a-b)/(a-c)$
2b	-1	$a > y \geq b > c$	$\eta = \pm[a - (a-b)\mathrm{sn}^2(x/g, k)]^{1/2}$	$2/\sqrt{a-c}$	$(a-b)/(a-c)$
3	$+1$	$y \geq a > b > c$	$\eta = \pm\left[\dfrac{(a-c) + c\,\mathrm{sn}^2(x/g, k)}{\mathrm{sn}^2(x/g, k)}\right]^{1/2}$	$2/\sqrt{a-c}$	$(b-c)/(a-c)$
4	-1	$a > b > c \geq y$	$\eta = \pm\left[\dfrac{a\,\mathrm{sn}^2(x/g, k) + (c-a)}{\mathrm{sn}^2(x/g, k)}\right]^{1/2}$	$2/\sqrt{a-c}$	$(a-b)/(a-c)$

Table 6.1 (Continued)

No.	ε	Root condition	Solution	g	k^2
5	+1	$y \geqslant a,$ $b, c \in C$	$\eta = \pm \left[\dfrac{(A-a)\operatorname{cn}(x/g,k)+(a+A)}{1-\operatorname{cn}(x/g,k)} \right]^{1/2}$	$1/\sqrt{A}$	$\dfrac{A+(\operatorname{Re} b)-a}{2A}$
			$A^2 = ((\operatorname{Re} b)-a)^2 + (\operatorname{Im} b)^2$		
6	-1	$y < a,$ $b, c \in C$	$\eta = \pm \left[\dfrac{(a+A)\operatorname{cn}(x/g,k)+(A-a)}{\operatorname{cn}(x/g,k)-1} \right]^{1/2}$	$1/\sqrt{A}$	$\dfrac{A-((\operatorname{Re} b)+a)}{2A}$
7	+1	$y \geqslant a = b = c$	$\eta = \pm \left[\dfrac{4+ax^2}{x^2} \right]^{1/2}$	—	—
8	+1	$y \geqslant a > b = c$	$\eta = \pm \left[a + (a-b)\tan\dfrac{x\sqrt{(a-b)}}{2} \right]^{1/2}$	—	—
9	+1	$c = b \geqslant y \geqslant a$	$\eta = \pm \left[a + (b-a)\tanh^2\left(\dfrac{x}{2}(b-a)\right) \right]^{1/2}$	—	—

where $\hat{\Delta}$ is related to the energy gap and $\hat{\varepsilon}_k$ is the energy of non-interacting electrons relative to the Fermi level. In this case,

$$\frac{d^2 E_k}{dk^2} = E_k^{-1} \left(\frac{d\hat{\varepsilon}_k}{dk} \right)^2 \left\{ \frac{\hat{\Delta}^2}{\left(\hat{\varepsilon}_k^2 + \hat{\Delta}^2\right)} + \frac{\hat{\varepsilon}_k d^2 \hat{\varepsilon}_k / dk^2}{(d\hat{\varepsilon}_k/dk)^2} \right\}. \tag{6.53}$$

It is quite conceivable that this could become negative when $d^2\hat{\varepsilon}_k/dk^2$ is negative and, simultaneously, $d\hat{\varepsilon}_k/dk$ is small enough in magnitude, as is often the case with optical phonons. In the vicinity of $T = T_c$, and especially so for high-T_c superconductors [5], we expect $\hat{\Delta}$ to be very small. Thus, the actual normal-to-superconducting phase transition could be envisaged as driven/controlled by the temperature dependence of the effective mass and one could then postulate that, close to $T = T_c$, $m^* \simeq m_0(T - T_c)$. The sign reversal of m^* would, therefore, lead to the changeover in the stability properties of the disordered (normal) and ordered (superconducting) phases. The mean field coefficient A, in the vicinity of T_c, might only be weakly temperature dependent in this case, while for standard superconductors, the opposite phenomenon is expected, i.e. m^* is virtually constant while A varies as $T - T_c$. Thus, the superconducting phase could be arrived at through short-range interactions in new superconductors, whereas it is via long-range interactions for standard superconductors.

To summarize, we can envisage two distinct possibilities for a normal-to-superconducting phase transition.

1. When $m^* > 0$, the transition is induced through long-range correlations, which are manifested by the mean field parameter $A \to 0$ as $T \to T_c$. The result is that the lowest energy solution is a mean field or a phase with an extremely long-range coherence length. Here, we may treat m^* as virtually constant with respect to temperature. This description appears quite adequate as a phenomenological description of standard superconductors.

2. The other possibility is that m^* is strongly temperature-dependent in the vicinity of T_c and it actually controls the transition which would occur at $T = T_c$ such that $m^* = 0$. The main difference between this case and the previous one is that the mean field is far from being the lowest energy solution. Indeed, modulational instability can be seen, in which the higher the frequency and the amplitude of an elliptic solution are, the lower is its energy. This modulational instability could be stabilized by introducing a cut-off which is physically motivated, namely by relating the lattice spacing with the solution's period.

In the next two sections we turn our attention to a completely different class of condensed-matter system, i.e. magnetic structures. The first section deals with the very interesting group of magnetic compounds called metamagnets. What will be new in the use of the MCS, as an effective tool, is that their Hamiltonians involve *only* spin degrees of freedom. Through a somewhat circuitous route, the MCS is shown to be a valid approach, nonetheless producing exact results.

6.3 MCS and metamagnetism

A number of compounds exhibit the property that we call *metamagnetism*. They are characterized by the fact that they undergo order–order magnetic phase transitions, the associated phases being of several types, such as ferromagnetic (F), antiferromagnetic (AF), ferrimagnetic (Fi), or helicoidal (AH). The antiferromagnetic phases can be of several different types, such as collinear (AF), canted (AC), declinational (AD), and several others. Metamagnetic compounds are sometimes designated as localized spin systems, e.g. Mn_3GaC, FeRh, $Mn_{2-x}Cr_xSb$, and $FeCl_2$ [6–7], but others involve itinerant electrons, as in the tertiary intermetallics, such as Cr(Sb–Te), Fe(Pd–Pt), or Tb(Cu–Zn) [8]. In general, these compounds can be divided into two groups:

(1) isotropic or weakly isotropic, which are characterized by the predominance of simple rotations of the local magnetic moments, and

(2) anisotropic, the main property of which is an abrupt spin reversal. In Fig. 6.6(a)–(c) we have shown how these properties are manifested in the typical phase diagrams for metamagnets.

In order to prevent the occurrence of spin-flip phases, the system must have sufficiently strong crystalline anisotropy. Furthermore, for a first-order phase transition to take place at low temperatures, there must be competition between two types of magnetic order, and commonly this takes the form of competing AF-type inter-sublattice interactions and F-type intra-sublattice interactions. These properties give insight into the underlying mechanisms that drive these compounds to phase transitions with or without external magnetic fields. Attempts at a theoretical understanding of these materials fall into several different categories. A large number of these approaches involve the phenomenological Landau expansions of the free energy in terms of uniform and staggered magnetization [8] or sublattice magnetization [9]. Other efforts include the role of critical

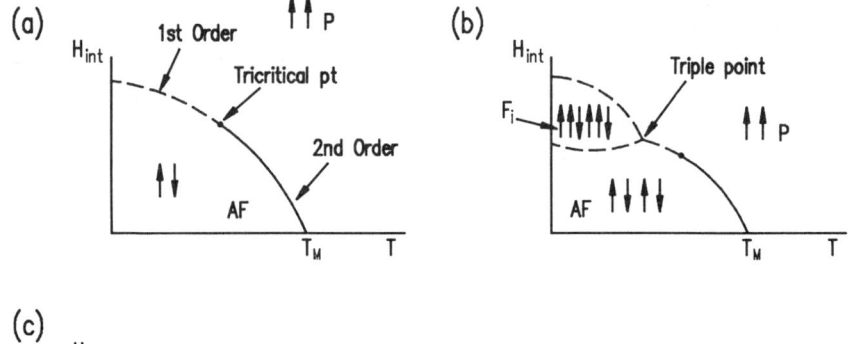
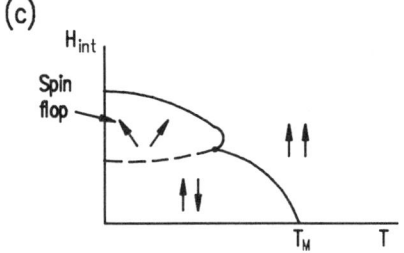

Fig. 6.6 Typical phase diagrams for metamagnets: (a) as in $FeCl_2$, $DyPO_4$; (b) as in $CoCl_2 \cdot 2H_2O$, $FeCl_2 \cdot 2H_2O$; (c) as in MnF_2, $GdAlO_3$.

fluctuations through Landau–Ginzburg methods [10] or microscopic spin-fluctuation calculations via a modified Hubbard Hamiltonian [11].

A model which does lead to phase diagrams for collinear metamagnetics and AF- to F-like transitions is of Heisenberg type [6, 7, 12]. It incorporates the interaction between an external magnetic field and spins of one-half as well as spin–spin interactions between nearest and next-nearest neighbour spins. These latter interactions are associated with exchange coupling constants J_1 and J_2, respectively. The phase behaviour depends on a parameter $\lambda = Z_2 J_2 / Z_1 J_1$, where Z_1 and Z_2 are the numbers of nearest and next-nearest neighbours, respectively. This latter approach was generalized by Niemeijer [12] to include possibly helical phases. Excluding the interaction with an external field, the Hamiltonian takes the form [70]

$$H = J_1 \sum_{i=1}^{N} S_i \cdot S_{i+1} + J_2 \sum_{i=1}^{N} S_i \cdot S_{i+2}, \qquad (6.54)$$

where we have assumed that there are N spins on the chain, and we assume periodic boundary conditions at the ends. The spins are again $S = \frac{1}{2}$, but spins lie in an easy-magnetization plane and may be arbitrarily orientated in going from plane to plane. It turns out that the zero temperature ground state energy of the Hamiltonian in eqn (6.54), taken classically, is

$$E_0 = \begin{cases} -S^2(J_1^2/8J_2 + J_2) & \text{if } J_2 > 0 \text{ and } |J_1| < 4|J_2|, \\ S^2(\pm J_1 + J_2) & \text{otherwise.} \end{cases}$$

The top expression corresponds to helical magnetic order with spins tilted by a constant angle with respect to the nearest neighbours. The bottom expression describes the antiferromagnetic ground state, with a minus sign and the ferromagnetic ground state for the plus sign. Niemeijer [12] used Hartree–Fock approximations that depended on the relative strengths of J_1 and J_2, and demonstrated numerically that ferromagnetic, antiferromagnetic, or helimagnetic ground states could be obtained at finite temperatures.

In this section we demonstrate how powerful the method of coherent structures is when we apply it to the spin system in eqn (6.54). The régimes in J_1 and J_2 which produce the phases above emerge automatically and fully confirm the earlier numerical work. As we have seen, our method is exact, but for the purposes of this exposition it is only necessary to go to zero order. In principle, the method could also be used to incorporate distance-dependent interactions and include spins further away than nearest and next-nearest neighbours.

Our first task is to transform eqn (6.54) into an appropriate second-quantized form. Initially, we follow Stephenson [13] and write

$$S_j^x = \tfrac{1}{2}(a_j^* + a_j), \qquad S_j^y = \frac{i}{2}(a_j - a_j^*), \qquad S_j^z = a_j^* a_j - \tfrac{1}{2}, \quad (6.55)$$

where the superscripts indicate the x, y, and z components and the annihilators and creators for each site j anti-commute, but commute between different sites. Thus, the operators a_j and a_j^* partly resemble Fermion annihilators (creators) on each site, i.e.

$$[a_j, a_j^*]_+ = a_j a_j^* + a_j^* a_j = 1, \qquad a_j^2 = (a_j^*)^2 = 0, \quad (6.56)$$

and partly Bose operators in that for different sites, $i \neq j$, all these operators commute. It is readily verified that, on a particular site, the spin components in eqn (6.55) satisfy angular momentum commutation rules

$$[S_j^x, S_j^y]_- = S_j^x S_j^y - S_j^y S_j^x = iS_j^z, \qquad [S_j^y, S_j^z]_- = iS_j^x,$$
$$[S_j^z, S_j^x]_- = iS_j^y. \quad (6.57)$$

It is easy to show that the Hamiltonian in eqn (6.54) may be written as

$$H = \frac{J_1}{4} \sum_{j=1}^{N} \{2a_j^* a_{j+1} + 2a_j a_{j+1}^* + 4a_j^* a_j a_{j+1}^* a_{j+1} - 2a_j^* a_j - 2a_{j+1}^* a_{j+1} + 1\}$$

$$+ \frac{J_2}{4} \sum_{j=1}^{N} \{2a_j^* a_{j+2} + 2a_j a_{j+2}^* + 4a_j^* a_j a_{j+2}^* a_{j+2} - 2a_j^* a_j$$

$$- 2a_{j+2}^* a_{j+2} + 1\}. \tag{6.58}$$

The operators a_j and a_j^* are awkward to work with, however, as they have some properties of Fermion and some of Boson annihilators and creators. Thus, we introduce Fermion operators c_j (c_j^*), satisfying

$$[c_i, c_j^*]_+ = \delta_{ij}. \tag{6.59}$$

After some algebra, which we present in Appendix D, we may verify that

$$c_j = \exp\left[i\pi \sum_{k=1}^{j-1} a_k^* a_k\right] a_j \tag{6.60}$$

and

$$c_j^* = a_j^* \exp\left[-i\pi \sum_{k=1}^{j-1} a_k^* a_k\right] \tag{6.61}$$

satisfy eqn (6.59) and are the operators that we are looking for. Inverting eqns (6.60) and (6.61) we find that

$$a_j = \exp\left[-i\pi \sum_{k=1}^{j-1} c_k^* c_k\right] c_j \tag{6.62}$$

and

$$a_j^* = c_j^* \exp\left[+i\pi \sum_{k=1}^{j-1} c_k^* c_k\right]. \tag{6.63}$$

By substituting eqns (6.62) and (6.63) into eqn (6.58), it is straightforward to show that the Hamiltonian becomes (also see Appendix D for the method to obtain expressions for $\exp[-i\pi c_j^* c_j]$)

$$H = \frac{J_1}{4} \sum_{j=1}^{N} \{2c_j^* c_{j+1} + 2c_{j+1}^* c_j + 4c_j^* c_j c_{j+1}^* c_{j+1} - 2c_j^* c_j - 2c_{j+1}^* c_{j+1}\}$$

$$+ \frac{N}{4}(J_1 + J_2)$$

$$+ \frac{J_2}{4} \sum_{j=1}^{N} \{2c_j^* c_{j+2} - 4c_j^* c_{j+1}^* c_{j+1} c_{j+2} + 2c_{j+2}^* c_j - 4c_{j+2}^* c_{j+1}^* c_{j+1} c_j$$

$$+ 4c_j^* c_j c_{j+2}^* c_{j+2} - 2c_j^* c_j - 2c_{j+2}^* c_{j+2}\}, \tag{6.64}$$

where terms have been written in such a way as to make the Hermitian nature of eqn (6.64) obvious. To illustrate the method, we use the fact that

$$\exp[-i\pi c_j^* c_j] = 1 - 2c_j^* c_j. \tag{6.65}$$

Thus,

$$a_j^* a_{j+1} = c_j^* \exp\left[+i\pi \sum_{k=1}^{j-1} c_k^* c_k\right] \exp\left[-i\pi \sum_{k=1}^{j} c_k^* c_k\right] c_{j+1}$$

$$= c_j^* \{\exp[-i\pi c_j^* c_j]\} c_{j+1}$$

$$= c_j^*(1 - 2c_j^* c_j) c_{j+1} = c_j^* c_{j+1}. \tag{6.66}$$

However, for pairs of operators connecting sites two neighbours away we have, similarly,

$$a_j^* a_{j+2} = c_j^* \exp[-i\pi c_j^* c_j] \exp[-i\pi c_{j+1}^* c_{j+1}] c_{j+2}$$

$$= c_j^* [1 - 2c_{j+1}^* c_{j+1}] c_{j+2}. \tag{6.67}$$

A further transformation is now required to remove all the site labels j. We do this by expressing each annihilator c_j and creator c_j^* as a Fourier series. Thus

$$c_j = N^{-1/2} \sum_{k=1}^{N} \eta_k \exp[+i\phi_k j], \quad c_j^* = N^{-1/2} \sum_{k=1}^{N} \eta_k^* \exp[-i\phi_k j], \tag{6.68}$$

where the angles ϕ_k are defined by

$$\phi_k = 2\pi k/N \quad \text{for} \quad k = 1, 2, \ldots, N, \tag{6.69}$$

and the operators η_k and η_k^* satisfy Fermi-commutation rules. We now substitute c_j and c_j^* from eqn (6.68) into eqn (6.64) and utilize the relation

$$\sum_{j=1}^{N} \exp[\mp i\phi_j(k-l)] = \sum_{j=1}^{N} \exp[\mp ij(\phi_k - \phi_l)] = N, \tag{6.70}$$

if $\phi_k - \phi_l$ is a multiple of 2π (the only case is $\phi_k = \phi_l$) and is zero otherwise. This gives

$$H = \frac{N}{4}(J_1 + J_2) + \sum_{k=1}^{N} \{J_1(\cos\phi_k - 1) + J_2[\cos(2\phi_k) - 1]\} \eta_k^* \eta_k$$

$$+ \frac{1}{N} \sum_{k_1 k_2 k_3 k_4} \{J_1 \cos(\phi_{k_1} - \phi_{k_4}) - 2J_2 \cos(\phi_{k_1} + \phi_{k_4})$$

$$+ J_2 \cos 2(\phi_{k_1} - \phi_{k_4})\}$$

$$\times \Delta(\phi_{k_1} + \phi_{k_2} - \phi_{k_3} - \phi_{k_4}) \eta_{k_1}^* \eta_{k_2}^* \eta_{k_3} \eta_{k_4}, \tag{6.71}$$

where $\Delta(\phi) = 1$ when $\phi = 2\pi n$ with n an integer or zero, and zero otherwise.

We have thereby transformed the original *spin* Hamiltonian (in an extended Heisenberg form) to the one in eqn (6.71) which is still exact—with *no* explicit spin dependence. We can now take this as a starting point in the MCS formalism. Note that eqn (6.71) contains the familiar two-legged one-body term and the four-legged two-body coupling or interaction term. The Hamiltonian coefficients, in J_1 and J_2, appear in the 'one-body' term! As usual in the MCS, we first calculate the operator's time evolution using Heisenberg's equations.

From the commutation rules we have

$$[\eta_k^* \eta_k, \eta_{k'}]_- = -\delta_{kk'} \eta_k. \tag{6.72}$$

Similarly,

$$[\eta_{k_1}^* \eta_{k_2}^* \eta_{k_3} \eta_{k_4}, \eta_{k'}]_- = \delta_{k',k_2} \eta_{k_1}^* \eta_{k_3} \eta_{k_4} - \delta_{k',k_1} \eta_{k_2}^* \eta_{k_3} \eta_{k_4}. \tag{6.73}$$

Thus,

$$i\hbar \partial_t \eta_{k'} = \omega_k \delta_{k,k'} \eta_k - \sum_{k,l,m} \frac{\Delta_{klm}}{N}$$
$$\times \{\delta_{k',l} \eta_k^* \eta_m \eta_{l+k-m} - \delta_{k',k} \eta_l^* \eta_m \eta_{l+k-m}\}, \tag{6.74}$$

where we have assumed that the dominant term in the interaction is that for which $k_1 + k_2 - k_3 - k_4 = 0$.

For simplicity, we have defined

$$\omega_k = J_1(\cos \phi_k - 1) + J_2(\cos(2\phi_k) - 1) \tag{6.75}$$

and

$$\Delta_{klm} = [J_1 \cos(\phi_m - \phi_l) - 2J_2 \cos(2\phi_k + \phi_l - \phi_m)$$
$$+ 2J_2 \cos 2(\phi_m - \phi_l)]. \tag{6.76}$$

Both sides of eqn (6.74) are multiplied by $\exp(-i\bar{k}'x)$, divided by $\sqrt{N}$, and summed over k' from $k' = 1$ to $k' = N$. At the same time, we define a quantum field operator ψ by the usual relationship in one dimension,

$$\psi(x) = N^{-1/2} \sum_k \eta_k(t) \exp(-i\bar{k}x), \tag{6.77}$$

where $\bar{k}$ is defined by $\bar{k} = 2\pi k/N$. We are using a one-dimensional chain of spins, so we define the field using $\exp[-i\bar{k}x]$ as opposed to the general situation in which the exponential would be of the form $\exp[-i\bar{\mathbf{k}} \cdot \mathbf{r}]$. Furthermore, $\bar{k}$ is regarded as a C-number relative to η_k.

The procedure now in MCS is to take only the zeroth-order interaction terms. That is, we expand $\Delta_{k',k,m}$ about a critical point in reciprocal space (noting that we may write effectively that $\Delta_{klm} = -\Delta_{lkm}$) and retain only the zeroth-order term. We make the specific assumption that, at this critical point, $\phi_k = \phi_{k'} = \phi_m = \phi_0$. We make the further assumption that ϕ_0 takes the value $\phi_0 = 0$ for a ferromagnetic ground state, $\phi_0 = \pi$ for the antiferromagnetic case, and $0 < \phi_0 < \pi$ for the helicoidal structure. In other words, we identify the ground states of the various phases with a particular value of ϕ_k, for any k, namely ϕ_0. The Δ's may be therefore taken out of the summation and, using the definition of the quantum field ψ in eqn (6.77) and performing the summations, we find that the interaction term gives $\mu_3 \psi^* \psi \psi$, where

$$\mu_3 = 2(J_1 + J_2 - 2J_1 \cos(\phi_0) - 2J_2 \cos(2\phi_0)). \tag{6.78}$$

In 'zeroth' order, ω_k must be expanded to quadratic deviations from the critical point. The first and second derivatives of ω_k with respect to k are

$$d\omega_k/dk = (2\pi/N)\{-J_1 \sin \phi_k - 2J_2 \sin 2\phi_k\} \tag{6.79}$$

and

$$d^2\omega_k/dk^2 = (2\pi/N)^2\{-J_1 \cos \phi_k - 4J_2 \cos 2\phi_k\}. \tag{6.80}$$

Thus,

$$\omega_k \simeq \mu_0 + \mu_1(k - k_0) + \mu_2(k - k_0)^2, \tag{6.81}$$

where

$$\mu_0 = J_1(\cos \phi_0 - 1) + J_2(\cos 2\phi_0 - 1), \tag{6.82}$$
$$\mu_1 = (2\pi/N)\{-J_1 \sin \phi_0 - 2J_2 \sin 2\phi_0\}, \tag{6.83}$$

and

$$\mu_2 = \tfrac{1}{2}(2\pi/N)^2\{-J_1 \cos \phi_0 + J_2 \cos 2\phi_0\}. \tag{6.84}$$

Hence, using the above relationships, our equation of motion becomes

$$i\hbar \partial_t \psi = +\hat{\mu}_0 \psi + i\hat{\mu}_1 \partial_x \psi + \hat{\mu}_2 \partial_{xx} \psi + \hat{\mu}_3 \psi^* \psi \psi, \tag{6.85}$$

where the coefficients are defined by

$$\hat{\mu}_0 = J_1(\cos \phi_0 - 1) + J_2(\cos 2\phi_0 - 1) + \phi_0(J_1 \sin \phi_0 + 2J_2 \sin 2\phi_0)$$
$$- \frac{\phi_0^2}{2}(J_1 \cos \phi_0 + 4J_2 \cos 2\phi_0), \tag{6.86}$$

$$\hat{\mu}_1 = \phi_0(J_1 \cos \phi_0 + 4J_2 \cos 2\phi_0) - (J_1 \sin \phi_0 + 2J_2 \sin 2\phi_0), \tag{6.87}$$
$$\hat{\mu}_2 = \tfrac{1}{2}(J_1 \cos \phi_0 + 4J_2 \cos \phi_0), \quad \text{and} \quad \hat{\mu}_3 = \mu_3. \tag{6.88, 6.89}$$

Equation (6.85) is a nonlinear Schrödinger equation (NLS) which has been extensively studied in the past [14], and its solutions have been thoroughly investigated, since it has been found to be completely integrable in $(1+1)$ dimensions. Both the term in $\hat{\mu}_0$ and that in $\hat{\mu}_1$ can be easily removed by two simple transformations. Putting

$$\psi = \overline{\psi} \exp\left[-\frac{i}{\hbar}\hat{\mu}_0 t\right]$$

reduces eqn (6.85) to

$$i\hbar\, \partial_t \overline{\psi} = +i\hat{\mu}_1 \partial_x \overline{\psi} + \hat{\mu}_2 \partial_{xx}^2 \overline{\psi} + \hat{\mu}_3 \overline{\psi}^* \overline{\psi}\overline{\psi} \tag{6.90}$$

and removes the $\hat{\mu}_0$ term. To remove the term in $\partial_x \psi$ we change the independent variable from x to x', where

$$x' = x + vt, \qquad t' = t. \tag{6.91}$$

That is, we choose a moving co-ordinate system with the 'drift' velocity $v = +\hat{\mu}_1/\hbar$ to find

$$i\hbar\, \partial_{t'} \overline{\psi} = \hat{\mu}_2 \partial_{x'x'}^2 \overline{\psi} + \hat{\mu}_3 \overline{\psi}^* \overline{\psi}\overline{\psi}. \tag{6.92}$$

This equation may, of course, be transformed into an irreducible form in which the coefficients have modulus unity, by introducing new dependent and independent variables as

$$\overline{\psi} = \lambda\tilde{\phi} \quad \text{and} \quad x' = \alpha x'', \quad \text{with } t' = \hbar t'', \tag{6.93}$$

to give

$$i\partial_{t''} \tilde{\phi} = \varepsilon_1 \partial_{x''x''}^2 \tilde{\phi} + \varepsilon_2 \tilde{\phi}^* \tilde{\phi}\tilde{\phi}, \tag{6.94}$$

by choosing $\alpha = \sqrt{|\hat{\mu}_2|}$ and $\lambda = 1/\sqrt{|\hat{\mu}_3|}$. The signatures ε_1 and ε_2 remaining are given by

$$\varepsilon_1 = \text{sgn}(\hat{\mu}_2), \qquad \varepsilon_2 = \text{sgn}(\hat{\mu}_3).$$

The various solutions of eqn (6.94) include the celebrated sech-solitons and elliptic waves of several kinds, as well as singular solutions. For a detailed discussion of these solutions and their properties, we refer the reader to the book by Dodd, Eilbeck, Gibbon, and Morris [14], and also to Chapter 7.

We have examined expressions such as eqn (6.92) earlier when discussing superconductivity, where it was found that if the sign of the $\hat{\mu}_2$ coefficient were to change from positive to negative, constant solutions (mean fields) would become unstable with respect to fluctuations. This behaviour is the so-called modulational instability and occurs when $\hat{\mu}_2$ tends to zero. From eqns (6.82)–(6.84) it is easy to see that for the F phase we must have $J_1 = -4J_2$, but in the AF phase $J_1 = 4J_2$. In the AH-structure phase, on the other hand,

$$\cos \phi_0 = \left[-1 \pm (1 + 8\alpha^2)^{1/2}\right]/4\alpha, \qquad (6.95)$$

where $\alpha = 4J_2/J_1$. For the AH case, $-1 \leqslant \cos \phi_0 \leqslant 1$, so its limiting values occur when either $4J_2 = \pm J_1$ or $J_2 = 0$.

We can analyse the regions of stability fairly readily by calculating the corresponding Hamiltonian density H_{LG}. This has the familiar Landau–Ginzburg form

$$H_{\mathrm{LG}} = -\mu_0 |\psi|^2 + \frac{\mu_3}{2} |\psi|^4 - \mu_2 |\partial_x \psi|^2, \qquad (6.96)$$

since its Euler–Lagrange equation takes the form

$$\frac{\partial H_{\mathrm{LG}}}{\partial \psi^*} - \frac{\partial}{\partial x} \frac{\partial H_{\mathrm{LG}}}{\partial(\partial_x \psi^*)} = 0. \qquad (6.97)$$

It can be shown that the stability of the energy functional formed with H_{LG} agrees exactly with the earlier numerical results [12]. That is, the F phase is stable when $J_1 < 0$ and $|J_1| > 4J_2$, but the AF phase is stable when $J_1 > 0$ and $|J_1| < 4J_2$. These latter two phases may become destabilized as a result of $|J_1|$ approaching $4J_2$, producing the helical structure. An AH phase itself may lose stability when J_2 tends to zero. In this case the sign of J_2 is the deciding factor as to the choice of the other two phases which represent the ground state. Solutions of eqn (6.94), which may be written in terms of Jacobi elliptic functions, e.g. sn, cn, dn, etc., describe spin-wave excitations in the magnetic system. We point to a very important fact about the sech-soliton and elliptic-wave solutions, in that their energies are known exactly [15]. These could, therefore, be used to find the finite temperature properties of this system through a reconstruction of the partition function.

Note in this connection that, depending on the sign of μ_2, two groups of nonsingular analytical solutions may be found. For $\mu_2 > 0$ one obtains a localized sech-type solitary wave and two branches of elliptic cn-waves (depending on whether $\mu_0 > 0$ or $\mu_0 < 0$). On the other hand, for $\mu_2 < 0$ a localized tanh-type solitary wave (domain wall) is found in addition to elliptic sn- and dn-waves. In Fig. 6.7 we have illustrated graphically their space dependence. We have also calculated the energies of the two types of localized solutions (tanh and sech) and compared them with the energies of the constant solutions of eqn (6.97), i.e. $\mathrm{const}_1 = 0$ (disordered

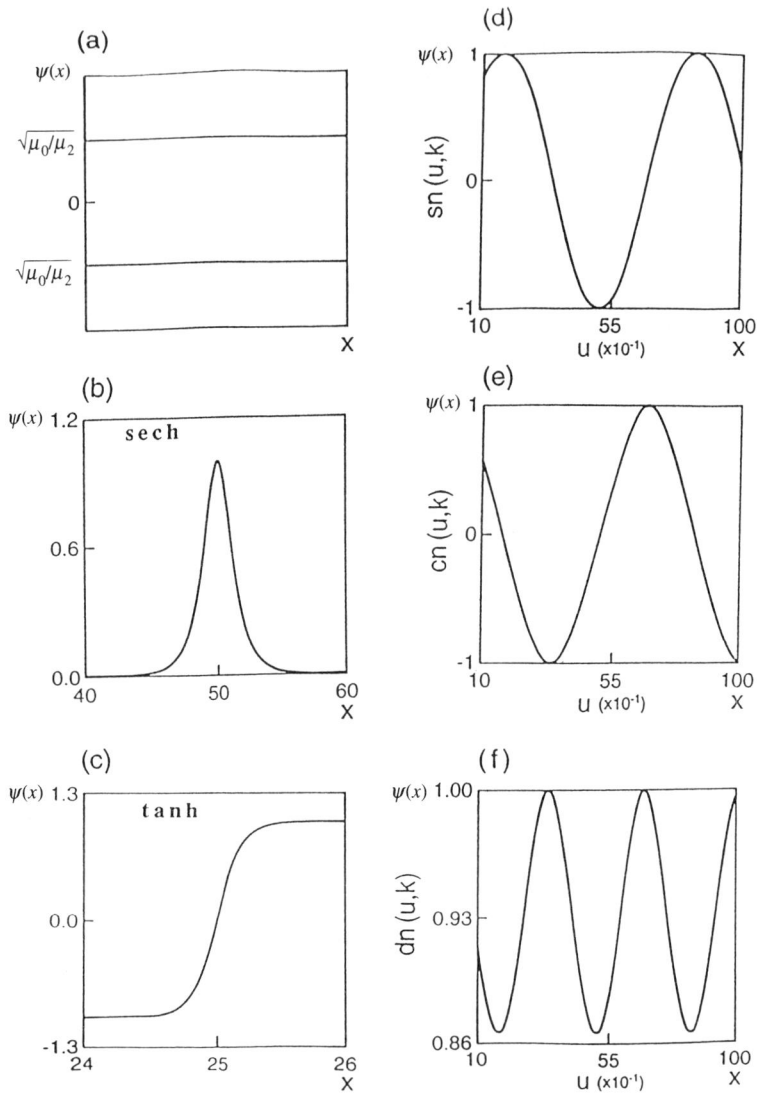

Fig. 6.7 Plots of the space dependence of the nonsingular solutions to eqn (6.97).

phase) and $\text{const}_2 = \sqrt{\mu_0/\mu_3}$ (perfectly ordered phase). This is shown in Fig. 6.8.

A comment is in order regarding the range of validity of our calculations. First of all, the method adopted relies on a zeroth-order approxima-

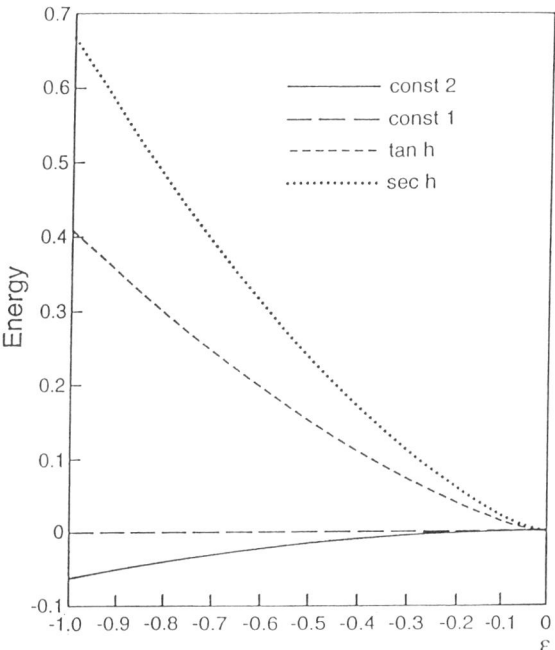

Fig. 6.8 Plots of scaled energies for the two constant and two localized solutions of eqn (6.97), presented as a function of the reduced temperature $\varepsilon = (T - T_c)/T_c$.

tion for the interaction term, i.e. no dispersive nonlinear terms have been included. This means that we ignore mode–mode coupling or scattering of nonlinear spin-waves off one another. Second, the expansion in k-space for the coefficient ω_k (the dispersion relation) has been carried out to second order, which should be quite accurate. Finally, in our stability analysis a classical approximation was used which neglects the role of quantum fluctuations in the vicinity of a transition point. Since the resultant energy functional takes a Landau–Ginzburg form (see eqn (6.96)), restrictions placed on its applicability are similar to those imposed on Landau–Ginzburg models in general. We thus conclude that our results are valid relatively close to transition temperatures and for temperatures sufficiently removed from 0 K, where quantum effects become increasingly significant. Despite a number of approximations made, however, our method has retained the most important nonlinear terms and thus should properly describe the main physical features of the model.

Regarding the question of fully quantum-mechanical treatments of spin-half Heisenberg or Ising metamagnetism, the reader is referred to reference [16] for an up-to-date review. However, it should be said that it has been known for a number of years now that the inclusion of next-nearest neighbours changes the nature of the ground state dramatically. While the quantum ground state is disordered in the nearest-neighbour case, a gap opens in the excitation spectrum of the next-nearest neighbour model at a critical value of $j = J_2/J_1$. Simultaneously, ground-state correlations change from those governed by a power law to an exponential decay type [17, 18]. The ground state, which is often referred to as the dimer phase, is characterized by the domination of short-range, two-spin singlets, i.e. resonating valence bonds (RVB). A comparison with the ground state of the classical Heisenberg model indicates that the RVB state appears in the helical phase [19]. According to the fully quantum-mechanical calculations of Pimpinelli [19], the RVB ground state remains the ground state even when the couplings are extended to $2n$ nearest neighbours, provided that the relations $J_1 = 2n$, $J_2 = 2n - 1, \ldots J_{2n} = 1$ hold.

Exact, fully quantum-mechanical calculations were, in the past, performed for several types of special cases. For example, for $J_2 = 0$ and $J_2 = J_1/2$, the obtained results [20] indicate the absence of Néel long-range order. Instead, an ultrashort-range RVB has been found, as mentioned above. This RVB-type ground state seems to be also a characteristic feature of quantum Hamiltonians with bilinear and biquadratic exchange terms [21]. It therefore appears that quantum fluctuations destroy long-range order arising due to the inherent nonlinearity of these problems. The extent to which this competition is resolved in favour of the disordering role of quantum fluctuations is not entirely clear and thus our results might be of use to this end. In fact, a number of papers have been published [22–24] which attack the problem starting from the classical limit (which is especially appropriate for spins larger than 1/2) and then superimposing quantum corrections. It has been demonstrated that the emerging picture with thermally activated solitons works well in the low-temperature régime. A very recent series of papers [25–27] applied a method very similar to ours to study nonlinear excitations in the antiferromagnetic CeAs. These authors employed a coherent state representation and the results obtained there are entirely consistent with ours.

This particular example of the application of the MCS has shown that the criteria which are obtained for phase stability of the phases of the $S = \frac{1}{2}$ spin chain agree completely with earlier estimates based on numerical minimization calculations performed by Niemeijer [12]. The spin chain, therefore, provides yet another physical situation in which the power of the MCS is quite clear. As we have pointed out earlier, the MCS is also capable, because of known exact solutions, of being utilized in finite-temperature calculations. Finally, it should be apparent from the above development that it could be used readily for more sophisticated spin systems; for example, where distance-dependent interactions are involved and in which spins further away than nearest and next-nearest neighbours are incorporated.

In this example we have only used the classical part of the MCS formalism, since our interests concentrated around the phase diagrams for metamagnets. In the next section we investigate anti-ferromagnetic spin chains but their quantum energy spectra will become all important as we try to elucidate the Haldane gap problem.

6.4 The Haldane gap problem for quantum spin chains

In 1983, Haldane [28] suggested that antiferromagnetic quantum spin chains, where each spin is integer, have a finite energy gap in their spectrum, whereas chains made up of half-odd-integer spins are gapless. Numerical studies were undertaken to try and confirm that there was indeed a gap for chains with spins $S = 1$ [29]. There were, however, at least initially, numerical convergence problems that clouded the issue, and Bethe–ansatz-like approaches seemed to indicate that there might not be a gap when the spins were integral. In this latter approach, however, the interactions between the spins were not of the same form as those discussed by Haldane, so possibly like systems were not being compared with like. It should be mentioned that, two decades earlier, Lieb, Schultz, and Mattis [30] had provided a rigorous proof that there was no gap for half-odd-integer spins, but the method was shown to fail for integer spins. Spin-wave theory for the simplest antiferromagnetic systems was developed long ago by Anderson [31], Ziman [32], and Kubo [33], and the situation was reviewed by Nagamiya *et al.* [34].

In this section we wish to discuss the problem of quantum chains using the MCS. The Hamiltonian that we shall use takes the form [35] [71]

$$H = J_1 \sum_{i=1}^{N} S_i \cdot S_{i+1} + J_2 \sum_{i=1}^{N} S_i \cdot S_{i+2}. \quad (6.98)$$

That is, we have only incorporated interactions between nearest and next-nearest sites labelled i, J_1 and J_2 being the corresponding coupling coefficients. We assume that there are N spins on the chain and we apply periodic boundary conditions at the ends, i.e. $i - j$ is only a multiple of N when $i = j$.

We wish to point out that this Hamiltonian is identical to the Hamiltonian in eqn (6.54) in the last section. Here, however, our main focus is on quantum excitations on the chain. Furthermore, in the metamagnetism example, the spin was taken to be always half, while here we do not presuppose any particular spin value. What we wish to do, in order to use the MCS, is to express eqn (6.93) in second-quantized form. We could opt to use Fermion operators but, other than in the spin-half case, this seems to be unduly complicated for general half-integral spins. For example, we might write the spin component operators on each site i as

$$S_{zi} = \sum_m m a_{im}^+ a_{im}, \qquad S_{+i} = \sum_m \sqrt{S(S+1) - m(m-1)}\, a_{im}^+ a_{i(m-1)},$$

$$S_{-i} = \sum_m \sqrt{S(S+1) - m(m+1)}\, a_{im}^+ a_{i(m+1)}, \qquad (6.99)$$

where the sum over m is over all the spin components, $2S+1$ in number, on one site [36]. If we adopted this approach, it would become necessary to introduce a field for each component and, furthermore, even though they satisfy SU(2) commutation rules on-site, they commute between sites, so that a further transformation to other operators would become necessary to make them true Fermion operators. Hence, we shall use Boson operators defined by the Holstein and Primakov [37] transformation, which is best understood as a particular case of a coupled Schwinger Boson representation [38]. Thus, we use

$$S_{zi} = S - b_i^+ b_i, \qquad (6.100)$$

$$S_{+i} = S_{xi} + iS_{yi} = \sqrt{2S}\left(1 - \frac{b_i^+ b_i}{2S}\right)^{1/2} b_i, \qquad (6.101)$$

and

$$S_{-i} = S_{xi} - iS_{yi} = \sqrt{2S}\, b_i^+ \left(1 - \frac{b_i^+ b_i}{2S}\right)^{1/2}. \qquad (6.102)$$

In eqns (6.100)–(6.102), S is the total spin on each site and the z, x, and y components of spin are denoted by S_{zi}, S_{xi}, and S_{yi} for the spin on site i, respectively. The operators on each site, b_i and b_i^+, are Bose annihilators and creators which satisfy the commutation rules

$$[b_i, b_j^+]_- = \delta_{ij}, \qquad [b_i, b_j]_- = [b_i^+, b_j^+]_- = 0. \qquad (6.103)$$

From eqn (6.103), these operators already commute between sites and it is not now necessary to have labels for different components of spin, i.e. each operator is a creator or annihilator labelled with a site only. There are, however, disadvantages in that the Bose operators operate in a large-dimensional Hilbert space, whereas the spin operators act only in a

($2S + 1$)-dimensional space. In principle, therefore, we ought to introduce projectors so that the new operators are confined to the smaller of the two spaces (in fact, it is clear that when the eigenvalues of $b_i^+ b_i$ exceed $2S$ the root will become imaginary). It is usually unnecessary to do this, as at low temperatures 'spin deviations' (see reference [39], for example) are slight and $b_i^+ b_i / 2S$ may be considered small. We shall, therefore, ignore this problem.

It should be mentioned that two other representations could be used. The first is the coupled Fermion–Drone System due to Jordan and Wigner [40], which has been discussed by Mattis [39]. Unfortunately, this would lead to two nonlinear coupled field equations and would be unnecessarily complicated. A second representation introduces Dyson–Maleev operators [41] $a_i(a_i^+)$ (Bosons), so that

$$S_i^+ = \sqrt{2S}\,(1 - (a_i^+ a_i / 2S))a_i, \qquad S_i^- = \sqrt{2S}\, a_i^+, \qquad S_i^z = S - a_i^+ a_i. \tag{6.104}$$

These operators overcome the problem of the square root which we have in a Holstein–Primakov representation, but this is at the expense of the Hamiltonian becoming non-Hermitian and the transformation nonunitary, although the operators in eqn (6.104) do satisfy the correct commutation rules. Thus, we shall adhere to the operators in eqns (6.100)–(6.102) and note that the spin operators do satisfy the correct commutation relations

$$[S_{zi}, S_{+j}]_- = \delta_{ij} S_{+i}, \qquad [S_{zi}, S_{-j}]_- = -\delta_{ij} S_{-i}, \qquad [S_{+i}, S_{-j}]_- = \delta_{ij} 2 S_{zi}. \tag{6.105}$$

It may also be verified that the operators in eqns (6.100)–(6.102) also satisfy

$$\mathbf{S}_i \cdot \mathbf{S}_i = S_{zi} S_{zi} + \tfrac{1}{2} S_{-i} S_{+i} + \tfrac{1}{2} S_{+i} S_{-i} = S(S+1). \tag{6.106}$$

Substituting the relations in eqns (6.100)–(6.102) in the Hamiltonian of eqn (6.98), we obtain

$$H = J_1 \sum_{i=1}^{N} \Bigg[(S - b_i^+ b_i)(S - b_{i+1}^+ b_{i+1})$$

$$+ S b_i^+ \left(1 - \frac{b_i^+ b_i}{2S}\right)^{1/2} \left(1 - \frac{b_{i+1}^+ b_{i+1}}{2S}\right)^{1/2} b_{i+1}$$

$$+ S\left(1 - \frac{b_i^+ b_i}{2S}\right)^{1/2} b_i b_{i+1}^+ \left(1 - \frac{b_{i+1}^+ b_{i+1}}{2S}\right)^{1/2} \Bigg]$$

$$+ \text{next-nearest neighbour interactions}. \tag{6.107}$$

Note that eqn (6.107) holds even for half-integral spins, S, despite the fact that these are Fermions and we have a Bosonic second-quantized representation. The next problem is to tackle the awkward square roots that appear in eqn (6.107). In order to use MCS in its simplest form, we wish to retain only one-body and two-body-like interaction terms, i.e. two-legged or four-legged operators only, respectively, since we want to retain only the first and simplest nonlinear term. There are at least three ways of doing this:

1. Simply expand each square root and retain only the second term. Such a procedure seems a little naive, since one would be truncating an infinite series with little hope of seeing how good retaining the second term would be.

2. We could perform a so-called Maleev similarity transformation to rationalize the square roots. It does preserve the commutation rules and the on-site number operators [38], but because of its nonunitary character we will not adopt this.

3. Suppose that we denote the number operator $b_i^+ b_i$ by n and consider the eigenvalues of n for each value of S. From eqn (6.100) we see that if $S = \frac{1}{2}$ then the eigenvalues of S_{zi} are $+\frac{1}{2}$ and $-\frac{1}{2}$, so the eigenvalues of n are those of $S - S_{zi}$, namely $n = 0$ or $n = 1$. Similarly, if $S = 1$ the eigenvalues of n are 2, 1, and 0, whereas when $S = 3/2$ they will be 3, 2, 1, and 0. It may be shown that, for any finite total spin S, the root operator may be written, for the eigenvalues at least, as a polynomial of order $2S$. That is,

$$\sqrt{1 - \frac{n}{2s}} = a_0 + a_1 n + a_2 n^2 + \cdots + a_{2s} n^{2S}, \qquad (6.108)$$

where the above is understood to be a relation for the eigenvalues of n. For example, if the total spin $S = 1$, then

$$\sqrt{1 - \frac{n}{2}} = 1 + \left(-\frac{3}{2} + \frac{2}{\sqrt{2}}\right) n + \left(\frac{1}{2} - \frac{1}{\sqrt{2}}\right) n^2. \qquad (6.109)$$

The first coefficient, a_0, is always unity, but all of the coefficients $a_0, a_1, a_2, \ldots, a_{2S}$ may be found from a set of $2S + 1$ linear equations, one for each eigenvalue, there being $2S + 1$ coefficients $a_0, a_1, \ldots, a_{2S}$ to evaluate.

We shall take the coefficients appearing in eqn (6.108), for each spin, to be those in an expansion of each root, n now being replaced, for example in the nearest-neighbour interaction, by $b_i^+ b_i$ or $b_{i+1}^+ b_{i+1}$ as appropriate.

One advantage of this approach is that there are a finite number of terms, namely $2S + 1$, for each total spin S, so it is much easier to see whether remainder terms are small relative to those retained. Furthermore, we eventually have to solve *classical* field equations, so to insist that the eigenvalues of the operators from the roots are correct is not at all unreasonable. In Table 6.2 we list the coefficients $a_1, a_2, \ldots, a_{2S}$ for spins $S = \frac{1}{2}, 1, \frac{3}{2}, 2, \ldots, 10$. The one noticeable point to make about them is that a_1 for integral and half-integral spins is always negative. It is also obvious from Table 6.2 that the coefficients a_i rapidly decrease with increasing i, a useful feature in itself.

Using eqn (6.108) and substituting into the Hamiltonian of eqn (6.107), we find that

$$\begin{aligned}
H = J_1 \sum_{i=1}^{N} \{ & S^2 - S(b_i^+ b_i + b_{i+1}^+ b_{i+1} - b_i^+ b_{i+1} - b_{i+1}^+ b_i) \\
& + (b_i^+ b_i b_{i+1}^+ b_{i+1} + Sa_1 b_i^+ b_{i+1}^+ b_{i+1} b_{i+1} \\
& + Sa_1 b_{i+1}^+ b_{i+1}^+ b_{i+1} b_i \\
& + Sa_1 b_i^+ b_i^+ b_i b_{i+1} + Sa_1 b_{i+1}^+ b_i^+ b_i b_i) \\
& + (Sa_2 b_i^+ b_{i+1}^+ b_{i+1} b_{i+1}^+ b_{i+1} b_{i+1} \\
& + Sa_2 b_{i+1}^+ b_{i+1}^+ b_{i+1} b_{i+1}^+ b_{i+1} b_i \\
& + Sa_1^2 b_i^+ b_i^+ b_i b_{i+1}^+ b_{i+1} b_{i+1} + Sa_1^2 b_{i+1}^+ b_{i+1}^+ b_{i+1} b_i^+ b_i b_i \\
& + Sa_2 b_i^+ b_i^+ b_i b_i^+ b_i b_{i+1} + Sa_2 b_{i+1}^+ b_i^+ b_i b_i^+ b_i b_i) \} \\
& + \text{next-nearest-neighbour term.} \qquad (6.110)
\end{aligned}$$

We have retained the six-legged operators in eqn (6.110). Some of these latter terms, as well as contributions from additional terms, that we have apparently dropped are *not* negligible. We shall assume at this stage that the parameter a_1 appearing in the four-legged parts of eqn (6.110) is an effective one. That we can view this parameter in this way, without affecting the form of the operators, becomes clear at a later stage. Clearly, the next-nearest-neighbour term may be obtained from that for the nearest-neighbour interaction, given explicitly in eqn (6.108), by the following transformation:

$$J_1 \to J_2, \quad b_i(b_i^+) \to b_i(b_i^+), \quad b_{i+1}(b_{i+1}^+) \to b_{i+2}(b_{i+2}^+). \quad (6.111)$$

Realizing that, by eqn (6.103), Bose operators on different sites commute, we have used this to arrange the terms in eqn (6.110) into pairs, which are obviously the Hermitian conjugates of each other.

Table 6.2 The values of the coefficients a_i in eqn (6.108).

Spin, S	a_1	a_2	a_3	a_4	a_5	a_6	a_7	a_8	a_9	a_{10}	Sa_1
0.5	−1.0										−0.5
1.0	−0.08579	−0.20711									−0.08579
1.5	−0.27506	+0.13931	−0.04775								−0.41259
2.0	−0.11779	−0.02377	+0.01195	−0.00436							−0.23559
2.5	−0.11938	+0.03097	−0.02593	+0.01071	−0.00194						−0.29845
3.0	−0.09378	+0.01593	−0.01440	+0.00670	−0.00178	+0.00020					−0.28133
3.5	−0.08036	+0.01561	−0.01698	+0.01155	−0.00524	+0.00141	−0.00017				−0.28125
4.0	−0.06946	+0.01277	−0.01525	+0.01224	−0.00703	+0.00271	−0.00062	+0.00007			−0.27785
4.5	−0.06115	+0.01070	−0.01393	+0.01294	−0.00906	+0.00457	−0.00156	+0.00032	−0.00003		−0.27515
5.0	−0.05454	+0.00894	−0.01236	+0.01271	−0.01023	+0.00621	−0.00274	+0.00083	−0.00015	+0.00001	−0.27270
5.5	−0.04919	+0.00749	−0.01078	+0.01185	−0.01045	+0.00720	−0.00377	+0.00144	−0.00038	+0.00006	−0.27057
6.0	−0.04479	+0.00629	−0.00921	+0.01038	−0.00946	+0.00682	−0.00380	+0.00159	−0.00048	+0.00010	−0.26872
6.5	−0.04109	+0.00529	−0.00768	+0.00837	−0.00706	−0.00430	−0.00158	+0.00002	+0.00037	−0.00024	−0.26711
7.0	−0.03796	+0.00446	−0.00621	+0.00588	−0.00307	−0.00122	+0.00463	−0.00547	+0.00411	−0.00216	−0.26571
7.5	−0.03527	+0.00376	−0.00479	+0.00295	+0.00267	−0.01068	+0.01708	−0.01834	+0.01447	−0.00858	−0.26449
8.0	−0.03293	+0.00317	−0.00343	−0.00040	+0.01033	−0.02509	+0.03860	−0.04367	+0.03792	−0.02559	−0.26342
8.5	−0.03088	+0.00266	−0.00212	−0.00413	+0.02003	−0.04554	+0.07264	−0.08848	+0.08477	−0.06451	−0.26248
9.0	−0.02907	+0.00222	−0.00087	−0.00822	+0.03194	−0.07320	+0.12332	−0.16223	+0.17066	−0.14489	−0.26164
9.5	−0.02746	+0.00184	+0.00035	−0.01269	+0.04625	−0.10950	+0.19580	−0.27762	+0.31871	−0.29896	−0.26089
10.0	−0.02602	+0.00151	+0.00154	−0.01762	+0.06350	−0.15676	+0.29775	−0.45363	+0.56509	−0.58092	−0.26022

Table 6.2 (*Continued*)

Spin	a_{11}	a_{12}	a_{13}	a_{14}	a_{15}	a_{16}	a_{17}	a_{18}	a_{19}	a_{20}
5.5	0.00000									
6.0	−0.00001	0.0000								
6.5	+0.00008	+0.00002	0.00000							
7.0	+0.00080	−0.00020	+0.00003	0.00000						
7.5	+0.00380	−0.00122	+0.00027	−0.00004	0.00000					
8.0	+0.01337	−0.00532	+0.00156	−0.00032	+0.00004	0.00000				
8.5	+0.03894	−0.01846	+0.00674	−0.00183	+0.00035	−0.00004	0.00000			
9.0	+0.09935	−0.05469	+0.02386	−0.00807	+0.00204	−0.00036	+0.00004	0.00000		
9.5	+0.22966	−0.14402	+0.07312	−0.02962	+0.00935	−0.00222	+0.00037	−0.00004	0.00000	
10.0	+0.49448	−0.34814	+0.20167	−0.09518	+0.03603	−0.01068	+0.00239	−0.00038	+0.00004	0.00000

The next step in the analysis is to Fourier transform each annihilator and creator to remove the site dependence. Thus, we put

$$b_j = N^{-1/2} \sum_{k=1}^{N} b_k \exp[+i\phi_k j], \qquad b_j^+ = N^{-1/2} \sum_{k=1}^{N} b_k^+ \exp[-i\phi_k j], \tag{6.112}$$

where the angles ϕ_k are defined by

$$\phi_k = 2\pi k/N \qquad \text{for } k = 1, 2, \ldots, N, \tag{6.113}$$

and the operators b_k and b_k^+ satisfy Bose-commutation rules.

Using eqn (6.112), we see that the truncated Hamiltonian becomes

$$H = S^2 N(J_1 + J_2) + 2S \sum_{k=1}^{N} \{J_1(\cos(\phi_k) - 1) + J_2(\cos(2\phi_k) - 1)\} b_k^+ b_k$$

$$+ \frac{1}{N} \sum_{k_1 k_2 k_3 k_4} \{J_1 \cos(\phi_{k_1} - \phi_{k_4}) + J_2 \cos 2(\phi_{k_1} - \phi_{k_4})$$

$$+ 2SJ_1 a_1 \left[\cos(\phi_{k_2} - \phi_{k_3} - \phi_{k_4}) + \cos(\phi_{k_1} + \phi_{k_2} - \phi_{k_3}) \right]$$

$$+ 2SJ_2 a_1 \left[\cos(2(\phi_{k_2} - \phi_{k_3} - \phi_{k_4})) + \cos(2(\phi_{k_1} + \phi_{k_2} - \phi_{k_3})) \right]\}$$

$$\times \Delta(\phi_{k_1} + \phi_{k_2} - \phi_{k_3} - \phi_{k_4}) b_{k_1}^+ b_{k_2}^+ b_{k_3} b_{k_4}. \tag{6.114}$$

It should be carefully remembered that a_1 above is a function of spin, i.e. $a_1 = a_1(S)$. Defining

$$\omega_k = [J_1(\cos(\phi_k) - 1) + J_2(\cos(2\phi_k) - 1)]2S, \tag{6.115}$$

and denoting the summand in eqn (6.114), i.e. $\Delta\{\quad\}$ in the two-body term, by $\Omega_{k_1, k_2, k_3, k_1+k_2-k_3}$, we can readily find the equation of motion of a particular Bose operator b_n by using Heisenberg's equation of motion. For simplicity of exposition, we write $k_1 = k$, $k_2 = l$, $k_3 = m$, and $\Omega_{k_1, k_2, k_3, k_1+k_2-k_3} = \Omega_{klm}$, so the Hamiltonian H becomes

$$H = S^2 N(J_1 + J_2) + \sum_k \omega_k b_k^+ b_k + \sum_{k,l,m} \frac{\Omega_{klm}}{N} b_k^+ b_l^+ b_m b_{k+l-m}. \tag{6.116}$$

This Hamiltonian is written in an exactly identical form to the Hamiltonian for metamagnets discussed in Section 6.3 for the two-body term. Obviously, the coefficients are different, since the spin is not necessarily $S = \frac{1}{2}$. We therefore take advantage of the structural similarity and define the field variable as

$$\psi(x) = N^{-1/2} \sum_k b_k \exp(-i\bar{k}x), \tag{6.117}$$

where $\bar{k} = 2\pi k/N$. The subsequent steps simply retrace those made in Section 6.3 and we take a short-cut to the final equation of motion for ψ, which is completely analogous to eqn (6.85) but with modified coefficients. Thus, we have

$$i\hbar \partial_t \psi = \hat{\mu}_0 \psi + i\hat{\mu}_1 \partial_x \psi + \hat{\mu}_2 \partial_{xx} \psi + \hat{\mu}_3 \psi^+ \psi\psi, \qquad (6.118)$$

where the coefficients are

$$\hat{\mu}_0 = 2S[J_1(\cos\phi_0 - 1) + J_2(\cos 2\phi_0 - 1)]$$
$$+ 2S\phi_0(J_1 \sin\phi_0 + 2J_2 \sin 2\phi_0)$$
$$- S\phi_0^2(J_1 \cos\phi_0 + 4J_2 \cos 2\phi_0), \qquad (6.119)$$

$$\hat{\mu}_1 = 2S\{-J_1 \sin\phi_0 - 2J_2 \sin 2\phi_0\} - 2S\{-J_1 \cos\phi_0 - 4J_2 \cos 2\phi_0\}\phi_0, \qquad (6.120)$$

$$\hat{\mu}_2 = S\{+J_1 \cos\phi_0 + 4J_2 \cos 2\phi_0\}, \qquad (6.121)$$

and

$$\hat{\mu}_3 = 2\{J_1 + J_2 + 4Sa_1 J_1 \cos\phi_0 + 4Sa_1 J_2 \cos 2\phi_0\}. \qquad (6.122)$$

As commented earlier, we cannot simply drop all six-legged terms or additional terms when square roots are expanded. To see this we examine the six-legged terms which arise from nearest neighbour couplings in eqn (6.110). By using the commutation relations for the annihilators and creators, it is obviously possible to write them in the form $b^+b^+b^+bbb$, the standard form of a second-quantized three-body operator. *However*, we find that when we do this two-body operators of the form b^+b^+bb also appear, and these will contribute to the two-body operators that we already have in eqn (6.110), and *also* will be of the *same order of magnitude* as retained after the Fourier transformation in eqn (6.112). The associated matrix elements for the three-body operators remaining are at least a factor of $1/N$ smaller, so we can safely neglect them. From eqn (6.110) it is clear we shall obtain the additional contributions

$$Sa_2 b_i^+ b_{i+1}^+ b_{i+1} b_{i+1} + Sa_2 b_{i+1}^+ b_{i+1}^+ b_{i+1} b_i$$
$$+ Sa_2 b_i^+ b_i^+ b_i b_{i+1} + Sa_2 b_{i+1}^+ b_i^+ b_i b_i,$$

the terms in a_1^2 being already in the three-body form, since annihilators and creators on different sites commute. By comparing the two-body terms of eqn (6.110) with the four immediately above, we see that each type of operator's coefficient is modified by the addition of a_2 to a_1 to form an effective a_1. In fact, if we consider all other-legged operators we find that the effective a_1 becomes $a_{1\text{eff}}$, where

$$a_{1\text{eff}} = a_1 + a_2 + a_3 + a_4 + \cdots. \qquad (6.123)$$

It is worth pointing out that the three one-body coefficients, $\hat{\mu}_0$, $\hat{\mu}_1$, and $\hat{\mu}_2$, degenerate into the metamagnetism case when $S = \frac{1}{2}$.

To summarize, we have so far been able to obtain the NLS equation in eqn (6.118) for the effective field. What still needs to be done is to find out under what circumstances the spectrum of quantum excitations, which may be formed over the classical envelope, will be gapless.

To see how this may be done, we use the MCS method and assume that the dominant part of the quantum field is classical in nature, but that its quantum character may be recovered later through a semiclassical approximation by linearizing about the classical field. Initially, we remove the first gradient term in eqn (6.139) by transforming to a moving frame of reference and choose a new independent variable so that $x \to x + (\hat{\mu}_1/\hbar)t$. In one dimension, the equation of motion becomes

$$i\hbar \partial_t \psi = \hat{\mu}_0 \psi + \hat{\mu}_2 \frac{\partial^2 \psi}{\partial x^2} + \hat{\mu}_3 \psi^+ \psi \psi. \quad (6.124)$$

Next, we adopt a modulus–argument form for the field, η, and introduce time dependence in the phase with angular frequency ω, so that

$$\psi(r,t) = \exp(i\omega t)\eta(r)\exp(i\phi(r)). \quad (6.125)$$

When eqn (6.125) is substituted into eqn (6.124), two coupled equations result for the real and imaginary parts, giving

Re, $\quad (\hat{\mu}_0 + \hbar\omega)\eta + \hat{\mu}_3\eta^3 + \hat{\mu}_2\nabla^2\eta - \hat{\mu}_2\eta(\nabla\phi)^2 = 0; \quad (6.126)$

Im, $\quad 2(\nabla\eta)\cdot\nabla\phi + \eta\nabla^2\phi = 0. \quad (6.127)$

Equation (6.127) can be thought of as a continuity equation, since it is equivalent to

$$\frac{1}{\eta}\nabla\cdot[\eta^2\nabla\phi] = 0. \quad (6.128)$$

It is easy to see that eqn (6.128) is readily integrated to give

$$\eta^2\nabla\phi = C_0 + \nabla\times F \equiv j, \quad (6.129)$$

where C_0 is an arbitrary vector integration constant, the second term is the curl of an arbitrary vector function, F, and the whole of the right-hand side has been equated with a phase current density, j. This latter quantity can be defined to be

$$j = \tfrac{1}{2}\text{Im}(\psi^*\nabla\psi - \psi\nabla\psi^*). \quad (6.130)$$

Using eqn (6.129) in eqn (6.126) for the field envelope η, we obtain an autonomous, effectively decoupled, differential equation in the form

$$(\hat{\mu}_0 + \hbar\omega)\eta + \hat{\mu}_3\eta^3 + \hat{\mu}_2\nabla^2\eta - \hat{\mu}_2 j^2/\eta^3 = 0. \tag{6.131}$$

Before we analyse eqn (6.131), we consider the time-dependent oscillatory part of the field $\eta(r,t)$. The underlying Hamiltonian density from which the equation of motion in eqn (6.124) may be derived variationally is

$$H_{LG} = \int dx \left[-\hat{\mu}_2 |\nabla\psi|^2 + \hat{\mu}_0 |\psi|^2 + \frac{\hat{\mu}_3}{2} |\psi|^4 \right] \tag{6.132}$$

and is of Landau–Ginzburg form. With this functional, the mean field, and higher localized solutions, which we obtain later, it is straightforwardly shown that their energy increases as ω increases. Thus, to obtain the lowest lying solution we shall henceforth put $\omega = 0$, so that stationary solutions will be investigated in the first instance.

Going back to eqn (6.131) (with $\omega = 0$) we first consider the case in which there are no currents present, i.e. $j = 0$. In one dimension it reduces to a stationary nonlinear Klein–Gordon (NLKG) equation of the form

$$\hat{\mu}_0\eta + \hat{\mu}_3\eta^3 + \hat{\mu}_2 d^2\eta/dx^2 = 0. \tag{6.133}$$

Equation (6.133) may be readily integrated to yield

$$\left(\frac{d\eta}{dx}\right)^2 = \frac{-\hat{\mu}_3}{2\hat{\mu}_2}(\eta^2 - \eta_1^2)(\eta^2 - \eta_2^2), \tag{6.134}$$

where $\mp\eta_1$ and $\mp\eta_2$ are the (possibly complex) roots of the quartic polynomial

$$\eta^4 + \frac{2\hat{\mu}_0}{\hat{\mu}_3}\eta^2 + C_0 = P(\eta), \tag{6.135}$$

in which C_0 is an integration constant setting the energy scale. Three types of real nonsingular solution can be readily found for eqn (6.134). First, we can clearly obtain constant solutions given by

$$\eta = +\eta_0 = \pm(-\hat{\mu}_0/\hat{\mu}_3)^{1/2}, \tag{6.136}$$

which correspond to an antiferromagnetic ground state since the magnitude of the order parameter, ψ, can be interpreted as the sublattice magnetization. Second, elliptic wave solutions of the form

$$\eta = \eta_1 \operatorname{sn}\left[\pm\sqrt{\Delta}\ \eta_2(x - x_0), k\right] \tag{6.137}$$

also exist, where sn denotes a Jacobi elliptic function (see Appendix B), $\Delta = -\hat{\mu}_3/2\hat{\mu}_2$, and k is the Jacobi modulus, given by $k^2 = \eta_1/\eta_2$. Third, a degenerate form of eqn (6.137) exists when $k \to 1$ and is the well-known localized form

$$\eta = \pm \eta_1 \tanh\left[\sqrt{\Delta}\ \eta_1(x - x_0)\right], \tag{6.138}$$

where x_0 is an arbitrary integration constant. This last solution will be of central importance in what follows.

It is apparent from eqn (6.133) that, when the cubic term is small, i.e. $\hat{\mu}_3 \to 0$, for positive values of $\hat{\mu}_0/\hat{\mu}_2$, oscillatory solutions exist—or from eqn (6.137) above when $k \to 0$ (see also Appendix B). Since $-\hat{\mu}_3/\hat{\mu}_2$ is negative, the cubic term will have the effect of lowering the energy of the oscillatory solutions. When $k \to 0$ one can demonstrate that the elliptic wave solutions above have higher energy than the solitonic form when $k \to 1$. That solitonic solutions can coexist with spin waves is very strongly supported by the recent work of Takeno and Kawasaki [42], who used a Dyson–Maleev transformation from spins to second-quantized operators, a coherent state ansatz, and—from a time-dependent variational principle—deduced appropriate equations of motion. They showed that there were two varieties of intrinsic self-localized modes, symmetric and antisymmetric, below the magnon frequency band, which arose from the nonlinearity in magnon systems in one-dimensional Heisenberg antiferromagnets.

Before we consider quantization of the field, let us turn to the case in which a nonzero current density j is present. In this situation, eqn (6.131) will be satisfied and $\omega = 0$ again for stationary solutions. It may be readily integrated to give

$$(d\eta/dx)^2 = \Delta[\eta^4 + \alpha\eta^2 + \beta + \gamma/\eta^2] = R(\eta), \tag{6.139}$$

where $\alpha = 2\hat{\mu}_0/\hat{\mu}_3$, β is an arbitrary constant, and $\gamma = 2(\hat{\mu}_2/\hat{\mu}_3)j^2$. The solutions of eqn (6.139) can be analysed (see Appendix A) by plotting $(d\eta/dx)^2$ as a function of η. As will be seen from the latter appendix, it is crucial to determine the number and location of the extrema of the polynomial $R(\eta)$. The derivative of $R(\eta)$ with respect to η, when equated to zero and substituting $y = \eta^2$, reduces the problem to the solutions of the cubic equation $2y^3 + \alpha y^2 - \gamma = 0$, the discriminant of which is proportional to $(\gamma^2/16)(-1 + \alpha^3/27\gamma)$. Elementary considerations show that when

$$j < j_c = \sqrt{\tfrac{4}{27}\left[\hat{\mu}_0^3/\hat{\mu}_2\hat{\mu}_3\right]} \tag{6.140}$$

the polynomial $R(\eta)$ may have up to six real roots depending on the integration constant β. If, on the other hand, $j \geq j_c$, the cubic equation has only *one* real solution and, consequently, $R(\eta)$ has only *two* real roots. Unlike the $j < j_c$ case, when $j \geq j_c$ no nonsingular real solutions exist.

Having analysed the classical solutions, we now quantize semiclassically by linearizing about the classical solution, as we have described earlier in Chapter 5. As this case is considerably more complicated, because of the presence of currents, we draw the reader's attention to the second part of Appendix D. There we show that for a classical field of the form

$$\psi_c = \exp(i(\phi(r) + \omega t))\eta(r), \qquad (6.141)$$

the quantum field, dropping Fock operators for convenience (see the earlier part of Appendix D), may be taken to be

$$\psi = \exp(i\phi(r) + \omega t + \varepsilon f(r,t))(\eta(r) + \varepsilon \Lambda(r,t)) \qquad (6.142)$$

where $f(r,t)$ and $\Lambda(r,t)$ are real space- and time-dependent functions and ε is a symbol, later dropped, to indicate the 'small' quantum correction. We find, subject to certain approximations, that the Schrödinger equation satisfied by a component of the spatial part of the quantum field (which has been written as $\Lambda_2 \Lambda_3 = \Lambda_1$—see Appendix D) is

$$\varepsilon \Lambda_2 \simeq \hat{\mu}_0 \Lambda_2 + 3\hat{\mu}_3 \eta^2 \Lambda_2 + \hat{\mu}_2 d^2 \Lambda_2/dx^2, \qquad (6.143)$$

where $\varepsilon = -\hbar(\omega + \alpha)$. Thus we see that the constant α denotes the shift in energy away from the energy of the classical field, eqn (6.143) being a one-dimensional Schrödinger equation with an effective potential proportional to η^2 as a result of a nonlinear classical envelope providing a binding potential. Note that although the currents should appear in this equation as well as the classical equation, their effect in the quantum equation is dominated by the $3\hat{\mu}_3 \eta^2 \Lambda_2$ contribution to the potential energy (see Appendix D for further details.) For a particular classical field, $\hbar \alpha$ denotes, in general, a series of quantum energies. Let us now examine what form these energies take by considering the insertion of the three types of classical solution into eqn (6.143). First, constant classical solutions lead to a continuum of plane waves in free space as solutions of eqn (6.143) and no energy gaps are formed. If η is a singular solution scattering states will be formed, but again no bound states with energy gaps will arise. The third class of solution discussed above were elliptic waves. For periodic solutions η the Schrödinger eqn (6.143) becomes a Lamé equation. The characteristic feature here is that energies associated with the solutions of this latter equation exhibit band formation. Although gaps may exist between bands, within a given band energies are distributed continuously. The only possibility remaining is when η describes a solitary wave proportional to the hyperbolic tanh or sech function. In both these cases, when the elliptic modulus $k \to 1$, the Schrödinger eqn (6.143) can be reduced to the same general form, namely

$$-\frac{d^2\Lambda}{d\xi^2} + \left[(L^2 - \omega_n^2) - L(L+1)\text{sech}^2\xi\right]\Lambda = 0, \quad (6.144)$$

with the independent variable ξ and the other coefficients being transformed into the form in eqn (6.144) for convenience. In both cases the parameter $L = 2$. In the language of Morse and Feshbach [43], $L(L+1) = v$ and $\omega_n^2 - L^2 = \varepsilon - v$. With these parameters, the bound energies are given by

$$\varepsilon = \varepsilon_n = v - \left[\sqrt{v + \tfrac{1}{4}} - (n + \tfrac{1}{2})\right]^2, \quad (6.145)$$

where

$$n = 0, 1, \cdots < \sqrt{v + \tfrac{1}{4}} - \tfrac{1}{2}. \quad (6.146)$$

As $L = 2$ there are only *two* quantum bound states [43, 44] corresponding to $n = 1$ and $n = 2$ separated from the above continuum. The picture that then emerges is that the existence of localised sech or tanh solutions is a prerequisite for the presence of an energy gap conjectured by Haldane.

In order to possess such a tanh-soliton, the asymptotic states $\pm\psi_0$ which correspond to the nonlinear potential minima between which the soliton interpolates must be physically realizable. This means that ψ_0 solves *the equation of state*

$$\hat{\mu}_0 \psi_0 + \hat{\mu}_3 \psi_0^+ \psi_0 \psi_0 = 0, \quad (6.147)$$

but it simultaneously *must be* compatible with the simple physical fact that the value of spin projection ranges from $-S$ to $+S$. Therefore, the field variable squared, i.e. $\psi_0^+ \psi_0$, must not exceed $2S$, which corresponds to a complete spin reversal from $-S$ to $+S$ on a single site. We therefore conclude that the following inequality must be satisfied if the two quantum levels are of physical consequence: that is,

$$|\psi_0|^2 = |\hat{\mu}_0/\hat{\mu}_3| \leq 2S, \quad (6.148)$$

if the chain is to have a gap. To enable the reader to understand this, in Fig. 6.9 we have schematically compared the two situations that arise.

Let us now be more specific and interpret eqn (6.148) in terms of model parameters. Substituting $\hat{\mu}_0$ and $\hat{\mu}_3$ and assuming the antiferromagnetic ground state, i.e. $\phi_0 = \pi$ (see Section 6.3) we obtain the following criterion:

$$\left|\frac{(4 - \pi^2) + 4\pi^2 r_0}{2\{(1 - 4Sa_1) + r_0(1 + 4Sa_1)\}}\right| \leq 2, \quad (6.149)$$

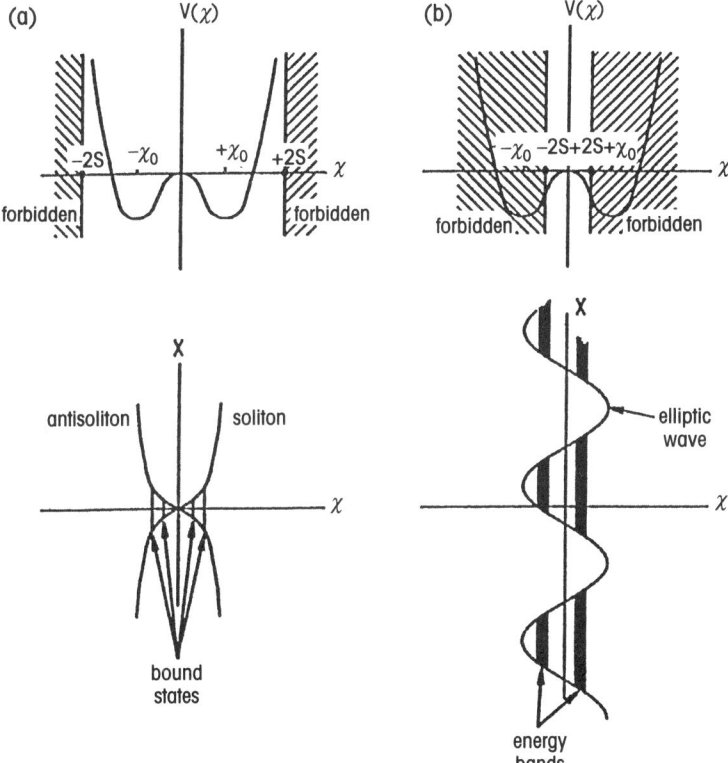

Fig. 6.9 A schematic illustration of the two possibilities in eqn (6.118). (a) When $|\psi_0|^2 < 2S$ a soliton state is allowed which has two quantum bound states (b) when $|\psi|^2 > 2S$ the kink soliton state is no longer physically admissible (see Appendix A) and a periodic solution with a continuum of quantum states within a band replaces the soliton.

where $r_0 = J_2/J_1$. First, in all the experimentally investigated materials the value of r_0 is very small indeed, $r_0 \ll 1$, and hence the appropriate terms in eqn (6.149) can be ignored. Second, using the values of Sa_1 from Table 6.2 we find that the above condition is satisfied in the asymptotic limit of large spin, S, and on this basis we would expect the Haldane gap to be present there. Furthermore, all the lower spin values satisfy the above criterion. In the special case of $S = 1$ the result of this analysis strongly depends on the value of r_0. It turns out that for $r_0 > 0.01$ the condition is satisfied and a gap exists. This latter result, however, has incorporated only a_1 in the inequality from Table 6.2 and has not included additional components to an $a_{1\text{eff}}$ which arise from six-legged and higher-legged operators when the square root operators are expanded in eqn (6.107). When the $a_{1\text{eff}}$ is used from the above expression, the condition that, when $S = 1$, $r_0 > 0.01$ is no longer applicable and eqn (6.149) holds for all spin values, S, including $S = 1$. It is interesting to note that even numerical computations have suffered from difficulties in the $S = 1$ case.

On the basis of our analysis so far, we could not account for the difference in behaviour between integer and half-integer spin cases. However, an important distinction has to be taken into account when dealing with the half-integer cases. Here, the local spin state is not invariant with respect to a 2π rotation but results in a sign reversal, unlike in the integer spin cases. Since a 2π rotation is not an invariant for the ground state of half-integral spins, it induces a degeneracy which has to be accommodated by the system if time-reversal symmetry is not to be violated. Suppose that the antiferromagnetic ground state of the spin chain begins at one end with a spin-up projection; then a completely equivalent arrangement beginning on a spin-down projection is allowed to exist. In order not to violate the time-reversal symmetry, the spin chain will oscillate in time between these two equivalent ground states. This can be viewed as a flow of 'phase current' backwards and forwards along the chain. The mathematical implications of the equation of motion are very simple to understand and we can make use of an analogous derivation which we have made earlier in this chapter for superconductors (see Section 6.2). A steady-state solution to the first integral of eqn (6.118) satisfies

$$\hat{\mu}_0 \eta + \hat{\mu}_2 \nabla^2 \eta + \hat{\mu}_3 \eta^3 - \frac{\hat{\mu}_2 j^2}{\eta^3} = 0, \qquad (6.150)$$

where the classical field ψ in eqn (6.118) has been written in modulus–argument form as $\psi = \eta \exp(i\phi)$ and the current density j is defined as

$$j = \eta^2 \nabla \phi. \qquad (6.151)$$

A very similar argument to that presented for superconductors can now be used to show that there exists a critical value of the current j_c for which all nonsingular solutions disappear (including the localized one). This value is given by the following condition (see eqn (6.48) and Fig. 6.5):

$$-\hat{\mu}_2 j_c^2 = \tfrac{4}{27} \hat{\mu}_0^3 / \hat{\mu}_3^2, \qquad (6.152)$$

where $j_c = 2S\pi$, since there exists a phase shift of π between adjacent

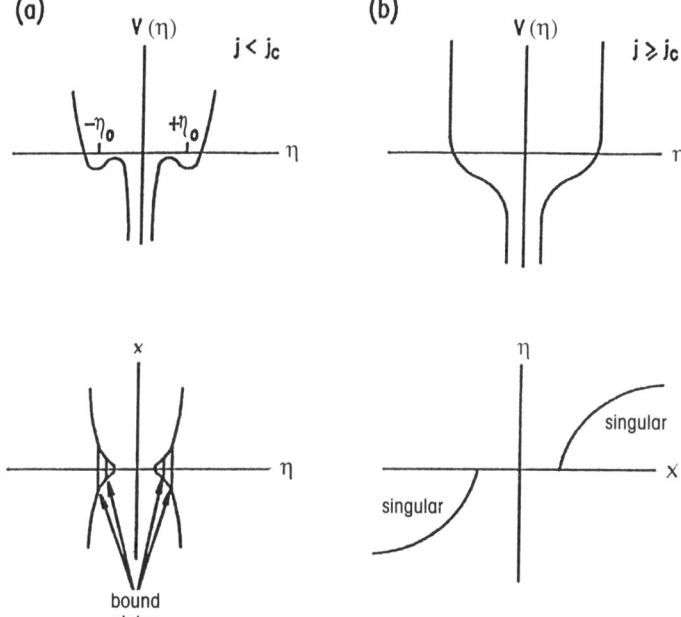

Fig. 6.10 A graphical illustration of the two possibilities in eqn (6.150), depending on the values of the current density: (a) $j < j_c$ supports a localized envelope and two bound states within it; (b) $j \geqslant j_c$ causes the envelope to become delocalized (in fact, singular) and no bound states can be found.

spins in an antiferromagnetic state and $\eta^2 \simeq 2S$ under the same conditions. The criterion in eqn (6.152) translates into

$$\pi^2 = \frac{(-4 - \pi^2(4r_0 - 1))^3}{4 \times 27(1 - 4r_0)\{1 + r_0 + 4Sa_1(r_0 - 1)\}^2}, \qquad (6.153)$$

which takes precedence over eqn (6.149) for half-integer spin chains.

In Fig. 6.10 we have graphically illustrated the two generic situations for half-integer spins in regard to the value of the critical current density; i.e. (a) $j < j_c$, implying the existence of localized envelopes with two bound states, and (b) $j \geqslant j_c$, where not only the bound states disappear but the classical envelope becomes delocalized.

Taking the asymptotic limit for Sa_1 to be -0.25, for simplicity, we can solve the cubic in eqn (6.153) for r_0 to obtain the following approximate results for the roots of the cubic polynomial:

$$r_0 \simeq +0.25, \quad +0.61, \quad -0.42.$$

We can, therefore, conclude that:

(1) the gap disappears whenever

$$-0.42 \leqslant J_2/J_1 = r_0 \leqslant +0.25 \quad \text{or} \quad J_2/J_1 > +0.61;$$

(2) there may be a gap present (if the current density is less than j_c) for either

$$+0.61 > J_2/J_1 = r_0 > 0.25 \quad \text{or} \quad J_2/J_1 < -0.42.$$

It can easily be demonstrated that similar conclusions hold for any spin. These conclusions are strongly supported by inserting values for $\hat{\mu}_0$, $\hat{\mu}_2$, and $\hat{\mu}_3$ from the exact treatment that we have presented for the $S = \frac{1}{2}$ case in the metamagnetism section. With these values in eqn (6.153) there is only one real root which is negative and well away from the origin. The minimum in the cubic curve is above but close to the horizontal axis at $r_0 \simeq +0.26$. As we already mentioned, all the known experimental cases involve r_0 the value of which is much less than one. Indeed, we recall from Section 6.3, that in the spin $S = \frac{1}{2}$ chain one of the criteria for antiferromagnetism was that r_0 be less than 0.25. This then strongly indicates that the energy gap that appeared to exist based on eqn (6.149) becomes forbidden as a result of the presence of phase currents, the densities of which exceed the critical value for all practical purposes. Thus, we conclude that our analysis supports the Haldane theorem and we restate our findings in the following summary:

1. For all integer spins there is an energy gap. This conclusion is without qualification and the apparent problem for $S = 1$ may be eliminated when $Sa_{1\text{eff}}$ is used instead of Sa_1.

2. For all half-integer spins the energy gap is absent as a result of phase currents which can be seen as oscillations of the ground state in time between the equally allowed spin projections. A qualification we have to make here is that when J_2/J_1 exceeds the value of $+0.25$ but is less than 0.61, a gap may re-emerge. This, however, is most unlikely but we nevertheless encourage experimental studies in this direction.

In the two sections which follow we shift our attention to a completely different class of problem, i.e. we will attempt to convince the reader that a multi-electron atom can be viewed as a strongly nonlinear localized structure in which nonlinearity results from Coulomb repulsion between electrons. The attractive potential due to the nucleus treated here in a one-body approximation results in the electronic field's localized character and stability but, as we have known for close to a century, its internal structure is a highly complex ladder of quantum energy levels. An explanation of the latter property is, therefore, our main challenge using the MCS. The first section that follows is intended to treat the atom in a hydrogenic approximation, ignoring the Coulomb repulsion between electrons. This is shown to result in a *linear* field equation. Nevertheless, we do this to see whether a quantum field approach to the problem is sensible at all.

The next two sections of this chapter are concerned with the modelling of the internal structures of atoms. It is therefore in order to elaborate on the emergence and development of atomistic ideas throughout the history of science.

6.5 Noninteracting electron systems: the hydrogen atom

It appears that the first evidence of atomistic thought can be traced back to Indian philosophers at least a millennium before the Christian era. In ancient Greece the views on this subject were divided. Anaxagorus and Aristotle believed that matter was continuous, while Democritus and Epicurus were strong proponents of the granularity of matter. This latter view was expounded by the Roman poet Lucretius. However, the Aristotelian view prevailed throughout the Middle Ages. A return to an atomistic philosophy of matter was initiated by Isaac Newton, Robert Boyle, René Descartes, and Galileo Galilei, and other eminent scientists of the time. Little progress was made in practical applications of this idea until the turn of the eighteenth and nineteenth centuries, when a connection was made between the concept of an atom and chemical compositional processes. In particular, John Dalton made these ideas more quantitative. Avogadro formulated his famous hypothesis and Prout observed that atomic weights are approximately integral multiples of the hydrogenic mass. Michael Faraday formulated his laws of electrolysis and linked atomicity to the elementary charge of an electron. An epoch-making breakthrough was made by Mendeleev in 1834 with his periodic classification of elements. Modern views concerning atomic structure stem from the discoveries made at the turn of the last century. In particular, in 1895 Röentgen discovered X-rays. In 1896 Becquerel made the first observations of radioactivity, and a year later J. J. Thomson measured the mass of an electron. Contemporary ideas of atomic structure have been largely inspired by Ernest Rutherford's experimental results, which demonstrated the existence of a positively charged nucleus surrounded by electrons whose total charge balance the nuclear charge. Detailed investigations of atomic structure were at the centre of the scientific revolution that resulted in the formulation of quantum mechanics. Quantized absorption and emission spectra, obtained initially for simple gases, especially hydrogen, provided direct proof that, at a microscopic level, energies are discrete. The various spectral series (Balmer, Paschen, Lyman, Brackett, etc.) found their physical interpretation in the semiclassical theory of atomic structure formulated by Bohr, which was later refined by Sommerfeld with elliptical orbits. The results of the Bohr–Sommerfeld investigations were later verified by full quantum-mechanical calculations. Over the course of some 50 years which followed, calculations of atomic energy levels were performed for virtually all of the elements of the periodic table, and they necessitated the inclusion of an even-increasing number of corrections and additional interactions such as spin–orbit coupling and other relativistic effects, the hyperfine splitting, the Lamb shift, and spin–spin interactions, to mention but a few. Simultaneously, a large computational effort was afforded by the development of high-speed computers, and techniques such as the Hartree and later the Dirac–Hartree–Fock schemes were routinely used. Nevertheless, serious discrepancies persisted between theoretical and experimental results, especially for multi-electron atoms starting from the transition series.

We have argued throughout this book that many-particle systems supporting significant two-body interactions inevitably lead to nonlinearities in their theoretical description. Up until fairly recently, these nonlinearities have been de-emphasized, largely due to lack of adequate analytical and numerical tools. Clearly, multi-electron atoms fall into the category of strongly interacting many-body systems, but the number of particles involved is finite, which causes additional complications. It is, therefore, extremely challenging to attempt the formulation of a nonlinear theory of the many-electron atom. On the other hand, these systems have been very thoroughly investigated over the years and a wealth of experimental and theoretical data is available for comparison. In the two sections that follow we intend to demonstrate that nonlinear analysis can provide an accurate, fascinating description of the age-old problem of the structure of the atom.

Before embarking on the study of interacting atom electron systems it is instructive to see how hydrogenic-like spectra arise using a field description by employing a Hamiltonian density in terms of the field. As we have seen earlier (see Chapter 5), for Euclidean signatures, i.e. $\varepsilon_1 = \varepsilon_2 = \varepsilon_3 = 1$, the equation of motion for the field ψ, when the nonlinear interaction term vanishes, takes the form

$$i\hbar \, \partial_t \psi = \nu_0 \psi + i\boldsymbol{\nu}_1 \cdot (\nabla \psi) - \tfrac{1}{2} \nabla^2 \psi. \qquad (6.154)$$

The first step is to remove the gradient term involving the constant vector

THE HYDROGEN ATOM   353

v_1. This is readily performed with a Galilean transformation which may be written as

$$t' = t, \quad r' = r - vt, \tag{6.155}$$

where the velocity v is to be chosen to eliminate the term in v_1. The chain rule in the following two forms,

$$\partial_x \psi = \partial_{x'} \psi \partial_x x' + \partial_{y'} \psi \partial_x y' + \partial_{z'} \psi \partial_x z' + \partial_{t'} \psi \partial_x t' \tag{6.156}$$

and

$$\partial_t \psi = \partial_{t'} \psi \partial_t t' + \partial_{x'} \psi \partial_t x' + \partial_{y'} \psi \partial_t y' + \partial_{z'} \psi \partial_t z', \tag{6.157}$$

with similar forms for derivatives with respect to y and z, enables us to obtain

$$\nabla \psi = \nabla' \psi \quad \text{and} \quad \partial_t \psi = \partial_{t'} \psi - \mathbf{v} \cdot \nabla' \psi, \tag{6.158, 6.159}$$

where the superscript primes on the deltas denote gradients with respect to the primed co-ordinates. Thus, if we choose

$$-\hbar v = v_1, \tag{6.160}$$

then eqn (6.154) becomes

$$i\hbar \partial_{t'} \psi = v_0 \psi - \tfrac{1}{2}(\nabla')^2 \psi. \tag{6.161}$$

Equation (6.161) is, of course, linear and can be solved easily by putting

$$\psi = f(r') \exp[-it'\lambda/\hbar]. \tag{6.162}$$

Inserting eqn (6.162) into eqn (6.161) and rearranging gives

$$(\nabla')^2 f + 2(\lambda - v_0) f = 0. \tag{6.163}$$

We can now either use the method of separation of variables and plod through the algebra or, to shorten the exposition, we can simply put

$$f = R(r') Y_l^m(\theta', \varphi'), \tag{6.164}$$

where Y_l^m denotes a spherical harmonic. The notes for Rotenberg's Tables [45] are now used, in particular equation (1.31), to enable us to write

$$(\nabla')^2 Y_l^m(\theta', \varphi') R(r') = Y_l^m(\theta', \varphi') \left[\frac{1}{r'} \frac{d^2}{d(r')^2} (r') - \frac{l(l+1)}{(r')^2} \right] R(r'). \tag{6.165}$$

When eqn (6.165) is inserted into eqn (6.163), the resulting equation may be multiplied through by $Y_{l'}^{m'*}(\theta', \varphi')$ and integrated over angles using the relation

$$\int_0^{2\pi}\int_0^{\pi} Y_{l'}^{m'*}(\theta', \varphi')Y_l^m(\theta', \varphi')\sin\theta'\,d\theta'\,d\phi' = \delta_{l,l'}\delta_{m,m'}, \quad (6.166)$$

to give

$$\frac{d^2 R}{d(r')^2} + \frac{2}{r'}\frac{dR}{dr'} + \left(-\frac{l(l+1)}{(r')^2} + \Lambda\right)R = 0, \quad (6.167)$$

where we have defined Λ by

$$\Lambda = 2(\lambda - \nu_0). \quad (6.168)$$

We now compare eqn (6.167) with that in Gradshteyn and Ryzhik [46],

$$u'' + \frac{(1-2\alpha)}{z}u' + \left[(\beta\gamma z^{\gamma-1})^2 + \frac{\alpha^2 - \delta^2\gamma^2}{z^2}\right]u = 0, \quad (6.169)$$

which has one solution

$$u = z^\alpha Z_\delta(\beta z^\gamma), \quad (6.170)$$

$Z_\delta(z)$ being an arbitrary Bessel function at this stage, but we will particularize and use a specific one, $J_\delta(z)$. By comparing eqn (6.169) with eqn (6.167), we have

$$\alpha = -\tfrac{1}{2}, \quad \gamma = 1, \quad \beta = \mp\sqrt{\Lambda}, \quad \delta^2 = \tfrac{1}{4} + l(l+1). \quad (6.171)$$

Thus, a solution of eqn (6.167) is

$$R = (r')^{-1/2} J_\delta(\sqrt{\Lambda}\,r'), \quad (6.172)$$

and the field ψ itself is given by

$$\psi = N_0(r')^{-1/2} J_\mu(\sqrt{\Lambda}\,r')Y_l^m(\theta', \phi')\exp[-it'\lambda/\hbar], \quad (6.173)$$

where $\mu = l + \tfrac{1}{2}$ and N_0 is some constant, with $N_0 \neq 0$.

When there are no interactions present we cannot take ψ to be its classical counterpart and then expand about this linearly to obtain a Schrödinger equation for the quantum field, as we have done previously. The equation for the quantum field will, in fact, be identical (because eqn (6.163) is linear) to the classical field and the effective potential zero. Thus, we must look for an alternative way of calculating the quantum energies. This may be done by using the field calculated in eqn (6.173) and the fact that in the Hamiltonian itself the leading term is

$$H \simeq \frac{\nu_0}{\Omega}\int_\Omega \psi^+(r)\psi(r)d^3r, \quad (6.174)$$

THE HYDROGEN ATOM 355

where the integral is over the volume Ω. The reader should perhaps consult the part of Chapter 5 that is concerned with the form of the Hamiltonian itself in terms of the field ψ and bear in mind that the current-like term in λ has been transformed away by the Galilean transformation. To make our calculation more definite we will assume that the boundary of the volume, Ω, is the sphere of radius $r = R_0$. This is convenient, since the noninteracting Hamiltonian, H_0, is

$$H_0 = -\frac{\hbar^2}{2m}\nabla^2 - \frac{Ze^2}{r}, \tag{6.175}$$

which is invariant under rotations. The shape of the boundary should not affect our results provided that the volume it encloses is large enough.

Before we calculate the energy, we need three preliminary results.

Relations between N_0, the magnitude of the wave vector at the critical point k_0 and the orientation of the position of this 'point', k_0, in reciprocal space

Consider the definition of the field ψ and invert it to give

$$q_k^+ = \frac{1}{\sqrt{\Omega}} \int_\Omega \psi^+(r) \exp[-i\mathbf{k}\cdot\mathbf{r}] \, d^3r, \tag{6.176}$$

and similarly for its complex conjugate. Close to the critical point suppose that $\psi \to \psi_0$ and $k \to k_0$. Then, if $|0\rangle$ denotes the vacuum state, we have

$$|k_0\rangle = \frac{1}{\sqrt{\Omega}} \int_\Omega \psi_c^+(r) \exp[-i\mathbf{k}_0\cdot\mathbf{r}] \, d^3r \, |0\rangle. \tag{6.177}$$

We also have that

$$\langle k_0 | k_0 \rangle = 1 = \frac{1}{\Omega} \int_\Omega \psi_c(r)\exp[+i\mathbf{k}_0\cdot\mathbf{r}]d^3r \int_\Omega \psi_c^+(r'')\exp[-i\mathbf{k}_0\cdot\mathbf{r}'']d^3r''. \tag{6.178}$$

Consider one of the integrals above and insert eqn (6.173) (taking out the time dependency) into it for ψ_c. We obtain

$$I = \frac{N_0}{\sqrt{\Omega}} \int_\Omega \frac{1}{r^{1/2}} J_{l+1/2}(\sqrt{\Lambda}\, r) Y_l^m(\theta, \phi) \exp[+i\mathbf{k}_0\cdot\mathbf{r}] d^3r. \tag{6.179}$$

Note that in eqn (6.178) this argument could also be used for any other vector near this point, so that we have $k_0 = l_0 = \cdots$, etc. The integral I is readily evaluated by using the relation [46]

$$\exp[i\mathbf{k}_0\cdot\mathbf{r}] = \sum_{LM} 4\pi i^L j_L(k_0 r) Y_L^{M*}(\theta, \phi) Y_L^M(\theta_0, \phi_0), \tag{6.180}$$

where j_L is a spherical Bessel function, which may be expressed as

$$j_L(k_0 r) = \sqrt{\frac{\pi}{2k_0 r}} J_{L+1/2}(k_0 r), \qquad (6.181)$$

and (θ_0, ϕ_0) is the angular orientation of the 'vector' k_0. When eqn (6.181) is inserted into eqn (6.180), and thence into eqn (6.179), the angular integrals performed using eqn (6.166), making the substitution $x = r/R_0$, and using the fact that [46]

$$\int_0^1 x J_\nu(\alpha x) J_\nu(\beta x)\, dx = 0 \qquad [\alpha \neq \beta]$$

$$= \tfrac{1}{2}\{J_{\nu+1}(\alpha)\}^2 \qquad [\alpha = \beta], \qquad (6.182)$$

we find that we must have $k_0 = \sqrt{\Lambda}$ and

$$I = \frac{N_0}{\sqrt{\Omega}} i^l Y_l^m(\theta_0, \phi_0) 4\pi \sqrt{\frac{\pi}{2\sqrt{\Lambda}}} R_0^{\frac{1}{2}} \left[J_{l+3/2}(\sqrt{\Lambda}\, R_0) \right]^2. \qquad (6.183)$$

In fact, we also require that $J_\nu(\alpha) = J_\nu(\beta) = 0$ for eqn (6.182) to be satisfied, but as we shall see later for $\nu = l + \tfrac{1}{2}$ and $\alpha = \sqrt{\Lambda}\, R_0$, this is equivalent to a boundary condition on the sphere $r = R_0$.

The arguments of our Bessel functions are all real so we can now use, assuming that R_0 is large, the asymptotic expression [47]

$$J_\nu(z) \approx \sqrt{\frac{2}{\pi z}} \cos\left[z - \frac{\pi}{2}(\nu + \tfrac{1}{2}) \right], \qquad (6.184)$$

for $z = |z| e^{i\phi}$ with $-\pi/2 < \phi < \pi/2$. Hence, when R_0 is large,

$$1 = |I|^2 \simeq \frac{N_0^2}{\Omega} |Y_l^m(\theta_0, \phi_0)|^2 \frac{8\pi R_0^2}{\Lambda^{3/2}} \cos^4\left[\sqrt{\Lambda}\, R_0 - \frac{\pi}{2}(l+2) \right]. \qquad (6.185)$$

Boundary condition on the surface of the sphere, $r = R_0$

We take as our boundary condition that, when $r = R_0$, the classical field ψ_0 in eqn (6.173) vanishes. Using eqn (6.184), this is equivalent to saying that when $z = \sqrt{\Lambda}\, R_0$ and $\nu = l + \tfrac{1}{2}$, the argument of the cosine should be a multiple of π, i.e.

$$\sqrt{\Lambda}\, R_0 - \tfrac{1}{2}\pi(l+1) = (2n-1)\frac{\pi}{2} \quad \text{or} \quad \sqrt{\Lambda}\, R_0 = \pi\left(n + \frac{l}{2} \right). \qquad (6.186)$$

Note that when eqn (6.186) is inserted into eqn (6.185), $\cos^4[\ ]$ becomes unity.

THE HYDROGEN ATOM

The evaluation of $\omega_{k,l}$ near the critical point

The piece of information that we require here is an expression for ω_{k_0,l_0}. From eqn (6.182), ω_{k_0,l_0} must be diagonal, so we need

$$J = \left\langle k_0 \left| -\frac{\hbar^2}{2m}\nabla^2 - \frac{Ze^2}{r} \right| k_0 \right\rangle. \tag{6.187}$$

The kinetic energy part of eqn (6.187) is trivial, giving $\hbar^2 k_0^2/2m$. The potential energy is

$$-\frac{Ze^2}{\Omega} \int_0^{R_0} \frac{1}{r} r^2 \, dr 4\pi = -\frac{3Ze^2}{2R_0}. \tag{6.188}$$

The calculation of the energy is now straightforward and is equal to

$$E = \frac{\omega_{k_0,k_0}}{\Omega} \int_\Omega \psi_c^+(r)\psi_c(r) \, d^3r. \tag{6.189}$$

Substituting in eqn (6.189) from eqns (6.187), (6.188), and (6.173), we find that

$$E = \left(\frac{\hbar^2 k_0^2}{2m} - \frac{3Ze^2}{2R_0} \right) \frac{1}{\Omega} N_0^2 \int_0^{R_0} J_{l+1/2}^2(\sqrt{\Lambda}\, r) r \, dr. \tag{6.190}$$

The substitution $x = r/R_0$ again, the relation in eqns (6.182) and (6.183), and conditions (6.186) and (6.185) reduce eqn (6.190), after some algebra, to

$$E = \frac{\Lambda^2}{16\pi^2 |Y_l^m(\theta_0 \phi_0)|^2} \left[\frac{\hbar^2 k_0^2}{\sqrt{\Lambda}\, m\pi(n+l/2)} - \frac{3Ze^2}{\pi^2(n+l/2)^2} \right]. \tag{6.191}$$

If λ is close to ν_0, then, from eqn (6.168), Λ will be small and the kinetic energy contribution in (6.191) relatively small compared to the Coulomb attraction (since $k_0 = \sqrt{\Lambda}$) and a *hydrogenic-like spectrum of bound energies* results!

Although this exercise has proved to be more involved than might be expected, the final result is very encouraging, since we have recovered the standard hydrogenic energy spectrum of the quantum theory of atoms, where $2n + l$ corresponds to the principal quantum number!

What we attempt to show in the final section of this chapter is far less conventional and much more spectacular, as we make full use of two-body interactions (linked to the dreaded correlations in standard theory) and incorporate them in the nonlinear terms of the field equation.

6.6 A nonlinear theory of multi-electron atoms

The problem of an adequate description of multi-electron atoms has been grappled with by physicists since the turn of the century. The discovery of spectral lines precipitated a scientific revolution that led to the formulation of quantum mechanics. Although early quantum mechanics initially made many in-roads, a full development of atomic physics has proved to be a huge undertaking. The many complexities involved in describing the energy spectra of multi-electron atoms appeared to be almost insurmountable until the advent of high-speed computers. However, even today we lack proper analytical methods to deal with this classic many-body problem. What we intend to do in this section is perhaps imperfect in many ways, but our intention is to introduce a new conceptual framework to discuss this problem. We intend to treat this as a problem in nonlinear physics and thus rekindle the idea, somewhat underutilized, that was introduced by Thomas and Fermi some 65 years ago and quoted by numerous authors [48].

6.6.1 Preliminaries

Our approach to this intriguing problem, as mentioned earlier, will be an application of the method of coherent structures (MCS). However, due to the specific nature of the atomic system several new features will be introduced. The main new characteristic is the finite number of interacting particles, N, which must be conserved. This requires an extra constraint on the electronic field ψ, which may be expressed by [72, 73]

$$\left\langle \int \psi^+ \psi \, \mathrm{d}^3 r \right\rangle = N, \tag{6.192}$$

where $\langle \; \rangle$ indicates a suitable average and the integral is either over the whole of space or over an appropriate volume associated with the system. In the MCS we suppose that ψ is adequately described by a large classical component, ψ_c, plus a small quantum component so that

$$\int \psi_c^+ \psi_c \, \mathrm{d}^3 r \simeq N. \tag{6.193}$$

In eqn (6.193) the integrand $\psi_c^+(r)\psi_c(r)$ is therefore playing the role of a number density. We shall refer to this condition as the normalization requirement. Thus, in regions in which particle numbers are zero or very small, we expect the modulus of ψ_c to be also very small. We use this idea and imagine a physical system made up of a very large number of identical atoms. The number of particles will be very large in this system and, therefore, all our truncation procedures (see Chapter 5) should go through as before. In addition, we now visualize the atoms as being very well separated, i.e. by much larger separations than those we find in solids. In view of our interpretation above of $|\psi_c|^2$ as a particle density, we expect that midway between a pair of atoms, the modulus of ψ_c will be extremely small, particularly if we view the atoms as located on a very large cubic net. We are not incorporating any ideas of temperature in this picture, and any zero point motion of the atoms is ignored. Thus, for the system as a whole, the classical field may be written as

$$\psi_c(r) = \sum_s \psi_{cs}(r), \tag{6.194}$$

where s labels each site at which an atom is situated. As the atoms are well separated,

$$\psi_{cs}\psi_{cs'} \simeq 0 \quad \text{for } s \neq s'. \tag{6.195}$$

That is, between *different* sites the overlap of the classical field components associated with different sites is effectively zero. Our equations of motion will therefore reduce to the sum of equations of motion for the fields on each site, but—and this is important—there will still remain a highly nonlinear part of this equation for each site separately. Thus, our equation of motion is essentially a sum of identical equations of motion for each site, and therefore we need only to consider that associated with one such site and drop the s label on the classical field.

Another question will readily spring to the mind of the reader, and that is what will happen when the atoms come closer and closer together and effectively become no longer noninteracting. This scenario falls into the category of what is termed a 'coupled field' system, also discussed in Chapter 5, which is of greater complexity than the one-field system, but the MCS may still be used effectively. We will not elaborate on this in detail, but concentrate solely on the one-atom case.

6.6.2 Initial specifics

Our starting point is the configuration space Hamiltonian

$$H = -\frac{\hbar^2}{2m} \sum_i \nabla_i^2 - \sum_i \frac{Ze^2}{r_i} + \tfrac{1}{2} \sum_{\substack{i,j \\ i \neq j}} \frac{e^2}{|r_i - r_j|}, \tag{6.196}$$

the first term representing the kinetic energy of the electrons, the second the attraction of each electron to the nucleus, with charge $Z|e|$, and the third the repulsion between the pairs of electrons. When H is second-quantized using a plane-wave basis (perhaps a strange one to use, but recall that equations of motion obtained in MCS are of the same form independent of the basis used, as we have shown in Chapter 5), it takes the form

$$H = \sum_{k,l} \Lambda_{k,l} q_k^+ q_l + \sum_{k,l} \varepsilon_{k,l} q_k^+ q_l$$
$$+ \sum_{k,l,m} \Delta_{k,l,m} q_k^+ q_l^+ q_m q_{k+l-m}, \qquad (6.197)$$

where

$$\Lambda_{k,l} = \left\langle k \left| -\frac{\hbar^2}{2m} \nabla^2 \right| l \right\rangle, \quad \varepsilon_{k,l} = \left\langle k \left| -\frac{Ze^2}{r} \right| l \right\rangle, \qquad (6.198)$$

and

$$\Delta_{k,l,m} = \left\langle k, l \left| \frac{e^2}{2|r_1 - r_2|} \right| k+l-m, m \right\rangle, \qquad (6.199)$$

and the annihilators and creators satisfy Fermi-commutation rules. From Heisenberg's equations of motion, for the Fermion operators we have

$$i\hbar \, \partial_t q_\eta = \sum_k \Lambda_{\eta,k} q_k + \sum_k \varepsilon_{\eta,k} q_k$$
$$+ 2 \sum_{k,m} \Delta_{\eta,k,m} q_k^+ q_m q_{k+\eta-m}. \qquad (6.200)$$

We now follow the usual procedure (see Chapter 5) and define the quantum field by

$$\psi = V^{-1/2} \sum_k \exp(-i k \cdot r) q_k, \qquad (6.201)$$

where V is the normalization volume. Equation (6.200) is now multiplied through by $\exp(-i\eta \cdot r)$, divided by $\sqrt{V}$, and summed over η. The result is

$$i\hbar \, \partial_t \psi = \sum_{k,\eta} \Lambda_{\eta,k} \frac{\exp(-i\eta \cdot r)}{\sqrt{V}} q_k \delta_{k,\eta}$$
$$+ \sum_{k,\eta} \varepsilon_{\eta,k} \frac{\exp(-i\eta \cdot r)}{\sqrt{V}} q_k$$
$$+ 2V \sum_{k,m,\eta} \Delta_{\eta,k,m} \frac{\exp(+i k \cdot r)}{\sqrt{V}} q_k^+$$
$$\times \frac{\exp(-i m \cdot r)}{\sqrt{V}} q_m \frac{\exp(-i(k+\eta-m) \cdot r)}{\sqrt{V}} q_{k+\eta-m}. \qquad (6.202)$$

As explained in Section 5.8, we *could* assume $\eta = k$ in the second term on the right of eqn (6.202). The result would be simpler but, as a consequence, when redressing is considered (again see Chapter 5), all other linear terms, including the kinetic energy, become modified. Basically, this is attempting to expand a $1/r$-like potential as a series of plane waves. In this particular instance, because of the form of the original attractive potential, this is highly inconvenient and physically implausible, as so many plane waves would be required. Hence, we follow the alternative procedure and use the result in eqn (5.175). Thus, converting eqn (6.202) into fields, using the definition in eqn (6.201) we obtain

$$i\hbar \partial_t \psi = -\frac{\hbar^2}{2m}\nabla^2\psi - \frac{Ze^2}{r}\psi + \omega_0 \psi^+\psi\psi, \qquad (6.203)$$

where $\omega_0 = 2V\Delta$ and we have used the fact that the kinetic energy matrix element in eqn (6.198) is diagonal. It is evident from eqn (6.203) that we have included the Coulomb repulsion between electrons in the cubic nonlinear term, i.e. to the lowest nontrivial order only. Notice that spin has nowhere made an appearance. The reason for this is simply that, as we have explained elsewhere, if spin-dependent fields are included, the *form* of the field equations is the same provided that the interactions themselves do not depend explicitly on spin. In the above, all of the Coulomb terms depend only on the distance between charges, so we need not insert spin directly. In eqn (6.203) we have that ω_0 is positive, since l_0 and m_0 are the values of l and m at the critical point,

$$\omega_0 = \frac{e^2}{V}\iint \frac{\exp(i(l_0 - m_0)\cdot(r_1 - r_2))}{|r_1 - r_2|} d^3r_1\, d^3r_2$$

$$= \frac{e^2}{V}\cdot\frac{4\pi}{|l_0 - m_0|^2} \quad \text{if } l_0 \neq m_0. \qquad (6.204)$$

Here, we have used the fact that the wave vectors η, k, and m are defined as appropriate to a cubic box of volume V which is used to normalize the plane waves (see reference [49], pp. 60 and 61, for details). When $l_0 = m_0$ the quantity Δ will again be positive because of its positive integrand—in fact, it is proportional to an electronic self potential energy of one electronic charge distributed through the volume V.

6.6.3 Classical and quantum field equations

To solve eqn (6.203) we follow Section 5.71 and Appendix C and write

$$\psi = M\exp(iN), \qquad (6.205)$$

where M and N are real functions of space and time, insert eqn (6.205) into eqn (6.203), and equate real and imaginary parts. As can be seen in Appendix C, quantization is afforded by writing the classical parts of M and N as ϕ_c and $-E_0 t/\hbar$, respectively. The resulting equation of motion for the classical field is (C12) with $\omega = -Ze^2/r$. That is,

$$E_0 \phi_c = -\frac{\hbar^2}{2m} \nabla^2 \phi_c - \frac{Ze^2}{r} \phi_c + \omega_0 \phi_c^3, \qquad (6.206)$$

and assume that ϕ_c is only radial. Then, to simplify eqn (6.206) we scale the dependent and independent variables as follows:

$$\phi_c(r) = \phi_c(r), \quad r = a_0 x, \quad \text{and} \quad \phi_c = \beta y, \qquad (6.207)$$

where a_0 is the radius of the first Bohr orbit. We find that eqn (6.206) becomes

$$\frac{d^2 y}{dx^2} + \frac{2}{x} \frac{dy}{dx} + \left(\frac{2Z'}{x} + \Gamma \right) y - 2\mu y^3 = 0, \qquad (6.208)$$

where

$$Z' = \frac{ma_0}{\hbar^2} Ze^2, \quad \Gamma = E_0 \frac{2ma_0^2}{\hbar^2}, \quad \mu = \frac{ma_0^2}{\hbar^2} \omega_0 \beta^2. \qquad (6.209)$$

We choose as our atom neutral zinc with $Z = 30$ and demand that the total electronic charge on the atom is $Z|e|$: that is,

$$4\pi \int_0^\infty \phi_c^2 r^2 \, dr = Z. \qquad (6.210)$$

Our supposition that the atom is neutral is not only for simplicity but that any possible divergences in matrix elements for small wave vectors can be assumed to cancel out or, in other words, the problems due to the long-range character of the Coulomb interaction are eliminated [2, 49]. Our assumption that ϕ_c is purely radial is also consistent with the fact that in its ground state, the Zn atom has electrons in closed orbitals. By expanding ϕ_c in the radial eigenfunctions of hydrogen and using eqn (6.210), we find that Γ may only take the approximate form

$$\Gamma = -(Z')^2 \left[1 - (Z')^2 \mu \right]. \qquad (6.211)$$

In numerical calculations we find this to be very accurate, so we adopt it here. We also find, intriguingly, that in order for the normalization condition in eqn (6.210) to hold, the parameter μ can only have a *discrete* spectrum of values given by

$$\mu = \mu_0 (1 - 1/n^2), \qquad (6.212)$$

where $\mu_0 = 1/(Z')^2$. The reader should remember that eqn (6.208) is a *classical* equation and yet a discretization has appeared due to the fact that its solutions are required to satisfy proper normalization as expressed by eqn (6.210). Furthermore, as the value of the integer, n, increases from $n = 2$ to infinity, the number of nodes of the classical field solution ϕ_c increases correspondingly from one to infinity. Obviously, the $n = 1$ case results in a linear form of eqn (6.208) which is hydrogenic and there is no effect from Coulomb repulsion between electrons. As $n \to \infty$, $\mu \to \mu_0$, which is a critical point of this parameter, and it manifests a bifurcation from regular to divergent behaviour of ϕ_c. This can be well illustrated on a phase diagram, on which we plot dy/dx versus y. For $\mu < \mu_0$ the solutions are represented by curves that spiral down to a focus point at $y = 0$, $dy/dx = 0$. In fact, it takes an infinite number of turns of the spiral to reach this asymptotic focus point. As already mentioned, within the continuum of values $0 < \mu < \mu_0$, only a countable set given by eqn (6.212) satisfy the normalization condition of eqn (6.210) and the phase space spirals wind down on to the focus point rapidly enough (see Fig. 6.11(a)). Otherwise, i.e. when eqn (6.212) is not satisfied, we still see spirals in the phase space, but the rate of winding down is much slower and normalization cannot be met. When $\mu > \mu_0$ the phase portrait suddenly changes, exhibiting a bifurcation phenomenon and the focus point, although still present, becomes unstable. After a finite number of turns the solution shoots away from the vicinity of the focus point (see Fig. 6.11(b)).

In summary, the physical interpretation of this effect is that three distinct régimes of behaviour exist for the classical solutions. When $\mu > \mu_0$, corresponding to an excess of electronic charge, only a small fraction of the charge, represented by the few oscillations about the focus point, remain attracted to the nucleus, while the remainder escapes to the outer reaches of space. On the other hand, when $\mu < \mu_0$, and does not correspond to the discrete values given by eqn (6.212), the classical solution exhibits a damped oscillatory behaviour which could be characterized as charged 'ring' waves. The total amount of charge corresponding to these solutions is still infinite, i.e. the solutions are not normalizable. Although this and the previous class of solutions appear unphysical in the atomic application, they may be quite important in plasma physics problems, with the presence of localized attraction centres. The remaining class of solution is normalizable and represents the various physically acceptable classical charge distribution functions. It is truly amazing that their character is almost identical to the quantum hydrogenic wave functions that will be later obtained as perturbations, i.e. this structure is certainly self-similar. The classical field solutions are radially symmetric, localized around the centre, and develop an increasing number of nodes as the parameter,

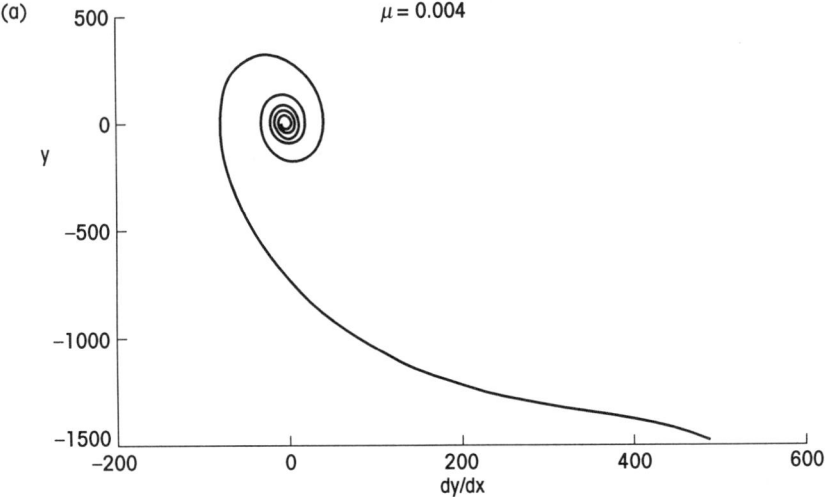

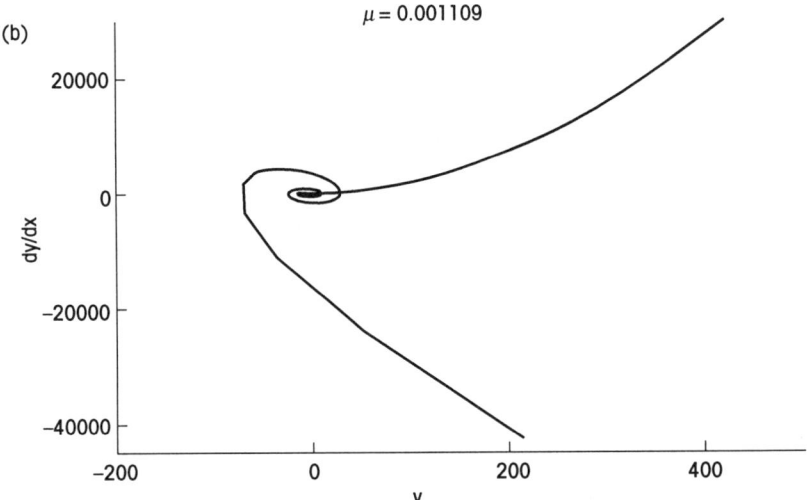

Fig. 6.11 The phase portraits for eqn (6.208) when (a) $0 < \mu < \mu_0$, and (b) $\mu \geqslant \mu_0$.

n, increases. (This latter parameter, however, should *not* be mistaken for the familiar principal quantum number.) Furthermore, as n increases the electronic charge distribution gradually but steadily expands concentrically away from the nucleus in the form of concentric spheres corresponding to the peaks of the charge distribution. Asymptotically, as $n \to \infty$ the electronic charge transfer away from the centre becomes so large that the nucleus loses its binding power and the process of ionization takes place. It is also important to emphasize that the *amplitude of μ is very small, and that this also is a result of the normalization requirement* and *not* an additional assumption.

Having analysed the classical solutions, we now turn to quantization. To quantize we write the amplitude of the field ψ, i.e. M, as

$$M = \phi_c + \varepsilon u \qquad (6.213)$$

and its phase as

$$N = -E_0 t/\hbar + \varepsilon \Omega. \qquad (6.214)$$

Inserting eqns (6.214) and (6.213) into eqn (6.203), we retain terms of order ε only (see Appendix C for details) and obtain

$$E_0 u - \hbar \Omega_t \phi_c = -\frac{\hbar^2}{2m} \nabla^2 u - \frac{Ze^2}{r} u + 3\omega_0 \phi_c^2 u \qquad (6.215)$$

and

$$\hbar u_t = -\frac{\hbar^2}{2m} \{2\nabla\Omega \cdot \nabla\phi_c + \phi_c \nabla^2 \Omega\}. \qquad (6.216)$$

Using a separation of variables argument and a further transformation, these latter two equations become

$$-\hbar K_1 K \omega_1 V = -\frac{\hbar^2}{2m} \nabla^2 W + \left(-\frac{Ze^2}{r} - E_0\right) W + 3\omega_0 \phi_c^2 W \qquad (6.217)$$

and

$$+\hbar \frac{K_2}{K} \omega_1 W = -\frac{\hbar^2}{2m} \nabla^2 V + \left(-\frac{Ze^2}{r} - E_0\right) V + \omega_0 \phi_c^2 V \qquad (6.218)$$

where K_1 and K_2 are separation constants, ω_1 is a constant angular frequency, and V is related to the spatial part of Ω, namely $\Omega_1(\mathbf{r})$, by $\Omega_1 \phi_c = KV$ (K being a constant). The operator nature of u and Ω will be dropped (see Appendix C) since it does not affect the form of the above equations. To obtain an approximate form of solution of the coupled equations (6.217) and (6.218) we note that ω_0 is proportional to the parameter μ introduced earlier, and hence is small in some sense. We solve them numerically later to confirm the results we obtain by perturbative means. To simplify eqns (6.217) and (6.218), we first put

$$W = \xi V, \qquad (6.219)$$

Table 6.3 Allowed signs of K_1, K_2, K, and ξ in eqns (6.217) and (6.218) due to the consistency requirement.

Case	K_1	K_2	K	ξ
A	+	−	+	+
B	−	+	+	+
C	+	−	−	+
D	−	+	−	+
E	+	−	+	−
F	−	+	+	−
G	+	−	−	−
H	−	+	−	−

where ξ is a constant. Inserting this latter expression into both eqns (6.217) and (6.218) we find, for consistency, that

$$\xi^2 = -\frac{K_1}{K_2} K^2 \qquad (6.220)$$

(the factor μ merely has the effect of sending $K^2 \to \mu^2 K^2$). Clearly, if ξ is real, K_1/K_2 must be negative, if K is also real. Thus, eight different cases arise depending on the signs of K_1, K_2, K, and ξ, and are exhibited in Table 6.3.

In all cases, the transformed equations (6.217) and (6.218) become identical and hydrogenic with a nuclear charge $Z|e|$. In cases B, C, E, and H, the associated energies, in Rydbergs (units of $e^2/2a_0$), are

$$\hbar\omega_1 \sqrt{|K_1||K_2|} = -Z^2/s^2 - E_0, \qquad (6.221)$$

where s is an integer and, although E_0 has discrete values from the classical field, only one of its values appears here corresponding to one classical charge distribution. In these latter cases all positive values of s are allowed, which makes the right-hand side of (6.221) positive. However, in cases A, D, F, and G, the energy equation is

$$-\hbar\omega_1 \sqrt{|K_1||K_2|} = -Z^2/s^2 - E_0, \qquad (6.222)$$

and encompasses those energies not allowed in eqn (6.221). When the product of K_1 and K_2 is zero, the time-dependent components become constant (there then appears a linear dependence on time and the only physically plausible part of this is constant). When $K_1 K_2 \neq 0$, if for example the time dependence of the quantum component of the amplitude, M, is

$$T = T_0 \cos\left[\sqrt{|K_1||K_2|}\,\omega_1 t\right], \qquad (6.223)$$

then the quantum component of the phase, N, is

$$\Omega_2 = -T_0 \frac{\sqrt{|K_1||K_2|}}{K_2 \mu_2} \mu_1 \sin\left[\sqrt{|K_1||K_2|}\, \omega_1 t\right], \qquad (6.224)$$

where the factor μ_1/μ_2 arises from the operator character of u and Ω. That is, T and Ω_2 are $\pi/2$ out of phase.

Another approach, of course, is to solve eqn (6.217) for V and substitute in eqn (6.218) to form a fourth-order eigenvalue equation, the eigenvalue now being $\hbar^2 \omega_1^2 |K_1||K_2|$. That hydrogenic eigenvalues emerge when $\omega_0 = 0$ is readily confirmed numerically.

6.6.4 A perturbative approach

When μ or $\omega_0 \neq 0$ we can, in view of the smallness of μ (this, remember, is implied by the normalization and is *not* an assumption) use a perturbative method. For nontrivial solutions the angular dependence of V and W can be taken to be the same. Furthermore, as ϕ_c^2 is purely radial, it will not have matrix elements between degenerate hydrogenic zero order-states, so nondegenerate perturbation theory may be adopted.

One way to proceed is to use the fourth-order equation for W obtained by substituting V from eqn (6.217) into eqn (6.218), and obtain

$$\tilde{E}W = \left[L + \omega_0 \phi_c^2\right]\left[L + 3\omega_0 \phi_c^2\right]W, \qquad (6.225)$$

where $\tilde{E} = -\hbar^2 K_1 K_2 \omega_1^2$ and L is the operator

$$L = -\frac{\hbar^2}{2m} \nabla^2 + \left(-\frac{Ze^2}{r} - E_0\right), \qquad (6.226)$$

and

$$E_0 = -Z^2(1 - \tilde{\mu}Z^2) \text{ (in Rydbergs)}, \quad \tilde{\mu} \equiv \frac{1}{Z^2}(1 - 1/n^2). \qquad (6.227)$$

Now we know that the eigenvalues of L are, in Rydbergs,

$$E = -Z^2/s^2 + Z^2(1 - \tilde{\mu}Z^2). \qquad (6.228)$$

Note that we cannot include the term in $\tilde{\mu}$ in eqn (6.228) in the perturbation because of its size. Thus, writing

$$W = W_0 + \sum_{s \neq 0} a_s W_s, \qquad (6.229)$$

where W_0 is a zero-order eigenfunction, and inserting into eqn (6.225), we have

$$(E^2 + \delta E)\left(W_0 + \sum_{s \neq 0} a_s W_s\right) = L^2\left(W_0 + \sum_{s \neq 0} a_s W_s\right)$$
$$+ 3\omega_0 L \phi_c^2\left(W_0 + \sum_{s \neq 0} a_s W_s\right)$$
$$+ \omega_0 \phi_c^2 L\left(W_0 + \sum_{s \neq 0} a_s W_s\right) + O(\omega_0^2),$$
(6.230)

where ω_0 is the magnitude of the perturbation and the coefficients a_n with $n \neq 0$ are of the same size or smaller. Multiplying eqn (6.230) through by W_0^* and integrating over all space we find that

$$\bar{E} = -\hbar^2 K_1 K_2 \omega_1^2 = E^2 + 4\omega_0 E \langle \phi_c^2 \rangle, \qquad (6.231)$$

where $\langle \phi_c^2 \rangle$ is the average of ϕ_c^2 for the state W_0. Hence, using one of the cases B, C, E, or H, the associated energies that we require are

$$\hbar \omega_1 \sqrt{|K_1 K_2|} = -Z^2/s^2 + Z^2(1 - \mu Z^2) + 2\omega_0 \langle \phi_c^2 \rangle_s, \qquad (6.232)$$

where

$$\langle \phi_c^2 \rangle_s = \int W_{0s}^* \phi_c^2 W_{0s} \, d^3 r \qquad (6.233)$$

and W_{0s} is the sth zero-order hydrogenic eigenfunction.

Note that in eqn (6.232) there is an ambiguity in the value of ω_0 in relation to the dimensionless parameter μ introduced in eqn (6.209). Thus, we treat the coefficient $2\omega_0$ in eqn (6.232) as part of an adjustable parameter x. The choice of μ is dictated by the number of nodes that one might expect in ϕ_c^2. These in turn provide local potential minima within which electronic charge is bound.

In fact, if $|n, l\rangle$ is a zero-order hydrogenic state [50], we know that

$$\langle n, l | r | n, l \rangle = \tfrac{1}{2}[3n^2 - l(l+1)], \qquad (6.234)$$

so in a central field approximation we might expect classically that charge would accumulate in states with a given principal quantum number, n, since the effect of l from eqn (6.234) is small. For *neutral* zinc with a configuration of the form

$$1s^2 2s^2 2p^6 3s^2 3p^6 3d^{10} 4s^2, \qquad (6.235)$$

MULTI-ELECTRON ATOMS 369

there are four such groups of orbitals: namely, 1s; 2s and 2p; 3s, 3p and 3d; and fourthly 4s. Thus, we may associate the classical part of the field with a 5s-like function having four nodes. To a very good degree of approximation, this will have energy $-Z^2/25$ Rydbergs with $Z = 30$ and be normalized to Z, the number of electrons bound. This identification of the classical field may be viewed as the reintroduction of the particle statistics since, although the *form* of the equation of motion is independent of the statistics—as one would expect for a *classical* field—the allowed charge distribution is dependent on the statistics of the particles and a lesser number of nodes in the classical field would imply that all electrons, in the ground state configuration, occupy three or less shells. By Pauli's Principle, this, of course, is impossible.

A more direct argument, based on the anticommutation rules for the radial quantum field, reaches the same conclusion to the one above based on charge density considerations and also restores the particle statistics. The idea is to use the anticommutation relation in one radial dimension (since we have assumed that the classical field is radial),

$$\psi(r)\psi^+(r') + \psi^+(r')\psi(r) = \delta(r - r'), \quad (6.236)$$

and consider the case in which $r = r'$. Let us suppose that there is a radial position in real space, say when $r = r_0$, which corresponds to the position of the critical point $k = k_0$ in reciprocal space. Close to this radial position we expect $\psi(r) = \psi(r)$ to be, to a very good approximation, a classical c-function. Thus, from eqn (6.260), we have

$$\int_{r_0-\delta}^{r_0+\delta} (\psi(r)\psi^+(r) + \psi^+(r)\psi(r))\,\mathrm{d}r = \int_{r_0-\delta}^{r_0+\delta} \delta(0)\,\mathrm{d}r. \quad (6.237)$$

Provided that δ is small enough, $\psi(r)$ becomes $\phi_c(r)\hat{I}$. If the critical transition is related to the process of ionization, then we expect $r = r_0$ to be the 'ionic radius', i.e. some position outside the last orbital to be filled—in our case 4s. As δ is very small but $\delta \neq 0$, eqn (6.237) becomes, since r_0 is not close to $r = 0$,

$$4\delta\phi_c^2(r_0) = 0. \quad (6.238)$$

We interpret eqn (6.238) as saying that $\phi_c(r)$ must have a zero with a radial position, r, greater than any of the zeros (in particular, the one with the greatest r) of the last occupied orbital (here 4s). At the same time, since we are concerned with the atom in its ground state, ϕ_c must have the lowest energy consistent with this latter condition. These two conditions imply that ϕ_c must be 5s-like. In fact, these two conditions uniquely specify ϕ_c for the atom in its ground state.

An approximate hydrogenic radial form for this classical function is given by

$$\phi_c = \frac{Z^2}{(5a_0)^{3/2}\sqrt{\pi}} \left\{ 1 - 4\left(\frac{Zr}{5a_0}\right) + 4\left(\frac{Zr}{5a_0}\right)^2 - \frac{4}{3}\left(\frac{Zr}{5a_0}\right)^3 \right.$$

$$\left. + \frac{2}{15}\left(\frac{Zr}{5a_0}\right)^4 \right\} \exp\left(\frac{-Zr}{5a_0}\right). \tag{6.239}$$

The above form for ϕ_c appears to be a fairly good approximation to a function obtained numerically, especially because of its value and gradient close to $r = 0$, the number of nodes, asymptotic behaviour, and normalization to $Z = 30$.

The zero-order hydrogenic eigenfunctions, normalized to unity for a nuclear charge Z, are as follows:

$$R_{1s} = (Z/a_0)^{3/2} 2\exp(-Zr/a_0), \tag{6.240}$$

$$R_{2s} = (Z/(2a_0))^{3/2} 2(1 - (Zr/2a_0))\exp(-(Zr/(2a_0))), \tag{6.241}$$

$$R_{2p} = (Z/(2a_0))^{3/2}(Zr/\sqrt{3}\,a_0)\exp(-(Zr/(2a_0))), \tag{6.242}$$

$$R_{3s} = (Z/(3a_0))^{3/2} 2\left[1 - (2/3)(Zr/a_0) + (2/27)(Zr/a_0)^2\right]$$
$$\times \exp(-(Zr/(3a_0))), \tag{6.243}$$

$$R_{3p} = (Z/(3a_0))^{3/2}\left(8Zr/(9a_0\sqrt{2})\right)[1 - (Zr/6a_0)]$$
$$\times \exp(-(Zr/(3a_0))), \tag{6.244}$$

$$R_{3d} = (4/(3\sqrt{10}))(Z/(3a_0))^{3/2}(Zr/(3a_0))^2$$
$$\times \exp(-(Zr/(3a_0))), \tag{6.245}$$

$$R_{4s} = 2(Z/(4a_0))^{3/2}\left[1 - 3(Zr/(4a_0)) + 2(Zr/(4a_0))^2 \right.$$
$$\left. - (1/3)(Zr/(4a_0))^3\right]\exp(-(Zr/(4a_0))). \tag{6.246}$$

The matrix elements of ϕ_c^2 may now be readily calculated using the above eigenfunctions and the square of ϕ_c from eqn (6.239). Thus, for example,

$$\langle 1s|\phi_c^2|1s\rangle = \langle n=1, l=0|\phi_c^2|n=1, l=0\rangle$$
$$= \langle 1,0|\phi_c^2|1,0\rangle = \int_0^\infty R_{1s}^2 \phi_c^2 r^2\, dr. \tag{6.247}$$

After evaluation of the appropriate integrals, we find that

$$\langle 1,0| \phi_c^2 |1,0\rangle = 2.40 \times 10^{-4}, \quad \langle 2,0| \phi_c^2 |2,0\rangle = 3.77 \times 10^{-5}, \qquad (6.248, 6.249)$$

$$\langle 2,1| \phi_c^2 |2,1\rangle = 2.09 \times 10^{-5}, \quad \langle 3,0|\phi_c^2 |3,0\rangle = 1.44 \times 10^{-5}, \qquad (6.250, 6.251)$$

$$\langle 3,1| \phi_c^2 |3,1\rangle = 7.48 \times 10^{-6}, \quad \langle 3,2|\phi_c^2 |3,2\rangle = 6.20 \times 10^{-6}, \qquad (6.252, 6.253)$$

and

$$\langle 4,0|\phi_c^2 |4,0\rangle = 6.34 \times 10^{-6}. \qquad (6.254)$$

All of the above matrix elements are expressed as multiples of Z^4/a_0^3 corrected to two decimal places. Great care has had to be exercised in the evaluation of those integrals with a high value of n, particularly the average value of ϕ_c^2 for the 4s orbital, because of rounding errors. The total energies of these orbitals, including that of the classical field denoted by $E(n,l)$, in Rydbergs, are found to be

$$E(1,0) = -900 + 36 + 2.40\bar{x} = -864 + 2.40\bar{x}, \qquad (6.255)$$

$$E(2,0) = -225 + 36 + 0.377\bar{x} = -189 + 0.377\bar{x}, \qquad (6.256)$$

$$E(2,1) = -189 + 0.209\bar{x}, \quad E(3,0) = -64 + 0.144\bar{x}, \qquad (6.257, 6.258)$$

$$E(3,1) = -64 + 0.0748\bar{x}, \quad E(3,2) = -64 + 0.0620\bar{x}, \qquad (6.259, 6.260)$$

and

$$E(4,0) = -20.25 + 0.0634\bar{x}, \qquad (6.261)$$

where in eqns (6.255) and (6.256), in the centre, we have separated first the hydrogenic energy in Rydbergs, then the approximate constant energy of the classical field and, third, the additional potential energy in the matrix element of ϕ_c^2 representing the effect of the Coulomb interaction between electrons. The parameter $\bar{x}$ is defined by

$$\bar{x} = 2\omega_0(Z^4/a_0^3) \times 10^{-4}. \qquad (6.262)$$

From the contributions in $\bar{x}$ it is clear that the '1s' energy is modified most by repulsion, with the shift for '2s' being a factor of just over six smaller.

6.6.5 The Hartree–Fock method

The reason why we have put 1s and 2s in inverted commas at the end of the last section is to remind us that the quantum energies in eqns (6.255)–(6.261) are *not* one-electron energies but energies for the whole atom, as we stressed in Chapter 5. To compare these energies with those obtained by more conventional techniques, we must interpret them in terms of concepts which are not only possibly more familiar to our readers, but are calculated for the whole atom and not for an individual electron. The most obvious results to compare with are those which are calculated using a Hartree–Fock approach. To do this we shall utilize the effective one-electron energies calculated by Herman and Skillman [51], given in Rydbergs, listed below:

$$\varepsilon(1,0) = -698.40, \quad \varepsilon(2,0) = -85.12, \quad \varepsilon(2,1) = -75.55,$$
$$\varepsilon(3,0) = -9.79, \quad \varepsilon(3,1) = -6.66, \quad \varepsilon(3,2) = -1.29,$$
$$\varepsilon(4,0) = -0.62. \tag{6.263}$$

One of the earliest formulations of a self-consistent method of tackling quantum-mechanical many-body systems was due to Hartree [52–55], and was later extended by Fock [56] to take the exchange interaction into account. For a nonrelativistic Hamiltonian, H, which we have considered in this book, a mean energy W is first calculated of the form

$$W = \frac{\int \phi^* H \phi \, d\tau}{\int \phi^* \phi \, d\tau}, \tag{6.264}$$

where

$$H = \sum_i f(i) + \sum_{i<j} g(i,j). \tag{6.265}$$

The sums are over electrons i, j and, for atoms, $f(i)$ and $g(i,j)$ are defined by

$$f(i) = p_i^2/2m - Ze^2/r_i, \tag{6.266}$$

with

$$g(i,j) = e^2/r_{ij}. \tag{6.267}$$

The function ϕ is usually a normalized determinantal state composed of individual functions $\phi_\alpha(r_i)$, each of which is a function of one electronic co-ordinate r_i, $d\tau = d^3r_1 d^3r_2 \cdots d^3r_N$, where N is the number of electrons bound and α denotes a set of one-electron quantum numbers. By performing a functional minimization $\delta W = 0$, the corresponding Euler–Lagrange equations result in N coupled integro-differential Hartree–Fock equations for the associated orbitals $\psi_\alpha(r_i)$, the variations $\delta\psi_\alpha^*$ and $\delta\psi_\alpha$ being treated as independent. It is to be understood that a description of the spin-function associated with a particular orbital is incorporated in the label α. There are also constraints on the orbitals $\psi_\alpha(r_i)$ to ensure that they are orthonormal and hence, in a standard way, introduce a series of Lagrange multipliers which it is not necessary to dwell on here. Labelling the orbitals $a, b, c, \ldots, n$ as a shorthand (these will be taken as labels for the associated quantum numbers n_a, l_a, m_{la} and m_{sa}), the mean energy W takes the well-known form

$$W = \sum_a I(a) + \sum_{\text{pairs}} [J(a,b) - K(a,b)], \quad (6.268)$$

where each pair is counted once only, with $a \neq b$ and the sums are over occupied orbitals in ϕ. In eqn (6.268), $I(a)$, $J(a,b)$, and $K(a,b)$ are defined by

$$I(a) = \langle a|f|a\rangle, \quad J(a,b) = \langle a,b|e^2/r_{12}|a,b\rangle, \quad (6.269, 6.270)$$

and

$$K(a,b) = \langle a,b|e^2/r_{12}|b,a\rangle, \quad (6.271)$$

where the matrix elements in eqns (6.269), (6.270), and (6.271) may be defined alternatively, so that there is no uncertainty, by

$$I(a) = \int \psi_a^*(r_1) f(1) \psi_a(r_1) d^3r_1, \quad (6.272)$$

$$J(a,b) = \iint |\psi_a(r_1)|^2 g(1,2) |\psi_b(r_2)|^2 d^3r_1 d^3r_2, \quad (6.273)$$

and

$$K(a,b) = \iint \psi_a^*(r_1) \psi_b^*(r_2) g(1,2) \psi_b(r_1) \psi_a(r_2) d^3r_1 d^3r_2. \quad (6.274)$$

Assuming that the spin dependence of each $\psi_a(r_1)$ is through a factor $\delta(m_{sa}, s_1)$, where m_{sa} is the component of spin for orbital a and s_1 its value, then the spin summations that arise in the I and J integrals produce a factor of unity. For the K integral, however, a factor $\delta(m_{sa}, m_{sb})$ is produced.

When the angular integrations are performed in $J(a,b)$ and $K(a,b)$ (see, for example, reference [50]), the following well-known results are obtained:

$$J(a,b) = \sum_k a^k(a,b) F^k(a,b) \quad (6.275)$$

and

$$K(a,b) = \sum_k b^k(a,b) G^k(a,b) \delta(m_{sa}, m_{sb}), \quad (6.276)$$

where a^k and b^k are factors involving the angular momenta l_a and l_b of the orbitals a and b and their components m_a and m_b [50]. The functions F^k and G^k are radial integrals defined by

$$F^k(a,b) = e^2 R^k(a,b;a,b) \quad \text{and} \quad G^k(a,b) = e^2 R^k(a,b;b,a),$$
(6.277, 6.278)

with

$$R^k(a,b;c,d) = \int_0^\infty \int_0^\infty \frac{r_<^k}{r_>^{k+1}} R_a(r_1) R_b(r_2) R_c(r_1) R_d(r_2) r_1^2 r_2^2 \, dr_1 \, dr_2,$$
(6.279)

and where $r_<$ denotes the lesser of r_1 and r_2 and $r_>$ the greater. The function $R_a(r_1)$ is the radial part of the 'a' orbital the spatial component of which, without the inclusion of spin, is assumed to have the form

$$\psi_a(r_1) = R_a(r_1) Y_{l_a}^{m_a}.$$
(6.280)

As we shall later be concerned only with the most dominant contribution from each of eqns (6.275) and (6.276), we shall put $k = 0$ and note that $a^0 = 1$ and $b^0 = \delta(m_a, m_b)\delta(m_{sa}, m_{sb})\delta(l_a, l_b)$. It is clear from its definition that the integral F^k is positive. Furthermore, F^k is a decreasing function of k. The exchange integrals G^k are also positive, but in this case $(2k+1)G^k$ is a decreasing function of k. It may also be shown [50], when the spin components, m_{sa}, and orbital angular momentum components are summed over, that

$$\sum_{m_{sa}, m_a} J(a,b) = 2(2l_a + 1)F^0(a,b).$$
(6.281)

When $k = 0$ we also have

$$\sum_{m_{sa}, m_a} K(a,b) = (2l_a + 1)G^0(a,b)b^0.$$
(6.282)

It is also obvious, from the definitions of F^k and G^k, that

$$F^k(a,b) = F^k(b,a) \quad \text{and} \quad G^k(a,b) = G^k(b,a). \quad (6.283, 6.284)$$

We have presented the values of the direct and exchange integrals, in Rydbergs, in Tables 6.4, 6.5, and 6.6, where their values have been computed with hydrogenic radial functions by Condon and Odabasi [50]. The appropriate value required for a particular Z ($Z = 30$ in our case) is simply the entry in the table multiplied by Z. From eqns (6.283) and (6.284) it is clear that we need only tabulate values on or above the

Table 6.4 Coulomb interaction integrals using hydrogenic ($Z = 1$) radial functions: direct integrals, $F^0(a, b)$ (in **Rydbergs**: to be multiplied by Z).

a/b	1s	2s	2p	3s	3p	3d	4s	4p
1s	1.25000	0.41975	0.48560	0.19898	0.21765	0.22205	0.11527	0.12303
2s		0.30078	0.32422	0.16823	0.18002	0.20716	0.10270	0.10799
2p			0.36328	0.17351	0.18789	0.21266	0.10480	0.11104
3s				0.13281	0.13759	0.14627	0.08980	0.09302
3p					0.14374	0.15386	0.09139	0.09499
3d						0.08605	0.09433	0.09825
4s							0.03728	0.03804
4p								0.03894

Table 6.5 Coulomb interaction integrals using hydrogenic ($Z = 1$) radial functions: direct integrals (in Rydbergs: to be multiplied by Z) $F^k(a, b)$.

k	(a, b)	(a, b)	(a, b)
2	(2p, 2p) = 0.17578	(3p, 3p) = 0.07198	(4p, 4p) = 0.01992
		(3d, 3d) = 0.04542	
4		(3d, 3d) = 0.02962	

diagonal of the tables. In the above, assumptions have been made about the nature of the one-electron orbitals, namely that their spatial part is of the form in eqn (6.280) and the spin through a factor $\delta(m_{sa}, s_1)$. The central field approximation, which underlies the major part of the theory of atomic energy levels, also assumes that the radial component, $R_a(r_1)$, of the spatial part of each orbital depends only on n_a and l_a (where n_a is the principal quantum number of the 'a' orbital), and not on (m_a, m_{sa}). In fact, the associated spherical symmetry is possessed only by configurations of closed shells [57], but the fields depart from this to some extent in other cases, so the validity of the central field approximation depends on the smallness of the departure from spherical symmetry. We shall use the integrals F^k and G^k tabulated above with caution because, as we shall later see, they are often larger than Hartree–Fock values which, in turn, are larger than those found by fitting to experiment. It is also well known

Table 6.6 Coulomb interaction integrals using hydrogenic ($Z = 1$) radial functions: exchange integrals, $G^k(a, b)$ (in Rydbergs: to be multiplied by Z) [$k = l_a + l_b$ except where two values are shown, the value of k then being written as a bracketed superscript].

a/b	1s	2s	2p	3s	3p	3d	4s	4p
1s	0.04390	0.10242	0.01154	0.02719	0.00247		0.00468	0.01107
2s		0.17578	0.01495	0.01699	0.07304		0.00479	0.00587
2p			0.01885	0.01982$^{(0)}$	0.07474$^{(1)}$		0.00648	0.00632$^{(0)}$
				0.02265$^{(2)}$	0.04360$^{(3)}$			0.00801$^{(2)}$
3s					0.08464	0.04557	0.00755	0.00854
3p						0.06836$^{(1)}$	0.00995	0.00886$^{(0)}$
						0.04811$^{(3)}$		0.01061$^{(2)}$
3d							0.00835	0.00925$^{(1)}$
								0.00960$^{(3)}$
4s								0.02446

that the Hartree–Fock method fails to produce term splittings which agree with experiment. The reason for this, amongst other things, is that it ignores the effects of interactions with other configurations. Despite this defect, calculations using the Hartree–Fock method do serve a useful purpose in providing estimates of electrostatic interaction parameters. As pointed out above, the parameters so obtained are mostly too large, but their variation across periods or isoelectronic sequences is uniform, as is also their ratio to values obtained directly from experiment. We may therefore use Hartree–Fock calculations, suitably scaled, as estimates of these parameters. In this account we shall explicitly assume that the hydrogenic values of two-body Coulomb integrals may also be suitably scaled in the same way to obtain the Hartree–Fock values required.

We have digressed somewhat in giving a very brief account of the Hartree–Fock technique. What we now do is to use it to obtain an approximate set of energies to compare with those from the method of coherent structures obtained in eqns (6.255)–(6.261). To reach our goal we interpret the 'whole-atom' energies, $E(n,l)$, as the difference in energy between the energies of two determinantal states. The first is the ground configuration described by the determinantal state

$$|G\rangle = \{1\overset{+}{s}, 1\overset{-}{s}, 2\overset{+}{s}, 2\overset{-}{s}, 2\overset{+}{p}(m=1), 2\overset{+}{p}(m=0), 2\overset{+}{p}(m=-1),$$

$$2\overset{-}{p}(m=1), 2\overset{-}{p}(m=0), 2\overset{-}{p}(m=-1),$$

$$3\overset{+}{s}, 3\overset{-}{s}, 3\overset{+}{p}(m=1), 3\overset{+}{p}(m=0), 3\overset{+}{p}(m=-1), 3\overset{-}{p}(m=1),$$

$$3\overset{-}{p}(m=0), 3\overset{-}{p}(m=-1),$$

$$3\overset{+}{d}(m=2), 3\overset{+}{d}(m=1), 3\overset{+}{d}(m=0), 3\overset{+}{d}(m=-1),$$

$$3\overset{+}{d}(m=-2), 3\overset{-}{d}(m=2), 3\overset{-}{d}(m=1), 3\overset{-}{d}(m=0),$$

$$3\overset{-}{d}(m=-1), 3\overset{-}{d}(m=-2), 4\overset{+}{s}, 4\overset{-}{s}\},$$

where '+' and '−', above a particular orbital, denote $m_s = 1/2$ and $m_s = -1/2$ components of spin, respectively.

The second determinantal state is obtained from $|G\rangle$ by removing an electron from the n,l orbital, with a particular component of orbital angular momentum, m_l, and spin component, m_s, and putting it into a 4p orbital ($n = 4$, $l = 1$), with particular values for the angular and spin components $m_{l'}$ and $m_{s'}$. As we see later, our approximate results do not depend too critically on the particular values of $m_{l'}$ and $m_{s'}$, since changes brought about by different values appear principally in the exchange integrals, and as these are somewhat smaller than the direct ones we do not dwell too much on these small differences, particularly bearing in mind the approximations that we have already made and will make. We denote this second determinant by $|EX: nl; 4p\, m_{l'}\, m_{s'}\rangle$. As we are concerned with energy 'differences' we shall feel free to add and subtract constant energies which remain constant for each of the four different shells denoted by n, l.

To relate our definition or interpretation of $E(n, l)$ to Hartree–Fock-like quantities, we follow reference [50] by Condon and Odabasi [CO, 485], where the number following the initials of the authors denotes a specific page. First, eqn (6.268) may be written in a slightly different equivalent form, as

$$W = \sum_a{}' q_a I(a) + \sum_{\text{pairs}} [J(a,b) - K(a,b)], \qquad (6.285)$$

where W is again the total energy of a particular determinant state, and the prime is put on the first summation to indicate that the sum is now over only (nl) values, and that spin and angular components do not appear explicitly in the sum. The number of electrons in the shell $a \equiv nl$ has been denoted by q_a. The energy of an electron in the $(nl) \equiv a$ orbital is

$$\omega(a) = I(a) + \sum_b{}' (q_b - \delta_{a,b})[J(a,b) - K(a,b)]. \qquad (6.286)$$

The integral $K(a, b)$ cannot, of course, appear as $K(a, a)$ in eqn (6.286), since if the spins of a and b are the same, it will cancel with $J(a, a)$ and vanish if the spin is different. The negative of $\omega(a)$ is interpreted as the binding energy of an electron in orbital a provided that, when the electron in a is removed, the spin orbitals of the other electrons in the system remain unchanged. The authors of reference [50] point out that $\omega(a) \neq \varepsilon(a)$, where the $\varepsilon(a)$ are the electron energy parameters in the Hartree–Fock Scheme (HFS). They arise as eigenvalues of the equation

$$\left[-\tfrac{1}{2}\nabla^2 + V(r)\right] P(a; r) = \varepsilon(a) P(a; r), \qquad (6.287)$$

where the self-consistent potential $V(r)$ is defined by [CO, 484] as

$$V(r) = \frac{-Z}{r} + \frac{1}{r} \int_0^r \sum_a q_a |P(a;s)|^2 \, ds + \int_r^\infty \frac{1}{s} \sum_a q_a |P(a;s)|^2 \, ds$$

$$- k_x \left[\frac{3}{4\pi^2 r^2} \sum_a q_a |P(a;r)|^2\right]^{1/3}, \qquad (6.288)$$

and the $P(a; r)$ are radial solutions to the one-electron HFS equation

(6.287). The last term is a local potential, introduced by Slater [58] to incorporate and approximate the nonlocal part of the potential. The energies $\varepsilon(a)$ may be identified with the one-electron energies found by Herman and Skillman [51]. As $\omega(a) \neq \varepsilon(a)$, a correction to the energy parameter $\varepsilon(a)$ is defined by

$$\omega(a) = \varepsilon(a) + \delta\varepsilon(a), \qquad (6.289)$$

where $\delta\varepsilon(a)$ is given by

$$\delta\varepsilon(a) = -\left\langle a \left| \frac{Z}{r} - V(r) \right| a \right\rangle + \sum_{b}{}' (q_b - \delta_{ab})[J(a,b) - K(a,b)]. \qquad (6.290)$$

That is, $\delta\varepsilon(a)$ is a balance between a self-consistent potential and the Coulomb interaction energies. This is usually small, and it is this fact that we make full use of. In fact, as stated in reference [50], the total energy of a particular configuration (determinantal state) may be written in the form

$$W = W' + \tfrac{1}{2} \sum_{a}{}' q_a \delta\varepsilon(a), \qquad (6.291)$$

with

$$W' = \sum_{a}{}' q_a \varepsilon(a) - \tfrac{1}{2} \sum_{a}{}' q_a \left\langle a \left| \frac{Z}{r} + V(r) \right| a \right\rangle. \qquad (6.292)$$

It has been demonstrated in reference [50] that the magnitude of the correction, the second term in eqn (6.291), is very small, so small in fact that we shall neglect it here and use eqn (6.292) as a good approximation to the determinantal energy. The relationship in eqn (6.292) may be verified by inserting it into eqn (6.291), using eqns (6.290) and (6.286) to produce eqn (6.286) with a different but equivalent form for the pair interaction. We note also that as the Σ'_b in eqn (6.290) is positive—see Tables 6.4, 6.5, and 6.6—and $\delta\varepsilon(a)$ is small, the matrix element $\langle a| Z/r + V(r) |a\rangle$ is also positive, so that we have approximately

$$W' \simeq \sum_{a}{}' q_a \varepsilon(a) - \sum_{a}{}' q_a \left\{ \tfrac{1}{2} \sum_{b}{}' (q_b - \delta_{ab})[J(a,b) - K(a,b)] \right\}. \qquad (6.293)$$

To shorten our notation, we write

$$\sum_{b}{}' (q_b - \delta_{ab})[J(a,b) - K(a,b)] = \sum_{b}{}' (a). \qquad (6.294)$$

We also denote constant energy contributions to the two determinants, defined above, by C. Thus the approximate energies of the two determinants are as follows. For $|G\rangle$,

$$W' \simeq \sum_{n,l}{}' \left[q_{nl} \varepsilon(nl) - \tfrac{1}{2} q_{nl} \sum_{b}{}' (nl) \right] + C. \qquad (6.295)$$

For $|EX: nl; 4p, m_l, m_{s'}\rangle$,

$$W' \simeq \sum_{n,l}{}' \left[q_{nl} \varepsilon(n,l) - \tfrac{1}{2} q_{nl} \sum_b{}'(nl) \right]$$
$$- \varepsilon(n,l) + \tfrac{1}{2} \sum_b{}'(nl) - \tfrac{1}{2} \sum_b{}'(4p) + C. \quad (6.296)$$

Thus, apart from constant shifts in energy, we have

$$E(nl) \simeq \varepsilon(nl) - \tfrac{1}{2} \sum_b{}'(nl), \quad (6.297)$$

where we have dropped $\tfrac{1}{2} \sum_b'(4p)$ because this is an energy common to all of the $E(nl)$.

6.6.6 A comparison of Hartree–Fock with the method of coherent structures

In this section we use the result of the earlier calculations above to make a comparison of energies—strictly, energy differences—derived using the quantization procedure of the method of coherent structures (MCS) and an approximate analytical Hartree–Fock calculation. Energies from MCS will be determined in two ways. The first is perturbative and described in earlier sections, and the second uses a computer for verification. To see exactly how the calculations are performed, we refer the reader to Table 6.7. On the left-hand side of this table we list the orbitals of Zn, with $Z = 30$, the energies of which are to be determined. In the middle column

Table 6.7 Performing the calculation.

Configuration	Hartree–Fock	Nonlinear
1s:	$-698.4 - \dfrac{\alpha}{2}(271.1328)$	$-864 + 2.3963\bar{x} + d$
2s:	$-85.12 - \dfrac{\alpha}{2}(199.557)$	$-189 + 0.37659\bar{x} + d$
2p:	$-75.55 - \dfrac{\alpha}{2}(213.8304)$	$-189 + 0.20871\bar{x} + d$
3s:	$-9.793 - \dfrac{\alpha}{2}(129.2413)$	$-64 + 0.14374\bar{x} + d$
3p:	$-6.661 - \dfrac{\alpha}{2}(135.5706)$	$-64 + 0.074769\bar{x} + d$
3d:	$-1.2882 - \dfrac{\alpha}{2}(129.3957)$	$-64 + 0.062039\bar{x} + d$
4s:	$-0.6185 - \dfrac{\alpha}{2}(82.1766)$	$-20.25 + 0.063384\bar{x} + d$

are the one-electron energies of Herman and Skillman (see eqn (2.263)), and the third column denotes $-\Sigma'_b(nl)/2$. Thus, the second and third columns are the first and second terms of eqn (6.297). The third column has been calculated using Tables 6.4, 6.5, and 6.6 with $Z = 30$ and the parameter α has been inserted to scale these hydrogenic calculations to produce a result close to Hartree–Fock. In Table 6.7 under 'Nonlinear' we present the results of the MCS, namely the results in eqns (6.255)–(6.261). In addition, we have added a constant shift in energy, d, to be determined.

We have chosen a plausible range for α, from $\alpha = 0.7$ to $\alpha = 1.30$, the latter value possibly being too large to accord with reality, but this brings out the trend in our calculated energies as α is increased.

In Table 6.8 we display, in the left-hand column, a value of α, and for a specific choice of α we have determined $\bar{x}$ (the Coulomb repulsion parameter) and the shift d so that the 1s and 2s values of $E(n, l)$ agree in the second and third columns of Table 6.7. That is, once α is chosen, $\bar{x}$ and d are fixed so that Hartree–Fock and MCS values for the $E(n, l)$ of 1s and 2s agree. The second and third columns of Table 6.8 record the values of $\bar{x}$ and d, respectively, that correspond to α in the first column. For a given α, the fourth to the tenth columns of Table 6.8 display $E(n, l)$ values, the top row being from a Hartree–Fock approach and the second row from MCS. We notice that for $\alpha = 0.7$ the 4s energy from MCS is just unbound—corresponding to a positive $E(n, l)$—and as α increases this orbital becomes more and more bound. Now, as pointed out in [CO, 473], $-\varepsilon(a)$ is always larger than the theoretical ionization potential. This means that the effect of $\delta\varepsilon(a)$ on each orbital is positive and hence when the correction $(1/2)\Sigma_a(q_a)\delta\varepsilon(a)$ is added to W', this energy becomes more unbound. Thus, when this correction is incorporated we expect the energies in the top row, for a particular α, to be more positive, and the 'effective' α would be lower in Table 6.8 with a larger magnitude. Hence, it is useful to see the effect of increasing α above unity. The agreement between HF and MCS is fairly good at $\alpha = 1$ and becomes much better as we go down Table 6.8.

As α determines both $\bar{x}$ and the shift d, it is interesting to see whether we can get a rough idea of what value of α to take. One way to do this is to compare the value of $F^0(2s, 2p)$ from the top row of Table 4^9a of [CO, 513], bearing in mind that values are those given in (a.u) [1 a.u. = 2 Ryd.], with values determined from Table 6.4 by varying Z. In Table 6.9 we display the ratio of HF to hydrogenic ones, the result being displayed graphically in Fig. 6.12. If we attempt to fit this dependence of R on Z by fitting the points with $Z = 6, 12$, and 16 to a quadratic form, we find that

$$Z = 84.3979 - 250.0000R + 198.1890R^2. \qquad (6.298)$$

Table 6.8 A comparison of Hartree–Fock energies, $E(n, l)$ (top row) with energies obtained from nonlinear analytical treatment (second row) for $\alpha = 0.70, 0.85, 1.00, 1.15,$ and 1.30, in Rydbergs; the computer results at the bottom are for $p = 0.08$.

α	$\bar{x}$	d	1s	2s	2p	3s	3p	3d	4s
0.70	+18.16	+27.20	−793.30	−154.97	−150.39	−55.03	−54.11	−46.58	−29.38
			−793.30	−154.97	−158.01	−34.19	−35.44	−35.68	+8.10
0.85	+15.50	+13.23	−813.63	−169.93	−166.43	−64.72	−64.28	−56.28	−35.54
			−813.63	−169.93	−172.53	−48.54	−49.61	−49.81	−6.04
1.00	+12.84	−0.73	−833.97	−184.90	−182.47	−74.41	−74.47	−65.99	−41.71
			−833.97	−184.90	−187.05	−62.89	−63.45	−63.94	−20.17
1.15	+10.18	−14.70	−854.30	−199.87	−198.50	−84.11	−84.61	−75.69	−47.87
			−854.30	−199.87	−201.57	−77.24	−77.94	−78.07	−34.30
1.30	+7.52	−28.67	−874.64	−214.83	−214.54	−93.80	−94.78	−85.40	−54.03
			−874.64	−214.83	−216.10	−91.58	−92.10	−92.20	−48.44
Computer Results $p = 0.083$		−3.12	−833.97	−184.90	−188.16	−65.05	−65.31	−65.99	−22.99

MULTI-ELECTRON ATOMS 383

Table 6.9 The ratio of HF to hydrogenic $F^0(2s, 2p)$ values, in Rydbergs.

Element	Z	HF	Hydrogenic	Ratio, R
C	6	1.13448	1.94532	0.5832
N	7	1.48876	2.26954	0.6560
O	8	1.82944	2.59376	0.7053
F	9	2.16448	2.91798	0.7418
Ne	10	2.49634	3.24220	0.7700
Na	11	2.8263	3.56642	0.7925
Mg	12	3.1550	3.89064	0.8109
Al	13	3.48286	4.21486	0.8263
Si	14	3.81006	4.53908	0.8394
P	15	4.13676	4.86330	0.8506
S	16	4.46306	5.18752	0.8603

Thus, when $Z = 30$ in eqn (6.298), one root is $R = 0.9819$, i.e. an appropriate value of α to use in Table 6.8 is $\alpha \simeq 1$, and possibly even larger when the correction in $\delta\varepsilon(a)$ is taken into account. The quite good agreement obtained in Table 6.8, bearing in mind all of the approximations that we have used, between HF and MCS is reassuring and confirms our interpretation of $E(n, l)$ energy values as representing, to within a constant energy shift, the difference in energy between the two determinantal states $|G\rangle$ and $|EX; nl; 4p, m_{l'}, m_{s'}\rangle$. The reader may perhaps more readily appreciate the agreement by comparing the second and third columns of energies of Fig. 6.13, where values of the energies $E(n, l)$ have been taken from

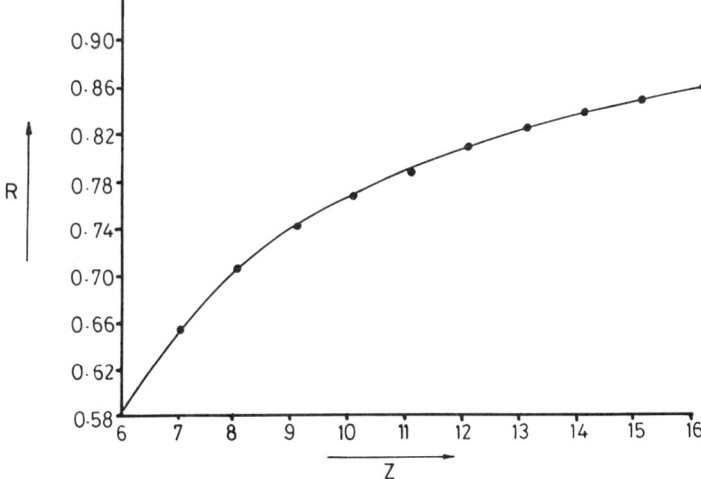

Fig. 6.12 A graphical display of the ratio of HF to hydrogenic $F^0(2s, 2p)$ values.

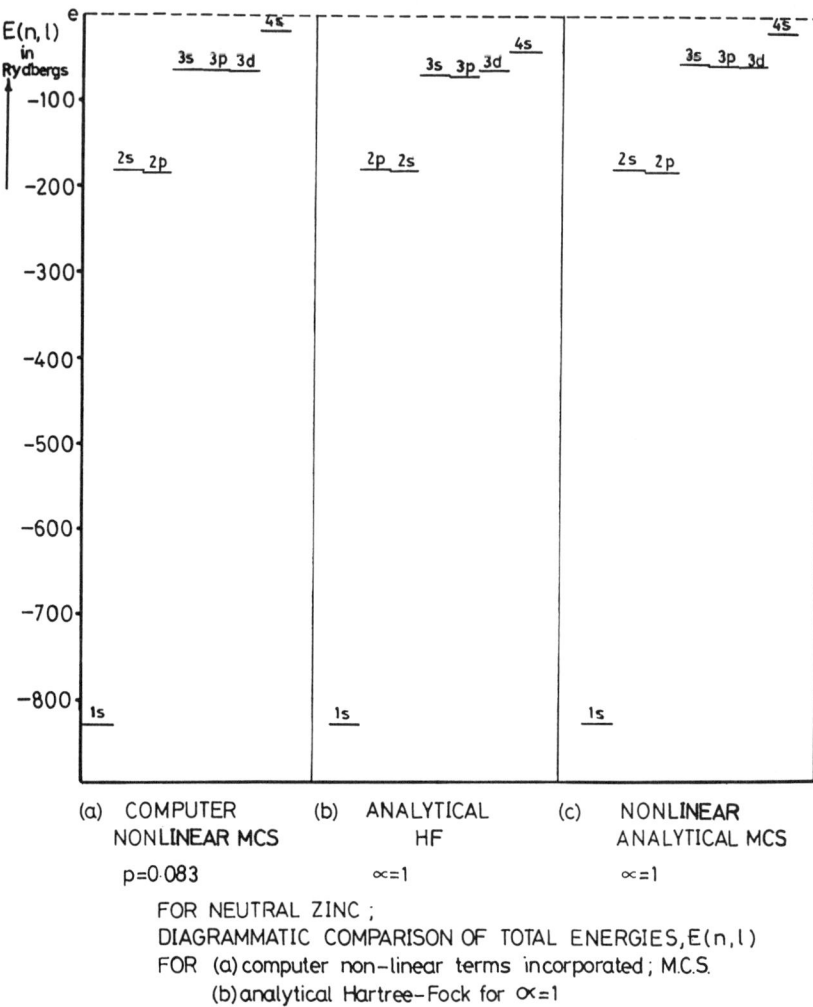

Fig. 6.13 A diagrammatic comparison of the total energies of neutral zinc, $E(n,l)$, for: (a) the computer nonlinear terms incorporated, MCS; (b) analytical Hartree–Fock for $\alpha = 1$; (c) nonlinear analytical method for $\alpha = 1$, MCS.

Table 6.8. It is interesting, but perhaps not altogether surprising in view of approximations made, that when $\alpha = 1$ in Fig. 6.13(c), the ordering of the energies of orbitals within a shell is modified. In fact, the discrepancy, although well within the errors due to approximations, decreases as α increases.

Another method of determining α or $\bar{x}$ (since once α is determined, so are $\bar{x}$ and d) is to use *Moseley's Law*, which concerns the X-ray emission lines from atoms. The number of elements examined by Moseley (the metals from aluminum to gold) was limited, but the series has been extended in both directions. He found that

$$\sqrt{\nu} = k(Z - \sigma), \quad (6.299)$$

where ν is the frequency of the emission, Z is the atomic number, and k and σ are quantities which are the same for any particular line on the Z versus $\sqrt{\nu}$ plot. It was found that a very good rule was to take the screening constant, σ, as the number of extra-nuclear electrons *inside* the orbit to which the electron falls in radiating plus $(n-1)/2$, where n is the number of electrons in that orbit. If we consider transitions down into the 1s level, then there are no other electrons inside this orbit and $n = 2$. Consequently, $\sigma = 1/2$. If we identify this transition as being from the $n = 2$ shell (taking $n = 2, l = 0$ as an example), we have from Table 6.7 that

$$h\nu = \tfrac{3}{4}Z^2 + 2.0198\bar{x} = E(2,0) - E(1,0). \quad (6.300)$$

Hence, comparing eqns (6.300) and (6.299) we find, assuming that the terms in $\bar{x}$ are small relative to those in Z^2,

$$\sigma = [2 \times 2.0197\bar{x}]/3Z. \quad (6.301)$$

From Table 6.8 when $\alpha = 1$ or $\bar{x} = 12.84$ ($Z = 30$), we have $\sigma = 0.5763$ and when $\alpha = 1.15$ or $\bar{x} = 10.18$, $\sigma = 0.4569$. Using the $E(2,1)$ level instead gives values $\sigma = 0.6242$ for $\bar{x} = 12.84$ and $\sigma = 0.4949$ when $\bar{x} = 10.18$. Thus, α should be very close to unity, as we had previously concluded. Moseley's law, therefore, provides strong collaborative evidence for the values of $\bar{x}$ that we have taken earlier. Indeed there are now *no* free parameters for the determination of the seven levels.

We should also note that ϕ_c^2 will have nondiagonal matrix elements among the hydrogenic basis, e.g. $\langle 1s| \phi_c^2 |2s\rangle$ and $\langle 2s| \phi_c^2 |3s\rangle$, even though the zero-order shifted energy differences will be larger. Thus, we have also determined the energies using a computer to check our results.

To see how this was done without solving a fourth-order differential equation, we go back to eqns (6.217) and (6.218). For simplicity, we put

$$\lambda_1 = -\hbar K_1 K \omega_1 \quad \text{and} \quad \lambda_2 = \hbar K_2 \omega_1/K. \quad (6.302)$$

The idea is to now write

$$V = S(r)W, \quad (6.303)$$

where $S(r)$ is a function of r to be determined. Equation (6.217) now becomes an eigenvalue equation, with $-\lambda_1 S(r)$ forming part of the potential energy. In a similar way, W in eqn (6.218) is replaced by $S^{-1}V$. As the structure of the two eigenvalue equations takes the same form, one in W and the other in V, the simplest way to ensure that they are consistent in the sense that their eigenvalues are the same is to arrange that the potential energy in each is the same. Strictly speaking, the two Hamiltonians should differ by a unitary transformation, but here this simple idea suffices. Putting the two potential energies equal determines $S(r)$ and we find that

$$S(r) = \left\{ + \omega_0 \phi_c^2 \mp \left[\omega_0^2 \phi_c^4 + \lambda_1 \lambda_2 \right]^{1/2} \right\} / \lambda_1. \quad (6.304)$$

Both eigenvalue equations then take the form

$$\frac{-\hbar^2}{2m} \nabla^2 W + \left(\frac{-Ze^2}{r} - E_0 + 2\omega_0 \phi_c^2 \mp \left[\omega_0^2 \phi_c^4 + E^2 \right]^{1/2} \right) W = 0, \quad (6.305)$$

where

$$E^2 = \hbar^2 \omega_1^2 |K_1||K_2|. \quad (6.306)$$

By scaling, eqn (6.305) may be transformed into

$$\frac{d^2 y}{dx^2} + \frac{2}{x} \frac{dy}{dx} - \frac{l(l+1)}{x^2} y + \left(\frac{2Z}{x} - \frac{Z^2}{25} \right) y$$
$$+ \left\{ -2py_c^2 \pm \left[E^2 + p^2 y_c^4 \right]^{1/2} \right\} y = 0, \quad (6.307)$$

where E_0 and E have been written in Rydbergs, p is a parameter which may be defined by ω_0/a_0^3, $x = r/a_0$, and y_c is defined from eqn (6.239) by replacing r/a_0 by x and extracting $a_0^{-3/2}$. The computational technique now adopted is usually referred to as the Numerov method [59] and solves equations of the form $y'' = F(x, y)$. This simply means removing the first gradient term from eqn (6.307) and following this method. The results are displayed graphically in Fig. 6.14. In Table 6.8 we have extracted one particular situation to compare with HF results, the MCS case in which $p = 0.083$ agreeing favourably with the $\alpha = 1$ situation for the Hartree–Fock calculation.

At this point a few specific remarks concerning how the MCS results compare with Hartree–Fock are in order. First of all, it is very important to stress that the energies $E(n, l)$ are *not* the one-electron energies $\varepsilon(n, l)$. In fact, the last bound one-electron energy is $\varepsilon(4s)$ but, from Fig. 6.14, where the energies $E(n, l)$ are plotted, there are additional levels such as 4p and 4d. The reason for this can be understood by reference to Table 6.7, the last row of which is for 4s.

However, even when the one-electron energies become zero, above 4s,

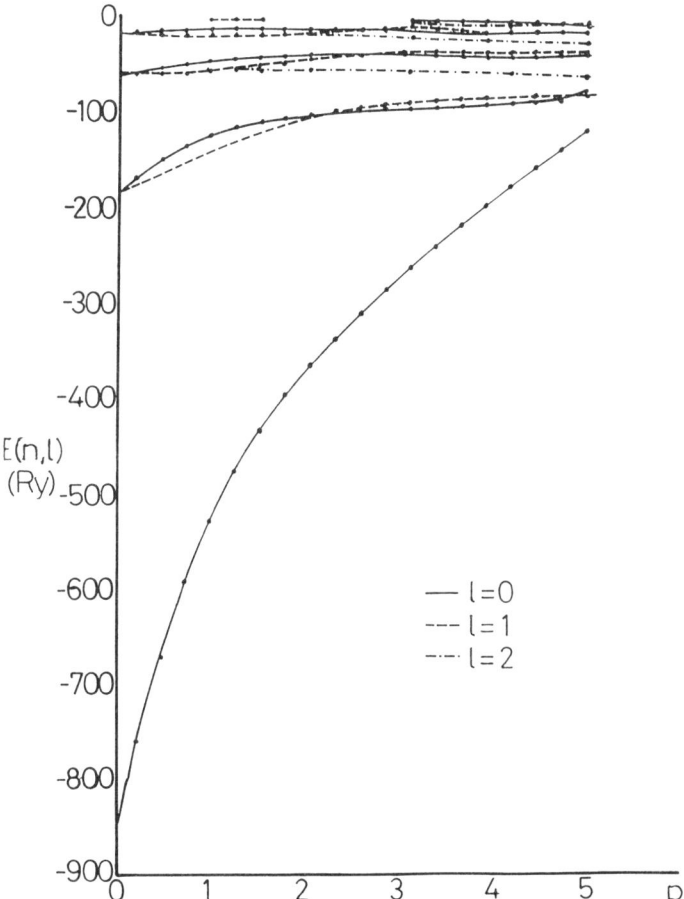

Fig. 6.14 $E(n, l)$ versus p for neutral zinc from the computer (in Rydbergs).

the terms in α do not. For example, $\Sigma'(4p) \simeq 87.17$, very similar to $\Sigma'(4s)$, for the configuration

$$1s^2 2s^2 2p^6 3s^2 3p^6 3p^{10} 4s^0 4p^2.$$

What we have to bear in mind is that as long as we have four sets of values of n or four shells, then the same classical function ϕ_c that we have utilized above may be used. However, removing two 1s electrons and putting them in the $n = 4$ shell is not allowed. We notice from Fig. 6.14 that the additional levels, other than 4d and 4p, arise in régimes of p in which HF cannot be fitted, usually high values of p. We conclude that the additional levels are due to rearrangement of electrons—the two 4s in the ground configuration—within the $n = 4$ shell. Another point to be borne in mind is that one *cannot* make a smooth transition, within MCS, to the one-electron case, simply because MCS does not apply then, no nonlinear terms arise, and in any case the inter-repulsive terms $J(a,b)$ and $K(a,b)$ are nonzero. We also comment that as $p = \omega_0/a_0^3$, $\bar{x}$ and p are related, both representing the strength of the Coulomb repulsion. For $p = 0.083$ from the computer and using eqn (6.262), we find $\bar{x} = 13.45$ for $Z = 30$, which is very close to the value of 12.84 in Table 6.8 for $\alpha = 1$. The value should be slightly higher than that in Table 6.8, since α is just a little lower than unity, as we have seen. In Table 6.8 itself, we see that agreement becomes better as we go down the table, or α increases. Comparing the computer result in Table 6.8 with the HF for $\alpha = 1$, we see also that, with the exception of 2p, energies from the computer are greater than or equal to HF.

The problem of level ordering within a shell is a separate issue and can be easily explained using higher-order nonlinear terms in the equation of motion. For further details, the reader should consult Appendix F.

The final comment that we wish to make in this section is regarding Kato's Theorem, which concerns the electron density at small values of r and the form of the density for large r. According to March [60] there are two properties of atoms which are of particular importance because they can be established from first principles. First, when the radius, r, is very small, as we see from Fig. 6.15, the hydrogenic-like 1s eigenfunction is dominant in magnitude. When $r \to 0$, the potential energy $-Ze^2/r$ dominates that from the repulsion described by ϕ_c^2 and hence we expect quantum eigenfunctions, W, close to $r = 0$ to be 1s-like, i.e.

$$W \sim \exp(-Zr/a_0). \quad (6.308)$$

Therefore, the electron density for $r \to 0$ will behave as $\rho \sim \exp(-2Zr/a_0)$:

$$(\partial \rho/\partial r)_{r=0} = (-2Z/a_0)\rho(0). \quad (6.309)$$

This, of course, is Kato's Theorem, and we expect our quantum eigenfunctions to satisfy it very closely. When r is large, we expect that ϕ_c^2 will become smaller and, again, the long-range part of the Coulomb potential energy will prevail. Thus, the electron density will follow the outermost occupied orbital and as this, for large r, behaves as

$$W \sim r^n \exp(-Zr/na_0), \quad (6.310)$$

the density ρ will have the asymptotic form

$$\rho \sim r^{2n} \exp(-2Zr/na_0). \quad (6.311)$$

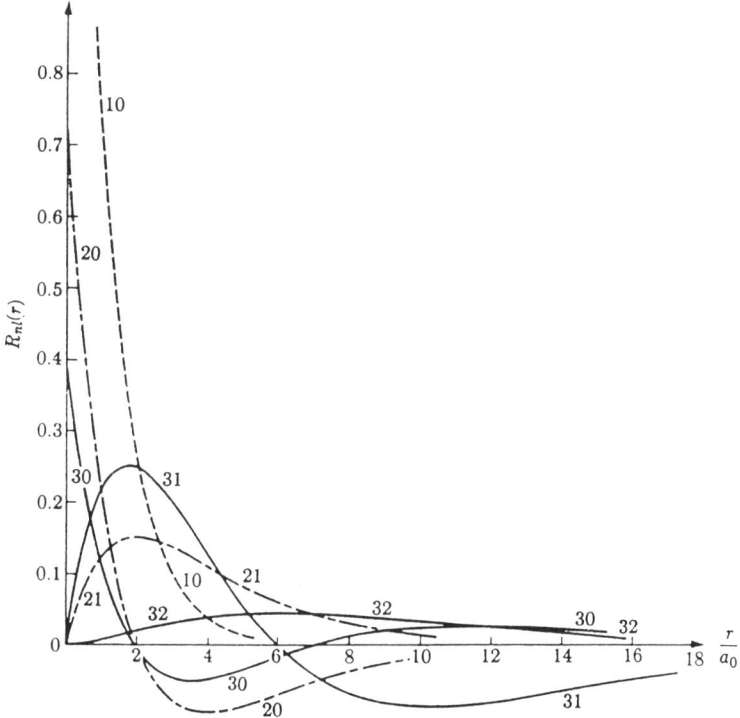

Fig. 6.15 The radial wave function $R_{nl}(r)$ for hydrogenic atoms, for $n = 1, 2, 3$. Each curve is labelled with two integers, representing the corresponding n and l values.

Again, this is consistent with one of the two main requirements listed by March [60].

Thus we conclude that quite good agreement with HFS studies is obtained for the quantum one-electron energies of neutral zinc. The parameters found which represent the magnitude of the repulsion between electrons and energy shifts agree well with the behaviour of $F^0(2s, 2p)$ as the atomic number, Z, is varied. Furthermore, these parameters are in quite good agreement with Moseley's law and consistent with variables obtained from the computer. We find that the associated classical field is only normalizable for certain discrete small values of μ, thus determining the scale parameter β. In fact, when μ is less than a critical value, μ_0, but does *not* correspond to one of the discrete values, the classical field is *not* normalizable but still apparently exhibits damped oscillatory behaviour. We describe these solutions as charged 'ring waves', and the charge in phase space still homes in on the stability point, but the rate of winding of the spirals is much less. When $\mu > \mu_0$ after a finite number of turns, the solution shoots away from the focus point and charge escapes from the attracting nucleus. Agreement is also obtained with two 'first principles' properties as outlined by March. The first is Kato's Theorem, which indicates how the charge density behaves for small r, and the second is its asymptotic behaviour. Energy orderings within a shell are a result of a delicate balancing of first- and second-order contributions to the effective potential generated by the classical field, from the $\psi^+\psi\psi$ part and additional components from higher order $(\nabla\psi^+)\psi(\nabla\psi)$ and $(\nabla^2\psi^+)\psi\psi + \psi\nabla^2\psi\psi$ terms in the quantum field equation.

6.7 The Hubbard Hamiltonian

A very general Hamiltonian which describes a system of N electrons on a periodic lattice can be written in configuration space as [61]

$$H = \sum_{k,\sigma} \varepsilon(k) c_{k\sigma}^+ c_{k\sigma} + \tilde{U} \sum_{k,l,m} c_{k-}^+ c_{l+}^+ c_{m+} c_{(l-m+k)-}. \tag{6.312}$$

'+' here denotes the $m = +1/2$ spin component and '−' the $m = -1/2$ part. The Hamiltonian above is commonly known as the Hubbard Hamiltonian [62], and has found numerous applications in the theory of strongly correlated electrons [16], including recent investigations of high-temperature superconductors [63].

The first step in analysing the Hubbard Hamiltonian using the MCS is to use Heisenberg's equations of motion for a particular annihilator $c_{\eta\sigma}$, say; namely, when $\sigma = +1/2$,

$$i\hbar \partial_t c_{\eta+} = \varepsilon(\eta) c_{\eta+} - \tilde{U} \sum_{k,m} c_{k-}^+ c_{m+} c_{k+\eta-m,-}. \tag{6.313}$$

We then define the spin component fields by

$$\psi_+ = \frac{1}{\sqrt{\Omega}} \sum_k e^{-i k \cdot r} c_{k+}, \quad \psi_- = \frac{1}{\sqrt{\Omega}} \sum_k e^{i k \cdot r} c_{k-}. \tag{6.314}$$

We now multiply both sides of eqn (6.313) by $\exp(-i\eta \cdot r)/\sqrt{\Omega}$, sum over η, and use the definitions in eqn (6.314) where possible. The interaction term in $\tilde{U}$ can clearly be expressed in terms of the fields ψ_+ and ψ_- by construction, but in order to represent the one-body contribution also in terms of these fields we must perform an expansion about some point $\eta = \eta_0$ in reciprocal space. That is

$$\varepsilon(\eta) = \omega_0 + \eta \cdot (\nabla\varepsilon)_0 + \frac{\eta^2}{2!}(\nabla^2\varepsilon)_0 + \cdots, \tag{6.315}$$

where ω_0 is the value of $\varepsilon(\boldsymbol{\eta})$ at $\boldsymbol{\eta} = \boldsymbol{\eta}_0$, $\boldsymbol{\omega}_1 = (\nabla \varepsilon)_0$ is the value of $\nabla \varepsilon$ at $\boldsymbol{\eta} = \boldsymbol{\eta}_0$ and, similarly, for $(\nabla^2 \varepsilon)_0$. Hence, using eqn (6.315), we obtain

$$i\hbar \partial_t \psi_+ = \omega_0 \psi_+ + i\boldsymbol{\omega}_1 \cdot \nabla \psi_+ - \omega_2 \nabla^2 \psi_+ \cdots + \overline{U} \psi_-^+ \psi_- \psi_+, \quad (6.316)$$

where $\overline{U} = \tilde{U}\Omega$ and $\omega_2 = (\nabla^2 \varepsilon)_0/2$.

In the method of coherent structures, $\varepsilon(\boldsymbol{\eta})$ is expanded about a critical point of the system, so that both ψ_+ and ψ_- will be predominantly classical in nature close to $\boldsymbol{\eta} = \boldsymbol{\eta}_0$ and, because of the nature of the expansion, the deviations of $\boldsymbol{\eta}$ away from $\boldsymbol{\eta}_0$ are very small, the expansion is terminated at the ∇^2 term. Similarly,

$$i\hbar \partial_t \psi_- = \omega_0 \psi_- + i\boldsymbol{\omega}_1 \cdot \nabla \psi_- - \omega_2 \nabla^2 \psi_- \cdots + \overline{U} \psi_+^+ \psi_+ \psi_- \quad (6.317)$$

for the other spin component field, ψ_-. Henceforth, we assume that ψ_+ and ψ_- are predominantly classical and that their quantum components can be reintroduced at a later stage as a correction. Furthermore, charge conservation implies that

$$\int (\psi_-^+ \psi_- + \psi_+^+ \psi_+) \, d^3\mathbf{r} = n, \quad (6.318)$$

where n is the total number of mobile electrons on the lattice. Before we investigate the nature of the solutions to eqns (6.316) and (6.317), we simplify the problem somewhat. To do this we transform away the gradient terms by shifting the spatial variable to a moving frame, by writing $t = t'$ with $x' = x - vt$ and simply choose $-\hbar v = \omega_1$. After this transformation, we go back to the notation (x,t) for convenience. The time derivative may also be eliminated by the representation

$$\psi_{\mp} = \phi_{\mp} \exp(-iE_0 t/\hbar), \quad (6.319)$$

to give

$$E_0 \phi_+ = \omega_0 \phi_+ - \omega_2 \nabla^2 \phi_+ + \overline{U} \phi_-^+ \phi_- \phi_+. \quad (6.320)$$

Similarly, for ϕ_- we obtain

$$E_0 \phi_- = \omega_0 \phi_- - \omega_2 \nabla^2 \phi_- + \overline{U} \phi_+^+ \phi_+ \phi_-. \quad (6.321)$$

In both of these equations, we assume that the time-dependent phase is the same.

The first step towards solving the field equations of motion is to represent each of the coupled fields in modulus–argument form, i.e.

$$\phi_+ = \eta \exp(i\xi), \qquad \phi_- = \mu \exp(i\theta). \quad (6.322)$$

Separating real and imaginary parts for each equation produces

$$\text{Re,} \quad 0 = \overline{\omega}_0 \eta - \omega_0 \left[\nabla^2 \eta - \eta(\nabla \xi)^2 \right] + \overline{U} \mu^2 \eta \quad (6.323)$$

and

$$\text{Im,} \quad 0 = -\omega_2 [2\nabla \xi \cdot \nabla \eta + \eta \nabla^2 \xi] \quad (6.324)$$

for ϕ_+ and, similarly, for ϕ_-.

In the above equation, we have combined ω_0 with E_0 so that $\bar{\omega}_0 = \omega_0 - E_0$. Equations for the imaginary parts may be readily integrated to give $K_1 = \eta^2 \nabla \xi$ and $K_2 = \mu^2 \nabla \theta$, where K_1 and K_2 are constant vector fields and, in principle, may include the curl of an arbitrary vector field. If K_1 and K_2 are assumed to be constant vectors, this situation corresponds to the presence of unidirectional currents; although we shall assume, for simplicity, that both are zero. This reduces the problem to solving two nonlinearly coupled differential equations on the amplitudes of the field variables which are displayed below. We have assumed that $\omega_2 \neq 0$.

For computational simplicity, we consider now only the *one-dimensional* case and to make the equations more symmetrical we scale the dependent variables so that $\eta = \varepsilon X$ and $\mu = \varepsilon Y$. If we write $\bar{\omega}_0/\omega_2 = \Lambda$ and choose ε so that $\varepsilon^2 \bar{U}/\omega_2 = \mp 1$, then the field equations reduce to the two coupled nonlinear differential equations below:

$$X'' = \Lambda X \mp Y^2 X \quad \text{and} \quad Y'' = \Lambda Y \mp X^2 Y. \quad (6.325, 6.326)$$

The value of the parameter Λ above is, *a priori*, arbitrary in both sign and magnitude, but the conditions in eqn (6.318) will impose restrictions through the solutions X and Y on the physically admissible values of Λ. Furthermore, the positive value of $\bar{U}$ leads to the plus signs in eqns (6.325) and (6.326) and vice versa. It can be readily recognized that the right-hand sides of these equations can be identified with appropriate derivatives of an effective field-dependent potential $V(X, Y)$:

$$V = -\frac{\Lambda}{2}\{X^2 + Y^2\} + \frac{\varepsilon}{2} X^2 Y^2. \quad (6.327)$$

The three physically interesting sign possibilities for the potential in eqn (6.327) are illustrated in Fig. 6.16.

Clearly, the case with $\Lambda > 0$ and the minus sign in the second term ($\varepsilon = -1$) results in an unstable potential and hence no physical (nonsingular) solutions can be expected in this case. When $\Lambda < 0$ and the plus sign ($\varepsilon = +1$) is chosen, the potential is stable in the entire range of field amplitudes. The remaining two cases are most interesting, however, since they exhibit extrema of V other than $X = Y = 0$; namely, $X^2 = Y^2 = |\Lambda|$, which are local maxima when there is a minus sign ($\varepsilon = -1$) in eqn (6.327) and represent local minima when the plus sign ($\varepsilon = +1$) is adopted. The topology of the potential surface is all important when it comes to analysing the resulting dynamics. This is done numerically using Poincaré sections in the phase space and the results obtained in the physically acceptable cases are summarised in Figs 6.17 and 6.18.

Although this has not been shown graphically, the case of the inverted potential well when $\Lambda > 0$ and $\varepsilon = -1$ leads only to unbounded orbits,

(a) $$V(X,Y) = \frac{1}{2}(X^2 + Y^2) + \frac{X^2Y^2}{2}$$

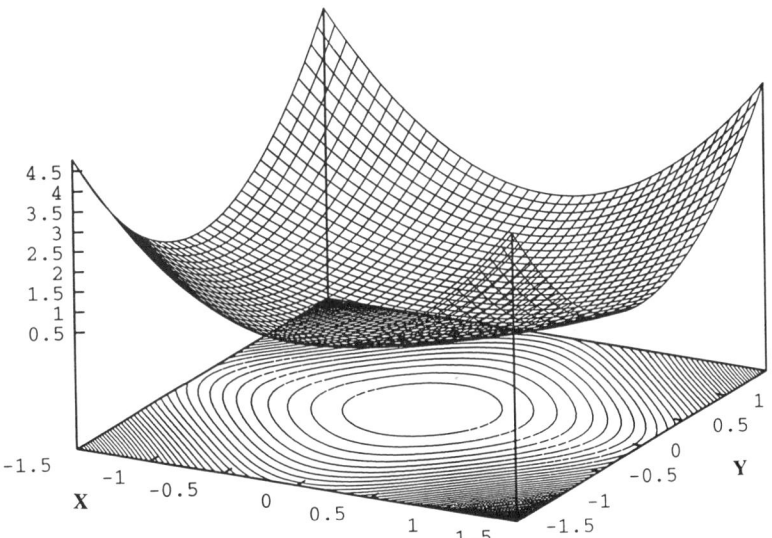

(b) $$V(X,Y) = -\frac{1}{2}(X^2 + Y^2) + \frac{X^2Y^2}{2}$$

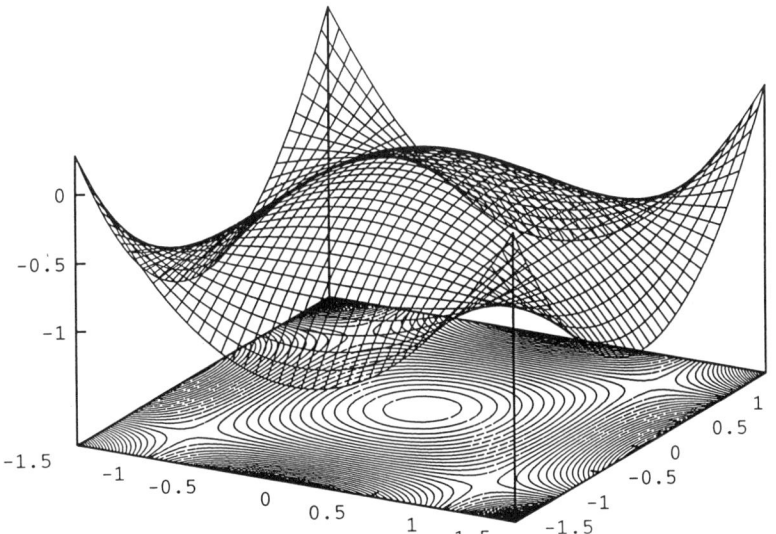

(c) $V(X,Y) = \frac{1}{2}(X^2 + Y^2) - \frac{X^2 Y^2}{2}$

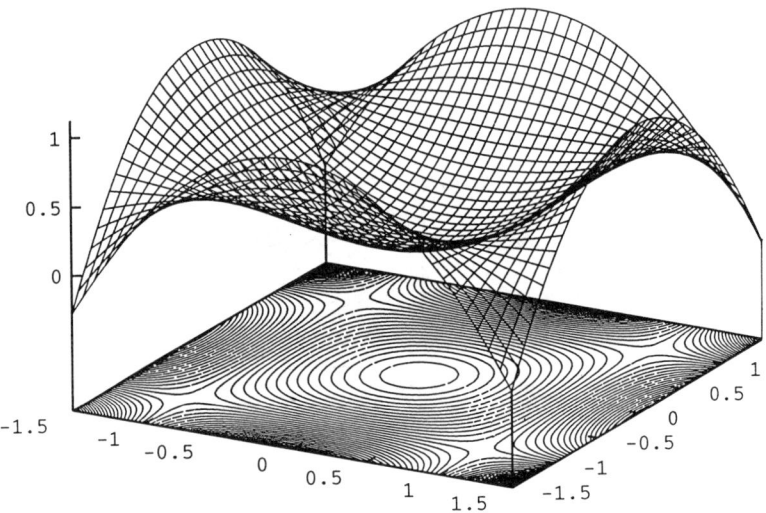

Fig. 6.16 An illustration of $V(X,Y)$ of eqn (6.327) for (a) $\Lambda = -1$, $\varepsilon = +1$ (b) $\Lambda = +1$, $\varepsilon = +1$, and (c) $\Lambda = -1$, $\varepsilon = -1$.

since the potential is globally unstable. On the other hand, when $\Lambda < 0$ and $\varepsilon = +1$, describing a single potential well in the X-Y space, all of the orbits are regular and centre around the only stable focus point, i.e. $X = Y = 0$. This is not very illuminating, and hence we have not plotted the obtained numerical results here either.

However, the two remaining cases are most interesting and we present them here in some detail. When $\Lambda > 0$ and $\varepsilon = +1$, the potential surface has a local maximum in the centre, around $X = Y = 0$. Due to the fact that the potential is asymptotically decreasing along the X- and Y-axes, a vast majority of the orbits that we found were unbounded. However, under two general types of conditions, periodic orbits were also found. The first type of periodic orbit represents straight intervals $Y = \pm X$ (see Fig. 6.17(a) and its time dependence is shown in Fig. 6.17(b). The second type of periodic orbit is found when the initial velocity is tangential to the potential surface. The trajectory in X-Y space has the shape of a cushion (see Fig. 6.17(c)) and the corresponding time dependence is shown in Fig. 6.17(d). When the initial conditions are selected differently, the phase-space behaviour that results is of unbounded character.

The last case that we considered is for $\Lambda < 0$ and $\varepsilon = -1$. Here, an

THE HUBBARD HAMILTONIAN

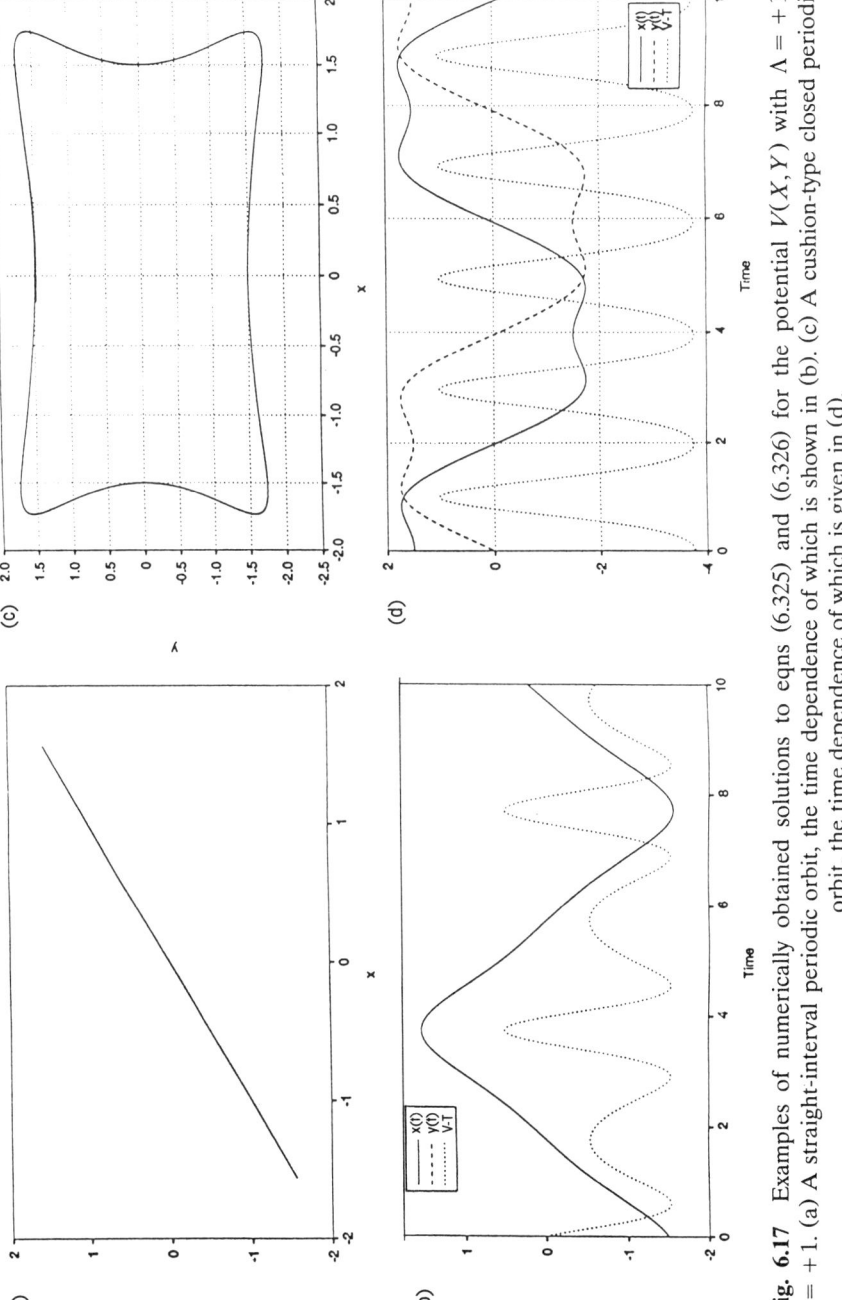

Fig. 6.17 Examples of numerically obtained solutions to eqns (6.325) and (6.326) for the potential $V(X, Y)$ with $\Lambda = +1$, $\varepsilon = +1$. (a) A straight-interval periodic orbit, the time dependence of which is shown in (b). (c) A cushion-type closed periodic orbit, the time dependence of which is given in (d).

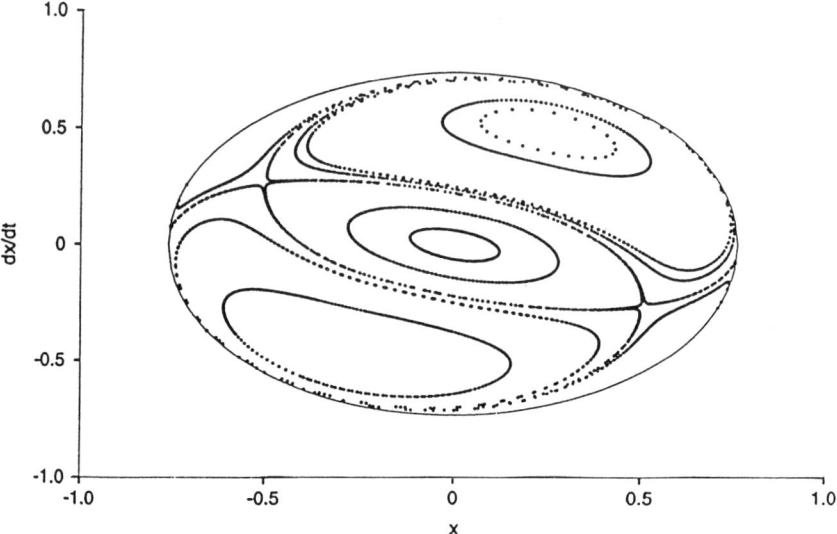

Fig. 6.18 A representative Poincaré section for the solution set of eqns (6.325) and (6.326) using the potential $V(X,Y)$ with $\Lambda = -1$ and $\varepsilon = -1$, and obtained at the energy level $E = 0.3$.

entirely different behaviour has been found from those presented above. In addition to the centrally located minimum of the potential surface, four symmetrically placed local minima also exist. As a result, Poincaré sections obtained when eqns (6.325) and (6.326) are solved numerically exhibit phase-space regions filled with closed (periodic) orbits which are separated by special orbits, commonly referred to as separatrices. Behaviour in the vicinity of separatrix motion has characteristics of chaos. A typical Poincaré section is shown in Fig. 6.18.

There exist two sets of analytical solutions to eqns (6.325) and (6.326), and they provide an additional insight into this problem. The first set is obtained by setting either $X = 0$ or $Y = 0$, i.e. the solution is restricted to only one field axis. In this case when $\Lambda > 0$ the solution diverges exponentially on one side and decays exponentially on the other. For $\Lambda < 0$, on the other hand, the solution is a sine wave as the problem reduces to a classical harmonic oscillator where the effective potential is quadratic in the field variable, which results in confining the motion.

The second set of analytical solutions is somewhat more interesting, as it leads to elliptic functions which have been encountered earlier in Chapters 5 and 6 (see also Appendix B). Here, we assume that $Y = \mp X$, i.e. that the motion will be confined to the diagonals of the potential surface $V(X,Y)$.

The reduced equation becomes

$$X'' = \Lambda X + \varepsilon_0 X^3, \qquad (6.328)$$

where $\varepsilon_0 = \mp 1$. This is obviously a standard elliptic equation [64], with the following types of solution depending on the signs above:

1. When $\Lambda > 0$ and $\varepsilon_0 = +1$, only $X = 0$ is a nonsingular solution.

2. When $\Lambda < 0$ and $\varepsilon_0 + 1$, in addition to the constant solutions of eqn (6.328), $X \sim \tanh$ and $X \sim \mathrm{sn}$ can be found for a range of integration constants. The former hyperbolic solution corresponds to tunnelling between the two oppositely located potential maxima. The latter elliptic wave oscillations represent solutions with a lower energy.

3. When $\Lambda > 0$ and $\varepsilon_0 = -1$ in addition to the constant solutions of eqn (6.328), $X \sim \mathrm{sech}$, $X \sim \mathrm{cn}$, and $X \sim \mathrm{dn}$ elliptic wave solutions exist, depending on the integration constants. The first hyperbolic solution corresponds to an asymptotic and marginally stable motion to the top of the centrally located potential maximum. The cn-wave solutions describe large-amplitude oscillations about both nonzero minima, while dn-function solutions represent small-amplitude oscillations about a single potential minimum.

4. When $\Lambda < 0$ and $\varepsilon_0 = -1$, the only type of solution present is $X \sim \mathrm{cn}$ and it describes arbitrary amplitude oscillations within this unbounded unistable potential, V.

It is important to make a connection between the signs of the reduced coefficients used above and those of the original Hubbard Hamiltonian. Since $\Lambda = (\omega_0 - E_0)/\omega_2$ and $\omega_2 = t \sum_R R^2 e^{i\eta_0 \cdot R}$, the sign of Λ is determined by several factors, one of which is t and another of which is the relative energy of ω_0. The value of ε_0 is solely dependent on the magnitude and sign of $\bar{U}/\omega_2$. It has been recently emphasized that the Hubbard model with a negative value of $\bar{U}$ exhibits qualitative features that render it very useful in modelling high-temperature superconductivity [65–68]. We have found that the case with $\Lambda < 0$ and $\varepsilon = -1$ exhibits broken symmetry and that nonzero values of the field variables minimize the effective potential. We believe that this latter case, although one-dimensional, may provide useful insights in applications to high-temperature superconductivity.

In this chapter we have considered a number of examples to illustrate how the MCS may be used directly to gain physical insight into a variety of phenomena. These examples fall into two distinct types. In the first type we have considered bound particles which are highly localized in space.

This, of course, is the atomic case, in which there are effectively relatively few particles and yet the MCS may be used. The problem here is to define fields in the first place, because the number of particles is low. We overcame this conflict by considering a very large number of essentially noninteracting atoms, the field for the whole ensemble being made up of components from each nuclear site separately. In the atomic case, as Hartree–Fock theory works quite well, the associated correlation energies are small relative to the other examples (the correlation energy here being defined as the difference between the actual energy of the system of particles and a corresponding Hartree–Fock energy). We found for an atom that nearly all higher-order terms in the equation of motion have to be used to obtain good agreement with experiment. This arises because, as the system is so localized, régimes in which the radius, r, is small are physically important, and so it is necessary to take into account deviations away from a critical point to high order in reciprocal space. For the particular system of a zinc atom it would also appear that the effects of renormalization may be neglected.

In all of the other examples, the number of particles is large and their spatial extent is very much greater than the atomic situation. For these cases, e.g. the Haldane problem and metamagnetism, only a few terms in the equation of motion are found to be necessary—we took only leading cubic components. We may observe that larger spatial distances become important in these cases, so we expect deviations away from the critical point, in reciprocal space, to be small. This was exactly what we found. In these latter systems also, we do not expect a Hartree–Fock approach to be appropriate, e.g. in a solid where we have a very large number of particles, N, as we quoted earlier, $\langle \psi_{HF} | \psi_{EXACT} \rangle \to 0$ as $N \to \infty$, where ψ_{HF} and ψ_{EXACT} are the Hartree–Fock wave function and the corresponding exact wave function, respectively. Thus correlation energies become paramount in these systems.

References

1. M. L. A. Nip, unpub. M.Sc. thesis, The University of Alberta, Edmonton (1991).
2. P. L. Taylor, *A Quantum Approach to the Solid State*, Prentice Hall, New York (1970).
3. K. Vos, Ph.D. thesis, University of Alberta (1992).
4. L. P. Gor'kov, *Sov. Phys. JETP*, **9**, 1364 (1959).
5. J. G. Bednorz and K. A. Müller, *Z. Phys. B*, **64**, 188 (1988).
6. E. Stryjewski and N. Giordano, *Adv. Phys.*, **26**, 487 (1977).
7. J. M. Kincaid and E. G. D. Cohen, *Phys. Lett. A* **50**, 317 (1974).
8. M. Shimizu, *J. Magn. Magn. Mat.*, **50**, 319 (1985).

REFERENCES

9. J. A. Tuszyński, *Phys. Lett. A*, **85**, 175 (1981).
10. J. A. Tuszyński and A. P. Smith, *J. Appl. Phys.* **64**, 5633 (1988).
11. M. Isoda, *J. Magn. Magn. Mat.* **27**, 235 (1982).
12. T. Niemeijer, *J. Math. Phys.* **12**, 1487 (1971).
13. J. Stephenson, Department of Physics, University of Alberta, private communication.
14. R. K. Dodd, J. C. Eilbeck, J. D. Gibbon, and H. C. Morris, *Solitons and Non-linear Wave Equations*, Academic Press, London (1982).
15. P. Winternitz, A. M. Grundland, and J. A. Tuszyński, *J. Phys. C*, **21**, 4931 (1988).
16. E. Fradkin, *Field Theories of Condensed Matter Systems*, Addison-Wesley, Redwood City, CA (1991).
17. F. D. M. Haldane, *Phys. Rev. B*, **25**, 4925 (1982).
18. K. Kuboki and H. Fukuyama, *J. Phys. Soc. Japan*, **56**, 3126 (1987).
19. A. Pimpinelli, *J. Phys. Cond. Matter*, **3**, 445 (1991).
20. C. K. Majumdar and D. K. Gosh, *J. Math. Phys.*, **10**, 1399 (1969).
21. I. Affleck, T. Kennedy, E. H. Lieb, and H. Tasaki, *Phys. Rev. Lett.*, **59**, 799 (1987).
22. H. J. Mikeska, B. da Costa, and H. C. Fogedby, *Z. Phys. B*, **77**, 119 (1989).
23. H. J. Mikeska, *J. Phys. C*, **11**, L29 (1978).
24. J. P. Boucher, L. P. Regnault, and H. Benner, in *Nonlinearity in Condensed Matter* (ed. A. R. Bishop, D. K. Campbell, P. Kumar, and S. E. Trullinger), Springer-Verlag, Berlin (1987), p. 24.
25. W.-M. Lin and B.-L. Zhou, *Phys. Lett. A*, **184**, 487 (1994).
26. W.-M. Lin, X.-M. Zhang, D. Zhang, S.-H. Li, and B.-L. Zhou, *Phys Stat. Sol.* (*b*), **182**, 437 (1994).
27. W.-M. Lin, H. Wang, F.-X. Bai, and B.-L. Zhou, *Phys. Stat. Sol.* (*b*), **182**, 445 (1994).
28. F. D. M. Haldane, *Phys. Lett.*, **93A**, 464 (1983); *Phys. Rev. Lett.*, **50**, 1153 (1983); *J. Appl. Phys.*, **57**, 3359 (1985).
29. R. Botet and R. Jullien, *Phys. Rev. B*, **27**, 613 (1983).
30. E. H. Lieb, T. Schultz, and D. J. Mattis, *Ann. Phys.*, *NY*, **16**, 407 (1961).
31. P. W. Anderson, *Phys. Rev. B*, **86**, 694 (1952).
32. J. M. Ziman, *Proc. Phys. Soc. A*, **65**, 540; 548 (1952).
33. R. Kubo, *Phys. Rev.*, **86**, 929 (1952); ibid., **87**, 568 (1952); *Rev. Mod. Phys.*, **25**, 344 (1953).
34. T. Nagamiya, K. Yosida, and R. Kubo, *Adv. Phys.*, **4**, 2 (1955).
35. I. Affleck, *J. Phys. Condens. Matter*, **1**, 3047 (1989).
36. B. R. Judd, *Second Quantisation and Atomic Spectrometry. G. S. Dieke Memorial Lectures*, Johns Hopkins Press, Baltimore (1967).
37. T. Holstein and H. Primakoff, *Phys. Rev.*, **58**, 1048 (1940).
38. J. Schwinger, *On Angular Momentum*, U.S. Atomic Energy Commission Rpt. NYO-3071 (1952); reprinted in L. Biedenharn and H. Van Dam (eds) *Quantum Theory of Angular Momentum*, Academic Press, New York (1965).
39. D. C. Mattis, *The Theory of Magnetism. I. Statics and Dynamics, Springer Series in Solid-State Sciences* **17**, Springer-Verlag, Berlin (1981).
40. P. Jordan and E. Wigner, *Z. Phys.*, **47**, 631 (1928).

41. M. Shimizu and E. P. Wohlfarth, *Phys. Lett.*, **11**, 108 (1964).
42. S. Takeno and K. Kawasaki, *Phys. Rev. B*, **45**, 9, 5083 (1992).
43. P. M. Morse and H. Feshbach, *Methods of Theoretical Physics, Part I*, McGraw-Hill, New York (1953).
44. F. M. Arscott, *Periodic Differential Equations*, Pergamon Press, London (1964).
45. M. Rotenberg, R. Bivins, N. Metropolis, and J. K. Wooten, Jr, *The 3j- and 6j-Symbols*, Technology Press, MIT, Cambridge, MA (1959).
46. I. S. Gradshteyn and I. M. Ryzhik, *Table of Integrals, Series and Products— Corrected and Enlarged Edition*, Academic Press, New York (1980).
47. P. M. Morse and H. Feshbach, *Methods of Theoretical Physics*, McGraw-Hill, New York (1953).
48. E. U. Condon and G. H. Shortley, *The Theory of Atomic Spectra*, Cambridge University Press, Cambridge (1964).
49. S. Raimes, *Many Electron Theory*, North-Holland, Amsterdam (1972).
50. E. U. Condon and H. Odabasi, *Atomic Structure*, Cambridge University Press, Cambridge (1980).
51. F. Herman and S. Skillman, *Atomic Structure Calculations*, Prentice Hall, Englewood Cliffs, NJ (1963).
52. D. R. Hartree, *Proc. Camb. Phil. Soc.* **24**, 89; 111; 426 (1927).
53. D. R. Hartree, W. Hartree, and B. Swirles, *Phil. Trans. R. Soc. Lond.*, **A238**, 229 (1939).
54. D. R. Hartree, *Rep. Progr. Phys.* **11**, 113 (1948).
55. D. R. Hartree, *The Calculation of Atomic Structure*, John Wiley, New York (1957).
56. V. Fock, *Z. Phys.* **61**, 126; **62**, 795 (1930).
57. M. Delbrück, *Proc. R. Soc. Lond.* **A129**, 686 (1930).
58. J. C. Slater, *Phys. Rev.*, **81**, 385 (1951).
59. F. Scheid, *Theory and Problems of Numerical Analysis, Schaum Outline Series*, McGraw-Hill, New York (1968).
60. N. H. March, *Phys. Lett.*, **113A**, 66 (1985).
61. A. P. Balachandran, E. Ercolessi, G. Morandi, and A. M. Srivastava, *Hubbard Model and Anyon Superconductivity*, Lecture Notes in Physics, Vol. 38, World Scientific, Singapore (1990).
62. J. Hubbard, *Proc. R. Soc. Lond.* **A281**, 410 (1964).
63. V. J. Emery, *Phys. Rev. Lett.* **58**, 2794 (1987).
64. P. F. Byrd and M. D. Friedman, *Handbook of Elliptical Integrals for Engineers and Scientists*, Springer-Verlag, New York (1971).
65. H. E. Castillo and C. A. Balseiro, *Phys. Rev. B*, **45**, 10549 (1992).
66. C. F. Chen, *Phys. Rev. B* **47**, 2861 (1993).
67. A. Taraphder, *Europhys. Letts.*, **21**, 79 (1993).
68. K. Kuroki and H. Aoki, *Phys. Rev. Letts*, **69**, 3820 (1992).
69. J. A. Tuszyński and J. M. Dixon, *Physica C*, **161**, 687 (1989).
70. J. A. Tuszyński and J. M. Dixon, to be published in *Int. J. Mod. Phys. B* (1997).
71. J. M. Dixon and J. A. Tuszyński, *Phys. Rev. E*, **51**, 5318 (1995).
72. J. M. Dixon, J. A. Tuszyński and M. L. A. Nip, *Phys. Letts. A*, **221**, 56 (1996).
73. J. M. Dixon, J. A. Tuszyński and M. L. A. Nip, to be published in *Int. J. Mod. Phys. B* (1997).

7 MATHEMATICAL METHODS AND RESULTS

The significant problems we have cannot be solved at the same level of thinking with which we created them.

Albert Einstein

7.1 Symmetry analysis of partial differential equations

In earlier chapters we have alluded frequently to the recent progress made in analysing nonlinear differential equations using both exact and approximate methods. This chapter is intended as a practical overview of the main developments in the field of nonlinear analysis. We have striven to make the exposition lucid, so that even an uninitiated reader can apply the methods presented to the physical problems at hand. Furthermore, many concrete examples have been provided, especially those which pertain to applications in condensed-matter systems described in Chapters 5 and 6. This chapter is replete with many references so that, as particular issues are discussed, additional information can be easily retrieved from the published literature.

7.1.1 Introduction

In this section we discuss applications of symmetry analysis (sometimes called similarity analysis) to find symmetry reductions and exact analytical solutions of nonlinear PDEs. Essentially all the fundamental equations of physics are nonlinear and, in general, such nonlinear equations are often very difficult to solve explicitly. Consequently, perturbation, asymptotic, and numerical methods are often used (with much success) to obtain *approximate* solutions of these equations; however, there is also much current interest in obtaining *exact* analytical solutions of nonlinear equations. Symmetry group techniques provide one method for obtaining such solutions and, furthermore, they do not depend upon whether or not the equation is 'integrable' (in any sense of the word).

Symmetry groups and associated reductions and exact solutions have several different applications in the context of nonlinear differential equations:

1. *Derive new solutions from old solutions.* Applying the symmetry group of a differential equation to a known solution yields a family of new

solutions. Quite often, interesting solutions can be obtained from trivial ones.

2. *Integration of ODEs.* Symmetry groups of ODEs can be used to reduce the order of the equation; for example, to reduce a second-order equation to a first-order one.

3. *Reductions of PDEs.* Symmetry groups of PDEs are used to reduce the total number of dependent and independent variables; for example, from a PDE with two independent and one dependent variable to an ODE.

4. *Linearization of PDEs.* Symmetry groups can be used to discover whether or not a PDE can be linearized and to construct an explicit linearization when one exists (cf. references [1–3]).

5. *Classification of equations.* Symmetry groups can be used to classify differential equations into equivalence classes and to choose simple representatives of such classes (cf. references [1, 4–7]).

6. *Asymptotics of solutions of PDEs.* It is known that as solutions of PDEs asymptotically tend to solutions of lower-dimensional equations obtained by symmetry reduction, some of these special solutions will illustrate important physical phenomena. In particular, exact solutions arising from symmetry methods can often be effectively used to study properties such as asymptotics and 'blow up' (cf. references [8–11]).

7. *Numerical methods and testing computer coding.* Symmetry groups and exact solutions of physically relevant PDEs are used in the design, testing, and evaluation of numerical algorithms; these solutions provide an important practical check on the accuracy and reliability of such integrators (cf. references [12–15]).

8. *Conservation laws.* The application of symmetries to conservation laws dates back to the work of Noether [16], who proved the remarkable result that for systems arising from a variational principle, every conservation law of the system comes from a corresponding symmetry property (cf. references [4, 17]).

9. *Further applications.* There are several other important applications of symmetry groups, including bifurcation theory (cf. references [18–21]), control theory (cf. references [22, 23]), special function theory (cf. reference [24]), boundary value problems (cf. references [1, 25]), and free boundary problems (cf. reference [26]).

In the mid-nineteenth century, Sophus Lie was searching for a general theory for solving differential equations. He made the profound and far-reaching discovery that the special methods, such as separable equations, homogeneous equations, exact equations, and integrating factors, for first order ODEs, were, in fact, all special cases of a general integration method based on the invariance of differential equations under a continuous group of symmetries, now known as a *Lie group* (a group which depends upon a continuous parameter—see Definition 7.2 below). Lie developed a theory for symmetry groups of differential equations which is highly algorithmic. This method is now commonly referred to as the classical Lie method for finding group-invariant solutions; this was first described in full generality by Lie (cf. references [28-31]). For a modern description, see for example the books by Ames [32, 33], Bluman and Cole [34], Bluman and Kumei [1], Hill [35], Ibragimov [17], Olver [4], Ovsiannikov [36], Rogers and Ames [14], and Stephani [37].

Subsequently, there have been many applications in numerous areas of mathematics, physics, chemistry, engineering, etc. However, a major difficulty in applying the theory to differential equations is that the task of finding the symmetry group of a given system of differential equations is often exceedingly cumbersome. Despite the fact that the method is entirely algorithmic, it usually involves a large amount of tedious algebra and the associated calculations (sometimes involving hundreds or even thousands of equations), can be virtually unmanageable if attempted manually. Recently, symbolic manipulations programs have been developed, for example in MACSYMA [38-40], MAPLE [41-44], MATHEMATICA [45-47], MUMATH [48], and REDUCE [49-55], to facilitate the calculations. An excellent survey of the different packages presently available and a discussion of their strengths and applications is given by Hereman [56].

First, we make some definitions:

DEFINITION 7.1. A *symmetry group $\mathscr{G}$ of a System of differential equations* is a group of transformations that maps any solution to another solution of the system.

DEFINITION 7.2. A *Lie group of transformations* is a group of transformations that depend continuously upon a parameter.

DEFINITION 7.3. A *Lie point transformation* is a transformation that depends only on the dependent and independent variables (and not on the derivatives of the dependent variables).

We shall restrict our attention to local Lie point transformations, which have the form

$$x^* = X(x, u; \mathscr{G}), \qquad u^* = U(x, u; \mathscr{G}),$$

where $x = (x_1, x_2, \ldots, x_p) \in \mathbb{C}^p$ and $u = (u_1, u_2, \ldots, u_q) \in \mathbb{C}^q$ are the independent and dependent variables and $\mathscr{G}$ is some symmetry group. Note that the functions X and U depend neither on the derivatives $\partial u_i / \partial x_j$ nor on higher-order derivatives. A generalized transformation that depends on the derivatives as well as the dependent and independent variables is called a *contact* or *Lie–Bäcklund transformation*. (For further details of such transformations see, for example, the monograph by Anderson and Ibragimov [57].)

Our primary application will be the use of symmetry groups to reduce a given PDE, or system of PDEs, to a lower-dimensional PDE, in particular to ODEs. However, it is still necessary to solve the lower-dimensional equations in order to obtain exact solutions. Frequently it is ascertained whether the resulting ODE is of Painlevé type (i.e. its solutions have no movable singularities other than poles). It appears to be the case that if the ODE is of Painlevé type then it can be solved explicitly, yielding exact solutions to the original PDE, while if it is not, then usually it is not solvable analytically. We shall discuss Painlevé analysis in Section 7.2.

Before discussing the general theory for determining symmetry groups of differential equations, we first consider some simple examples.

EXAMPLE 7.4. Consider the homogeneous first-order differential equation

$$\frac{dy}{dx} = \frac{x^2 + y^2}{xy}. \tag{7.1}$$

This is usually solved by first making the substitution $y = xv$, which yields

$$x \frac{dv}{dx} = \frac{1}{v}.$$

Then, by separating variables and integrating, one obtains the general solution of eqn (7.1),

$$y(x) = x\sqrt{2(\ln x + c)},$$

in which c is the constant of integration.

Why does the substitution $y = xv$ lead to a separable equation? Applying the transformation $x^* = xe^\alpha$, $y^* = ye^\beta$, with α and β real parameters, to eqn (7.1) yields

$$e^{\alpha - \beta} \frac{dy^*}{dx^*} = \frac{x^{*2} e^{2\beta} + y^{*2} e^{2\alpha}}{x^* y^* e^{\alpha + \beta}}. \tag{7.2}$$

Therefore, if $\alpha = \beta$ then we see that eqns (7.1) and (7.2) are the same, and so eqn (7.1) is said to be *invariant* under the transformation

$$x^* = xe^\alpha, \qquad y^* = ye^\alpha, \tag{7.3}$$

for *any* α. These transformations form a group, the so-called *group of scaling transformations*. This is a consequence of the fact that the transformation (7.3) possesses the following properties:

(1) $\alpha = 0$ gives the *identity* transformation $x^* = x$, $y^* = y$;
(2) $-\alpha$ characterizes the *inverse* transformation, $x = x^* e^{-\alpha}$, $y = y^* e^{-\alpha}$;
(3) if $x^{**} = x^* e^{\beta}$ and $y^{**} = y^* e^{\beta}$, then $x^{**} = x e^{\alpha + \beta}$ and $y^{**} = y e^{\alpha + \beta}$, which is a form of eqn (7.3) characterized by the parameter $\alpha + \beta$; and
(4) if also $x^{***} = x^{**} e^{\gamma}$ and $y^{***} = y^{**} e^{\gamma}$, then $x^{***} = x e^{\alpha + \beta + \gamma}$ and $y^{***} = y e^{\alpha + \beta + \gamma}$, from which associativity follows.

A transformation satisfying these four properties is said to be a *one-parameter group of transformations* (see Definition 7.8).

In Example 7.4, the substitution $v = y/x$ led to a separable equation, since $v(x, y)$ is an *invariant* of the transformation (7.3), i.e.

$$v(x^*, y^*) = y^*/x^* = y/x = v(x, y),$$

and it is this property which results in a simplification of eqn (7.1). In general, if an ODE is invariant under a one-parameter group of transformations, then the group leads to a simplification of the equation. If the equation is of first order then it becomes separable, while if it is of higher order, then the use of an invariant permits a reduction in the order of the equation by one.

EXAMPLE 7.5. The second-order equation for a simple pendulum (see Chapter 1 for a physical discussion) is

$$d^2 y/dx^2 + \sin y = 0, \qquad (7.4)$$

which is invariant under the transformation

$$x^* = x + \alpha, \qquad y^* = y,$$

with α a real parameter, for any α. These mappings form a continuous group, the so-called *group of translations* along the x-axis, and thus setting

$$p = dy/dx = dy^*/dx^*,$$

reduces eqn (7.4) to the first-order equation

$$p \frac{dp}{dy} + \sin y = 0.$$

This is easily integrated to give

$$\tfrac{1}{2} p^2 = \cos y + C,$$

with C an arbitrary constant, and so we obtain the first integral of eqn (7.4), given by

$$\tfrac{1}{2}(dy/dx)^2 = \cos y + C,$$

which can be further integrated in terms of elliptic functions using quadratures (see Section 1.4 for further discussion).

In an analogous manner, symmetries can be used to reduce the number of equations in a system.

EXAMPLE 7.6. Consider the system of equations (which often arises in applications such as those discussed in Section 1.6)

$$dx/dt = yz, \quad dy/dt = zx, \quad dz/dt = xy. \tag{7.5}$$

Since these coupled equations are autonomous, they are equivalent to the system

$$dy/dx = x/y, \quad dz/dx = x/z, \tag{7.6}$$

which have the respective solutions

$$y^2 = x^2 - c_1, \quad z^2 = x^2 - c_2, \tag{7.7}$$

where c_1 and c_2 are arbitrary constants. Substituting these into the first equation of eqn (7.5) yields

$$(dx/dt)^2 = (x^2 - c_1)(x^2 - c_2), \tag{7.8}$$

which is solvable in terms of Jacobi elliptic functions (see Appendix B and applications in Section 6.2).

For further details on their applications of Lie groups to ODEs, see, for example, the books by Ames [58], Bluman and Cole [34], Bluman and Kumei [1], Hill [35], Olver [4], or Stephani [37].

Henceforth we shall be concerned with applications of symmetry groups to PDEs for which a one-parameter group of transformations permits a reduction in the total number of dependent and independent variables.

EXAMPLE 7.7. Korteweg–de Vries equation. The Korteweg–de Vries (KdV) equation,

$$u_t + 6uu_x + u_{xxx} = 0, \tag{7.9}$$

is the prototypic example of a soliton equation solvable by the inverse scattering method due to Gardner, Greene, Kruskal, and Mirua [59]; we discuss inverse scattering for the KdV equation in Section 7.3. Furthermore, the KdV equation has arisen in several important physical contexts, including small-amplitude surface gravity waves in shallow water, collision-free hydromagnetic waves, stratified internal waves, ion-acoustic waves, plasma physics, lattice dynamics, etc. (see, for example, the book by Ablowitz and Segur [60] and the references therein, as well as our discussion in Section 1.7).

In this example we demonstrate how simple one-parameter transformation groups can be used to reduce the KdV equation to an ODE.

Case 7.7(i): travelling wave solution. Consider the transformation

$$x^* = x + x_0, \qquad t^* = t + t_0,$$

where x_0 and t_0 are arbitrary (nonzero) constants. Then it is easily seen that $x^* - ct^* = x - ct$, provided that $c = x_0/t_0$. Therefore, the KdV equation (7.9) is invariant under the group of translations

$$x^* = x + c\varepsilon, \qquad t^* = t + \varepsilon, \qquad u^* = u, \qquad (7.10)$$

where ε is the group parameter and c is an arbitrary constant, for which the associated invariants are u and $x - ct$, i.e.

$$x - ct = x^* - ct^* \qquad \text{and} \qquad u = u^*.$$

Setting these invariants to be the new independent and dependent variables, z and w, respectively, yields the similarity reduction

$$u(x,t) = w(z), \qquad z = x - ct, \qquad (7.11)$$

where $w(z)$ satisfies

$$\frac{d^3 w}{dz^3} + 6w \frac{dw}{dz} - c \frac{dw}{dz} = 0.$$

Integrating this with respect to z, multiplying by dw/dz, and integrating again yields

$$(dw/dz)^2 + 2w^3 - cw^2 = 2Aw + B, \qquad (7.12)$$

where A and B are arbitrary constants, which is solvable either in terms of elementary or Weierstrass elliptic functions. In particular, if $A = B = 0$, then the general solution is

$$w(z) = \tfrac{1}{2}c \operatorname{sech}^2\{\tfrac{1}{2}\sqrt{c}\,(z + \delta_0)\},$$

with δ_0 an arbitrary constant, which, on setting $c = 4k^2$, gives the well-known soliton solution of the KdV equation (7.9):

$$u(x,t) = 2k^2 \operatorname{sech}^2\{k(x - 4k^2 t + \delta_0)\}.$$

Furthermore, we remark that if
$$u(x,t) = u^*(x^*, t^*), \quad x^* = x + c\varepsilon, \quad t^* = t + \varepsilon,$$
then
$$u_t + 6uu_x + u_{xxx} = u^*_{t^*} + 6u^*u^*_{x^*} + u^*_{x^*x^*x^*}.$$

Therefore, if $u(x,t)$ is *any* solution of the KdV equation (7.9), then so also is $u^*(x^*, t^*)$, where $x^* = x + c\varepsilon$ and $t^* = t + \varepsilon$. The transformation (7.10) is a symmetry group which maps the set of solutions of the KdV equation into itself.

Case 7.7(ii): scaling reduction. If we apply the transformation
$$x^* = \alpha x, \quad t^* = \beta t, \quad u^* = \gamma u,$$
to the KdV equation (7.9), then
$$\frac{\beta}{\gamma} u^*_{t^*} + 6 \frac{\alpha}{\gamma^2} u^* u^*_{x^*} + \frac{\alpha^3}{\gamma} u^*_{x^*x^*x^*} = 0.$$

This is the same equation as eqn (7.9) if and only if $\beta = \alpha^3$ and $\gamma = 1/\alpha^2$. Therefore, setting $\alpha = e^\varepsilon$ we see that the KdV equation (7.9) is invariant under the group of scaling transformations
$$x^* = xe^\varepsilon, \quad t^* = te^{3\varepsilon}, \quad u^* = ue^{-2\varepsilon}, \tag{7.13}$$
where ε is the group parameter, for which the associated invariants are $x/t^{1/3}$ and $ut^{2/3}$, i.e.
$$x/t^{1/3} = x^*/t^{*1/3} \quad \text{and} \quad ut^{2/3} = u^*t^{*2/3}.$$
These yield the similarity reduction
$$u(x,t) = t^{-2/3} w(z), \quad z = x/t^{1/3}, \tag{7.14}$$
where $w(z)$ satisfies
$$\frac{d^3w}{dz^3} + 6w \frac{dw}{dz} - \tfrac{2}{3}w - \tfrac{1}{3}z \frac{dw}{dz} = 0, \tag{7.15}$$
which is obtained by substituting eqn (7.14) into eqn (7.9). This equation is solvable in terms of the second Painlevé equation [61]
$$d^2 y/dz^2 = 2y^3 + zy + a, \qquad \text{PII}$$
where a is an arbitrary constant (cf. reference [62]). We note that if
$$u(x,t) = e^{2\varepsilon} u^*(x^*, t^*), \quad x^* = xe^\varepsilon, \quad t^* = te^{3\varepsilon},$$
then
$$u_t + 6uu_x + u_{xxx} = e^{5\varepsilon}(u^*_{t^*} + 6u^*u^*_{x^*} + u^*_{x^*x^*x^*}).$$

Hence if $u(x,t)$ is a solution of the KdV equation (7.9), then so is $e^{2\varepsilon}u^*(x^*,t^*)$, where $x^* = xe^{\varepsilon}$ and $t^* = te^{3\varepsilon}$, and so the transformation (7.13) is a symmetry group, since it maps the set of solutions of the KdV equation into itself.

For linear PDEs, exact solutions are often derived using the well-known method of separation of variables (cf. reference [24]). This method can also be applied to nonlinear PDEs; for example, to the Improved Boussinesq equation [63–67].

In the remainder of this section we shall concentrate upon the classical Lie group method of infinitesimal transformations for determining symmetry reductions. Emphasis will be placed on explicit computational algorithms to discover symmetries admitted by differential equations and the construction of solutions arising from these symmetries.

7.1.2 One-parameter groups of transformations

DEFINITION 7.8. In the (x, y) plane, the transformation

$$x^* = f(x, y; \varepsilon), \qquad y^* = g(x, y; \varepsilon), \qquad (7.16)$$

where $\varepsilon \in \mathbb{C}$ is a *one-parameter group of transformations* if the following properties hold:

(i) the value $\varepsilon = 0$ characterizes the *identity* transformation, i.e.

$$x = f(x, y; 0), \qquad y = g(x, y; 0);$$

(ii) the parameter $-\varepsilon$ characterizes the *inverse* transformation, i.e.

$$x = f(x^*, y^*; -\varepsilon), \qquad y = g(x^*, y^*; -\varepsilon);$$

(iii) if $x^{**} = f(x^*, y^*; \delta)$, $y^{**} = g(x^*, y^*; \delta)$, then the product of the two transformations is also a member of the set of transformations (7.16) characterized by the parameter $\varepsilon + \delta$, i.e.

$$x^{**} = f(x, y; \varepsilon + \delta), \qquad y^{**} = g(x, y; \varepsilon + \delta),$$

so there is closure under composition; and

(iv) the associativity law for groups holds; this property follows from the formula given in (iii) since addition in $\mathbb{C}$ is associative.

EXAMPLE 7.9. Examples of groups of transformations are given in the following table.

Table 7.1 One-parameter groups of transformations

	x^*	y^*	
(a)	$x + \alpha\varepsilon$	$y + \beta\varepsilon$	Group of translations
(b)	$x \exp(\alpha\varepsilon)$	$y \exp(\beta\varepsilon)$	Group of scalings/dilations
(c)	$x \cos \varepsilon - y \sin \varepsilon$	$x \sin \varepsilon + y \cos \varepsilon$	Group of rotations
(d)	$\dfrac{x}{1-\varepsilon x}$	$\dfrac{y}{1-\varepsilon x}$	Projective group

ε is the group parameter and α and β are arbitrary parameters.

Expanding eqn (7.16) about (the identity) $\varepsilon = 0$ yields

$$x^* = x + \varepsilon\xi(x,y) + O(\varepsilon^2), \qquad y^* = y + \varepsilon\phi(x,y) + O(\varepsilon^2), \quad (7.17\text{a,b})$$

where

$$\xi(x,y) = \left.\frac{df}{d\varepsilon}\right|_{\varepsilon=0}, \qquad \phi(x,y) = \left.\frac{dg}{d\varepsilon}\right|_{\varepsilon=0}, \qquad (7.18)$$

since $f(x,y;0) = x$ and $g(x,y;0) = y$. Equation (7.16) is referred to as the *global* form of the group and eqn (7.17) as the *infinitesimal* form. The crucial property of a one-parameter group of transformations is that given the infinitesimal form, one can deduce the global form by integrating

$$dx^*/d\varepsilon = \xi(x^*, y^*), \qquad dy^*/d\varepsilon = \phi(x^*, y^*), \qquad (7.19)$$

subject to the initial conditions

$$x^* = x, \quad y^* = y, \qquad \text{at } \varepsilon = 0. \qquad (7.20)$$

This fundamental result is due to Lie.

EXAMPLE 7.10. In this example we shall derive the infinitesimal form of the scaling group (b) in Table 7.1 and then, by integration of eqn (7.19), deduce the global form of the group. For the scaling group we have

$$dx^*/d\varepsilon = \alpha x \exp(\alpha\varepsilon), \qquad dy^*/d\varepsilon = \beta y \exp(\beta\varepsilon),$$

and so setting $\varepsilon = 0$ from eqn (7.18) yields

$$\xi(x,y) = \alpha x, \qquad \phi(x,y) = \beta y.$$

Therefore we need to integrate

$$dx^*/d\varepsilon = \alpha x^*, \qquad dy^*/d\varepsilon = \beta y^*,$$

ANALYSIS OF DIFFERENTIAL EQUATIONS

subject to the initial conditions given in eqn (7.20). Hence it is easily seen that we recover the global form.

7.1.3 The classical Lie group method for second-order PDEs

Consider the second-order PDE

$$\Delta^{(2)}(u_{xx}, u_{xt}, u_{tt}, u_x, u_t, u, x, t) = 0, \qquad (7.21)$$

where u is the dependent variable and x and t are the independent variables. Suppose that $u = \theta(x, t)$ is a solution of eqn (7.21). Consider a Lie group of transformations

$$x^* = X(x, t, u; \varepsilon), \qquad t^* = T(x, t, u; \varepsilon), \qquad u^* = U(x, t, u; \varepsilon), \qquad (7.22\text{a–c})$$

where ε is the group parameter and $\varepsilon = 0$ is the identity. Expanding the transformation (7.22) about the identity $\varepsilon = 0$ yields

$$x^* = x + \varepsilon \xi(x, t, u) + O(\varepsilon^2), \qquad (7.23\text{a})$$
$$t^* = t + \varepsilon \tau(x, t, u) + O(\varepsilon^2), \qquad (7.23\text{b})$$
$$u^* = u + \varepsilon \phi(x, t, u) + O(\varepsilon^2), \qquad (7.23\text{c})$$

where

$$\xi(x, t, u) = \left.\frac{dx^*}{d\varepsilon}\right|_{\varepsilon=0}, \qquad \tau(x, t, u) = \left.\frac{dt^*}{d\varepsilon}\right|_{\varepsilon=0}, \qquad \phi(x, t, u) = \left.\frac{du^*}{d\varepsilon}\right|_{\varepsilon=0},$$

are called the *infinitesimals*. As discussed in Section 7.1.2, the important property of the one-parameter Lie groups under consideration is that one can deduce the global form of the group from the infinitesimal form. Thus, given $\xi(x, t, u)$, $\tau(x, t, u)$, and $\phi(x, t, u)$, one obtains $X(x, t, u; \varepsilon)$, $T(x, t, u, \varepsilon)$, and $U(x, t, u; \varepsilon)$ by integrating

$$dx^*/d\varepsilon = \xi(x^*, t^*, u^*), \qquad dt^*/d\varepsilon = \tau(x^*, t^*, u^*),$$
$$du^*/d\varepsilon = \eta(x^*, t^*, u^*),$$

subject to the initial conditions

$$x^*|_{\varepsilon=0} = x, \qquad t^*|_{\varepsilon=0} = t, \qquad u^*|_{\varepsilon=0} = u,$$

which follow by virtue of the group identity property. Thus the problem of finding the full invariance group is converted to the problem of determining the infinitesimals $\xi(x, t, u)$, $\tau(x, t, u)$, and $\phi(x, t, u)$.

Now consider the equation

$$\Delta^{(2)}(v_{x^*x^*}, v_{x^*t^*}, v_{t^*t^*}, v_{x^*}, v_{t^*}, v, x^*, t^*) = 0, \qquad (7.21^*)$$

which is eqn (7.21) with u replaced by v, x by $x^* = X(x, t, u; \varepsilon)$, and t by $t^* = T(x, t, u; \varepsilon)$, respectively. Evidently, eqn (7.21*) has the solution $v = \theta(x^*, t^*)$. The transformation (7.22) is said to leave eqn (7.21) *invariant* if and only if $v = U^*(x, t, \theta(x, t); \varepsilon)$ satisfies eqn (7.21*) whenever $u = \theta(x, t)$ is a solution of eqn (7.21) for all values of the parameter ε.

Now let us consider the problem of finding the group of transformations that leave a PDE invariant. Since it is the infinitesimal of a group which is used to reduce the number of variables in the PDE, we seek the infinitesimal (7.23) leaving invariant eqn (7.21). In order to find the infinitesimal (7.23) we need to extend the group to calculate how the derivatives transform. The first step is to determine $\partial u^*/\partial x^*$ and $\partial u^*/\partial t^*$. By the chain rule, we have

$$\frac{\partial u^*}{\partial x^*} = \frac{\partial u^*}{\partial x}\frac{\partial x}{\partial x^*} + \frac{\partial u^*}{\partial t}\frac{\partial t}{\partial x^*}, \qquad \frac{\partial u^*}{\partial t^*} = \frac{\partial u^*}{\partial x}\frac{\partial x}{\partial t^*} + \frac{\partial u^*}{\partial t}\frac{\partial t}{\partial t^*},$$

where u^* is considered to be a function of x and t. Hence we need to calculate

$$\frac{\partial x}{\partial x^*}, \quad \frac{\partial x}{\partial t^*}, \quad \frac{\partial t}{\partial x^*}, \quad \frac{\partial t}{\partial t^*},$$

where by $\partial x/\partial x^*$, we understand that only t^* is fixed. Therefore,

$$\frac{\partial x}{\partial x^*} = \frac{\partial}{\partial x^*}[x^* - \varepsilon\xi(x, t, u(x, t)) + O(\varepsilon^2)],$$

$$= 1 - \varepsilon\left(\frac{\partial \xi}{\partial x} + \frac{\partial \xi}{\partial u}\frac{\partial u}{\partial x}\right)\frac{\partial x}{\partial x^*} + O(\varepsilon^2)$$

$$= 1 - \varepsilon\left(\frac{\partial \xi}{\partial x} + \frac{\partial \xi}{\partial u}\frac{\partial u}{\partial x}\right) + O(\varepsilon^2), \qquad (7.24a)$$

by recursion. Similarly, it can be shown that

$$\frac{\partial t}{\partial x^*} = -\varepsilon\left(\frac{\partial \tau}{\partial x} + \frac{\partial \tau}{\partial u}\frac{\partial u}{\partial x}\right) + O(\varepsilon^2), \qquad (7.24b)$$

$$\frac{\partial x}{\partial t^*} = -\varepsilon\left(\frac{\partial \xi}{\partial t} + \frac{\partial \xi}{\partial u}\frac{\partial u}{\partial t}\right) + O(\varepsilon^2), \qquad (7.24c)$$

$$\frac{\partial t}{\partial t^*} = 1 - \varepsilon\left(\frac{\partial \tau}{\partial t} + \frac{\partial \tau}{\partial u}\frac{\partial u}{\partial t}\right) + O(\varepsilon^2). \qquad (7.24d)$$

Therefore,

$$\begin{aligned}\frac{\partial u^*}{\partial x^*} &= \frac{\partial}{\partial x^*}[u(x,t) + \varepsilon\phi(x,t,u(x,t))] + O(\varepsilon^2) \\ &= \frac{\partial}{\partial x}[u(x,t) + \varepsilon\phi(x,t,u(x,t))]\frac{\partial x}{\partial x^*} + \frac{\partial u}{\partial t}\frac{\partial t}{\partial x^*} + O(\varepsilon^2) \\ &= \frac{\partial u}{\partial x} + \varepsilon\left[\frac{\partial\phi}{\partial x} + \left(\frac{\partial\phi}{\partial u} - \frac{\partial\xi}{\partial x}\right)\frac{\partial u}{\partial x} - \frac{\partial\xi}{\partial u}\left(\frac{\partial u}{\partial x}\right)^2 \right. \\ &\quad \left. - \frac{\partial\tau}{\partial x}\frac{\partial u}{\partial t} - \frac{\partial\tau}{\partial u}\frac{\partial u}{\partial x}\frac{\partial u}{\partial t}\right] + O(\varepsilon^2),\end{aligned}$$

and, consequently,

$$\frac{\partial u^*}{\partial x^*} = \frac{\partial u}{\partial x} + \varepsilon\phi^{[x]} + O(\varepsilon^2), \qquad (7.25a)$$

where $\phi^{[x]}$ is the associated infinitesimal for $\partial u^*/\partial x^*$, given by

$$\begin{aligned}\phi^{[x]} &= \frac{\partial\phi}{\partial x} + \left(\frac{\partial\phi}{\partial u} - \frac{\partial\xi}{\partial x}\right)\frac{\partial u}{\partial x} - \frac{\partial\xi}{\partial u}\left(\frac{\partial u}{\partial x}\right)^2 \\ &\quad - \frac{\partial\tau}{\partial x}\frac{\partial u}{\partial t} - \frac{\partial\tau}{\partial u}\frac{\partial u}{\partial x}\frac{\partial u}{\partial t}.\end{aligned} \qquad (7.25b)$$

Similarly,

$$\frac{\partial u^*}{\partial t^*} = \frac{\partial u}{\partial t} + \varepsilon\phi^{[t]} + O(\varepsilon^2), \qquad (7.26a)$$

where $\phi^{[t]}$ is the associated infinitesimal for $\partial u^*/\partial t^*$, given by

$$\begin{aligned}\phi^{[t]} &= \frac{\partial\phi}{\partial t} + \left(\frac{\partial\phi}{\partial u} - \frac{\partial\tau}{\partial t}\right)\frac{\partial u}{\partial t} - \frac{\partial\tau}{\partial u}\left(\frac{\partial u}{\partial t}\right)^2 \\ &\quad - \frac{\partial\xi}{\partial t}\frac{\partial u}{\partial x} - \frac{\partial\xi}{\partial u}\frac{\partial u}{\partial x}\frac{\partial u}{\partial t}.\end{aligned} \qquad (7.26b)$$

In an analogous manner, reverting to using subscript notation, it can be shown that

$$u^*_{x^*x^*} = u_{xx} + \varepsilon\phi^{[xx]} + O(\varepsilon^2), \quad u^*_{x^*t^*} = u_{xt} + \varepsilon\phi^{[xt]} + O(\varepsilon^2),$$
$$u^*_{t^*t^*} = u_{tt} + \varepsilon\phi^{[tt]} + O(\varepsilon^2), \qquad (7.27a\text{--}c)$$

where $\phi^{[xx]}$, $\phi^{[xt]}$, and $\phi^{[tt]}$ are the associated infinitesimals for $u^*_{x^*x^*}$, $u^*_{x^*t^*}$, and $u^*_{t^*t^*}$ respectively, given by

$$\phi^{[xx]} = \phi_{xx} + (2\phi_{xu} - \xi_{xx})u_x - \tau_{xx}u_t + (\phi_{uu} - 2\xi_{xu})u_x^2$$
$$- 2\tau_{xu}u_xu_t - \xi_{uu}u_x^3 - \tau_{uu}u_x^2u_t$$
$$+ (\phi_u - 2\xi_x)u_{xx} - 2\tau_xu_{xt} - 3\xi_uu_xu_{xx} - \tau_uu_{xx}u_t - 2\tau_uu_{xt}u_x, \tag{7.27d}$$

$$\phi^{[xt]} = \phi_{xt} + (\phi_{xu} - \tau_{xt})u_t + (\phi_{tu} - \xi_{xt})u_x$$
$$+ (\phi_{uu} - \xi_{xu} - \tau_{tu})u_xu_t - \tau_{xu}u_t^2 - \xi_{tu}u_x^2$$
$$- \tau_{uu}u_xu_t^2 - \xi_{uu}u_tu_x^2 + (\phi_u - \xi_x - \tau_t)u_{xt} - \tau_xu_{tt} - \xi_tu_{xx}$$
$$- 2\tau_uu_tu_{xt} - 2\xi_uu_xu_{xt} - \tau_uu_xu_{tt} - \xi_uu_tu_{xx}, \tag{7.27e}$$

$$\phi^{[tt]} = \phi_{tt} + (2\phi_{tu} - \tau_{tt})u_t - \xi_{tt}u_x + (\phi_{uu} - 2\tau_{tu})u_t^2$$
$$- 2\xi_{tu}u_tu_x - \tau_{uu}u_t^3 - \xi_{uu}u_t^2u_x$$
$$+ (\phi_u - 2\tau_t)u_{tt} - 2\xi_tu_{tx} - 3\tau_uu_tu_{tt} - \xi_uu_{tt}u_x - 2\xi_uu_{tx}u_t. \tag{7.27f}$$

Suppose that $u(x,t)$ satisfies eqn (7.21) and $u^*(x^*,t^*)$ satisfies

$$\Delta^{(2)}(u^*_{x^*x^*}, u^*_{x^*t^*}, u^*_{t^*t^*}, u^*_{x^*}, u^*_{t^*}, u^*, x^*, t^*) = 0. \tag{7.28}$$

Then a necessary and sufficient condition for eqn (7.21) to be invariant under the transformation (7.22) is that the infinitesimal (i.e. $O(\varepsilon)$) term of eqn (7.28) is identically zero, i.e.

$$\phi^{[xx]}\frac{\partial \Delta^{(2)}}{\partial u_{xx}} + \phi^{[xt]}\frac{\partial \Delta^{(2)}}{\partial u_{xt}} + \phi^{[tt]}\frac{\partial \Delta^{(2)}}{\partial u_{tt}} + \phi^{[x]}\frac{\partial \Delta^{(2)}}{\partial u_x} + \phi^{[t]}\frac{\partial \Delta^{(2)}}{\partial u_t}$$
$$+ \phi\frac{\partial \Delta^{(2)}}{\partial u} + \xi\frac{\partial \Delta^{(2)}}{\partial x} + \tau\frac{\partial \Delta^{(2)}}{\partial t} = 0. \tag{7.29}$$

Hence the transformation (7.22) leaves eqn (7.21) invariant if and only if, given any $u(x,t)$ satisfying eqn (7.21), $u(x,t)$ also satisfies eqn (7.29), which contains the infinitesimals ξ, τ, and ϕ to be determined.

To find the infinitesimals $\xi(x,t,u)$, $\tau(x,t,u)$, and $\phi(x,t,u)$, we next substitute the original PDE (7.21) into eqn (7.28) (we normally eliminate the highest derivative term). The resulting equation is treated as a form in the (partial) derivatives of u, the coefficients of which depend upon (x,t,u) and the unknowns (ξ, τ, ϕ) that we wish to determine. Collecting coefficients of like derivatives of u and setting them all equal to zero (since the derivatives u are independent except for the fact that they satisfy eqn (7.21)) yields the *determining equations*. These are a *linear, homogeneous system* of PDEs for ξ, τ, ϕ, which are in general solvable.

Consider the *infinitesimal generator* associated with the transformation (7.22) given by

$$v \equiv \xi(x,t,u)\frac{\partial}{\partial x} + \tau(x,t,u)\frac{\partial}{\partial t} + \phi(x,t,u)\frac{\partial}{\partial u}. \qquad (7.30)$$

Then the operator

$$\mathrm{pr}^{(2)}v \equiv \phi^{[xx]}\frac{\partial}{\partial u_{xx}} + \phi^{[xt]}\frac{\partial}{\partial u_{xt}} + \phi^{[tt]}\frac{\partial}{\partial u_{tt}} + \phi^{[x]}\frac{\partial}{\partial u_x} + \phi^{[t]}\frac{\partial}{\partial u_t} + v, \qquad (7.31)$$

is known as the *second prolongation* (or *second extension*) of the infinitesimal operator (7.30). Hence the requirement that eqn (7.21) is invariant under the transformation (7.22) can be written as

$$\mathrm{pr}^{(2)}v(\Delta^{(2)})|_{\Delta^{(2)}=0} = 0.$$

Suppose that $u = \theta(x,t)$ is a solution surface of the equation (7.21). Then, setting $\eta = u - \theta(x,t)$, it follows that $u = \theta(x,t)$ is a *group-invariant* solution, provided that

$$v(\eta)|_{\eta=0} = \phi - \xi\theta_x - \tau\theta_t = 0. \qquad (7.32)$$

Hence we obtain the *invariant surface condition*,

$$\xi(x,t,\theta)\theta_x + \tau(x,t,\theta)\theta_t = \phi(x,t,\theta). \qquad (7.33)$$

The method of solving such quasi-linear first-order PDEs is discussed in many standard texts on PDEs (cf. reference [68]), and is usually obtained by solving the characteristic equations

$$\frac{dx}{\xi(x,t,\theta)} = \frac{dt}{\tau(x,t,\theta)} = \frac{d\theta}{\phi(x,t,\theta)}. \qquad (7.34)$$

In principle, the general solution of these equations can be found. This involves two constants (which are the associated invariants of the transformation), one of which becomes the independent variable $z(x,t,\theta)$, called the *similarity variable*, while the other plays the role of the dependent variable $w(z)$. Hence we obtain the *similarity form*,

$$u = \theta(x,t) = F(x,t,w(z)), \qquad (7.35)$$

where the dependence of F on x, t, and w, and that of z on x, t, and θ, is explicitly known. Substituting eqn (7.35) into eqn (7.21) then yields an

ODE for $w(z)$. Commonly, $\xi/\tau = \xi(x,t)$, i.e. the ratio of ξ and τ is independent of θ. In this case the similarity variable z is also independent of θ.

Therefore, given a group of transformations (7.22) leaving eqn (7.21) invariant, we can use the infinitesimal (7.23) of the group to reduce eqn (7.21) to an ODE. We make the following remarks: (i) only the infinitesimals ξ, τ, and ϕ appear to be needed in order to obtain the similarity form (7.35); (ii) the order of the equation, which can be nonlinear, is irrelevant; (iii) the transformations (7.22) considered are functions of both the dependent and independent variables; (iv) boundary conditions play no role; and (v) the determining equations for the infinitesimals turn out to be a *linear, homogeneous* system of equations.

Hence we have the following definition.

DEFINITION 7.11. Suppose that the PDE (7.21) admits a one-parameter Lie group of transformations with associated invariant surface condition (7.33). Then a solution $u = \theta(x, t)$ is an *invariant solution* of eqn (7.21) associated with eqn (7.33) if and only if $u = \theta(x, t)$ satisfies both eqns (7.21) and (7.33), i.e.

$$\Delta^{(2)}(\theta_{xx}, \theta_{xt}, \theta_{tt}, \theta_x, \theta_t, \theta, x, t) = 0$$

and

$$\xi(x, t, \theta)\theta_x + \tau(x, t, \theta)\theta_t = \phi(x, t, \theta).$$

Thus far we have shown that, given a group of transformations (7.22) that leave invariant a PDE (7.21), we use the infinitesimal form of the transformation (7.23) to reduce eqn (7.21) to an ODE. The classical symmetry analysis of a second-order PDE (7.21) involves the following steps:

1. Apply the second prolongation $\mathrm{pr}^{(2)}v$ to obtain the infinitesimal equation (7.28).

2. Use the fact that u satisfies the original PDE (7.21) to eliminate, say, the highest derivative term of u in (7.28).

3. Equate coefficients of like derivatives of u to zero to generate the determining equations for ξ, τ, and ϕ. This yields a linear homogeneous system of PDEs in ξ, τ, and ϕ.

4. Solve the determining equations for the infinitesimals ξ, τ, and ϕ.

5. Given the infinitesimals ξ, τ, and ϕ, solve the invariant surface condition (7.33) to obtain the associated symmetry reduction.

ANALYSIS OF DIFFERENTIAL EQUATIONS

EXAMPLE 7.12. *Burgers' equation.* In this example we shall derive the classical symmetry reductions for Burgers' equation [69]

$$u_t + uu_x = u_{xx}, \tag{7.36}$$

which can be mapped into the linear heat equation via the Cole–Hopf transformation [70, 71]. Burgers' equation (7.36) also arises in several physical applications, including turbulent flows [72–74], shock waves [75], and nonlinear acoustics [76, 77]; for a comprehensive survey of applications and properties of Burgers' equation (7.36), see the book by Sachdev [78].

Step 1. Applying the second prolongation $\text{pr}^{(2)}v$ to this equation yields the infinitesimal equation

$$\phi^{[t]} + \phi u_x + u\phi^{[x]} = \phi^{[xx]}.$$

Step 2. Substituting for $\phi^{[x]}$, $\phi^{[t]}$, and $\phi^{[xx]}$ and using the fact that u satisfies Burgers' equation (7.36) (i.e. replacing u_{xx} by $u_t + uu_x$) yields

$$\phi_t + (\phi_u - \tau_t)u_t - \tau_u u_t^2 - \xi_t u_x - \xi_u u_x u_t + \phi u_x + u$$
$$\times \left[\phi_x + (\phi_u - \xi_x)u_x - \xi_u u_x^2 - \tau_x u_t - \tau_u u_x u_t \right]$$
$$= \phi_{xx} + (2\phi_{xu} - \xi_{xx})u_x - \tau_{xx} u_t + (\phi_{uu} - 2\xi_{xu})u_x^2$$
$$- 2\tau_{xu} u_x u_t - \xi_{uu} u_x^3 - \tau_{uu} u_x^2 u_t + (\phi_u - 2\xi_x)(u_t + uu_x)$$
$$- 2\tau_x u_{xt} - 3\xi_u u_x(u_t + uu_x) - \tau_u u_t(u_t + uu_x) - 2\tau_u u_{xt} u_x.$$

Step 3. Equate the coefficients of derivatives of u to zero:

Monomial	Coefficient	
u_{xt}	$\tau_x = 0$	(7.37i)
$u_x u_{xt}$	$\tau_u = 0$	(7.37ii)
u_x^3	$\xi_{uu} = 0$	(7.37iii)
$u_x u_t$	$\xi_u = 0$	(7.37iv)
u_x^2	$\phi_{uu} = 0$	(7.37v)
u_t	$\tau_t - 2\xi_x = 0$	(7.37vi)
u_x	$2\phi_{xu} - \xi_{xx} - u\xi_x + \xi_t - \phi = 0$	(7.37vii)
1	$\phi_t + u\phi_x - \phi_{xx} = 0$	(7.37viii)

Step 4. Solve the determining equations (7.37) for the infinitesimals $\xi(x,t,u)$, $\tau(x,t,u)$, and $\phi(x,t,u)$. From eqns (7.37i,ii) and (7.37iv), we see that $\tau = \tau(t)$ and $\xi(x,t)$, respectively. Equation (7.37v) implies that

$$\phi(x,t,u) = u\psi_1(x,t) + \psi_0(x,t), \tag{7.38}$$

where $\psi_1(x,t)$ and $\psi_0(x,t)$ are to be determined. Substituting this into eqn (7.37vii) and equating coefficients of powers of u yields

$$\psi_1 + \xi_x = 0, \qquad \psi_0 - \xi_t - 2\psi_{1,x} = 0. \qquad (7.39a,b)$$

Since $\xi_{xx} = 0$, from eqns (7.37i,vi), then eqn (7.39a) yields $\psi_{1,x} = 0$ and thus from eqn (7.39b), we have $\psi_0 = \xi_t$. Substituting eqn (7.38) into eqn (7.37viii) and equating coefficients of powers of u yields

$$\psi_{1,t} + \psi_{0,x} = 0, \qquad \psi_{0,t} - \psi_{0,xx} = 0. \qquad (7.40a,b)$$

Since $\psi_1 = -\xi_x$ and $\psi_0 = \xi_t$, then eqn (7.40a) is identically satisfied and $\psi_{0,t} = \psi_{0,xx} = 0$. Thus $\xi_{tt} = 0$, which in conjunction with $\xi_{xx} = 0$ and $\xi_u = 0$ implies that

$$\xi = \alpha x + \beta t + \gamma xt + \delta,$$

where α, β, γ, and δ are arbitrary constants. Since, from eqn (7.37vi),

$$\tau_t = 2\xi_x = 2(\alpha + \gamma t),$$

then

$$\tau = 2\alpha t + \gamma t^2 + \kappa,$$

where κ is an arbitrary constant. Finally,

$$\psi_1 = -\xi_x = -(\alpha + \gamma t), \qquad \psi_0 = \xi_t = \beta + \gamma x,$$

and thus

$$\phi = \beta + \gamma x - (\alpha + \gamma t)u.$$

Hence we obtain the following five-parameter family of infinitesimals for Burger's equation (7.36):

$$\xi = \alpha x + \beta t + \gamma xt + \delta, \quad \tau = 2\alpha t + \gamma t^2 + \kappa, \quad \phi = \beta + \gamma x - (\alpha + \gamma t)u. \qquad (7.41)$$

Step 5. Having determined the infinitesimals, we next solve the characteristic equations

$$\frac{dx}{\xi} = \frac{dt}{\tau} = \frac{du}{\phi},$$

with ξ, τ, and ϕ as given by eqn (7.41), to yield the associated symmetry reductions. Essentially, we can split this analysis into several (canonical) cases.

Case 7.12(i): $\alpha = \beta = \gamma = 0$, $\kappa = 1$. In this case we have to solve

$$\frac{dx}{\delta} = \frac{dt}{1} = \frac{du}{0},$$

ANALYSIS OF DIFFERENTIAL EQUATIONS 419

and so we obtain the travelling wave reduction

$$u(x,t) = w(z), \qquad z = x - \delta t, \tag{7.42}$$

where $w(z)$ satisfies

$$w'' = ww' - \delta w',$$

with $' \equiv d/dz$. This can be integrated once to yield

$$w' = \tfrac{1}{2}w^2 - \delta w + A,$$

where A is an arbitrary constant, which is solvable in terms of elementary functions.

Case 7.12(ii): $\alpha = \gamma = 0$, $\beta \neq 0$, $\kappa = 1$. In this case we have to solve

$$\frac{dx}{\beta t + \delta} = \frac{dt}{1} = \frac{du}{\beta},$$

and so we obtain the accelerating wave reduction

$$u(x,t) = w(z) + \beta t, \qquad z = x - \tfrac{1}{2}\beta t^2 - \delta t, \tag{7.43}$$

where $w(z)$ satisfies

$$w'' = ww' - \delta w' + \beta.$$

This can be integrated once to yield

$$w' = \tfrac{1}{2}w^2 - \delta w + \beta z + A,$$

where A is an arbitrary constant. Then, making the transformation $w = -2\psi'/\psi$, and setting $\psi = \Psi \exp(-\tfrac{1}{2}\delta z)$ yields

$$\Psi'' + \left[\tfrac{1}{2}(\beta z + A) - \tfrac{1}{4}\delta^2\right]\Psi = 0.$$

This equation is solvable in terms of the Airy functions $\mathrm{Ai}(z)$ and $\mathrm{Bi}(z)$, which satisfy

$$\frac{d^2 F}{dz^2} - zF = 0, \tag{7.44}$$

and the asymptotic conditions

$$\mathrm{Ai}(z) \sim \tfrac{1}{2}\pi^{-1/2} z^{-1/4} \exp\left(-\tfrac{2}{3} z^{3/2}\right), \qquad \text{as } z \to \infty,$$

$$\mathrm{Bi}(z) \sim \pi^{-1/2} z^{-1/4} \exp\left(\tfrac{2}{3} z^{3/2}\right), \qquad \text{as } z \to \infty.$$

Case 7.12(iii): $\alpha = \beta = \kappa = 0$, $\gamma = 1$. In this case we have to solve

$$\frac{dx}{xt+\delta} = \frac{dt}{t^2} = \frac{du}{x-tu},$$

and so we obtain the reduction

$$u(x,t) = \frac{w(z)}{t} + \frac{x}{t} + \frac{\delta}{t^2}, \qquad z = \frac{x}{t} + \frac{\delta}{2t^2}, \qquad (7.45)$$

where $w(z)$ satisfies

$$w'' = ww' - \delta.$$

This can be integrated once to yield

$$w' = \tfrac{1}{2}w^2 - \delta z + A,$$

where A is an arbitrary constant. Then making the transformation $w = -2\psi'/\psi$ yields

$$\psi'' - \tfrac{1}{2}(\delta z - A)\psi = 0,$$

which is solvable in terms of Airy functions (if $\delta \neq 0$) or elementary functions (if $\delta = 0$).

Case 7.12(iv): $\alpha = 1$, $\gamma = \delta = \kappa = 0$. In this case we have to solve

$$\frac{dx}{x+\beta t} = \frac{dt}{2t} = \frac{du}{-u+\beta},$$

and so we obtain the scaling reduction

$$u(x,t) = \frac{w(z)}{\sqrt{t}} + \beta, \qquad z = \frac{x-\beta t}{\sqrt{t}} \qquad (7.46)$$

(if $\beta = 0$, then we obtain the pure scaling reduction), where $w(z)$ satisfies

$$w'' = ww' - \frac{zw'}{2} - \frac{w}{2}.$$

This can be integrated once to yield

$$w' = \frac{w^2}{2} - \frac{1}{2}zw + A,$$

where A is an arbitrary constant. Then making the transformation $w = -2\psi'/\psi$ and setting $\psi = \eta(\xi)$ with $\xi = -\tfrac{1}{2}z^2$ yields

$$\xi\eta'' + \left(\frac{1}{2} - \frac{\xi}{2}\right)\eta' - \tfrac{1}{4}A\eta = 0.$$

This equation is solvable in terms of the confluent hypergeometric function $_1F_1(\alpha, \beta; \gamma, z)$, which satisfies

$$z(1-z)\frac{d^2 F}{dz^2} + (\gamma - z)\frac{dF}{dz} - \alpha F = 0. \tag{7.47}$$

Case 7.12(v): $\beta = \delta = \kappa = 0$, $\gamma = 1$. In this case we have to solve

$$\frac{dx}{x(t+\alpha)} = \frac{dt}{t^2 + 2\alpha t} = \frac{du}{x - (t+\alpha)u},$$

and so we obtain the reduction

$$u(x,t) = \frac{w(z)}{\sqrt{t^2 + 2\alpha t}} + \frac{x}{t + 2\alpha}, \quad z = \frac{x}{\sqrt{t^2 + 2\alpha t}}, \tag{7.48}$$

where $w(z)$ satisfies

$$w'' = ww' - \alpha(zw' + w).$$

This can be integrated once to yield

$$w' = \tfrac{1}{2}w^2 - \alpha zw + A,$$

where A is an arbitrary constant. Then, making the transformation $w = -2\psi'/\psi$ and setting $\psi = \eta(\xi)$ with $\xi = -\tfrac{1}{2}\alpha z^2$ yields

$$\xi\eta'' + (\tfrac{1}{2} - \xi)\eta' - \frac{A}{4\alpha}\eta = 0,$$

which is solvable in terms of confluent hypergeometric functions.

Case 7.12(vi): *all constants nonzero*. In this case we have to solve

$$\frac{dx}{\alpha x + \beta t + \gamma xt + \delta} = \frac{dt}{2\alpha t + \gamma t^2 + \kappa} = \frac{du}{\beta + \gamma x - (\alpha + \gamma t)u},$$

and so we obtain the reduction

$$u(x,t) = \frac{w(z) + \gamma tz}{(\gamma t^2 + 2\alpha t + \kappa)^{1/2}} + \frac{\alpha\beta - \gamma\delta}{\alpha^2 - \gamma\kappa},$$

$$z = \frac{x(\alpha^2 - \gamma\kappa) - t(\alpha\beta - \gamma\delta) - (\beta\kappa - \alpha\delta)}{(\alpha^2 - \gamma\kappa)(\gamma t^2 + 2\alpha t + \kappa)^{1/2}}, \tag{7.49}$$

where $w(z)$ satisfies

$$w'' = ww' - \alpha(zw' + w) + \gamma\kappa z.$$

This can be integrated once to yield

$$w' - \tfrac{1}{2}w^2 - \alpha z w + \tfrac{1}{2}\gamma\kappa z^2 + \Lambda,$$

where A is an arbitrary constant. Then, making the transformation $w = -2\psi'/\psi$ and setting $\psi = \eta(\xi)\exp(\mu z^2)$ with $\xi = -\tfrac{1}{2}(\alpha^2 - \gamma\kappa)^{1/2}z^2$ and $\mu = \tfrac{1}{4}[-\alpha + (\alpha^2 - \gamma\kappa)^{1/2}]$ yields

$$\xi\eta'' + (\tfrac{1}{2} - \xi)\eta' + \left\{ \frac{\alpha - (\alpha^2 - \gamma\kappa)^{1/2} - A}{4(\alpha^2 - \gamma\kappa)^{1/2}} \right\}\eta = 0,$$

which is solvable in terms of confluent hypergeometric functions. The infinitesimal operator (or vector field) associated with eqn (7.41) is

$$\begin{aligned}v &= (\alpha x + \beta t + \gamma x t + \delta)\partial_x + (2\alpha t + \gamma t^2 + \kappa)\partial_t \\ &\quad + [\beta + \gamma x - (\alpha + \gamma t)u]\partial_u \\ &= \alpha(x\partial_x + 2t\partial_t - u\partial_u) + \beta(t\partial_x + \partial_u) \\ &\quad + \gamma\left[xt\partial_x + t^2\partial_t + (x - tu)\partial_u\right] + \delta\partial_x + \kappa\partial_t,\end{aligned}$$

where α, β, γ, and δ are arbitrary constants. The generators of this are

$v_1 = \partial_x,$	space translation,
$v_2 = \partial_t,$	time translation,
$v_3 = x\partial_x + 2t\partial_t - u\partial_u,$	scaling or dilation,
$v_4 = t\partial_x + \partial_u,$	Galilean boost,
$v_5 = xt\partial_x + t^2\partial_t + (x - tu)\partial_u,$	conformal point symmetry.

These generators are associated with the following transformations which map the set of solutions of Burgers' equation into itself

$v_1:$	$x_1 = x + \varepsilon,$	$t_1 = t,$	$u_1 = u,$
$v_2:$	$x_2 = x,$	$t_2 = t + \varepsilon,$	$u_2 = u,$
$v_3:$	$x_3 = x + e^\varepsilon,$	$t_3 = te^{2\varepsilon},$	$u_3 = ue^{-\varepsilon}$
$v_4:$	$x_4 = x + \varepsilon t,$	$t_4 = t,$	$u_4 = u + \varepsilon,$
$v_5:$	$x_5 = \dfrac{x}{1 - \varepsilon t},$	$t_5 = \dfrac{t}{1 - \varepsilon t},$	$u_5 = u + \varepsilon(x - tu).$

The generators of the vector field for the Burger's equation form a Lie algebra.

DEFINITION 7.13. A *Lie algebra* $\mathfrak{g}$ is a vector space over a field $\mathbb{F}$ (usually

ANALYSIS OF DIFFERENTIAL EQUATIONS 423

either $\mathbb{R}$ or $\mathbb{C}$), on which a bilinear product [,] called the *Lie bracket* (or *commutator*) is defined, with the following properties:

1. If $v_i, v_j \in \mathfrak{g}$, then $[v_i, v_j] \in \mathfrak{g}$.
2. If $\alpha, \beta \in \mathbb{F}$ and $v_i, v_j, v_k \in \mathfrak{g}$, then $[v_i, \alpha v_j + \beta v_k] = \alpha [v_i, v_j] + \beta [v_i, v_k]$.
3. If $v_i, v_j \in \mathfrak{g}$, then $[v_i, v_j] = -[v_j, v_i]$ (skew-symmetry).
4. If $v_i, v_j, v_k \in \mathfrak{g}$, then $[v_i, [v_j, v_k]] + [v_j, [v_k, v_i]] + [v_k, [v_i, v_j]] = 0$ (Jacobi identity).

It is easily seen that generators of the vector field for the Burger's equation form a five-dimensional Lie algebra with commutator bracket $[v_i, v_j] = v_i v_j - v_j v_i$, since

$$[v_1, v_2] = \partial_x \partial_t - \partial_t \partial_x \equiv 0,$$
$$[v_1, v_3] = \partial_x (x \partial_x + 2t \partial_t - u \partial_u) - (x \partial_x + 2t \partial_t - u \partial_u) \partial_x$$
$$\equiv \partial_x = v_1,$$
$$[v_1, v_4] = \partial_x (t \partial_x + \partial_u) - (t \partial_x + \partial_u) \partial_x \equiv 0,$$
$$[v_1, v_5] = \partial_x \left[xt \partial_x + t^2 \partial_t + (x - tu) \partial_u \right] - \left[xt \partial_x + t^2 \partial_t + (x - tu) \partial_u \right] \partial_x$$
$$\equiv t \partial_x + \partial_u = v_4,$$
$$[v_2, v_3] = \partial_t (x \partial_x + 2t \partial_t - u \partial_u) - (x \partial_x + 2t \partial_t - u \partial_u) \partial_t$$
$$\equiv 2 \partial_t = 2 v_2,$$
$$[v_2, v_4] = \partial_t (t \partial_x + \partial_u) - (t \partial_x + \partial_u) \partial_t$$
$$\equiv \partial_x = v_1,$$
$$[v_2, v_5] = \partial_t \left[xt \partial_x + t^2 \partial t + (x - tu) \partial_u \right] - \left[xt \partial_x + t^2 \partial_t + (x - tu) \partial_u \right] \partial_t$$
$$\equiv x \partial_x + 2t \partial_t - u \partial_u = v_3,$$
$$[v_3, v_4] = (x \partial_x + 2t \partial_t - u \partial_u)(t \partial_x + \partial_u) - (t \partial_x + \partial_u)(x \partial_x + 2t \partial_t - u \partial_u)$$
$$\equiv t \partial_x + \partial_u = v_4,$$
$$[v_3, v_5] = (x \partial_x + 2t \partial_t - u \partial_u)\left[xt \partial_x + t^2 \partial_t + (x - tu) \partial_u \right]$$
$$- \left[xt \partial_x + t^2 \partial_t + (x - tu) \partial_u \right](x \partial_x + 2t \partial_t - u \partial_u)$$
$$\equiv 2 \left[xt \partial_x + t^2 \partial_t + (x - tu) \partial_u \right] = 2 v_5,$$
$$[v_4, v_5] = (t \partial_x + \partial_u)\left[xt \partial_x + t^2 \partial_t + (x - tu) \partial_u \right]$$
$$- \left[xt \partial_x + t^2 \partial_t + (x - tu) \partial_u \right](t \partial_x + \partial_u) \equiv 0.$$

7.1.4 Higher-dimensional PDEs

In this subsection we extend the Lie method to the general case of multi-dimensional PDEs. This embraces both multi-component dependent variables (e.g. order parameters) and multi-dimensional physical space–times.

Let $x = (x_1, x_2, \ldots, x_p) \in \mathbb{C}^p$ be the independent variables, $u = (u_1, u_2, \ldots, u_q) \in \mathbb{C}^q$ the dependent variables, and let $u^{(k)}(x)$ denote the set of all partial derivatives of $u_1, u_2, \ldots, u_q$ of order k with respect to $x_1, x_2, \ldots, x_p$. To apply the classical method to the Nth order system of PDEs

$$\Delta(x, u, u^{(1)}(x), \ldots, u^{(N)}(x)) = 0, \qquad (7.50)$$

where $\Delta = (\Delta_1, \Delta_2, \ldots, \Delta_q)$, we consider the one-parameter Lie group of infinitesimal transformations in (x, u) given by

$$x^* = x + \varepsilon \xi(x, u) + O(\varepsilon^2), \qquad u^* = u + \varepsilon \phi(x, u) + O(\varepsilon^2), \qquad (7.51\text{a,b})$$

where $\xi = (\xi_1, \xi_2, \ldots, \xi_p)$ and $\phi = (\phi_1, \phi_2, \ldots, \phi_q)$, ε is the group parameter, and $\varepsilon = 0$ corresponds to the identity transformation. The associated infinitesimals for derivatives are given by

$$\frac{\partial u_m^*}{\partial x_j^*} = \frac{\partial u_m}{\partial x_j} + \varepsilon \phi_m^{[j]}(x, u, u^{(1)}) + O(\varepsilon^2), \qquad (7.52\text{a})$$

for $m = 1, 2, \ldots, q$ and $j = 1, 2, \ldots, p$, where the *first extension* $\phi_m^{[j]}(x, u, u^{(1)})$ is given by

$$\phi_m^{[j]}(x, u, u^{(1)}) = \frac{D \phi_m}{D x_j} - \sum_{l=1}^{p} \frac{D \xi_l}{D x_j} \frac{\partial u_m}{\partial x_l}, \qquad (7.52\text{b})$$

and

$$\frac{D}{D x_j} \equiv \frac{\partial}{\partial x_j} + \sum_{\lambda=1}^{q} \frac{\partial u_\lambda}{\partial x_j} \frac{\partial}{\partial u_\lambda}, \qquad (7.52\text{c})$$

is the *total derivative operator*. Note that if $\Phi = \Phi(x, u, u^{(1)}, \ldots, u^{(n)})$, then the *total derivative* of Φ is given by

$$\frac{D\Phi}{Dx_k} \equiv \frac{\partial \Phi}{\partial x_k} + \sum_{\lambda=1}^{q} \frac{\partial u_\lambda}{\partial x_k} \frac{\partial \Phi}{\partial u_\lambda} + \sum_{\lambda=1}^{q} \sum_{j=1}^{p} \frac{\partial^2 u_\lambda}{\partial x_j \partial x_k} \frac{\partial \Phi}{\partial u_{\lambda,j}}$$

$$+ \sum_{\lambda=1}^{q} \sum_{j_1 \leq j_2 = 1}^{p} \frac{\partial^3 u_\lambda}{\partial x_{j_1} \partial x_{j_2} \partial x_k} \frac{\partial \Phi}{\partial u_{\lambda, j_1 j_2}}$$

$$+ \cdots + \sum_{\lambda=1}^{q} \sum_{j_1 \leq j_2 \leq \cdots \leq j_N = 1}^{p} \frac{\partial^{n+1} u_\lambda}{\partial x_{j_1} \partial x_{j_2} \cdots \partial x_{j_n} \partial x_k} \frac{\partial \Phi}{\partial u_{\lambda, j_1 j_2 \cdots j_n}}, \qquad (7.53)$$

where

$$u_{\lambda,j} \equiv \frac{\partial u_\lambda}{\partial x_j}, \quad u_{\lambda,j_1 j_2} \equiv \frac{\partial^2 u_\lambda}{\partial x_{j_1} \partial x_{j_2}}, \quad \ldots, \quad u_{\lambda,j_1 j_2 \cdots j_n} \equiv \frac{\partial^n u_\lambda}{\partial x_{j_1} \partial x_{j_2} \cdots \partial x_{j_n}}.$$

In a similar way,

$$\frac{\partial^2 u_m^*}{\partial x_j^* x_k^*} = \frac{\partial^2 u_m}{\partial x_j x_k} + \varepsilon \phi_m^{[jk]}(x, u, u^{(1)}, u^{(2)}) + O(\varepsilon^2), \quad (7.54a)$$

for $m = 1, 2, \ldots, q$ and $j, k = 1, 2, \ldots, p$, where the *second extension* $\phi_m^{[jk]}(x, u, u^{(1)}, u^{(2)})$ is given by the recursion formula

$$\phi_m^{[jk]}(x, u, u^{(1)}, u^{(2)}) = \frac{D \phi_m^{[j]}}{D x_k} - \sum_{l=1}^{p} \frac{D \xi_l}{D x_k} \frac{\partial^2 u_m}{\partial x_j x_l}. \quad (7.54b)$$

In general, the nth order partial derivatives transform according to

$$\frac{\partial^n u_m^*}{\partial x_{j_1}^* x_{j_2}^* \cdots x_{j_n}^*} = \frac{\partial^n u_m}{\partial x_{j_1} \partial x_{j_2} \cdots \partial x_{j_n}} + \varepsilon \phi_m^{[j_1 j_2 \cdots j_n]}(x, u, u^{(1)}, \ldots, u^{(n)}) + O(\varepsilon^2), \quad (7.55a)$$

for $m = 1, 2, \ldots, q$ and $j_r = 1, 2, \ldots, p$, $r = 1, 2, \ldots, n$, where the nth extension $\phi_m^{[j_1 j_2 \cdots j_n]}(x, u, u^{(1)}, \ldots, u^{(n)})$ is given by the recursion formula

$$\phi_m^{[j_1 j_2 \cdots j_n]}(x, u, u^{(1)}, \ldots, u^{(n)}) = \frac{D \phi_m^{[j_1 j_2 \cdots j_{n-1}]}}{D x_{j_n}}$$

$$- \sum_{l=1}^{p} \frac{D \xi_l}{D x_{j_n}} \frac{\partial^n u_m}{\partial x_{j_1} \partial x_{j_2} \cdots \partial x_{j_{n-1}} \partial x_l}. \quad (7.55b)$$

Now consider the system of equations

$$\Delta(x^*, u^*, u^{*(1)}(x^*), \ldots, u^{*(N)}(x^*)) = 0, \quad (7.56)$$

which is eqn (7.50) with u replaced by u^* and x by x^*. It is easily seen that

$$\Delta(x^*, u^*, u^{*(1)}(x^*), \ldots, u^{*(N)}(x^*))$$
$$= \Delta(x, u, u^{(1)}(x), \ldots, u^{(N)}(x)) + \varepsilon \operatorname{pr}^{(N)} \mathbf{v}(\Delta) + O(\varepsilon^2),$$

where

$$\operatorname{pr}^{(N)} \mathbf{v} \equiv \sum_{j=1}^{p} \xi_j \frac{\partial}{\partial x_j} + \sum_{m=1}^{q} \phi_m \frac{\partial}{\partial u_m}$$

$$+ \sum_{m=1}^{q} \sum_{j=1}^{p} \phi_m^{[j]} \frac{\partial}{\partial u_{m,j}} + \sum_{m=1}^{q} \sum_{j_1 \leq j_2 = 1}^{p} \phi_m^{[j_1 j_2]} \frac{\partial}{\partial u_{m,j_1 j_2}}$$

$$+ \cdots + \sum_{m=1}^{q} \sum_{j_1 \leq j_2 \leq \cdots \leq j_N = 1}^{p} \phi_m^{[j_1 j_2 \cdots j_N]} \frac{\partial}{\partial u_{m,j_1 j_2 \cdots j_N}}, \quad (7.57)$$

is known as the *N*th *prolongation* (or *N*th *extension*) of the infinitesimal operator

$$\mathbf{v} \equiv \sum_{j=1}^{p} \xi_j(x,u)\frac{\partial}{\partial x_j} + \sum_{m=1}^{q} \phi_m(x,u)\frac{\partial}{\partial u_m}, \qquad (7.58)$$

and

$$u_{m,j_1j_2\cdots j_n} \equiv \frac{\partial^n u_m}{\partial x_{j_1}\partial x_{j_2}\cdots \partial x_{j_n}}.$$

The expression for $\mathrm{pr}^{(N)}\mathbf{v}$ is long and cumbersome. However, it is entirely algorithmic and explicit; moreover, it is the basic tool in the calculation of symmetry groups.

Requiring that eqn (7.50) is invariant under the transformation (7.51), i.e.

$$\mathrm{pr}^{(N)}\mathbf{v}(\Delta)|_{\Delta=0} = 0, \qquad (7.59)$$

yields an overdetermined, linear system of equations for the infinitesimals $\xi(x,u)$ and $\phi(x,u)$. This means that the Nth prolongation of Δ_m, $m = 1, 2, \ldots, q$, is zero whenever u is a solution of the original PDE (7.50). The symmetry solution is then obtained by solving the q invariant surface conditions

$$\sum_{j=1}^{p} \xi_j(x,u)\frac{\partial u_m}{\partial x_j} = \phi_m(x,u), \qquad m = 1,2,\ldots,q. \qquad (7.60)$$

or, equivalently, by solving the associated characteristic equations

$$\mathrm{d}x^*/\mathrm{d}\varepsilon = \xi(x^*,u^*), \qquad \mathrm{d}u^*/\mathrm{d}\varepsilon = \phi(x^*,u^*),$$

subject to the initial conditions $x^*|_{\varepsilon=0} = x$ and $u^*|_{\varepsilon=0} = u$.

EXAMPLE 7.14. *Boussinesq equation.* In this example we derive the classical symmetry reductions of the Boussinesq equation

$$u_{tt} + uu_{xx} + u_x^2 + u_{xxxx} = 0, \qquad (7.61)$$

which is a soliton equation solvable by inverse scattering [79–83] and arises in several physical applications including propagation of long waves in shallow water [75, 84–86], one-dimensional nonlinear lattice-waves [87, 88], vibrations in a nonlinear string [83] and ion sound waves in a plasma [89, 90].

To apply the classical Lie method to the Boussinesq equation (7.61), we require that the set of solutions of the equation is invariant under the transformation

$$x^* = x + \varepsilon \xi(x,t,u) + O(\varepsilon^2),$$
$$t^* = t + \varepsilon \tau(x,t,u) + O(\varepsilon^2),$$
$$u^* = u + \varepsilon \phi(x,t,u) + O(\varepsilon^2).$$

This yields a system of 12 determining equations for the infinitesimals which have the general solution

$$\xi = \alpha x + \beta, \qquad \tau = 2\alpha t + \gamma, \qquad \phi = -2\alpha u, \tag{7.62}$$

where α, β, and γ are arbitrary constants [91, 92] and the associated vector field is

$$\mathbf{v} = (\alpha x + \beta)\partial_x + (2\alpha t + \gamma)\partial_t - 2\alpha u \partial_u$$
$$= \alpha(x\partial_x + 2t\partial_t - 2u\partial_u) + \beta\partial_x + \gamma\partial_t.$$

The generators of this are

$\mathbf{v}_1 = \partial_x,$ space translational,
$\mathbf{v}_2 = \partial_t,$ time translational,
$\mathbf{v}_3 = x\partial_x + 2t\partial_t - 2u\partial_u,$ scaling or dilational.

Consequently, there are two canonical (classical) symmetry reductions.

Case 7.14(i): $\alpha = 0$. In this case we set $\gamma = 1$ and obtain the travelling wave reduction

$$u(x,t) = w(z), \qquad z = x - \beta t, \tag{7.63}$$

where $w(z)$ satisfies

$$w'''' + ww'' + (w')^2 + \beta^2 w'' = 0, \tag{7.64}$$

with $' := d/dz$. Integrating this with respect to z twice yields

$$w'' + \beta^2 w + \tfrac{1}{2}w^2 = Az + B,$$

with A and B arbitrary constants, which is solvable in terms of the first Painlevé equation PI (if $A \neq 0$) and elliptic or elementary functions (if $A = 0$).

Case 7.14(ii): $\alpha = 1$. In this case we set $\beta = \gamma = 0$ and obtain the scaling reduction

$$u(x,t) = t^{-1}w(z), \qquad z = x/t^{1/2}, \tag{7.65}$$

where $w(z)$ satisfies

$$w'''' + ww'' + (w')^2 + \tfrac{1}{4}z^2 w'' + \tfrac{7}{4}zw' + 2w = 0, \tag{7.66}$$

This equation is solvable in terms of the fourth Painlevé equation PIV [93].

EXAMPLE 7.15. *Nonlinear Schrödinger equation.* In this example we derive the classical symmetry reductions of the nonlinear Schrödinger equation

$$iu_t + u_{xx} = 2|u|^2 u, \qquad (7.67)$$

which was the second soliton equation to be solved by the inverse scattering method [99]. The nonlinear Schrödinger equation (7.67) arises in several physical applications, including: the evolution of an envelope of weakly nonlinear dispersive water waves [60, 89, 95]; the model of the coupling of Langmuir oscillations to ion acoustic waves in a plasma [100, 101]; the propagation of pulses in optical fibres [102–106]; the description of energy transport along alpha-helix proteins [107]; and quantum field theories [108, 109]. In the context of the MCS developed in Chapter 5, the NLS equation figures prominently as the field's equation of motion at the zeroth order. It appeared in a number of specific applications of the MCS developed in Chapter 6. The reader is also referred to Section 1.7 for a more general introduction.

Due to the nonanalyticity of the nonlinear term in the nonlinear Schrödinger equation (7.67), we set $v = \bar{u}$, the formal complex conjugate of u, and consider the coupled system

$$iu_t + u_{xx} = 2u^2 v, \qquad -iv_t + v_{xx} = 2v^2 u. \qquad (7.68a,b)$$

Alternatively, we could have set $u = p + iq$ and considered the coupled system

$$-q_t + p_{xx} = 2(p^2 + q^2)p, \qquad p_t + q_{xx} = 2(p^2 + q^2)q.$$

Requiring that the system (7.68) is invariant under the infinitesimal transformation

$$x^* = x + \varepsilon \xi(x,t,u,v) + O(\varepsilon^2), \qquad t^* = t + \varepsilon \tau(x,t,u,v) + O(\varepsilon^2),$$
$$u^* = u + \varepsilon \eta(x,t,u,v) + O(\varepsilon^2), \qquad v^* = v + \varepsilon \phi(x,t,u,v) + O(\varepsilon^2),$$

yields a system of 14 determining equations for the infinitesimals ξ, τ, η, and ϕ. The solution of this system is

$$\xi = \alpha x + 2i\beta t + \gamma, \qquad \tau = 2\alpha t + \delta, \qquad \eta = (\beta x - \alpha + \kappa)u,$$
$$\phi = -(\beta x + \alpha + \kappa)v, \qquad (7.69)$$

ANALYSIS OF DIFFERENTIAL EQUATIONS 429

and the associated vector fields are

$\mathbf{v}_1 = \partial_x,$ space translation,
$\mathbf{v}_2 = \partial_t,$ time translation,
$\mathbf{v}_3 = x\partial_x + 2t\partial_t - u\partial_u - v\partial_v,$ scaling or dilation,
$\mathbf{v}_4 = 2it\partial_x + xu\partial_u - xv\partial_v,$ Galilean boost,
$\mathbf{v}_5 = u\partial_u - v\partial_v,$ constant change of phase

(cf. references [52, 110]).

Case 7.15(i): $\alpha = \beta = \kappa = 0$, $\delta = 1$. In this case we obtain the travelling wave reduction

$$u(x,t) = f(z), \qquad v(x,t) = g(z), \qquad z = x - \gamma t, \qquad (7.70)$$

where $f(z)$ and $g(z)$ satisfy

$$f'' = 2f^2 g + i\gamma f', \qquad g'' = 2fg^2 - i\gamma g'. \qquad (7.71a,b)$$

Setting $f(z) = R(z)\exp\{i\theta(z)\}$ and $g(z) = R(z)\exp\{-i\theta(z)\}$ and equating real and imaginary parts yields

$$R'' - R(\theta')^2 = 2R^3 - \gamma R\theta', \qquad 2R'\theta' + R\theta'' = \gamma R'. \qquad (7.72a,b)$$

Multiplying eqn (7.72b) by R and integrating yields $\theta'(z) = \tfrac{1}{2}\gamma + C/R^2$, where C is an arbitrary constant. Substituting this into eqn (7.72a) and setting $R = w^{1/2}$ yields

$$w'' = \frac{(w')^2}{2w} + 4w^2 - \tfrac{1}{2}\gamma^2 w + \frac{2C^2}{w}. \qquad (7.73)$$

After rescaling w and z, this equation is equation XXX in Ince [61, Chapter 14] which is solvable in terms of elliptic functions.

Case 7.15(ii): $\alpha = 1$, $\beta = \gamma = \delta = 0$, $\kappa = -i\mu$. In this case we obtain the scaling reduction

$$u(x,t) = t^{-1/2}f(z)\exp(-\tfrac{1}{2}i\mu \ln t), \qquad v(x,t) = t^{-1/2}g(z)\exp(\tfrac{1}{2}i\mu \ln t),$$

$$z = x/t^{1/2}, \qquad (7.74)$$

where $f(z)$ and $g(z)$ satisfy

$$f'' = 2f^2 g + \tfrac{1}{2}(i - \mu)f + \tfrac{1}{2}izf', \qquad g'' = 2fg^2 - \tfrac{1}{2}(i + \mu)g - \tfrac{1}{2}izg'. \qquad (7.75a,b)$$

Setting $f(z) = R(z)\exp\{i\theta(z)\}$ and $g(z) = R(z)\exp\{-i\theta(z)\}$ and equating real and imaginary parts yields

$$R'' - (\theta')^2 R + \tfrac{1}{2}z\theta'R + \tfrac{1}{2}\mu R = 2R^3, \qquad 2R'\theta' + R\theta'' = \tfrac{1}{2}zR' + \tfrac{1}{2}R. \tag{7.76a,b}$$

Multiplying eqn (7.76b) by R and integrating yields

$$\theta' = \tfrac{1}{4}z + \frac{C}{R^2} + \frac{1}{4R^2}\int^z R^2(z_1)\,dz_1,$$

where C is an arbitrary constant. Substituting this into eqn (7.76a) and setting $W(z) = \int^z R^2(z_1)\,dz_1$ yields the third-order equation

$$2W'W''' - (W'')^2 + \left(\tfrac{1}{4}z^2 + 2\mu\right)(W')^2 - \tfrac{1}{4}W^2 - 8(W')^3 - 4C^2 - 2CW = 0. \tag{7.77}$$

This equation is solvable in terms of the fourth Painlevé equation PIV [111] (see also references [112, 113]).

Case 7.15(iii): $\alpha = 0$, $\beta = -i\mu \neq 0$, $\kappa = \gamma = 0$, $\delta = 1$. In this case we obtain the accelerating wave reduction

$$u(x,t) = \exp\{i\mu[zt + \tfrac{1}{3}\mu t^3]\}f(z), \tag{7.78a}$$

$$v(x,t) = \exp\{-i\mu[zt + \tfrac{1}{3}\mu t^3]\}g(z), \tag{7.78b}$$

$$z = x - \mu t^2, \tag{7.78c}$$

where $f(z)$ and $g(z)$ satisfy

$$f'' = 2f^2 g + \mu z f, \qquad g'' = 2fg^2 + \mu z g. \tag{7.79a,b}$$

Setting $f(z) = R(z)\exp\{i\theta(z)\}$ and $g(z) = R(z)\exp\{-i\theta(z)\}$ and equating real and imaginary parts yields

$$R'' - R(\theta')^2 = 2R^3 + \mu z R, \qquad 2R'\theta' + R\theta'' = 0. \tag{7.80a,b}$$

Multiplying eqn (7.80b) by R and integrating yields $\theta' = C/R^2$, where C is an arbitrary constant. Substituting this into eqn (7.80a) and setting $R = w^{1/2}$ yields

$$w'' = \frac{(w')^2}{2w} + 4w^2 + 2\mu z w + \frac{2C^2}{w}. \tag{7.81}$$

After rescaling w and z, this equation is equation XXXIV in Ince [61, Chapter 14] which is solvable in terms of the second Painlevé equation PII.

EXAMPLE 7.16. *Generalized nonlinear Schrödinger equation.* In this example we discuss symmetry reductions of the generalized nonlinear Schrödinger (GNLS) equation

$$iu_t + u_{xx} + (a_1 + ia_2)(|u|^2 u)_x + (b_1 + ib_2)u(|u|^2)_x + c|u|^4 u + d|u|^2 u = 0, \quad (7.82)$$

where a_1, a_2, b_1, b_2, c, and d are real constants. There are several special cases of this equation which have been studied previously: (i) if $a_1 = a_2 = b_1 = b_2 = c = 0$, then eqn (7.82) is the cubic nonlinear Schrödinger equation which is solvable by inverse scattering [99]; (ii) if $a_1 = a_2 = b_1 = b_2 = 0$, $c \neq 0$, then eqn (7.82) is the quintic nonlinear Schrödinger equation, which appears to be nonintegrable [114]; (iii) if $a_1 = b_1 = d = 0$, then eqn (7.82) is a generalized derivative nonlinear Schrödinger equation, which is completely integrable if and only if $c = \frac{1}{4}b_2(a_2 + b_2)$ [114], when it is known as the derivative nonlinear Schrödinger equation and is solvable by inverse scattering [115]; (iv) if $a_1 = b_1 = b_2 = c = 0$, then eqn (7.82) is the mixed nonlinear Schrödinger equation, which also is solvable by inverse scattering [116, 117]; (v) if $a_1 = a_2 = b_2 = d = 0$, $c = \frac{1}{2}b_1^2$, then eqn (7.82) is the Eckhaus equation, which is linearizable [118, 119].

The GNLS equation (7.82) also arises in several physical applications including: quantum field theory [120–122]; weakly nonlinear dispersive water waves [123–125]; and nonlinear optics [102–105, 126, 127]. We note that a number of subcases of eqn (7.82) appear in the first order of the MCS (see Section 5.6.3).

To determine symmetry reductions of eqn (7.82), we write it as the system

$$iu_t + u_{xx} + (a_1 + ia_2)(u^2 v)_x + (b_1 + ib_2)u(uv)_x + cu^3 v^2 + du^2 v = 0, \quad (7.83a)$$

$$-iv_t + v_{xx} + (a_1 - ia_2)(uv^2)_x + (b_1 - ib_2)v(uv)_x + cu^2 v^3 + duv^2 = 0. \quad (7.83b)$$

To apply the classical Lie group method we require that eqn (7.83) is invariant under the infinitesimal transformation

$$x^* = x + \varepsilon\xi(x,t,u,v) + O(\varepsilon^2), \quad t^* = t + \varepsilon\tau(x,t,u,v) + O(\varepsilon^2),$$
$$u^* = u + \varepsilon\eta(x,t,u,v) + O(\varepsilon^2), \quad v^* = v + \varepsilon\phi(x,t,u,v) + O(\varepsilon^2).$$

This yields a system of 16 determining equations, the solution of which yields the infinitesimals

$$\xi = 4\alpha_6 xt - 2i\alpha_5 t + 2\alpha_4 x + \alpha_1, \qquad \tau = 4\alpha_6 t^2 + 4\alpha_4 t + \alpha_2,$$

$$\eta = \left[-\alpha_6(2t - ix^2) + \alpha_5 x - \alpha_4 + \alpha_3\right]u,$$

$$\phi = -\left[\alpha_6(2t + ix^2) + \alpha_5 x + \alpha_4 + \alpha_3\right]v,$$

where $\alpha_1, \alpha_2, \ldots, \alpha_6$ are constants such that $d\alpha_4 = 0$, $\alpha_5 a_1 = \alpha_5 a_2 = 0$, and $\alpha_6 a_1 = \alpha_6 a_2 = \alpha_6 d = 0$ [128, 129]. The associated vector fields are

$\mathbf{v}_1 = \partial_x,$ \qquad x-translation,

$\mathbf{v}_2 = \partial_t,$ \qquad t-translation,

$\mathbf{v}_3 = u\partial_u - v\partial_v,$ \qquad constant change of phase,

$\mathbf{v}_4 = 2x\partial_x + 4t\partial_t - u\partial_u - v\partial_v,$ \qquad scaling (for $d = 0$),

$\mathbf{v}_5 = -2it\partial_x + xu\partial_u - xv\partial_v,$ \qquad Galilean boost (for $a_1 = a_2 = 0$),

$\mathbf{v}_6 = 4xt\partial_x + 4t^2\partial_t - (2t - ix^2)u\partial_u - (2t + ix^2)v\partial_v,$

conformal point symmetry (for $a_1 = a_2 = d = 0$).

These yield six types of classical symmetry reductions.

Case 7.16(i): $\alpha_4 = \alpha_5 = \alpha_6 = 0$. In this case we set $\alpha_1 = \lambda$, $\alpha_2 = -1$, and $\alpha_3 = i(\mu - \tfrac{1}{2}\lambda^2)$, without loss of generality, and obtain the travelling wave reduction

$$u(x,t) = R(z)\exp\{i[\theta(z) - (\tfrac{1}{2}\lambda x + \mu t)]\}, \qquad (7.84a)$$

$$v(x,t) = R(z)\exp\{-i[\theta(z) - (\tfrac{1}{2}\lambda x + \mu t)]\}, \qquad (7.84b)$$

where $z = x + \lambda t$, and $R(z)$ and $\theta(z)$ satisfy

$$R'' + (2b_1 + 3a_1)R^2 R' + cR^5 + (d + \tfrac{1}{2}\lambda a_2)R^3$$
$$- R(\theta')^2 - a_2 R^3\theta' + (\mu - \tfrac{1}{4}\lambda^2)R = 0, \qquad (7.85a)$$

$$R\theta'' + 2R'\theta' + (2b_2 + 3a_2)R^2 R' + a_1 R^3\theta' - \tfrac{1}{2}\lambda a_1 R^3 = 0. \qquad (7.85b)$$

If $a_1 = 0$, then multiplying eqn (7.85b) by R and integrating gives

$$\theta' = -(\tfrac{1}{2}b_2 + \tfrac{3}{4}a_2)R^2 - \gamma/R^2,$$

where γ is an arbitrary constant. Substituting this into eqn (7.85a) and setting $R = \sqrt{w}$ yields

$$ww'' = \tfrac{1}{2}(w')^2 - 2b_1 w^2 w' + \left(\tfrac{1}{2}b_2^2 + \tfrac{1}{2}b_2 a_2 - \tfrac{3}{8}a_2^2 - 2c\right)w^4$$
$$- (2d + \lambda a_2)w^3 - \left[2\mu - \tfrac{1}{2}\lambda^2 - (2b_2 + a_2)\gamma\right]w^2 + 2\gamma^2. \quad (7.86)$$

If $b_1 \neq 0$, $c = \tfrac{1}{4}(b_1^2 + b_2^2 + a_2 b_2) - \tfrac{3}{16}a_2^2$, and $d = -\tfrac{1}{2}\lambda a_2$, then eqn (7.86) is linearizable; if $b_1 \neq 0$, $c = \tfrac{1}{4}b_2(a_2 + b_2) - \tfrac{3}{16}a_2^2 - b_1^2$, $d = -\tfrac{1}{2}\lambda a_2$, $\mu = \tfrac{1}{4}\lambda^2$, and $\gamma = 0$, then it is solvable in terms of Weierstrass elliptic functions; while if $b_1 = 0$, then it is solvable in terms of Jacobi and Weierstrass elliptic functions. Otherwise, it is not of Painlevé type.

Case 7.16(ii): $a_1 = a_2 = 0$, $\alpha_4 = \alpha_6 = 0$, $\alpha_5 \neq 0$. In this case we set $\alpha_1 = \alpha_2 = 0$, $\alpha_3 = i\mu$ and $\alpha_5 = i\kappa$, without loss of generality, and obtain the accelerating wave reduction

$$u(x,t) = R(z)\exp\{i[\theta(z) - (\kappa xt + \tfrac{2}{3}\kappa^2 t^3 + \mu t)]\}, \quad (7.87a)$$
$$v(x,t) = R(z)\exp\{-i[\theta(z) - (\kappa xt + \tfrac{2}{3}\kappa^2 t^3 + \mu t)]\}, \quad (7.87b)$$

where $z = x + \kappa t^2$, $\kappa \neq 0$ is an arbitrary constant and $R(z)$ and $\theta(z)$ satisfy

$$R'' + 2b_1 R^2 R' + cR^5 + dR^3 - R(\theta')^2 + (\kappa z + \mu)R = 0, \quad (7.88a)$$
$$R\theta'' + 2R'\theta' + 2b_2 R^2 R' = 0. \quad (7.88b)$$

Multiplying eqn (7.88b) by R and integrating gives

$$\theta' = -\tfrac{1}{2}b_2 R^2 - \gamma/R^2,$$

where γ is an arbitrary constant. Substituting this into eqn (7.88a) and setting $R = \sqrt{w}$ yields

$$ww'' = \tfrac{1}{2}(w')^2 - 2b_1 w^2 w' + \left(\tfrac{1}{2}b_2^2 - 2c\right)w^4$$
$$- 2dw^3 - [2\kappa z + 2\mu - 2b_2\gamma]w^2 + 2\gamma^2. \quad (7.89)$$

If $b_1 = 0$, $c = \tfrac{1}{4}b_2^2 \neq 0$, and $d \neq 0$, then eqn (7.89) is solvable in terms of the second Painlevé equation PII, while if $b_1 \neq 0$, $c = \tfrac{1}{4}(b_1^2 + b_2^2)$, and $d = 0$ it is solvable in terms of a linear equation. Otherwise, it is not of Painlevé type.

Case 7.16(iii): $d = 0$, $\alpha_4 \neq 0$, $\alpha_6 = 0$. In this case we set $\alpha_1 = \alpha_2 = 0$,

$\alpha_3 = -4i\mu$, $\alpha_4 = 1$, and $\alpha_5 = 0$, without loss of generality, and obtain the scaling reduction

$$u(x,t) = t^{-1/4} R(z) \exp\{i[\theta(z) + \tfrac{1}{8}z^2 + \mu \ln t]\}, \qquad (7.90a)$$

$$v(x,t) = t^{-1/4} R(z) \exp\{-i[\theta(z) + \tfrac{1}{8}z^2 + \mu \ln t]\}, \qquad (7.90b)$$

where $z = x/t^{1/2}$, and $R(z)$ and $\theta(z)$ satisfy

$$R'' + (2b_1 + 3a_1)R^2 R' + cR^5 - R(\theta')^2$$
$$\quad - a_2 R^3(\theta' + \tfrac{1}{4}z) + (\mu - \tfrac{1}{16}z^2)R = 0, \qquad (7.91a)$$

$$R\theta'' + 2R'\theta' + (2b_2 + 3a_2)R^2 R' + a_1 R^3(\theta' + \tfrac{1}{4}z) = 0. \qquad (7.91b)$$

If $a_1 = 0$, then multiplying eqn (7.91b) by R and integrating gives

$$\theta' = -(\tfrac{1}{2}b_2 + \tfrac{3}{4}a_2)R^2 - \gamma/R^2,$$

where γ is an arbitrary constant. Substituting this into eqn (7.91a) and setting $R = \sqrt{w}$ yields

$$ww'' = \tfrac{1}{2}(w')^2 - 2b_1 w^2 w' + \left(\tfrac{1}{2}b_2^2 + \tfrac{1}{2}b_2 a_2 - \tfrac{3}{8}a_2^2 - 2c\right)w^4 + \tfrac{1}{2}a_2 z w^3$$
$$\quad + \left[2\mu + (2b_2 + a_2)\gamma - \tfrac{1}{8}z^2\right]w^2 + 2\gamma^2. \qquad (7.92)$$

If $b_1 = 0$ and $c = \tfrac{1}{4}b_2(a_2 + b_2)$, then eqn (7.92) is solvable in terms of the fourth Painlevé equation PIV, while if $b_1 \neq 0$, $a_2 = 0$, and $c = \tfrac{1}{4}(b_1^2 + b_2^2)$, then it is solvable in terms of a linear equation. Otherwise, it is not of Painlevé type.

Case 7.16(iv): $a_1 = a_2 = d = 0$, $\alpha_2 = 0$, $\alpha_6 \neq 0$. In this case we set $\alpha_1 = -2\beta$, $\alpha_3 = \gamma$, $\alpha_4 = 0$, $\alpha_5 = 0$, and $\alpha_6 = \tfrac{1}{4}$, and obtain the reduction

$$u(x,t) = t^{-1/2} R(z) \exp\{i[\theta(z) + \Phi(x,t)]\}, \qquad (7.93a)$$

$$v(x,t) = t^{-1/2} R(z) \exp\{-i[\theta(z) + \Phi(x,t)]\}, \qquad (7.93b)$$

where

$$z = \frac{xt - \beta}{t^2}, \qquad \Phi(x,t) = \tfrac{1}{4}z^2 t - \frac{\beta z + 2\gamma}{2t} - \frac{\beta^2}{12t^3}, \qquad (7.93c)$$

and $R(z)$ and $\theta(z)$ satisfy

$$R'' + 2b_1 R^2 R' + cR^5 - R(\theta')^2 - (\beta z + \gamma)R = 0, \qquad (7.94a)$$

$$R\theta'' + 2R'\theta' + 2b_2 R^2 R' = 0. \qquad (7.94b)$$

Multiplying eqn (7.94b) by R and integrating yields
$$\theta' = -\tfrac{1}{2}b_2 R^2 - \delta/R^2,$$
where δ is an arbitrary constant. Substituting this into eqn (7.94a) and setting $R = \sqrt{w}$ yields
$$ww'' = \tfrac{1}{2}(w')^2 - 2b_1 w^2 w' - 2(c - \tfrac{1}{4}b_2^2)w^4 + 2(\beta z + \gamma + b_2\delta)w^2 + 2\delta^2. \quad (7.95)$$

If $b_1 \neq 0$ and $c = \tfrac{1}{4}(b_1^2 + b_2^2)$, then eqn (7.95) is linearizable; if $b_1 \neq 0$, $c = \tfrac{1}{4}b_2^2 - b_1^2$, and $\beta = \gamma = \delta = 0$, then it is solvable in terms of Weierstrass elliptic functions; while if $b_1 = 0$, $c = \tfrac{1}{4}b_2^2$, and $\delta = 0$, then $R = \sqrt{w}$ satisfies the Airy equation (for $\beta \neq 0$)
$$R'' - (\beta z + \gamma)R = 0.$$
Otherwise, it is not of Painlevé type.

Case 7.16(v): $a_1 = a_2 = d = 0$, $\alpha_5 \neq 0$, $\alpha_6 = 0$. In this case we set $\alpha_1 = \alpha_2 = \alpha_3 = 0$, $\alpha_4 = 1$, and $\alpha_5 = i\lambda$, and obtain the reduction
$$u(x,t) = t^{-1/4} R(z)\exp\{i[\theta(z) + \Phi(x,t)]\}, \quad (7.96a)$$
$$v(x,t) = t^{-1/4} R(z)\exp\{-i[\theta(z) + \Phi(x,t)]\}, \quad (7.96b)$$
where
$$z = (x + \lambda t)/t^{1/2}, \quad \Phi(x,t) = \tfrac{1}{8}z^2 - \tfrac{1}{2}\lambda z t^{1/2} + \tfrac{1}{4}\lambda^2 t + \gamma \ln t. \quad (7.96c)$$
Note that if $\lambda = 0$ then this reduces to Case 7.16(iii) above; in fact, the case $\lambda \neq 0$ can be obtained using a Galilean boost of the case when $\lambda = 0$.

Case 7.16(vi): $a_1 = a_2 = d = 0$, $\alpha_2 \neq 0$, $\alpha_6 \neq 0$. In this case we set $\alpha_1 = -\lambda$, $\alpha_2 = 1$, $\alpha_3 = \gamma$, $\alpha_4 = 0$, $\alpha_5 = 0$, and $\alpha_6 = \tfrac{1}{4}$, and obtain the reduction
$$u(x,t) = (t^2 + 1)^{-1/4} R(z)\exp\{i[\theta(z) + \Phi(x,t)]\}, \quad (7.97a)$$
$$v(x,t) = (t^2 + 1)^{-1/4} R(z)\exp\{-i[\theta(z) + \Phi(x,t)]\}, \quad (7.97b)$$
where
$$z = \frac{x + \lambda t}{(t^2 + 1)^{1/2}}, \quad \Phi(x,t) = \tfrac{1}{4}(z^2 + \lambda^2)t - \tfrac{1}{2}\lambda z(t^2 + 1)^{1/2} + \gamma \tan^{-1} t. \quad (7.97c)$$

In Cases 7.16(v) and 7.16(vi), $R(z)$ and $\theta(z)$ satisfy
$$R'' + 2b_1 R^2 R' + cR^5 - R(\theta')^2 - (\alpha z^2 + \gamma)R = 0, \quad (7.98a)$$
$$R\theta'' + 2R'\theta' + 2b_2 R^2 R' = 0. \quad (7.98b)$$

with $\alpha = -\tfrac{1}{16}$ for Case 7.16(ii), and $\alpha = \tfrac{1}{4}$ for Case 7.16(iii). Multiplying eqn (7.98b) by R and integrating yields

$$\theta' = -\tfrac{1}{2}b_2 R^2 - \delta/R^2,$$

where δ is an arbitrary constant. Substituting this into eqn (7.98a) and setting $R = \sqrt{w}$ yields

$$ww'' = \tfrac{1}{2}(w')^2 - 2b_1 w^2 w' - 2(c - \tfrac{1}{4}b_2^2)w^4 + 2(\alpha z^2 + \gamma - b_2 \delta)w^2 + 2\delta^2, \tag{7.99}$$

where δ is an arbitrary constant. If $b_1 \neq 0$ and $c = \tfrac{1}{4}(b_1^2 + b_2^2)$, then eqn (7.98) is linearizable, while if $b_1 = 0$, $c = \tfrac{1}{4}b_2^2$, and $\delta = 0$, then $R = \sqrt{w}$ satisfies

$$R'' - (\alpha z^2 + \gamma)R = 0. \tag{7.100}$$

Otherwise, it is not of Painlevé type. Equation (7.100) is solvable in terms of the parabolic cylinder function equation $D_\nu(z)$, which is defined to be the solution of

$$\frac{d^2 D_\nu}{dz^2} = \left(\tfrac{1}{4}z^2 - \nu - \tfrac{1}{2}\right)D_\nu(z),$$

satisfying

$$D_\nu(z) \sim z^\nu \exp\left(-\tfrac{1}{4}z^2\right), \qquad \text{as } z \to +\infty,$$

$$D_\nu(z) \sim \frac{\sqrt{2\pi}}{\Gamma(-\nu)} e^{i\pi\nu} z^{-\nu-1} \exp\left(\tfrac{1}{4}z^2\right), \qquad \text{as } z \to -\infty,$$

provided that ν is not an integer (cf. reference [30]). Approximate solutions of eqn (7.98) with $b_1 = 0$, $\delta = 0$, and $c \neq \tfrac{1}{4}b_2^2$ are obtained by Florjańczyk and Gagnon [129] using the Rayleigh–Ritz variational method, and have been employed in the elucidation of self-focusing phenomena in optical fibres (see also Section 1.8).

REMARKS

1. The vector fields associated with classical symmetry reductions for the higher-order nonlinear Schrödinger equation

$$iu_t + u_{xx} + \kappa |u|^{2n} u = 0, \tag{7.101}$$

provided that $n \neq 2$ and $n \neq 0$, are

$\mathbf{v}_1 = \partial_x,$ space translation,
$\mathbf{v}_2 = \partial_t,$ time translation,
$\mathbf{v}_3 = nx\partial_x + 2nt\partial_t - u\partial_u - v\partial_v,$ scaling or dilation,
$\mathbf{v}_4 = 2it\partial_x - xu\partial_u + xv\partial_v,$ Galilean boost,
$\mathbf{v}_5 = u\partial_u - v\partial_v,$ constant change of phase.

If $n = 2$ there is one additional vector field,

$$\mathbf{v}_6 = xt\partial_x + t^2\partial_t + \left(\tfrac{1}{2}t + \tfrac{1}{4}ix^2\right)u\partial_u + \left(\tfrac{1}{2}t - \tfrac{1}{4}ix^2\right)v\partial_v, \qquad (7.102)$$

which corresponds to the conformal point symmetry. The case $n = 2$ is the so-called 'critical dimension' for eqn (7.101). A short discussion on other properties of this equation can be found in Section 1.7.

2. The associated one-parameter transformation group with eqn (7.102) is

$$\hat{x} = \frac{x}{1 - \varepsilon t}, \qquad \hat{u} = u(1 - \varepsilon t)^{1/2} \exp\left\{\frac{i\varepsilon x^2}{4(1 - \varepsilon t)}\right\},$$

$$\hat{t} = \frac{t}{1 - \varepsilon t}, \qquad \hat{v} = v(1 - \varepsilon t)^{1/2} \exp\left\{-\frac{i\varepsilon x^2}{4(1 - \varepsilon t)}\right\},$$

which is known as the *Talanov lens transformation* [131]. We further remark that the Talanov lens transformation (or conformal point symmetry) has been discussed by several authors in connection with reductions of the $(2 + 1)$-dimensional, cubic nonlinear Schrödinger equation [132, 133],

$$iu_t + u_{xx} + u_{yy} + \kappa |u|^2 u = 0,$$

and more generally for the $(n + 1)$-dimensional, nonlinear Schrödinger equation,

$$iu_t + \sum_{j=1}^{n} \frac{\partial^2 u}{\partial x_j^2} + \kappa |u|^{2\nu} u = 0,$$

in the so-called critical dimension when $n\nu = 2$ [134–138].

3. The symmetry group, associated reductions, and exact solutions of the $(3 + 1)$-dimensional, nonlinear Schrödinger equation,

$$iu_t + u_{xx} + u_{yy} + u_{zz} = a_0 u + a_1 |u|^2 u + a_2 |u|^4 u,$$

where a_0, a_1, and a_2 are constants, are extensively studied in [111, 139–142]. An extensive discussion on this particular case and its applicability within the MCS method can be found in Section 5.6.2.

7.1.5 Extensions of the classical Lie group method

There have been several generalizations of the classical Lie group method for symmetry reductions. It has been known for several years that there do exist PDEs which possess symmetry reductions that are not obtained using the classical Lie group method. For example, as noted by several authors [91, 92, 143–145], the Boussinesq equation,

$$u_{tt} + uu_{xx} + u_x^2 + u_{xxxx} = 0, \qquad (7.103)$$

possesses the accelerating wave solution

$$u(x,t) = w(z) - 4\mu^2 t^2, \qquad z = x + \mu t^2. \qquad (7.104)$$

where μ is an arbitrary constant and $w(z)$ satisfies

$$w''' + ww' + 2\mu w = 8\mu^2 z + A,$$

with A an arbitrary constant, which is solvable in terms of the second Painlevé equation PII. Associated infinitesimals for this reduction are $\xi = 2\mu t$, $\tau = -1$, and $\phi = 8\mu^2 t$, which are clearly *not* a special case of the classical infinitesimals (recall eqn (7.62)), with associated vector field $\mathbf{v} = 2\mu t \partial_x - \partial_t + 8\mu^2 t \partial_u$.

Further examples are noted in references [146–148], although in these papers there was no attempt to derive the additional reductions using a generalization of the classical Lie group method.

Nonclassical method

Bluman and Cole [149], in their study of symmetry reductions of the linear heat equation, proposed the so-called nonclassical method of group-invariant solutions, in the sequel referred to as the *nonclassical method*, which we discuss further below (see also references [33, §2.10] and [150]). This technique is also known as the 'method of conditional symmetries' (cf. references [151, 152]) and the 'method of partial symmetries of the first type' (cf. reference [157]). This method involves considerably more algebra and associated calculations then the classical Lie method.

Suppose that we wish to apply the nonclassical method to the general second-order PDE

$$\Delta^{(2)}(u_{xx}, u_{xt}, u_{tt}, u_x, u_t, u, x, t) = 0, \qquad (7.105)$$

where u is the dependent variable and x and t the independent variables. Consider the one-parameter Lie group of infinitesimal transformations in (x,t,u) given by

$$x^* = x + \varepsilon \xi(x,t,u) + O(\varepsilon^2), \qquad (7.106a)$$
$$t^* = t + \varepsilon \tau(x,t,u) + O(\varepsilon^2), \qquad (7.106b)$$
$$u^* = u + \varepsilon \phi(x,t,u) + O(\varepsilon^2), \qquad (7.106c)$$

ANALYSIS OF DIFFERENTIAL EQUATIONS

where ε is the group parameter, with the associated invariant surface condition

$$\psi \equiv \xi(x,t,u)u_x + \tau(x,t,u)u_t - \phi(x,t,u) = 0 \quad (7.107)$$

and infinitesimal generator

$$\mathbf{v} \equiv \xi(x,t,u)\frac{\partial}{\partial x} + \tau(x,t,u)\frac{\partial}{\partial t} + \phi(x,t,u)\frac{\partial}{\partial u}. \quad (7.108)$$

In the classical method it is required that the infinitesimal transformation (7.106) leaves the set of solutions

$$\mathcal{S} = \{u(x,t): \Delta^{(2)}(u) = 0\} \quad (7.109)$$

invariant. The associated symmetry analysis for the second-order PDE (7.105) involves applying the prolongation $\mathrm{pr}^{(2)}\mathbf{v}$ as given by eqn (7.31) to eqn (7.105) and requiring that the resulting expression vanishes for $u \in \mathcal{S}$, i.e.

$$\mathrm{pr}^{(2)}\mathbf{v}(\Delta^{(2)})|_{\Delta^{(2)}=0} = 0. \quad (7.110)$$

Equating coefficients of derivatives of u to zero generates the determining equations.

In the nonclassical method it is required that the infinitesimal transformation (7.106) leaves invariant the set of simultaneous solutions of the PDE (7.105) and the invariant surface condition (7.107), where ξ, τ, and ϕ are the same as in the transformation (7.106), That is, the method requires that the subset of $\mathcal{S}$ given by

$$\mathcal{S}_\psi = \{u(x,t): \Delta^{(2)}(u) = 0, \psi(u) = 0\} \quad (7.111)$$

is invariant under the transformation (7.106). Thus 'nonclassical symmetries', or 'conditional symmetries' as they are sometimes called, of a system of PDEs $\Delta^{(2)}$ are transformations that leave only the subset S_ψ of the solution set S of the system invariant. Other solutions of S that are not in the subset of S_ψ are *not* necessarily transformed to the set S.

The standard method of applying the nonclassical method (e.g. as described in reference [152]) to the second-order PDE (7.105) involves applying the prolongation $\mathrm{pr}^{(2)}\mathbf{v}$ to the system of equations given by the PDE (7.105) and the invariant surface condition (7.107) and requiring that the resulting expressions vanish for $u \in \mathcal{S}_\psi$, i.e.

$$\mathrm{pr}^{(2)}\mathbf{v}(\Delta^{(2)})|_{\Delta^{(2)}=0,\,\psi=0} = 0, \quad \mathrm{pr}^{(2)}\mathbf{v}(\psi)|_{\Delta^{(2)}=0,\,\psi=0} = 0. \quad (7.112, 7.113)$$

It is easily shown that

$$\mathrm{pr}^{(2)}\mathbf{v}(\psi) = -(\xi_u u_x + \tau_u u_t - \phi_u)\psi,$$

which vanishes identically when $\psi = 0$, without imposing any conditions upon the functions ξ, τ, and ϕ. However, as shown by Clarkson and Mansfield [158], this procedure for applying the nonclassical method can create difficulties, in particular in the implementation of symbolic manipulation programs. These difficulties often arise for PDEs such as the linear wave equation $u_{tt} = u_{xx}$, which require the use of differential consequences of the invariant surface condition (eqn (7.107)). In reference [158], Clarkson and Mansfield developed an algorithm for calculating determining equations associated with the nonclassical method which requires significantly less computational time than that standardly used and avoids many of the difficulties commonly encountered.

There are two cases to consider: (i) $\tau \neq 0$; and (ii) $\tau = 0$ and $\xi \neq 0$.

Case (i): $\tau \neq 0$. In this case we set $\tau = 1$, without loss of generality, and then from the invariant surface condition (7.107) obtain

$$u_t = \phi - \xi u_x,$$
$$u_{xt} = \phi_x + \phi_u u_x - (\xi_x u_x + \xi_u u_x^2 + \xi u_{xx}),$$
$$u_{tt} = \phi_t + \phi_u u_t - \xi_t u_x - \xi_u u_x u_t - \xi u_{xt}$$
$$= \phi_t + \phi_u(\phi - \xi u_x) - \xi_t u_x - \xi_u u_x(\phi - \xi u_x)$$
$$- \xi[\phi_x + \phi_u u_x - (\xi_x u_x + \xi_u u_x^2 + \xi u_{xx})].$$

Hence, eliminating u_t, u_{xt}, and u_{tt} in eqn (7.105) yields

$$\tilde{\Delta}^{(2)}(u_{xx}, u_x, u, x, t; \xi, \phi) = 0, \tag{7.114}$$

which is essentially an ODE for u, with t a parameter. Now we apply the classical Lie algorithm to this equation, i.e. we require that it is invariant under the transformation (7.106) with $\tau = 1$, and then use eqn (7.114) to eliminate u_{xx}. The determining equations are obtained by equating coefficients of powers of u_x to zero.

The requirement that both the PDE (7.105) and invariant surface condition (7.107) are invariant under the transformation (7.106) yields a *nonlinear* overdetermined system of equations for the infinitesimals $\xi(x, t, u)$, $\tau(x, t, u)$ and $\phi(x, t, u)$, which appear in both the transformation (7.106) and the supplementary condition (7.107); this is in contrast to the classical method in which the system of determining equations is linear. The set of solutions is larger than for the classical method, since the number of determining equations is smaller. Furthermore, it should be noted that u_t and u_x are no longer treated as being independent, since they are related by eqn (7.107). Also, it should be emphasized that the

ANALYSIS OF DIFFERENTIAL EQUATIONS

associated vector fields do *not* form a vector space, still less a Lie algebra, since the invariant surface condition (7.107) depends upon the particular reduction. Olver and Rosenau [143, 144] suggest that, for some PDEs, the determining equations for the nonclassical method might be actually too difficult to solve explicitly. Furthermore, it is known that for some equations such as the linear heat equation, the nonclassical method does not yield any additional symmetry reductions to those obtained using the classical Lie method, though there are PDEs which possess symmetry reductions that are *not* obtained using the classical Lie group method.

Case (ii): $\tau = 0$, $\xi \neq 0$. In this case we set $\xi = 1$, without loss of generality. The invariant surface condition (7.107) simplifies to

$$u_x = \phi(x, t, u),$$

and so

$$u_{xt} = \phi_t + \phi_u u_t, \qquad u_{xx} = \phi_x + \phi_u u_x = \phi_x + \phi \phi_u.$$

Hence, eliminating u_x, u_{xt}, and u_{xx} in eqn (7.105) yields

$$\hat{\Delta}^{(2)}(u_{tt}, u_t, u, x, t; \xi, \phi) = 0, \qquad (7.115)$$

which essentially is an ODE for u, with x a parameter. Now we apply the classical Lie algorithm to this equation, i.e. we require that it is invariant under the transformation (7.106) with $\tau = 0$ and $\xi = 1$, and then use eqn (7.115) to eliminate u_{tt}. The determining equations are obtained by equating coefficients of powers of u_t to zero.

EXAMPLE 7.18. *Nonlinear heat equation*. In this example, we discuss nonclassical symmetry reductions of the nonlinear heat equation

$$u_t = u_{xx} + f(u), \qquad (7.116)$$

which has been referred to in earlier chapters as the time-dependent Landau–Ginzburg equation in critical phenomena applications (see Chapter 2, for example). This equation arises in several other important physical applications including microwave heating, where $f(u)$ is the rate of absorption of microwave energy (cf. references [159, 160]), in the theory of chemical reactions, where $f(u)$ is the temperature dependent reaction rate (cf. references [32, 161, 162]), and in mathematical biology, where $f(u)$ represents the reaction kinetics in a diffusion process (cf. reference [163]). In Section 7.4 we show the results of the symmetry analysis for the $(3 + 1)$-dimensional analogue of this equation.

To apply the nonlinear heat equation (7.116) to the nonclassical method we require that the infinitesimal transformation leaves invariant the set of simultaneous solutions of eqn (7.116) and the invariant surface condition (eqn (7.107)), where τ, ξ, ϕ are as yet unspecified functions of x, t, and u.

In the case in which $\tau \neq 0$, we set $\tau \equiv 1$. Eliminating u_t in eqn (7.116) using eqn (7.107) yields

$$\psi - \xi u_x = u_{xx} + f(u). \tag{7.117}$$

Applying the classical algorithm to this equation and eliminating u_{xx} using it yields the following determining equations [164]:

$$\xi_{uu} = 0, \qquad \phi_{uu} - 2\xi_{xu} + 2\xi\xi_u = 0, \tag{7.118a,b}$$

$$2\phi_{xu} - 2\phi\xi_u + 3\xi_u f + \xi_t - \xi_{xx} + 2\xi\xi_x = 0, \tag{7.118c}$$

$$\phi_t - \phi_{xx} + \phi_u f - f_u \phi + 2\phi\xi_x - 2\xi_x f = 0. \tag{7.118d}$$

Case 7.18(i): $f(u) = \sigma(u + b/3)^3$, $\sigma = \pm 1$. In this case we obtain two sets of infinitesimals:

(a) $$\xi = \frac{x + c_1}{2t + c_2}, \qquad \phi = -\frac{u + b/3}{2t + c_2},$$

which generate the (classical) scaling reduction, and

(b) $$\xi = -\frac{3}{x + c_1}, \qquad \phi = -\frac{u + b/3}{(x + c_1)^2},$$

which generate the nonclassical reduction

$$u(x,t) = (x + c_1)w(z) - b/3, \qquad z = \tfrac{1}{2}x^2 + c_1 x + 3t,$$

where $w(z)$ satisfies

$$w'' + \sigma w^3 = 0. \tag{7.119}$$

The solution of this is

$$w(z) = \begin{cases} \tfrac{1}{2}\sqrt{2} \operatorname{sd}(z; \tfrac{1}{2}\sqrt{2}), & \text{if } \sigma = 1, \\ \sqrt{2} \operatorname{ds}(z; \tfrac{1}{2}\sqrt{2}), & \text{if } \sigma = -1, \end{cases}$$

where $\operatorname{sd}(z; k)$ and $\operatorname{ds}(z; k)$ are the Jacobi elliptic functions satisfying

$$(d\eta/dz)^2 = 1 + (2k^2 - 1)\eta^2 + k^2(k^2 - 1)\eta^4,$$

and

$$(d\eta/dz)^2 = k^2(k^2 - 1) + (2k^2 - 1)\eta^2 + \eta^4,$$

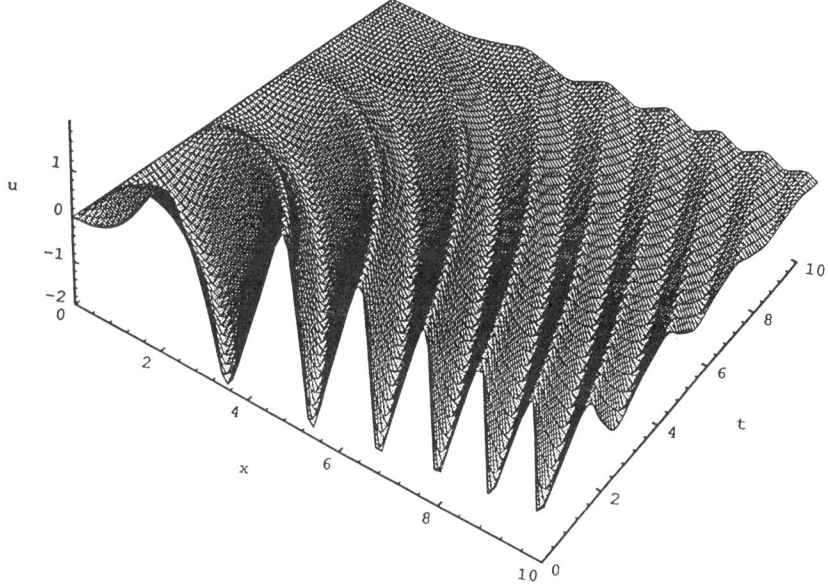

Fig. 7.1 The elliptic function solution (7.120) with $c_1 = 0$.

respectively. Hence we obtain the exact solution of

$$u_t = u_{xx} + \sigma u^3,$$

given by

$$u(x,t) = \tfrac{1}{2}\sqrt{2}\,(x+c_1)\operatorname{sd}\!\left(\tfrac{1}{2}x^2 + c_1 x + 3t; \tfrac{1}{2}\sqrt{2}\right) - \tfrac{1}{3}b, \qquad \text{if} \quad \sigma = 1, \tag{7.120}$$

$$u(x,t) = \sqrt{2}\,(x+c_1)\operatorname{ds}\!\left(\tfrac{1}{2}x^2 + c_1 x + 3t; \tfrac{1}{2}\sqrt{2}\right) - \tfrac{1}{3}b, \qquad \text{if} \quad \sigma = -1. \tag{7.121}$$

A plot of the elliptic function solution (7.120) for $c_1 = 0$ is given in Fig. 7.1.

Case 7.18(ii): $f(u) = \sigma[u^3 + bu^2 + cu + b(9c - 2b^2)/27]$, $\sigma = \pm 1$. In this case we obtain the infinitesimals

$$\xi = 3\mu \tan(\mu x), \qquad \phi = -\mu(3u + b)\sec^2(\mu x),$$

where $\mu^2 = \tfrac{1}{3}(b^2 - 3c)$. These generate the nonclassical reduction

$$u(x,t) = \mu \sin(\mu x)\exp(-3\mu^2 t)w(z) - b/3, \qquad z = \cos(\mu x)\exp(-3\mu^2 t),$$

where $w(z)$ satisfies eqn (7.119). The roots of $u^3 + bu^2 + cu + b(9c - 2b^2)/27 = 0$ are $-\tfrac{1}{3}b$, $-\tfrac{1}{3}b + \mu$, and $-\tfrac{1}{3}b - \mu$. Therefore they are collinear and the distances from the outer roots to the inner one are equal. Hence we obtain the exact solutions of

$$u_t = u_{xx} + [u^3 + bu^2 + cu + b(9c - 2b^2)/27], \quad (7.122)$$

given by

$$u(x,t) = \tfrac{1}{2}\sqrt{2}\,\mu \sin(\mu x)\exp(-3\mu^2 t)\operatorname{sd}\!\left[\cos(\mu x)\exp(-3\mu^2 t); \tfrac{1}{2}\sqrt{2}\right] - \tfrac{1}{3}b,$$

$$\text{if } b^2 > 3c, \quad (7.123)$$

$$u(x,t) = \tfrac{1}{2}\sqrt{2}\,\lambda \sinh(\lambda x)\exp(3\lambda^2 t)\operatorname{sd}\!\left[\cosh(\lambda x)\exp(3\lambda^2 t); \tfrac{1}{2}\sqrt{2}\right] - \tfrac{1}{3}b,$$

$$\text{if } b^2 < 3c, \quad (7.124)$$

where $\mu^2 = \tfrac{1}{6}(b^2 - 3c)$ and $\lambda^2 = \tfrac{1}{6}(3c - b^2)$, and the exact solutions of

$$u_t = u_{xx} - [u^3 + bu^2 + cu + b(9c - 2b^2)/27], \quad (7.125)$$

given by

$$u(x,t) = \sqrt{2}\,\mu \sinh(\mu x)\exp(3\mu^2 t)\operatorname{ds}\!\left[\cosh(\mu x)\exp(3\mu^2 t); \tfrac{1}{2}\sqrt{2}\right] - \tfrac{1}{3}b,$$

$$\text{if } b^2 > 3c, \quad (7.126)$$

$$u(x,t) = \tfrac{1}{2}\sqrt{2}\,\lambda \sin(\lambda x)\exp(-3\lambda^2 t)\operatorname{ds}\!\left[\cos(\lambda x)\exp(-3\lambda^2 t); \tfrac{1}{2}\sqrt{2}\right] - \tfrac{1}{3}b,$$

$$\text{if } b^2 < 3c. \quad (7.127)$$

A plot of the elliptic function solution (7.123), with $\mu = \tfrac{1}{2}$, is given in Fig. 7.2.

Case 7.18(iii): $f(u) = \sigma(u^3 + bu^2 + cu + d)$, $\sigma = \pm 1$. In this case we obtain the infinitesimals

$$\xi = \tfrac{3}{2}\sqrt{2}\,(u + b/3), \quad \phi = \tfrac{3}{2}(u^3 + bu^2 + cu + d), \quad \text{if } \sigma = 1,$$

$$\xi = \tfrac{3}{2}\sqrt{2}\,(u + b/3), \quad \phi = -\tfrac{3}{2}(u^3 + bu^2 + cu + d), \quad \text{if } \sigma = -1.$$

In the case $\sigma = -1$, the associated invariant surface condition is

$$\tfrac{3}{2}\sqrt{2}\,(u + b/3)u_x + u_t + \tfrac{3}{2}(u^3 + bu^2 + cu + d) = 0,$$

which is solvable as follows. Eliminate u_t using the original PDE, which gives

$$u_{xx} + \tfrac{3}{2}\sqrt{2}\,(u + b/3)u_x + \tfrac{1}{2}(u^3 + bu^2 + cu + d) = 0,$$

which can be linearized by the transformation $u = \sqrt{2}\,\Psi_x/\Psi$, yielding

$$2\sqrt{2}\,\Psi_{xxx} + 2b\Psi_{xx} + \sqrt{2}\,c\Psi_x + d\Psi = 0.$$

(a)
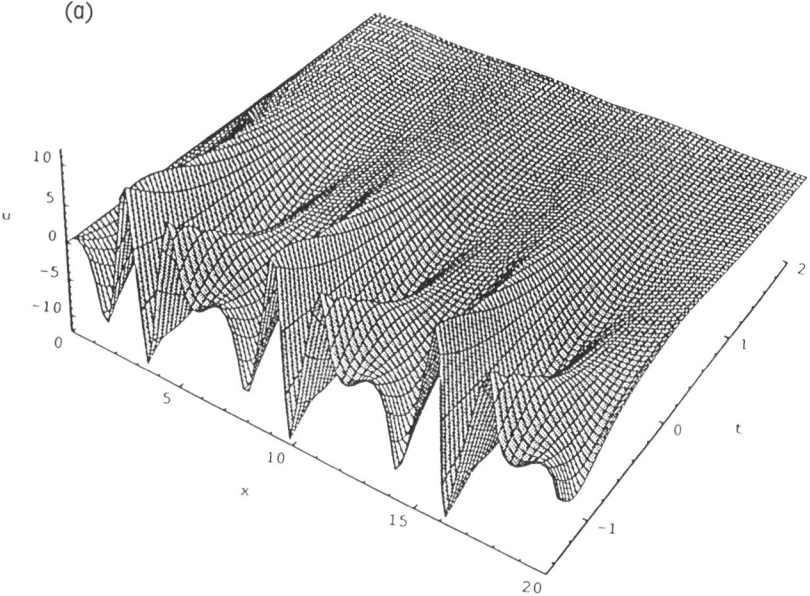

(b)
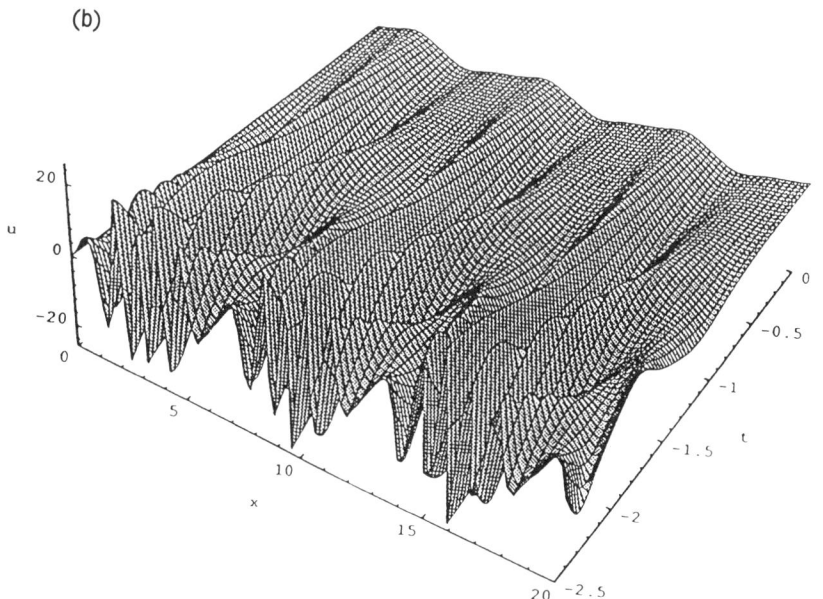

Fig. 7.2 The elliptic function solution (7.123) with $\mu = \frac{1}{2}$.

Suppose that the roots of the cubic

$$u^3 + bu^2 + cu + d = 0 \tag{7.128}$$

are m_1, m_2, and m_3. Then the roots of

$$2\sqrt{2}\, p^3 + 2bp^2 + \sqrt{2}\, cp + d = 0$$

are $\tfrac{1}{2}\sqrt{2}\, m_1$, $\tfrac{1}{2}\sqrt{2}\, m_2$, and $\tfrac{1}{2}\sqrt{2}\, m_3$.

(a) *Distinct real roots.* In the case in which m_1, m_2, and m_3 are distinct and real, we obtain the exact solution of

$$u_t = u_{xx} - (u - m_1)(u - m_2)(u - m_3), \tag{7.129}$$

given by

$$u(x,t) = \frac{c_1 m_1 \Psi_1 + c_2 m_2 \Psi_2 + c_3 m_3 \Psi_3}{c_1 \Psi_1 + c_2 \Psi_2 + c_3 \Psi_3},$$

where

$$\Psi_j(x,t) = \exp\{\tfrac{1}{2}\sqrt{2}\, m_j x - m_j(m_1 + m_2 + m_3 - \tfrac{3}{2} m_j)t\}, \quad j = 1,2,3, \tag{7.130}$$

where c_1, c_2, and c_3 are arbitrary constants. In particular, setting $m_1 = a$, $m_2 = 1$, and $m_3 = 0$ in eqn (7.130), with $a \neq 0$ and $a \neq 1$, which we may without loss of generality for the general cubic with real and distinct roots, yields the exact solution of the Fitzhugh–Nagumo equation,

$$u_t = u_{xx} + u(1-u)(u-a), \tag{7.131}$$

which models the transmission of nerve impulses [153, 154] and arises in population genetics [155, 156], given by

$$u(x,t) = \frac{a\kappa_1 \exp\{\tfrac{1}{2}(\pm\sqrt{2}\, ax + a^2 t)\} + \kappa_2 \exp\{\tfrac{1}{2}(\pm\sqrt{2}\, x + t)\}}{\kappa_1 \exp\{\tfrac{1}{2}(\pm\sqrt{2}\, ax + a^2 t)\} + \kappa_2 \exp\{\tfrac{1}{2}(\pm\sqrt{2}\, x + t)\} + \kappa_3 \exp(at)}. \tag{7.132}$$

This solution was also obtained by Vorob'ev [157] (who calls the associated symmetry a 'partial symmetry of the first type'), Kawahara and Tanaka [165] (using Hirota's bilinear method [166]), and Hereman [167] (using the truncated Painlevé expansion method [168, 169]). Plots of the exponential solution (7.132) with (a) $a = 0.4$ and (b) $a = 0.7$ are given in Fig. 7.3.

ANALYSIS OF DIFFERENTIAL EQUATIONS

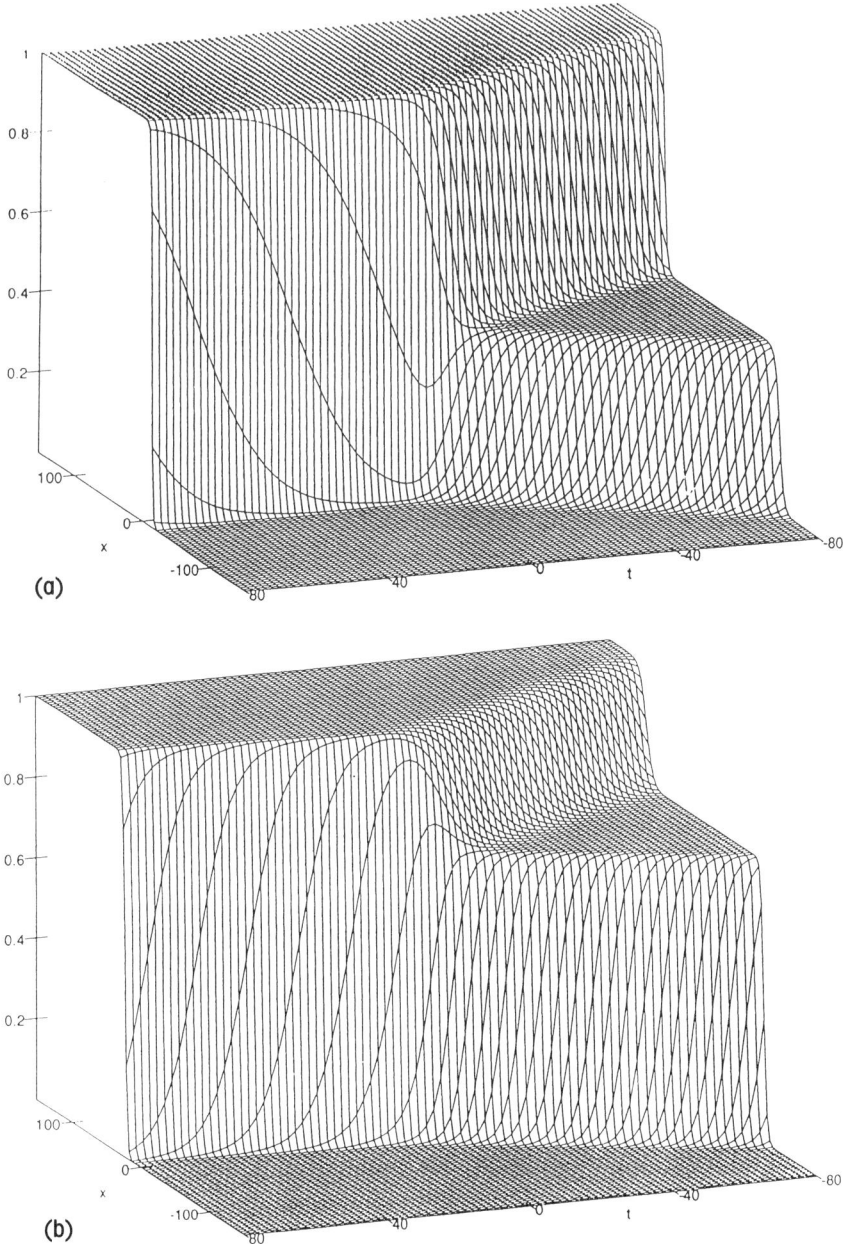

Fig. 7.3 The exponential solution (7.132) with (a) $a = 0.4$ and (b) $a = 0.7$.

(b) *Complex roots.* In the case in which the roots of eqn (7.128) are m (real), and $\alpha \pm i\beta$, we obtain the exact solution of

$$u_t = u_{xx} - u^3 + (m + 2\alpha)u^2 - (\alpha^2 + \beta^2 + 2\alpha m)u + (\alpha^2 + \beta^2)m, \tag{7.133}$$

given by

$$u(x,t) = \frac{c_1 m \exp\{\tfrac{1}{2}\sqrt{2}\,\mu x + [\tfrac{1}{2}\mu^2 + \tfrac{3}{2}\beta^2]t\} + c_2\sqrt{\alpha^2 + \beta^2}\,\sin\{\tfrac{1}{2}\sqrt{2}\,\beta x - \mu\beta t + \delta_0 + \theta_0\}}{c_1 \exp\{\tfrac{1}{2}\sqrt{2}\,\mu x + [\tfrac{1}{2}\mu^2 + \tfrac{3}{2}\beta^2]t\} + c_2 \cos\{\tfrac{1}{2}\sqrt{2}\,\beta x - \mu\beta t + \delta_0\}},$$

where $\mu = m - \alpha$, $\theta_0 = \tan^{-1}(\beta/\alpha)$, and c_1, c_2, and δ_0 are arbitrary constants. Setting $m = \alpha = 0$ and $\beta = 1$ in eqn (7.133) yields the exact solution of

$$u_t = u_{xx} - u(u^2 + 1),$$

given by

$$u(x,t) = \frac{c_2 \sin(\tfrac{1}{2}\sqrt{2}\,x)}{c_1 \exp(\tfrac{3}{2}t) + c_2 \cos(\tfrac{1}{2}\sqrt{2}\,x)}. \tag{7.134}$$

A plot of the exponential solution (7.134) with $c_1 = 1$ and $c_2 = 1.1$ is given in Fig. 7.4.

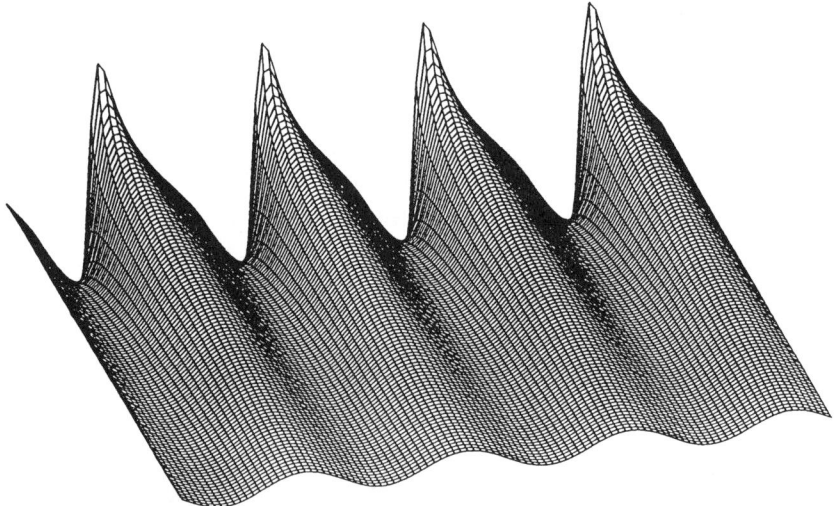

Fig. 7.4 The exponential solution (7.134) with $c_1 = 1.1$ and $c_2 = 1$.

ANALYSIS OF DIFFERENTIAL EQUATIONS 449

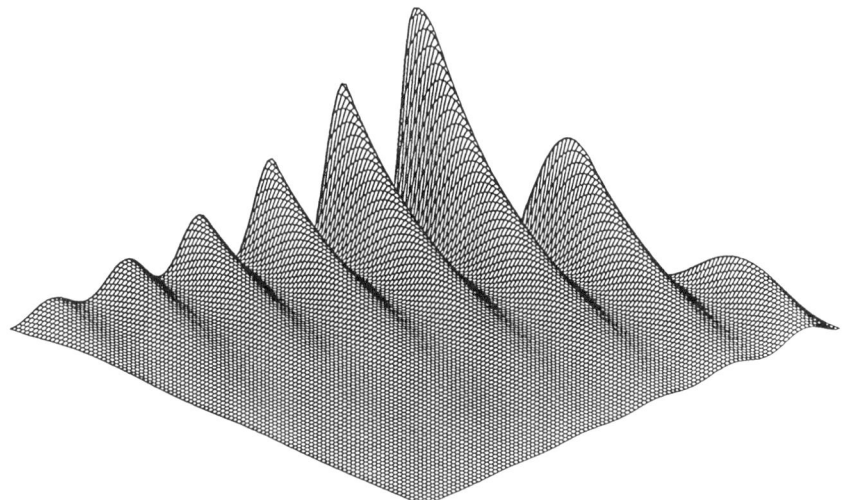

Fig. 7.5 The exponential solution (7.135) with $c_1 = 1$ and $c_2 = 1.1$, multiplied by $\exp\{\tfrac{9}{10}(\tfrac{1}{2}\sqrt{2}\,x + 2t)\}$.

Setting $m = 0$ and $\alpha = \beta = 1$ in eqn (7.133) yields the exact solution of
$$u_t = u_{xx} - u(u^2 - 2u + 2),$$
given by
$$u(x,t) = \frac{c_2 \cos\left(\tfrac{1}{2}\sqrt{2}\,x - t\right) + c_2 \sin\left(\tfrac{1}{2}\sqrt{2}\,x - t\right)}{c_1 \exp\left(\tfrac{1}{2}\sqrt{2}\,x + 2t\right) + c_2 \cos\left(\tfrac{1}{2}\sqrt{2}\,x - t\right)}. \tag{7.135}$$

A plot of the exponential solution (7.135) with $c_1 = 1$ and $c_2 = 1.1$, multiplied by $\exp\{\tfrac{9}{10}(\tfrac{1}{2}\sqrt{2}\,x + 2t)\}$ is given in Fig. 7.5.

(c) *Two equal roots.* Suppose that the cubic (7.128) has a single root m_1 and a double root m_2. Then we obtain the exact solution of
$$u_t = u_{xx} - (u - m_1)(u - m_2)^2,$$
given by
$$u(x,t) = \frac{m_1 c_1 \exp\{\tfrac{1}{2}\sqrt{2}\,\beta x + \tfrac{1}{2}\beta^2 t\} + c_2\left[m_2(x - \sqrt{2}\,\beta t) + \sqrt{2}\right]}{c_1 \exp\{\tfrac{1}{2}\sqrt{2}\,\beta x + \tfrac{1}{2}\beta^2 t\} + c_2(x - \sqrt{2}\,\beta t)},$$
where $\beta = m_2 - m_1$ and c_1, c_2, and c_3 are arbitrary constants. In particular, an exact solution of the Huxley equation
$$u_t = u_{xx} + u^2(1 - u),$$

is given by

$$u(x,t) = \frac{\exp\{\frac{1}{2}(-\sqrt{2}\,x+t)\} - \frac{1}{2}\sqrt{2}}{\exp\{\frac{1}{2}(-\sqrt{2}\,x+t)\} + \frac{1}{2}(x+\sqrt{2}\,t)}. \qquad (7.136)$$

A plot of the exponential solution (7.136) with (a) $b = -1$ and (b) $b = 1$ is given in Fig. 7.6.

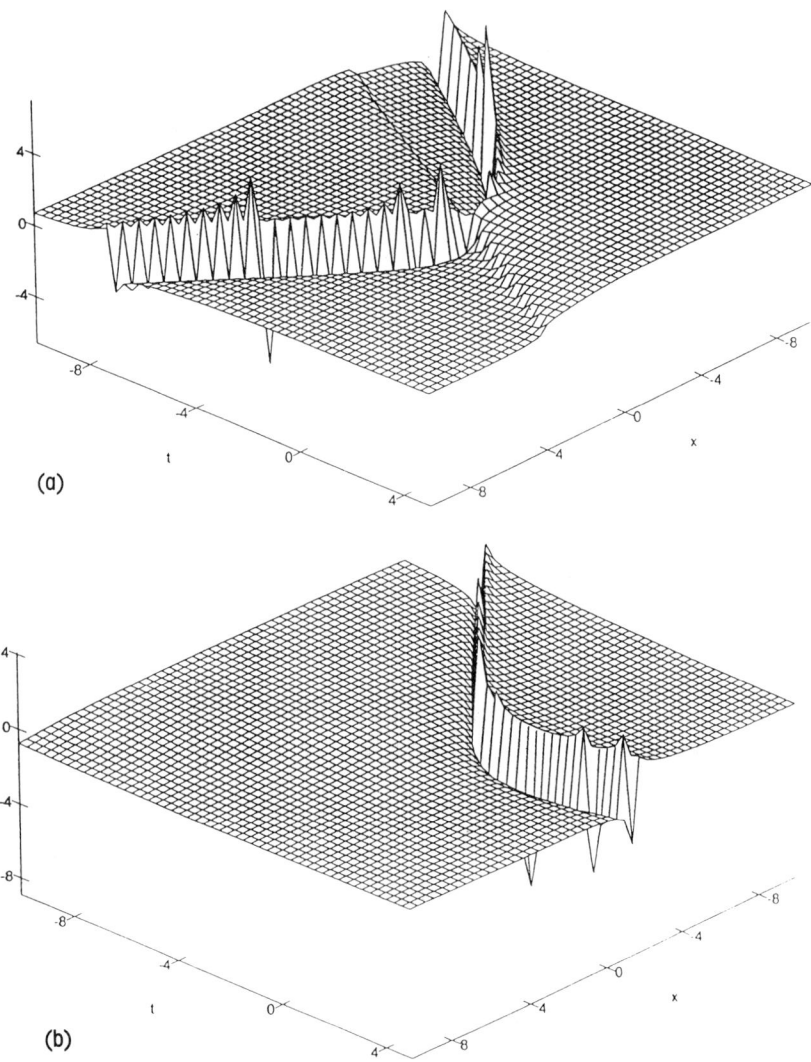

Fig. 7.6 The exponential solution (7.136) with (a) $b = -1$ and (b) $b = 1$.

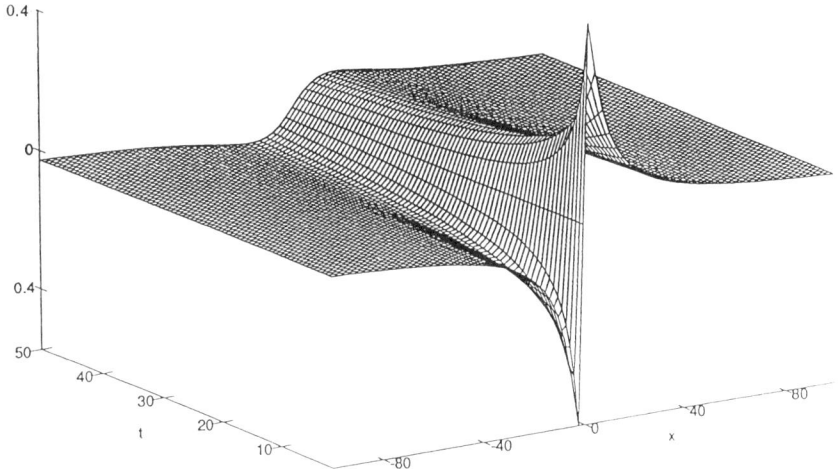

Fig. 7.7 The rational solution (7.137) with $c_1 = c_2 = 0$.

(d) *Three equal roots.* In this case we obtain the exact solution of

$$u_t = u_{xx} - u^3,$$

given by

$$u(x,t) = \frac{\sqrt{2}\,(2x + c_1)}{x^2 + c_1 x + 6t + c_2}. \tag{7.137}$$

A plot of the rational solution (7.137) with $c_1 = c_2 = 0$ is given in Fig. 7.7.

The nonclassical method was further generalized by Olver and Rosenau [143, 144] to include 'weak symmetries' and, even more generally, 'side conditions' or 'differential constraints', and they concluded that 'the unifying theme behind finding special solutions of PDEs is not, as is commonly supposed, group theory, but rather the more analytic subject of overdetermined systems of PDEs'.

Note that the Boussinesq equation

$$u_{tt} + u u_{xx} + u_x^2 + u_{xxxx} = 0 \tag{7.138}$$

possesses the accelerating wave solution

$$u(x,t) = w(z) - 4\mu^2 t^2, \qquad z = x + \mu t^2. \tag{7.139}$$

This reduction was first written down by Nakamoto and Tajiri [91] and Quispel, Nijhoff and Capel [145], and subsequently Olver, Rosenau and Schwarzmeier [92, 143, 144] pointed out that the reduction is not obtainable using the classical Lie method. Motivated by this observation, Clarkson and Kruskal [93] developed an algorithmic and direct method for finding symmetry reductions (hereafter referred to as the *direct method*) in an attempt to understand the origin of the symmetry reduction (7.139) and derive it systematically (the previous derivations had been by seemingly *ad hoc* techniques). However, using the direct method in addition to deriving eqn (7.139), Clarkson and Kruskal [93] also obtained several previously unknown symmetry reductions of the Boussinesq equation (7.138).

The basic idea of the direct method is to seek a solution of a PDE such as eqn (7.138) in the form

$$u(x,t) = F(x,t,w(z(w,t))), \qquad (7.140)$$

and require that $w(z)$ satisfies an ODE. This imposes conditions upon $F(x,t,w)$, $z(x,t)$, and their derivatives in the form of an overdetermined system of equations, the solution of which yields the desired reductions. The novel characteristic about the direct method, in comparison to the others mentioned above, such as the classical and nonclassical methods, is that it involves no use of group theory. We remark that the direct method has certain resemblances to the so-called 'method of free parameter analysis' (cf. reference [170]); although in this latter method, the boundary conditions are crucially used in the determination of the reduction, whereas they are not used in the direct method.

Levi and Winternitz [152] gave a group theoretical explanation of Clarkson and Kruskal's results by showing that all the new symmetry reductions of the Boussinesq equation (7.138) derived in [93] using the direct method can also be obtained using the nonclassical method. Clarkson and Winternitz [185] proved a similar result for the Kadomtsev–Petviashvili equation by showing that the direct and nonclassical methods gave the same reduction. The results in [152, 185] suggested that perhaps the direct and nonclassical methods were equivalent, i.e. they yield the same reductions. Indeed, Clarkson and Kruskal [93] posed the question on the relationship between these two methods in the conclusion of their original paper on the direct method.

This question of the relationship between these two methods was investigated by Nucci and Clarkson [186], who applied both the direct and nonclassical methods to the Fitzhugh–Nagumo equation (7.131) and showed that the exact solution (7.132) is *not* obtainable using the direct method, although it is using the nonclassical method. This demonstrated that the nonclassical method is more general than the direct method, at least as it was originally formulated. Estévez [187] has subsequently shown that the exact solution (7.132) can be obtained using a generalization of the direct method which uses ideas from the singular manifold method [168, 169].

More generally, the singular manifold method [168, 169] has proven to be an effective method for generating exact solutions of several PDEs. In particular, we mention the work of Cariello and Tabor [188, 189] and Estévez [187, 190], who have shown that the so-called singularity method plays the role of a similarity variable, and that the singular manifold method is capable of yielding solutions corresponding to both classical and nonclassical symmetries. The relationship between nonclassical symmetries and the singular manifold method is considered in depth by Estévez and Gordoa [191].

The ansatz $u(x, t) = F(x, t, w(z))$ with $z = z(x, t)$ used in the direct method assumes that the symmetry variable z does not depend upon u. Consequently, it is implicitly assumed that the ratio of infinitesimal ξ/τ is independent of u. For the exact solution (7.132), this ratio is dependent upon u. Recently, Olver [192] has shown that the direct method is equivalent to the nonclassical method when this ratio of infinitesimals ξ/τ is independent of u, thus generating a group of so-called 'fibre-preserving transformations' (see also references [193–197]).

In the past few years the direct and nonclassical methods have been used to generate many new symmetry reductions and exact solutions for several physically significant PDEs, which represents important progress; in addition to the aforementioned references, see also references [198–221].

Since the original development of the direct method, it has been known that there exist exact solutions of PDEs which are not obtainable using the direct method. For example, the two-soliton solution of the Boussinesq equation (7.138) given by

$$u(x,t) = 12 \frac{\partial^2}{\partial x^2} [1 + \exp(\eta_1) + \exp(\eta_2) + A_{12} \exp(\eta_1 + \eta_2)] - 1,$$

(7.141a)

where

$$\eta_1 = \kappa_1(x + \mu_1 t) + \delta_1, \qquad \eta_2 = \kappa_2(x + \mu_2 t) + \delta_2,$$

$$A_{12} = \frac{\mu_1 \mu_2 + 2\kappa_1^2 - 3\kappa_1 \kappa_2 + 2\kappa_2^2 - 1}{\mu_1 \mu_2 + 2\kappa_1^2 + 3\kappa_1 \kappa_2 + 2\kappa_2^2 - 1},$$

(7.141b)

with $\mu_1 = (1 - \kappa_1^2)^{1/2}$, $\mu_2 = (1 - \kappa_2^2)^{1/2}$, and δ_1 and δ_2 arbitrary constants. The two-soliton solution (7.141), and more generally the N-soliton solution, of completely integrable equations such as the Boussinesq equation (7.138) are standardly obtainable using generalized or Lie–Bäcklund symmetries (cf. reference [57]). However, recently, Clarkson and Mansfield [158, 207] have shown that some special 'two-soliton' solutions arise from nonclassical symmetries. We remark that the special case of the solution

(7.141) with $A_{12} = 0$ can be obtained using the truncated Painlevé expansion (singular manifold) method. Hirota and Ito [222] refer to this special case of the general soliton solution as being the 'resonant state' in which either two solitons fuse together after colliding with each other or a single soliton splits into two solitons. Additionally, the solution (7.132) of the Fitzhugh–Nagumo equation (7.131) has certain similarities to a 'two-soliton' solution and, as mentioned above, it has also been obtained using Hirota's bilinear method [165]; since the Fitzhugh–Nagumo equation (7.131) is thought to be nonintegrable, the solution (7.132) is not strictly a 'two-soliton'.

These results pose the following important open question: For which PDEs does the nonclassical method yield more symmetry reductions than the direct method? Furthermore, it remains an open question to determine *a priori* which PDEs possess symmetry reductions that are not obtainable using the classical Lie group approach.

Other methods

The direct and nonclassical methods have generated many reductions and exact solutions for a number of physically important PDEs. What further generalizations are possible? Galaktionov's method of 'nonlinear separation' [8–9] and Olver and Rosenau's method of differential constraints [143, 144] are very general methods. In fact, the method of differential constraints can be viewed as the ultimate generalization, since it includes (almost) all the known methods for determining special solutions of PDEs such as classical group-invariant solutions, nonclassical and weak symmetry reductions, partially invariant solutions, separation of variables solutions etc. (see also the earlier work of Yanenko [223] and Meleshko [224]). They do have a group-theoretic interpretation as 'generalized weak symmetries' although, as Olver [225] admits, this does *not* appear to be an overly useful observation. The main difficulty with this approach is that it appears to be *too general* to be of practical use. The question is how to determine *a priori* which constraints are useful and which ansätze will yield computable solutions.

Ovsiannikov [36] developed the concept of *partially invariant solutions* and introduced a complicated algorithm for calculating them. The partially invariant solution method has not yet, to date, been extensively applied as other symmetry methods, first because the algorithm is quite complicated and, second, because in many cases the associated solutions could have been determined more easily through the classical Lie group method. However, in the past few years there has been some renewed interest in the method [226–229]; in particular, the recent work of Ondich [229] has made some important advances.

Potential symmetries (sometimes known as *nonlocal symmetries*) are another type of symmetries which have received some attention recently (cf. references [1, 230–236]). The existence of potential symmetries has led to the construction of corresponding invariant solutions as well as to the linearization of nonlinear PDEs through noninvertible mappings [1, 237]. In this approach, a PDE is written as a system of PDEs. For example, one can write the Boussinesq equation (7.138) as the system

$$v_x = -u_t, \qquad v_t = u_{xxx} + uu_x. \qquad (7.142\mathrm{i, ii})$$

It can be shown that this system has the same classical symmetries as the Boussinesq equation (7.138), although this is not always the situation.

Anderson, Kamran, and Olver [238] introduced the concept of an *internal symmetry* for systems of differential equations, in particular underdetermined systems. These symmetries are also known as 'dynamical symmetries' and have received attention in the work of Vinogradov and collaborators [239]. For a wide class of systems of equations, every internal symmetry comes from a first-order generalized symmetry and, conversely, every first-order generalized symmetry satisfying certain conditions determines an internal symmetry. Anderson, Kamran, and Olver [238] deduced necessary conditions for a system of differential equations to admit a genuine internal symmetry.

Differential-difference and difference equations

There is much current interest in deriving symmetries of differential-difference equations such as the Toda lattice equation

$$\partial^2 u_n / \partial x \partial t = \exp\{u_{n-1} - u_n\} - \exp\{u_n - u_{n+1}\}, \qquad (7.143)$$

which is a completely integrable soliton equation [240–244]. Numerous physical applications of such equations have been outlined in Section 1.7. There have been two different approaches to the determination of classical Lie symmetry reductions of differential-difference equations, one proposed by Levi and Winternitz [245–247] and another by Quispel, Capel, and Sahadevan [248, 249]. Levi and Winternitz [246] have also introduced the concept of conditional symmetries for the Toda lattice equation (7.143). Essentially, the difference between these two approaches concerns the interpretation of the discrete variable n. Levi and Winternitz [245–247] essentially treat u_{n-1}, u_n, and u_{n+1} as different dependent variables. On the other hand, Quispel, Capel and Sahadevan [248, 249] treat n as an independent variable (in addition to x and t) and thus, in effect, treat the differential-difference equation as a differential-delay equation.

Symmetries for difference equations are discussed by Dorodnitsyn

[250, 251] (see also reference [13]), Maeda [252, 253], and Quispel and Sahadevan [254, 255]. As for symmetries of differential-difference equations, these three approaches have some important and fundamental differences.

7.2 The Painlevé tests

7.2.1 Singularities of ODEs

Singularities of ODEs are of two types: *fixed* and *movable*. We begin by reviewing some facts about *linear* ODEs (cf. reference [61, Chapter 15]). Consider the nth-order equation in the complex domain,

$$\frac{d^n w}{dz^n} + p_{n-1}(z)\frac{d^{n-1} w}{dz^{n-1}} + \cdots + p_1(z)\frac{dw}{dz} + p_0(z)w = 0. \quad (7.144)$$

If the n coefficients $p_0(z), \ldots, p_{n-1}(z)$ are all analytic in the neighbourhood of a point z_0, then z_0 is a *regular* point of the ODE, and there exists a unique solution of eqn (7.144) such that it and its first $n-1$ derivatives assume arbitrarily assigned values at $z = z_0$. This solution is expressible as a Taylor series in $(z - z_0)$, convergent at least within the circle centred at z_0 and passing through that singular point of the coefficients lying nearest to z_0. Therefore the singularities of solutions of the equation can be located only at singularities of the coefficients (which are either regular or irregular singular points). These singularities are said to be *fixed*, since their locations do not vary with the particular solution chosen but depend only on the equation. For example, solutions of the hypergeometric equation

$$z(1-z)\frac{d^2 w}{dz^2} + [c - (a+b+1)z]\frac{dw}{dz} - abw = 0,$$

can only have singularities where the coefficients do, namely at 0, 1, and ∞; the latter is classified by making the transformation $z = 1/\xi$ and then analysing the point $\xi = 0$.

It is a general property of linear ODEs that any singularities of their solutions are fixed. Nonlinear ODEs lose this property. Consider the very simple equation

$$dw/dz + w^2 = 0. \quad (7.145)$$

The general solution of this equation is

$$w(z) = (z - z_0)^{-1},$$

where z_0 is a constant of integration, which is also the location of the singularity. This singularity, a pole, is said to be *movable*, since its location depends on the constant of integration z_0. Nonlinear ODEs can have both movable and fixed singularities.

EXAMPLE 7.19. A solution of an ODE can have various kinds of singularities. Anything other than a pole, of whatever order, is called a critical point: however, this is not entirely correct and would exclude equations with solutions that have a single-valued movable essential singularity such as $\exp\{1/(z - z_0)\}$. Furthermore, Painlevé himself, in one of his definitive works [256], defines a critical point to be where branching, i.e. multi-valuedness, occurs (see also the discussions of this in references [257, 258]). Branch points may have a finite number (greater than or equal to two) of branches, or infinitely many. For example, consider the equations in the following table.

Table 7.2 Singularities of ordinary differential equations,

	Equation	Solution
(i)	$dw/dz + w^3 = 0$	$2^{-1/2}(z - z_0)^{-1/2}$
(ii)	$w\dfrac{d^2w}{dz^2} - \dfrac{dw}{dz} + 1 = 0$	$(z - z_0)\ln(z - z_0) + C(z - z_0)$
(iii)	$aw\dfrac{d^2w}{dz^2} + (1-a)\left(\dfrac{dw}{dz}\right)^2 = 0$	$C(z - z_0)^a$
(iv)	$\left[w\dfrac{d^2w}{dz^2} - \left(\dfrac{dw}{dz}\right)^2\right]^2 + 4w\left(\dfrac{dw}{dz}\right)^3 = 0$	$C\exp\{(z - z_0)^{-1}\}$
(v)	$(1 + w^2)\dfrac{d^2w}{dz^2} + (1 - 2w)\left(\dfrac{dw}{dz}\right)^2 = 0$	$\tan\{\ln C(z - z_0)\}$

C and z_0 are arbitrary constants. These singularities are, respectively, a movable algebraic branch point, a movable logarithmic branch point, a transcendental singular point unless a is rational, a movable isolated essential singularity, and a nonisolated movable essential singularity. In (v), $w(z)$ has no limit (even allowing infinity) as $z \to z_0$ unless z is taken along special paths. In fact, an infinite number of distinct branches originate from the point z_0, which is a limit point of poles, a branch point, and an essential singularity.

If the equation is put into normal form, i.e. solved for its highest derivative

$$\frac{d^n w}{dz^n} = F\left(\frac{d^{n-1}w}{dz^{n-1}}, \ldots, \frac{dw}{dz}, w, z\right),$$

then its singularities are those of F. Values of z at which F is singular independently of the other variables are fixed singularities, while the others are movable.

7.2.2 First-order ODEs

Painlevé [259–261] proved that for first-order equations of the form
$$G(dw/dz, w, z) = 0, \qquad (7.146)$$
with G a polynomial in dw/dz and w, and analytic in z, the movable singularities of the solutions are poles and/or algebraic branch points (see also references [61, 256, 262]). Fuchs [263] showed that the only equation of the form
$$\frac{dw}{dz} = F(w, z) = \frac{P(w, z)}{Q(w, z)}, \qquad (7.147)$$
where P and Q are polynomials in w with coefficients analytic in z, the critical points of the solutions of which are all fixed, is the generalized Riccati equation
$$dw/dz = p_2(z)w^2 + p_1(z)w + p_0(z), \qquad (7.148)$$
where $p_2(z)$, $p_1(z)$, and $p_0(z)$ are analytic functions (for proof, see references [262, p. 89] and [61, p. 293]). If $p_2(z) \equiv 0$, then eqn (7.148) is just a linear equation. Otherwise, one makes the transformation
$$w(z) = -\frac{1}{p_2(z)y} \frac{dy}{dz}, \qquad (7.149)$$
whereupon eqn (7.148) becomes the second-order linear equation
$$p_2(z)\frac{d^2y}{dz^2} - \left[\frac{dp_2}{dz} + p_1(z)p_2(z)\right]\frac{dy}{dz} + p_0(z)p_2^2(z)y = 0. \qquad (7.150)$$

Its general solution has the form
$$y(z) = \alpha_1 y_1(z) + \alpha_2 y_2(z),$$
where $y_1(z)$ and $y_2(z)$ are any two linearly independent solutions and α_1 and α_2 are arbitrary constants. Therefore, the general solution of the Riccati equation (7.148) is
$$w(z) = -\left[\alpha_1 \frac{dy_1}{dz} + \alpha_2 \frac{dy_2}{dz}\right] \bigg/ [p_2(z)(\alpha_1 y_1 + \alpha_2 y_2)].$$

We shall now consider equations of the form
$$(dw/dz)^n = p(w), \qquad (7.151)$$
where p is a polynomial in w of degree m with constant coefficients. (It is assumed that $p(w)$ is not the kth power of a polynomial with k a divisor of

THE PAINLEVÉ TESTS

n.) If solutions of eqn (7.151) have no movable critical points, then (necessarily) $m \leq 2n$ (for proof see references [262, p. 418] and [61, p. 312]). We shall determine all equations of the form (7.151) the solutions of which have no movable critical points for $n = 1$ and $n = 2$.

EXAMPLE 7.20. For $n = 1$, the only possibility is a Riccati equation, and there are three possible forms (up to a change of scale of z, and omitting the trivial case $dw/dz = a$) which, together with their general solutions, are given in the following table.

Table 7.3 First-order first-degree equations

	Equation	Solution
(i)	$dw/dz = w - a$	$a + \exp\{(z - z_0)\}$
(ii)	$dw/dz = (w - a)(w - b)$	$\dfrac{b \exp\{(a - b)(z - z_0)\} - a}{\exp\{(a - b)(z - z_0)\} - 1}$
(iii)	$dw/dz = (w - a)^2$	$a - (z - z_0)^{-1}$

z_0 is an arbitrary constant.

EXAMPLE 7.21. Similarly, for $n = 2$, there are eight cases; six of these lead to equations which are given in the following table and can be solved in terms of elementary functions.

Table 7.4 First-order second-degree equations

	Equation	Solution
(i)	$(dw/dz)^2 = w - a$	$\tfrac{1}{4}(z - z_0)^2 + a$
(ii)	$(dw/dz)^2 = (w - a)(w - b)$	$\tfrac{1}{2}(a + b) + \tfrac{1}{2}(a - b)\cosh(z - z_0)$
(iii)	$(dw/dz)^2 = (w - a)^3$	$a + 4(z - z_0)^{-2}$
(iv)	$(dw/dz)^2 = (w - a)^2(w - b)$	$a + (b - a)\sec^2(\tfrac{1}{2}\sqrt{b - a}\,(z - z_0))$
(v)	$(dw/dz)^2 = (w - a)^3(w - b)$	$a + \dfrac{4(a - b)}{(a - b)^2(z - z_0)^2 - 4}$
(vi)	$(dw/dz)^2 = (w - a)^2(w - b)(w - c)$	$a + \dfrac{2(b - a)(c - a)}{(b + c - 2a) + (b - c)\cosh\{\mu(z - z_0)\}}$

In Table 7.4, $\mu = \sqrt{(b - a)(c - a)}$.

The remaining two cases define elliptic functions. These are the cases in which $p(w)$ has three or four distinct roots.

If $p(w)$ has three distinct roots, then eqn (7.151) may be brought (by a linear transformation) into the form

$$(dw/dz)^2 = 4(w - a)(w - b)(w - c) = 4w^3 - g_2 w - g_3, \quad (7.152)$$

where $a + b + c = 0$, $4(ab + bc + ca) = -g_2$, and $4abc = g_3$. This is solved by the quadrature

$$t - z_0 = \int^w \frac{dy}{\sqrt{4y^3 - g_2 y - g_3}}, \qquad (7.153)$$

with z_0 an arbitrary constant. This elliptic integral defines z as a function of w; the inverse being given by the *Weierstrass elliptic function* $\wp(z)$ (cf. [264]), so the general solution of eqn (7.152) is

$$w(z) = \wp(z - z_0; g_2, g_3). \qquad (7.154)$$

If $p(w)$ has four distinct roots, then eqn (7.151) may be brought into the form

$$(dw/dz)^2 = (1 - w^2)(1 - k^2 w^2), \qquad k \neq \pm 1,$$

by means of a Möbius (bilinear rational) transformation. The general solution is the *Jacobian elliptic function* $w(z) = \mathrm{sn}(z - z_0; k)$, which is a doubly periodic meromorphic function of z with simple poles (see Whittaker and Watson [264, Chapter 22] for further details about the Jacobian elliptic function). A discussion of graphical methods to solve elliptic and hyperelliptic equations is given in Appendix A.

7.2.3 Second-order ODEs

We now consider ODEs of the form

$$d^2 w/dz^2 = F(dw/dz, w, z). \qquad (7.155)$$

where F is a rational function of dw/dz and w, and an analytic function of z. In 1887, Picard posed the problem of determining which equations of the form (7.155) have solutions the only movable singularities of which are poles, i.e. which solutions have *no movable critical points* so that the locations of singularities of any of the solutions other than poles are independent of the particular solution chosen and thus dependent only on the equation. This property is now known as the '*Painlevé property*', and ODEs that possess it are said to be of '*Painlevé type*'. The problem was solved by Painlevé and his colleagues: a comprehensive review of their work is given in Ince [61, Chapter 14]; see also [262, p. 439]). More recently, the problem has been reconsidered by Bureau [265] and Cosgrove [266] (see also [267]). Whereas for first-order equations of the form (7.146), the only possible movable singularities of solutions are poles and algebraic branch points, Painlevé showed that, for second-order equations of the form (7.155), the movable singularities may also be logarithmic branch points and essential singularities. The general method used to study this problem is known as the 'α-method' and is due to Painlevé; this is also called the 'theorem of stability' by Bureau [265].

Painlevé's α-method consists first of finding a set of necessary conditions for the absence of movable critical points and then verifying, by direct integration or otherwise, that these are also sufficient. The second part of the method is very tedious, since it involves considering a large number of cases and then showing that the equations in each different case are either reducible to equations the solutions of which are known to have no movable critical points or define new transcendents.

Painlevé et al. showed that out of all possible equations of the form (7.155), there are only 50 canonical types that have the property of having no movable critical points. Furthermore, they showed that of these 50 equations, 44 are either integrable in terms of previously known functions (such as elliptic functions or are equivalent to linear equations) or reducible to one of six new nonlinear ODEs. These 50 canonical types are generalizable by the Möbius transformation

$$W(Z) = \frac{a(z)w + b(z)}{c(z)w + d(z)}, \quad Z = \phi(z),$$

where a, b, c, d, and ϕ are analytic functions, and these equations contain all second-order ODEs of the form (7.155) the general solutions of which have no movable critical points. The most interesting of the 50 canonical equations are those which define new transcendental functions. These six equations are

$$d^2w/dz^2 = 6w^2 + z, \quad d^2w/dz^2 = 2w^3 + zw + \alpha, \qquad \text{PI, PII}$$

$$\frac{d^2w}{dz^2} = \frac{1}{w}\left(\frac{dw}{dz}\right)^2 - \frac{1}{z}\frac{dw}{dz} + \frac{1}{z}(\alpha w^2 + \beta) + \gamma w^3 + \frac{\delta}{w}, \qquad \text{PIII}$$

$$\frac{d^2w}{dz^2} = \frac{1}{2w}\left(\frac{dw}{dz}\right)^2 + \tfrac{3}{2}w^3 + 4zw^2 + 2(z^2 - \alpha)w + \frac{\beta}{w}, \qquad \text{PIV}$$

$$\frac{d^2w}{dz^2} = \left\{\frac{1}{2w} + \frac{1}{w-1}\right\}\left(\frac{dw}{dz}\right)^2 - \frac{1}{z}\frac{dw}{dz} + \frac{(w-1)^2}{z^2}\left(\alpha w + \frac{\beta}{w}\right)$$
$$+ \frac{\gamma w}{z} + \frac{\delta w(w+1)}{w-1}, \qquad \text{PV}$$

$$\frac{d^2w}{dz^2} = \tfrac{1}{2}\left\{\frac{1}{w} + \frac{1}{w-1} + \frac{1}{w-z}\right\}\left(\frac{dw}{dz}\right)^2 - \left\{\frac{1}{z} + \frac{1}{z-1} + \frac{1}{w-z}\right\}\frac{dw}{dz}$$
$$+ \frac{w(w-1)(w-z)}{z^2(z-1)^2}\left\{\alpha + \frac{\beta z}{w^2} + \frac{\gamma(z-1)}{(w-1)^2} + \frac{\delta z(z-1)}{(w-z)^2}\right\}, \qquad \text{PVI}$$

where α, β, γ, and δ are arbitrary constants, these are known as the *Painlevé equations* and they require the introduction of new transcendental functions for their solution. These functions are called *Painlevé transcendents* (the six equations are often referred to as the Painlevé transcendents, although we shall use this term solely for their solutions). The first three equations, PI, PII, and PIII, were discovered by Painlevé, the fourth and fifth, PIV and PV, were added by Gambier, and the general form of the sixth, PVI, was given by Fuchs [263]. The solutions of PI, PII, and PIV are meromorphic functions of x. If the substitution $z = e^t$ is made in PIII and PV (i.e. $z = 0$ is a fixed singular point), then the solutions become meromorphic functions of t. However, for PVI, 0, 1, and ∞ are fixed critical points (in fact, they are all essential singularities), and hence the general solution of PVI is *not* meromorphic throughout the finite complex plane; furthermore, in contrast to the case of PIII and PV above, since there are three critical points, there exists no transformation which removes all of them from the finite complex plane. In reality, equation PVI contains the other five, since these may be obtained from it by passage to the limit and coalescence [256] (see also references [61, 270]). This is illustrated in the following diagram:

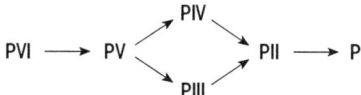

Since Painlevé proved that solutions of PI define new transcendental functions [271–275], then one expects that the solutions pf PII,...,PVI also do, except for special values of the constants α, β, γ, and δ. This is usually a difficult part of the proof that an equation possessed the Painlevé property (cf. reference [61, p. 347]), although in practice it is rarely done. However, recently, Joshi and Kruskal [276] have given a simple proof that solutions of the Painlevé equations have no movable essential singularities. Furthermore, Kruskal [258] has developed a method for detecting movable essential singularities which involves expansions in terms of large (or small) exponentials.

Although the Painlevé equations were first discovered from strictly mathematical considerations, they have recently appeared in several physical applications, including: asymptotics of nonlinear evolution equations [509, 277]; scattering of electromagnetic waves from a strip [278, 279]; time-dependent correlation functions of the one-dimensional X–Y model with spin $\frac{1}{2}$ [280, 281]; the spin–spin correlation function of the two-dimensional Ising Model [282, 283]; the one-particle reduced density matrix of the one-dimensional Bose gas [285, 286]; plasma physics [287]; resonant oscillations in shallow water [288]; natural convective flows with viscous dissipation [289–290]; Görtler vortices in boundary layers [291–293]; the study of polyelectrolytes [294]; quantum gravity [295–302]; general relativity [303–313]; and nonlinear and fibre optics [128, 314] (see also [267, 353, 315] and the references therein). Mason and Woodhouse [316] have recently shown that all six Painlevé equations arise as reductions of the self-dual Yang–Mills equations, a four-dimensional system having important applications in mathematical physics. The Painlevé equations may also be thought of as nonlinear analogues of the classical special functions.

7.2.4 Higher-degree and higher-order ODEs

Many physical problems are modelled by systems of first-order ODEs of the form (see Section 1.6 for a physical motivation)

$$d\mathbf{w}/dt = \mathbf{F}(\mathbf{w}, t), \qquad (7.156)$$

with $\mathbf{w} = (w_1, w_2, \ldots, w_n)$ and $\mathbf{F} = (F_1, F_2, \ldots, F_n)$ and where each F_k is analytic in t and rational with respect to its other arguments. Examples include the motion of stars in the galaxy, the three-body problem, nonlinear lattices, turbulence, chemically active media, competition of species in biology, wave interactions, and many other examples (cf. references [317, 318]). These systems describe the time evolution of a set of dynamical variables $w_k(t)$ and recently there has been considerable interest in determining when such systems are integrable, often using Painlevé analysis (see, for example, references [319–323]). Despite many extensive investigations, the precise relationship between the Painlevé property and integrability of nonlinear dynamical systems has yet to be rigorously established mathematically. However, there have been some notable results in recent years on the relationship between the Painlevé property and so-called algebraic complete integrability (cf. references [324–331]).

Systems of higher-order equations,

$$\frac{d^n w}{dz^n} = F\left(\frac{d^{n-1}w}{dz^{n-1}}, \ldots, \frac{dw}{dz}, w, z\right), \qquad (7.157)$$

can easily be rewritten in the form (7.156). Painlevé's α-method, which enabled a complete discussion of second-order equations of the form (7.155), can also be applied to equations of third (and higher) order of the form (7.157). As for the second-order case, the procedure divides into two parts; first the determination of necessary conditions for an equation to have no movable critical points, and then the proof that these conditions are also sufficient. The determination of many necessary conditions presents few difficulties: however, the proof that these conditions are also sufficient increases in difficulty with the order of the equations under investigation.

At present there are no comprehensive results for third- and higher-order equations despite several substantial investigations. Partial classifications have been given by Chazy [332], Garnier [333], Bureau [334], Lukashevich [335], and Martynov [336–339] for the third-order equation

$$\frac{d^3w}{dz^3} = F\left(\frac{d^2w}{dz^2}, \frac{dw}{dz}, w, z\right). \tag{7.158}$$

We remark that neither are there comprehensive results for the second-order, second-degree equation given by

$$\left(\frac{d^2w}{dz^2}\right)^2 = F\left(\frac{dw}{dz}, w, z\right)\frac{d^2w}{dz^2} + G\left(\frac{dw}{dz}, w, z\right), \tag{7.159}$$

where F and G are rational in dw/dz and w and analytic in z, which is effectively a subcase of eqn (7.158) (see references [340–342] and the references therein for some results). Cosgrove and Scoufis [343] have classified all equations of the special form of (7.159) with $F \equiv 0$ that are of Painlevé type. Despite the fact that the aforementioned papers on eqns (7.158) and (7.159) are of considerable length, the classification problem is far from complete. Furthermore, a curiosity is that, to date, no new equations have been discovered that require the introduction of new transcendental functions. All of the equations of the forms (7.158) and (7.159) investigated to date are solvable in terms of the six classical Painlevé transcendents, elliptic functions, elementary functions, or linear equations. Special cases of the second-order second-degree equation (7.159) have arisen in the study of exact solutions of Einstein's equations [344, 345], the theory of tau-functions [347, 348], and in investigations of transformation properties of Painlevé equations [62, 349].

An interesting feature of the second-order second-degree and third-order equations is the nature of the movable singularities that can occur. To date, these have been observed through exact solutions, for which considerable skill would be required in order to demonstrate their existence using Painlevé analysis.

For example, consider the second-order equation given by Painlevé [271, pp. 230–233] (see also references [61, p. 344] and [276]),

$$\left[\frac{d^2w}{dz^2} + \frac{w(1+k^2-2k^2w^2)}{(1-w^2)(1-k^2w^2)}\left(\frac{dw}{dz}\right)^2\right]^2 = \frac{1}{\lambda^2(1-w^2)(1-k^2w^2)}\left(\frac{dw}{dz}\right)^4,$$

which has the explicit solution

$$w(z) = \pm\operatorname{sn}[\lambda \ln(Az+B), k],$$

where A and B are arbitrary constants, and $\text{sn}(\xi; k)$ is the Jacobi elliptic function. The point $z = -B/A$ is a movable essential singularity, since it is an accumulation point of poles.

A major difficulty at third order is the occurrence of movable *natural boundaries*; a natural boundary being a line of singularities through which the solution cannot be analytically continued (cf. references [262, 350]). Chazy [332, 351, 352] studied the following ODEs:

$$\frac{d^3w}{dz^3} = 2w\frac{d^2w}{dz^2} - 3\left(\frac{dw}{dz}\right)^2 \tag{7.160}$$

and

$$\frac{d^3w}{dz^3} = 2w\frac{d^2w}{dz^2} - 3\left(\frac{dw}{dz}\right)^2 + \frac{4}{36-n^2}\left(6\frac{dw}{dz} - w^2\right)^2, \tag{7.161}$$

where n is an integer larger than 6. The general solution of eqn (7.161), and that of eqn (7.160) in the limit $n \to \infty$, is given parametrically by

$$w(z(s)) = \frac{6}{y_1}\frac{dy_1}{dz} = \frac{6}{y_1}\frac{dy_1}{ds}\frac{ds}{dz}, \quad z(s) = \frac{y_2(s)}{y_1(s)},$$

where $y_1(s)$ and $y_2(s)$ are two linearly independent solutions of the hypergeometric equation

$$s(1-s)\frac{d^2y}{ds^2} + (\tfrac{1}{2} - \tfrac{7}{6}s)\frac{dy}{ds} + \left(\frac{1}{4n^2} - \frac{1}{144}\right)y = 0.$$

The function $z(s)$ maps the upper half s-plane into the interior of a so-called spherical triangle with angles $\tfrac{1}{n}\pi$, $\tfrac{1}{2}\pi$, and $\tfrac{1}{3}\pi$ (cf. reference [350, p. 206]). The inverse function $s(z)$ is the Schwarzian triangle function $S(z; \tfrac{1}{n}, \tfrac{1}{2}, \tfrac{1}{3})$, which has a straight line or a circle as a natural boundary, and the fundamental triangle of which also has angles $\tfrac{1}{n}\pi$, $\tfrac{1}{2}\pi$, and $\tfrac{1}{3}\pi$. Consequently, every solution of the Chazy equation (7.160) is analytic either in a punctured plane or in a domain bounded by a straight line or a circle, the location of which is dependent on the constants of integration, and the solution cannot be analytically continued outside the domain (see, for example, references [353, §6.5] for further details about the Chazy equation).

We remark that recently there has been renewed interest in the Chazy equation (7.160). The Chazy equation is deeply connected to special automorphic functions (elliptic modular functions) which arise in various branches of mathematics; in particular, number theory (for further details, see, for example, reference [354]). A Painlevé analysis demonstrates that the Chazy equation also possesses three 'negative resonances' about which there is much current interest (cf. references [257, 355]). The Chazy equation (7.160) has assumed added importance, since it appears as a reduction of the self-dual Yang–Mills (SDYM) equations [356]. Ward [357] conjectured that in some sense all soliton equations arise as special cases of the SDYM equations. Subsequently, many of the well-known soliton equations, such as the Korteweg–de Vries, nonlinear Schrödinger, Sine–Gordon, Kadomtsev–Petviashvili, Davey–Stewartson, and Painlevé equations, have been discovered to be exact or asymptotic reductions of the SDYM equations (cf. reference [353, §6.5]). All these classical soliton equations arise when it is assumed that the Yang–Mills potentials take values in a finite-dimensional Lie algebra such as $\mathfrak{su}(2)$. By contrast, the Chazy equation arises when it is assumed that the Yang–Mills potentials take values in the infinite-dimensional Lie algebra $\mathfrak{sdiff}(SU(2))$ of all 'divergence-free' vector fields on SU(2) [356]. In fact, the Chazy equation is perhaps the simplest equation that arises from an infinite-dimensional Lie algebra. Therefore, the Chazy equation plays an important role in soliton theory and integrable systems.

7.2.5 The ARS algorithm

Ablowitz, Ramani, and Segur [358] give an algorithm, henceforth referred to as the ARS algorithm, for determining whether solutions of a given ODE (or system of ODEs) have movable branch points. This algorithm is similar to the method used by Kowalevski in her study of the motion of a rigid body rotating about a fixed point [359, 360].

In this section we shall use the ARS algorithm to demonstrate that solutions of

$$d^2w/dz^2 = 6w^2 + a(z), \quad (7.162)$$

where $a(z)$ is an analytic function, have no movable branch points. Recall that an ODE is said to be of *Painlevé type* if all solutions have no movable critical points; that is, the only movable singularities of its solutions in the finite complex plane are poles.

Step 1: find the dominant behaviour. The first step is to find the dominant behaviour of the solutions of eqn (7.162) in the neighbourhood of a movable singularity at $z = z_0$. We assume that

$$w(z) \sim c_0(z - z_0)^p \quad \text{as } z \to z_0, \quad (7.163)$$

where c_0 ($\neq 0$) and p are constants to be determined. Substituting eqn (7.163) into eqn (7.162), in order for the dominant terms to balance it is necessary that $p - 2 = 2p$, i.e. $p = -2$ and $c_0 = 1$. Hence

$$w(z) = (z - z_0)^{-2} + o\left((z - z_0)^{-2}\right) \quad \text{as } z \to z_0. \quad (7.164)$$

THE PAINLEVÉ TESTS   467

If p had not turned out to be an integer, the dominant behaviour would have been that of a movable algebraic branch point, so the equation would not have been of Painlevé type.

Step 2: Find the resonances. Since eqn (7.162) is of second order, its general solution has two constants of integration, and we expand eqn (7.164) in powers of $z - z_0$ until a second constant of integration appears (z_0 being the first). The second part of the algorithm is to find the power of $z - z_0$, called a *resonance*, at which this second constant appears. This resonance may be determined by substituting

$$w(z) = \xi^{-2} + \beta \xi^{r-2}, \tag{7.165}$$

with $\xi = z - z_0$, into the dominant terms of eqn (7.162) (d^2w/dz^2 and $6w^2$). To first order in β, this gives

$$\beta[(r-2)(r-3) - 12]\xi^{r-4} = 0, \tag{7.166}$$

which gives for r the algebraic equation

$$r^2 - 5r - 6 \equiv (r+1)(r-6) = 0. \tag{7.167}$$

The root $r = -1$ corresponds to the arbitrariness of z_0, while the root $r = 6$ indicates the location of the second arbitrary constant. If the second root had not been a real integer, it would have indicated the presence of a movable branch point at z_0.

Step 3: find the constants of integration. The final step is to substitute

$$w = \xi^{-2} + c_1 \xi^{-1} + c_2 + c_3 \xi + c_4 \xi^2 + c_5 \xi^3 + c_6 \xi^4 + O(\xi^5), \tag{7.168}$$

into eqn (7.162) and obtain a recurrence relation by equating powers of ξ [expanding $a(z)$ in a Taylor series about z_0]. There is little point in continuing the expansion (7.168) beyond the resonance (in higher-order equations, beyond the largest resonance), since the recurrence relations determines all the c_j for $j \geq 7$. Thus

$$\xi^{-3}: 2c_1 = 12c_1, \tag{7.169i}$$

$$\xi^{-2}: 0 = 12c_2 + 6c_1^2, \tag{7.169ii}$$

$$\xi^{-1}: 0 = 12c_3 + 12c_1 c_2, \tag{7.169iii}$$

$$\xi^0: 2c_4 = 12c_4 + 12c_1 c_3 + 6c_2^2 + a(z_0), \tag{7.169iv}$$

$$\xi^1: 6c_5 = 12c_5 + 12c_1 c_4 + 12c_2 c_3 + \frac{da}{dt}(z_0), \tag{7.169v}$$

$$\xi^2: 12c_6 = 12c_6 + 12c_1 c_5 + 12c_2 c_4 + 6c_3^2 + \tfrac{1}{2}\frac{d^2 a}{dz^2}(z_0). \tag{7.169vi}$$

Solving eqns (7.169i, ii, iii, iv) gives

$$c_1 = c_2 = c_3 = 0, \quad c_4 = -\tfrac{1}{10}a(z_0), \quad c_5 = -\tfrac{1}{6}\frac{da}{dt}(z_0). \quad (7.170)$$

Because eqn (7.169v) is a resonance equation, a_6 cancels out, leaving a *compatibility condition*, which must be identically satisfied for the series (7.168) to be the start of the Laurent series of the general solution in the neighbourhood of a movable pole. Using the known values of $c_1, \ldots, c_5$ we see that this compatibility condition amounts to

$$\frac{d^2 a}{dz^2}(z_0) = 0. \quad (7.171)$$

Since z_0 is arbitrary, it follows that eqn (7.162) is only free from movable branch points if $d^2 a/dz^2 = 0$, and hence has the linear form $a(z) = a_1 z + a_0$ where a_1 and a_0 are constants. If $d^2 a/dz^2 \neq 0$, then the compatibility condition is not identically satisfied and eqn (7.168) must be supplemented with logarithmic terms:

$$w(z) = \xi^{-2} + c_1 \xi^{-1} + c_2 + c_3 \xi + c_4 \xi^2 + c_5 \xi^3 + (c_6 + b_6 \ln \xi)\xi^4 + O(\xi^5). \quad (7.172)$$

Substituting this into eqn (7.162) and equating coefficients of powers of ξ, we find that $c_1, \ldots, c_5$ are as given in eqn (7.170) and $b_6 = -\tfrac{1}{7}(d^2 a/dz^2)(z_0)$ (higher terms contain higher powers of ξ and $\ln \xi$), and so we have a movable logarithmic branch point. In any case, c_6 is the second arbitrary constant.

Therefore, eqn (7.162) is of Painlevé type only if it has the form

$$d^2 w/dz^2 = 6w^2 + a_1 z + a_0. \quad (7.173)$$

To complete the proof, we now show that eqn (7.173) has no movable branch points. We prove that the expansion (7.168), (7.170) does indeed represent the beginning of a Laurent series of the general solution of eqn (7.173) in the neighbourhood of a movable pole. (In practice, such a calculation is rarely done—but should be for a rigorous proof.) For eqn (7.173) the expansion is

$$w(z) = \sum_{n=0}^{\infty} c_n (z - z_0)^{n-2}, \quad (7.174)$$

where

$$c_0 = 1, \quad c_1 = 0, \quad c_2 = 0, \quad c_3 = 0, \quad (7.175a)$$

$$c_4 = -\tfrac{1}{10}(a_1 z_0 + a_0), \quad c_5 = -\tfrac{1}{6}a_0, \quad c_6 \text{ is arbitrary}, \quad c_7 = 0, \quad (7.175b)$$

$$c_n = \frac{6}{(n+1)(n-6)} \sum_{j=4}^{n-4} c_j c_{n-j}, \quad \text{for } n \geq 8. \quad (7.175c)$$

To prove that this series is convergent in some neighbourhood of $z = z_0$, choose a positive number μ such that $|c_4| < \mu^4, |c_5| < \mu^5$, and $|c_6| < \mu^6$; then, from eqn (7.175),

$$|c_n| < \frac{6(n-7)\mu^n}{(n+1)(n-6)} < \mu^n, \quad \text{for } n \geq 8.$$

Therefore the right-hand side of eqn (7.174) converges in the punctured disc $0 < |z - z_0| < \mu^{-1}$, and obviously to the general solution of eqn (7.173).

This suggests that eqn (7.162) has no movable branch points if and only if it has the special form (7.173): however, to make a more convincing argument, we return to the original equation and solve by recursive asymptotics (we do not need to assume an expansion). We leave it to the interested reader to consult the proof in reference [61, p. 347] that eqn (7.173) also has no movable essential singularities.

The ARS algorithm is applicable to both higher-order ODEs and systems of ODEs. As in the above application, there are three basic steps: (i) find the dominant behaviour; (ii) find the resonances; and (iii) determine the constants of integration. For further discussion on the ARS algorithm, the reader should consult, for example, the seminal paper by Ablowitz, Ramani, and Segur [358], the book of Ablowitz and Segur [60], or the reviews [320–322].

7.2.6 The relationship between the Painlevé equations and inverse scattering

Ablowitz and Segur [361] have demonstrated a close connection between completely integrable PDEs and the Painlevé equations. Consider the following physically relevant examples.

EXAMPLE 7.22. *The Korteweg–de Vries equation.* The Korteweg–de Vries (KdV) equation

$$u_t + 6uu_x + u_{xxx} = 0, \tag{7.176}$$

is solvable by inverse scattering [59] (see Section 7.3). It possesses the symmetry reduction (a scaling or self-similar reduction)

$$u(x,t) = (3t)^{-2/3} v(z), \quad z = x/(3t)^{1/3},$$

where $v(z)$ satisfies

$$v''' + 6vv' - (2v + zv') = 0, \tag{7.177}$$

the solutions of which are expressible in terms of solutions of the second

Painlevé equation PII; there is a one-to-one correspondence between solutions of eqn (7.177) and those of PII (cf. reference [62]). The KdV equation (7.176) also has the symmetry reduction (an accelerating wave reduction)

$$u(x,t) = w(z) - \lambda t, \qquad z = x + 3\lambda t^2,$$

with λ an arbitrary constant, where $w(z)$ is solvable in terms of the first Painlevé equation PI (cf. references [4, 62]).

EXAMPLE 7.23. *The Sine–Gordon equation.* The Sine–Gordon equation,

$$u_{xt} = \sin u, \qquad (7.178)$$

is completely integrable [363] and has the scaling reduction

$$u(x,t) = v(z), \qquad z = xt. \qquad (7.179)$$

After making the transformation $w = e^{iv}$, $w(z)$ satisfies

$$z[ww'' - (w')^2] + ww' = \tfrac{1}{2}w(w^2 - 1), \qquad (7.180)$$

which is a special case of the third Painlevé equation PIII [361].

EXAMPLE 7.24. *The nonlinear Schrödinger equation.* The nonlinear Schrödinger equation

$$iu_t + u_{xx} = 2|u|^2 u, \qquad (7.181)$$

is completely integrable [99]. This equation has the scaling reduction

$$u(x,t) = (t)^{-1/2} w(z) \exp(i\mu \ln t), \qquad z = x/t^{1/2}, \qquad (7.182)$$

where $w(z)$ satisfies

$$w'' - \tfrac{1}{2}izw' = 2|w|^2 w + (\mu + \tfrac{1}{2}i)w, \qquad (7.183)$$

which is solvable in terms of the fourth Painlevé equation PIV (cf. reference [364]). It also possesses the accelerating wave reduction

$$u(x,t) = w(z)\exp\{i[\mu xt - \tfrac{2}{3}\mu^2 t^3]\}, \qquad z = x - \mu t^2, \qquad (7.184)$$

where $w(z)$ satisfies

$$w'' = 2|w|^2 w + \mu z w, \qquad (7.185)$$

which is solvable in terms of the second Painlevé equation PII.

For further examples of completely integrable PDEs which are reducible to the Painlevé equations through a symmetry reduction see, for example, references [60, 353, 358, 365, 366].

Inspired by the observation of Ablowitz and Segur [361], Ablowitz, Ramani, and Segur [358, 365] and Hastings and McLeod [368] formulated the *Painlevé Conjecture* or *Painlevé ODE test*:

Every ODE which arises as a symmetry reduction of a completely integrable PDE is of Painlevé type, perhaps after a transformation of variables.

This conjecture, if true, would provide a useful *necessary* condition to test whether a given PDE might be completely integrable. Ablowitz, Ramani, and Segur [367] and McLeod and Olver [369] have proved weakened versions of the Painlevé ODE test. These proofs are based on the fact that if a PDE is completely integrable, then its solutions are expressible in terms of the solutions of the Gel'fand–Levitan–Marchenko equation, which is a *linear* integral equation. Nevertheless, despite the absence of a proof of the Painlevé Conjecture, in a more general setting, there is considerable evidence which suggests that it is true (cf. references [60, 353, 358, 365, 366, 369]). Unfortunately, the Painlevé Conjecture has been widely misunderstood, with many authors interpreting it as a *sufficient* condition rather than a necessary condition (see the example below). Furthermore, the Painlevé Conjecture has a major limitation, in that it fails to draw any conclusion for PDEs that do not have any symmetry reductions.

7.2.7 The Painlevé ODE test

Since the formulation of the Painlevé ODE test, there has been considerable interest in using the Painlevé property as a means of determining whether given equations, both PDEs and ODEs, are integrable. The first person to use the Painlevé property as a method of identifying possible integrable equations was Sophie Kowalevski [359, 360], whose classic work was on the theory of the motion of a rigid body (a top) about a fixed point (see also references [267, 370]).

The Painlevé ODE test may be applied as follows: if a given PDE is reducible to an ODE which is *not* of Painlevé type (even after allowing for a possible change of variables), then the Painlevé ODE test predicts that the PDE is not completely integrable. An important point here is *not* whether a given PDE is reducible to one of the six Painlevé equations but, rather, whether an ODE arising as from symmetry reduction of the PDE is of Painlevé type. The one-parameter Lie group method of infinitesimal transformations which we described in Section 7.1 above is a systematic, algorithmic method for determining symmetry reductions of a given system of PDEs (although we recall that this method does *not* give all possible symmetry reductions for some PDEs, such as the Boussinesq equation (7.63) (cf. reference [93]). However, it is often straightforward to determine simple reductions, for example travelling wave solutions and self-similar or scaling solutions, by considering 'obvious' symmetries of a given PDE. Then one can determine whether the resulting ODE is of Painlevé type by using the ARS algorithm described in Section 7.2.6. We shall now consider some important examples to demonstrate the application of the Painlevé ODE test.

EXAMPLE 7.25. *Nonlinear Klein–Gordon equations.* Consider the nonlinear Klein–Gordon equation in characteristic co-ordinates,

$$u_{xt} = f(u), \tag{7.186}$$

where $f(u)$ is (i) a rational function, and (ii) a linear combination of exponentials $\exp(b_j u)$.

Case 7.25(i): $f(u)$, *a rational function*. In this case McLeod and Olver [369] showed that the only PDEs of the form (7.186) which are reducible to ODEs of Painlevé type by travelling wave solutions

$$u(x,t) = w(z), \qquad z = x - ct,$$

are those of the form

$$u_{xt} = a_3 u^3 + a_2 u^2 + a_1 u + a_0, \tag{7.187}$$

for some constants $a_3, a_2, a_1,$ and a_0. A PDE which is included in the form (7.187) is the ϕ^4 equation

$$u_{xt} = 2u^3 - u, \tag{7.188}$$

which is not thought to be completely integrable, since numerical evidence suggests that the interaction of solitary waves for eqn (7.188) is inelastic and so they do not behave like solitons [372].

Next we consider the symmetry reduction of eqn (7.187) given by

$$u(x,t) = w(z), \qquad z = xt;$$

then $w(z)$ satisfies

$$zw'' + w' = a_3 w^3 + a_2 w^2 + a_1 w + a_0,$$

with $' := d/dz$. It is easily shown that this equation is of Painlevé type if and only if $a_3 = a_2 = 0$. Hence the Painlevé ODE test predicts that if a PDE of the form (7.186) with $f(u) = \sum_{j=1}^{n} a_j u^j$ is completely integrable, then it is necessarily a linear equation.

Case 7.25(ii): $f(u) = \sum_{j=1}^{n} a_j \exp(b_j u)$. In this case McLeod and Olver [369] showed that the only PDEs of the form (7.168) which are reducible to ODEs of Painlevé type by travelling wave solutions

$$u(x,t) = w(z), \quad z = x - ct,$$

are those of the form

$$u_{xt} = a_2 e^{2\beta u} + a_1 e^{\beta u} + a_{-1} e^{-\beta u} + a_{-2} e^{-2\beta u}, \quad (7.189)$$

for some (possibly complex) constants β, a_2, a_1, a_{-1}, and a_{-2}. In fact, the singularities of u are not really poles, but rather 'pure logarithms' in the sense that u_x, u_t, and $\exp(\beta u)$ have only poles—this possibility is allowed by the final clause of the Painlevé ODE test given above. However, a PDE which is included in the form (7.189) is the double Sine–Gordon equation

$$u_{xt} = \sin u + \lambda \sin 2u, \quad (7.190)$$

with $\lambda \neq 0$ a constant, which is not thought to be completely integrable, since numerical evidence suggests that the interaction of solitary waves for eqn (7.190) is inelastic and so they do not behave like solitons [372].

Now suppose that we consider the symmetry reduction of eqn (7.189) given by

$$u(x,t) = v(z), \quad z = xt/\beta.$$

Then, after making the transformation $w = \exp(\beta v)$ (to put the equation into rational form), $w(z)$ satisfies

$$z[ww'' - (w')^2] + ww' = a_2 w^4 + a_1 w^3 + a_{-1} w + a_{-2},$$

with $' := d/dz$. This equation is of Painlevé type if and only if $a_2 a_1 = 0$ and $a_{-2} a_{-1} = 0$ [373]. Therefore the Painlevé ODE test predicts that if a PDE of the form (7.186) is completely integrable, then it is necessarily one of the three standard forms

$$u_{xt} = \sin u, \quad u_{xt} = e^u - e^{-2u}, \quad u_{xt} = e^u, \quad (7.191\text{–}7.193)$$

possibly after scaling and/or translation (which rules out the double Sine–Gordon equation (7.190)). Equation (7.191) is the Sine–Gordon equation, which is known to be completely integrable [363]. Equation (7.192) is also known to be completely integrable [240–242]. Equation (7.193) is the Liouville equation, and is reducible using a Bäcklund transformation to the linear PDE $v_{xt} = 0$ (cf. reference [373]). Therefore, we conclude from the Painlevé ODE test that the only PDEs of the form (7.186) that might be completely integrable are equivalent to one of the three equations (7.191)–(7.193).

Equation (7.192) also highlights an important aspect in the application of the Painlevé ODE test. It reduces under the symmetry reduction $u(x,t) = \ln w(z)$, with $z = xt$, to the equation

$$z[ww'' - (w')^2] + ww' = w^3 - 1. \tag{7.194}$$

This second-order equation is of Painlevé type and yet is *not* one of the 50 canonical equations on the list in Ince [61], which were given by Painlevé and his colleagues! However, if we make the transformation

$$w(z) = z^{-1/4}\eta(\xi), \qquad \xi = \tfrac{4}{3}z^{3/4},$$

then eqn (7.194) is transformed into a special case of the third Painlevé equation PIII. This shows the problem of using the list in Ince superficially; that is, without considering possible transformations. It also shows that it is often necessary to consider more than one symmetry reduction.

EXAMPLE 7.26: *The $(n+1)$-dimensional nonlinear Klein–Gordon equations.* In this example we consider the $(n+1)$-dimensional nonlinear Klein–Gordon equation of the types

$$\sum_{j=1}^{n} \frac{\partial^2 u}{\partial x_j^2} - \frac{\partial^2 u}{\partial t^2} = \begin{cases} 2\sinh u, \\ e^u, \\ e^u - e^{-2u}, \end{cases} \tag{7.195}$$

i.e. essentially the $(n+1)$-dimensional generalizations of eqns (7.191)–(7.193), all of which—as we have seen—are completely integrable if $n = 1$. We make the symmetry reduction

$$u(x_1, x_2, \ldots, x_n, t) = v(z), \qquad z = \tfrac{1}{2}(x_1^2 + x_2^2 + \cdots + x_n^2 - t^2). \tag{7.196}$$

Then, after making the transformation $w = e^v$, $w(z)$ satisfies

$$z[ww'' - (w')^2] + \tfrac{1}{2}(n+1)ww' = \begin{cases} w^3 - w, \\ w^3, \\ w^3 - 1, \end{cases} \tag{7.197}$$

respectively. In all three cases, it can be shown that, in the vicinity of a pole at z_0, as $z \to z_0$,

$$w(z) = \frac{2z_0}{(z-z_0)^2} - \frac{n-1}{z-z_0} + c_1 + \frac{n(n-1)}{3z_0}\ln(z-z_0) + o(1), \tag{7.198}$$

where c_1 is an arbitrary constant. Hence $w(z)$ has a movable logarithmic branch point and so eqn (7.197) is not of Painlevé type unless either $n = 0$

or $n = 1$. Therefore the Painlevé ODE test predicts that the $(n + 1)$-dimensional Klein–Gordon equations (7.195) are *not* completely integrable if $n \geq 2$. An overview of the results of symmetry analysis for the $(3 + 1)$-dimensional analogue of this equation is given in Section 7.5.

EXAMPLE 7.27. *The $(n + 1)$-dimensional nonlinear Schrödinger equation.* In this example we consider the $(n + 1)$-dimensional nonlinear Schrödinger equation

$$iu_t + \nabla^2 u = 2|u|^2 u, \tag{7.199}$$

which is the $(n + 1)$-dimensional generalization of the cubic nonlinear Schrödinger equation that is completely integrable if $n = 1$. Consider the symmetry reduction

$$u(x_1, x_2, \ldots, x_n, t) = w(z)e^{i\lambda t}, \quad z = (x_1^2 + x_2^2 + \cdots + x_n^2)^{1/2}, \tag{7.200}$$

where $w(z)$ satisfies

$$w'' + \frac{n-1}{z} w' - 2w^3 - \mu w = 0. \tag{7.201}$$

It can be shown that, in the vicinity of a pole at z_0, as $z \to z_0$,

$$w(z) = \frac{1}{\xi} - \frac{(n-1)}{6z_0} - \frac{(n^2 - 8n + 6\mu z_0^2 + 7)}{36 z_0^2} \xi$$

$$- \frac{(n-1)(4n^2 - 35n + 18\mu z_0^2 + 85)}{216 z_0^3} \xi^2$$

$$+ \left\{ c_4 + \frac{(n-1)(2n^3 - 21n^2 + 9\mu z_0^2 n + 72n - 18\mu z_0^2 - 80)}{135 z_0^4} \ln \xi \right\}$$

$$\times \xi^3 + o(\xi^3), \tag{7.202}$$

where $\xi = z - z_0$ and c_4 is an arbitrary constant. Hence $w(z)$ has a movable logarithmic branch point and so eqn (7.201) is not of Painlevé type unless either $n = 0$ or $n = 1$. Therefore the Painlevé ODE test predicts that the $(n + 1)$-dimensional Schrödinger equation (7.199) is *not* completely integrable if $n \geq 2$.

We remark that the crucial question is not whether a given PDE is reducible to one of the Painlevé equations, but rather whether it is reducible to an ODE which is of Painlevé type. If a given PDE is reducible to an ODE of Painlevé type, even to one of the Painlevé equations, then the Painlevé ODE test says nothing about whether the PDE might be integrable. The Painlevé ODE test requires that every reduction should give an ODE of Painlevé type: however, this is *not* sufficient.

For the example of the modified Benjamin–Bona–Mahoney equation,

$$u_t + u_x + u^2 u_x - u_{xxt} = 0, \qquad (7.203)$$

every ODE arising as a symmetry reduction of the equation obtained by the classical Lie group method is of Painlevé type [199]. However, numerical evidence suggests that the interaction of solitary waves for eqn (7.203) is inelastic and so they do not behave like solitons [374]. Whereas this does not contradict the assertion of the Painlevé ODE test, it strongly suggests that the Painlevé ODE test may *not* be generalized to provide a *necessary and sufficient* condition for a PDE to be completely integrable.

7.2.8 The Painlevé PDE test

The major drawback in applying the Painlevé ODE test to a given PDE is that one has to first reduce the PDE to an ODE (there may be several possible reductions) and then determine whether each resulting ODE is of Painlevé type (after allowing for possible transformations). Despite the existence of a systematic method of finding symmetry reductions (by exploiting the symmetries in the PDE, although, as we remarked, this method does not necessarily determine *all* possible symmetry solutions of a given PDE), the Painlevé ODE test is less effective for PDEs that have a limited number of such symmetries, such as the MBBM equation [199]; and if a PDE has no symmetries, then the Painlevé ODE test is ineffective. In principle, it would also be better if one could directly test the PDE.

Weiss, Tabor, and Carnevale [168] introduced the *Painlevé property for PDEs*, or *Painlevé PDE Test*, as a method of applying the Painlevé ODE test directly to a given PDE without having to reduce it to an ODE. A PDE is said to *possess the Painlevé property* if solutions of the PDE are 'single-valued' in the neighbourhood of noncharacteristic, movable singularity manifolds. Weiss, Tabor, and Carnevale [168] proposed a method of applying the Painlevé PDE Test to a given PDE, which is analogous to the ARS algorithm discussed above to determine whether a given PDE might be of Painlevé type, by seeking a solution of a given PDE in the form of a Laurent series

$$u(x_1, x_2, \ldots, x_n) = u(x) = \phi^{-p}(x) \sum_{j=0}^{\infty} u_j(x) \phi^j(x), \qquad (7.204)$$

where $u_j(x)$, $j = 0, 1, \ldots$, are analytic functions of $x = (x_1, x_2, \ldots, x_n)$, with

$u_0(x) \neq 0$, in the neighbourhood of a noncharacteristic, movable singularity manifold defined by $\phi(x) = 0$, where $\phi(x)$ is an analytic function of $x_1, x_2, \ldots, x_n$. Substituting eqn (7.204) into the equation and equating coefficients of like powers of ϕ determines p and defines recursion relations for u_n, for $n \geq 1$, of the form

$$(n - \beta_1)(n - \beta_2) \cdots (n - \beta_N) u_n = F_n(u_0, u_1, \ldots, u_{n-1}, \phi, x),$$

where N is the order of the equation, for some functional F_n. This defines u_n unless $n = \beta_j$ for some j, $1 \leq j \leq N$. The integers $n = \beta_1, \beta_2, \ldots, \beta_N$ are the *resonances* (commonly, $n = -1$ is a resonance and it is usually associated to the singularity manifold defined by $\phi = 0$ being arbitrary). For each positive integer resonance there is a compatibility condition (i.e. $F_\beta = 0$) which must be identically satisfied for the PDE to have a solution of the form (7.204), and then $u_\beta(x)$ is an arbitrary function. Essentially, in order for a given PDE to pass the Painlevé PDE test, it is required that p is an integer and there are $N - 1$ consistent recursion relations (i.e. all the compatibility conditions are satisfied), so that the series (7.204) contains the requisite number of arbitrary functions as required by the Cauchy–Kowalevski theorem ($\phi(x)$ is the Nth arbitrary function) and thus corresponds to the general solution of the equation. We remark that recently there have been studies into the role of negative resonances, suggesting that they are important (cf. references [257, 258, 355]).

Also, as for the Painlevé ODE test, it may be necessary to make a change of variables in order to apply the Painlevé PDE Test: for example, the Sine–Gordon equation,

$$u_{xt} = \sin u, \tag{7.205}$$

is known to be completely integrable [363], and in order to apply the Painlevé PDE Test we must first make the transformation $v = e^{iu}$ to put eqn (7.205) into rational form.

As pointed out by Ward [375], the singularity manifold must be noncharacteristic, since on a characteristic manifold *any* type of singularity may propagate (see also references [376–378]). For example, consider the linear wave equation

$$u_{tt} = u_{xx},$$

which has general solution

$$u(x,t) = f(x+t) + g(x-t), \tag{7.206}$$

where f and g are arbitrary, twice differentiable, functions. Clearly, f and g can possess any type of singularity, for example essential singularities, so the general solution (7.206) will also possess the same singularities on the characteristic manifolds. Such singularities clearly should not be permitted to have any bearing upon the Painlevé PDE Test. Characteristic manifolds for PDEs are the analogue of fixed singularities in ODEs, since they are determined by the PDE and not by the particular solution. We remark that Weiss [379] has suggested that *all* singularity manifolds, irrespective of whether or not they are characteristic, should be investigated in the application of the Painlevé PDE Test.

Weiss, Tabor, and Carnevale [168] did not attempt to prove any relationship between the Painlevé property and complete integrability: however, a 'proof' can be inferred from the 'proof' of the Painlevé ODE test given by McLeod and Olver [369]. As mentioned above, McLeod and Olver essentially proved that, under certain restrictions, the solution $u(x_1, x_2, \ldots, x_n)$ of a completely integrable PDE is meromorphic when $x_1, x_2, \ldots, x_n$ assume complex values.

Despite being by no means foolproof (cf. references [199, 258, 267, 353, 376, 380]), the Painlevé PDE test also provides a useful criterion for the identification of completely integrable PDEs. In addition to providing a valuable first test for whether a given PDE is completely integrable, other important information relating to completely integrable equations can be obtained by use of Painlevé analysis, including Bäcklund transformations, Lax pairs, Hirota's bilinear representation, and special and rational solutions (cf. references [322, 323, 376, 381–402]). Many of these results are obtained by seeking solutions of the PDE in the form of a *truncated Laurent series expansion*:

$$u(x) = u_0(x)\phi^{-p}(x) + u_1(x)\phi^{-p+1}(x) + \cdots + u_p(x). \quad (7.207)$$

We remark that equating the coefficient of each order of ϕ to zero can be too restrictive and a more general approach may be required (cf., [376]).

If a compatibility condition is not satisfied for arbitrary $\phi(x)$ (i.e. $F_\beta \neq 0$ for some β), then it is necessary to introduce terms of the form $\phi(x)^{\beta-p} \ln \phi(x)$ into the series (7.204) at this order to make the recursion relations consistent, thereby rendering it a so-called multivalued logarithmic psi series (i.e. a series of the form $\sum_{j=0}^{\infty} \sum_{k=0}^{j} u_{jk} \phi^j (\ln \phi^k)$. If, for special choices of $\phi(x)$, all of the compatibility conditions are identically satisfied (i.e. $\phi(x)$ satisfies a set of '*consistency conditions*' $F_\beta = 0$), then the equation is said to have the '*conditional Painlevé property*' [395]. In these cases useful information, such as special solutions, for nonintegrable equations can be obtained by using truncated Laurent series expansions (7.207) (cf. references [188, 189, 383, 384, 403–422]). Additionally, even if the equation possesses neither the Painlevé property nor the conditional Painlevé property for any choice of $\phi(x)$, then analysis of the associated logarithmic psi series can still yield valuable insights (cf. references [423, 424]).

Recently, Cosgrove [377, 378] has proposed the following definition of the Painlevé property for PDEs.

DEFINITION 7.28. A PDE is said to *possess the Painlevé property* if all the movable singularities (that is, the singularities the location of which involves one of the arbitrary functions of integration, provided that they are noncharacteristic), if any, of its general solutions are poles.

In two long papers, Cosgrove [377, 378] has classified all semilinear, second-order PDEs of the form

$$A(x,y)u_{xx} + B(x,y)u_{xy} + C(x,y)u_{yy} = F(x,y,u,u_x,u_y), \quad (7.208)$$

where A, B, and C are locally analytic in x and y, and F is rational in u, u_x, and u_y and locally analytic in x and y, that possess the Painlevé PDE property as defined in Definition (7.28). In the case of the semilinear PDE (7.208), Definition 7.28 requires that all singularities of $u(x,y)$ that lie on the curve $y = \phi(x)$, where ϕ is an arbitrary locally analytic function of integration, are poles. Cosgrove used a generalization of Painlevé's α-method to derive four sets of necessary conditions that the PDE must satisfy in order to possess the Painlevé property, and then used the Weiss, Tabor, and Carnevale Painlevé expansion technique to check whether these are sufficient as well. Hlavartý [425, 427] has obtained partial results on the classification of PDEs with the Painlevé property, although he imposes severe restrictions upon the class of PDEs considered.

7.2.9 Applications of the Painlevé PDE test

We shall now discuss an important physical example of the application of the Painlevé PDE test.

EXAMPLE 7.29. *A generalized nonlinear Schrödinger equation.* In this example, we consider the generalized nonlinear Schrödinger equation

$$iu_t + u_{xx} - 2|u|^2 u = a(x,t)u + b(x,t), \quad (7.209)$$

where $a(x,t)$ and $b(x,t)$ are analytic functions, which arises in several physical applications (cf. references [430, 115]), as has been outlined in Chapters 5 and 6 of this book. We now wish to determine the choices of the functions $a(x,t)$ and $b(x,t)$ such that eqn (7.209) passes the Painlevé PDE test, following reference [431]. Due to the nonanalyticity of the nonlinear term, we first complexify all variables and write eqn (7.209) as the system

$$iu_t + u_{xx} - 2u^2 v = a(x,t)u + b(x,t), \quad (7.210a)$$

$$-iv_t + v_{xx} - 2uv^2 = \bar{a}(x,t)v + \bar{b}(x,t), \quad (7.210b)$$

in which u and v are treated as *independent* complex functions of the (complex) variables x and t, and in which $\bar{a}(x,t)$ and $\bar{b}(x,t)$ are the formal complex conjugates of $a(x,t)$ and $b(x,t)$, respectively.

Now we seek solutions of eqn (7.210) in the form

$$u(x,t) = \phi^p(x,t) \sum_{j=0}^{\infty} u_j(t) \phi^j(x,t), \quad v(x,t) = \phi^q(x,t) \sum_{j=0}^{\infty} v_j(t) \phi^j(x,t),$$
$$(7.211)$$

with $\phi(x,t) = x + \psi(t)$, where $\psi(t)$ is an arbitrary analytic function and $u_j(t), v_j(t)$, $j = 0, 1, 2, \ldots$, are analytic functions, such that $u_0 \neq 0$, in the neighbourhood of a noncharacteristic movable singularity manifold defined by $\phi(x,t) = 0$. It is also necessary to expand the function $a(x,t)$ about the singularity manifold as follows:

$$a(x,t) = \sum_{j=0}^{\infty} a_j(t) \phi^j(x,t), \quad a_j(t) := \frac{1}{j!} \frac{\partial^j a}{\partial x^j}(x,t) \Big|_{x=-\psi(t)},$$

and similarly for $b(x,t)$, $\bar{a}(x,t)$ and $\bar{b}(x,t)$.

Leading order analysis shows that $p = q = -1$ and $u_0 v_0 = 1$. Equating the coefficients of powers of ϕ^{j-3} yields the general recursion relation

$$\mathbf{Q}(j) \begin{pmatrix} u_j \\ v_j \end{pmatrix} \equiv \begin{pmatrix} j^2 - 3j - 2 & -2u_0^2 \\ -2v_0^2 & j^2 - 3j - 2 \end{pmatrix} \begin{pmatrix} u_j \\ v_j \end{pmatrix} = \begin{pmatrix} F_j \\ G_j \end{pmatrix}, \quad (7.212a)$$

where

$$F_j = 2 \sum_{k=1}^{j-1} \sum_{l=0}^{k} u_l u_{k-l} v_{j-k} + 2v_0 \sum_{l=0}^{j-1} u_l u_{j-l} - i(j-2)u_{j-1} \frac{d\psi}{dt} - i \frac{du_{j-2}}{dt}$$

$$+ \sum_{k=0}^{j-2} a_k u_{j-k-2} + b_{j-3}, \quad (7.212b)$$

for $j \geq 1$ (define $u_j = 0$ and $v_j = 0$ for $j < 0$) and G_j is obtained from F_j by interchanging u_j and v_j and letting $i \to -i$. These recursion relations uniquely define u_j and v_j unless

$$\det \mathbf{Q}(j) = (j+1)j(j-3)(j-4) = 0.$$

Therefore the resonances are $j = -1, 0, 3, 4$. The resonances $j = -1$ and $j = 0$ correspond to the fact that $\psi(t)$ is arbitrary and that there is only one equation defining u_0 and v_0 (so one, say u_0, is arbitrary), respectively.

For $j = 1$ and $j = 2$, we find that

$$u_1 = -\tfrac{1}{2}iu_0\,d\psi/dt, \qquad v_1 = \tfrac{1}{2}iv_0\,d\psi/dt, \tag{7.213a}$$

$$u_2 = -\tfrac{1}{12}u_0(d\psi/dt)^2 - \tfrac{1}{12}u_0 w_0 + \tfrac{1}{6}u_0(\bar{a}_0 - 2a_0), \tag{7.213b}$$

$$v_2 = -\tfrac{1}{12}v_0(d\psi/dt)^2 + \tfrac{1}{12}v_0 w_0 + \tfrac{1}{6}u_0(a_0 - 2\bar{a}_0), \tag{7.213c}$$

where

$$w_0 = u_0 \frac{dv_0}{dt} - v_0 \frac{du_0}{dt}. \tag{7.213d}$$

Setting $j = 3$ in eqn (7.212), we see that

$$-2(v_0 u_3 + u_0 v_3) = F_3 v_0, \qquad -2(v_0 u_3 + u_0 v_3) = G_3 u_0,$$

therefore these are compatible if and only if $F_3 v_0 = G_3 u_0$, i.e. using eqns (7.212) and (7.213), if and only if

$$a_1 - \bar{a}_1 + v_0 b_0 - u_0 \bar{b}_0 = 0.$$

Since u_0 is arbitrary and $v_0 = 1/v_0$, then, necessarily,

$$a_1 = \bar{a}_1, \qquad b_0 = \bar{b}_0 \equiv 0.$$

By definition $b_0(t) = b(-\psi(t), t)$: therefore, since $\psi(t)$ is arbitrary, then necessarily

$$b(x, t) \equiv 0, \qquad \bar{b}(x, t) \equiv 0.$$

Similarly, the compatibility condition for $j = 4$ is

$$v_0 F_4 + u_0 G_4 = 0 = -\tfrac{1}{2}(a_0 - \bar{a}_0)^2 + \tfrac{1}{2}i\frac{d}{dt}(a_0 - \bar{a}_0) + a_2 + \bar{a}_2. \tag{7.214}$$

Then, by supposing that

$$a(x, t) = \alpha(x, t) + i\beta(x, t), \qquad \bar{a}(x, t) = \alpha(x, t) - i\beta(x, t),$$

it is easily seen that eqn (7.214) holds only if

$$\alpha(x, t) = x^2\left(\tfrac{1}{2}\frac{d\beta}{dt} - \beta^2\right) + x\alpha_1(t) + \alpha_0(t), \qquad \beta(x, t) = \beta(t),$$

where $\alpha_1(t)$, $\alpha_0(t)$, and $\beta(t)$ are arbitrary functions. Hence the Painlevé PDE test suggests that necessary conditions for the generalized nonlinear Schrödinger equation (7.209) to be completely integrable are that

$$a(x, t) = x^2\left(\tfrac{1}{2}\frac{d\beta}{dt} - \beta^2\right) + i\beta(t) + x\alpha_1(t) + \alpha_0(t), \qquad b(x, t) \equiv 0, \tag{7.215}$$

where $\alpha_1(t)$, $\alpha_0(t)$, and $\beta(t)$ are arbitrary functions (see reference [431] for further details). Unless $a(x,t)$ and $b(x,t)$ take these special forms (together with analogous ones for $\bar{a}(x,t)$ and $\bar{b}(x,t)$), then the series expansions (7.211) are inconsistent and logarithmic terms $\ln \phi$ are required at the order at which the compatibility condition is not satisfied; at higher powers of ϕ, higher and higher powers of $\ln \phi$ are required, an indication of non-Painlevé behaviour.

Additionally, it can be shown that eqns (7.215) are necessary and sufficient conditions for the generalized nonlinear Schrödinger equation (7.209) to be mapped into the standard nonlinear Schrödinger equation

$$i\eta_t + \eta_{\xi\xi} - 2|\eta|^2 \eta = 0$$

(see reference [431] for further details).

We remark that applying the Painlevé PDE test to the modified Benjamin–Bona–Mahoney equation (7.203) yields a negative result, in contrast to the fact that it satisfied the necessary conditions of the Painlevé ODE test [199]. This suggests that it is nonintegrable, which is in agreement with the numerical evidence that shows that the interaction of solitary waves for eqn (7.203) is inelastic, and so they do not behave like solitons [374].

As mentioned above, in the literature there has been much lively debate as to whether the Painlevé tests provide necessary and/or sufficient conditions for a given PDE to be completely integrable. As described above, the Painlevé tests are *necessary* conditions: however, some authors (e.g., Weiss, Tabor, and Carnevale [168]) interpret them as *sufficient* conditions.

A prime source of contention concerns quasi-linear PDEs such as the Dym equation [432],

$$u_t = (u^{-1/2})_{xxx}, \qquad (7.216)$$

which is solvable by inverse scattering [116]—see also reference [371]). The Dym equation can be transformed via hodograph transformations (i.e. transformations involving the interchange of dependent and independent variables) into both the KdV equation,

$$u_t + 6uu_x + u_{xxx} = 0$$

(cf. reference [433]), and the MKdV equation,

$$u_t - 6u^2 u_x + u_{xxx} = 0$$

(cf. reference [434]), both of which are solvable by inverse scattering and pass the Painlevé PDE test [168]. However, the Dym equation (7.216) does not *directly* pass the Painlevé PDE test, since it has an expansion of the form

$$u(x,t) = \phi^{-4/3}(x,t) \sum_{j=0}^{\infty} u_j(t) \phi^{j/3}(x,t),$$

with $\phi(x,t) = x + \psi(t)$, in the neighbourhood of an arbitrary noncharacteristic movable singularity manifold defined by $\phi = 0$ [169], and so it is 'weak Painlevé' (cf. references [435, 436]). Consequently, it might be conjectured that the 'weak-Painlevé' property would provide the requisite requirement, but this is not sufficient. For example, the higher KdV equation

$$u_t + u^3 u_x + u_{xxx} = 0$$

is also 'weak-Painlevé' [397]; yet it is thought not to be completely integrable since (i) it has only three independent polynomial conservation laws of a certain type [437], (ii) the interaction of solitary wave solutions is inelastic [438], and (iii) it appears not to be solvable by inverse scattering [369]. Therefore the 'weak-Painlevé' concept does not appear to distinguish between integrable and nonintegrable PDEs.

Recall that the Painlevé tests require that an integrable PDE possesses the Painlevé property possibly only after a transformation of variables, so that we may first have to make a change of variables before applying the tests. An open question remains as to what kind of transformations are allowable in the application of the Painlevé tests (i.e. Which transformations does one have to check?). It seems that completely integrable quasi-linear PDEs such as the Dym equation (7.216), which are 'weak-Painlevé' can be transformed into a PDE with the 'full-Painlevé' property through an appropriate hodograph transformation (for further examples, see references [114, 439]). Clarkson, Fokas, and Ablowitz [439] have developed an algorithmic method for transforming a quasi-linear PDE into a form seemingly more suitable for applying the Painlevé tests.

To summarize, the Painlevé ODE and PDE tests have been proven to provide useful criteria for determining whether a given PDE might be completely integrable. However, at present, they can only be regarded as 'rules of thumb', since there are several aspects of their application which are not satisfactorily resolved. One of the most important of these is the question of transformations. It is known that the Painlevé property is not invariant to a transformation, so how do we know if it is necessary to make a transformation in order to apply either of the Painlevé tests, and if so, what is the desired transformation? What transformations do we have to check?

Despite the 'success' of the Painlevé tests, the lack of rigour and absence of precise definitions are notable limitations in their effectiveness. For further discussion on the limitations and flexibility required in the application of the Painlevé tests, the reader is advised to consult, for example, references [258, 267, 353, 376, 440].

7.3 The inverse scattering method

7.3.1 Introduction

In this section we shall describe the inverse scattering method which was originally developed by Gardner, Greene, Kruskal, and Miura [59] in order to solve the Cauchy problem for the Korteweg–de Vries (KdV) equation. Subsequently, this was developed into a new method of mathematical physics, often referred to as the inverse scattering transform (I.S.T.), which has led to the solution of numerous nonlinear evolution equations, such as the nonlinear Schrödinger, Sine–Gordon, Modified KdV, and Boussinesq equations. Several of the nonlinear evolution equations solvable by inverse scattering, the so-called 'soliton equations', have arisen in many branches of physics, including water waves, stratified fluids, fibre optics, general relativity, plasma physics, statistical mechanics, and quantum field theory (see Chapter 1). Furthermore, during the past 20 years or so, soliton equations have attracted considerable interest from many branches of mathematics, including differential geometry, Hamiltonian mechanics, group theory, partial differential equations, and numerical analysis. Several books (see, e.g., references [60, 353, 371, 441–448]) and edited works (see, e.g., references [449–452]) have been written on various aspects of soliton theory. The description below may be viewed as a guide through the literature; there are many references given which give further details.

'Solitons' were first observed in 1834 by J. Scott Russell [453, 454] while riding on horseback beside the narrow Union canal near Edinburgh, Scotland. He described his observations as follows:

I was observing the motion of a boat which was rapidly drawn along a narrow channel by a pair of horses, when the boat suddenly stopped—not so the mass of water in the channel which it had put in motion; it accumulates round the prow of the vessel in a state of violent agitation, then suddenly leaving it behind, rolled forward with great velocity, assuming the form of a large solitary elevation, a rounded, smooth and well-defined heap of water, which continued its course along the channel apparently without change of form or diminution of speed. I followed it on horseback, and overtook it still rolling on at a rate of some eight or nine miles an hour, preserving its original figure some thirty feet long and a foot to a foot and a half in height. Its height gradually diminished, and after a chase of one or two miles I lost it in the windings of the channel. Such, in the month of August 1834, was my first chance interview with that rare and beautiful phenomenon which I have called the Wave of Translation....

Subsequently, Russell carried out extensive experiments in a laboratory scale wave tank in order to study this phenomenon more carefully. Russell observed solitary waves, which are long, shallow, water waves of permanent form, and hence he deduced that they *exist*; this is his most significant result. Second, he determined that the speed of propagation, c, of a solitary wave in a channel of uniform depth h is given by $c^2 = g(h + \eta)$, where η is the amplitude of the wave and g is the force due to gravity.

Subsequent investigations were undertaken by Airy [455], Stokes [456], Boussinesq [84, 85], and Rayleigh [457]. Boussinesq and Rayleigh independently obtained approximate descriptions of the solitary wave; Boussinesq derived a one-dimensional nonlinear evolution equation, which now bears his name,

$$u_{tt} = u_{xx} + (u^2)_{xx} + u_{xxxx}, \tag{7.217}$$

in order to obtain his result.

These investigations provoked much lively discussion and controversy as to whether the inviscid equations of water waves would possess such solitary wave solutions. The issue was finally resolved by Korteweg and de Vries [458]. They derived a nonlinear evolution equation governing long one-dimensional, small-amplitude, surface gravity waves propagating in a shallow channel of water,

$$\frac{\partial \eta}{\partial \tau} = \frac{3}{2}\sqrt{\frac{g}{h}} \frac{\partial}{\partial \xi}\left(\frac{1}{2}\eta^2 + \frac{2}{3}\alpha\eta + \frac{1}{3}\sigma \frac{\partial^2 \eta}{\partial \xi^2}\right), \qquad \sigma = \frac{1}{3}h^3 - \frac{Th}{\rho g}, \tag{7.218}$$

where η is the surface elevation of the wave, h is the equilibrium level, α is a small arbitrary constant related to the uniform motion of the liquid, g is the gravitational constant, T is the surface tension, and ρ is the density. The adjectives 'long' and 'small' are in comparison to the depth of the channel.

The controversy was now resolved, since eqn (7.218), now known as the Korteweg–de Vries (KdV) equation, has permanent wave solutions, including *solitary wave solutions* which we shall discuss later. Equation (7.218) may be brought into nondimensional form by making the transformation

$$t = \frac{1}{2}\sqrt{\frac{g}{h\sigma}}\,\tau, \qquad x = -\sigma^{-1/2}\xi, \qquad u = \frac{1}{2}\eta + \frac{1}{3}\alpha.$$

Hence, we obtain

$$u_t + 6uu_x + u_{xxx} = 0, \tag{7.219}$$

where subscripts denote partial differentiations and the constant coefficients may be modified arbitrarily by suitably scaling the variables.

Despite the early derivation of the KdV equation in 1895, it was not until 1960 that any new application of the equation was discovered. Gardner and Morikawa [459] rediscovered the KdV equation in the study of collision-free hydromagnetic waves. Subsequently, the KdV equation has arisen in a number of other physical contexts, stratified internal waves in the ocean, ion-acoustic waves, plasma physics, lattice dynamics, etc. (cf. reference [60] and the references therein).

It has been well known for a long time that the KdV equation (7.219) has a special kind of solitary wave solution,

$$u(x,t) = 2k^2 \operatorname{sech}^2\{k(x - 4k^2 t - \delta_0)\}, \qquad (7.220)$$

in which k and δ_0 are constants. Note that the velocity of this wave, $4k^2$, is proportional to its amplitude $2k^2$, so that taller solitary waves are narrower and faster than shorter ones. Solitary waves were first observed scientifically by John Scott Russell in 1834, yet the remarkable properties of solitary wave solutions of the KdV equation were unknown until Zabusky and Kruskal [460] solved eqn (7.219) numerically, and made the remarkable discovery that these solitary wave solutions have the property that the interaction of two solitary wave solutions is elastic. Zabusky and Kruskal were studying the Fermi–Pasta–Ulam (FPU) problem of a one-dimensional anharmonic lattice of equal masses coupled by nonlinear strings, which in the continuum limit reduces to the KdV equation. When two solitary wave solutions of the KdV equation are initially well separated, with the larger to the left (since they travel to the right), then the larger, faster one catches up with the smaller, slower one, and they overlap and interact nonlinearly. Zabusky and Kruskal [460] discovered that after the interaction, the waves separate, with the larger one on the right, having regained their initial amplitudes and velocity, and that the only effect of the interaction is a phase shift, so that the centres of the waves are at different positions from where they would have been without the interaction.

At the centre of these observations is the discovery that these nonlinear waves can interact elastically, and continue afterwards almost as if there had been no interaction at all. Because of the analogy with particles, Zabusky and Kruskal called these solitary waves *solitons*. Their remarkable numerical discovery demanded an analytical explanation and detailed mathematical study of the KdV equation. However, the KdV equation is nonlinear, and at that time no general method of solution for nonlinear equations existed.

Following the discovery of the soliton by Zabusky and Kruskal, it was found that the KdV equation has several other remarkable properties, including the possession of an infinite number of polynomial conservation laws. A conservation law (here) is an equation of the form

$$\partial T/\partial t + \partial X/\partial t = 0,$$

where T is the *conserved density* and X the *associated flux*. After studying these conservation laws and those associated with another equation, the modified KdV (MKdV) equation

$$v_t - 6v^2 v_x + v_{xxx} = 0, \qquad (7.221)$$

Miura [461] discovered that given a solution v of eqn (7.221),

$$u = -v^2 - v_x, \qquad (7.222)$$

is a solution of the KdV equation. This relationship, known as the *Miura transformation*, led to the development by Gardner, Greene, Kruskal, and Miura [59] of the inverse scattering method of solving the initial value problem for the KdV equation, subject to the initial condition

$$u(x,0) = f(x),$$

where $f(x)$ decays sufficiently rapidly at infinity.

The basic idea is to associate with the KdV equation the time-independent Schrödinger equation

$$L\psi := \psi_{xx} + u(x,t)\psi = \lambda\psi, \qquad (7.223)$$

which has been extensively studied by mathematicians and physicists. Equation (7.222) may be viewed as a Riccati equation for v in terms of u, and it is well known that it may be linearized by the transformation $v = \psi_x/\psi$, which yields

$$u + v^2 + v_x = u + \frac{\psi_x^2}{\psi^2} + \left(\frac{\psi_{xx}}{\psi} - \frac{\psi_x^2}{\psi^2} \right) = 0,$$

and so

$$\psi_{xx} + u\psi = 0. \qquad (7.224)$$

Since the KdV equation is Galilean invariant—that is, it is invariant under the transformation

$$(x, t, u(x,t)) \to (x - 6\lambda t, t, u(x,t) + \lambda),$$

where λ is some constant—then it is natural to consider eqn (7.223) rather than eqn (7.224), in which t plays the role of a parameter and $u(x,t)$ the potential.

For eqn (7.223), the eigenvalues and the behaviour of the eigenfunctions as functions of x determine the *scattering data*, $S(\lambda, t)$, which depends upon the potential $u(x, t)$. The *direct scattering problem* is to map the potential into the scattering data. The *inverse scattering problem* is to reconstruct the potential from the scattering data.

The time dependence of the eigenfunctions of eqn (7.223) is given by

$$\psi_t = (\gamma + u_x)\psi - (4\lambda + 2u)\psi_x, \tag{7.225}$$

where γ is an arbitrary constant. Assuming that $\gamma_t = 0$, then from eqns (7.223) and (7.225) we obtain

$$\psi_{txx} = [(\gamma + u_x)(\lambda - u) + u_{xxx} + 6uu_x]\psi - (4\lambda + 2u)(\lambda - u)\psi_x, \tag{7.226a}$$

$$\psi_{xxt} = [(\lambda - u)(\gamma + u_x) - u_t]\psi - (\lambda - u)(4\lambda + 2u)\psi_x. \tag{7.226b}$$

Therefore eqns (7.223) and (7.225) are compatible (i.e. $\psi_{xxt} = \psi_{txx}$), if and only if u satisfies the KdV equation (7.219). Similarly, if eqn (7.219) is satisfied, then necessarily the eigenvalues must be time independent (i.e. $\partial \lambda / \partial t = 0$).

The solution of the KdV equation (7.219) using the inverse scattering method may be summarized as follows (we shall give further details in Section 7.3.3).

Step 1: direct problem. At time $t = 0$, given $u(x, 0)$ we solve the direct scattering problem. The spectrum of the Schrödinger equation (7.223) consists of a finite number of discrete eigenvalues, $\lambda = \kappa_n^2$, $n = 1, 2, \ldots, N$, for $\lambda > 0$ and a continuum, $\lambda = -k^2$, for $\lambda < 0$. The eigenfunctions corresponding to these eigenvalues may be computed and their asymptotic behaviour written as follows. For $0 < \lambda = \kappa_n^2$,

$$\psi_n(x, 0) \sim c_n(0)\exp(-\kappa_n x), \quad \text{as } x \to -\infty, \tag{7.227}$$

with

$$\int_{-\infty}^{\infty} \psi_n^2(x, 0)\,dx = 1; \tag{7.228}$$

and for $0 > \lambda = -k^2$,

$$\psi(x, 0) \sim a(k, 0)e^{-ikx} + b(k, 0)e^{ikx}, \quad \text{as } x \to \infty, \tag{7.229a}$$

$$\psi(x, 0) \sim e^{-ikx}, \quad \text{as } x \to -\infty; \tag{7.229b}$$

where $\rho(k, 0) \equiv b(k, 0)/a(k, 0)$ is the *reflection coefficient* and $\tau(k, 0) \equiv 1/a(k, 0)$ the *transmission coefficient*. Therefore we have the scattering data at time $t = 0$:

$$S(\lambda, 0) = \left(\{\kappa_n, c_n(0)\}_{n=1}^{N}, \rho(k, 0), \tau(k, 0)\right).$$

THE INVERSE SCATTERING METHOD

Step 2: time evolution. From eqn (7.225) we can determine the time evolution of the scattering data. It may be shown that

$$\kappa_n = \text{constant}, \qquad n = 1, 2, \ldots, N, \qquad (7.230\text{a})$$
$$c_n(t) = c_n(0)\exp(4\kappa_n^3 t), \qquad n = 1, 2, \ldots, N, \qquad (7.230\text{b})$$
$$\tau(k, t) = \tau(k, 0), \qquad (7.230\text{c})$$
$$\rho(k, t) = \rho(k, 0)\exp(8ik^3 t), \qquad (7.230\text{d})$$

so we have the scattering data at time t:

$$S(\lambda, t) = \left(\{\kappa_n, c_n(t)\}_{n=1}^N, \rho(k, t), \tau(k, t) \right).$$

Step 3: inverse problem. Given the scattering data at the initial time, it is possible to determine its time evolution. The inverse scattering problem is to reconstruct from knowledge of the scattering $S(\lambda, t)$ data, the potential $u(x, t)$ which is the required solution of the KdV equation. This problem was considered by Gel'fand and Levitan [462]; see also references [463, 464].

The inverse scattering problem solution may be summarized as follows. First, using the scattering data in eqn (7.230), define the function

$$F(\xi, t) = \sum_{n=1}^N c_n^2(t)\exp(-\kappa_n \xi) + \frac{1}{2\pi} \int_{-\infty}^{\infty} \rho(k, t) e^{ik\xi} \, dk. \quad (7.231)$$

Then solve the linear integral equation

$$K(x, y; t) + F(x + y; t) + \int_x^{\infty} K(x, z, t) F(z + y, t) \, dz = 0, \quad (7.232)$$

known as the *Gel'fund–Levitan–Marchenko* equation. Finally, reconstruct the potential by the relation

$$u(x, t) = 2 \frac{\partial}{\partial x} [K(x, x; t)]. \qquad (7.233)$$

This method is therefore conceptually analogous in many ways to the Fourier transform method for solving linear equations, which we discuss in the following subsection, except, however, that the final step of solving the inverse scattering problem is highly nontrivial; in Section 7.3.3, we shall view this technique in a different, although equivalent, way; namely as a

Riemann–Hilbert boundary value problem. Schematically, the inverse scattering method may be written as

$$u(x,0) \xrightarrow{\text{direct scattering}} S(\lambda,0) = \left(\{\kappa_n, c_n(0)\}_{n=1}^{N}, \rho(k,0), \tau(k,0)\right)$$

$$\downarrow : \text{time evolution}$$

$$u(x,t) \xleftarrow{\text{inverse scattering}} S(\lambda,t) = \left(\{\kappa_n, c_n(t)\}_{n=1}^{N}, \rho(k,t), \tau(k,t)\right).$$

The scattering data plays the role of the Fourier transform and the inverse scattering problem the inverse Fourier transform (cf. section 7.3.2).

In the special case in which $\rho(k,0) = 0$, then $\rho(k,t) = 0$ for all t, and so the integral equation (7.232) is degenerate and is solvable in closed form [465, 466] (see below). In this case the potential $u(x,0)$ is said to be *reflectionless*, there is no continuous spectrum, and it gives rise to the special soliton solutions.

Shortly after the discovery of the inverse scattering method by Gardner, Greene, Kruskal, and Miura [59], Lax [467] showed that the KdV equation can be interpreted as the compatibility condition for two linear operators L and M. If L is the operator of the spectral problem (7.223) and M is the operator of the associated time evolution of the eigenfunction

$$\psi_t = M\psi := (\gamma + u_x)\psi - (4\lambda + 2u)\psi_x, \tag{7.234}$$

then L and M satisfy

$$L_t = [M, L] := ML - LM \tag{7.235}$$

(assuming that $\psi_t \neq 0$). For L and M as chosen here, eqn (7.235) implies that $u(x,t)$ satisfies the KdV equation (7.219). For any given PDE, such a commutator representation, if it exists, is called a *Lax representation* and the operators a *Lax pair*.

Zakharov and Shabat [99] extended Lax's ideas to solve the nonlinear Schrödinger equation

$$iq_t = q_{xx} \pm 2|q|^2 q. \tag{7.236}$$

Ablowitz, Kaup, Newell, and Segur [363] then generalized the technique to solve a wide class of physically interesting nonlinear PDEs, including the MKdV equation (7.221) and the Sine–Gordon equation

$$u_{xt} = \sin u$$

(see Section 7.3.5). Subsequently, many other physically interesting (1 + 1)-dimensional nonlinear PDEs—that is, PDEs in one spatial and one temporal dimensions—were solved by this method; for example, the Boussinesq equation (7.217), the three-wave interaction equation, and the massive Thirring model (see, for example, the books by Ablowitz and Clarkson [353] and Ablowitz and Segur [60] for further details).

The inverse scattering technique has also been extended to solve other types of equations.

EXAMPLE 7.30. *Ordinary differential equations.* The second Painlevé equation

$$d^2y/dx^2 = 2y^3 + xy + \alpha, \qquad (7.237)$$

in which α is a constant, was analysed by Ablowitz, Ramani, and Segur [367], Flaschka and Newell [468], and Fokas and Ablowitz [469].

EXAMPLE 7.31. *Differential-difference equations.* The Toda lattice equation

$$d^2u_n/dt^2 = \exp\{-(u_n - u_{n-1})\} - \exp\{-(u_{n+1} - u_n)\} \qquad (7.238)$$

was studied by Flaschka [470, 471] and the discrete nonlinear Schrödinger equation

$$i\,du_n/dt = (u_{n+1} + u_{n-1} - 2u_n) + \sigma\,|u_n|^2(u_{n+1} + u_{n-1}), \qquad (7.239)$$

with $\sigma = \pm 1$, by Ablowitz and Ladik [472–475].

EXAMPLE 7.32. *Integro-differential equations.* The Benjamin–Ono equation

$$u_t + 2uu_x + Hu_{xx} = 0, \qquad (7.240)$$

where Hu is the Hilbert transform

$$(Hf)(x) = \frac{1}{\pi}\int_{-\infty}^{\infty} \frac{f(y)}{y-x}\,dy,$$

in which $\int_{-\infty}^{\infty}$ is the Cauchy principal value integral, was investigated by Fokas and Ablowitz [476]. The Intermediate Long Wave equation

$$u_t + \delta^{-1}u_x + 2uu_x + Tu_{xx} = 0, \qquad (7.241)$$

where δ is a constant and Tu is the singular integral operator

$$(Tf)(x) = \frac{1}{2\delta}\int_{-\infty}^{\infty} \coth\left\{\frac{\pi}{2\delta}(y-x)\right\}f(y)\,dy$$

is studied by Kodama, Ablowitz, and Satsuma [477].

EXAMPLE 7.33. *(2 + 1)-dimensional equations.* The Kadomtsev–Petviashvili equation

$$(u_t + 6uu_x + u_{xxx})_x = -3\sigma^2 u_{yy}, \qquad (7.242)$$

with $\sigma^2 = \pm 1$ was investigated by Dryuma [478], Manakov [97], Ablowitz, BarYaacov, and Fokas [98], and Fokas and Ablowitz [96], and the Davey–Stewartson equation

$$iq_t - \sigma_1 q_{xx} + q_{yy} = 2\sigma_2 |q|^2 q + 2q\phi, \qquad (7.243a)$$

$$\phi_{xx} + \sigma_1 \phi_{yy} = -2\sigma_1 \sigma_2 (|q|^2)_{xx}, \qquad (7.243b)$$

with $\sigma_1 = \pm 1$ and $\sigma_2 = \pm 1$, by Ablowitz and Haberman [480] and Fokas and Ablowitz [479].

An equation that is solvable by some form of the inverse scattering method is often referred to as being *completely integrable*, this terminology arising from the interpretation of the KdV equation as a completely integrable Hamiltonian system [481].

Completely integrable equations all seem to possess several remarkable properties in common, including:

- the 'elastic' interaction of solitary waves, i.e. multi-soliton solutions
- Bäcklund transformations
- an infinite number of independent conservation laws
- a complete set of action-angle variables
- an underlying Hamiltonian formulation
- a Lax representation
- a bilinear representation *á la* Hirota
- an associated linear eigenvalue problem, the eigenvalues of which are constants of the motion

(cf. references [60, 353]). However, the precise interrelationships between all these properties has yet to be rigorously formulated. A fundamental open question is *What is it that really characterizes completely integrable equations?*

7.3.2 Fourier transforms

Prior to discussing the inverse scattering method in more detail, we shall discuss a general method for solving *linear* partial differential equations, namely the Fourier transform method.

Consider the PDE

$$u_t = \Delta(x, t, u(x, t)), \qquad (7.244)$$

where $\Delta(x, t, u)$ is a function of $u(x, t)$ and its spatial derivatives, with

$t \in \mathbb{R}$ and $x \in \mathbb{R}$, the temporal and spatial variables, respectively. A typical problem associated with eqn (7.244) is to solve it subject to the given initial condition

$$u(x,0) = f(x), \qquad (7.245)$$

provided that the problem is well posed for f in some Banach space $\mathcal{B}$ (usually either a space of functions vanishing sufficiently rapidly as $|x| \to \infty$ such as $f \in L^1(\mathbb{R})$, or a space of periodic functions). Given the linear PDE in the form

$$u_t = -i\omega(-i\partial_x)u, \qquad (7.246)$$

where $\partial_x \equiv \partial/\partial x$ and ω is a polynomial function with constant coefficients, referred to as the dispersion relation. A solution of eqn (7.246) is given by

$$u(x,t) = \exp\{i[kx - \omega(k)t]\}, \qquad (7.247)$$

as is easily verified. The solution of eqn (7.246), given the initial condition (7.245), is obtained by use of the Fourier transform pair

$$u(x,t) = \frac{1}{2\pi} \int_{-\infty}^{\infty} U(k,t)e^{ikx}\,dk, \qquad (7.248)$$

$$U(k,t) = \int_{-\infty}^{\infty} u(x,t)e^{-ikx}\,dx. \qquad (7.249)$$

Assuming the validity of the interchange of derivative and integral, by taking the Fourier transform of eqn (7.246) we obtain a linear ODE for $U(k,t)$

$$dU/dt = -i\omega(k)U, \qquad (7.250)$$

which has the general solution

$$U(k,t) = U(k,0)\exp\{-i\omega(k)t\}, \qquad (7.251)$$

where

$$U(k,0) = F(k) = \int_{-\infty}^{\infty} f(x)e^{-ikx}\,dx. \qquad (7.252)$$

Therefore, from eqn (7.248),

$$u(x,t) = \frac{1}{2\pi} \int_{-\infty}^{\infty} F(k)\exp\{i[kx - \omega(k)t]\}\,dk. \qquad (7.253)$$

Conceptually, the Fourier transform method may be viewed as follows:

1. The initial data is transformed into Fourier space by means of eqn (7.252).

2. The time evolution in Fourier space is particularly simple, since it satisfies eqn (7.250).

3. $u(x,t)$ is recovered from the inverse transform (7.248) and is given by eqn (7.253).

Schematically, this may be written as:

$$u(x,0) \xrightarrow{\text{Fourier transform}} U(k,0)$$

$$\downarrow \omega(k) \text{: dispersion relation}$$

$$u(x,t) \xleftarrow{\text{inverse Fourier transform}} U(k,t)$$

Although eqn (7.253) is a solution in quadrature form, useful information may be obtained from it in the asymptotic limit $t \to \infty$. If the system is conservative ($\omega(k)$ is real for real k) and dispersive ($\omega''(k) \neq 0$), then the initial data decays into wave packets which move with their group velocity $\omega'(k)$ and decay algebraically as $t \to \infty$. In general, asymptotic formulae can be given to describe this; see also below. Hence the asymptotic behaviour of these linear problems is relatively simple.

EXAMPLE 7.34. *The linearized Korteweg–de Vries equation*. In this example we consider the linearized KdV equation

$$u_t + u_{xxx} = 0, \qquad (7.254)$$

with initial condition $u(x,0) = f(x)$, where $f(x)$ decays sufficiently rapidly as $|x| \to \infty$. The dispersion relation is given by $\omega(k) = -k^3$ and so the Fourier transform solution of eqn (7.254) is

$$u(x,t) = \frac{1}{2\pi} \int_{-\infty}^{\infty} F(k) \exp\{i(kx + k^3 t)\} \, dk, \qquad (7.255a)$$

where

$$F(k) = \int_{-\infty}^{\infty} f(x) e^{-ikx} \, dx. \qquad (7.255b)$$

If the initial condition is given by $f(x) = \delta(x)$, the Dirac delta function, then $F(k) = 1$ and so

$$u(x,t) = \frac{1}{2\pi} \int_{-\infty}^{\infty} \exp\{i(kx + k^3 t)\} \, dk = (3t)^{-1/3} \text{Ai}\left(\frac{x}{(3t)^{1/3}}\right),$$

where $\text{Ai}(z)$ is the Airy function, defined by

$$\text{Ai}(z) = \frac{1}{2\pi} \int_{-\infty}^{\infty} \exp\{i(sz + \tfrac{1}{3}s^3)\} \, ds = \frac{1}{\pi} \int_{0}^{\infty} \cos\{i(sz + \tfrac{1}{3}s^3)\} \, ds.$$

If the initial condition is given by $f(x) = \exp(-x^2)$, then $F(k) = \sqrt{\pi}\exp(-\tfrac{1}{4}k^2)$ and so

$$u(x,t) = \frac{1}{2\sqrt{\pi}} \int_{-\infty}^{\infty} \exp\{i(kx + k^3 t) - \tfrac{1}{4}k^2\} dk.$$

7.3.3 The Korteweg–de Vries equation

The prototype example for inverse scattering in one spatial dimension is the Korteweg–de Vries (KdV) equation,

$$u_t + 6uu_x + u_{xxx} = 0. \tag{7.256}$$

Recall that the Lax pair is

$$L\psi = \psi_{xx} + [u(x,t) + \lambda]\psi, \qquad \psi_t = M\psi = (u_x + \gamma)\psi - (2u + \lambda)\psi_x, \tag{7.257a,b}$$

where γ is an arbitrary constant parameter and λ is the spectral parameter. We shall now outline the inverse scattering method for solving the KdV equation; for further details, see Ablowitz and Clarkson [353].

The direct scattering problem

Suppose that $\lambda = -k^2$. Then eqn (7.257) becomes

$$\psi_{xx} + \{u(x) + k^2\}\psi = 0, \qquad \psi_t = (u_x + \gamma)\psi + (4k^2 - 2u)\psi_x, \tag{7.258a,b}$$

where we have suppressed the time dependence in u for convenience. We shall further assume that $u(x)$ lies in the space P_μ, where

$$P_\mu = \left\{ u(x) : \int_{-\infty}^{\infty} (1 + |x|^\mu) |u(x)| dx < \infty \right\},$$

for either $\mu = 1$ or $\mu = 2$ (see reference [464] for the case $\mu = 2$; P_μ is sometimes referred to as L_μ^1). The case $\mu = 1$ was the original condition given by Faddeev [463] and has been shown to be valid by Marchenko [482].

Consider the eigenfunctions associated with eqn (7.258a), which are bounded for all values of x and have the following asymptotic behaviour:

$$\phi(x;k) \sim e^{-ikx}, \quad \overline{\phi}(x;k) \sim e^{ikx}, \qquad \text{as } x \to -\infty, \tag{7.259a}$$

$$\psi(x,k) \sim e^{ikx}, \quad \overline{\psi}(x;k) \sim e^{-ikx}, \qquad \text{as } x \to \infty. \tag{7.259b}$$

It turns out that it is more convenient to work with the (modified) eigenfunctions $M(x;k)$, $\overline{M}(x;k)$, $N(x;k)$, and $\overline{N}(x;k)$, defined by

$$M(x;k) := \phi(x;k)e^{ikx}, \qquad \overline{M}(x;k) := \overline{\phi}(x;k)e^{ikx}, \qquad (7.260a)$$
$$N(x;k) := \psi(x;k)e^{ikx}, \qquad \overline{N}(x;k) := \overline{\psi}(x;k)e^{ikx}, \qquad (7.260b)$$

then

$$M(x;k) \sim 1, \quad \overline{M}(x;k) \sim e^{2ikx}, \qquad \text{as } x \to -\infty, \qquad (7.261a)$$
$$N(x;k) \sim e^{2ikx}, \quad \overline{N}(x;k) \sim 1, \qquad \text{as } x \to \infty. \qquad (7.261b)$$

From completeness of these eigenfunctions (there can only be two linearly independent solutions of a second order equation), we obtain

$$M(x;k) = a(k)\overline{N}(x;k) + b(k)N(x;k), \qquad (7.262)$$

where $\tau(k) \equiv 1/a(k)$ and $\rho(k) \equiv b(k)/a(k)$ are called the *transmission* and *reflection* coefficients, and satisfy

$$|\rho(k)|^2 + |\tau(k)|^2 = 1.$$

Furthermore, we have the so-called symmetry condition $N(x;k) = \overline{N}(x;-k)e^{2ikx}$, which is easily derived using the differential equations satisfied by $N(x;k)$ and $\overline{N}(x;-k)$ together with their asymptotic behaviour. Therefore eqn (7.262) yields

$$\frac{M(x;k)}{a(k)} = \overline{N}(x;k) + \rho(k)e^{2ikx}\overline{N}(x;-k), \qquad (7.263)$$

which is the fundamental equation. This is equivalent to a *Riemann–Hilbert boundary value problem* due to the following analyticity properties:

(1) $M(x;k)$ and $a(k)$ can be analytically extended to the upper half k-plane and tend to unity as $|k| \to \infty$ for $\text{Im } k > 0$;

(2) $\overline{N}(x;k)$ can be analytically extended to the lower half k-plane and tends to unity as $|k| \to \infty$ for $\text{Im } k < 0$.

These properties are established using the following linear integral equations for $M(x;k)$ and $\overline{N}(x;k)$,

$$M(x;k) = 1 + \frac{1}{2ik}\int_{-\infty}^{x} \{1 - e^{2ik(x-\xi)}\}u(\xi)M(\xi,k)\,d\xi, \qquad (7.264a)$$

$$\overline{N}(x;k) = 1 - \frac{1}{2ik}\int_{x}^{\infty} \{1 - e^{2ik(x-\xi)}\}u(\xi)\overline{N}(\xi;k)\,d\xi, \qquad (7.264b)$$

together with following integral representations for $a(k)$ and $b(k)$,

$$a(k) = 1 + \frac{1}{2ik} \int_{-\infty}^{\infty} u(\xi) M(\xi; k) \, d\xi, \qquad (7.264c)$$

$$b(k) = -\frac{1}{2ik} \int_{-\infty}^{\infty} u(\xi) M(\xi; k) e^{-2ik\xi} \, d\xi. \qquad (7.264d)$$

It is straightforward to show that the function $a(k)$ has a finite number of simple zeros at $k_1, k_2, \ldots, k_N$, where $k_n = i\kappa_n$, $\kappa_n \in \mathbb{R}$, $n = 1, 2, \ldots, N$, so they all lie on the imaginary axis, in the upper half k-plane.

Hence, in general $M(x; k)/a(k)$ is a meromorphic function in the upper half k-plane with a finite number of poles at $k = i\kappa_1, i\kappa_2, \ldots, i\kappa_N$. Then set

$$\frac{M(x; k)}{a(k)} = \mu_+(x; k) + \sum_{n=1}^{N} \frac{A_n(x)}{k - i\kappa_n}, \qquad (7.265)$$

where $\mu_+(x; k)$ is analytic in the upper half k-plane. Integrating eqn (7.265) around $i\kappa_n$ and using eqn (7.263) shows that

$$A_n(x) = C_n \overline{N}(x; -i\kappa_n) \exp(-2\kappa_n x),$$

where C_n are the *normalization constants*. These constants satisfy

$$C_n = \frac{c_n}{a'(i\kappa_n)} = ic_n^2, \qquad (7.266)$$

where c_n are as defined in eqn (7.227). Hence, from eqn (7.263), we obtain

$$\mu_+(x; k) = \overline{N}(x; k) - \sum_{n=1}^{N} \frac{C_n}{k - i\kappa_n} \exp(-2\kappa_n x) \overline{N}(x; -i\kappa_n)$$

$$+ \rho(k) \exp(2ikx) \overline{N}(x; -k). \qquad (7.267)$$

Furthermore, $\mu_+(x; k) \to 1$ as $|k| \to \infty$ (for Im $k > 0$).

The inverse scattering problem

Equation (7.267) defines a Riemann–Hilbert problem in terms of the scattering data $S(\lambda) = \{(\kappa_n, C_n)_{n=1}^N, \rho(k), a(k)\}$. By applying a suitable projection operator to eqn (7.267), it can be shown that the integral equations for the eigenfunctions are given by

$$N(x;k) = e^{2ikx}\left\{1 - \sum_{n=1}^{N} \frac{C_n N_n(x)}{k+i\kappa_n} + \frac{1}{2\pi i}\int_{-\infty}^{\infty}\frac{\rho(\zeta)N(x;\zeta)}{\zeta+k+i0}d\zeta\right\}, \quad (7.268)$$

$$N_m(x) = \exp(-2\kappa_m x)$$
$$\times\left\{1 + i\sum_{n=1}^{N}\frac{C_n N_n(x)}{\kappa_m + \kappa_n} + \frac{1}{2\pi i}\int_{-\infty}^{\infty}\frac{\rho(\zeta)N(x;\zeta)}{\zeta+i\kappa_m}d\zeta\right\}, \quad (7.269)$$

where $N_m(x) := \overline{N}(x; -i\kappa_m)\exp(-2\kappa_m x)$, for $m = 1, 2, \ldots, N$. The potential is then reconstructed by the following expression:

$$u(x) = \frac{\partial}{\partial x}\left\{2i\sum_{n=1}^{N}C_n N_n(x) - \frac{1}{\pi}\int_{-\infty}^{\infty}\rho(k)N(x;k)\,dk\right\}. \quad (7.270)$$

Time dependence of the scattering data

The time evolution of the scattering data is obtained by analysing the asymptotic behaviour of the associated time evolution equation, i.e. eqn (7.258b). Making the transformation

$$v(x;k;t) = \phi(x;k;t) = M(x;k;t)e^{-ikx}$$

yields

$$M_t = (\gamma - 4ik^3 + u_x + 2iku)M + (4k^2 - 2u)M_x. \quad (7.271)$$

The asymptotic behaviour of $M(x;k;t)$ is given by

$$M(x;k;t) \to 1, \qquad \text{as } x \to -\infty, \quad (7.272a)$$
$$M(x;k;t) \to a(k,t) + b(k,t)e^{2ikx}, \qquad \text{as } x \to \infty, \quad (7.272b)$$

and consequently it can be shown that

$$\frac{\partial a}{\partial t}(k,t) = 0, \qquad \frac{\partial b}{\partial t}(k,t) = 8ik^3 b(k,t).$$

Therefore

$$a(k,t) = a(k,0), \qquad b(k,t) = b(k,0)\exp(8ik^3 t), \quad (7.273)$$

and so

$$\tau(k,t) \equiv \frac{1}{a(k,t)} = \tau(k,0) \quad \text{and} \quad \rho(k,t) \equiv \frac{b(k,t)}{a(k,t)} = \rho(k,0)\exp(8ik^3 t). \quad (7.274)$$

The discrete eigenvalues, i.e. the zeros of $a(k)$, satisfy

$$k_n = i\kappa_n = \text{constant}, \qquad n = 1, 2, \ldots, N, \qquad (7.275)$$

since we have a so-called isospectral problem. Similarly, the time dependence of the normalization constants is given by

$$C_n(t) = C_n(0)\exp(8ik_n^3 t) = C_n(0)\exp(8\kappa_n^3 t). \qquad (7.276)$$

Using eqns (7.273)–(7.275) we may now solve the initial value problem for the KdV equation as follows:

1. Given $u(x,0)$, solve eqn (7.264a) for $M(x;k;0)$ and hence, using eqns (7.264c, d), determine the initial scattering data κ_n, $C_n(0)$, and $\rho(k,0)$.

2. Use eqns (7.273)–(7.275) to compute $C_n(t)$ and $\rho(k,t)$.

3. Solve eqn (7.268) to obtain $N(x;k;t)$ and then construct $u(x,t)$ from eqn (7.270).

Schematically, this is given by

$$u(x,0) \xrightarrow{\text{direct scattering}} S(\lambda,0) = \left(\{\kappa_n, C_n(0)\}_{n=1}^N, \rho(k,0), a(k,0)\right)$$

$$\downarrow : \text{time evolution}$$

$$u(x,t) \xleftarrow{\text{inverse scattering}} S(\lambda,t) = \left(\{\kappa_n, C_n(t)\}_{n=1}^N, \rho(k,t), a(k,t)\right).$$

The Gel'fand–Levitan–Marchenko approach

The Gel'fand–Levitan–Marchenko (GLM) equation may be derived from this approach as follows. Assuming that $u \to 0$ sufficiently rapidly as $|x| \to \infty$, we seek a solution of eqn (7.269) in the 'triangular' form

$$N(x;k;t) = e^{2ikx}\left\{1 + \int_x^\infty K(x;s;t)e^{ik(s-x)}\,ds\right\}. \qquad (7.277)$$

Substituting this into eqn (7.269) and operating with $\mathscr{F}$, where

$$(\mathscr{F}f)(x,y) = \frac{1}{2\pi}\int_{-\infty}^\infty f(k)e^{ik(x-y)}\,dk, \qquad \text{for } y > x, \qquad (7.278)$$

i.e. taking the Fourier transform, it can be shown that

$$K(x,y;t) + F(x+y;t) + \int_x^\infty K(x,s;t)F(s+y;t)\,\mathrm{d}s = 0, \qquad y > x, \tag{7.279}$$

where

$$F(\xi;t) = -\mathrm{i}\sum_{n=1}^N C_n(0)\exp(8\kappa_n^3 t - \kappa_n \xi) + \frac{1}{2\pi}\int_{-\infty}^\infty \rho(k,t)\mathrm{e}^{\mathrm{i}k\xi}\,\mathrm{d}k. \tag{7.280}$$

Recall from eqn (7.266) that $C_n(0) = \mathrm{i}c_n^2(0)$, $n = 1,2,\ldots,N$ and so this agrees with eqn (7.231). The existence and uniqueness of solutions of linear integral equations such as eqn (7.279) is usually proven using the Fredholm alternative (cf., for example, reference [441, §4.1]). Substituting eqn (7.277) into eqn (7.270), we have

$$u(x,t) = 2\frac{\partial}{\partial x}\{K(x,x;t)\}. \tag{7.281}$$

This demonstrates that the GLM equation (7.279) arises as a direct consequence of the Riemann–Hilbert boundary value problem. Whereas there are not necessarily any disadvantages in using the GLM equation, it is not as fundamental as the Riemann–Hilbert boundary value problem, since some problems do not have an associated GLM equation.

7.3.4 Soliton solutions of the Korteweg–de Vries equation

Here we describe how the inverse scattering method can be used to solve an initial value problem for the Korteweg–de Vries equation

$$u_t + 6uu_x + u_{xxx} = 0. \tag{7.282}$$

Our objective is to solve eqn (7.282) subject to the initial condition $u(x,0) = f(x)$, where it is assumed that $f(x)$ decays sufficiently rapidly as $|x| \to \infty$. In particular, the inverse scattering method is well illustrated by the initial condition $f(x) = Q\,\mathrm{sech}^2 x$, where Q is a constant.

In this case the Schrödinger scattering problem is

$$\psi_{xx} + (Q\,\mathrm{sech}^2 x - \lambda)\psi = 0. \tag{7.283}$$

If we make the transformation $\psi(x) = \Psi(\xi)$, with $\xi = \tanh x$ (thus $-1 < \xi < 1$ for $-\infty < x < \infty$), then eqn (7.283) becomes

$$(1-\xi^2)\frac{\mathrm{d}^2\Psi}{\mathrm{d}\xi^2} - 2\xi\frac{\mathrm{d}\Psi}{\mathrm{d}\xi} + \left(Q - \frac{\lambda}{1-\xi^2}\right)\Psi = 0, \tag{7.284}$$

which is the *associated Legendre equation*.

Continuous spectrum

First we shall discuss the continuous spectrum. If we set $\lambda = -k^2$ and make the transformation

$$\Psi(\xi) = (1-\xi^2)^{-ik/2} F(z), \qquad z = \tfrac{1}{2}(1+\xi),$$

then the associated Legendre equation (7.284) becomes

$$z(1-z)\frac{d^2 F}{dz^2} + (1-ik)(1-2z)\frac{dF}{dz} + (Q+ik+k^2)F = 0, \quad (7.285)$$

which is the hypergeometric equation. Thus the solution of eqn (7.283) satisfying the asymptotic condition $\psi(x) \sim e^{-ikx}$ as $x \to -\infty$ is given by

$$\psi(x;k) = 2^{ik}(\operatorname{sech} x)^{-ik}\,{}_2F_1(\alpha,\beta;\gamma,z), \qquad z = \tfrac{1}{2}(1+\tanh x), \quad (7.286)$$

with

$$\alpha = \tfrac{1}{2} - ik + (Q+\tfrac{1}{4})^{1/2}, \qquad \beta = \tfrac{1}{2} - ik + (Q+\tfrac{1}{4})^{1/2}, \qquad \gamma = 1 - ik, \quad (7.287)$$

and where ${}_2F_1(\alpha,\beta;\gamma;z)$ is the hypergeometric function satisfying ${}_2F_1(\alpha,\beta;\gamma;0) = 1$.

To determine the transmission and reflection coefficients, we need to express eqn (7.286) in the form

$$\psi(x) \sim a(k)\exp(-ikx) + b(k)e^{ikx}, \qquad \text{as } z \to +\infty.$$

Using the following identity for hypergeometric functions,

$$\begin{aligned}{}_2F_1(\alpha,\beta;\gamma;z) &= \frac{\Gamma(\gamma)\Gamma(\gamma-\alpha-\beta)}{\Gamma(\gamma-\alpha)\Gamma(\gamma-\beta)}\,{}_2F_1(\alpha,\beta;\alpha+\beta-\gamma+1;1-z) \\ &+ (1-z)^{\gamma-\alpha-\beta}\frac{\Gamma(\gamma)\Gamma(\alpha+\beta-\gamma)}{\Gamma(\alpha)\Gamma(\beta)} \\ &\quad \times {}_2F_1(\gamma-\alpha,\gamma-\beta;1+\gamma-\alpha-\beta;1-z),\end{aligned}$$

it is straightforward to show that

$$\psi(x) \sim \frac{\Gamma(\gamma)\Gamma(\alpha+\beta-\gamma)}{\Gamma(\alpha)\Gamma(\beta)}e^{-ikx} + \frac{\Gamma(\gamma)\Gamma(\gamma-\alpha-\beta)}{\Gamma(\gamma-\alpha)\Gamma(\gamma-\beta)}e^{ikx}, \qquad \text{as } x \to \infty, \quad (7.288)$$

where α, β, and γ are as given in eqn (7.287) and $\Gamma(\nu)$ is the gamma function. Therefore we see that the scattering coefficients $a(k)$ and $b(k)$ are given by

$$a(k) = \frac{\Gamma(\gamma)\Gamma(\alpha+\beta-\gamma)}{\Gamma(\alpha)\Gamma(\beta)}, \quad b(k) = \frac{\Gamma(\gamma)\Gamma(\gamma-\alpha-\beta)}{\Gamma(\gamma-\alpha)\Gamma(\gamma-\beta)}. \quad (7.289)$$

Next we shall determine the values of k such that $b(k) = 0$, which corresponds to the case in which the potential $u(x,0)$ is reflectionless. Using the gamma function identity $\Gamma(1-\nu)\Gamma(\nu) = \pi \csc \pi\nu$, we see that

$$\Gamma(\gamma-\alpha)\Gamma(\gamma-\beta) = \Gamma\left(\tfrac{1}{2} - (Q+\tfrac{1}{4})^{1/2}\right)\Gamma\left(\tfrac{1}{2} + (Q+\tfrac{1}{4})^{1/2}\right)$$

$$= \pi \sec\left[\pi(Q+\tfrac{1}{4})^{1/2}\right].$$

Consequently, $b(k) = 0$ for all k if $Q = N(N+1)$, where N is a positive integer.

Furthermore, $a(k)$ has zeros at the poles of $\Gamma(\alpha)$ and $\Gamma(\beta)$. The gamma function $\Gamma(\mu)$ has poles at $\mu = -n$, for $n = 0, 1, 2, \ldots$, and so $\Gamma(\beta)$ has poles when

$$k = i\left[(Q+\tfrac{1}{4})^{1/2} - n - \tfrac{1}{2}\right]. \quad (7.290)$$

Therefore, there are a finite number of discrete eigenvalues (for finite Q) if $(Q+\tfrac{1}{4})^{1/2} > \tfrac{1}{2}$, i.e. if $Q > 0$. In the special case in which $Q = N(N+1)$, then eqn (7.290) simplifies to $k = i(N-n)$ and so the eigenvalues are $\kappa_n = n$, for $n = 1, 2, \ldots, N$.

Discrete spectrum

Now suppose that $Q = N(N+1)$, where N is a (strictly) positive integer. If $\lambda = \kappa^2$, then, as shown above, eqn (7.284) has N distinct eigenvalues $\kappa_n = n$, for $n = 1, 2, \ldots, N$ and associated bounded eigenfunctions $\psi_1(x), \psi_2(x), \ldots, \psi_N(x)$, given by

$$\psi_n(x) = \gamma_n P_N^n(\tanh x),$$

where γ_n is a normalizing constant and $P_N^n(\xi)$ is the associated Legendre polynomial, defined by

$$P_N^n(\xi) = (-1)^n (1-\xi^2)^{n/2} \frac{d^n}{d\xi^n}\left\{\frac{1}{2^N N!}\frac{d^N}{d\xi^N}\left[(\xi^2-1)^N\right]\right\}.$$

EXAMPLE 7.35. *One-soliton solution.* In the case in which $N = 1$, the initial condition is $u(x,0) = 2\,\text{sech}^2 x$. There is one discrete eigenvalue $\kappa_1 = 1$ and the associated eigenfunction is

$$\psi_1(x) = \gamma_1 P_1^1(\tanh x) = -\gamma_1 \,\text{sech}\, x.$$

THE INVERSE SCATTERING METHOD 503

Since $\int_{-\infty}^{\infty} \mathrm{sech}^2 x\,dx = 2$, then the normalized eigenfunction is $\psi_1(x) = \frac{1}{2}\sqrt{2}\,\mathrm{sech}\,x$ (the sign of ψ_1 is irrelevant). The asymptotic behaviour of this solution is given by

$$\psi_1(x) \sim \sqrt{2}\,e^{-x}, \quad \text{as } x \to +\infty.$$

Hence $c_1(0) = \lim_{x \to +\infty} \psi_1(x) e^x = \sqrt{2}$, and so $c_1(t) = c_1(0)e^{4t} = \sqrt{2}\,e^{4t}$. Therefore, from eqn (7.280), $F(\xi;t) = 2\exp(8t - \xi)$, since $C_1(t) = ic_1^2(t)$, and so the GLM equation (7.279) reduces to

$$K(x,y;t) + 2\exp\{8t - (x+y)\} + 2\int_x^\infty K(x,z;t)\exp\{8t - (x+z)\}\,dz = 0.$$

We seek a solution of this equation in the form $K(x,y;t) = L(x,t)e^{-y}$, and thus $L(x,t)$ satisfies the algebraic equation

$$L(x,t) + 2\exp(8t - x) + L(x,t)\exp(8t - 2x) = 0.$$

Therefore,

$$L(x,t) = -\frac{2\exp(8t - x)}{1 + \exp(8t - 2x)},$$

and hence we obtain the one-soliton solution of the KdV equation:

$$u(x,t) = 2\frac{\partial}{\partial x}K(x,x;t) = -2\frac{\partial}{\partial x}\left\{\frac{\exp(8t - 2x)}{1 + \exp(8t - 2x)}\right\} = 2\,\mathrm{sech}^2(x - 4t).$$

EXAMPLE 7.36. *Two-soliton solution*. In the case in which $N = 2$, the initial condition is $u(x,0) = 6\,\mathrm{sech}^2 x$. There are two discrete eigenvalues, $\kappa_1 = 1$ and $\kappa_2 = 2$, with associated normalized eigenfunctions

$$\psi_1(x) = \tfrac{1}{2}\sqrt{6}\,\tanh x\,\mathrm{sech}\,x, \qquad \psi_2(x) = \tfrac{1}{2}\sqrt{3}\,\mathrm{sech}^2 x.$$

The asymptotic behaviours of these solutions are

$$\psi_1(x) \sim \sqrt{6}\,e^{-x}, \quad \psi_2(x) \sim 2\sqrt{3}\,e^{-2x}, \quad \text{as } x \to +\infty.$$

Hence $c_1(0) = \sqrt{6}$ and $c_2(0) = 2\sqrt{3}$, and so $c_1(t) = \sqrt{6}\,e^{8t}$ and $c_2(t) = 2\sqrt{3}\,e^{16t}$. Therefore, from eqn (7.280),

$$F(\xi;t) = 6\exp(8t - \xi) + 12\exp(64t - 2\xi)$$

and so the GLM equation (7.279) becomes

$$K(x,y;t) + 6\exp\{8t - (x+y)\} + 12\exp\{64t - 2(x+y)\}$$
$$+ \int_x^\infty K(x,z;t)[6\exp\{8t - (x+z)\} + 12\exp\{64t - 2(x+z)\}]\,dz = 0.$$

(7.291)

Since $F(\xi;t)$ is separable, we seek a solution of this equation in the form

$$K(x,y;t) = L_1(x,t)e^{-y} + L_2(x,t)e^{-2y}.$$

Substituting this into eqn (7.291) and equating coefficients of powers of e^{-y} and e^{-2y} yields

$$L_1 + 6\exp(8t - x) + 3L_1\exp(8t - 2x) + 2L_2\exp(8t - 3x) = 0,$$
$$L_2 + 12\exp(64t - 2x) + 4L_1\exp(64t - 3x) + 3L_2\exp(64t - 4x) = 0,$$

which are easily solved to yield

$$L_1(x,t) = 6\left\{\frac{\exp(72t - 5x) - \exp(8t - x)}{1 + 3\exp(8t - 2x) + 3\exp(64t - 4x) + \exp(72t - 6x)}\right\},$$

$$L_2(x,t) = -12\left\{\frac{\exp(64t - 2x) + \exp(72t - 4x)}{1 + 3\exp(8t - 2x) + 3\exp(64t - 4x) + \exp(72t - 6x)}\right\}.$$

Hence we obtain the two-soliton solution of the KdV equation (7.282), given by

$$u(x,t) = 2\frac{\partial}{\partial x}K(x,x;t) = 2\frac{\partial}{\partial x}\left[L_1(x,t)e^{-x} + L_2(x,t)e^{-2x}\right]$$
$$= -12\frac{\partial}{\partial x}\left\{\frac{\exp(8t - 2x) + 2\exp(64t - 4x) + \exp(72t - 6x)}{1 + 3\exp(8t - 2x) + 3\exp(64t - 4x) + \exp(72t - 6x)}\right\},$$

which can be simplified to yield

$$u(x,t) = 12\left\{\frac{3 + 4\cosh(2x - 8t) + \cosh(4x - 64t)}{[3\cosh(x - 28t) + \cosh(3x - 36t)]^2}\right\}. \quad (7.292)$$

This solution is plotted in Fig. 7.8(a) and a contour plot of the solution is given in Fig. 7.8(b). Plots of the two-soliton solution (7.292) for (i) $t = 0$, (ii) $t = 0.05$, (iii) $t = 0.1$, (iv) $t = 0.15$, (v) $t = 0.2$, and (vi) $t = 0.3$ are given in Fig. 7.9. The initial condition is $u(x,0) = 6\operatorname{sech}^2 x$ and two solitons emerge, with amplitudes 2 and 8.

We shall now consider the asymptotic behaviour of eqn (7.292) as $t \to \pm\infty$. In particular, we show that eqn (7.292) asymptotically decomposes as $t \to \pm\infty$ into the sum of two (single) solitons, and we thus obtain the phase shifts which each of these undergo as a consequence of their interaction.

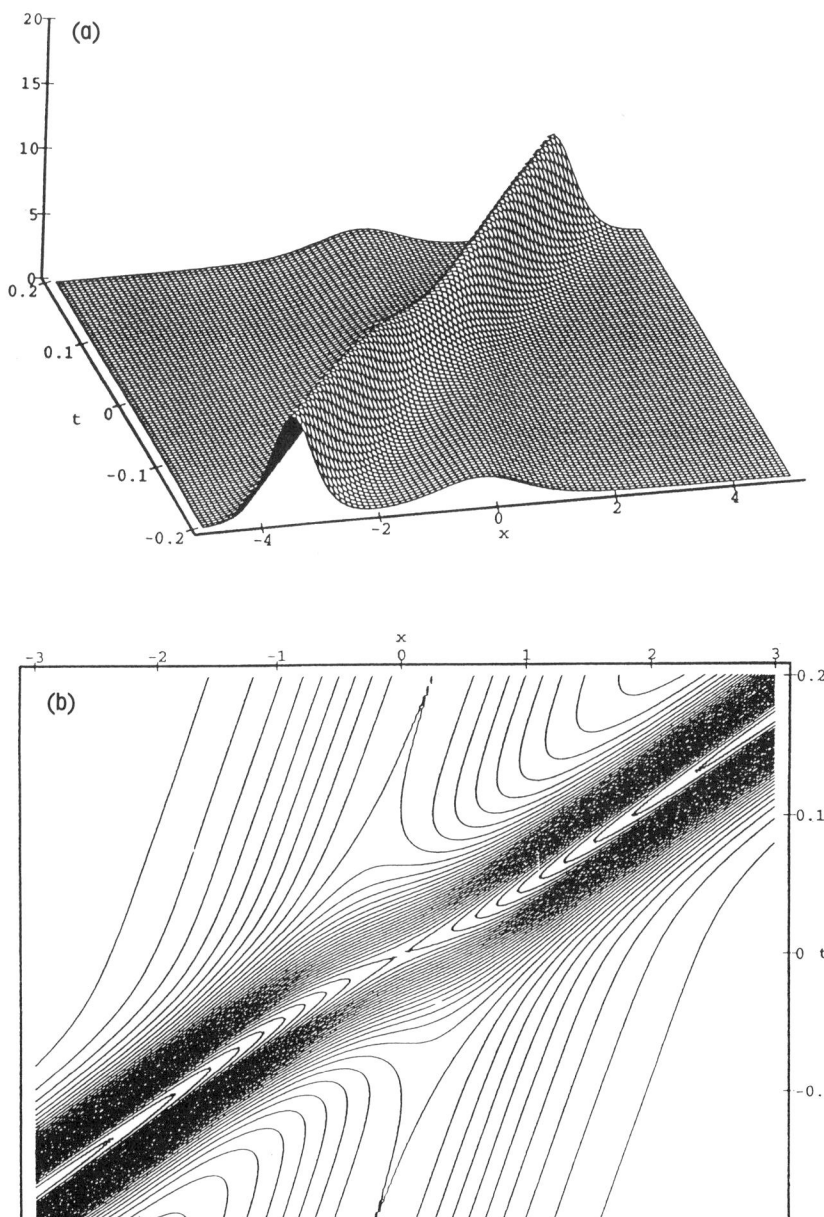

Fig. 7.8 The two-soliton solution (7.292) of the KdV equation: (a) a perspective view; (b) a contour plot.

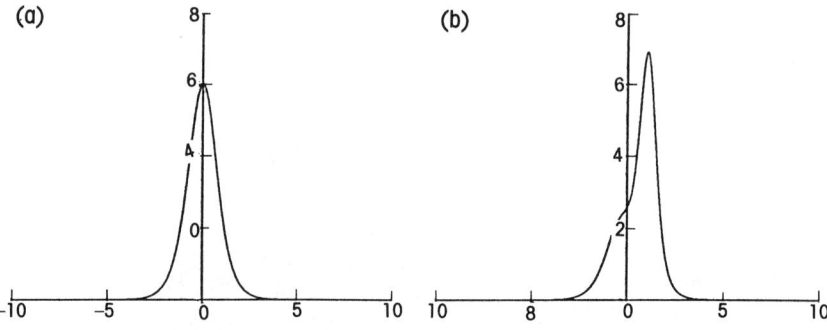

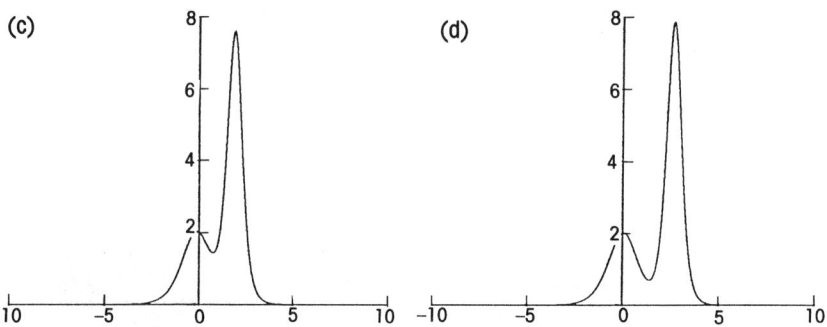

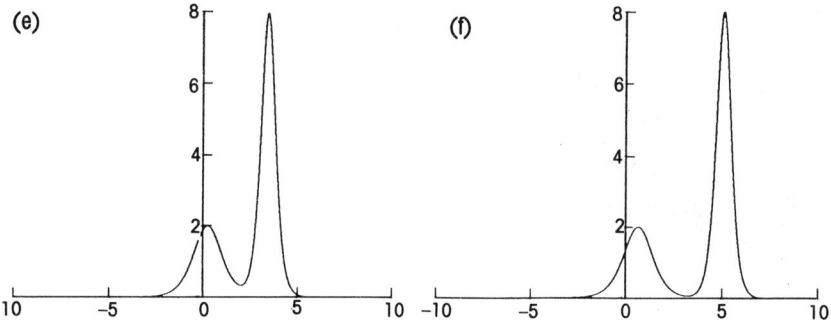

Fig. 7.9 The two-soliton solution (7.292) for (a) $t = 0$, (b) $t = 0.05$, (c) $t = 0.1$, (d) $t = 0.15$, (e) $t = 0.2$, and (f) $t = 0.3$.

If we introduce $\xi = x - 16t$ and $\eta = x - 4t$: then the two-soliton solution can be expressed as

$$u(x,t) = 12 \left\{ \frac{3 + 4\cosh(2\xi + 24t) + \cosh(4\xi)}{[3\cosh(\xi - 12t) + \cosh(3\xi + 12t)]^2} \right\} \quad (7.293)$$

and

$$u(x,t) = 12 \left\{ \frac{3 + 4\cosh(2\eta) + \cosh(4\eta - 48t)}{[3\cosh(\eta - 24t) + \cosh(3\eta - 24t)]^2} \right\}. \quad (7.294)$$

Then, keeping ξ fixed and taking the limit of eqn (7.293) as $t \to \pm\infty$ shows that

$$u(x,t) \sim 8\,\text{sech}^2(2\xi \mp \tfrac{1}{2}\ln 3), \quad \text{as } t \to \pm\infty. \quad (7.295)$$

Similarly, keeping η fixed and taking the limit of eqn (7.294) as $t \to \pm\infty$ yields

$$u(x,t) \sim 2\,\text{sech}^2(\eta \pm \tfrac{1}{2}\ln 3), \quad \text{as } t \to \pm\infty. \quad (7.296)$$

These two asymptotic expansions may be combined to give the uniformly valid asymptotic solution

$$u(x,t) \sim 2\,\text{sech}^2(\eta \pm \tfrac{1}{2}\ln 3) + 8\,\text{sech}^2(2\xi \mp \tfrac{1}{2}\ln 3), \quad \text{as } t \to \pm\infty, \quad (7.297)$$

since the error terms are exponentially small. Therefore the solution (7.294) effectively comprises of two solitons at infinity, with explicit phase shifts. The asymptotic expansion (7.297) shows that the taller, faster, soliton moves forward by $\tfrac{1}{2}\ln 3$ and the smaller, slower soliton moves backward by $\tfrac{1}{2}\ln 3$.

Reflectionless potentials

Now we shall derive the general N-soliton solution for the KdV equation. Suppose that the potential $u(x,0)$ is reflectionless and that the Schrödinger equation,

$$\psi_{xx} + [u(x,0) - \kappa^2]\psi = 0,$$

has N discrete eigenvalues κ_n, $n = 1, 2, \ldots, N$ such that $0 < \kappa_1 < \kappa_2 < \cdots < \kappa_N$. The associated eigenfunctions $\psi_n(x)$, $n = 1, 2, \ldots, N$, satisfy the condition $\int_{-\infty}^{\infty} \psi_n^2(x)\,dx = 1$, and the asymptotic limit $\lim_{x \to +\infty} \psi_n(x)\exp(\kappa_n x) = c_n(0)$, where $c_n(0)$ is the so-called 'normalization constant'. Furthermore, $c_n(t) = c_n(0)\exp(4\kappa_n^3 t)$.

If $u(x,0)$ is reflectionless, then $\rho(k,t) = 0$, and so the kernel and the inhomogeneous terms in the GLM integral equation (7.279) are reduced to finite sums over the discrete spectrum. This is solvable by separation of variables (cf. references [465, 466]). In this case in which $\rho(k,t) = 0$, then, from eqn (7.280), we have

$$F(\xi;t) = \sum_{n=1}^{N} c_n^2(0)\exp(8\kappa_n^3 t - \kappa_n \xi) = \sum_{n=1}^{N} c_n^2(t)\exp(-\kappa_n \xi),$$

and so the GLM equation (7.279) becomes

$$K(x,y;t) + \sum_{n=1}^{N} c_n^2(t)\exp\{-\kappa_n(x+y)\}$$

$$+ \int_x^\infty K(x,z;t) \sum_{n=1}^{N} c_n^2(t)\exp\{-\kappa_n(z+y)\} \, dz = 0. \quad (7.298)$$

Now we seek a solution of this equation in the form

$$K(x,y;t) = \sum_{n=1}^{N} L_n(x,t)\exp(-\kappa_n y),$$

and then the integral equation (7.298) reduces to the algebraic system

$$L_n(x,t) + \sum_{m=1}^{N} \frac{c_n^2(t)}{\kappa_m + \kappa_n} \exp\{-(\kappa_m + \kappa_n)x\} L_m(x,t)$$

$$+ c_n^2(t)\exp(-\kappa_n x) = 0,$$

for $n = 1, 2, \ldots, N$. Setting $L_n(x,t) = -c_n(t)w_n(x,t)$, for $n = 1, 2, \ldots, N$, yields a system which can be written in the form

$$(\mathbf{I} + \mathbf{B})w = f, \quad (7.299)$$

where $w = (w_1, w_2, \ldots, w_N)$, $f = (f_1, f_2, \ldots, f_N)$ with $f_n(x,t) = c_n(t)\exp(-\kappa_n x)$, for $n = 1, 2, \ldots, N$, $\mathbf{I}$ is the $N \times N$ identity matrix, and $\mathbf{B}$ is a symmetric $N \times N$ matrix, with entries

$$B_{mn} = \frac{c_m(t)c_n(t)}{\kappa_m + \kappa_n} \exp\{-(\kappa_m + \kappa_n)x\}, \quad m,n = 1,2,\ldots,N.$$

A sufficient condition for eqn (7.299) to have a unique solution is that $\mathbf{B}$ is positive definite. To demonstrate this, consider the quadratic form

$$v^T \mathbf{B} v = \sum_{m=1}^{N} \sum_{n=1}^{N} \frac{c_m(t)c_n(t)v_m v_n}{\kappa_m + \kappa_n} \exp\{-(\kappa_m + \kappa_n)x\}$$

$$= \int_x^\infty \left[\sum_{n=1}^{N} c_n(t)v_n \exp(-\kappa_n y)\right]^2 dy,$$

where $\boldsymbol{v} = (v_1, v_2, \ldots, v_N)$, which is clearly positive and equal to zero only if $\boldsymbol{v} = \boldsymbol{0}$, and so $\boldsymbol{B}$ is positive definite.

Set the solution of eqn (7.299) to be $\boldsymbol{w} = (\boldsymbol{I} + \boldsymbol{B})^{-1}\boldsymbol{f}$, so that

$$K(x, x; t) = -\sum_{n=1}^{N} c_n(t) w_n(x, t) \exp(-\kappa_n y) = -\sum_{n=1}^{N} f_n(x, t) w_n(x, t).$$

Then, using the observation,

$$\frac{\partial(\delta_{mn} + B_{mn})}{\partial x} = \frac{\partial B_{mn}}{\partial x} = -c_m(t) c_n(t) \exp\{-(\kappa_m + \kappa_n)x\}$$
$$= -f_m(x, t) f_n(x, t),$$

where δ_{mn} is the Kroneker delta, we set $\boldsymbol{A} = \boldsymbol{I} + \boldsymbol{B}$. Then,

$$K = -\sum_{n=1}^{N} f_n w_n = -\sum_{n=1}^{N} f_n \sum_{m=1}^{N} A_{mn}^{-1} f_m = \sum_{n=1}^{N} \sum_{m=1}^{N} A_{mn}^{-1} \frac{\partial A_{mn}}{\partial x}.$$

This can be written in the form

$$K(x, x; t) = \mathrm{tr}\left(\boldsymbol{A}^{-1} \frac{\partial \boldsymbol{A}}{\partial x}\right) = \frac{1}{\Delta} \frac{\partial \Delta}{\partial x} = \frac{\partial}{\partial x}(\ln \Delta),$$

where $\Delta = \det(\boldsymbol{A}) = \det(\boldsymbol{I} + \boldsymbol{B})$. Hence the unique solution of the KdV equation (7.282) in this case is

$$u(x, t) = 2 \frac{\partial^2}{\partial x^2} \{\ln \det(\boldsymbol{I} + \boldsymbol{B})\}. \tag{7.300}$$

This is the N-soliton solution which corresponds to N waves which asymptotically, as $t \to \pm\infty$, have the form

$$u_n(x, t) \sim 2 \sum_{n=1}^{N} \kappa_n^2 \mathrm{sech}^2 \{\kappa_n(x - 4\kappa_n^2 t + \delta_n^{\pm})\}, \tag{7.301}$$

where $\delta_n^{\pm}$ are constants. There is a one-to-one relationship between the number of discrete eigenvalues and the number of solitons which emerge asymptotically (see below). As mentioned previously above, these waves interact in such a way as to preserve their identities in the asymptotic limit. However, by using the inverse scattering method to solve the KdV equation, we are able mathematically to confirm the numerical observations of Zabusky and Kruskal.

The three-soliton solution of the KdV equation with $\kappa_1 = 1$, $\kappa_2 = 2$, and $\kappa_3 = 3$ is plotted in Fig. 7.10(a) and a contour plot of the solution is given in Fig. 7.10(b).

Plots of the three-soliton solution (7.292) for (i) $t = 0$, (ii) $t = 0.02$, (iii) $t = 0.04$, (iv) $t = 0.08$, (v) $t = 0.12$, and (vi) $t = 0.16$ are given in Fig. 7.11.

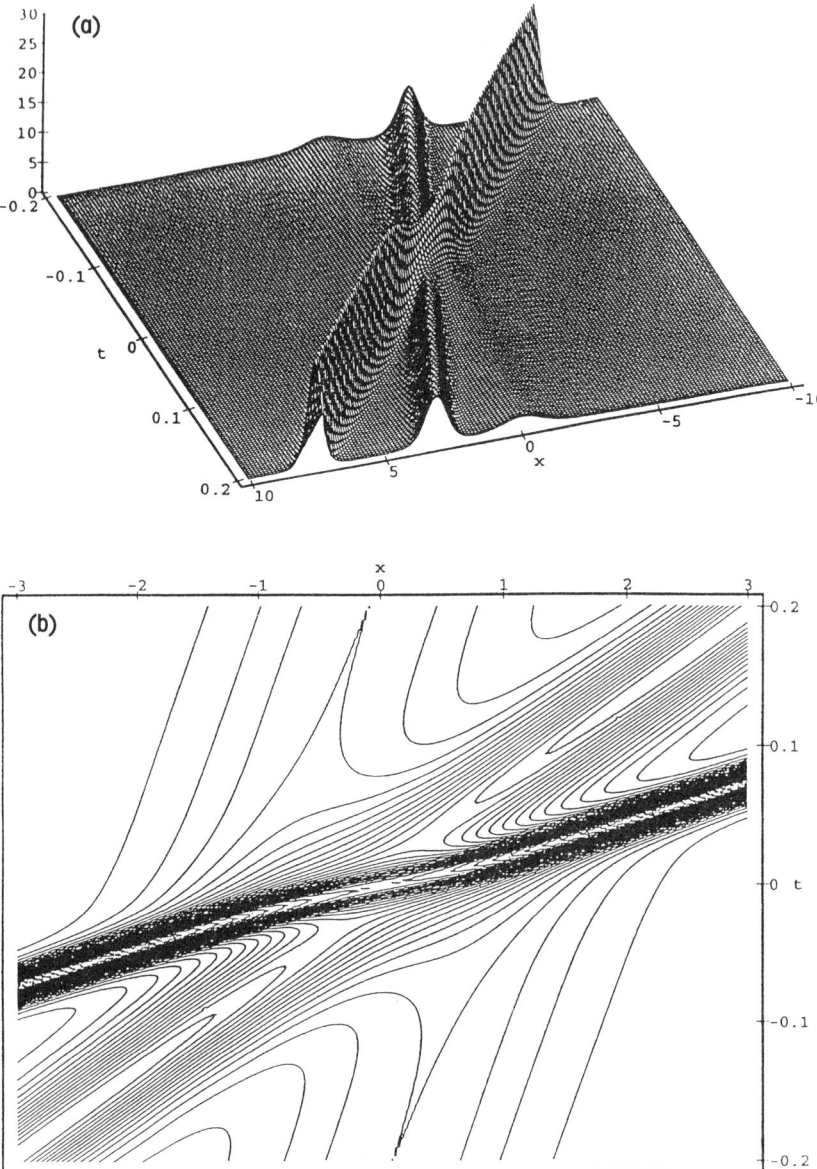

Fig. 7.10 The three-soliton solution of the KdV equation with $\kappa_1 = 1$, $\kappa_2 = 2$, and $\kappa_3 = 3$: (a) a perspective view; (b) a contour plot.

The initial condition is $u(x, 0) = 12 \operatorname{sech}^2 x$, and three solitons emerge with amplitudes 2, 8 and 18.

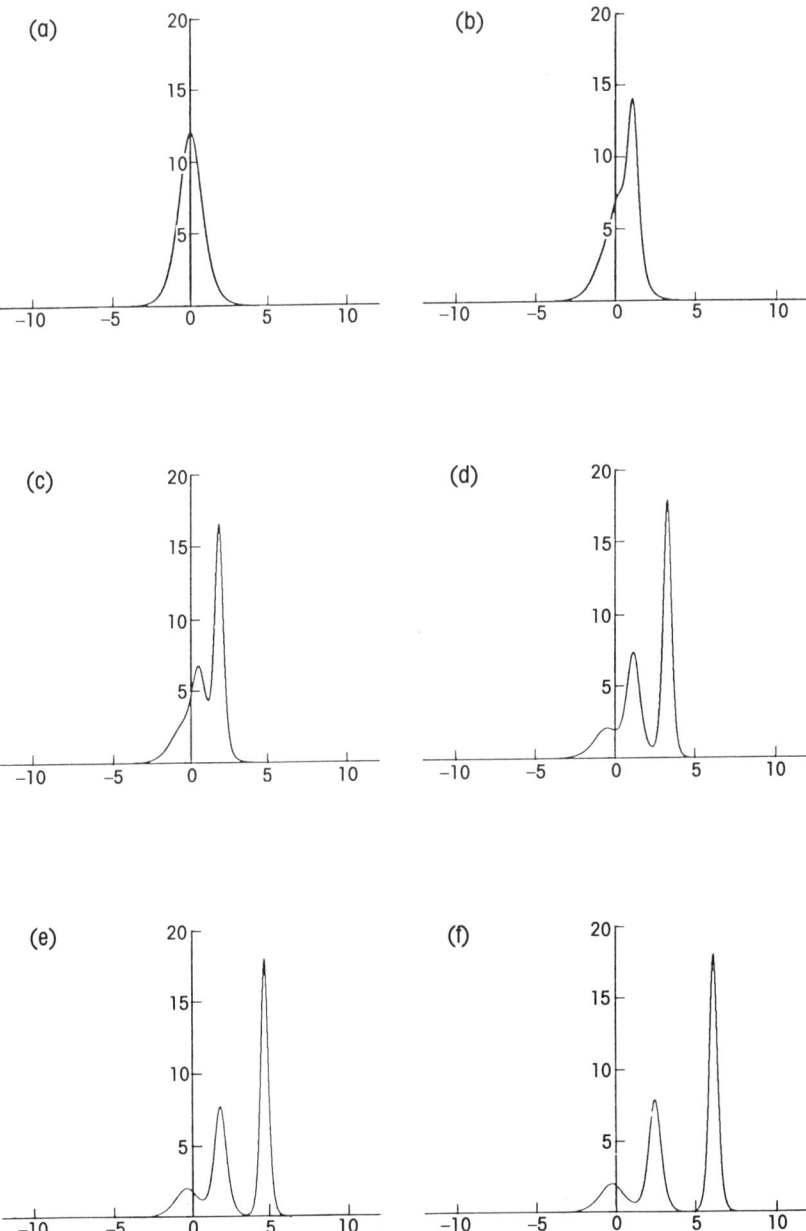

Fig. 7.11 The three-soliton solution for (a) $t = 0$, (b) $t = 0.02$, (c) $t = 0.04$, (d) $t = 0.08$, (e) $t = 0.12$, and (f) $t = 0.16$.

The difference $\delta_n = \delta_n^+ - \delta_n^-$ in eqn (7.301) represents the phase shift which the nth soliton undergoes during the interaction.

In the case in which $N = 2$, assuming that $\kappa_2 > \kappa_1 > 0$, then it is straightforward to show that, for fixed $\xi_1 = x - 4\kappa_1^2 t$,

$$u(x,t) \sim 2\kappa_1^2 \operatorname{sech}^2(\xi_1 + \delta_1^\pm), \quad \text{as } t \to \pm\infty, \qquad (7.302)$$

with $\delta_1^+ = 0$ and $\delta_1^- = \tfrac{1}{2} A_{12}$. Similarly, for fixed $\xi_2 = x - 4\kappa_2^2 t$,

$$u(x,t) \sim 2\kappa_2^2 \operatorname{sech}^2(\xi_2 + \delta_2^\pm), \quad \text{as } t \to \pm\infty, \qquad (7.303)$$

with $\delta_2^+ = \tfrac{1}{2} A_{12}$ and $\delta_2^- = 0$. For large negative time, the larger soliton lies to the left of the smaller soliton, and vice versa for large positive time, since the solitons are travelling from right to left. From eqns (7.302) and (7.303) it follows that, as a consequence of the interaction, the larger and smaller solitons undergo phase shifts given by

$$\delta_2 = \delta_2^+ - \delta_2^- = -\tfrac{1}{2} A_{12} = -\ln\left(\frac{\kappa_2 - \kappa_1}{\kappa_2 + \kappa_1}\right) > 0, \qquad (7.304\text{a})$$

$$\delta_1 = \delta_1^+ - \delta_1^- = \tfrac{1}{2} A_{12} = \ln\left(\frac{\kappa_2 - \kappa_1}{\kappa_2 + \kappa_1}\right) < 0, \qquad (7.304\text{b})$$

respectively. Therefore, after the interaction, the larger soliton has been shifted to the right of where it would have been had there been no interaction, and the smaller shifted to the left by the same amount.

A similar calculation can be done in the N-soliton case. Assuming that $0 < \kappa_1 < \kappa_2 < \cdots < \kappa_N$, then, for fixed ξ_n,

$$u(x,t) \sim 2\kappa_n^2 \operatorname{sech}^2(\xi_n + \delta_n^\pm), \quad \text{as } t \to \pm\infty,$$

where

$$\delta_n = \delta_n^+ - \delta_n^- = \sum_{m=n+1}^{N} \ln\left(\frac{\kappa_m - \kappa_n}{\kappa_n + \kappa_m}\right) - \sum_{m=1}^{n-1} \ln\left(\frac{\kappa_n - \kappa_m}{\kappa_n + \kappa_m}\right).$$

It is clear from this expression that the total phase shift experienced by the nth soliton is equivalent to the sum of phase shifts that arise from pairwise interactions with every other soliton.

Delta function initial profile

In the previous section we obtained pure soliton solutions for the KdV equation arising from reflectionless potentials. However, more general choices of the initial data $u(x,0)$ give rise to nonzero reflection coefficients; unfortunately, in this case it is generally not possible to obtain solutions in closed form.

THE INVERSE SCATTERING METHOD

As an example, we consider the delta function initial profile given by

$$u(x;0) = Q\delta(x), \tag{7.305}$$

where Q is a constant and $\delta(x)$ is the Dirac delta function. In this case the reflection and transmission coefficients are given by

$$\rho(k;0) = -\frac{Q}{Q+2ik}, \quad \tau(k;0) = \frac{2ik}{Q+2ik}.$$

If $Q > 0$, then there is a single discrete eigenvalue $\lambda = -\kappa_1^2$, and a normalization constant given by

$$\kappa_1 = \tfrac{1}{2}Q, \quad C_1(0) = \tfrac{1}{2}iQ = i\kappa_1 = \text{Res } \rho(i\kappa_1).$$

If $Q < 0$, then there are no discrete eigenvalues.

Hence eqns (7.268) and (7.269) become

$$N(x,t;k) = e^{2ikx}\left\{1 - \frac{i\kappa_1 N_1(x,t)\exp(8\kappa_1^3 t)}{k+i\kappa_1}\right.$$

$$\left. - \frac{\kappa_1}{2\pi i}\int_{-\infty}^{\infty}\frac{N(x,t;\zeta)\exp(8i\zeta^3 t)}{(\zeta+k+i0)(\kappa_1+ik)}d\zeta\right\}, \tag{7.306a}$$

$$N_1(x,t) = \exp(-2\kappa_1 x)\left\{1 - \tfrac{1}{2}N_1(x,t)\exp(8\kappa_1^3 t)\right.$$

$$\left. + \frac{\kappa_1}{2\pi}\int_{-\infty}^{\infty}\frac{N(x,t;\zeta)\exp(8i\zeta^3 t)}{\zeta^2+\kappa_1^2}d\zeta\right\}, \tag{7.306b}$$

and the potential $u(x,t)$ is given by

$$u(x,t) = \frac{\partial}{\partial x}$$

$$\times \left\{-2\kappa_1 N_1(x,t)\exp(8\kappa_1^3 t) + \frac{\kappa_1}{\pi}\int_{-\infty}^{\infty}\frac{N(x,t;k)\exp(8i\kappa^3 t)}{\kappa_1+i\kappa}dk\right\}. \tag{7.307}$$

Despite the fact that it is not possible to solve eqns (7.306) in closed form, it is possible to obtain useful information through asymptotic analysis. For example, it can be shown that the solution $u(x,t)$ incorporates the single soliton associated with the single discrete eigenvalue $\kappa_1 = \alpha$, and given by

$$u(x,t) \sim 2\kappa_1 \text{sech}^2\{\kappa_1(x - 4\kappa_1^2 t - \delta_1)\} = Q\text{sech}^2\{\tfrac{1}{2}Q(x - Q^2 t - \delta_1)\},$$

as $t \to +\infty$, with $x - Q^2 t$ fixed, and where $\exp(Q\delta_1) = \frac{1}{2}$. By considering the limit as $t \to +\infty$ for $x < 0$ (in this region of (x, t)-space the soliton is exponentially small), it can be shown that the solution is an oscillatory dispersive wave propagating to the left with an amplitude which decays as $t^{-1/3}$. If $Q < 0$, then there is no discrete eigenvalue. Consequently, there is no soliton and only the dispersive-wave train arises in the solution for $t > 0$.

7.3.5 Lax's generalization

Lax [467] put the inverse scattering method for solving the KdV equation into a more general framework, which subsequently paved the way to generalizations of the technique as a method for solving other partial differential equations. Consider two operators L and M, where L is the operator of the spectral problem and M is the operator governing the associated time evolution of the eigenfunctions

$$L\psi = \lambda \psi, \qquad \psi_t = M\psi. \qquad (7.308\text{a},\text{b})$$

Now take $\partial/\partial t$ of eqn (7.308a), giving

$$L_t \psi + L \psi_t = \lambda_t \psi + \lambda \psi_t;$$

hence, using eqn (7.308b),

$$L_t \psi + LM\psi = \lambda_t \psi + \lambda M \psi,$$
$$= \lambda_t \psi + M \lambda \psi,$$
$$= \lambda_t \psi + ML\psi.$$

Therefore, we obtain

$$[L_t + (LM - ML)]\psi = \lambda_t \psi,$$

and hence in order to solve for nontrivial eigenfunctions $\psi(x, t)$,

$$L_t + [L, M] = 0, \qquad (7.309)$$

where

$$[L, M] := LM - ML,$$

if and only if $\lambda_t = 0$. Equation (7.309) is called *Lax's equation*, and contains a nonlinear evolution equation for suitably chosen L and M. For example, if we take

$$L := \frac{\partial^2}{\partial x^2} + u, \qquad M := (\gamma + u_x) - (4\lambda + 2u)\frac{\partial}{\partial x}, \qquad (7.310\text{a},\text{b})$$

then L and M satisfy eqn (7.309), provided that u satisfies the KdV equation

$$u_t + 6uu_x + u_{xxx} = 0.$$

Therefore, the KdV equation may be thought of as the compatibility condition of the two linear operators given by eqn (7.310).

If a nonlinear partial differential equation arises as the compatibility condition of two such operators L and M, then eqn (7.309) is called the *Lax representation* of the partial differential equation and L and M are the *Lax pair*.

7.3.6 The AKNS method

Following the development of the method of inverse scattering to solve the initial value problem for the KdV equation by Gardner, Greene, Kruskal, and Miura [59], it was then of considerable interest to determine whether the method would be applicable to other physically important nonlinear evolution equations. The method of inverse scattering is highly nontrivial and was thought by some to be a fluke, an ingenious transformation analogous to the Cole–Hopf transformation [70, 71] which linearizes Burgers' equation:

$$u_t + 2uu_x - u_{xx} = 0.$$

If we make the transformation $u = -\phi_x/\phi$, then $\phi(x,t)$ satisfies the linear heat equation

$$\phi_t - \phi_{xx} = 0.$$

In 1972, Zakharov and Shabat [99] proved that the method was indeed no fluke by extending Lax's ideas in order to relate the nonlinear Schrödinger equation,

$$iu_t + u_{xx} + \kappa u^2 u^* = 0, \tag{7.311}$$

where u^* is the formal complex conjugate of u and κ is a constant, to a certain linear scattering problem. In particular, they showed that if

$$L := i\begin{pmatrix} 1+k & 0 \\ 0 & 1-k \end{pmatrix}\frac{\partial}{\partial x} + \begin{pmatrix} 0 & u^* \\ u & 0 \end{pmatrix}, \tag{7.312a}$$

$$M := ik\begin{pmatrix} 1 & 0 \\ 0 & 1 \end{pmatrix}\frac{\partial^2}{\partial x^2} + \begin{pmatrix} \dfrac{-iuu^*}{1+k} & u_x^* \\ -u_x & \dfrac{iuu^*}{1-k} \end{pmatrix}, \tag{7.312b}$$

with $\kappa = 2/(1-k^2)$, then L and M satisfy Lax's equation (7.309) if and only if $u(x,t)$ satisfies the nonlinear Schrödinger equation (7.311). Using the operators (7.312), Zakharov and Shabat were able to solve eqn (7.311), given initial data $u(x,0) = f(x)$, provided that $f(x)$ decays sufficiently rapidly as $|x| \to \infty$. Shortly thereafter, Wadati [362] gave the method of solution for the modified KdV equation,

$$u_t - 6u^2 u_x + u_{xxx} = 0,$$

and Ablowitz, Kaup, Newell, and Segur [483], motivated by several important observations by Kruskal, solved the Sine–Gordon equation

$$u_{xt} = \sin u.$$

Ablowitz, Kaup, Newell, and Segur [363] developed a procedure that showed that the initial value problem for a remarkably large class of physically interesting nonlinear evolution equations could be solved by the inverse scattering method. Because of the analogy between the Fourier transform method for solving the initial value problem for linear evolution equations and the inverse scattering method for solving the initial value problem for nonlinear evolution equations, they termed the inverse scattering technique the *Inverse Scattering Transform* (I.S.T.).

As an example, consider the 2×2 scattering problem given by

$$\psi_{1,x} = -ik\psi_1 + q(x,t)\psi_2, \qquad \psi_{2,x} = ik\psi_2 + r(x,t)\psi_1, \quad (7.313a,b)$$

and the most general linear time dependence, which is given by

$$\psi_{1,t} = A\psi_1 + B\psi_2, \qquad \psi_{2,t} = C\psi_1 + D\psi_2, \qquad (7.314a,b)$$

where A, B, C, and D are scalar functions of $q(x,t)$, $r(x,t)$ and k, and are independent of (ψ_1, ψ_2). Note that if there were any x-derivatives on the right-hand side of eqn (7.314) then they can be eliminated by use of eqn (7.313). Furthermore, when $r = -1$, then eqn (7.313) reduces to the Schrödinger scattering problem

$$\psi_{2,xx} + (k^2 + q)\psi_2 = 0. \tag{7.315}$$

It is interesting to note that the most physically interesting nonlinear evolution equations arise from this procedure when either $r = -1$ or $r = q^*$ (or $r = q$ if q is real).

This procedure provides a simple technique that allows us to find integrable nonlinear evolution equations. The compatibility of eqns (7.313 and 7.314), that is, requiring that $\psi_{j,xt} = \psi_{j,tx}$, for $j = 1, 2$, and assuming

THE INVERSE SCATTERING METHOD

that the eigenvalue k is time-independent, that is, $dk/dt = 0$, imposes a set of conditions that A, B, C, and D must satisfy. Therefore,

$$\psi_{1,xt} = -ik\psi_{1,t} + q_t\psi_2 + q\psi_{2,t}$$
$$= -ik(A\psi_1 + B\psi_2) + q_t\psi_2 + q(C\psi_1 + D\psi_2),$$
$$\psi_{1,tx} = A_x\psi_1 + A\psi_{1,x} + B_x\psi_2 + B\psi_{2,x}$$
$$= A_x\psi_1 + A(-ik\psi_1 + q\psi_2) + B_x\psi_2 + B(ik\psi_2 + r\psi_1).$$

Hence, by equating the coefficients of ψ_1 and ψ_2, we obtain

$$A_x = qC - rB, \qquad B_x + 2ikB = q_t - (A - D)q, \qquad (7.316a,b)$$

respectively. Similarly,

$$\psi_{2,xt} = ik\psi_{2,t} + r_t\psi_1 + r\psi_{1,t}$$
$$= ik(C\psi_1 + D\psi_2) + r_t\psi_1 + r(A\psi_1 + B\psi_2),$$
$$\psi_{2,tx} = C_x\psi_1 + C\psi_{1,x} + D_x\psi_2 + D\psi_{2,x}$$
$$= C_x\psi_1 + C(-ik\psi_1 + q\psi_2) + D_x\psi_2 + D(ik\psi_2 + r\psi_1),$$

and, equating the coefficients of ψ_1 and ψ_2, we obtain

$$C_x - 2ikC = r_t + (A - D)r, \qquad -D_x = qC - rB. \qquad (7.317a,b)$$

Therefore, from eqns (7.316a) and (7.317b), without loss of generality we may assume that $D = -A$, and hence it is seen that A, B, and C necessarily satisfy the compatibility conditions

$$A_x = qC - rB, \qquad B_x + 2ikB = q_t - 2Aq, \qquad C_x - 2ikC = r_t + 2Ar.$$
$$(7.318a-c)$$

We now solve eqns (7.318) for A, B, and C (thus ensuring that eqns (7.313) and (7.314) are compatible). In general, this can only be done if another condition on r and q is satisfied, this condition being the evolution equation. Since k, the eigenvalue, is a free parameter, we may find solvable evolution equations by seeking finite power series expansions for A, B, and C:

$$A = \sum_{j=0}^{n} A_j k^j, \qquad B = \sum_{j=0}^{n} B_j k^j, \qquad C = \sum_{j=0}^{n} C_j k^j. \qquad (7.319)$$

Substituting eqn (7.319) into eqn (7.318) and equating coefficients of powers of k, we obtain $3n + 5$ equations. There are $3n + 3$ unknowns, A_j, B_j, and C_j, $j = 0, 1, \ldots, n$, and so we also obtain two nonlinear evolution equations for r and q. Now let us consider some examples.

EXAMPLE 7.37. $n = 2$. Suppose that A, B, and C are quadratic in k; that is,

$$A = A_2 k^2 + A_1 k + A_0, \quad B = B_2 k^2 + B_1 k + B_0, \quad C = C_2 k^2 + C_1 k + C_0.$$

We substitute these into eqn (7.318) and then equate powers of k. The coefficients of k^3 immediately give $B_2 = C_2 = 0$. At order k^2, we obtain $A_2 = a$, a constant, $B_1 = iaq$, and $C_1 = iar$. At order k^1, we obtain $A_1 = b$, a constant. For simplicity, we set $b = 0$ (if $b \neq 0$, then a more general evolution equation is obtained); then $B_0 = -\frac{1}{2}aq_x$ and $C_0 = \frac{1}{2}ar_x$. Finally, at order k^0, we obtain $A_0 = \frac{1}{2}arq - c$, with c a constant (again, for simplicity we set $c = 0$). Therefore we obtain the following evolution equations:

$$-\tfrac{1}{2}aq_{xx} = q_t - aq^2 r, \quad \tfrac{1}{2}ar_{xx} = r_t + aqr^2. \qquad (7.320a,b)$$

If in eqn (7.320) we set $r = \mp q^*$ and $a = 2i$, then we obtain the nonlinear Schrödinger equation

$$iq_t = q_{xx} \pm 2q^2 q^*. \qquad (7.321)$$

In summary, given the scattering problem (7.313) and the associated time dependence (7.314), then these are compatible provided that eqn (7.318) holds. Hence the Lax pair for the nonlinear Schrödinger equation (7.321) is given by

$$\psi_x = \begin{pmatrix} -ik & q \\ \mp q^* & ik \end{pmatrix} \psi,$$

$$\psi_t = \begin{pmatrix} 2ik^2 \mp iqq^* & -2kq - iq_x \\ \pm 2kq^* \mp iq_x^* & -2ik^2 \pm iqq^* \end{pmatrix} \psi.$$

EXAMPLE 7.38. $n = 3$. If we substitute the third-order polynomials in k,

$$A = a_3 k^3 + a_2 k^2 + \tfrac{1}{2}(a_3 qr + a_1)k + \tfrac{1}{2}a_2 qr - \tfrac{1}{4}ia_3(qr_x - rq_x) + a_0,$$

$$B = ia_3 qk^2 + (ia_2 q - \tfrac{1}{2}a_3 q_x)k + [ia_1 q - \tfrac{1}{2}a_2 q_x + \tfrac{1}{4}ia_3(2q^2 r - q_{xx})],$$

$$C = ia_3 rk^2 + (ia_2 r + \tfrac{1}{2}a_3 r_x)k + [ia_1 r + \tfrac{1}{2}a_2 r_x + \tfrac{1}{4}ia_3(2r^2 q - r_{xx})],$$

in eqn (7.318), with a_3, a_2, a_1, and a_0 constants, then we find that $q(x,t)$ and $r(x,t)$ satisfy the evolution equations

$$q_t + \tfrac{1}{4}ia_3(q_{xxx} - 6qrq_x) + \tfrac{1}{2}a_2(q_{xx} - 2q^2 r) - ia_1 q_x - 2a_0 q = 0, \qquad (7.322a)$$

$$r_t + \tfrac{1}{4}ia_3(r_{xxx} - 6qrr_x) - \tfrac{1}{2}a_2(r_{xx} - 2qr^2) - ia_1 r_x + 2a_0 r = 0. \qquad (7.322b)$$

THE INVERSE SCATTERING METHOD

For special choices of the constants a_3, a_2, a_1, and a_0 in eqn (7.322), we find physically interesting evolution equations. If $a_0 = a_1 = a_2 = 0$, $a_3 = -4i$, and $r = -1$, then we obtain the KdV equation

$$q_t + 6qq_x + q_{xxx} = 0.$$

If $a_0 = a_1 = a_2 = 0$, $a_3 = -4i$, and $r = q$, then we obtain the MKdV equation

$$q_t - 6q^2 q_x + q_{xxx} = 0. \tag{7.323}$$

(Note that if $a_0 = a_1 = a_3 = 0$, $a_2 = -2i$, and $r = -q^*$, then we obtain the nonlinear Schrödinger equation (7.321).) Hence the Lax pair for the MKdV equation (7.323) is given by

$$\psi_x = \begin{pmatrix} -ik & q \\ q & ik \end{pmatrix} \psi,$$

$$\psi_t = \begin{pmatrix} -4ik^3 - 2ikq & 4k^2 q + 2iq_x + 2q^3 - q_{xx} \\ 4k^2 q - 2iq_x + 2q^3 - q_{xx} & 4ik^3 + 2ikq \end{pmatrix} \psi.$$

We can also consider expansions of A, B, and C in inverse powers of k.

EXAMPLE 7.39. $n = -1$. Suppose that

$$A = \frac{a(x,t)}{k}, \qquad B = \frac{b(x,t)}{k}, \qquad C = \frac{c(x,t)}{k}$$

then the compatibility conditions (7.318) are satisfied if

$$a_x = \tfrac{1}{2}(qr)_t, \qquad q_{xt} = -4iaq, \qquad r_{xt} = -4iar.$$

Special cases of these are: (i) $a = \tfrac{1}{4}i\cos u$, $b = c = \tfrac{1}{4}i\sin u$, and $q = -r = -\tfrac{1}{2}u_x$—then u satisfies the Sinh–Gordon equation

$$u_{xt} = \sin u; \tag{7.324}$$

and (ii) $a = \tfrac{1}{4}i\cosh u$, $b = c = \tfrac{1}{4}i\sinh u$, and $q = r = -\tfrac{1}{2}u_x$, where u satisfies the Sinh–Gordon equation

$$u_{xt} = \sinh u. \tag{7.325}$$

Hence the Lax pair for the Sinh–Gordon equation (7.324) is given by

$$\psi_x = \begin{pmatrix} -ik & -\tfrac{1}{2}u_x \\ \tfrac{1}{2}u_x & ik \end{pmatrix} \psi, \qquad \psi_t = \frac{i}{4k}\begin{pmatrix} \cos u & \sin u \\ \sin u & -\cos u \end{pmatrix} \psi.$$

The inverse scattering method associated with the scattering problem (7.313) is discussed in detail by Ablowitz and Segur [60] using the GLM integral equation approach and by Ablowitz and Clarkson [353] using the Riemann–Hilbert boundary value problem technique.

7.3.7 Other equations that are solvable by the inverse scattering method

In this section we briefly discuss scattering problems used to solve several other equations by an associated inverse scattering method; for further details see, for example, the book by Ablowitz and Clarkson [353].

(1 + 1)-dimensional equations

The 'natural' generalization of the Schrödinger scattering problem

$$d^2\psi/dx^2 + u(x)\psi = \lambda\psi,$$

which was used to solve the KdV equation, is the Nth order Sturm–Liouville scattering problem proposed by Gel'fand and Dikii [484],

$$\frac{d^N\psi}{dx^N} + \sum_{j=2}^{N} u_j(x)\frac{d^{N-j}\psi}{dx^{N-j}} = \lambda\psi. \tag{7.326}$$

The third-order scattering problem

$$\psi_{xxx} + Q(x)\psi_x + R(x)\psi = \lambda\psi, \tag{7.327}$$

where $Q(x)$ and $R(x)$ are potentials that vanish sufficiently rapidly as $|x| \to \infty$, is discussed by Kaup [485]. Included amongst the evolution equations associated scattering problem (7.327) are:

(1) the Boussinesq equation [79–83]

$$u_{tt} = u_{xx} + 3(u^2)_{xx} + u_{xxxx}, \tag{7.328}$$

with the associated scattering problem

$$4\psi_{xxx} + (1 + 6u)\psi_x + \left[3u_x - i\sqrt{3}\,\partial^{-1}(u_t)\right]\psi = \lambda\psi,$$
$$\psi_t = i\sqrt{3}\,(\psi_{xx} + u\psi);$$

(2) two fifth-order KdV-type equations, namely the Sawada–Kotera equation [486, 487]

$$u_t + u_{xxxxx} + 10uu_{xxx} + 10u_x u_{xx} + 20u^2 u_x = 0, \tag{7.329}$$

and the Kaup–Kuperschmidt equation [485, 488]

$$u_t + u_{xxxxx} + 10uu_{xxx} + 15u_x u_{xx} + 20u^2 u_x = 0, \tag{7.330}$$

THE INVERSE SCATTERING METHOD   521

with the associated scattering problems

$$\psi_{xxx} + 6u\psi_x = \lambda\psi,$$
$$\psi_t = (9\lambda - 18u_x)\psi_{xx} + 6(u_{xx} - 6u^2)\psi_x + 36\lambda u\psi,$$

and

$$\psi_{xxx} + 6u\psi_x + 3u_x\psi = \lambda\psi,$$
$$\psi_t = 9\lambda\psi_{xx} - 3(u_{xx} + 12u^2)\psi_x + 3(u_{xxx} + 12u\lambda + 24uu_x)\psi,$$

respectively. We remark that eqns (7.329) and (7.330) are related by a Miura type transformation (Fordy and Gibbons [489], Hirota and Ramani [490]).

The Nth order Sturm–Liouville scattering problem (7.326) has been discussed in detail by Beals [491] and in the monograph by Beals, Deift, and Tomei [492]. The study of the scattering problem (7.326) can be reduced to the study of $N \times N$ systems. Beals [491] and Beals, Deift, and Tomei [492] show that one can construct suitable sectionally meromorphic functions and hence obtain a Riemann–Hilbert problem.

There have been numerous applications and generalizations of the AKNS method. Ablowitz and Haberman [79] generalized the 2×2 scattering problem

$$\boldsymbol{\psi}_x = ik\begin{pmatrix} -1 & 0 \\ 0 & 1 \end{pmatrix}\boldsymbol{\psi} + \begin{pmatrix} 0 & q(x) \\ r(x) & 0 \end{pmatrix}\boldsymbol{\psi},$$

to the $N \times N$ scattering problem

$$\boldsymbol{\psi}_x = ik\mathbf{J}\boldsymbol{\psi} + \mathbf{Q}\boldsymbol{\psi}, \qquad (7.331)$$

where $\boldsymbol{\psi} = (\psi_1, \psi_2, \ldots, \psi_N)$, and $\mathbf{J}$ and $\mathbf{Q}$ are $N \times N$ matrices such that $\mathbf{J} = \mathrm{diag}(J_1, J_2, \ldots, J_N)$, with $J_i \neq J_j$ for $i \neq j$, and $Q_{ii} = 0$.

Ablowitz and Haberman [79] established that the scattering problem (7.331) with $N = 3$ can be used to solve several physically interesting equations, such as the three-wave interaction equations in one spatial dimension

$$\frac{\partial u_i}{\partial t} + c_i \frac{\partial u_i}{\partial x} = b_i u_j^* u_k^*,$$

where b_i are constants, $i, j, k = 1, 2, 3$ are cyclically permuted, and c_i are the three group velocities associated with the underlying waves (see also references [493–495]) and the Boussinesq equation (7.328) (see also references [80, 81]). Other examples include the coupled nonlinear Schrödinger equations

$$iu_t + u_{xx} + 2(|u|^2 \pm |v|^2)u = 0, \qquad iv_t \pm v_{xx} + 2(|v|^2 \pm |u|^2)v = 0,$$

[496, 497] and an equation modelling the interaction of Langmuir waves with ion-acoustic waves in a plasma [498]

$$iu_t + u_{xx} - uv = 0, \qquad v_t + v_x + (|u|^2)_x = 0.$$

The inverse problem associated with the scattering problem (7.331) in the general $N \times N$ case has been rigorously studied by Beals and Coifman [499–502] (see also references [81, 241, 242, 503]).

Another generalization of the AKNS method is that due to Wadati, Konno, and Ichikawa [116, 117]. Consider the scattering problem

$$\psi_x = \begin{pmatrix} f(k) & 0 \\ 0 & g(k) \end{pmatrix} \psi + \begin{pmatrix} 0 & q(x) \\ r(x) & 0 \end{pmatrix} \psi, \qquad (7.332)$$

where $f(k)$ and $g(k)$ are functions of the eigenvalue k, and the time dependence is given (as previously) by

$$\psi_t = \begin{pmatrix} A & B \\ C & -A \end{pmatrix} \psi. \qquad (7.333)$$

The compatibility of eqns (7.332) and (7.333) requires that A, B, and C satisfy

$$A_x = g(qC - rB), \qquad B_x + 2fB = gq_t - 2Agq, \qquad C_x - 2fC = gr_t + 2Agr. \qquad (7.334\text{a–c})$$

As earlier, postulating that A, B, C, f, and g have finite power series expansions in k (where the expansions for f and g have constant coefficients), then one obtains a variety of physically interesting evolution equations.

EXAMPLE 7.40. Choosing

$$f(k) = iak^2 - 2bk, \qquad g(k) = ak + ib,$$
$$A = -2ia^2k^4 + 8abk^3 + (8ib^2 - ia^2rq)k^2 + 2abrqk + ibrq,$$
$$B = 2a^2qk^3 + 6iabqk^2 + (-4b^2q + iaq_x + a^2rq^2)k + b(-q_x + iarq^2),$$
$$C = 2a^2rk^3 + 6iabrk^2 + (-4b^2r - iar_x + a^2r^2q)k + b(r_x + iar^2q),$$

where a and b are constants, then the compatibility conditions (7.334) are satisfied if r and q are solutions of the evolution equations

$$iq_t + q_{xx} - ia(rq^2)_x + 2b^2rq^2 = 0, \qquad (7.335\text{a})$$
$$-ir_t + r_{xx} + ia(r^2q)_x + 2b^2r^2q = 0. \qquad (7.335\text{b})$$

If in this equation we set $r = q^*$, then q satisfies the generalized nonlinear Schrödinger equation

$$iq_t + q_{xx} - ia(q^2 q^*)_x + 2b^2 q^2 q^* = 0, \qquad (7.336)$$

which reduces to the nonlinear Schrödinger equation if $a = 0$, and to the derivative nonlinear Schrödinger equation

$$iq_t + q_{xx} - ia(q^2 q^*)_x = 0, \qquad (7.337)$$

if $b = 0$, which was solved by inverse scattering in this way by Kaup and Newell [115]; in fact, if $a \neq 0$, then the generalized nonlinear Schrödinger equation (7.336) can be transformed into the derivative nonlinear Schrödinger equation (7.337) [114].

Ordinary differential equations

In this section we discuss inverse problems associated with the second Painlevé equation (PII)

$$d^2 \eta / d\xi^2 = 2\eta^3 + \xi\eta + \alpha, \qquad (7.338)$$

where α is an arbitrary constant. There are two approaches in order to develop inverse problems for the Painlevé equations; either through a GLM integral equation, or through the *isomonodromic deformation method*, which can be viewed as the nonlinear analogue of the Laplace transform method. Since PII arises through the symmetry (scaling) reduction

$$u(x, t) = (3t)^{-1/3} \eta(\xi), \qquad \xi = x/(3t)^{1/3}, \qquad (7.339)$$

of the modified KdV (mKdV) equation

$$u_t - 6u^2 u_x + u_{xxx} = 0, \qquad (7.340)$$

both approaches may be derived from the inverse scattering method for the mKdV equation (see, for example, reference [353] for details).

Ablowitz and Segur [361] characterized a one-parameter family of solutions for a special case of the second Painlevé equation given by

$$d^2 \eta / d\xi^2 = 2\eta^3 + \xi\eta, \qquad (7.341a)$$

satisfying

$$\eta(\xi) \to 0 \qquad \text{as } x \to +\infty, \qquad (7.341b)$$

in terms of the solution of the linear integral equation

$$K(\xi, \zeta) = k \, \text{Ai}\left(\frac{\xi + \zeta}{2}\right) + \tfrac{1}{4} k^2 \int_\xi^\infty \int_\xi^\infty K(\xi, s) \text{Ai}\left(\frac{s + t}{2}\right) \text{Ai}\left(\frac{t + \zeta}{2}\right) ds \, dt, \qquad (7.342)$$

where $\text{Ai}(x)$ denotes Airy's function, which is a GLM equation of Fredholm type. This integral equation can be derived from the associated integral equation that arises in the inverse scattering method of solution for the mKdV equation (7.340). If $\eta(\xi) = K(\xi, \xi)$, then it can be shown that $\eta(\xi)$ satisfies eqn (7.341) [361, 367]. Using the integral equation (7.342), it is possible to derive many properties of solutions of eqn (7.341), including global existence, uniqueness, and connection formulae, relating the asymptotic behaviour of the solution as $\xi \to +\infty$ to the asymptotic behaviour as $\xi \to -\infty$ (cf. references [358, 361, 367, 368, 505–509]).

The symmetry reduction (7.339) from the mKdV equation (7.340) to PII also leads the way towards finding compatible linear systems which form the basis of the monodromy method. Flaschka and Newell [468] expressed PII (7.338) as the compatibility condition of the following linear systems of equations:

$$\frac{\partial w}{\partial \xi} = \begin{pmatrix} -i\zeta & \eta \\ \eta & i\zeta \end{pmatrix} w, \qquad (7.343a)$$

$$\frac{\partial w}{\partial \zeta} = \begin{pmatrix} -i(4\zeta^2 + 2\eta^2 + \xi) & 4\zeta\eta + 2i\frac{d\eta}{d\xi} + \frac{\alpha}{\zeta} \\ 4\zeta\eta - 2i\frac{d\eta}{d\xi} + \frac{\alpha}{\zeta} & i(4\zeta^2 + 2\eta^2 + \xi) \end{pmatrix} w, \qquad (7.343b)$$

i.e., $w_{\xi\zeta} = w_{\zeta\xi}$ if and only if $\eta(\xi)$ satisfies PII (7.338). The linear system (7.343b) provides the key to solving PII (7.338). Flaschka and Newell [468] investigated eqn (7.343b) and showed that the monodromy data around the irregular singular point $\xi = 0$ and the irregular singular point $\xi = \infty$ are constant if and only if $\eta(\xi)$ satisfies PII (7.338). Therefore the second Painlevé equation (7.338) may be viewed as an *isomonodromy-preserving deformation* of an irregular monodromy problem.

Classically, Fuchs [268], Garnier [333], and Schlesinger [510] considered the Painlevé equations as the isomonodromic conditions for suitable linear systems with rational coefficients possessing regular and irregular singular points (see also reference [511]). Since the development of the inverse scattering method for solving PDEs, there has been renewed interest in expressing the Painlevé equations as isomonodromic conditions for suitable linear systems (see, for example, references [347, 348, 468, 512, 513]).

Using monodromy theory, Flaschka and Newell [468] obtained a formal system of singular integral equations from which the solutions of PII (7.338) may be found, although the highly nontrivial question of existence of solutions was left open. Subsequently, Fokas and Ablowitz [469] demonstrated that the inverse problem for PII can be formulated as a matrix, singular, discontinuous, homogeneous, Riemann–Hilbert boundary value problem, defined on a complicated contour (six semi-rays intersecting at the origin). They mapped this Riemann–Hilbert problem through a series of transformations to three different standard Riemann–Hilbert problems (which need to be solved sequentially). Each of these Riemann–Hilbert problems can be solved through a system of two Fredholm integral equations. In a special case, using these Riemann–Hilbert problems, Lebeau and Lochak [514] derived connection formulae for PII. Fokas and Zhou [515] have rigorously studied the Riemann–Hilbert problems associated with PII and showed that, in general, the Cauchy problem for PII possesses global solutions that are meromorphic (in ξ). The proof of the existence of meromorphic solutions is more transparent than the original proof of Painlevé. Also using this method, it is possible to find those monodromy data (and hence those initial data) for which the solution is free from poles. Recently, there has been considerable interest in the use of the monodromy theory to derive properties of solutions of the Painlevé equations (cf. references [296, 297, 512, 515, 516–531]).

Differential-difference equations

EXAMPLE 7.41. *The Toda lattice.* The Toda lattice equation [87, 532, 533],
$$d^2 u_n/dt^2 = \exp\{-(u_n - u_{n-1})\} - \exp\{-(u_{n+1} - u_n)\}, \quad (7.344)$$
is solvable by inverse scattering through the discrete Schrödinger scattering problem
$$\alpha_n \psi_{n+1} + \alpha_{n-1} \psi_{n-1} + \beta_n \psi_n = k \psi_n, \quad (7.345)$$
with
$$\alpha_n = \tfrac{1}{2} \exp\{-\tfrac{1}{2}(u_n - u_{n-1})\}, \quad \beta_n = -\tfrac{1}{2} du_{n-1}/dt. \quad (7.346)$$
The inverse scattering scheme for the Toda lattice (7.344) equation was developed by Flaschka [470, 471] and Manakov [97], and is conceptually very similar to that for the KdV equation discussed above.

The (2 + 1)-dimensional Toda equation
$$\frac{\partial^2 \theta_n}{\partial x^2} - \varepsilon^2 \frac{\partial^2 \theta_n}{\partial t^2} = 2\sigma^2 [\exp\{2(\theta_{n+1} - \theta_n)\} - \exp\{2(\theta_n - \theta_{n-1})\}], \quad (7.347)$$
where $\theta = \theta(x, n, t) \equiv \theta_n(x, t)$, in which $\varepsilon^2 = \pm 1$, x and t are continuous variables, n is a discrete variable (e.g. $n \in \mathbb{Z}$), and $\sigma^2 = \pm 1$, is solvable through the scattering problem
$$\psi_{n,x} + (\varepsilon \theta_{n,t} - \theta_{n,x}) \psi_n + i\sigma$$
$$\times \left\{ k \psi_{n+1} - \left(k + \frac{1}{k}\right) \psi_n + k^{-1} \exp\{2(\theta_n - \theta_{n-1})\} \psi_{n-1} \right\} = 0,$$
$$\psi_{n,t} + \left(\frac{1}{\varepsilon} \theta_{n,x} - \theta_{n,t}\right) \psi_n + i\sigma \varepsilon^{-1}$$
$$\times \left\{ \left(k - \frac{1}{k}\right) \psi_n + \frac{1}{k} \exp\{2(\theta_n - \theta_{n-1})\} \psi_{n-1} - k \psi_{n+1} \right\} = 0,$$

where k is the spectral parameter [240–242]. The inverse scattering method of solution for the $(2+1)$-dimensional Toda equation (7.347) is discussed in detail by Villarroel and Ablowitz [243, 244].

Integro-differential equations

EXAMPLE 7.42. *The Intermediate Long Wave equation*. The scattering problem for the Intermediate Long Wave (ILW) equation

$$u_t + \delta^{-1} u_x + 2uu_x + Tu_{xx} = 0, \qquad (7.348a)$$

where δ is a constant and Tu is the singular integral operator

$$(Tf)(x) := \frac{1}{2\delta} \int_{-\infty}^{\infty} \coth\left\{\frac{\pi}{2\delta}(y-x)\right\} f(y)\, dy \qquad (7.348b)$$

($\int_{-\infty}^{\infty}$ represents the Cauchy principal value integral), is given by

$$i\psi_x^+ + (u-\lambda)\psi^+ = \mu \psi^-, \qquad (7.349a)$$

$$i\psi_t^\pm + i(2\lambda + \delta^{-1})\psi_x^\pm + \psi_{xx}^\pm + (\pm i u_x - T(u_x) + \nu)\psi^\pm = 0, \qquad (7.349b)$$

where it is convenient to parameterize λ and μ as

$$\lambda(k) = -\tfrac{1}{2} k \coth(k\delta) \qquad \text{and} \qquad \mu(k) = \tfrac{1}{2} k \operatorname{cosech}(k\delta).$$

The constant k is interpreted as a spectral parameter and ν is an arbitrary constant. Given u, eqn (7.349a) defines a Riemann–Hilbert problem in a strip of the complex z-plane (where z is the complex extension of x) and $v^\pm(x)$ represent the boundary values of functions which are analytic in the horizontal strips between $\operatorname{Im}(z) = 0$ and $\operatorname{Im}(z) = 2\delta$ ($z = x + iy$), and periodically extended vertically (that is, $v^\pm(x) := \lim_{y \downarrow 0} v(x \pm iy)$; $v^-(x) = v^+(x + 2i\delta)$).

EXAMPLE 7.43. *The Benjamin–Ono equation*. The scattering problem for the Benjamin–Ono (BO) equation

$$u_t + 2uu_x + Hu_{xx} = 0, \qquad (7.350a)$$

where Hu is the Hilbert transform (formally letting $\delta \to \infty$ in eqn (7.348))

$$(Hf)(x) := \frac{1}{\pi} \int_{-\infty}^{\infty} \frac{f(y)}{y-x}\, dy, \qquad (7.350b)$$

is given by

$$i\psi_x^+ + k(\psi^+ - \psi^-) = -u\psi^+, \qquad (7.351a)$$

$$i\psi_t^\pm - 2ik\psi_x^\pm + \psi_{xx}^\pm - 2i[u]_x^\pm \psi^\pm = -\rho\psi^\pm, \qquad (7.351b)$$

with

$$[u]^\pm := \pm \tfrac{1}{2} u - \tfrac{1}{2} iH(u), \qquad (7.351c)$$

where k is a constant, interpreted as the spectral parameter, and p is an arbitrary constant. The quantities $w^{\pm}(x)$ represent the boundary values of functions which are analytic in the upper (+) and lower (−) half x-plane, that is, $w^{\pm}(x) = \lim_{y \downarrow 0} w(x \pm iy)$; similarly, $[u]^+$ and $[u]^-$ are analytic in the upper and lower half z-plane respectively (where z is the complex extension of x). Given u, eqn (7.351a) defines a differential Riemann–Hilbert problem in the complex x-plane.

The ILW equation (7.348) arises in the context of long internal gravity waves in a stratified fluid with finite depth, characterized by the parameter δ. We note that the ILW equation (7.348) reduces to the KdV equation

$$u_t + 2uu_x + \tfrac{1}{3}\delta u_{xxx} = 0, \tag{7.352}$$

as $\delta \to 0$ (shallow water limit), and to the BO equation (7.350), as $\delta \to \infty$ (deep water limit). The BO equation physically arises in the context of long internal gravity waves in a stratified fluid [534–536]. In addition to the significance of the ILW and BO equations as models of physically interesting phenomena, the inverse scattering schemes associated with these two equations are interesting from a mathematical point of view.

The limit of the inverse scattering scheme for the ILW equation as $\delta \to 0$ reduces to that for the KdV equation [477, 537], and the limit is rather straightforward. The limit of the inverse scattering scheme for the ILW equation as $\delta \to \infty$ to that for the BO equation was established by Santini, Ablowitz, and Fokas [538], although it is subtle and presents some difficulties. There are certain significant differences between the inverse scattering schemes for the ILW equation (for finite δ) and the BO equation. The inverse scattering scheme for the ILW (for finite δ) equation is conceptually similar to that for the KdV equation; however, the inverse scattering scheme for the BO equation is similar in many respects to that of some $(2 + 1)$-dimensional equations (for example, the Kadomtsev–Petviashvili equation—see below). Therefore the limit $\delta \to \infty$ in some sense provides a bridge between the inverse scattering schemes for $(1 + 1)$- and $(2 + 1)$-dimensional problems.

(2 + 1)-dimensional equations

EXAMPLE 7.44. *The Kadomstev–Petviashvili equation.* The scattering problem associated with the Kadomstev–Petviashvili (KP) equation [94]

$$(u_t + 6uu_x + u_{xxx})_x + 3\sigma^2 u_{yy} = 0, \tag{7.353}$$

where $\sigma^2 \pm 1$, is given by

$$\sigma \psi_y + \psi_{xx} + (u + \lambda)\psi = 0, \tag{7.354}$$

$$\psi_t + 4\psi_{xxx} + 6u\psi_x + 3u_x\psi - 3\sigma\left(\partial_x^{-1} u_y\right)\psi + \alpha\psi = 0, \tag{7.355}$$

where α is an arbitrary constant and $(\partial_x^{-1} f)(x) := \tfrac{1}{2}(\int_{-\infty}^x - \int_x^\infty) f(x')\,dx'$.

(Note that: (i) in the case when $\sigma = i$, eqn (7.354) becomes the 'time'-dependent Schrödinger equation; (ii) the definition of ∂_x^{-1} is not the usual one, i.e. $\int_{-\infty}^{x}$.)

The KP equation (7.353), which may be thought of as a two spatial dimensional analogue of the KdV equation, plays the role in 2 + 1 dimensions (two space, one time) that the KdV equation plays in 1 + 1 dimensions. It is one of the classical prototype problems in the field of exactly solvable equations and arises generically in physical contexts with both choices of the sign of σ^2, e.g. plasma physics [94] and surface water waves [95]. The choice of sign is critical with respect to the stability characteristics of one-dimensional (i.e. y independent) solitons subject to slow transverse perturbations (in the y-direction). For $\sigma^2 = -1$, the solitons are unstable, whereas for $\sigma^2 = +1$, they are stable. Furthermore, the sign of σ^2 is also critical in the development of the inverse scattering scheme for solving the initial value problem for the KP equation. In particular, for $\sigma^2 = -1$, there exist lump-type solitons, which decay as $r = \sqrt{x^2 + y^2} \to \infty$; while for $\sigma^2 = +1$, it is known that for appropriately decaying and sufficiently small initial data, lump-type solitons do not exist (it seems unlikely that lump-type solitons will exist even for large initial data, since they have neither been observed nor is there any mathematical evidence suggesting their existence).

The inverse scattering scheme for eqn (7.353) with $\sigma^2 = -1$, which we call KPI, was developed by Manakov [97], Segur [539], and Fokas and Ablowitz [96]. However, there are various significant differences in the solution of KPI by an inverse scattering scheme in comparison with the standard one-dimensional equations: (i) the inverse problem is formulated in terms of a nonlocal (as opposed to local) Riemann–Hilbert boundary value problem; and (ii) algebraically decaying solutions, the so-called lump solutions, arise in a rather novel way. (We remark that the BO equation (7.350) also possesses all the above features.) In Manakov's treatment of KPI [97], the initial value problem was solved for a restricted class of initial data in terms of a GLM equation. However, this GLM equation is defined in terms of a certain function $f(k, l)$. The main difficulty in the solution of the inverse problem for KPI is to relate $f(k, l)$ to the scattering data; this relationship is established by use of another GLM-type equation. The lump solutions were excluded in this formulation. Segur [539] obtains the scattering data in terms of Volterra integral equations, and proved the existence of solutions to the direct problem, but otherwise the development parallels that of [97]. Fokas and Ablowitz [96] obtained explicit expressions for the scattering data in terms of a newly introduced eigenfunction which is analytic neither in the upper nor the lower half spectral parameter plane (the development is analogous to that for the

Benjamin–Ono equation). Their analysis incorporated the lump solutions naturally.

However, for eqn (7.353) with $\sigma^2 = +1$, which we call KPII, the previously used ideas in scattering theory are no longer sufficient. The inverse problem cannot be formulated as a Riemann–Hilbert boundary value problem, since there exist eigenfunctions for the associated spectral problem which, although bounded, are nowhere analytic in the spectral parameter. In this case a generalization of the Riemann–Hilbert problem is required (the $\bar{\partial}$ 'DBAR' problem), and was developed for KPII by Ablowitz, BarYaacov, and Fokas [98]. The $\bar{\partial}$ problem concept was introduced by Beals and Coifman [499, 500] in connection with the inverse scattering scheme for certain first-order one-dimensional systems of differential equations (see also references [501, 502]). However, in these latter problems the $\bar{\partial}$ method is not crucial, since the inverse problem for ordinary differential equations can always be reduced to a Riemann–Hilbert problem. On the other hand, Ablowitz, BarYaacov, and Fokas [98] showed that the $\bar{\partial}$ approach is essential in the development of the inverse scattering scheme for KPII. This is significant, since it was the first case discovered in which the Riemann–Hilbert problem formulation of the inverse scattering scheme is inadequate.

EXAMPLE 7.45. *The Davey–Stewartson equation.* The scattering problem associated with the Davey–Stewartson equations

$$iq_t + \tfrac{1}{2}(\sigma^2 q_{xx} + q_{yy}) = (\phi - qr)q, \quad (7.356a)$$

$$-ir_t + \tfrac{1}{2}(\sigma^2 r_{xx} + r_{yy}) = (\phi - qr)r, \quad (7.356b)$$

$$\phi_{xx} - \sigma^2 \phi_{yy} = 2(qr)_{xx}, \quad (7.356c)$$

is given by

$$\psi_x = -\sigma \mathbf{J}\psi_y + \mathbf{Q}\psi, \quad (7.357a)$$

$$v_t = \mathbf{A}\psi - (i/\sigma)\mathbf{Q}\psi_y + i\mathbf{J}\psi_{yy}, \quad (7.357b)$$

with

$$\mathbf{J} = \mathrm{diag}(1, -1), \quad \mathbf{Q}\begin{pmatrix} 0 & q(x,y) \\ r(x,y) & 0 \end{pmatrix}, \quad \mathbf{A} = \begin{pmatrix} A^{11}(x,y) & A^{12}(x,y) \\ A^{21}(x,y) & A^{22}(x,y) \end{pmatrix},$$

and where the entries of $\mathbf{A}$ satisfy

$$A^{12} = \tfrac{1}{2}i\sigma^2(\partial_x - \sigma\partial_y)q,$$

$$A^{21} = -\tfrac{1}{2}i\sigma^2(\partial_x + \sigma\partial_y)r,$$

$$(\partial_x + \sigma\partial_y)A^{11} = -\tfrac{1}{2}i\sigma^2(\partial_x - \sigma\partial_y)(qr),$$

$$(\partial_x - \sigma\partial_y)A^{22} = \tfrac{1}{2}i\sigma^2(\partial_x + \sigma\partial_y)(qr),$$

with $\partial_x \equiv \partial/\partial x$ and $\partial_y \equiv \partial/\partial y$. The mean field ϕ in the Davey–Stewartson equation (7.356) is determined from

$$\phi = i(A^{11} - A^{22}) + qr.$$

Analogous to the situation for the KP equation (7.353), the inverse scattering scheme for eqn (7.356) with $\sigma^2 = +1$, which is known as DSI, is formulated in terms of a Riemann–Hilbert boundary value problem, and that for eqn (7.356) with $\sigma^2 = -1$, which is known as DSII, in terms of a $\bar{\partial}$ problem.

The Davey–Stewartson system (7.356) was derived to describe the evolution of progressive waves of slowly varying amplitude moving under gravity in water of finite depth [540, 541]. The discovery that DSI possesses exponentially localized coherent structures generated by the boundary conditions has aroused considerable interest in the system. These solutions were first discovered by Boiti, Leon, Martina, and Pimpinelli [542] using a Bäcklund transformation approach. Subsequently, Fokas and Santini [543, 544] developed a new inverse scattering method to incorporate the non-trivial boundary conditions and obtained a wider class of solutions that they called 'dromions'. Hietarinta and Hirota [545] and Jaulent, Manna, and Martinez-Alonso [546] obtained a broader class of solutions expressible in terms of Wronskian determinants and polynomials in exponentials, respectively, using the Hirota bilinear method. Recently, Gilson and Nimmo [547] and Gilson [548] developed an alternative method in which the bilinear approach is generalized to that using a direct approach involving Grammian determinants. They obtained three classes of localized special solutions for eqn (7.356): (i) solitons (plane waves), (ii) dromions (exponential lumps), and (iii) 'solitoffs' (which are localized in every direction except one and resemble solitons, except that they tend to a nonzero value in only one direction).

For further details on the solution of (2 + 1)-dimension PDEs and, more generally, multi-dimensional inverse scattering in higher dimensions, see, for example, the books by Ablowitz and Clarkson [353] and Konopelchenko [444, 445] and the review papers [502, 549–551].

7.3.8 Bäcklund transformations

Over one hundred years ago, Bäcklund investigated transformation properties of pseudospherical surfaces, that is surfaces of constant negative curvature [552–555]. (A surface of constant curvature is one with the same total curvature κ, the product of the principal curvatures, at every point. A pseudospherical surface is one with $\kappa < 0$.) As a consequence of this classical study of surfaces $M^2 \in \mathbb{R}^3$, Bäcklund derived a transformation which generates a new solution of the Sine–Gordon equation

$$u_{xt} = \sin u, \qquad (7.358)$$

from a given solution

DEFINITION 7.46. A *Bäcklund transformation* is essentially defined as being a system of equations relating the solution of a given equation either to another solution of the same equation (when it is often called an *auto-Bäcklund transformation*), or to a solution of another equation. (See Ablowitz and Segur [60, p. 154] for a precise definition.)

EXAMPLE 7.47. *Sine–Gordon equation.* The Bäcklund transformation for the Sine–Gordon equation (7.358) is given by

$$u_x + v_x = 2\beta \sin\left(\frac{u-v}{2}\right), \quad u_t - v_t = \frac{2}{\beta} \sin\left(\frac{u+v}{2}\right), \qquad (7.359)$$

where β is an arbitrary constant, the so-called 'Bäcklund' parameter. By cross-differentiation, we see that

$$(u+v)_{xt} = 2\beta(u-v)_t \cos\left(\frac{u-v}{2}\right) = 2\sin\left(\frac{u+v}{2}\right)\cos\left(\frac{u-v}{2}\right)$$

and

$$(u-v)_{tx} = \frac{2}{\beta}(u+v)_x \cos\left(\frac{u+v}{2}\right) = 2\sin\left(\frac{u-v}{2}\right)\cos\left(\frac{u+v}{2}\right).$$

Hence we see that eqns (7.359) are consistent if and only if $u(x,t)$ and $v(x,t)$ satisfy eqn (7.358). Thus (7.359) are an auto-Bäcklund transformation for the Sine–Gordon equation. Later, Bianchi [556, 557] obtained a permutability theorem for surfaces which provides superposition formulae for the Sine–Gordon equation.

The Sine–Gordon equation (7.358) clearly possesses the trivial solution $u = 0$. We can use this to generate a nontrivial solution. Setting $v = 0$ in the Bäcklund transformation (7.359) yields

$$u_x = 2\beta \sin(\tfrac{1}{2}u), \quad u_t = \frac{2}{\beta}\sin(\tfrac{1}{2}u). \qquad (7.360)$$

Hence, in this case,

$$\beta^{-1} u_x - \beta u_t = 0,$$

and so $u(x,t) = w(z)$, with $z = \beta x + t/\beta$. Substituting this into eqn (7.360) yields

$$w' = 2\sin(\tfrac{1}{2}w),$$

with $' \equiv d/dz$. This is easily solved by making the transformation $w(z) = 4\tan^{-1}\theta(z)$, yielding $\theta' = \theta$, which is readily integrated. Hence we obtain the one-soliton solution of the Sine–Gordon equation (7.358), given by

$$u(x,t) = 4\tan^{-1}[C\exp(\beta^2 x + t/\beta)],$$

where C is an arbitrary constant.

This procedure may now be used to generate an infinite sequence of multi-soliton solutions of the Sine–Gordon equation (7.358). Suppose that v is a known solution of the Sine–Gordon equation (7.358). Then, from the Bäcklund transformation (7.359), we see that

$$\beta^{-1}u_x + \beta u_t + \beta^{-1}v_x - \beta v_t = 4\sin(u/2)\cos(v/2),$$

which is a quasi-linear first-order equation for u, which can be solved by the method of characteristics. However, while this procedure does generate a hierarchy of solutions of the Sine–Gordon equation (7.358), it also involves an integration at each stage that becomes increasingly more cumbersome at each iteration and the algebraic expressions for the solutions more complex.

There is an alternative procedure for generating solutions which only involves *algebraic manipulations*. Suppose that u_1 and u_2 are solutions of eqn (7.358) obtained by application of the Bäcklund transformation (7.359) to a known solution u_0 with different Bäcklund parameters β_1 and β_2 respectively. Now apply the Bäcklund transformation (7.359) with parameter β_2 to u_1, giving u_{12}, and with parameter β_1 to u_2, giving u_{21}. Eliminating $u_{1,x}$, $u_{1,t}$, $u_{2,x}$, and $u_{2,t}$ from the resulting expressions yields

$$(u_{12} - u_0)_x = 2\beta_2 \sin\left(\frac{u_{12} - u_1}{2}\right) - 2\beta_1 \sin\left(\frac{u_1 - u_0}{2}\right), \quad (7.361a)$$

$$(u_{12} - u_0)_t = \frac{2}{\beta_2}\sin\left(\frac{u_{12} + u_1}{2}\right) - \frac{2}{\beta_1}\sin\left(\frac{u_1 + u_0}{2}\right), \quad (7.361b)$$

$$(u_{21} - u_0)_x = 2\beta_2 \sin\left(\frac{u_{21} - u_2}{2}\right) - 2\beta_1 \sin\left(\frac{u_2 - u_0}{2}\right), \quad (7.361c)$$

$$(u_{21} - u_0)_t = \frac{2}{\beta_2}\sin\left(\frac{u_{21} + u_2}{2}\right) - \frac{2}{\beta_1}\sin\left(\frac{u_2 + u_0}{2}\right). \quad (7.361d)$$

From Bianchi's theorem of permutability (cf. reference [446, p. 247]), it follows that $u_{12} = u_{21} = u$, and so eqns (7.361) imply that

$$\beta_2 \sin\left(\frac{u_{12} - u_1}{2}\right) - \beta_1 \sin\left(\frac{u_1 - u_0}{2}\right)$$
$$= \beta_2 \sin\left(\frac{u_{21} - u_2}{2}\right) - \beta_1 \sin\left(\frac{u_2 - u_0}{2}\right), \quad (7.362a)$$

$$\frac{1}{\beta_2} \sin\left(\frac{u_{12} + u_1}{2}\right) - \frac{1}{\beta_1} \sin\left(\frac{u_1 + u_0}{2}\right)$$
$$= \frac{1}{\beta_2} \sin\left(\frac{u_{21} + u_2}{2}\right) - \frac{1}{\beta_1} \sin\left(\frac{u_2 + u_0}{2}\right). \quad (7.362b)$$

If we now define

$$\theta_1 = \tfrac{1}{4}(u + u_2 + u_1 + u_0), \qquad \theta_2 = \tfrac{1}{4}(u - u_2 - u_1 + u_0),$$
$$\theta_3 = \tfrac{1}{4}(u + u_2 - u_1 - u_0), \qquad \theta_4 = \tfrac{1}{4}(u - u_2 + u_1 - u_0),$$

then the conditions (7.362) may be written as

$$\beta_1 \sin(\theta_1 - \theta_3) + \beta_2 \sin(\theta_1 + \theta_4) = \beta_1 \sin(\theta_1 + \theta_3) + \beta_2 \sin(\theta_1 - \theta_4),$$
$$\beta_1 \sin(\theta_2 - \theta_3) + \beta_2 \sin(\theta_2 + \theta_4) = \beta_1 \sin(\theta_2 + \theta_3) + \beta_2 \sin(\theta_2 - \theta_4).$$

Using the standard trigonometric formulae, these equations reduce to

$$(\beta_1 \sin\theta_3 - \beta_2 \sin\theta_4)\cos\theta_1 = 0, \qquad (\beta_1 \sin\theta_3 - \beta_2 \sin\theta_4)\cos\theta_2 = 0.$$

Hence $\beta_1 \sin\theta_3 = \beta_2 \sin\theta_4$, and so we obtain

$$u = 4\tan^{-1}\left\{\left(\frac{\beta_2 + \beta_1}{\beta_2 - \beta_1}\right)\tan\left(\frac{u_2 - u_1}{4}\right)\right\} + u_0, \quad (7.363)$$

(see the monograph by Rogers and Shadwick [558] for further details). This is known as a *nonlinear superposition principle*, since it allows the construction of higher solutions using purely algebraic means.

In particular, if $u_0 = 0$, then using eqn (7.359) yields

$$u_1 = 4\tan^{-1}[\exp(\beta_1 x + t/\beta_1)], \qquad u_2 = 4\tan^{-1}[\exp(\beta_2 x + t/\beta_2)],$$

and so from eqn (7.363) we obtain the two-soliton solution

$$u(x,t) = 4\tan^{-1}\left\{\left(\frac{\beta_2 + \beta_1}{\beta_2 - \beta_1}\right)\tan\left(\frac{u_2 - u_1}{4}\right)\right\}. \quad (7.364)$$

This soliton may be written in the form

$$u(x,t) = 4\tan^{-1}\left\{\left(\frac{\beta_2 + \beta_1}{\beta_2 - \beta_1}\right)\frac{\exp(z_2) - \exp(z_1)}{1 + \exp(z_1 + z_2)}\right\}, \quad (7.365)$$

where $z_1 = \beta_1 x + t/\beta_1$ and $z_2 = \beta_2 x + t/\beta_2$. It is easily shown that the solution (7.365) is a kink–kink interaction if $\beta_2 > -\beta_1 > 0$, and a kink–

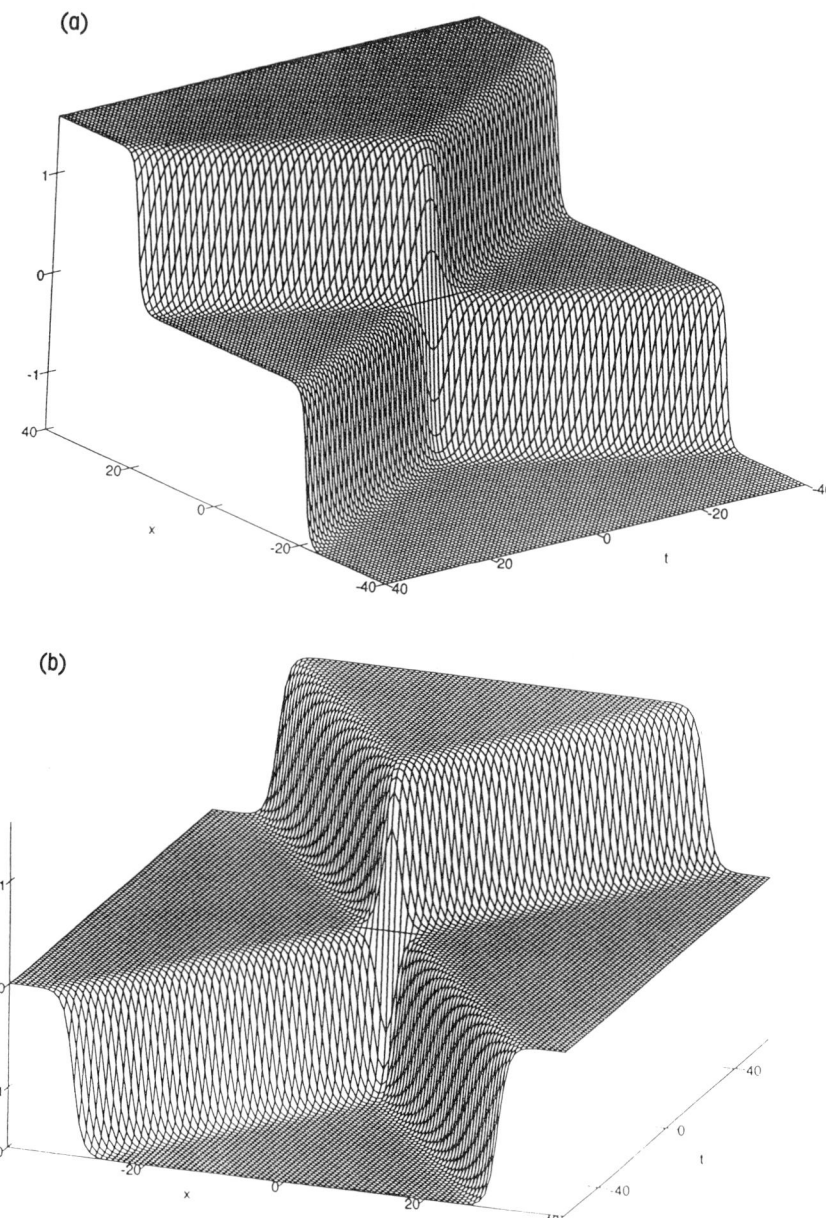

Fig. 7.12 The (a) kink–kink and (b) kink–antikink solutions of the Sine–Gordon equation.

antikink interaction if $\beta_2 > \beta_1 > 0$. Furthermore, setting $\beta_1 = \lambda - i\mu$ and $\beta_2 = \lambda + i\mu$ in eqn (7.365) yields the breather solution

$$u(x,t) = 4\tan^{-1}\left\{\frac{\lambda}{\mu}\sin\left(\mu x - \frac{\mu t}{\lambda^2 + \mu^2}\right)\text{sech}\left(\lambda x + \frac{\lambda t}{\lambda^2 + \mu^2}\right)\right\}. \quad (7.366)$$

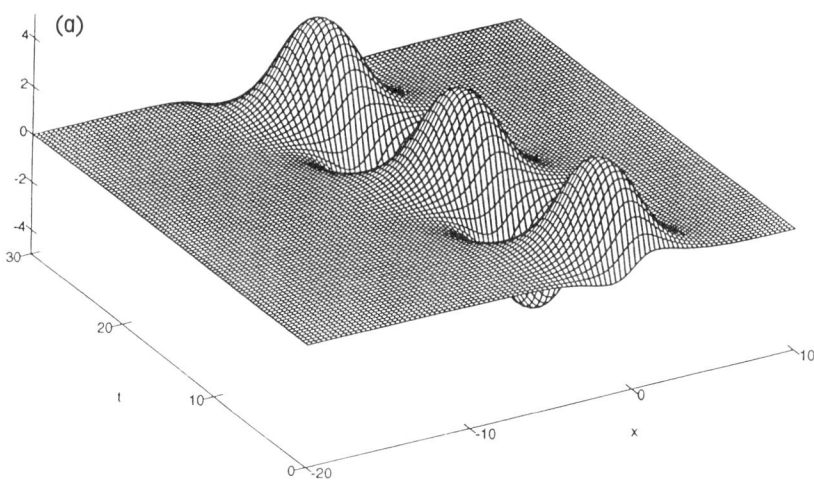

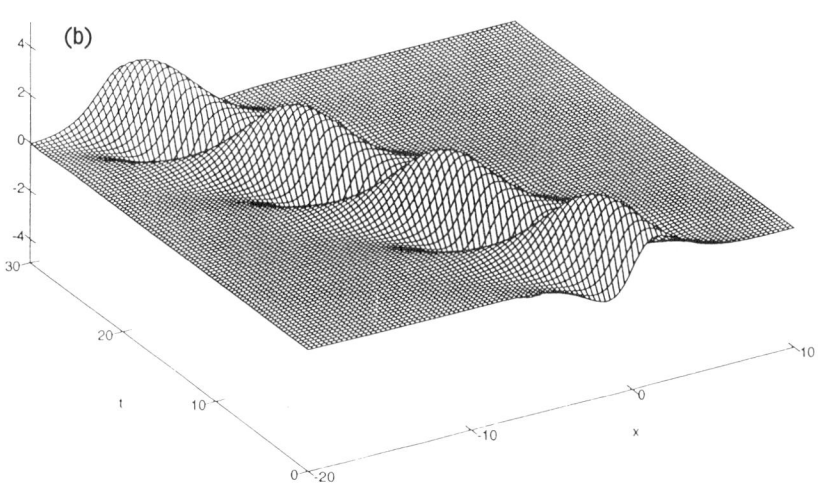

Fig. 7.13 Two breathers solutions (7.366) of the Sine–Gordon equation with (a) $\lambda = 1$ and $\mu = 0.7$ and (b) $\lambda = 1$ and $\mu = 1.3$.

The kink–kink and kink–antikink solutions of the Sine–Gordon equation are plotted in Fig. 7.12. Two breather solutions (7.366) of the Sine–Gordon equation, with (i) $\lambda = 1$ and $\mu = 0.7$ and (ii) $\lambda = 1$ and $\mu = 1.3$, are plotted in Fig. 7.13.

References

1. G. W. Bluman and S. Kumei, *Symmetries and Differential Equations*, Appl. Math. Sci., **81**, Springer-Verlag, Berlin (1989).
2. G. W. Bluman and S. Kumei, *Europ. J. Appl. Math.*, **1**, 189–216 (1990).
3. S. Kumei and G. W. Bluman, *SIAM. J. Appl. Math.*, **42**, 1157–73 (1982).
4. P. J. Olver, *Applications of Lie Groups to Differential Equations*, Graduate Texts Math., **107**, Springer-Verlag, New York (1986).
5. I. G. Lisle, 'Equivalence transformations for classes of differential equations', Ph.D. thesis, Department of Mathematics, University of British Columbia, Vancouver, Canada (1992).
6. P. Winternitz, in *Partially Integrable Evolution Equations in Physics* (ed. R. Conte and N. Boccara), NATO ASI Series C, **310**, Kluwer, Dordrecht (1990), pp. 515–67.
7. P. Winternitz, in *Integrable Systems, Quantum Groups, and Quantum Field Theories* (ed. L. A. Ibort and M. A. Rodriguez), NATO ASI Series C, **409**, Kluwer, Dordrecht (1993), pp. 425–95.
8. V. A. Galaktionov, *Diff. Int. Eqns*, **3**, 863–74 (1990).
9. V. A. Galaktionov, *Nonlinear Anal., Theory, Meth., Appl.*, **23**, 1595–621 (1994).
10. V. A. Galaktionov, *Proc. R. Soc. Edinb.*, **125A**, 225–46 (1995).
11. V. A. Galaktionov, V. A. Dorodnytzin, G. G. Elenin, S. P. Kurdjumov, and A. A. Samarskii, *J. Sov. Math.*, **41**, 1222–92 (1988).
12. W. F. Ames, F. V. Postell, and E. Adams, *Appl. Num. Math.*, **10**, 235–59 (1992).
13. N. H. Ibragimov (ed.), *CRC Handbook of Lie Groups Analysis of Differential Equations. I. Symmetries, Exact Solutions and Conservation Laws*, CRC Press, Boca Raton (1993).
14. C. Rogers and W. F. Ames, *Nonlinear Boundary Value Problems in Science and Engineering*, Academic Press, Boston (1989).
15. Yu. I. Shokin, *The Method of Differential Approximation*, Springer-Verlag, New York (1983).
16. E. Noether, *Nachr. König Gesell. Wissen. Göttingen, Math.-phys. Kl.*, 235–257 (1918). [*Transport Theory Stat. Phys.*, **1** (1971), 186–207.]
17. N. H. Ibragimov, *Transformation Groups Applied to Mathematical Physics*. D. Reidel, Boston (1985).
18. G. Gaeta, *Phys. Rep.*, **189**, 1–87 (1990).
19. G. Gaeta, *Acta Appl. Math.*, **28**, 43–68 (1992).
20. M. Golubitsky, I. Stewart, and D. Schaeffer, *Singularity and Groups in Bifurcation Theory. II*, Springer-Verlag, Berlin (1988).

21. D. H. Sattinger, *Group-theoretic Methods in Bifurcation Theory*, Lect. Notes Math., **762**, Springer-Verlag, New York (1972).
22. A. J. van der Schaft, *SIAM J. Control Optimization*, **25**, 245–59 (1987).
23. V. Ramakrishnan and H. Schaettler, *J. Math. Syst. Est. Control*, **1**, 263–78 (1991).
24. W. Miller, *Symmetry and Separation of Variables*, Addison-Wesley, Reading, MA (1977).
25. R. Seshadri and T. Y. Na, *Group Invariance in Engineering Boundary Value Problems*, Springer-Verlag, New York (1985).
26. T. B. Benjamin and P. J. Olver, *J. Fluid Mech.*, **125**, 137–85 (1982).
27. S. Lie, *Theorie der Transformationgruppen. I*, Teubner, Leipzig (1888). (Reprinted by Chelsea, New York (1970).)
28. S. Lie, *Math. Ann.*, **32**, 213–81 (1888).
29. S. Lie, *Theorie der Transformationgruppen. II*, Teubner, Leipzig (1890). (Reprinted by Chelsea, New York (1970).)
30. S. Lie, *Theorie der Transformationgruppen. III*, Teubner, Leipzig (1893). (Reprinted by Chelsea, New York (1970).)
31. S. Lie, *Leipz. Berich.*, **3**, 342–57 (1895).
32. W. F. Ames, *Nonlinear Partial Differential Equations in Engineering. I*, Academic Press, New York (1967).
33. W. F. Ames, *Nonlinear Partial Differential Equations in Engineering. II*, Academic Press, New York (1972).
34. G. W. Bluman and J. D. Cole, *Similarity Methods for Differential Equations*, Appl. Math. Sci., **13**, Springer-Verlag, Berlin (1974).
35. J. M. Hill, *Differential Equations and Group Methods for Scientists and Engineers*, CRC Press, Boca Raton (1992).
36. L. V. Ovsiannikov, *Group Analysis of Differential Equation* (Trans. W. F. Ames), Academic Press, New York (1982).
37. H. Stephani, *Differential Equations, their Solution using Symmetries* (ed. M. MacCallum), Cambridge University Press, Cambridge (1989).
38. B. Champagne, W. Hereman, and P. Winternitz, *Comput. Phys. Communs*, **66**, 319–40 (1991).
39. P. Vafeades, *MACSYMA Newsletter*, **9**(1), 1–15 (1992).
40. P. Vafeades, *MACSYMA Newsletter*, **9**(2–4), 5–20 (1992).
41. J. Carminati, J. S. Devitt and G. J. Fee, *J. Symbol. Comput.*, **14**, 103–20 (1992).
42. G. J. Reid, *J. Phys. A: Math. Gen.*, **23**, L853–9 (1990).
43. G. J. Reid, *Europ. J. Appl. Math.*, **2**, 293–318 (1991).
44. G. J. Reid, *Europ. J. Appl. Math.*, **2**, 319–40 (1991).
45. G. Baumann, Lie symmetries of differential equations, preprint, Universität Ulm, Germany (1992).
46. D. Bérubé and M. de Montigny, preprint CRM-1822, Centre de Researches Mathématiques, Université de Montréal (1992).
47. S. Herod, MathSym: a MATHEMATICA program for computing Lie symmetries, preprint, Program in Applied Mathematics, University of Colorado at Boulder (1992).

48. A. Head, *Comput. Phys. Communs*, **77**, 241–8 (1993).
49. P. H. M. Kersten, *Infinitesimal Symmetries: a Computational Approach*, CWI Tract, **34**, Amsterdam (1987).
50. M. C. Nucci, Interactive REDUCE programs for calculating classical, non-classical and Lie–Bäcklund symmetries of differential equations, preprint GT Math: 062090-051, School of Mathematics, Georgia Institute of Technology, Atlanta (1990).
51. F. Schwarz, *Computing*, **34**, 91–106 (1985).
52. F. Schwarz, *SIAM Rev.*, **30**, 450–81 (1988).
53. F. Schwarz, *Computing*, **49**, 95–115 (1992).
54. J. Sherring, Symmetry determination and linear differential equation packages, preprint, LaTrobe University, Australia (1992).
55. T. Wolf, and A. Brand, in *Modern Group Analysis: Advanced Analytical and Computational Methods in Mathematical Physics* (ed. N. H. Ibragimov, M. Torrisi, and A. Valenti), Kluwer, Dordrecht (1993), pp. 377–85.
56. W. Hereman, *Euromath Bull.*, **2**(1), 45–79 (1993).
57. R. L. Anderson and N. H. Ibragimov, *Lie–Bäcklund Transformations in Applications*, SIAM, Philadelphia (1979).
58. W. F. Ames, *Nonlinear Ordinary Differential Equations in Transport Processes*, Academic Press, New York (1968).
59. C. S. Gardner, J. M. Greene, M. D. Kruskal, and R. M. Miura, *Phys. Rev. Lett.*, **19**, 1095–7 (1967).
60. M. J. Ablowitz and H. Segur, *Solitons and the Inverse Scattering Transform*, SIAM, Philadelphia (1981).
61. E. L. Ince, *Ordinary Differential Equations*, Dover, New York (1956).
62. A. S. Fokas and M. J. Ablowitz, *J. Math. Phys.*, **23**, 2033–42 (1983).
63. P. A. Clarkson, R. J. LeVeque, and R. Saxton, *Stud. Appl. Math.*, **75**, 95–121 (1987).
64. M. P. Soerensen, P. L. Christiansen, and P. S. Lomdahl, *J. Acoust. Soc. Am.*, **76**, 871–9 (1984).
65. M. P. Soerensen, P. L. Christiansen, P. S. Lomdahl, and O. Skovgaard, *J. Acoust. Soc. Am.*, **81**, 1718–22 (1987).
66. P. L. Christiansen and V. Muto, *Physica*, **68D**, 93–6 (1993).
67. V. Muto, J. Halding, P. L. Christiansen, and A. C. Scott, *J. Biomol. Struct. Dyn.*, **4**, 873–94 (1988).
68. R. Courant and D. Hilbert, *Methods of Mathematical Physics. II. Partial Differential Equations*, John Wiley, New York (1962).
69. J. M. Burgers, *Adv. Appl. Mech.*, **1**, 177–99 (1948).
70. J. D. Cole, *Q. Appl. Math.*, **9**, 225–36 (1951).
71. E. Hopf, *Communs Pure Appl. Math.*, **3**, 201–30 (1950).
72. J. M. Burgers, *The Nonlinear Diffusion Equation*, Reidel, Dordrecht (1974).
73. S. N. Gurbatov, A. I. Saichev, and I. G. Yakushkin, *Sov. Phys. Usp.*, **26**, 857–76 (1983).
74. J. Qian, Phys. Fluids, **27**, 1957–65 (1984).
75. G. B. Whitham, *Linear and Nonlinear Waves*, John Wiley, New York (1974).

76. D. G. Crighton, *Ann. Fluid Mech.*, **11**, 11–33 (1979).
77. D. G. Crighton and J. F. Scott, *Phil. Trans. R. Soc. Lond. A*, **292**, 107–34 (1979).
78. P. L. Sachdev, *Nonlinear Diffusive Waves*, Cambridge University Press, Cambridge (1987).
79. M. J. Ablowitz and R. Haberman, *J. Math. Phys.*, **16**, 2301–5 (1975).
80. P. J. Caudrey, *Phys. Lett.*, **79A**, 264–8 (1980).
81. P. J. Caudrey, *Physica*, **6D**, 51–66 (1982).
82. P. Deift, C. Tomei, and E. Trubowitz, *Communs Pure Appl. Math.*, **35**, 567–628 (1982).
83. V. E. Zakharov, *Sov. Phys. JETP*, **38**, 108–10 (1974).
84. J. Boussinesq, *C.R. Acad. Sci. Paris*, **72**, 755–9 (1871).
85. J. Boussinesq, *J. Pure Appl.*, **7**, 55–108 (1872).
86. F. Ursell, *Proc. Camb. Phil. Soc.*, **49**, 685–94 (1953).
87. M. Toda, *Phys. Rep.*, **8**, 1–125 (1975).
88. N. J. Zabusky, in *Nonlinear Partial Differential Equations* (ed. W. F. Ames), Academic Press, New York (1967), pp. 233–58.
89. E. Infeld and G. W. Rowlands, *Nonlinear Waves, Solitons and Chaos*, Cambridge University Press, Cambridge (1990).
90. A. C. Scott, in *Bäcklund Transformations* (ed. R. M. Miura), *Lect. Notes Math.*, **515**, Springer-Verlag, Berlin (1975), pp. 80–105.
91. T. Nishitani and M. Tajiri, *Phys. Lett.*, **89A**, 379–80 (1982).
92. P. Rosenau and J. L. Schwarzmeier, *Phys. Lett.*, **115A**, 75–7 (1986).
93. P. A. Clarkson and M. D. Kruskal, *J. Math. Phys.*, **30**, 2201–13 (1989).
94. V. V. Kadomtsev and V. I. Petviashvili, *Sov. Phys. Dokl.*, **15**, 539–41 (1970).
95. M. J. Ablowitz and H. Segur, *J. Fluid Mech.*, **92**, 691–715 (1979).
96. A. S. Fokas and M. J. Ablowitz, *Stud. Appl. Math.*, **69**, 211–28 (1983).
97. S. V. Manakov, *Physica*, **3D**, 420–7 (1981).
98. M. J. Ablowitz, D. BarYaacov, and A. S. Fokas, *Stud. Appl. Math.*, **69**, 135–43 (1983).
99. V. E. Zakharov and A. B. Shabat, *Sov. Phys. JETP*, **34**, 62–9 (1972).
100. V. E. Zakharov, *Sov. Phys. JETP*, **38**, 108–10 (1972).
101. D. R. Nicholson and M. V. Goldman, *Phys. Rev. Lett.*, **41**, 406–10 (1978).
102. A. Hasegawa, *Optical Solitons in Fibres*, 2nd edn, Springer-Verlag, Berlin (1990).
103. A. Hasegawa and Y. Kodama, *Proc. IEEE*, **69**, 1145–50 (1981).
104. Y. Kodama, *J. Stat. Phys.*, **39**, 597–614 (1985).
105. Y. Kodama and A. Hasegawa, *IEEE J. Quantum Elect.*, **QE-23**, 510–24 (1987).
106. L. F. Mollenauer, *Phil. Trans. R. Soc. Lond. A*, **315A**, 437–50 (1985).
107. J. M. Hyman, D. W. McLaughlin, and A. C. Scott, *Physica*, **3D**, 23–44 (1981).
108. R. Jackiw, *Rev. Mod. Phys.*, **49**, 681–706 (1977).
109. R. Rajaraman, *Solitons and Instantons*, North-Holland, Amsterdam (1982).
110. M. Tajiri, *J. Phys. Soc. Japan*, **52**, 1908–17 (1983).
111. L. Gagnon, B. Grammaticos, A. Ramani, and P. Winternitz, *J. Phys. A: Math. Gen.*, **22**, 499–509 (1989).

112. M. Boiti and F. Pempinelli, *Nuovo Cim.*, **59B**, 40–58 (1980).
113. M. Can, *Nuovo Cim.*, **59B**, 205–7 (1991).
114. P. A. Clarkson and C. M. Cosgrove, *J. Phys. A: Math. Gen.*, **20**, 2003–24 (1987).
115. D. J. Kaup and A. C. Newell, *J. Math. Phys.*, **19**, 798–801 (1978).
116. M. Wadati, K. Konno, and Y.-H. Ichikawa, *J. Phys. Soc. Japan*, **46**, 1965–6 (1979).
117. M. Wadati, K. Konno, and Y.-H. Ichikawa, *J. Phys. Soc. Japan*, **47**, 1698–700 (1979).
118. F. Calogero and S. De Lillo, *Inverse Problems*, **3**, 633–81 (1987).
119. F. Calogero and S. De Lillo, *Inverse Problems*, **4**, L33–7 (1988).
120. P. A. Clarkson and J. A. Tuszyński, *J. Phys. A: Math. Gen.*, **23**, 4269–88 (1990).
121. J. M. Dixon and J. A. Tuszyński, *J. Phys. A: Math. Gen.*, **22**, 4895–920 (1989).
122. J. A. Tuszyński and J. M. Dixon, *J. Phys. A: Math. Gen.*, **22**, 4877–94 (1989).
123. R. S. Johnson, *Proc. R. Soc. London A*, **357**, 131–41 (1977).
124. T. Kakutani and K. Michihiro, *J. Phys. Soc. Japan*, **52**, 4129–37 (1983).
125. E. J. Parkes, *J. Phys. A: Math. Gen.*, **20**, 2025–36 (1987).
126. L. Gagnon and P. A. Bélanger, *Opt. Lett.*, **15**, 466–8 (1990).
127. L. Gagnon and P. A. Bélanger, *Phys. Rev.*, **43A**, 6187–93 (1991).
128. M. Florjańczyk and L. Gagnon, *Phys. Rev. A*, **41**, 4478–85 (1990).
129. M. Florjańczyk and L. Gagnon, *Phys. Rev. A*, **45**, 6881–3 (1992).
130. H. Buchholz, *The Confluent Hypergeometric Equation*, Springer-Verlag, Berlin (1969).
131. U. I. Talanov, *JETP Lett.*, **11**, 199–201 (1970).
132. L. Gagnon, *J. Opt. Soc. Am. B*, **7**, 1098–102 (1990).
133. L. Gagnon and C. Paré, *J. Opt. Soc. Am. A*, **8**, 601–7 (1991).
134. E. A. Kuznetsov and S. K. Turitsyn, *Phys. Lett.*, **112A**, 273–5 (1985).
135. B. J. LeMesurier, G. Papanicolaou, C. Sulem, and P. L. Sulem, *Physica*, **31D**, 78–102 (1988).
136. J. J. Rasmussen and K. Rypdal, *Phys. Scr.*, **33**, 481–97 (1985).
137. K. Rypdal and J. J. Rasmussen, *Phys. Scr.*, **33**, 498–504 (1985).
138. K. Rypdal, J. J. Rasmussen, and K. Thomsen, *Physica*, **16D**, 339–57 (1985).
139. L. Gagnon and P. Winternitz, *J. Phys. A: Math. Gen.*, **21**, 1493–511 (1988).
140. L. Gagnon and P. Winternitz, *Phys. Lett.*, **134A**, 276–81 (1989).
141. L. Gagnon and P. Winternitz, *J. Phys. A: Math. Gen.*, **22**, 469–97 (1989).
142. L. Gagnon and P. Winternitz, *Phys. Rev.*, **39A**, 296–306 (1989).
143. P. J. Olver and P. Rosenau, *Phys. Lett.*, **114A**, 107–12 (1986).
144. P. J. Olver and P. Rosenau, *SIAM J. Appl. Math.*, **47**, 263–75 (1987).
145. G. R. W. Quispel, F. W. Nijhoff, and H. W. Capel, *Phys. Lett.*, **91A**, 143–5 (1982).
146. R. E. Boisvert, W. F. Ames, and U. N. Srivastava, *J. Eng. Math.*, **17**, 203–21 (1983).
147. A. Oron and P. Rosenau, *Phys. Lett.*, **118A**, 172–6 (1986).
148. G. R. W. Quispel and H. W. Capel, *Physica*, **117A**, 76–102 (1983).

149. G. W. Bluman and J. D. Cole, *J. Math. Mech.*, **18**, 1025–42 (1969).
150. B. K. Harrison and F. B. Estabrook, *J. Math. Phys.* **12**, 653–66 (1970).
151. G. Gaeta, *J. Phys. A: Math. Gen.*, **23**, 3643–5 (1990).
152. D. Levi and P. Winternitz, *J. Phys. A: Math. Gen.*, **22**, 2915–24 (1989).
153. R. Fitzhugh, *Biophys. J.*, **1**, 445–66 (1961).
154. J. S. Nagumo, S. Arimoto, and S. Yoshizawa, *Proc. IRE*, **50**, 2061–70 (1962).
155. D. G. Aronson and H. F. Weinberger, in *Partial Differential Equations and Related Topics* (ed. J. A. Goldstein), *Lect. Notes Math.*, **446**, Springer-Verlag, Berlin (1975), pp. 5–49.
156. D. G. Aronson and H. F. Weinberger, *Adv. Math.*, **30**, 33–76 (1978).
157. E. M. Vorob'ev, *Acta Appl. Math.*, **24**, 1–24 (1991).
158. P. A. Clarkson and E. L. Mansfield, *SIAM J. Appl. Math.*, **54**, 1693–719 (1994).
159. A. H. Pincombe and N. F. Smyth, *Proc. R. Soc. Lond. A*, **433**, 479–98 (1991).
160. N. F. Smyth, *J. Austral. Math. Soc., Ser. B*, **33**, 403–13 (1992).
161. R. Aris, *The Mathematical Theory of Diffusion and Reaction in Permeable Catalysts*, Vols I and II, Oxford University Press, Oxford (1975).
162. D. A. Frank-Kamenetskii, *Diffusion and Heat Exchange in Chemical Kinetics*, Princeton University Press, Princeton (1955).
163. J. D. Murray, *Mathematical Biology*, Springer-Verlag, New York (1989).
164. P. A. Clarkson and E. L. Mansfield, *Physica*, **70D**, 250–88 (1993).
165. T. Kawahara and M. Tanaka, *Phys. Lett.*, **97A**, 311–14 (1983).
166. R. Hirota, in *Solitons* (eds R. K. Bullough and P. J. Caudrey), *Topics in Current Physics*, **17**, Springer-Verlag, Berlin (1980), pp. 157–76.
167. W. Hereman, in *Partially Integrable Evolution Equations in Physics* (ed. R. Conte and N. Boccara), *NATO ASI Series C*, **310**, Kluwer, Dordrecht (1990), pp. 585–6.
168. J. Weiss, M. Tabor, and G. Carnevale, *J. Math. Phys.*, **24**, 522–6 (1983).
169. J. Weiss, *J. Math. Phys.*, **24**, 1405–13 (1983).
170. A. G. Hansen, *Similarity Analyses of Boundary Value Problems in Engineering*, Prentice-Hall, Englewood Cliffs (1964).
171. P. A. Clarkson, *Europ. J. Appl. Math.*, **1**, 279–300 (1990).
172. H. Airault, *Stud. Appl. Math.*, **61**, 31–53 (1979).
173. A. P. Bassom, P. A. Clarkson, A. C. Hicks, and J. B. McLeod, *Proc. R. Soc. Lond. A*, **437**, 1–24 (1992).
174. A. P. Bassom, P. A. Clarkson, and A. C. Hicks, *IMA J. Appl. Math.*, **437**, 1–24 (1993).
175. V. I. Gromak, *Diff. Eqns*, **14**, 1510–13 (1977).
176. V. I. Gromak, *Diff. Eqns*, **23**, 506–13 (1987).
177. N. A. Lukashevich, *Diff. Eqns*, **1**, 561–4 (1965).
178. N. A. Lukashevich, *Diff. Eqns*, **3**, 395–9 (1967).
179. N. A. Lukashevich, *Diff. Eqns*, **7**, 853–4 (1971).
180. Y. Murato, *Funk. Ekvacioj*, **28**, 1–32 (1985).
181. K. Okamoto, *Math. Ann.*, **275**, 222–55 (1986).
182. S.-Y. Lou, *Phys. Lett.*, **151A**, 133–5 (1990).

183. M. Abramowitz and I. A. Stegun, *Handbook of Mathematical Functions*, Dover, New York (1965).
184. A. Erdélyi, W. Magnus, F. Oberhettinger, and F. G. Tricomi, *Higher Transcendental Functions, II*, McGraw-Hill, New York (1953).
185. P. A. Clarkson and P. Winternitz, *Physica*, **49D**, 257–72 (1991).
186. M. C. Nucci and P. A. Clarkson, *Phys. Lett.*, **164A**, 49–56 (1992).
187. P. G. Estévez, *Phys. Lett.*, **171A**, 259–61 (1992).
188. F. Cariello and M. Tabor, *Physica*, **39D**, 77–94 (1989).
189. F. Cariello and M. Tabor, *Physica*, **53D**, 59–70 (1991).
190. P. G. Estévez and P. R. Gordoa, *Theoret. Math. Phys.*, **99**, 562–6 (1994).
191. P. G. Estévez and P. R. Gordoa, *Stud. Appl. Math.*, **95**, 73–113 (1995).
192. P. J. Olver, *Proc. R. Soc. Lond. A*, **444**, 509–23 (1994).
193. D. Arrigo, P. Broadbridge, and J. M. Hill, *J. Math. Phys.*, **34**, 4692–703 (1993).
194. P. A. Clarkson and S. Hood, *Europ. J. Appl. Math.*, **3**, 381–414 (1992).
195. E. Pucci, *J. Phys. A: Math. Gen.*, **25**, 2631–40 (1992).
196. S. Zidowitz, Nichtklassische Symmetrierduktionen hydrodynamischer Plasmagleichungen, Master's thesis, Institut für Mathematische Physik, Technische Universität Carolo-Wilhelmina zu Branschweig, Germany (1992).
197. S. Zidowitz, in *Modern Group Analysis: Advanced Analytical and Computational Methods in Mathematical Physics* (ed. N. H. Ibragimov, M. Torrisi, and A. Valenti), Kluwer, Dordrecht (1993), pp. 387–93.
198. P. A. Clarkson, *J. Phys. A: Math. Gen.*, **22**, 2355–67 (1989).
199. P. A. Clarkson, *J. Phys. A: Math. Gen.*, **22**, 3821–48 (1989).
200. P. A. Clarkson, *Nonlinearity*, **5**, 453–72 (1992).
201. P. A. Clarkson, *Math. Comp. Model.*, **18**, 45–68 (1993).
202. P. A. Clarkson and S. Hood, *J. Phys. A: Math. Gen.*, **26**, 133–50 (1993).
203. P. A. Clarkson and S. Hood, *J. Math. Phys.*, **34**, 255–83 (1993).
204. P. A. Clarkson and S. Hood, New symmetry reductions and exact solutions of the Davey–Stewartson equation. II. Reductions to partial differential equations, preprint 170, Program in Applied Math., University of Colorado, Boulder (1993).
205. P. A. Clarkson and D. K. Ludlow, *J. Math. Analysis Appl.*, **186**, 132–55 (1994).
206. P. A. Clarkson and E. L. Mansfield, in *Applications of Analytic and Geometric Methods to Nonlinear Differential Equations* (ed. P. A. Clarkson) *NATO ASI Series C*, **413**, Kluwer, Dordrecht (1993), pp. 375–89.
207. P. A. Clarkson and E. L. Mansfield, *Nonlinearity*, **7**, 975–1000 (1994).
208. P. A. Clarkson and E. L. Mansfield, *Acta Appl. Math.*, **39**, 245–76 (1995).
209. W. I. Fushchich, *Ukrain. Mat. Zh.*, **43**, 1456–70 (1991).
210. W. I. Fushchich and A. G. Nikitin, *Symmetries of Maxwell's Equations*, D. Reidel, Dordrecht (1987).
211. W. I. Fushchich and R. Z. Zhdanov, *Phys. Rep.*, **172**, 123–74 (1989).
212. S.-Y. Lou, *J. Phys. A: Math. Gen.*, **23**, L649–54 (1990).
213. S.-Y. Lou, *Sci. China, Ser. A*, **34**, 1098–108 (1991).
214. S.-Y. Lou, *J. Math. Phys.*, **33**, 4300–5 (1992).

REFERENCES

215. S.-Y. Lou and G.-J. Ni, *Communs Theoret Phys.*, **15**, 465–472 (1991).
216. S.-Y. Lou, H.-Y. Ryan, D.-F. Chen, and W.-Z. Chen, *J. Phys. A: Math. Gen.*, **24**, 1455–67 (1991).
217. N. Manganaro and D. F. Parker, *J. Phys. A: Math. Gen.*, **26**, 4093–106 (1993).
218. M. C. Nucci, *J. Math. Analysis Appl.*, **178**, 294–300 (1993).
219. M. C. Nucci and W. F. Ames, *J. Math. Analysis Appl.*, **178**, 584–581 (1993).
220. E. Pucci and G. Saccomandi, *J. Math. Analysis Appl.*, **163**, 588–98 (1992).
221. P. Winternitz, Conditional symmetries and conditional integrability for nonlinear systems, preprint CRM-1709, Centre de Researches Mathématiques, Université de Montréal (1990).
222. R. Hirota and M. Ito, *J. Phys. Soc. Japan*, **52**, 744–8 (1983).
223. N. N. Yanenko, in *Proceedings of the Fourth All-Union Mathematica Congress*, Leningrad, pp. 247–59 [in Russian] (1964).
224. S. V. Meleshko, *Sov. Math. Dokl.*, **28**, 37–41 (1983).
225. P. J. Olver, *Appl. Num. Math.*, **10**, 307–24 (1992).
226. K. A. Dunn, D. R. K. S. Rao, and C. C. A. Sastri, *J. Math. Phys.*, **28**, 1473–76 (1987).
227. L. Martina, G. Soliani, and P. Winternitz, *J. Phys. A: Math. Gen.*, **25**, 4425–35 (1992).
228. L. Martina and P. Winternitz, *J. Math. Phys.*, **33**, 2718–27 (1992).
229. J. Ondich, *Europ. J. Appl. Math.*, **6**, 631–7 (1995).
230. G. W. Bluman and S. Kumei, *J. Math. Phys.*, **28**, 307–18 (1987).
231. G. W. Bluman, S. Kumei, and G. J. Reid, *J. Math. Phys.*, **29**, 806–11 (1988).
232. G. W. Bluman, in *Applications of Analytic and Geometric Methods to Nonlinear Differential Equations* (ed. P. A. Clarkson), *NATO ASI Series C*, **413**, Kluwer, Dordrecht (1993), pp. 363–73.
233. G. W. Bluman, *Math. Comp. Model.*, **18**, 1–14 (1993).
234. G. W. Bluman, in *Exploiting Symmetry in Applied and Numerical Analysis* (ed. E. Allgower, K. Georg, and R. Miranda), *Lect. Appl. Math.*, **29**, A.M.S., Providence (1993), pp. 97–109.
235. G. W. Bluman, in *Modern Group Analysis: Advanced Analytical and Computational Methods in Mathematical Physics* (ed. N. H. Ibragimov, M. Torrisi, and A. Valenti), Kluwer, Dordrecht (1993), pp. 71–84.
236. E. Pucci and G. Saccomandi, *J. Phys. A: Math. Gen.*, **26**, 681–90 (1993).
237. G. W. Bluman and S. Kumei, *Europ. J. Appl. Math.*, **1**, 217–23 (1990).
238. R. L. Anderson, N. Kamran, and P. J. Olver, *Adv. Math.*, **100**, 53–100 (1993).
239. I. S. Krasil'shchik, V. V. Lychagin, and A. M. Vinogradov, *Geometry of Jet Spaces and Nonlinear Partial Differential Equations*, Gordon and Breach, New York (1986).
240. A. P. Fordy and J. Gibbons, *Communs Math. Phys.*, **77**, 21–30 (1980).
241. A. V. Mikhailov, *Sov. Phys. JETP Lett.*, **30**, 414–18 (1979).
242. A. V. Mikhailov, *Physica*, **3D**, 73–117 (1981).
243. J. Villarroel and M. J. Ablowitz, *Phys. Lett.*, **77**, 293–8 (1992).
244. J. Villarroel and M. J. Ablowitz, *Physica*, **65D**, 48–70 (1993).
245. D. Levi and P. Winternitz, *Phys. Lett.*, **152A**, 335–8 (1992).

246. D. Levi and P. Winternitz, *J. Math. Phys.*, **34**, 3713–40 (1993).
247. D. Levi and P. Winternitz, in *Applications of Analytic and Geometric Methods to Nonlinear Differential Equations* (ed. P. A. Clarkson), *NATO ASI Series C*, **413**, Kluwer, Dordrecht (1993), pp. 405–13.
248. G. R. W. Quispel, H. W. Capel, and R. Sahadevan, *Phys. Lett.*, **170A**, 379–83 (1992).
249. G. R. W. Quispel, H. W. Capel, and R. Sahadevan, in *Applications of Analytic and Geometric Methods to Nonlinear Differential Equations* (ed. P. A. Clarkson), *NATO ASI Series C*, **413**, Kluwer, Dordrecht (1993), pp. 431–9.
250. V. A. Dorodnytzin, *J. Sov. Math.*, **55**, 1490–517 (1991).
251. V. A. Dorodnytzin, in *Modern Group Analysis: Advanced Analytical and Computational Methods in Mathematical Physics* (ed. N. H. Ibragimov, M. Torrisi, and A. Valenti), Kluwer, Dordrecht (1993), pp. 191–201.
252. S. Maeda, *Math. Japonica*, **25**, 405–20 (1980).
253. S. Maeda, *IMA J. Appl. Math.*, **35**, 129–34 (1987).
254. G. R. W. Quispel and R. Sahadevan, in *Modern Group Analysis: Advanced Analytical and Computational Methods in Mathematical Physics* (ed. N. H. Ibragimov, M. Torrisi, and A. Valenti), Kluwer, Dordrecht (1993), pp. 299–302.
255. G. R. W. Quispel and R. Sahadevan, *Phys. Lett.*, **184A**, 64–70 (1993).
256. P. Painlevé, *Acta Math.*, 1–85 (1902).
257. R. Conte, A. P. Fordy, and A. Pickering, *Physica*, **69D**, 33–58 (1993).
258. M. Kruskal, in *Painlevé Transcendents. Their Asymptotics and Physical Applications* (ed. D. Levi and P. Winternitz), *NATO ASI Series B: Physics*, **278**, Plenum Press, New York (1991), pp. 187–95.
259. P. Painlevé, *C.R. Acad. Sci. Paris*, **107**, 221–4 (1888).
260. P. Painlevé, *C.R. Acad. Sci. Paris*, **107**, 320–3 (1888).
261. P. Painlevé, *C.R. Acad. Sci. Paris*, **107**, 724–6 (1888).
262. E. Hille, *Ordinary Differential Equations in the Complex Domain*, John Wiley, New York (1976).
263. L. Fuchs, *Sitz. Akad. Wiss.*, Berlin, **32**, 669–720 (1884).
264. E. T. Whittaker and G. N. Watson, *A Course of Modern Analysis*, 4th edn, Cambridge University Press, Cambridge (1927).
265. F. Bureau, *Ann. Mat. Pura. Appl. (IV)*, **64**, 229–364 (1964).
266. C. M. Cosgrove, *Stud. Appl. Math.*, **90**, 119–87 (1993).
267. M. D. Kruskal and P. A. Clarkson, *Stud. Appl. Math.*, **86**, 87–165 (1992).
268. R. Fuchs, *Math. Ann.*, **63**, 301–321 (1907).
269. P. Painlevé, *C.R. Acad. Sci. Paris*, **143**, 1111–7 (1906).
270. K. Iwasaki, H. Kimura, S. Shimomura, and M. Yoshida, *From Gauss to Painlevé: a Modern Theory of Special Functions*, Aspects of Maths, **E16**, Viewag, Wiesbaden (1991).
271. P. Painlevé, *Bull. Soc. Math. France*, **28**, 201–261 (1900).
272. P. Painlevé, *C.R. Acad. Sci. Paris*, **135**, 411–15 (1900).
273. P. Painlevé, *C.R. Acad. Sci. Paris*, **135**, 641–7 (1902).
274. P. Painlevé, *C.R. Acad. Sci. Paris*, **135**, 1020–5 (1902).
275. P. Painlevé, *C.R. Acad. Sci. Paris*, **135**, 757–61 (1902).

276. N. Joshi and M. D. Kruskal, *Stud. Appl. Math.*, **86**, 315–76 (1992).
277. M. J. Ablowitz and H. Segur, *Stud. Appl. Math.*, **57**, 13–44 (1977).
278. J. Myers, *J. Math. Phys.* **6**, 1839–46 (1965).
279. J. Myers, *Physica*, **11**, 51–89 (1984).
280. B. M. McCoy, J. H. H. Perk, and R. E. Shrock, *Nucl. Phys.*, **B220[FS8]**, 35–47 (1983).
281. B. M. McCoy, J. H. H. Perk, and R. E. Shrock, *Nucl. Phys.*, **B220[FS8]**, 269–82 (1983).
282. E. Barouch, B. M. McCoy, and T. T. Wu, *Phys. Rev. Lett.*, **31**, 1409–11 (1973).
283. T. T. Wu, B. M. McCoy, C. A. Tracy, and E. Barouch, *Phys. Rev.*, **B13**, 316–74 (1976).
284. D. B. Creamer, H. B. Thacker, and D. Wilkinson, *Phys. Rev.*, **23D**, 3081–4 (1981).
285. D. B. Creamer, H. B. Thacker, and D. Wilkinson, *Physica*, **20D**, 155–86 (1986).
286. M. Jimbo, T. Miwa, Y. Mori, and M. Sato, *Physica*, **1D**, 80–158 (1980).
287. P. C. T. de Boer and L. S. S. Ludford, *Plasma Phys.*, **17**, 29–43 (1975).
288. R. N. Franklin and J. R. Ockendon, *J. Plasma Phys.*, **4**, 371–82 (1970).
289. P. J. Holmes and D. A. Spence, *Q. J. Mech. Appl. Math.*, **37**, 525–38 (1984).
290. D. L. Turcotte, D. A. Spence, and H. H. Bau, *Int. J. Heat Mass Transfer*, **25**, 699–706 (1982).
291. A. P. Bassom and P. Hall, *Stud. Appl. Math.*, **81**, 185–219 (1989).
292. P. Hall, *IMA J. Appl. Math.*, **29**, 173–96 (1982).
293. P. Hall and W. D. Lakin, *Proc. R. Soc. Lond. A*, **415**, 421–44 (1988).
294. J. S. McCaskill and E. D. Fackerell, *J. Chem. Soc., Faraday Trans. 2*, **84**, 161–79 (1988).
295. E. Brézin and V. A. Kazakov, *Phys. Lett.*, **236B**, 144–50 (1990).
296. A. S. Fokas, A. R. Its, and A. V. Kitaev, *Communs Math. Phys.*, **142**, 313–44 (1991).
297. A. S. Fokas, A. R. Its, and A. V. Kitaev, *Communs Math. Phys.*, **147**, 395–430 (1992).
298. D. J. Gross and A. A. Migdal, *Phys. Rev. Lett.*, **64**, 127–30 (1990).
299. D. J. Gross and A. A. Migdal, *Phys. Rev. Lett.*, **64**, 717–20 (1990).
300. D. J. Gross and A. A. Migdal, *Nucl. Phys.*, **B340**, 333–65 (1990).
301. G. Moore, *Communs Math. Phys.*, **133**, 261–304 (1990).
302. G. Moore, *Progr. Theoret. Phys. Suppl.*, **102**, 255–85 (1990).
303. S. Chandrasekhar, *Proc. R. Soc. Lond. A*, **408**, 209–32 (1986).
304. B. Léauté and G. Marcilhacy, *Ann. Inst. Henri Poincaré*, **31**, 363–75 (1979).
305. B. Léauté and G. Marcilhacy, *Phys. Lett.*, **87A**, 159–61 (1982).
306. B. Léauté and G. Marcilhacy, *Phys. Lett.*, **93A**, 394–8 (1983).
307. B. Léauté and G. Marcilhacy, *J. Math. Phys.* **26**, 1938–41 (1985).
308. B. Léauté and G. Marcilhacy, *J. Math. Phys.*, **27**, 703–6 (1986).
309. B. Léauté and G. Marcilhacy, *J. Math. Phys.*, **28**, 774–6 (1987).
310. G. Marcilhacy, *Phys. Lett.*, **73A**, 157–8 (1979).
311. S. Persides and B. C. Xanthopoulos, *J. Math. Phys.*, **29**, 674–80 (1988).

312. P. Wils, *Phys. Lett.*, **135A**, 425-7 (1989).
313. P. Wils, *Class. Quantum Grav.*, **6**, 1231-41 (1989).
314. J. A. Giannini and R. I. Joseph, *Phys. Lett.*, **141**, 417 9 (1989).
315. P. Winternitz and D. Levi (eds), *Painlevé Transcendents, their Asymptotics and Physical Applications*, NATO ASI Series B: Physics, **278**, Plenum Press, New York (1991).
316. L. J. Mason and N. M. J. Woodhouse, *Inverse Problems*, **6**, 569-81 (1993).
317. J. Guckenheimer and P. Holmes, *Nonlinear Oscillations, Dynamical Systems, and Bifurcations of Vector Fields*, Springer-Verlag, New York (1983).
318. R. S. MacKay and J. D. Meiss (eds), *Hamiltonian Dynamical Systems*, Adam Hilger, Bristol (1987).
319. J. Hietarinta, *Phys. Rev. Lett.*, **52**, 1057-60 (1984).
320. M. D. Kruskal, A. Ramani, and B. Grammaticos, in *Partially Integrable Evolution Equations in Physics* (ed. R. Conte and N. Boccara), NATO ASI Series C, **310**, Kluwer, Dordrecht (1991), pp. 321-72.
321. M. Lakshmanan and R. Sahadevan, *Phys. Rep.*, **224**, 1-93 (1992).
322. A. Ramani, B. Grammaticos, and T. Bountis, *Phys. Rep.*, **180**, 159-245 (1989).
323. W.-H. Steeb and N. Euler, *Nonlinear Evolution Equations and Painlevé Test*, World Scientific, Singapore (1988).
324. M. Adler and P. van Moerbeke, *Communs Math. Phys.*, **83**, 83-106 (1982).
325. M. Adler and P. van Moerbeke, *Invent. Math.*, **67**, 297-326 (1982).
326. M. Adler and P. van Moerbeke, *Math. Ann.*, **279**, 25-85 (1987).
327. M. Adler and P. van Moerbeke, *Communs Math. Phys.*, **113**, 659-700 (1988).
328. M. Adler and P. van Moerbeke, *Invent. Math.*, **97**, 3-51 (1989).
329. N. Ercolani and E. Siggia, *Phys. Lett.*, **119A**, 112-16 (1986).
330. N. Ercolani and E. Siggia, *Physica*, **34D**, 303-46 (1989).
331. L. Haine, *Communs Math. Phys.*, **94**, 271-87 (1984).
332. J. Chazy, *Acta Math.*, **34**, 317-85 (1911).
333. R. Garnier, *Ann. Sci. Ecole Norm. Sup.*, **29**, 1-126 (1912).
334. F. Bureau, *Ann. Mat. Pura Appl.* (*IV*), **66**, 1-116 (1964).
335. N. A. Lukashevich, *Diff. Eqns*, **18**, 557-63 (1982).
336. I. P. Martynov, *Diff. Eqns*, **17**, 154-8 (1981).
337. I. P. Martynov, *Diff. Eqns*, **18**, 568-78 (1982).
338. I. P. Martynov, *Diff. Eqns*, **21**, 512-17 (1985).
339. I. P. Martynov, *Diff. Eqns*, **21**, 623-30 (1985).
340. F. Bureau, *Ann. Mat. Pura Appl.* (*IV*), **91**, 163-281 (1972).
341. F. Bureau, A Garcet, and J. Goffar, *Ann. Mat. Pura Appl.* (*IV*), **92**, 177-91 (1972).
342. R. Chalkley, *J. Diff. Eqns*, **68**, 72-117 (1987).
343. C. M. Cosgrove and G. Scoufis, *Stud. Appl. Math.*, **88**, 25-87 (1993).
344. C. M. Cosgrove, *J. Phys. A: Math. Gen.*, **10**, 2093-105 (1977).
345. C. M. Cosgrove, *J. Phys. A: Math. Gen.*, **11**, 2405-30 (1978).
346. M. Jimbo, *Publ. RIMS, Kyoto Univ.*, **18**, 1137-61 (1982).
347. M. Jimbo and T. Miwa, *Physica*, **2D**, 407-488 (1981).
348. M. Jimbo and T. Miwa, *Physica*, **4D**, 26-46 (1981).

349. A. S. Fokas and Y. C. Yortsos, *Lett. Nuovo Cim.*, **30**, 539–44 (1981).
350. Z. Nehari, *Conformal Mapping*, McGraw-Hill, New York (1952).
351. J. Chazy, *C.R. Acad. Sci. Paris*, **149**, 563–5 (1909).
352. J. Chazy, *C.R. Acad. Sci. Paris*, **150**, 456–8 (1910).
353. M. J. Ablowitz and P. A. Clarkson, *Solitons, Nonlinear Evolution Equations and Inverse Scattering*, *L.M.S. Lect. Note Series*, **149**, Cambridge University Press, Cambridge (1991).
354. L. A. Takhtajan, *Theoret. Math. Phys.*, **93**, 1308–17 (1993).
355. A. P. Fordy and A. Pickering, *Phys. Lett.*, **160A**, 347–54 (1991).
356. S. Chakravarty, M. J. Ablowitz, and P. A. Clarkson, *Phys. Rev. Lett.*, **65**, 1085–7 (1990).
357. R. S. Ward, *Phil. Trans. R. Soc. Lond.*, **A315**, 451–7 (1985).
358. M. J. Ablowitz, A. Ramani, and H. Segur, *J. Math. Phys.*, **21**, 715–21 (1980).
359. S. Kowalevski, *Acta Math.*, **12**, 177–232 (1889).
360. S. Kowalevski, *Acta Math.*, **14**, 81–93 (1889).
361. M. J. Ablowitz and H. Segur, *Phys. Rev. Lett.*, **38**, 1103–7 (1977).
362. M. Wadati, *J. Phys. Soc. Japan*, **32**, 1681 (1972).
363. M. J. Ablowitz, D. J. Kaup, A. C. Newell, and H. Segur, *Stud. Appl. Math.*, **53**, 249–315 (1974).
364. M. Boiti and F. Pempinelli, *Nuovo Cim.*, **59B**, 40–58 (1980).
365. M. J. Ablowitz, A. Ramani, and H. Segur, *Lett. Nuovo Cim.*, **23**, 333–8 (1978).
366. M. Lakshmanan and P. Kaliappan, *J. Math. Phys.*, **24**, 795–806 (1983).
367. M. J. Ablowitz, A. Ramani, and H. Segur, *J. Math. Phys.*, **23**, 1006–15 (1980).
368. S. P. Hastings and J. B. McLeod, *Arch. Rat. Mech. Anal.*, **73**, 31–51 (1980).
369. J. B. McLeod and P. J. Olver, *SIAM J. Math. Anal.*, **14**, 488–506 (1983).
370. V. V. Golubov, *Lectures on the Integration of the Equations of Motion of a Rigid Body about a Fixed Point*, State Publishing House, Moscow (1953).
371. F. Calogero and A. Degasperis, *Spectral Transform and Solitons. I*, North-Holland, Amsterdam (1982).
372. M. J. Ablowitz, M. D. Kruskal, and J. F. Ladik, *SIAM J. Appl. Math.*, **36**, 428–37 (1979).
373. P. A. Clarkson, J. B. McLeod, P. J. Olver, and A. Ramani, *SIAM J. Appl. Math.*, **17**, 798–802 (1986).
374. V. G. Makhankov, *Phys. Rep.*, **35**, 1–128 (1978).
375. R. S. Ward, *Phys. Lett.*, **102A**, 279–82 (1984).
376. A. C. Newell, M. Tabor, and Y. B. Zeng, *Physica*, **29D**, 1–68 (1987).
377. C. M. Cosgrove, *Stud. Appl. Math.*, **89**, 1–61 (1993).
378. C. M. Cosgrove, *Stud. Appl. Math.*, **89**, 95–151 (1993).
379. J. Weiss, in *Solitons in Physics, Mathematics and Nonlinear Optics* (ed. P. J. Olver and D. H. Sattinger), *IMA Series*, **25**, Springer-Verlag, Berlin (1990), pp. 175–202.
380. A. K. Pogrebkov, *Inverse Problems*, **5**, L7–10 (1989).
381. S. Carillo and B. Fuchssteiner, *J. Math. Phys.*, **30**, 1606–13 (1989).
382. D. V. Chudnovsky, G. V. Chudnovsky, and M. Tabor, *Phys. Lett.*, **97A**, 268–74 (1983).

383. R. Conte, *Phys. Lett.*. **134A**, 100–4 (1988).
384. R. Conte, *Phys. Lett.*, **140A**, 383–90 (1989).
385. R. Conte and M. Musette, *J. Math. Phys.*, **32**, 1450–7 (1991).
386. P. G. Estévez, P. R. Gordoa, L. M. Alonso, and E. M. Reus, *J. Phys. A: Math. Gen.*, **26**, 1915–25 (1993).
387. H. Flaschka, A. C. Newell, and M. Tabor, in *What is Integrability?* (ed. V. E. Zakharov), Springer-Verlag, New York (1991), pp. 73–114.
388. B. Fuchssteiner and S. Carillo, *Physica*, **154A**, 467–510 (1989).
389. J. D. Gibbon, P. Radmore, M. Tabor, and D. Wood, *Stud. Appl. Math.*, **72**, 39–63 (1985).
390. J. D. Gibbon, A. C. Newell, M. Tabor, and Y. B. Zeng, *Nonliearity*, **1**, 481–90 (1988).
391. M. Musette and R. Conte, *J. Math. Phys.*, **32**, 1450–7 (1991).
392. A. Pickering, *J. Math. Phys.*, **35**, 821–33 (1994).
393. R. L. Sachs, *Physica*, **30D**, 1–27 (1988).
394. J. Weiss, *J. Math. Phys.*, **25**, 13–24 (1984).
395. J. Weiss, *J. Math. Phys.*, **25**, 2226–35 (1984).
396. J. Weiss, *J. Math. Phys.*, **26**, 258–69 (1985).
397. J. Weiss, *J. Math. Phys.*, **27**, 1293–305 (1986).
398. J. Weiss, *J. Math. Phys.*, **27**, 2647–56 (1986).
399. J. Weiss, *J. Math. Phys.*, **28**, 2025–39 (1987).
400. J. Weiss, in *Partially Integrable Evolution Equations in Physics* (ed. R. Conte and N. Boccara), *NATO ASI Series C*, **310**, Kluwer, Dordrecht (1990), pp. 375–411.
401. J. Weiss, in *Lie Theory, Differential Equations and Representation Theory* (ed. V. Hussin), Les Publications de Centre de Recherches Mathématiques, Montréal (1900), pp. 405–28.
402. J. Weiss, in *Painlevé Transcendents, their Asymptotics and Physical Applications* (ed. P. Winternitz and D. Levi), *NATO ASI Series B: Physics*, **278**, Plenum Press, New York (1991), pp. 225–47.
403. Z. X. Chen and B. Y. Guo, *J. Phys. A: Math. Gen.*, **22**, 5187–94 (1989).
404. Z. X. Chen and B. Y. Guo, *IMA J. Appl. Math.*, **48**, 107–15 (1992).
405. Z. X. Chen, B. Y. Guo, and L. W. Xiang, *J. Math. Phys.*, **31**, 2851–5 (1990).
406. R. Conte and M. Musette, *J. Phys. A: Math. Gen.*, **22**, 169–77 (1989).
407. P. G. Estévez and P. R. Gordoa, *J. Phys. A: Math. Gen.*, **23**, 4831–7 (1990).
408. N. A. Kudryashov, *Phys. Lett.*, **155A**, 269–75 (1991).
409. N. A. Kudryashov, *Phys. Lett.*, **169A**, 237–42 (1992).
410. N. A. Kudryashov, *Phys. Lett.*, **178A**, 99–104 (1993).
411. N. A. Kudryashov, *Phys. Lett.*, **182A**, 356–62 (1993).
412. A. L. Larsen, *Phys. Lett.*, **179A**, 284–90 (1993).
413. K. Nozaki, *J. Phys. Soc. Japan*, **56**, 3052–4 (1987).
414. A. Pickering, *J. Phys. A: Math. Gen.*, **26**, 4395–405 (1993).
415. J. A. Powell and P. K. Jacobsen, *Physica*, **64D**, 132–52 (1993).
416. J. A. Powell, A. C. Newell, and C. K. R. T. Jones, *Phys. Rev. A*, **44**, 3636–52 (1991).

417. J. A. Powell and M. Tabor, *J. Phys. A: Math. Gen.*, **25**, 3773–93 (1991).
418. G. M. Webb, *Physica*, **41D**, 208–18 (1990).
419. G. M. Webb and G. P. Zank, *Phys. Lett.*, **150A**, 14–22 (1990).
420. G. M. Webb and G. P. Zank, *J. Phys. A: Math. Gen.*, **23**, 5465–77 (1990).
421. G. M. Webb and G. P. Zank, *Nonlinear Anal.-Theo. Meth. Appl.*, **19**, 167–76 (1992).
422. G. M. Webb and G. P. Zank, *Astrophys. J.*, **396**, 549–74 (1992).
423. J. D. Fournier, G. Levine, and M. Tabor, *J. Phys. A: Math. Gen.*, **21**, 33–54 (1988).
424. G. Levine and M. Tabor, *Physica*, **33D**, 189–210 (1988).
425. L. Hlavatý, *J. Math. Phys.*, **31**, 605–9 (1990).
426. L. Hlavatý, *J. Math. Phys.*, **33**, 888–94 (1992).
427. L. Hlavatý, *Czech. J. Phys.*, **42**, 765–81 (1992).
428. R. Grimshaw, *Proc. R. Soc. Lond. A*, **368**, 359–75 (1979).
429. M. Jimbo, M. D. Kruskal, and T. Miwa, *Phys. Lett.*, **92A**, 59–60 (1982).
430. R. Grimshaw, *Proc. R. Soc. Lond. A*, **368**, 377–88 (1979).
431. P. A. Clarkson, *Proc. R. Soc. Edinb.*, **109A**, 109–26 (1988).
432. M. D. Kruskal, in *Dynamical Systems Theory and Applications* (ed. J. Moser), *Lect. Notes Phys.*, **38**, Springer-Verlag, New York (1975), pp. 310–54.
433. D. Levi, O. Ragnisco, and A. Sym, *Phys. Lett.*, **100A**, 7–10 (1984).
434. S. Kawamoto, *J. Phys. Soc. Japan*, **54**, 2055–6 (1985).
435. A. Ramani, B. Dorizzi, and B. Grammaticos, *Phys. Rev. Lett.*, **49**, 1539–41 (1982).
436. A. F. Ranada, A. Ramani, B. Dorizzi, and B. Grammaticos, *J. Math. Phys.*, **26**, 708–10 (1985).
437. R. M. Miura, *SIAM Rev.*, **18**, 412–59 (1976).
438. B. Fornberg and G. B. Whitham, *Phil. Trans. R. Soc. Lond. A*, **289**, 373–404 (1978).
439. P. A. Clarkson, A. S. Fokas, and M. J. Ablowitz, *SIAM J. Appl. Math.*, **49**, 1188–209 (1989).
440. M. D. Kruskal and N. Joshi, in *Chaos and Order* (ed. N. Joshi and R. L. Dewar), World Scientific, Singapore (1991), pp. 82–96.
441. R. K. Dodd, J. C. Eilbeck, J. D. Gibbon, and H. C. Morris, *Solitons and Nonlinear Wave Equations*, Academic Press, London (1982).
442. P. G. Drazin and R. S. Johnson, *Solitons: an Introduction*, Cambridge University Press, Cambridge (1989).
443. G. Eilenberger, *Solitons, Mathematical Methods for Physicists*, Springer Series in Solid-State Sciences, **19**, Springer-Verlag, Berlin (1981).
444. B. G. Konopelchenko, *Introduction to Multidimensional Integrable Equations*, Plenum Press, New York (1992).
445. B. G. Konopelchenko, *Soliton in Multidimensions*, World Scientific, Singapore (1993).
446. G. L. Lamb, *Elements of Soliton Theory*, John Wiley, New York (1980).
447. A. C. Newell, *Solitons in Mathematics and Physics*, SIAM, Philadelphia (1985).

448. S. P. Novikov, S. V. Manakov, L. P. Pitaevskii, and V. E. Zakharov, *Theory of Solitons: The Inverse Scattering Method*, Plenum Press, New York (1984).
449. R. K. Bullough and P. J. Caudrey (eds), *Solitons, Topics in Current Physics*, **17**, Springer-Verlag, Berlin (1980).
450. A. S. Fokas and V. E. Zakharov (eds), *Important Developments in Soliton Theory*, Springer Series in Nonlinear Dynamics, Springer-Verlag, Berlin (1993).
451. A. P. Fordy (ed.), *Soliton Theory: a Survey of Results*, Manchester University Press, Manchester (1990).
452. V. E. Zakharov (ed.), *What is Integrability?* Springer Series in Nonlinear Dynamics, Springer-Verlag, Berlin (1991).
453. J. Scott Russell, Report of the 7th Meeting of the British Association for the Advancement of Science, Liverpool (1838), pp. 417–96.
454. J. Scott Russell, Report of the 14th Meeting of the British Association for the Advancement of Science, John Murray, London (1844), pp. 311–90.
455. G. B. Airy, Tides and waves, *Encyclopaedia Metropolotana*, **5**, 241–396, London (1845).
456. G. Stokes, *Trans. Camb. Phil. Soc.*, **8**, 441–55 (1847).
457. Lord Rayleigh, *Phil. Mag. Ser. (5)*, **1**, 257–79 (1876).
458. D. J. Korteweg and G. DeVries, *Phil. Mag. Ser. (5)*, **39**, 422–43 (1895).
459. C. S. Gardner and G. K. Morikawa, Similarity in the asymptotic behaviour of collision free hydrodynamic waves and water waves, Courant Inst. Math. Sci. Res. Rep. NYO-9082, New York University, New York (1960).
460. V. E. Zakharov and E. A. Kuznetzov, *Physica*, **18D**, 455–63 (1986).
461. R. M. Miura, *J. Math. Phys.*, **9**, 1202–4 (1968).
462. I. M. Gel'fand and B. M. Levitan, *Am. Math. Soc. Transl., Ser. 2*, **1**, 253–304 (1955).
463. L. D. Faddeev, *J. Math. Phys.*, **4**, 72–104 (1963).
464. P. Deift and E. Trubowitz, *Communs Pure Appl. Math.*, **32**, 121–251 (1979).
465. I. Kay and H. E. Moses, *J. Appl. Phys.*, **27**, 1503–8 (1956).
466. C. S. Gardner, J. M. Greene, M. D. Kruskal, and R. M. Miura, *Communs Pure Appl. Math.*, **27**, 97–133 (1974).
467. P. D. Lax, *Communs Pure Appl. Math.*, **21**, 467–490 (1968).
468. H. Flaschka and A. C. Newell, *Communs Math. Phys.*, **76**, 65–116 (1980).
469. A. S. Fokas and M. J. Ablowitz, *Communs Math. Phys.*, **91**, 381–403 (1983).
470. H. Flaschka, *Phys. Rev. B*, **9**, 1924–5 (1974).
471. H. Flaschka. *Progr. Theoret. Phys.*, **51**, 703–16 (1974).
472. M. J. Ablowitz and J. F. Ladik, *J. Math. Phys.*, **16**, 598–603 (1975).
473. M. J. Ablowitz and J. F. Ladik, *J. Math. Phys.*, **17**, 1011–18 (1976).
474. M. J. Ablowitz and J. F. Ladik, *Stud. Appl. Math.*, **55**, 213–29 (1976).
475. M. J. Ablowitz and J. F. Ladik, *Stud. Appl. Math.*, **57**, 1–12 (1977).
476. A. S. Fokas and M. J. Ablowitz, *Stud. Appl. Math.*, **68**, 1–10 (1983).
477. Y. Kodama, M. J. Ablowitz, and J. Satsuma, *J. Math. Phys.*, **23**, 564–76 (1982).
478. V. Dryuma, *Sov. Phys. JETP Lett.*, **19**, 381–8 (1974).
479. A. S. Fokas and M. J. Ablowitz, *J. Math. Phys.*, **25**, 2494–505 (1984).

REFERENCES

480. M. J. Ablowitz and R. Haberman, *Phys. Rev. Lett.*, **35**, 1185–8 (1975).
481. C. S. Gardner, *J. Math. Phys.*, **12**, 1548–51 (1971).
482. V. A. Marchenko, *Sturm–Liouville Operators and Applications*, Birkhauser-Verlag, Basel (1986).
483. M. J. Ablowitz, D. J. Kaup, A. C. Newell, and H. Segur, *Phys. Rev. Lett.*, **30**, 1262–4 (1973).
484. I. M. Gel'fand and L. A. Dikii, *Func. Anal. Appl.* **11**, 93–104 (1977).
485. D. J. Kaup, *Stud. Appl. Math.*, **62**, 189–216 (1980).
486. S. Sawada and T. Kotera, *Progr. Theoret. Phys.*, **51**, 1355–67 (1974).
487. P. J. Caudrey, R. K. Dodd, and J. D. Gibbon, *Proc. R. Soc. Lond. A*, **351**, 407–22 (1976).
488. J. D. Gibbon, *Phil. Trans. R. Soc. Lond. A*, **315**, 335–65 (1985).
489. A. P. Fordy and J. Gibbons, *Phys. Lett.*, **75A**, 325 (1980).
490. R. Hirota and A. Ramani, *Phys. Lett.*, **76A**, 95–6 (1980).
491. R. Beals, *Am. J. Math.*, **107**, 281–366 (1985).
492. R. Beals, P. Deift, and C. Tomei, *Direct and Inverse Scattering on the Line*, Math. Surv. Mono., **28**, American Mathematical Society, Providence, R.I. (1988).
493. V. E. Zakharov and S. V. Manakov, *Sov. Phys. JETP Lett.*, **18**, 243–5 (1973).
494. V. E. Zakharov and S. V. Manakov, *Sov. Phys. JETP*, **42**, 842–50 (1976).
495. D. J. Kaup, *Stud. Appl. Math.*, **55**, 9–44 (1976).
496. S. V. Manakov, *Physica*, **3D**, 420–7 (1981).
497. V. E. Zakharov and E. I. Schulman, *Physica*, **4D**, 270–4 (1982).
498. N. Yajima and M. Oikawa, *Progr. Theoret. Phys.*, **56**, 1719–39 (1976).
499. R. Beals and R. R. Coifman, *Communs Pure Appl. Math.*, **37**, 39–90 (1984).
500. R. Beals and R. R. Coifman, *Communs Pure Appl. Math.*, **38**, 29–42 (1985).
501. R. Beals and R. R. Coifman, *Inverse Problems*, **3**, 577–93 (1987).
502. R. Beals and R. R. Coifman, *Inverse Problems*, **5**, 87–130 (1989).
503. X. Zhou, *Communs Pure Appl. Math.*, **42**, 895–938 (1989).
504. K. Konno and A. Jeffrey, in *Advances in Nonlinear Waves* (ed. L. Debnath), Res. Notes Math., **95**, Pitman, London (1984), pp. 162–83.
505. P. A. Clarkson and J. B. McLeod, *Arch. Rat. Mech. Anal.*, **103**, 97–138 (1988).
506. J. W. Miles, *Proc. R. Soc. Lond.*, **A361**, 277–291 (1978).
507. J. W. Miles, *Mechanics Today*, **5**, 297–313, Pergamon Press, Oxford (1980).
508. R. R. Rosales, *Phys. Rev. Lett.*, **A361**, 265–275 (1978).
509. H. Segur and M. J. Ablowitz, *Physica*, **3D**, 165–84 (1981).
510. L. Schlesinger, *J. für Math.*, **141**, 96–145 (1912).
511. K. Okamoto, *J. Fac. Sci. Univ. Tokyo*, **33**, 575–618 (1986).
512. A. R. Its and V. Yu. Novokshenov, *The Isomonodromic Deformation Method in the Theory of Painlevé Equations*, Lect. Notes Math., **1191**, Springer-Verlag, Berlin (1986).
513. M. Jimbo, T. Miwa, and K. Ueno, *Physica*, **2D**, 306–52 (1981).
514. G. Lebaeu and P. Lochak, *J. Diff. Eqns*, **68**, 344–72 (1987).

515. A. S. Fokas and X. Zhou, *Communs Math. Phys.*, **144**, 601–22 (1992).
516. R. Beals and D. H. Sattinger, *Physica*, **65D**, 17–47 (1983).
517. P. A. Deift and X. Zhou, *Ann. Math.*, 295–370 (1993).
518. A. S. Fokas and A. R. Its, in *Important Developments in Soliton Theory* (ed. A. S. Fokas and V. E. Zakharov), Springer Series in Nonlinear Dynamics, Springer-Verlag, Berlin (1993), pp. 99–122.
519. A. F. Fokas, U. Mugan, and M. J. Ablowitz, *Physica*, **30D**, 247 (1988).
520. A. S. Fokas, U. Mugan, and X. Zhou, *Inverse Problems*, **8**, 757–85 (1992).
521. A. R. Its and A. A. Kapaev, *Math. USSR Izvest.*, **31**, 193–207 (1988).
522. A. A. Kapaev, *Diff. Eqns*, **24**, 1107–1115 (1989).
523. A. A. Kapaev, *Theoret. Math. Phys.*, **77**, 1227–1234 (1989).
524. A. A. Kapaev, *Phys. Lett.*, **167A**, 356–62 (1992).
525. A. A. Kapaev and V. Yu. Novokshenov, *Sov. Phys. Dokl.*, **31**, 719–21 (1986).
526. A. A. Kitaev, *Theoret. Math. Phys.*, **64**, 878–94 (1985).
527. A. A. Kitaev, *Math. USSR Sbornik*, **62**, 421–43 (1989).
528. U. Mugan and A. S. Fokas, *J. Math. Phys.*, **33**, 2031–45 (1992).
529. V. Yu. Novokshenov, *Math. USSR Izvest.*, **54**, 587–609 (1990).
530. V. Yu. Novokshenov, *Physica*, **63D**, 1–7 (1991).
531. B. I. Suleimanov, *Diff. Eqns*, **23**, 569–76 (1987).
532. M. Toda, *J. Phys. Soc. Japan*, **22**, 431–6 (1967).
533. M. Toda, *Suppl. Progr. Theoret. Phys.*, **59**, 1–161 (1976).
534. T. B. Benjamin, *J. Fluid Mech.*, **29**, 559–92 (1967).
535. R. E. Davies and A. Acrivos, *J. Fluid Mech.*, **29**, 593–607 (1967).
536. H. Ono, *J. Phys. Soc. Japan*, **39**, 1082–91 (1975).
537. J. Satsuma, M. J. Ablowitz, and Y. Kodama, *Phys. Lett.*, **73A**, 283–6 (1979).
538. P. M. Santini, M. J. Ablowitz, and A. S. Fokas, *J. Math. Phys.*, **25**, 892–99 (1984).
539. H. Segur, in *Mathematical Methods in Hydrodynamics and Integrability in Dynamical Systems* (ed. M. Tabor and Y. M. Treve), *AIP Conf. Proc.*, **88**, pp. 211–28 (1982).
540. A. Davey and K. Stewartson, *Proc. R. Soc. Lond. A*, **338**, 101–10 (1974).
541. D. J. Benney and G. J. Roskes, *Stud. Appl. Math.*, **48**, 377–85 (1969).
542. M. Boiti, J. J.-P. Leon, L. Martina, and F. Pempinelli, *Phys. Lett.*, **132A**, 432–9 (1988).
543. A. S. Fokas and P. M. Santini, *Phys. Rev. Lett.*, **63**, 1329–33 (1989).
544. A. S. Fokas and P. M. Santini, *Physica*, **44D**, 99–130 (1990).
545. J. Hietarinta and R. Hirota, *Phys. Lett.*, **145A**, 237 (1990).
546. M. Jaulent, M. Manna, and L. Martinez-Alonso, *Phys. Lett.*, **151A**, 303 (1990).
547. C. R. Gilson and J. J. C. Nimmo, *Proc. R. Soc. Lond. A*, **435**, 339 (1991).
548. C. R. Gilson, *Phys. Lett.*, **161A**, 423 (1992).
549. M. J. Ablowitz and A. I. Nachman, *Physica*, **18D**, 223–41 (1986).
550. R. R. Coifman and A. S. Fokas, in *Important Developments in Soliton Theory* (eds A. S. Fokas and V. E. Zakharov), Springer Series in Nonlinear Dynamics, Springer-Verlag, Berlin (1993), pp. 58–85.

551. A. S. Fokas and L.-Y. Sung, **8**, *Inverse Problems*, 673–708 (1992).
552. A. V. Bäcklund, *Math. Ann.*, **9**, 207–320 (1876).
553. A. V. Bäcklund, *Math. Ann.*, **17**, 285–328 (1880).
554. A. V. Bäcklund, *Math. Ann.*, **19**, 387–422 (1882).
555. A. V. Bäcklund, *Concerning Surfaces with Constant Negative Curvature* (Trans. E. M. Coddington), New Ear, Lancaster, PA (1905).
556. L. Bianchi, *Lezioni sulla Teoria dei Gruppi Continui Finiti di Transformazioni*, Enrico Spoerri, Pisa (1918).
557. L. Bianchi, *Lezioni di Geometria Differenziale*, Vol. I, Enrico Spoerri, Pisa (1922).
558. C. Rogers and W. F. Shadwick, *Bäcklund Transformations and their Applications*, Academic Press, New York (1982).
559. H. D. Wahlquist and F. B. Estabrook, *Phys. Rev. Lett.*, **23**, 1386–9 (1973).

8 CONCLUSIONS

What we observe is not nature itself, but nature exposed to our method of questioning.

Werner Heisenberg

It is the customary fate of new truths to begin as heresies and to end as superstitions.

Thomas Henry Huxley

The motivation for this book was to emphasize the importance of nonlinearity in many-body systems, especially in condensed-matter physics. This approach is based on the belief that interactions between particles play a dominant role in solid state phenomena and that the constituent particles, in most cases, may not be treated as nearly independent entities. While the introduction of quasi-particles, representing collective modes of behaviour of the solid, goes a long way towards understanding the physics of such systems, there are always limitations to this approach. The first limitation is the stability of the underlying structure, e.g. the crystal lattice, which cannot be recovered using independent particle approaches. A related limitation is that, in most cases, collective modes interact, which may lead to the emergence of localized excitation modes that cannot be derived from a perturbative approach. Our strategy or philosophy, therefore, has been to complement the traditional methods of linearizing physical problems and instead explore new possibilities that nonlinear physics affords.

We recognize and acknowledge the fact that most of the developments in the physical sciences from the time of Newton up until very recently were permeated by the idea of linearity. This conceptual framework was influenced by linear differential equations and their properties. An important example of this is the Superposition Principle, which states that by adding any two solutions of a linear equation another valid solution is produced. Its physical consequence is that a multi-constituent entity may be broken down into a collection of weakly interacting subsystems, the properties of which are independent, so that this resembles clockwork arrangements in which the knowledge of its parts determines the behaviour of the whole system. Another attendant property of a linear system is that an effect is always proportional to its cause. Thus, a small perturbation always results in a small or minor change in the system's behaviour, while a large disturbance is expected to give a much larger outcome. As we have argued throughout the book, there exists a growing number of natural phenomena which do not entirely conform to this physical picture. One such case is the so-called 'Butterfly effect' exhibited by weather systems, as illustrated by the Lorenz equations. Conversely, it is believed by the practitioners of the science of complexity that most dynamical systems are robustly stable with a tendency to remain at the edge of chaos. This is well illustrated by sand pile behaviour, called self-organized criticality, where, independently of the amount of added sand, the system evolves towards self-similarity. Thus the linear cause–effect relationship may not be universally true.

We admit that 'linear physics' has played a very important and valuable role in the past 300 years by providing a very systematic and internally consistent way of describing natural phenomena. Its remarkable successes, such as the development of the main physical disciplines of classical mechanics (Newton, Hamilton, Lagrange, and Jacobi), electromagnetism (Maxwell), thermodynamics (Gibbs, Boltzmann), quantum mechanics (Planck, Heisenberg, Schrödinger, and Bohr) and special relativity (Einstein) have led not only to an unprecedented growth of knowledge about the Universe but also spearheaded a technological revolution, the greatest such advancement in the history of mankind. However, starting with the doubts expressed by Poincaré at the turn of the century about the general validity of linear approaches to investigate nature, subsequent developments in physics have gradually undermined universal applicability of the linear principle. For example, Einstein's general relativity is expressed in terms of nonlinear differential equations, where mass causes the curvature of space in which it moves. Many problems regarding the foundations of quantum mechanics, particularly related to the process of measurement, remain unresolved. Several of the creators of quantum mechanics, including Einstein, Pauli, Planck, Heisenberg, and Dirac, have expressed dissatisfaction with their own brainchild towards the end of their lives. Einstein in particular advocated an effort to make quantum mechanics a nonlinear field theory. The successor to quantum mechanics, quantum field theory, has been developing in the direction of accommodating more and more nonlinear concepts in its foundations, as has been exemplified by the incorporation of solitons, instantons, Goldstone modes, and Higgs Bosons, all of which are nonperturbative effects. A serious difficulty still remains, however, and that is how simultaneously to make a theory nonlinear and nonclassical, i.e. a nonlinear quantum theory. It appears that the price for ignoring nonlinearity through perturbative schemes may be the emergence of divergences, as is possibly best demonstrated in Feynman's quantum electrodynamics. In one of his last lectures, Dirac strongly objected to the ease with which modern physics has been accepting such ideas as discarding infinities.

Condensed-matter physics has been making great strides in accepting nonlinearity and its manifestations. This is perhaps largely due to its more utilitarian emphasis and the less pedantic approach. Beginning with the models of phase transitions, this area of physics has embraced bifurcations, catastrophes, and instabilities into its theoretical framework. The concept of scaling and self-similarity has been very fruitfully employed by Kadanoff and Wilson, later to become instrumental in the formulation of chaos theory by Feigenbaum and of fractality by Mandelbrot. Numerous effects have been observed experimentally which have pointed to the importance of global behaviour, in contrast to a more traditional approach that focused on the system's components. Many examples of self-organization and co-operative phenomena spring to mind. In this context we wish to mention several spectacular examples, such as laser action, pattern formation in convective flows, crystal growth phenomena, and superconductivity.

At present, nonlinear science is one of the most dynamically developing areas of intellectual activity and is rapidly gaining respect and recognition by the community of scientists at large. Much of the credit should be given to such individuals as Mandelbrot, Feigenbaum, or Lorenz, who developed so many of these new and original ideas. It is true, however, that even today nonlinear physics is not a unified scientific discipline but is made up of rather loosely connected endeavours, the common aspect of which is to look beyond linearity.

Viewing the situation historically, we can see many parallels between the present situation in physics and the turn of the previous century. At that time, before quantum mechanics and relativity appeared on the scene, a common conviction among scientists was that the edifice of physics was very nearly completed. This is well illustrated by Lord Kelvin's address to the Royal Society in which he argued that 'the age of discovery in physics was coming to an end'. What precipitated the greatest breakthrough in the understanding of nature by mankind were initially only thought to be minor problems. Difficulties were encountered when trying to explain black-body radiation characteristics, atomic spectra, and the photoelectric effect. Today, many scientists are echoing the same comments. In 1980, S. Hawking gave a lecture entitled 'Is the end in sight for theoretical physics'. In it, he said that 'we already know the physical laws that govern everything we experience in everyday life'. If history is to be taken seriously, there is much cause for optimism, and we believe that physics will be as vigorous 20 years from now as it was in the 1920s.

One of the reasons for writing this book was to present the reader with an accessible overview of the developments in nonlinear science, especially as it pertains to condensed-matter systems. We have also attempted to present a logical sequence of steps in extracting nonlinear behaviour from the basics of many-body physics. Therefore, our discussion was initially based on Landau's idea of an order parameter and its bifurcations. This was assumed to be at first an homogeneous and time-independent quantity, and subsequent generalizations led us to consider order parameter fields, both classical and quantum in nature. Before a quantum field theory analogue of an order parameter was properly introduced, we presented a required background knowledge involving elementary excitations and quantum fields. We believe that the two chapters on the Method of Coherent Structures can be viewed as a synthesis that combines elements of critical phenomena and quantum field theory, as well as nonlinear dynamics. We have tried to demonstrate that a unified nonlinear quantum field theory formalism can be used to describe interacting many-particle systems of diverse character. Moreover, at a relatively simple level of approximation, the Landau–Ginzburg Hamiltonian has emerged naturally in the description. A characteristic feature of the method proposed is the inherent nonlinearity of the equations of motion for the field. The availability of both analytical and numerical means of tackling these equations made this development possible. Indeed, the previous chapter of our monograph is solely devoted to an up-to-date presentation of the mathematical developments currently at our disposal. In order for this physical theory to make contact with experiment and to satisfy our own curiosity, we have provided the reader with practical examples of the use of MCS. Hence, a chapter was included discussing various applications. The rather challenging examples that we selected include spin chains and their phase boundaries as well as quantum excitations, the phenomenon of superconductivity, and multi-electron atoms and their spectra. Although many other examples have been contemplated by the authors, the limited size of the book and the time available for its completion prevented us from extending this list of examples.

Appendix A: SOLVING $\ddot{x} = g(x)$ GRAPHICALLY

The autonomous second-order differential equation

$$\ddot{x} = g(x) \tag{A1}$$

has appeared in this book in a number of contexts. When the right-hand side is simple enough, e.g. a cubic polynomial, the equation may be solved analytically. Fortunately, in this and all other situations a graphical method can be used that provides a means of analysing the types of solution of eqn (A1) as well as the conditions on their existence. We integrate eqn (A1) once to find

$$\tfrac{1}{2}\dot{x}^2 = \int dx\, g(x) + c_0, \tag{A2}$$

where c_0 is an integration constant, and we define the potential function, $G(x)$, as

$$G(x) \equiv \int dx\, g(x). \tag{A3}$$

The graphical method that we present here is based on the plot of $\dot{x}^2$ versus x (i.e. plotting $G(x)$ for a number of choices of the integration constant c_0). A general plot of this type is illustrated in Fig. A.1.

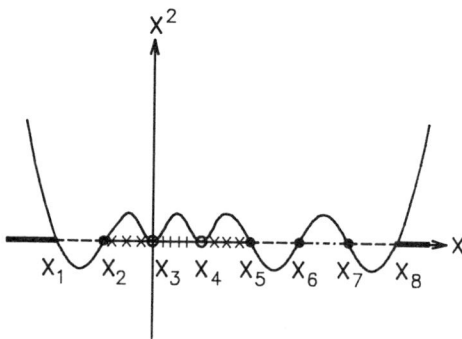

Fig. A.1 A plot of $\dot{x}^2$ versus x for eqn (A2). ●, Single root; ○, double root; ———, real singular solution; – – –, complex solution; ++++, topological solitary wave; —×—×—, nontopological solitary wave; —·—·—, real periodic solution (nonsingular).

APPENDIX A 559

First of all, for a given choice of the integration constant c_0, the resultant graph represents as many possible solutions as there are regions on the x-axis defined by the real roots of the function $G(x) + c_0$. In our case, the roots x_i are defined by

$$G(x) + c_0 = 0 \quad \text{for } x = x_i \quad (i = 1, \ldots, 8). \tag{A4}$$

Second, whether a given solution is real or complex depends on the value of $G(x) + c_0$ for all x within a single region. This follows from eqn (A2), since it can be integrated formally to yield

$$\pm \int_{x_0}^{x} \frac{\mathrm{d}x}{\sqrt{2(G(x) + c_0)}} = t - t_0. \tag{A5}$$

Thus, whenever $G(x) + c_0 \geq 0$ for $x_i \leq x \leq x_{i+1}$, the solution is *real*. Here, x_i and x_{i+1} denote two neighbouring roots of $G(x) + c_0$. For example, real solutions exist for $x < x_1$, $x_2 \leq x \leq x_3$; $x_3 \leq x \leq x_4$; $x_4 \leq x \leq x_5$; $x_6 \leq x \leq x_7$, and $x \geq x_8$. On the other hand, if $G(x) + c_0 \neq 0$ between any two neighbouring roots, the corresponding solution is *complex*, which follows directly from eqn (A5). Thus, using Fig. A.1, we conclude that the only complex solutions are for $x_1 \leq x \leq x_2$, $x_5 \leq x \leq x_6$, and $x_7 \leq x \leq x_8$.

Having determined which solutions are real, we now investigate the nature of real solutions. In general, we can distinguish four types of real solutions: (a) singular; (b) periodic nonsingular; (c) topological solitary waves; and (d) nontopological solitary waves. Below, we give the criteria for the existence of each of these types.

First, the distinction between *singular* and *regular* solutions is made solely on the basis of whether one or two real roots of $G(x) + c_0$ exist in the neighbourhood of the initial value of x at $t = 0$, i.e. $x(0)$. If two real roots exist, the solution is always bounded by these two roots, i.e.

$$x_i \leq x(t) \leq x_{i+1} \quad \text{for } -\infty \leq t \leq +\infty, \tag{A6}$$

such as is the case for all but the two external regions shown in Fig. A.1. Conversely, if only one real root separates the initial point $x(0)$ between the other solutions and $+\infty$ or $-\infty$, the solution that is found is singular, i.e.

$$-\infty \leq x(t) \leq x_{\min} \quad \text{for } -\infty \leq t \leq \infty$$

or

$$x_{\max} \leq x(t) \leq +\infty \quad \text{for } -\infty \leq t \leq +\infty, \tag{A7}$$

This is the situation involving the region below x_1 and above x_8, as shown in Fig. A.1.

The remaining cases involve two real roots on both sides of the initial point $x(0)$. If both of these roots are single, the resultant solution, $x(t)$, is periodic, since in the vicinity of a single root, say x_i, we approximate eqn (A5) by

$$\pm \int_{x_i}^{x} \frac{dx}{\sqrt{(x-x_i)}} \simeq \alpha(t-t_0), \tag{A8}$$

where α is approximately constant close to x_i. The integral on the left-hand side of eqn (A8) is *nonsingular*, meaning that it takes a *finite* amount of time to reach x_i. On the other hand, if x_i is a multiple root with the degree of multiplicity k ($=2,3,\ldots$), the same level of approximation for $x(t)$ in its vicinity requires that

$$\pm \int_{x_i}^{x} \frac{dx}{\sqrt{(x-x_i)^k}} \simeq \beta(t-t_0). \tag{A9}$$

Clearly, for all $k \geq 2$ the left-hand side of eqn (A9) is singular and it takes an infinite amount of time to reach the root x_i by the solution $x(t)$. Thus, the asymptotics are governed by

$$x \simeq x_i + \exp[\pm\beta |t-t_0|], \tag{A10}$$

as $t \to \pm\infty$, for $k = 2$, while

$$x \simeq x_i + \left\{[\beta(t-t_0)]\left(1-\frac{k}{2}\right)\right\}^{1/(1-k/2)}, \tag{A11}$$

for $k \geq 3$. As a consequence, if a solution $x(t)$ is bounded by two real roots, one of which is *single* and the other *multiple*, it will be in the form of a *nontopological solitary wave* (a bump or a well). If, on the other hand, it is bounded by *two-multiple roots*, $x(t)$ will have the shape of a *topological solitary wave* (a kink or an antikink). The reader is referred to Chapter 1 for graphical illustrations. In Fig. A.1 a nonsingular periodic solution is found for $x_6 < x < x_7$, and two bump-like solutions for $x_4 \leq x \leq x_5$ and $x_2 \leq x \leq x_3$, while a kink-type solution is given for $x_3 \leq x \leq x_4$.

Appendix B: THE ELLIPTIC FUNCTIONS OF JACOBI

In the text we have utilized these functions without definition and have not described their properties. Their use is essential for the solution of some of the simplest nonlinear ordinary differential equations as they encapsulate succinctly, sometimes with complicated arguments, the main features of the mathematical structure of the solutions. Knowing something of their properties also enables us to see, almost at a glance, what physical features they represent, so they are extremely beneficial assets. Despite the fact that historically they were defined independently by Jacobi and Abel, many of their properties had been investigated almost twenty years earlier by Gauss. Elliptic functions are fascinating in their own right but here, inevitably, we cannot delve into all their attributes, but only dwell on those we believe essential in our discourse on coherent structures.

Generalities

We begin by defining what we mean by an analytic function of a complex variable Z. Suppose that a function is single-valued and differentiable at every point of some domain, let us call it D, in the complex plane, except possibly for a finite number of points. Then we say it is analytic in D. These latter exceptional points are called singularities of the function and are of a variety of types including, for example, simple poles, multiple poles, removable singularities, and isolated essential singularities. Our function F may, of course, have no singularities in the domain, in which case we shall say that it is *regular* there. A function F is periodic if there exists a constant 2Λ such that

$$F(Z + 2\Lambda) = F(Z). \tag{B1}$$

In physics and mathematics there are many functions which are periodic, like the trigonometric and hyperbolic functions, which have periods of 2π and $2\pi i$, respectively. We call 2Λ a period of F, so $2n\Lambda$ is also a period, where n is an integer. Here, we shall define a *fundamental period* as a period no submultiple of which is also a period, and we call a function that has only one fundamental period *simply periodic*, as are the two examples cited above. Bearing in mind that one of our examples has an imaginary period leads us naturally to ask whether there exist analytical functions, regular except at poles that possess *two* fundamental periods, let us say $2\Lambda_1$ and $2\Lambda_2$, such that $2\Lambda_1/2\Lambda_2$ is real. Indeed, could such a function have more than two fundamental periods? To our great benefit, it was Jacobi in 1829, in his magnum opus entitled *Fundamenta Nova Theoriae Functionum Ellipticarum*, who showed that such functions must be constant.

Primitive periods, poles, and zeros

Now suppose that our function F is not a constant and possesses fundamental periods the ratio of which is not real. Then it may be demonstrated that such a function is necessarily doubly periodic. If we designate its periods by $2\Lambda_1$ and $2\Lambda_2$, then if every other period is a sum of multiples of these, we say these form a pair of *primitive periods*. If F had no poles but the ratio of its primitive periods was real, then mathematicians have demonstrated that it will again be a constant. Thus, such a function *must* have poles if it is not a constant. Note here that although the cos z, sin z, cosh z, and sinh z functions have zeros in the complex plane, and are periodic, they do *not* have poles. An *elliptic function* is defined to be analytic, to be doubly periodic, and to have only singularities which are poles in the finite complex plane. We can visualize the complex plane as divided up into rectangles, the length of each, along one of the axes (Im axis or Re axis), defines a primitive period. We illustrate this in Fig. B.1, writing a general period as

$$P_{m,n} = 2\Lambda_1 m + 2\Lambda_2 ni,$$

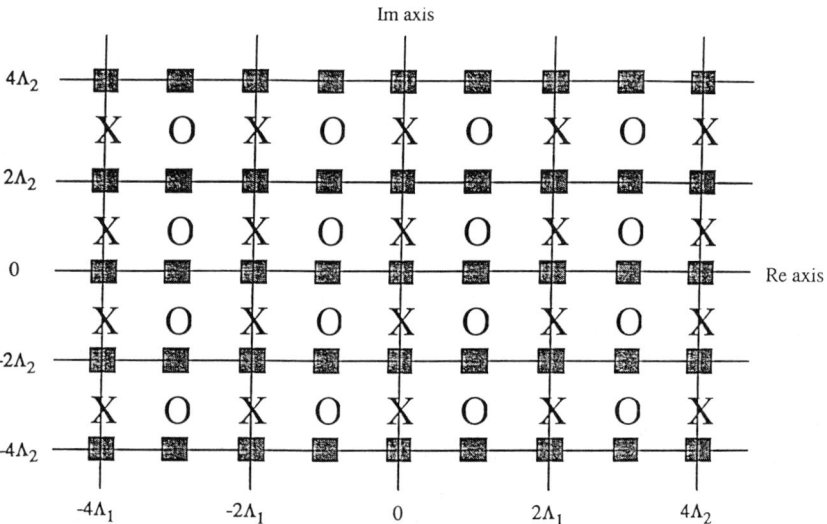

Fig. B.1 Period parallelograms of an elliptic function.

where $m, n = 0, \mp 1, \mp 2, \ldots$. The points $P_{m,n}$, $P_{m+1,n}$, $P_{m,n+1}$, and $P_{m+1,n+1}$ define a mesh, the whole of the complex plane being covered by nonoverlapping meshes. Within any one mesh, there are only a finite number of poles and a finite number of zeros for an elliptic function.

Jacobi elliptic functions

Definitions

An integral of the type

$$x = \int_0^\phi \frac{d\theta}{\sqrt{1 - k^2 \sin^2\theta}} \tag{B2}$$

is called an elliptic integral of the first kind. There are two other canonical types, those of the second kind,

$$y = \int_0^\phi \sqrt{1 - k^2 \sin^2\theta}\, d\theta, \tag{B3}$$

and a third kind,

$$z = \int_0^\phi \frac{d\theta}{(1 - \alpha^2 \sin^2\theta)\sqrt{1 - k^2 \sin^2\theta}}. \tag{B4}$$

In these integrals, k is called the *elliptic modulus* and may take any real or imaginary value, in principle, and α^2 is a real constant. In physics k, when real, is usually assumed to lie within the range $0 \leq k \leq 1$ but, even if it is not, a variety of transformations may be used to find an equivalent modulus in this range. The *complementary elliptic modulus* is defined in a standard way, as used in the literature, by

$$k' = \sqrt{1 - k^2}. \tag{B5}$$

If we consider the integral for x, we may clearly represent it as a function of ϕ and the complex parameter k by $x = F(\phi, k)$. If F is an analytical function of ϕ, regular in a neighbourhood of a given point, z_0, in the complex plane, and takes the value x_0 there, then the necessary and sufficient condition that $x = F(\phi, k)$ should have a unique inverse, ϕ, there, i.e. $\phi = G(x, k)$, which is regular in a neighbourhood of x_0, is that the derivative of F with respect to ϕ at z_0 should not vanish [1]. By a neighbourhood of a point z_0 we mean a set of points z such that $|z - z_0| < r$, where r is the radius of the neighbourhood. Thus, the expression for x, as its derivative will not vanish at any point, if $0 \leq k \leq 1$, may be inverted to find ϕ or, indeed, $\sin \phi$, so we define the simplest elliptic functions sn, cn, dn, am, and tn, by

$$\text{sn}(x,k) = \sin \phi, \qquad \text{cn}(x,k) = \cos \phi, \qquad \text{(B6, B7)}$$

$$\text{dn}(x,k) = \sqrt{1 - k^2 \sin^2 \phi}, \qquad \text{(B8)}$$

$$\text{am}(x,k) = \phi, \qquad \text{tn}(x,k) = \tan \phi. \qquad \text{(B9, B10)}$$

It is convenient to define the complete elliptic integral of the first kind by

$$K = F(\pi/2, k) \qquad \text{(B11)}$$

and the corresponding complementary complete elliptic integral as

$$K' = F(\pi/2, k'). \qquad \text{(B12)}$$

The function $\text{sn}(x,k)$ is doubly periodic with periods $2\Lambda_1 = 4K$ and $2\Lambda_2 = i2K'$, so referring back to Fig. B.1 the behaviour of this function within each rectangle is identical via its periodicity properties. At the points $4mK + i(2n + 1)K'$, the function $\text{sn}(x,k)$ has simple poles (denoted by crosses in Fig. B.1), and also at $(4m + 2)K + i(2n + 1)K'$ (indicated by circles). This same function has zeros at the points $2mK + i2nK'$, where $m, n = 0, \mp 1, \mp 2$ (in Fig. B.1 these are denoted by squares). Close to one of the poles (of the cross variety), e.g. at $x = iK'$,

$$\text{sn}(x,k) = \frac{1}{k(x - iK')} + \overline{\phi}, \qquad \text{(B13)}$$

where $\overline{\phi}$ is regular at $x = iK'$, i.e. this type of pole has a residue of $+1/k$. The other poles (circles) have residue $-1/k$ at every point congruent to $2K + iK'$ (shifted from it by a period). In Table B1 we have listed the magnitudes of periods, zeros, poles, and residues for $\text{sn}(x,k)$, $\text{cn}(x,k)$, and $\text{dn}(x,k)$ following Davis [2], all such points in the complex plane being reached from the pair in this figure by translations $4Km + i2nK'$.

Table B1 Magnitudes of periods, zeros, poles, and residues for elliptic functions.

	$\text{sn}(x,k)$	$\text{cn}(x,k)$	$\text{dn}(x,k)$
Periods	$K, 2iK'$	$4K, 2K + 2iK'$	$2K, 4iK'$
Zeros	$0, 2K$	$K, 3K$	$K + iK', K + 3iK'$
Poles	$iK', 2K + iK'$	$iK', 2K + iK'$	$iK', 3iK'$
Residues	$1/k, -1/k$	$-i/k, i/k$	$-i, i$

Notation for quotients of Jacobi elliptic functions

In the literature it is common to find the following definitions for reciprocals and quotients of Jacobi elliptic functions:

$$1/\text{sn } x = \text{ns } x, \quad \text{sn } x/\text{cn } x = \text{sc } x, \quad \text{cn } x/\text{sn } x = \text{cs } x, \quad \text{(B14)}$$
$$1/\text{cn } x = \text{nc } x, \quad \text{sn } x/\text{dn } x = \text{sd } x, \quad \text{dn } x/\text{sn } x = \text{ds } x, \quad \text{(B15)}$$
$$1/\text{dn } x = \text{nd } x, \quad \text{cn } x/\text{dn } x = \text{cd } x, \quad \text{dn }x/\text{cn } x = \text{dc } x. \quad \text{(B16)}$$

Elliptic functions expressed as series

Maclaurin series about the point $x = 0$ may be obtained in a straightforward manner by successively evaluating appropriate derivatives at this point. The most common of these are as given below:

$$\text{sn}(x, k) = x - (1 + k^2)\frac{x^3}{3!} + (1 + 14k^2 + k^4)\frac{x^5}{5!}$$
$$- (1 + 135k^2 + 135k^4 + k^6)\frac{x^7}{7!} \cdots, \quad \text{(B17)}$$

$$\text{cn}(x, k) = 1 - \frac{x^2}{2!} + (1 + 4k^2)\frac{x^4}{4!} - (1 + 44k^2 + 16k^4)\frac{x^6}{6!}$$
$$+ (1 + 408k^2 + 912k^4 + 64k^6)\frac{x^8}{8!} \cdots, \quad \text{(B18)}$$

$$\text{dn}(x, k) = 1 - k^2\frac{x^2}{2!} + (4 + k^2)k^2\frac{x^4}{4!} - (16 + 44k^2 + k^4)k^2\frac{x^6}{6!}$$
$$+ (64 + 912k^2 + 408k^4 + k^6)k^2\frac{x^8}{8!} \cdots, \quad \text{(B19)}$$

and

$$\text{am}(x, k) = x - k^2\frac{x^3}{3!} + (4 + k^2)k^2\frac{x^5}{5!} - (16 + 44k^2 + k^4)k^2\frac{x^7}{7!}. \quad \text{(B20)}$$

These series are taken from Byrd and Friedman [3], where many other useful series and relationships may be found. Each of these functions may also be expressed as a Fourier series, which are often useful in practice.

Derivatives and integrals of elliptic functions with respect to the argument x

Strictly speaking, each of the Jacobi functions $\text{sn}(x,k)$, $\text{cn}(x,k)$, and $\text{dn}(x,k)$ are functions of *both* x and the modulus k. (Note that x *also* depends on k!) In many applications, however, particularly when k is a fixed constant, the argument k is dropped and, for example, $\text{sn}(x,k)$ becomes $\text{sn}(x)$. Although this is conventional, it should always be remembered when taking derivatives because when k is fixed, derivatives with respect to x are written as if they were total derivatives. Thus, when differentiated, $\text{sn}(x)$ is written as

$$\frac{d}{dx}\text{sn}(x) = \text{cn}(x)\text{dn}(x), \tag{B21}$$

although, strictly speaking, the left-hand side is

$$\frac{\partial}{\partial x}\text{sn}(x,k).$$

With this convention, we list below some common derivatives with respect to x:

$$\frac{d}{dx}\text{cn}(x) = -\text{sn}\,x\,\text{dn}\,x, \quad \frac{d}{dx}\text{dn}(x) = -k^2\,\text{sn}\,x\,\text{cn}\,x, \quad \frac{d}{dx}\text{am}\,x = \text{dn}\,x. \tag{B22}$$

Derivatives with respect to the elliptic modulus are not often used, and so we omit them here, but the interested reader may consult, for example, Byrd and Friedman [3].

The Jacobi functions may also be integrated indefinitely with respect to x to give the following results, which are readily verified by differentiation:

$$\int \text{sn}\,x\,dx = \frac{1}{k}\log(\text{dn}\,x - k\,\text{cn}\,x), \tag{B23}$$

$$\int \text{cn}\,x\,dx = \frac{1}{k}\arccos(\text{dn}\,x), \tag{B24}$$

and

$$\int \text{dn}\,x\,dx = \arcsin(\text{sn}\,x). \tag{B25}$$

Here, arbitrary constants have been omitted for simplicity. There are a whole series of exact indefinite integrals like this, which we need from time to time, with integrands which are rational functions of the Jacobi functions, but we have only presented the most commonly used above.

Approximate relations when the elliptic modulus is close to zero or unity

In physical applications it is often very useful to have approximation formulae for $k \approx 0$ (when Jacobi elliptic functions are approximately trigonometric) or $k \approx 1$ (when these functions become closely hyperbolic). Below, we give some of the most common relationships: others may be found by the reader in Byrd and Friedman [3] and Davis [2]:

(1) $k \simeq 0$:

$$\operatorname{sn}(x,k) \approx \sin x - \frac{k^2}{4} \cos x (x - \sin x \cos x), \qquad \text{(B26)}$$

$$\operatorname{cn}(x,k) \approx \cos x + \frac{k^2}{4} \sin x (x - \sin x \cos x), \qquad \text{(B27)}$$

$$\operatorname{dn}(x,k) \approx 1 - \frac{k^2}{2} \sin^2 x, \qquad \text{(B28)}$$

$$\operatorname{am}(x,k) \approx x - \frac{k^2}{4}(x - \sin x \cos x). \qquad \text{(B29)}$$

(2) $k \simeq 1$:

$$\operatorname{sn}(x,k) \approx \tanh x + \frac{(k')^2}{4} \operatorname{sech}^2 x (\sinh x \cosh x - x), \qquad \text{(B30)}$$

$$\operatorname{cn}(x,k) \approx \operatorname{sech} x - \frac{(k')^2}{3} \tanh x \operatorname{sech} x (\sinh x \cosh x - x), \qquad \text{(B31)}$$

$$\operatorname{dn}(x,k) \approx \operatorname{sech} x + \frac{(k')^2}{4} \tanh x \operatorname{sech} x (\sinh x \cosh x + x), \qquad \text{(B32)}$$

$$\operatorname{am}(x,k) \approx \sin^{-1}(\tanh x) + \frac{(k')^2}{4} \operatorname{sech} x (\sinh x \cosh x - x). \qquad \text{(B33)}$$

Differential equations and identities

1. The most common equations that we use a great deal are the following

(where conventions on derivatives are the ones discussed above):

$$\left[\frac{d}{dx}(\text{sn } x)\right]^2 = (1 - \text{sn}^2 x)(1 - k^2 \text{sn}^2 x), \tag{B34}$$

$$\left[\frac{d}{dx}(\text{cn } x)\right]^2 = (1 - \text{cn}^2 x)((k')^2 + k^2 \text{cn}^2 x), \tag{B35}$$

$$\left[\frac{d}{dx}(\text{dn } x)\right]^2 = (1 - \text{dn}^2 x)(\text{dn}^2 x - (k')^2), \tag{B36}$$

and

$$\left[\frac{d}{dx}(\text{tn } x)\right]^2 = (1 + \text{tn}^2 x)(1 + (k')^2 \text{tn}^2 x). \tag{B37}$$

2. *Identities.* There are so many identities and relationships among the Jacobi elliptic functions that we restrict ourselves to those that we most commonly use. In fact, those that we do give are often regarded as fundamental:

(a) *Fundamental relations:*

$$\text{sn}^2 x + \text{cn}^2 x = 1, \tag{B38}$$

$$k^2 \text{sn}^2 x + \text{dn}^2 x = 1, \tag{B39}$$

$$\text{dn}^2 x - k^2 \text{cn}^2 x = (k')^2, \tag{B40}$$

$$(k')^2 \text{sn}^2 x + \text{cn}^2 x = \text{dn}^2 x. \tag{B41}$$

(b) *Double argument formulae and additional theorems.* These are considerably more complex than the corresponding trigonometric equations, but we give only three below:

$$\text{sn } 2x = (2 \text{ sn } x \text{ cn } x \text{ dn } x)/(1 - k^2 \text{sn}^4 x), \tag{B42}$$

$$\text{cn } 2x = (\text{cn}^2 x - \text{sn}^2 x \text{ dn}^2 x)/(1 - k^2 \text{sn}^4 x), \tag{B43}$$

and

$$\text{dn } 2x = (\text{dn}^2 x - k^2 \text{sn}^2 x \text{ cn}^2 x)/(1 - k^2 \text{sn}^4 x). \tag{B44}$$

Half-argument formulae may also be derived. We do not give them here, but draw the attention of the reader to Davis [2]. These latter relations may be expressed in numerous forms, which may be derived from those given and other equations such as the fundamental relations. Some particular forms that result in useful connections are the addition formulae that we give below:

$$\text{sn}(x+y) = (\text{sn } x \text{ cn } y \text{ dn } y + \text{cn } x \text{ sn } y \text{ dn } x)/(1 - k^2 \text{sn}^2 x \text{ sn}^2 y), \quad \text{(B45)}$$

$$\text{cn}(x+y) = (\text{cn } x \text{ cn } y - \text{sn } x \text{ sn } y \text{ dn } x \text{ dn } y)/(1 - k^2 \text{sn}^2 x \text{ sn}^2 y), \quad \text{(B46)}$$

and

$$\text{dn}(x+y) = (\text{dn } x \text{ dn } y - k^2 \text{ sn } x \text{ sn } y \text{ cn } x \text{ cn } y)/(1 - k^2 \text{sn}^2 x \text{ sn}^2 y). \quad \text{(B47)}$$

Jacobi elliptic functions of differences may easily be obtained from the above by the transformation $x \to x$, $y \to -y$, and by using the parities of cn x, sn x, and dn x, so we do not give them explicitly here.

(c) *Complex and imaginary arguments.* By making use of the relationships

$$\text{sn}(ix, k) = i \text{tn}(x, k'), \quad \text{(B48)}$$

$$\text{cn}(ix, k) = \text{nc}(x, k'), \quad \text{(B49)}$$

$$\text{dn}(ix, k) = \text{dc}(x, k'), \quad \text{(B50)}$$

and

$$\text{tn}(ix, k) = i \text{sn}(x, k'), \quad \text{(B51)}$$

the additional formulae above, together with the fundamental relations, one can obtain expressions for the Jacobi elliptic functions when the argument is complex. For example,

$$\text{sn}(x + iy, k) = \frac{\text{sn}(x, k)\text{dn}(y, k') + i\text{cn}(x, k)\text{dn}(x, k)\text{sn}(y, k')\text{cn}(y, k')}{1 - \text{sn}^2(y, k')\text{dn}^2(x, k)}. \quad \text{(B52)}$$

Such equations for $\text{cn}(x + iy, k)$ and $(\text{dn}(x + iy, k)$ may be derived in a similar way.

(d) *Reciprocal modulus transformations.* If the elliptic modulus is greater than one, these may be used to obtain elliptic functions the modulus of which does lie in the range $0 \leqslant k \leqslant 1$. These transformations are useful in their own right, so we give below those which are most likely to prove useful to us:

$$\left.\begin{aligned} \operatorname{sn}(kx, k_1) &= k\,\operatorname{sn}(x, k) \\ \operatorname{cn}(kx, k_1) &= \operatorname{dn}(x, k) \\ \operatorname{dn}(kx, k_1) &= \operatorname{cn}(x, k) \\ \operatorname{tn}(kx, k_1) &= k\,\operatorname{sd}(x, k) \end{aligned}\right\} \quad \text{where } k_1 = 1/k. \quad (B53)$$

Other transformations that modify both the argument x and the elliptic modulus k, e.g. Landen's and Gauss's transformation [3], may be used, particularly if applied successively to obtain numerical values from tables the range of which for x and k is not that which is required.

References

1. E. T. Copson, *An Introduction to the Theory of Functions of a Complex Variable*, Clarendon Press, Oxford (1962).
2. J. H. T. Davis, *Introduction to Nonlinear Differential and Integral Equations*, Dover, New York (1962).
3. P. F. Byrd and M. E. Friedman, *Handbook of Elliptic Integrals for Engineers and Scientists*, Springer-Verlag, Berlin (1971).

Appendix C: AN EXTENDED PRESENTATION OF SEMICLASSICAL QUANTIZATION FOR CLASSICAL TIME-DEPENDENT FIELDS

The main idea of the MCS is to treat the quantum field ψ as being made up of a large classical field ϕ_c plus a relatively small quantum component. Thus, one first ignores the quantum component, obtaining an equation for the classical field. This side of the procedure has been thoroughly examined in Chapter 5. To find a systematic method for establishing the correspondence between the classical solutions of a given field theory and its quantum version is a nontrivial matter, and during the mid-1970s many theorists developed a number of approaches, including canonical operator methods, semiclassical methods for Fermion fields, and functional integral techniques. Rajaraman [1] has given a very clear exposition of the central idea, in which he begins with a scalar field, $\phi(r,t)$, governed by an appropriate Lagrangian. The equation for the classical field is simply obtained, in the standard way, as an Euler–Lagrange equation from the functional. In Chapter 5, our classical field equation can also be derived in this way, although we proceeded via Heisenberg's equations of motion for the annihilators and then recouped the classical field as the predominant part of the quantum field near a critical point. Our initial equation of motion was a quantum one to begin with, and so it is to this we return. Rajaraman [1] introduces the quantum oscillations, at least initially, by performing a functional Taylor expansion about one minimum, say $\phi_0(r)$, of a classical static field, but this is strictly limited, as he states, to the 'weak coupling approximation'. In the MCS we use the opposite limit and go back to the quantum equation of motion which in zeroth order is (as the simplest case containing nonlinearity):

$$i\hbar \partial_t \psi = \frac{-\hbar^2}{2m} \nabla^2 \psi + \omega \psi + \omega_0 \psi^+ \psi \psi. \tag{C1}$$

These are three main cases that we wish to consider.

Case 1: when $\psi = \eta \exp(-iEt/\hbar)$, where η is real and space-dependent only

Inserting this form into eqn (C1) gives

$$0 = \frac{-\hbar^2}{2m} \nabla^2 \eta + (\omega - E)\eta + \omega_0 \eta^3. \tag{C2}$$

To quantize, we write $\eta = \phi_c \hat{I} + \varepsilon \hat{u}$, where ϕ_c is the classical field, $\hat{u}$ is a small quantum component, and ε is a symbol, which we later drop, to indicate that the quantum part is small. Here $\hat{I}$ is the unit operator in Fock space and the circumflex above u is to indicate its operator character. In addition, we put $E = E_0 + E_1$, where E_0 is the energy of the classical field and E_1 the quantum deviation in energy away from it. When these latter forms for η and E are inserted into eqn (C2) and only terms of order ε are retained, we find that

$$E_1 u = \frac{-\hbar^2}{2m} \nabla^2 u + (\omega - E_0) u + \omega_0 3 \phi_c^2 u, \tag{C3}$$

where u is the expectation value of $\hat{u}$. This is a Schrödinger equation, with an effective potential proportional to $3\omega_0 \phi_c^2$, for the quantum expectation value, u, with energy E_1. Here we have explicitly assumed that the quantum energies, E_1, are of the same order of magnitude as the energy, E_0, of the classical field.

Case 2: when ψ is complex and the classical field's time dependence is $\exp(-iE_0 t/\hbar)$ only

In this case we write

$$\psi = M \exp(iN), \tag{C4}$$

where both M and N are real and, in general, are operators and functions of time and space variables. To ensure that space and time derivatives of N commute with N itself, we assume that N and hence $\exp(iN)$ can be expanded as a linear combination of products of functions of r, t, and operators in the space in which the a_k act. Thus, writing

$$N = \sum_n u_n(r, t) \hat{V}_n, \tag{C5}$$

we see that

$$\left. \begin{array}{l} [\nabla N, N]_- = \sum_{n,n'} \left[\hat{V}_n, \hat{V}_{n'} \right]_- (\nabla u_n) u_{n'} \simeq 0, \\ \text{and} \\ [N_t, N]_- = \sum_{n,n'} \left[\hat{V}_n, \hat{V}_{n'} \right]_- (u_{nt}) u_{n'} \simeq 0. \end{array} \right\} \tag{C6}$$

where $\hat{V}_n$ are the associated operators. This follows since the leading component of each $\hat{V}_n$ is the identity operator $\hat{I}$, as we see later, so corrections to the right-hand side of eqn (C6) are $O(\varepsilon^2)$. Making the general assumption that M has the same form as N results in

$$[M, e^{iN}]_- = 0. \tag{C7}$$

APPENDIX C

Inserting eqn (C4) into eqn (C1), equating real and imaginary parts, and utilizing eqns (C5), (C6), and (C7), we find that

Re, $$-\hbar N_t M = \frac{-\hbar^2}{2m}(\nabla^2 M - M(\nabla N)^2) + \omega M + \omega_0 M^3,\qquad (C8)$$

and

Im, $$+\hbar M_t = \frac{-\hbar^2}{2m}(2\nabla N \cdot \nabla M + M\nabla^2 N),\qquad (C9)$$

since clearly N and M commute, as do ∇N and ∇M. Bearing in mind that what we try to do is expand about the classical solution, we write

$$M = \phi_c \hat{I} + \varepsilon \hat{u} \qquad (C10)$$

and

$$N = \frac{-E_0 t}{\hbar}\hat{I} + \varepsilon \hat{\Omega}, \qquad (C11)$$

where ε is simply inserted to indicate that the quantum components of the modulus and argument, which we denote by $\hat{u}$ and $\hat{\Omega}$, respectively, are small. $\hat{I}$ is the unit operator in Fock space. Equations (C10) and (C11) are now inserted into eqns (C8) and (C9) and only terms of order unity and ε are retained. We subsequently find that

$$E_0 \phi_c = \frac{-\hbar^2}{2m}\nabla^2 \phi_c + \omega \phi_c + \omega_0 \phi_c^3, \qquad (C12)$$

$$-\hbar\left[\frac{-E_0}{\hbar}u + \Omega_t \phi_c\right] = \frac{-\hbar^2}{2m}\nabla^2 u + \omega u + 3\omega_0 \phi_c^2 u, \qquad (C13)$$

and

$$\hbar u_t = \frac{-\hbar^2}{2m}\{2\nabla\Omega \cdot \nabla\phi_c + \phi_c \nabla^2 \Omega\}. \qquad (C14)$$

It is to be understood that, in eqn (C12), an expectation value over $\hat{I}$ has been taken, and over $\hat{u}$ and $\hat{\Omega}$ in eqns (C13) and (C14), so that $\langle \hat{u} \rangle = u$ and $\langle \hat{\Omega} \rangle = \Omega$. Equation (C12) is clearly our earlier equation for the classical field ϕ_c which is satisfied by definition, and eqns (C13) and (C14) are two coupled equations for Ω and u. The easiest way to proceed is to write each of u and Ω as products of a function of time and a space function, so we put

$$u = W(r)T(t) \quad \text{and} \quad \Omega = \Omega_1(r)\Omega_2(t), \qquad (C15, C16)$$

Putting eqns (C16) and (C15) into eqns (C13) and (C14) clearly requires using a separation of variables argument, and we find that

$$\Omega_{2t}(t)/T(t) = K_1 \omega_1 \quad \text{and} \quad T_t(t)/\Omega_2(t) = K_2 \omega_1, \quad \text{(C17, C18)}$$

where ω_1 is some constant frequency. Hence,

$$T_{tt} = K_2 K_1 \omega_1^2 T \quad \text{and} \quad \Omega_{2tt} = K_2 K_1 \omega_1^2 \Omega_2, \quad \text{(C19, C20)}$$

where T_t and T_{tt} indicate total time derivatives, T and Ω_2 being dimensionless.

In a similar way, we may show that

$$-K_1 \hbar \Omega_1 \phi_c \omega_1 = \frac{-\hbar^2}{2m} \nabla^2 W + (\omega - E_0) W + 3 \omega_0 \phi_c^2 W \quad \text{(C21)}$$

and

$$K_2 \hbar W \omega_1 = \frac{-\hbar^2}{2m} \{ 2 \nabla \Omega_1 \cdot \nabla \phi_c + \phi_c \nabla^2 \Omega_1 \}. \quad \text{(C22)}$$

By changing the function Ω_1 to V, where $\Omega_1 \phi_c = KV$, K being a constant, eqns (C21) and (C22) become transformed into a more symmetric form, as

$$-\hbar K_1 K \omega_1 V = \frac{-\hbar^2}{2m} \nabla^2 W + (\omega - E_0) W + 3 \omega_0 \phi_c^2 W \quad \text{(C23)}$$

and

$$+(\hbar K_2 \omega_1 / K) W = \frac{-\hbar^2}{2m} \nabla^2 V + (\omega - E_0) V + \omega_0 \phi_c^2 V, \quad \text{(C24)}$$

where, in eqn (C24), eqn (C12) has been used to attain eqn (C24) in the above form.

Equations (C23) and (C24) can be made equivalent in a very simple way, so as to ensure that their eigenvalues are equal. This can be done by putting $V = SW$ into eqn (C23) and $W = S^{-1}V$ into eqn (C24), where S is a spatially varying function, and arranging that the potential energies are equal. One finds that

$$S = \left(-\omega_0 \phi_c^2 \mp \left[\omega_0^2 \phi_c^4 - \hbar^2 K_1 K_2 \omega_1^2 \right]^{1/2} \right) / (K_1 K \omega_1 \hbar), \quad \text{(C25)}$$

and that

$$\frac{-\hbar}{2m} \nabla^2 W + (\omega - E_0) W + \left\{ 2 \omega_0 \phi_c^2 \mp \left[\omega_0^2 \phi_c^4 - \hbar^2 \omega_1^2 K_1 K_2 \right]^{1/2} \right\} W = 0.$$

$$\text{(C26)}$$

APPENDIX C   575

Another way to proceed is to assume that the gradient of the spatial dependence of V is very weak, i.e. that the gradient of V does not depend very greatly on the space variables; then $\nabla^2 V \sim 0$. Hence, on substituting from eqn (C24), for V, into eqn (C23) we obtain

$$\frac{-\hbar^2}{2m}\nabla^2 W + \left[(\omega - E_0) + 3\omega_0\phi_c^2 + \frac{\hbar^2 K_1 K_2 \omega_1^2}{[(\omega - E_0) + \omega_0\phi_c^2]}\right]W \simeq 0. \quad (C27)$$

If one supposes that $\omega - E_0 \gg \omega_0\phi_c^2$ in eqn (C27), then an effective Schrödinger equation results, namely

$$-\frac{\hbar^2}{2m}\nabla^2 W + \left[(\omega - E_0) + \omega_0\phi_c^2\left(3 - \frac{\hbar^2 K_1 K_2 \omega_1^2}{[(\omega - E_0)^2]}\right)\right]W$$

$$\simeq \frac{-\hbar^2 K_1 K_2 \omega_1^2}{[(\omega - E_0)]}W, \quad (C28)$$

where $-\hbar^2 K_1 K_2 \omega_1^2/(\omega - E_0)$ is playing the role of an eigenvalue (remember that K_1 and K_2 are arbitrary separation constants up to this stage and ω and E_0 are fixed). We may expect the constants in the second rounded bracket in eqn (C28) to be close to 3, as we have taken it to be in the main text.

In principle, eqns (C23) and (C24) result in a fourth-order eigenvalue equation for W, the most general result, the eigenvalue there being proportional to $K_1 K_2$, as is obvious if one puts W from eqn (C24) into eqn (C23). The reader may wonder whether the assumption of a weak gradient for V is a consistent picture to emerge from eqns (C23) and (C24). This consistency is readily demonstrated by inserting the approximation

$$W \simeq \{(\omega - E_0) + \omega_0\phi_c^2\}(K/\hbar K_2 \omega_1)V, \quad (C29)$$

obtained from eqn (C24) by making the weak gradient approximation, into eqn (C23), to give

$$-\hbar^2 K_1 K_2 \omega_1^2 V = \frac{-\hbar^2}{2m}\{\omega_0 V \nabla^2(\phi_c^2) + 2\omega_0 \nabla(\phi_c^2)\cdot\nabla V$$
$$+ [(\omega - E_0) + \omega_0\phi_c^2]\nabla^2 V\} + (\omega - E_0)$$
$$\times [(\omega - E_0) + \omega_0\phi_c^2]V + 3\phi_c^2\omega_0[(\omega - E_0) + \omega_0\phi_c^2]V,$$
$$(C30)$$

where ω has been assumed not to vary with spatial variables. The right-hand side of eqn (C30) may be considerably simplified by using the equation satisfied by the classical field, ϕ_c, namely eqn (C12), to give

$$\text{R.H.S.} = \left\{ -\frac{\hbar^2}{2m} 2(\nabla\phi_c)^2 + 2\phi_c^2(\omega - E_0) + \omega_0 \phi_c^4 \right\} \omega_0 V$$
$$+ (\omega - E_0)^2 V + \text{terms in } \nabla V \text{ and } \nabla^2 V. \tag{C31}$$

With the aid of eqn (C12), it is easy to demonstrate that the gradient of the term in the {} braces of eqn (C31) vanishes provided that $\nabla\phi_c \neq 0$, so one possibility is that the terms within the braces may be replaced by an arbitrary constant, D. As ω and E_0 are fixed and $\omega_1^2 K_1 K_2$ is proportional to the square of the quantum energy, the constant may be adjusted for each quantum energy level, so that the terms in ∇V and $\nabla^2 V$ vanish, this being consistent with our earlier approximation.

When ω is a spatial function the above argument *cannot* be used, since eqn (C31) contains an extra term of the form

$$-\hbar^2/2m(\nabla^2\omega)$$

and the coefficient of ∇V contains an additional $\nabla\omega$ component. The term $(\omega - E_0)^2$, of course, is now a function of r also, so that the coupled equations (C23) and (C24) must now be solved.

Case 3: when ψ is complex with no restrictions on the space dependence of the phase of the classical field: the presence of currents

This is an even more general case, which we have used in the main text, where the complexity of the classical field arises not only through the time-dependent phase factor, $\exp(-iE_0 t/\hbar)$, but by a space-dependent phase. As we saw in Chapter 6, when describing the Haldane problem, this means that current densities are present. The classical field is now written as

$$\psi_c = \exp(i(\phi(r) + \bar{\omega}t))\eta(r), \tag{C32}$$

where $\bar{\omega}$ is a constant angular frequency ($\bar{\omega} = -E_0/\hbar$) and $\eta(r)$ is the real space-dependent amplitude of the field. We quantize about ψ_c by writing

$$\psi = \exp(i(\phi(r) + \bar{\omega}t + \varepsilon f(r,t)))(\eta(r) + \varepsilon \Lambda(r,t)), \tag{C33}$$

where we have dropped the Fock space designations since we can treat them in a similar manner to that in case 2 which we discussed earlier. In eqn (C33) the functions $f(r,t)$ and $\Lambda(r,t)$ are *real* functions of space and

APPENDIX C 577

time and the symbol ε has its usual meaning. There are now *four* equations which correspond to eqns (C12), (C13), and (C14) in case 2, and are as follows:

Re: ε^0:
$$-\hbar\bar{\omega}\eta = \frac{-\hbar^2}{2m}\{\nabla^2\eta - (\nabla\phi)^2\eta\} + \omega\eta + \omega_0\eta^3, \qquad (C34)$$

Im: ε^0:
$$0 = \frac{-\hbar^2}{2m}\{(\nabla^2\phi)\eta + 2\nabla\phi\cdot\nabla\eta\}, \qquad (C35)$$

Re: ε^1: $-\hbar(\bar{\omega}\Lambda + \eta f_t) = \frac{-\hbar^2}{2m}\{\nabla^2\Lambda - (\nabla\phi)^2\Lambda - 2(\nabla\phi)\cdot(\nabla f)\eta\}$
$$+ \omega\Lambda + \omega_0 3\eta^2\Lambda, \qquad (C36)$$

Im: ε^1:
$$+\hbar\Lambda_t = \frac{-\hbar^2}{2m}\{(\nabla^2\phi)\Lambda + (\nabla^2 f)\eta + 2(\nabla\phi\cdot\nabla\Lambda + \nabla f\cdot\nabla\eta)\}. \qquad (C37)$$

Clearly, eqn (C35) may be integrated directly to give
$$\eta^2\nabla\phi = C_0 + \nabla\times F, \qquad (C38)$$

where C_0 and F are arbitrary vectors. In the main text we identified the contribution $\eta^2\nabla\phi$ with a current density, j, in the system, and showed that the nonsingular solutions of eqn (C34) are destroyed by sufficiently strong current densities. What we will show here is that eqns (C36) and (C37) produce a quantum equation in which η^2 again provides an effective potential (*but* currents are included since the new η, in the presence of currents, must be a solution of eqn (C34)). However, *there are no other potential energy terms which involve* $\nabla\phi$ *explicitly*. This would appear to be very reminiscent of the effect of choosing a particular gauge when electromagnetic fields are introduced in a quantum context.

We first make eqns (C36) and (C37) a little more symmetrical by putting $f = V\eta^{-1}$ and utilize eqns (C34) and (C35) to simplify the result. Equations (C36) and (C37) now become

$$-\hbar V_t = \frac{-\hbar^2}{2m}\{\nabla^2\Lambda - (\nabla\phi)^2\Lambda\} + (\omega + \hbar\bar{\omega})\Lambda + 3\omega_0\eta^2\Lambda$$
$$+ \frac{\hbar^2}{2m}2\nabla\phi\cdot[\nabla V - \eta^{-1}V\nabla\eta], \qquad (C39)$$

$$+\hbar\Lambda_t = \frac{-\hbar^2}{2m}\{\nabla^2 V - (\nabla\phi)^2 V\} + (\omega + \hbar\bar{\omega})V + \omega_0\eta^2 V$$
$$+ \frac{\hbar^2}{2m}2\nabla\phi\cdot[-\nabla\Lambda + \eta^{-1}\Lambda\nabla\eta]. \qquad (C40)$$

As we shall later wish to remove single gradient terms by transformation, these two equations are still in an inconvenient form, since each involves the ∇^2 of one dependent variable and the gradient of the other. Thus we define two new dependent variables, $S = V - i\Lambda$ and $T = V + i\Lambda$. Equations (C39) and (C40) now become

$$-\hbar S_t = \frac{-\hbar^2}{2m} i\left[\nabla^2 S - (\nabla\phi)^2 S\right] + i(\omega + \hbar\bar{\omega})S$$

$$+ \frac{\hbar^2}{2m} 2\nabla\phi \cdot \{\nabla S - \eta^{-1} S \nabla\eta\} - i\omega_0 \eta^2 T + 2\eta^2 \omega_0 iS, \quad \text{(C41)}$$

$$-\hbar T_t = \frac{+\hbar^2}{2m} i\left[\nabla^2 T - (\nabla\phi)^2 T\right] - i(\omega + \hbar\bar{\omega})T$$

$$+ \frac{\hbar^2}{2m} 2\nabla\phi \cdot \{\nabla T - \eta^{-1} T \nabla\eta\} + i\omega_0 \eta^2 S - 2\eta^2 \omega_0 iT. \quad \text{(C42)}$$

We are now in a position to remove the single gradient terms in eqns (C41) and (C42) by writing $S = \hat{S} S_2$ and $T = \hat{T} T_2$, where S_2 and T_2 are only spatially dependent, and are chosen so that

$$\nabla S_2 + i S_2 (\nabla\phi) = 0 \quad \text{and} \quad \nabla T_2 - i T_2 (\nabla\phi) = 0, \quad \text{(C43)}$$

which are obviously satisfied by

$$S_2 = \exp(-i(\phi + c)) \quad \text{and} \quad T_2 = \exp(+i(\phi + d)), \quad \text{(C44)}$$

where c and d are arbitrary integration constants. The result of these transformations modifies eqns (C41) and (C42) to

$$+i\hbar\hat{S}_t = \frac{-\hbar^2}{2m} \nabla^2 \hat{S} + (\omega + \hbar\bar{\omega})\hat{S} - \omega_0 \eta^2 \hat{T} \exp(i(2\phi + d + c)) + 2\eta^2 \omega_0 \hat{S},$$

(C45)

$$-i\hbar\hat{T}_t = \frac{-\hbar^2}{2m} \nabla^2 \hat{T} + (\omega + \hbar\bar{\omega})\hat{T} - \omega_0 \eta^2 \hat{S} \exp(-i(2\phi + c + d)) + 2\eta^2 \omega_0 \hat{T},$$

(C46)

where we have made use of eqn (C35).

To make further progress we put

$$\hat{S} = \tilde{S} \exp(-iEt/\hbar) \quad \text{and} \quad \hat{T} = \tilde{T} \exp(-iEt/\hbar), \quad \text{(C47)}$$

where $\tilde{S}$ and $\tilde{T}$ are only spatially dependent. The resulting two equations are

$$E\tilde{S} = \frac{-\hbar^2}{2m}\nabla^2\tilde{S} + (\omega + \hbar\bar{\omega})\tilde{S} - \omega_0\eta^2\tilde{T}\exp(i(2\phi + d + c)) + 2\eta^2\omega_0\tilde{S}, \quad (C48)$$

$$-E\tilde{T} = \frac{-\hbar^2}{2m}\nabla^2\tilde{T} + (\omega + \hbar\bar{\omega})\tilde{T} - \omega_0\eta^2\tilde{S}\exp(-i(2\phi + c + d)) + 2\eta^2\omega_0\tilde{T}. \quad (C49)$$

Equations (C48) and (C49) are again two coupled Schrödinger equations, so we use a similar procedure as in case 2 and make $\tilde{T} = \bar{S}(r)\tilde{S}$ in such a way that the potential energies in these equations is the same. We find that we require

$$\bar{S} = \left\{\frac{-E \mp \left[E^2 + \omega_0^2\eta^4\right]^{1/2}}{\omega_0\eta^2}\right\}\exp(-i(2\phi + c + d)). \quad (C50)$$

Inserting $\tilde{T}$, and using eqn (C50) back into eqn (C48) gives

$$0 = \frac{-\hbar^2}{2m}\nabla^2\tilde{S} + \left\{(\omega + \hbar\bar{\omega}) \pm \left[E^2 + \omega_0^2\eta^4\right]^{1/2} + 2\eta^2\omega_0\right\}\tilde{S}, \quad (C51)$$

which is very similar to eqn (C26) when no currents were present. Thus the form of eqn (C51) is identical to eqn (C26) except that η is now a solution of (C34) *with* currents present. *However, no other terms in eqn (C51) occur which involve the currents (or $\nabla\phi$) explicitly.* This means that when currents are present *in the classical equation* they only appear in the effective quantum potential through η, and no other explicit current terms give rise to interference with those in η^2.

Reference

1. R. Rajaraman, *Solitons and Instantons: an introduction to Solitons and Instantons in Quantum Field Theory*, North-Holland, Amsterdam (1987).

Appendix D: CONVERSION OF ANNIHILATORS AND CREATORS WITH MIXED STATISTICS TO FERMIONS FOR THE SPIN CHAIN

In Section 6.3, operators a_j (a_j^*) were introduced, which acted as Fermion operators on the sites j, but operators from different sites commuted. The 'sites', j, were on a chain, there being N in number, and periodic boundary conditions were satisfied at the ends. It was stated in this latter section that new operators c_j and c_j^*, defined by

$$c_j = \exp\left[+i\pi \sum_{k=1}^{j-1} a_k^* a_k\right] a_j \tag{D1}$$

and

$$c_j^* = a_j^* \exp\left[-i\pi \sum_{k=1}^{j-1} a_k^* a_k\right], \tag{D2}$$

satisfy

$$[c_i, c_j^*]_+ = \delta_{ij}. \tag{D3}$$

The case when $i = j$

From the definition in eqns (D1) and (D2), we find that

$$c_i c_j^* = \exp\left[+i\pi \sum_{k=1}^{i-1} a_k^* a_k\right] a_i a_j^* \exp\left[-i\pi \sum_{k=1}^{j-1} a_k^* a_k\right]$$

$$= \exp\left[+i\pi \sum_{k=1}^{j-1} a_k^* a_k\right](1 - a_j^* a_j)\exp\left[-i\pi \sum_{k=1}^{j-1} a_k^* a_k\right]. \tag{D4}$$

From the on-site commutation rules for the a_j's and using the fact that a_k operators for different sites commute, eqn (D4) becomes

$$c_j c_j^* = 1 - c_j^* c_j. \tag{D5}$$

APPENDIX D

The case when $i \neq j$

(1) $$c_i^* c_j = a_i^* \exp\left[-i\pi \sum_{k=1}^{i-1} a_k^* a_k\right]\exp\left[+i\pi \sum_{k=1}^{j-1} a_k^* a_k\right]a_j. \tag{D6}$$

If $i > j$,

$$\text{R.H.S. of eqn (D6)} = a_i^* \exp\left[-i\pi \sum_{k=j}^{i-1} a_k^* a_k\right]a_j. \tag{D7}$$

If $i < j$,

$$\text{R.H.S. of eqn (D6)} = a_i^* \exp\left[+i\pi \sum_{k=i}^{j-1} a_k^* a_k\right]a_j. \tag{D8}$$

(2) $$c_j c_i^* = \exp\left[+i\pi \sum_{k=1}^{j-1} a_k^* a_k\right]a_j a_i^* \exp\left[-i\pi \sum_{k=1}^{i-1} a_k^* a_k\right]. \tag{D9}$$

If $i > j$,

$$\text{R.H.S. of eqn (D9)} = a_i^* \exp\left[-i\pi \sum_{k=1}^{i-1} a_k^* a_k\right]$$
$$\times \exp\left[+i\pi \sum_{k=1}^{j-1} a_k^* a_k\right](-a_j) = -c_i^* c_j, \tag{D10}$$

since a_i^* commutes with all operators to its left in eqn (D9) and a_j commutes with all operators to the right in eqn (D9) except $\exp[-i\pi a_j^* a_j]$. In this latter case,

$$a_j \exp(-i\pi a_j^* a_j) = a_j(1 - 2a_j^* a_j) = a_j - 2a_j(1 - a_j a_j^*) = -a_j, \tag{D11}$$

since

$$\exp(-i\pi a_j^* a_j) = 1 - i\pi a_j^* a_j + \tfrac{1}{2}(-i\pi)^2(a_j^* a_j)^2 + \ldots$$
$$= 1 + a_j^* a_j(e^{-i\pi} - 1). \tag{D12}$$

This follows because

$$a_j^* a_j a_j^* a_j = a_j^*(1 - a_j^* a_j)a_j = a_j^* a_j, \tag{D13}$$

since $a_j^2 = 0$. Thus

$$\exp(-i\pi a_j^* a_j) = 1 + (e^{-i\pi} - 1)a_j^* a_j = 1 - 2a_j^* a_j. \tag{D14}$$

The right-hand side of eqn (D10) follows from eqn (D7).
If $i < j$,

$$\text{R.H.S. of eqn (D9)} = (-a_i^*)\exp\left[+i\pi \sum_{k=1}^{j-1} a_k^* a_k\right]$$

$$\times \exp\left[-i\pi \sum_{k=1}^{i-1} a_k^* a_k\right] a_j = -c_i^* c_j. \quad (D15)$$

In eqn (D15), a_j commutes with all operators in eqn (D9) to the right, but a_i^* commutes with all terms to its left except $\exp[+i\pi a_i^* a_i]$. However,

$$\exp[+i\pi a_i^* a_i]a_i^* = (1 - 2a_i^* a_i)a_i^* = a_i^* - 2a_i^*(1 - a_i^* a_i) = -a_i^*, \quad (D16)$$

since $(a_i^*)^2 = 0$. The far right side of eqn (D15) follows from eqn (D8). Thus, from eqn (D15) and $i < j$, we have $c_j c_i^* = -c_i^* c_j$.

If $i > j$ we have, from eqns (D9) and (D10), $c_j c_i^* = -c_i^* c_j$. Hence, in all cases we have the Fermi–Dirac commutation rule

$$[c_i, c_j^*]_+ = \delta_{ij}. \quad (D17)$$

Appendix E: A DERIVATION OF THE FORM OF THE EQUATION OF MOTION FOR THE CLASSICAL FIELD FROM THE CORRESPONDING EQUATION OF MOTION FOR THE QUANTUM FIELD OPERATOR

In this appendix we demonstrate one method of obtaining the equation of motion for the classical field from that for the quantum field. We shall only consider Fermion fields as this appears to be the most complicated situation. In order to do this we make full use of the transformation used in Appendix D from annihilators and creators with mixed statistics to operators obeying Fermi–Dirac commutation rules. The identification of combinations of mixed statistical operators to form spin operators will also be utilized (see Section 6.3).

We begin by observing that the commutation relations for Boson fields imply that, in the classical limit when $\hbar \to 0$, the Fock operators used in their formation must commute. However, for Fermion fields the Fock operators become Grassman numbers (C_k, C_k^*), and always mutually anti-commute and never satisfy the Fermi–Dirac relation $C_k C_k^* + C_k^* C_k = 1$. The quantum field, ψ, is a sum of products of c-numbers, varying with position r, and time-dependent Fock operators,

$$q_k(t) = \exp(iHt/\hbar) q_k \exp(-iHt/\hbar),$$

where H is the Hamiltonian. On the other hand, the classical field, ϕ_c, is simply a function of space and time. Clearly, to obtain ϕ_c from ψ an expectation value of ψ is required with a state of some kind, and the obvious choice to make is a coherent state. However, the conventional way this is done is to use Grassman numbers, but if we use Appendix D this is most inconvenient and awkward because there we must use Fermion operators. Therefore, we do not follow this route but define our own 'coherent state' using operators that satisfy Fermi–Dirac statistics.

To recapitulate, one of the properties of a coherent state, defined in a conventional manner by

$$|\xi\rangle = \exp\left(\sum_k C_k q_k + C_k^* q_k^+\right)|0\rangle, \tag{E1}$$

is that it is an eigenstate of a time-independent annihilator, q_k, with eigenvalue C_k^* (see eqn (4.53)), so that

$$q_k |\xi\rangle = C_k^* |\xi\rangle. \tag{E2}$$

APPENDIX E

In this appendix we wish to find another 'coherent state' with a property analogous to that in eqn (E2) but defined using Fermi–Dirac operators (η_k, η_k^*). We do this by defining this state, $|\xi'\rangle$, by

$$|\xi'\rangle = f\left(\sum_k \eta_k q_k + \eta_k^* q_k^+\right)|0\rangle \tag{E3}$$

for some function f to be found, so that both η_k and η_k^* commute with q_k and q_k^+ and

$$[\eta_k, \eta_{k'}^*]_+ = \delta_{k,k'}. \tag{E4}$$

We seek the function f so that

$$q_k|\xi'\rangle = \eta_k^*|\xi'\rangle. \tag{E5}$$

For simplicity, we define an operator X by

$$X = \sum_k (\eta_k q_k + \eta_k^* q_k^+), \tag{E6}$$

so that the functional operator f can be written as

$$f = \sum_{n=0}^{\infty} a_n X^n, \tag{E7}$$

where the a_n are coefficients to be determined. Using the requirement in eqn (E5) and the commutation relations for the q_k's, one finds—after some algebra—that

$$\left.\begin{aligned} q_k \cdot \{a_0 + a_1 X + a_2 X^2 + a_3 X^3 + \ldots\}|0\rangle \\ = \eta_k^* \{a_1 + 2a_2 X + a_3(3X^2 - 2) + a_4(4X^3 - 8X) + \cdots\}|0\rangle \\ = \eta_k^* f(X)|0\rangle \end{aligned}\right\}. \tag{E8}$$

It is found that the functions, f_n, of X which multiply each coefficient, a_n, in the central part of eqn (E8) satisfy the recurrence relation

$$Xf_{n-1} - (n-1)f_{n-2} + X^{n-1} = f_n \quad \text{with } f_0 = 0. \tag{E9}$$

We do not try to find the individual coefficients, a_n, but find the whole function f by noting that the last equality in eqn (E8) is equivalent to

$$f = \frac{df}{dX} - \frac{1}{3}\frac{d^3 f}{dX^3} + \frac{1}{15}\frac{d^5 f}{dX^5} - \frac{1}{105}\frac{d^7 f}{dX^7} + \ldots. \tag{E10}$$

The auxiliary equation for the ODE in eqn (E10) is

$$1 = m - \frac{m^3}{3!!} + \frac{m^5}{5!!} - \frac{m^7}{7!!} + \ldots, \tag{E11}$$

where the semifactorials are defined by $3!! = 3 \times 1$, $5!! = 5 \times 3 \times 1$, $7!! = 7 \times 5 \times 3 \times 1,\ldots$. One solution of eqn (E10) may be written as $f = A\exp(mX)$, where m is a solution of eqn (E11) and A is an arbitrary constant. Equation (E11) may be re-expressed in several different ways, one of which is

$$1 = \sqrt{2}\,\text{daw}(m/\sqrt{2}) \tag{E12}$$

where 'daw' is Dawson's integral [1]. Interestingly, if we define z by $z = m/\sqrt{2}$, eqn (E11) may be cast into the form

$$W(z) - \exp(-z^2) = \sqrt{\frac{2}{\pi}}\,\text{i} = \frac{W(z) - W(-z)}{2}, \tag{E13}$$

where $W(z)$ is the complex error function, the altitude chart of which has been provided by Abramowitz and Stegun [2], as well as tables of its value as a function of z. A more familiar way, perhaps, of expressing eqn (E11) is

$$1 = \exp(-m^2/2)\int_0^m \exp(+t^2/2)dt, \tag{E14}$$

from which a value of m may be found which satisfies it, namely $m = 1.2007 + 0.7167\text{i}$, where the real and imaginary parts of m have been given to four decimal places.

At this point, having shown that there is a state $|\xi'\rangle$ which is an eigenstate of $q_{k'}$ the eigenvalue of which is another Fermion operator, $\eta_{k'}^*$, we define a further state by

$$|G\rangle = |\xi'(t)\rangle|S\rangle, \tag{E15}$$

where $|S\rangle$ is a spin state given by

$$|S\rangle = \prod_k (u_k|+\rangle_k + v_k|-\rangle_k). \tag{E16}$$

The u_k and v_k are constants for a particular k and we have associated with each k a spin of one half. We have written $|+\rangle$ for the component $m_s = +\tfrac{1}{2}$ and $|-\rangle$ for $m_s = -\tfrac{1}{2}$. A spin operator is now defined for each k, in terms of annihilators and creators, a_k and a_k^*, such that they commute between different values of k, i.e. $[a_k, a_{k'}]_- = [a_k^+, a_{k'}^+]_- = [a_k, a_{k'}^+]_- = 0$ for $k \neq k'$, whereas for the same k they behave like Fermi–Dirac operators. Thus

$$S_k^x = \tfrac{1}{2}(a_k^+ + a_k), \quad S_k^y = \frac{\text{i}}{2}(a_k - a_k^+), \quad S_k^z = a_k^+ a_k - \tfrac{1}{2}. \tag{E17}$$

From the properties of the a_k operators, spin operators for different k's commute (see Appendix D). The idea, as in Appendix D, is to associate the η_k, once ordered, in an appropriate way (using the fact that they are countable) with the c_j in eqn (D1), so that

$$\eta_{k_j} = \exp\left(i\pi \sum_{s=1}^{j-1} \left(\tfrac{1}{2} + S_{k_s}^z\right)\right) S_{k_j-}. \tag{E18}$$

Henceforth we shall drop the j on the wave vectors k, the ordering being understood. We define the state $|\xi'(t)\rangle$ in eqn (E15) to be the state in eqn (E3) with each annihilator and creator, q_k and q_k^+, replaced by their Heisenberg time-dependent representatives, e.g.

$$q_k \to \exp(iHt/\hbar) q_k \exp(-iHt/\hbar) = q_k(t).$$

Clearly, an operator equation such as eqn (E5) will be valid when q_k is replaced by $q_k(t)$ and $\xi'(t)$. That is,

$$q_k(t)|\xi'(t)\rangle = \eta_k^*|\xi'(t)\rangle, \tag{E19}$$

since the result in eqn (E5) is derived using only the commutation rules for the annihilators and creators and the fact that $q_{k'}|0\rangle = 0$. This follows, since

$$\left.\begin{array}{r}[q_k(t), q_{k'}^+(t)]_+ = \delta_{k,k'} \\ q_k(t)|0\rangle = 0 \\ H|0\rangle = 0\end{array}\right\}, \tag{E20}$$

where H is the Hamiltonian. For convenience, we shall now put $A = 1$, although this is not necessary for the rest of the development. It may now be established, by examining the terms in the series for $|\xi'\rangle$ and using the Fermion commutation rules for η_k's and η_k^*'s, that

$$\eta_k^*|\xi'\rangle = \frac{(e + e^{-1})}{2} |\bar{\xi}'\rangle \eta_k^*. \tag{E21}$$

In eqn (E21), the state $|\bar{\xi}'\rangle$ is defined by

$$|\bar{\xi}'\rangle = \exp(-mX). \tag{E22}$$

Since the η_k^* commute with H and the q_k's, following eqn (E21) we must have

$$\eta_k^*|\xi'(t)\rangle = \frac{(e + e^{-1})}{2} |\bar{\xi}'(t)\rangle \eta_k^*. \tag{E23}$$

We are now in a position to define the classical field, ϕ_0, in terms of an expectation value using $|G\rangle$. Thus we write

$$\phi_c(r, t) = \exp(-iEt/\hbar)\langle G|\psi|G\rangle, \tag{E24}$$

where E is a constant energy that we determine later. Going back to the equation of motion for the quantum field, ψ, in eqn (5.56), we take an expectation value of both sides with $|G\rangle$ and multiply both sides by $\exp(-iEt/\hbar)$. Thus we obtain

$$\exp(-iEt/\hbar)\langle G|i\hbar\partial_t\psi|G\rangle = \exp(-iEt/\hbar)$$
$$\times \{\lambda_0\langle\psi\rangle + i\mathbf{v}_1\cdot\nabla_\varepsilon\langle\psi\rangle - \tfrac{1}{2}\nabla_\varepsilon^2\langle\psi\rangle$$
$$+ \Omega f(\boldsymbol{\eta}_0,\mathbf{k}_0,\mathbf{m}_0)\langle\psi^+\psi\psi\rangle + \ldots\}, \quad (E25)$$

where

$$\langle\psi\rangle = \langle G|\psi|G\rangle \quad (E26)$$

and

$$\langle\psi^+\psi\psi\rangle = \langle G|\psi^+\psi\psi|G\rangle. \quad (E27)$$

The right-hand side of eqn (E26) becomes

$$\langle G|\sum_k \frac{1}{\sqrt{\Omega}}\exp(-i\mathbf{k}\cdot\mathbf{r})q_k(t)|G\rangle$$
$$= \sum_k \frac{\exp(-i\mathbf{k}\cdot\mathbf{r})}{\sqrt{\Omega}}\langle S|\langle\xi'(t)|q_k(t)|\xi'(t)\rangle|S\rangle. \quad (E28)$$

We simplify the expectation value in eqn (E28) using eqns (E19) and (E23), to give

$$\langle S|\langle\xi'(t)|q_k(t)|\xi'(t)\rangle|S\rangle = \langle S|\langle\xi'(t)|\eta_k^*|\xi'(t)\rangle|S\rangle$$
$$= \langle S|\langle\xi'(t)|\frac{(e+e^{-1})}{2}|\bar{\xi}'(t)\rangle\eta_k^*|S\rangle$$
$$= \frac{(e+e^{-1})}{2}\langle S|\eta_k^*|S\rangle. \quad (E29)$$

The expectation value in eqn (E28) is thus reduced to a spin matrix element. Using eqn (E18),

$$\text{R.H.S. of eqn (E29)} = \frac{(e+e^{-1})}{2}\langle S|S_{k_+}\exp\left(-i\pi\sum_{s=1}^{j-1}(\tfrac{1}{2}+S_s^z)\right)|S\rangle, \quad (E30)$$

where j in the upper limit of the sum in eqn (E30) refers to the position of $\mathbf{k}$ when the $\mathbf{k}$'s are properly ordered, and the sum is up to the $\mathbf{k}$ vector which is one less in the order. The spin operators now mutually commute for different $\mathbf{k}$'s and an elementary calculation reduces eqn (E30) to

$$\frac{(e+e^{-1})}{2}u_k^* v_k \prod_{i=1}^{j-1}(-u_{k_i}^* u_{k_i} + v_{k_i}^* v_{k_i}) = \Lambda_k. \quad (E31)$$

Thus from eqns (E24), (E26), and (E30),

$$\phi_c(r,t) = \exp(-iEt/\hbar) \sum_k \frac{\exp(-i k \cdot r)}{\sqrt{\Omega}} \Lambda_k, \quad (E32)$$

which is merely a time phase multiplied by a Fourier series for the spatial part of ϕ_c. For all the linear terms on the right-hand side of eqn (E25), the above procedure is straightforward, but now we must examine the cubic component on the right-hand side and the left-hand side more carefully.

We examine the left-hand side of eqn (E25) first and choose E so that

$$\exp(-iEt/\hbar)\langle G|i\hbar\partial_t\psi|G\rangle = i\hbar\partial_t(\exp(-iEt/\hbar)\langle G|\psi|G\rangle). \quad (E33)$$

Already we have seen that $\langle G|\psi|G\rangle$ is independent of time, so the right-hand side of eqn (E33) only involves the time-dependent phase. Inside the expectation value on the left-hand side we can apply Heisenberg's equation, so these two features reduce eqn (E33) to

$$-\langle S|\langle\xi'(t)|\sum_k \frac{1}{\sqrt{\Omega}} \exp(-i k \cdot r)[H, q_k(t)]_-|\xi'(t)\rangle|S\rangle$$

$$= E\langle G|\psi|G\rangle$$

$$= E\langle S|\langle\xi'(t)|\sum_k \frac{1}{\sqrt{\Omega}} \exp(-i k \cdot r) q_k(t)|\xi'(t)\rangle|S\rangle. \quad (E34)$$

We may now use the fact that

$$|\xi'(t)\rangle = \exp(iHt/\hbar)\exp(mX)\exp(-iHt/\hbar)|0\rangle$$

$$= \exp(iHt/\hbar)\exp(mX)|0\rangle, \quad (E35)$$

since $H|0\rangle = 0$. Hence eqn (E34) becomes

$$-\langle S|\langle\xi'|\sum_k \exp(-i k \cdot r)[H, q_k]_-|\xi'\rangle|S\rangle$$

$$= E\langle S|\langle\xi'|\sum_k \exp(-i k \cdot r) q_k|\xi'\rangle|S\rangle, \quad (E36)$$

where all the operators and states in eqn (E36) now contain no time factors. This equation serves to define the energy, E, and each of the matrix elements may be evaluated, in principle, using eqn (E23) or its conjugate and eqn (E5).

As a result of our manoeuvres so far, the equation of motion in eqn (E25) becomes, because of our choice of E,

$$i\hbar\partial_t\phi_c = \lambda_0\phi_c + i\mathbf{v}_1 \cdot \nabla_\varepsilon\phi_c - \tfrac{1}{2}\nabla_\varepsilon^2\phi_c$$

$$+ \Omega f(\eta_0, k_0, m_0)\exp(-iEt/\hbar)\langle\psi^+\psi\psi\rangle. \quad (E37)$$

We now follow the above procedure for the cubic term, and we find that

$$\langle \psi^+ \psi \psi \rangle = \langle S | \sum_k d_k^* \eta_k \sum_{k'} d_{k'} \eta_{k'}^* \sum_{k''} d_{k''} \eta_{k''}^* | S \rangle, \qquad (E38)$$

where

$$d_k = \frac{1}{\sqrt{\Omega}} \tfrac{1}{2}(e + e^{-1}) \exp(-i \mathbf{k} \cdot \mathbf{r}). \qquad (E39)$$

Changing the operators η_k and η_k^* in eqn (E38) to spin operators using eqn (E18), the triple summand, excluding the factors d_k, becomes

$$\langle S | \exp\left(+i\pi \sum_{s=1}^{j-1} (\tfrac{1}{2} + S_s^z) \right) S_{k-} S_{k'+}$$

$$\times \exp\left(-i\pi \sum_{s=1}^{j'-1} (\tfrac{1}{2} + S_s^z) \right) S_{k''+} \exp\left(-i\pi \sum_{s=1}^{j''-1} (\tfrac{1}{2} + S_s^z) \right) | S \rangle, \qquad (E40)$$

where j, j', and j'' denote the positions of $\mathbf{k}$, $\mathbf{k}'$, and $\mathbf{k}''$, respectively, when ordered to be in correspondence with the positive integers. When evaluated, expression (E40) does *not* give $\Lambda_k^* \Lambda_{k'} \Lambda_{k''}$ in general, as we would require if $\langle \psi^+ \psi \psi \rangle$ simply reduced to $\phi_c^* \phi_c \phi_c$. For a given set of constants, u_{k_i} and v_{k_i}, eqn (E40) always produces a factor of $v_k^* u_k u_{k'}^* v_{k'} u_{k''}^* v_{k''}$ modulated by a constant involving some or all of the constants u_{k_i} and v_{k_i}, but is different for different sets of the three wave vectors $\mathbf{k}$, $\mathbf{k}'$, and $\mathbf{k}''$. Thus we can write eqn (E40) as

$$F(\mathbf{k}, \mathbf{k}', \mathbf{k}'') v_k^* u_k u_{k'}^* v_{k'} u_{k''}^* v_{k''}, \qquad (E41)$$

where F is some function of the three wavevectors $\mathbf{k}$, $\mathbf{k}'$, and $\mathbf{k}''$. The form in eqn (E41) may be written as

$$G(\mathbf{k}, \mathbf{k}', \mathbf{k}'') \Lambda_k^* \Lambda_{k'} \Lambda_{k''}, \qquad (E42)$$

where

$$G(\mathbf{k}, \mathbf{k}', \mathbf{k}'') = F(\mathbf{k}, \mathbf{k}', \mathbf{k}'') v_k^* u_k u_{k'}^* v_{k'} u_{k''}^* v_{k''} / (\Lambda_k^* \Lambda_{k'} \Lambda_{k''}), \qquad (E43)$$

and none of Λ_k^*, $\Lambda_{k'}$, and $\Lambda_{k''}$ are zero. Assuming that G may be represented as a Taylor expansion about $\mathbf{k} = \mathbf{k}_e$, $\mathbf{k}' = \mathbf{k}'_e$, and $\mathbf{k}'' = \mathbf{k}''_e$, we have

$$G = a_0 + \mathbf{a}_1 \cdot (\mathbf{k} - \mathbf{k}_e) + \mathbf{b}_1 \cdot (\mathbf{k}' - \mathbf{k}'_e) + \mathbf{c}_1 \cdot (\mathbf{k}'' - \mathbf{k}''_e) + \dots \qquad (E44)$$

In eqn (E44), a_0 is the value of G at $\mathbf{k}_e$, $\mathbf{k}'_e$, $\mathbf{k}''_e$, where these wave vectors are constant, as are a_1, b_1, and c_1. Equation (F37) therefore takes the form

$$i\hbar \partial_t \phi_c = \lambda_0 \phi_c + i\mathbf{v}_1 \cdot \nabla_\varepsilon \phi_c - \tfrac{1}{2}\nabla_\varepsilon^2 \phi_c + \Omega f(\eta_0, k_0, m_0)$$
$$\times \{\tilde{a}_0 \phi_c^* \phi_c \phi_c - i\mathbf{a}_1 \cdot \nabla \phi_c^* \phi_c \phi_c + i\mathbf{b}_1 \cdot \phi_c^* (\nabla \phi_c) \phi_c$$
$$+ i\mathbf{c}_1 \cdot \phi_c^* \phi_c \nabla \phi_c + \ldots \}, \tag{E45}$$

where $\tilde{a}_0 = a_0 - \mathbf{a}_1 \cdot \mathbf{k}_e - \mathbf{b}_1 \cdot \mathbf{k}_e' - \mathbf{c}_1 \cdot \mathbf{k}_e'' + \ldots$.

The form of the terms in the braces { } of eqn (E45) are already present in the expansion of the interaction in the equation of motion, but where ϕ_c is replaced by ψ. In a similar way, if we had begun with another nonlinear interaction term in the equation of motion, for example $\langle (\nabla \psi^+) \psi \psi \rangle$, this too would reduce to c-number-like components like those in { } of eqn (E45). Thus, on going to the classical limit, with the equation of motion for the quantum field, merely results in a classical PDE for ϕ_c which is of exactly the same form as the equation of motion for the quantum field, but the coefficients will be modified.

References

1. J. Spanier and K. B. Oldham, *An Atlas of Functions*, Hemisphere, London (1987).
2. M. Abramowitz and I. A. Stegun (eds), *Handbook of Mathematical Functions*, National Bureau of Standards (Applied Mathematics Series No. 55), Washington, DC (1964).

Appendix F: INTRA-SHELL ORDERING OF ENERGY LEVELS

At this point we investigate the position of levels within a shell, since in the study so far it is quite clear (see, for example, Table 6.8) that for a particular shell, e.g. 2s, 2p, the level ordering is reversed in the MCS case compared with HF. We have made two simplifications earlier; namely to ignore all F^k and G^k terms other than those for $k = 0$ and, also, to ignore the correction from $\delta\varepsilon$ terms. As we saw in Chapter 6, the $\delta\varepsilon$ terms have a tendency to make levels more unbound but, in any case, subsidiary calculations appear to indicate that their effect is too small significantly to affect reordering within a shell. The F^k contributions, for $k > 0$, can act in the opposite direction to make states more bound, but again their effect has been estimated and found to be too small.

It is quite clear from Fig. 6.13 that for physically realistic values of p (a measure of the strength of the Coulomb repulsion) this level inversion, e.g. for 2s and 2p energies, occurs as a result of the computer calculation, so that continuing to higher order with the perturbation method, e.g. to order ω_0^2, will never reverse these energy level positions. In fact, a simple second-order calculation shows that 2s is shifted only by -0.7 Ryd and 2p by -0.05 Ryd, thus confirming this observation. However, what we shall now show is that these levels *are* reversed by other terms in the quantum field equation (5.59) where, so far, we have only included $\psi^+\psi\psi$ contributions from the repulsion.

Let us now examine other components in the quantum field equation (5.59), drop the residual term in R_{ij}, as we have spherical symmetry, and assume that the part in μ_1 has already been removed by transforming to a moving frame of reference. Taking all terms into account in eqn (5.59), the *classical* field equation is obtained by replacing the quantum field, ψ, by ψ_c. Subsequently, to obtain the quantum equation we linearize about the classical field, ψ_c. After this latter process, if u is the quantum probability amplitude, e.g. for 2p or 2s, we expect to find:

(1) from $(\nabla_\varepsilon \psi^+)\psi(\nabla_\varepsilon \psi)$,

$$\text{(a) } (\nabla\phi_c)^2 u, \quad \text{(b) } \phi_c \nabla\phi_c \cdot \nabla u; \tag{F1}$$

(2) from $\psi^+\psi(\mu_4 \cdot \nabla_\varepsilon)\psi + \psi^+((\mu_4 \cdot \nabla_\varepsilon)\psi)\psi$,

$$\text{(c) } \mu_4 \cdot \phi_c^2 \nabla u, \quad \text{(d) } \mu_4 \cdot \phi_c \nabla \phi_c u; \tag{F2}$$

(3) from $(\nabla_\varepsilon^2 \psi^+)\psi\psi + \psi^+ \psi \nabla_\varepsilon^2 \psi$,

$$(\text{e}) \ \phi_c^2 \nabla^2 u, \quad (\text{f}) \ \phi_c \nabla^2 \phi_c u. \quad (\text{F3})$$

The coefficients of these terms must be determined from the Coulomb repulsion (see eqn (F13) later).

Let us now investigate the possibilities arising in eqns (F1), (F2), and (F3), and examine first those in eqn (F2) as they represent contributions which arise through linear deviations in wave vectors away from the critical point. If we ignore higher-order derivatives for the time being and follow the quantization procedure in Section 6.6.3 and Appendix C, we find that now there is a component of order ε^0 from the equation corresponding to eqn (C9) and this takes the form $2\phi_c^2 \boldsymbol{\mu}_4 \cdot \nabla \phi_c = 0$. As ϕ_c is, by construction, purely radial, this can only mean that only the angular parts of ∇ are operative, with the consequence that (d) of eqn (F2) vanishes. For case (c) we need matrix elements of the type

$$\int v \phi_c^2 \nabla u \, \mathrm{d}\tau,$$

where v and u are either 2p or 2s and the integral is over all space. Using eqn (1.32) of Rotenberg et al. [1] it is straightforward to show that diagonal elements, i.e. $v = 2s$, $u = 2s$ or $v = 2p$, $u = 2p$, vanish *but*, in general, nondiagonal elements are not zero, being the same whichever component of 2p we take. Unfortunately, diagonalizing these matrix elements within the whole 2s, 2p shell would lead to a splitting of the 2p energy level. This is clearly inconsistent with spherical symmetry, so contributions for case (c) must vanish and, similarly, matrix elements in case (b) will also vanish for the same reason.

Before we tackle components in eqns (F1) and (F3), which involve quadratic deviations from the critical point, we need to appreciate exactly how this particular part of the equation of motion for the quantum field, in eqn (5.59), is obtained. Simply performing a Taylor expansion about the critical point results in terms of the form

$$(\partial_x \ \partial_y \ \partial_z) \begin{pmatrix} a & b & b \\ b & a & b \\ b & b & a \end{pmatrix} \begin{pmatrix} \partial_x \psi \\ \partial_y \psi \\ \partial_z \psi \end{pmatrix}, \quad (\text{F4})$$

premultiplied by ψ^+ and postmultiplied by ψ, where a and b are constants. A simple transformation of the independent variables brings the matrix in eqn (F4) into diagonal form, two eigenvalues being $a - b$ and one being $a - b + (3b)$, where the contribution in round brackets is a component of the residual term. For the operators appearing in eqn (F4) we find a value for $a - b$ which is positive. A similar set of terms of the form

$$(\partial_x \psi^+ \quad \partial_y \psi^+ \quad \partial_z \psi^+) \begin{pmatrix} a' & b' & b' \\ b' & a' & b' \\ b' & b' & a' \end{pmatrix} \begin{pmatrix} \partial_x \psi \\ \partial_y \psi \\ \partial_z \psi \end{pmatrix} \psi \quad \text{(F5)}$$

also appear. The same transformation as that applied to eqn (F4) reduces eqn (F5) to diagonal form, but now $a' - b' = 2(a - b) > 0$. This latter transformation, of course, will not affect the terms in $\nabla^2 \psi$ or $\psi^+ \psi \psi$.

In eqn (F3), cases (e) and (f) are fairly easy to handle because $\nabla^2 \phi_c$ can be replaced using the equation for the classical field, namely eqn (6.206), and $\nabla^2 u$ can be replaced using the corresponding hydrogenic quantum equation for u. We find that the net result for (e) and (f) is given by

$$\frac{4\pi e^2}{|m_0 - k_0|^4} \left(\frac{-2m}{\hbar^2}\right) \int \left[(E_u + 3E_c)\phi_c^2 + 4\frac{Ze^2}{r}\phi_c^2 - 3\omega_0 \phi_c^4\right] u^* u \, d\tau, \quad \text{(F6)}$$

where the integral is over all space, E_u is the energy of the u orbital, E_c is the energy of the classical field, and we have used the same quantization procedure as in Appendix C, assuming that Ω_1 and W are purely radial.

For the interaction of type (a) in eqn (F1) we need to study integrals over all space of the form

$$\int u^* (\nabla \phi_c)^2 u \, d\tau, \quad \text{(F7)}$$

where only a diagonal element is required because $(\nabla \phi_c)^2$ is purely radial. This can be done by observing that the divergence of $\phi_c (d\phi_c/dr)\hat{r} u u^*$ is given by

$$\nabla \cdot \left(\phi_c \frac{d\phi_c}{dr}\hat{r} u u^*\right) = \phi_c \frac{d\phi_c}{dr} u u^* \nabla \cdot \hat{r} + \hat{r} \cdot \left\{\hat{r}\left(\frac{d\phi_c}{dr}\right)^2 u u^* + \hat{r}\phi_c \frac{d^2\phi_c}{dr^2} u u^* \right.$$
$$\left. + \phi_c \frac{d\phi_c}{dr}((\nabla u)u^* + u\nabla u^*)\right\} \quad \text{(F8)}$$

where $\hat{r}$ is a radial unit vector. Applying Gauss's Divergence Theorem, the surface integral vanishes, since $u \to 0$ as $r \to \infty$, and the last two terms on the right-hand side of eqn (F8) in ∇u and ∇u^* also vanish for the same reason, which we used above for cases (b) and (c). As $\nabla \cdot \hat{r} = 2/r$ we have, from eqn (F8),

$$\int (d\phi_c/dr)^2 uu^* \, d\tau = -\int \phi_c \nabla^2 \phi_c uu^* \, d\tau. \tag{F9}$$

Thus we may again use the classical field of eqn (6.206) and combine the result in eqn (F9) with that in eqn (F6) to give the net result for all terms, (a) to (f), given by

$$\frac{4\pi e^2}{|m_0 - k_0|^4} \left(\frac{-2m}{\hbar^2}\right) \int \left[(E_u + 5E_c)\phi_c^2 + 6\,\frac{Ze^2}{r}\,\phi_c^2 - 5\omega_0 \phi_c^4 \right] u^*u \, d\tau, \tag{F10}$$

where the contributions in eqns (F9) and (F6) add in this way because the component on the left-hand side of eqn (F9) has the opposite sign to those terms in eqn (F6) from the quantum field equation. All of the contributions in eqn (F10) are calculated below, the terms in $E_u + 5E_c$ and $\omega_0 \phi_c^4$ being relatively small compared with the central term. What is quite apparent is that these shifts are *always negative* and hence make both states more bound.

We have calculated the matrix elements of ϕ_c^2/r for 2s and 2p orbitals, and we find that

$$\left. \begin{array}{l} \langle \phi_c^2/r \rangle_{2s} = 0.21 \times 10^{-3} \\ \langle \phi_c^2/r \rangle_{2p} = 0.26 \times 10^{-4} \end{array} \right\} \tag{F11}$$

both in units of $(Z/\pi)(Z/a_0)^4$. Thus the shift in 2p is some ten times smaller than that for 2s. We do not know where the critical point is nor how renormalization will modify the size of the interaction repulsion, but we can show, for quite reasonable values of $1/|m_0 - k_0|^4$, that the shifts are quite sufficient to reverse the energy ordering of 2s and 2p orbitals. Writing this factor as $a_0^4((2\pi)^4\lambda^4)$ we evaluate all the terms of eqn (F10) as a function of the parameter λ. In this way, we can find a value of λ giving the correct reordered energies relative to Hartree–Fock energies. We can then use this value of λ to show that we obtain very similar values of x, which we needed to obtain agreement with Hartree–Fock earlier. We carry out this procedure using only 2s, since the shift in 2p is much smaller, and use the fact that we need a shift of $(4.58 - 0.7)$ Ryd to agree with Hartree–Fock energies (if we take the case in which $\alpha = 1$, taking out the second order part -0.7). We find that we require $\lambda = 6.89$.

To show that the above value of λ agrees with our earlier estimated values of x, we note that

$$2\Omega \Delta_{\eta, k, m} = f(\eta, k, m) = 4\pi e^2/|m - k|^2. \tag{F12}$$

Up to second-order deviations, we can straightforwardly show that

$$\omega_0 = \left\{ f(\boldsymbol{\eta}_0, \boldsymbol{k}_0, \boldsymbol{m}_0) - \boldsymbol{k}_0 \cdot (\nabla_k f) - \boldsymbol{m}_0 \cdot (\nabla_m f)_0 \right.$$

$$+ \frac{1}{2!} \left[\sum_{i,j=x,y,z} m_{0i} m_{0j} \left(\frac{\partial^2 f}{\partial m_i \partial m_j} \right)_0 + \sum_{i,j=x,y,z} k_{0i} k_{0j} \left(\frac{\partial^2 f}{\partial k_i \partial k_j} \right)_0 \right.$$

$$\left. \left. + 2 \sum_{i,j=x,y,z} k_{0i} m_{0j} \left(\frac{\partial^2 f}{\partial k_i \partial m_j} \right)_0 \right] \right\}. \quad \text{(F13)}$$

From eqn (F12) we can calculate all the partial derivatives in eqn (F13), and we obtain

$$\omega_0 = 24\pi e^2 / |\boldsymbol{m}_0 - \boldsymbol{k}_0|^2. \quad \text{(F14)}$$

Using eqn (F13) and the value of $\lambda = 6.89$ above, we obtain

$$x = 2\omega_0 (Z^4/a_0^3) \times 10^{-4} = 13.05 \text{ Ryds} \quad (1 \text{ Ryd} = e^2/2a_0),$$

which is very close to our $x = 12.84$ obtained earlier. Bearing in mind uncertainties in the calculation and the dependencies of energies displayed in Table 6.8, this agreement is quite good.

Reference

1. M. Rotenberg, R. Bivins, N. Metropolis, and J. K. Wooten, Jr, *The 3j- and 6j-Symbols*, Technology Press, MIT, Cambridge, MA (1959).

INDEX

Abel, N. H. 561
Ablowitz, M. J. 38, 407, 466, 469, 471, 483, 490–2, 495, 516, 519–21, 523–4, 526–31
Ablowitz–Ladik equation 28, 204
Abramowitz, M. 585
Abrikosov lattice 169–70
acoustic
  mode 153
  resonator 27, 29
Airy 419, 435, 485, 523
Amit, D. J. 225, 233
Anaxagorus 351
Anderson, P. W. vii, 92, 95, 333
  –Higgs mechanism 196–7
Anderson, R. L. 404, 455
angular
  frequency 43, 142, 146, 153, 164, 342
  momentum 101, 139, 211, 323, 374, 377
  velocity 16, 55
anharmonic oscillator 25, 28
  motion vii, 10, 12–13
  potential 155
annihilator 143–7, 149, 161–2, 177–8, 184, 203, 205, 223, 226–7, 229, 272, 277, 279–80, 282–3, 285, 287–8, 323–5, 334, 340–1, 390, 580, 583, 585–6
anticommutation rule 149, 187–8, 228, 369
antiferroelectric 78
antiferromagnet 49, 81–2, 93–4, 97, 104, 111, 300, 321, 323, 332–3, 346, 348–50
antikink 35–6, 534–6, 560
antisoliton 167, 314
Aristotle 351
asymptotic analysis 40
atomic structure ix, 351–2
attractor 57–8, 60, 65
Austin, S. 222
Avogadro, A. 351
  number 2, 92

Bäcklund transform 473, 478, 492, 530–2
Balmer series 352
Banach space 493
Bar Yaacov, D. 492, 529
Bardeen–Cooper–Schrieffer (BCS) theory 85, 180, 221, 224, 300, 306–8
Bargmann criterion 270
basin of attraction 18–19, 24
Beals, R. 521–2, 529
Becquerel, A. H. 351

Bednorz, J. G. 84
Belousov–Zhabotinskii system 49
  reaction 58
Bénard cell 93
Benjamin–Bona–Mahoney equation 476, 482
Benjamin–Ono equation 39, 491, 526–7, 529
Berezin, F.A. 217
Bernoulli equation 261
Bessel function 120, 244, 354, 356
Bethe ansatz 333
Bethe–Peierls approximation 106–7
Bianchi, L. 531, 533
bifurcation vii, viii, 1–3, 5, 6, 8, 21–2, 24, 46, 54, 58–9, 363, 402, 556–7
  cascade 15
  diagram 6, 10, 14, 40–1, 55, 61, 85, 209
  pitchfork 21–2, 57, 60–1
  point 20, 57
  subcritical 4, 5, 22, 60
  supercritical 4, 21–2, 60
  transcritical 7, 21–2
binary alloy 20, 81, 84, 97
binary fluid 20, 43, 86
Bloch theory 194
Bluman, G. W. 403, 406, 438
Bogolyubov transformation 176, 180–1
Bogolyubov–Valatin transformation 307
Bohr, N. 300, 352, 555
  magneton 101
  model 211
  orbit 362
  theory 217
Boiti, M. 530
Boltzmann, L. 3, 8, 106, 130, 157, 212, 555
Born, M. 151, 154
Born–Infeld equation 39
Born–Oppenheimer approximation 151
Bose condensation 85, 179–180, 183–4, 215, 219
Bose–Einstein 146, 161, 176, 178, 186–7, 218, 226–8, 283–4
Boson 85, 176, 178–9, 182, 186, 196, 203, 206, 221, 227–8, 234, 271, 279–81, 284, 288–9, 314, 324, 334–5, 583
Boussinesq equation 39, 409, 426, 438, 451–3, 455, 472, 484–5, 491, 520–1
Boyle, R. 351
Brackett series 352

INDEX

Bragg–Williams approximation 106–7
Bravais lattice 74–5, 77
breather 31–2, 36, 191, 535
Breit–Coulomb term 291
Breit interaction 290
Brillouin
  function 101
  zone 152–3
broken symmetry viii, ix, 2, 38, 80, 85, 89–91, 94, 175, 196–8, 200, 219, 276, 292, 397
Brusselator 23, 58
Bureau, F. 460, 464
Burgers' equation 39, 417–18, 422–3, 515
butterfly catastrophe 4–7, 116
Byrd, P. F. 266, 565–7

Cahn–Hilliard equation 131
canonical ensemble 104
Capel, H. W. 451, 455
Cariello, F. 453
Carnevale, G. 476, 478–9, 482
catastrophe vii, 3, 4, 116, 556
Cauchy, A. L.
  integral 491
  problem 484, 525
Cauchy–Kovalevski theorem 477
cellular automaton 65–8
central field 1, 212, 368, 376
chaos vii, viii, 15, 17, 20, 25, 41, 46, 54–61, 65, 68, 207–9, 294, 556
chaotic vii, 2, 18, 20, 23, 45, 55–7, 63, 289, 303, 555
  map 3
  region 2, 60, 303
charge density wave (CDW) viii, 68, 86, 94, 168
Chazy, J. 464–5
  equation 465–6
chemical
  instability 48
  kinetics 3, 20, 33
  potential 47, 103, 179, 181
  reaction 49, 441
Chiao, R. Y. 294
chirality 86
cholesteric 86, 94
Clarkson, P. A. 38, 440, 451–3, 483, 491, 495, 520, 530
Clausius–Clapeyron equation 89
coexistence manifold 88
coherence length 169, 309, 314, 320
coherent
  light 206, 208
  state ix, 182–5, 189, 217–18, 295–6, 332, 344, 583–4
  structure 41, 54, 189, 244, 275–6, 300–1, 530, 561

Coifman, R. R. 522, 529
Cole, J. D. 403, 406, 438
Cole–Hopf transform 515
Coleman, S. 194
collective viii, 128
  coordinate method 192
  excitation 108, 136, 141, 157
  mode 92, 554
collision 27, 31, 37
commensurate–incommensurate transition 48, 173
commutation rule 142, 144–5, 188, 204, 206, 226, 228, 278–80, 283–4, 287–8, 291, 323, 325–6, 334–5, 340–1, 360, 580, 582–4, 586
commutator 185, 226–8, 277
complexity 59
compressibility 96, 99, 119
condensate 38, 85, 93, 179, 309, 313–14
condensation 93, 179, 189, 191
Condon, E. U. 374, 378
conformal symmetry 422, 432, 437
conservation laws 40
continuity equation 310, 342
continuum
  approximation 28
  limit 35, 40, 166, 172, 189
  medium 40
control parameter 3–5, 9, 20, 22, 51, 54–60, 65, 67, 112, 182, 233
convection 43, 62
  instability 60, 132
Cooper pair 85, 93, 181, 281, 289, 309
correlation
  function 96, 108, 119, 121, 123, 462
  length 67, 74, 107, 119–122, 128, 136, 233
correspondence principle 211, 286
Cosgrove, C. M. 460, 464, 479
Coulomb, C. 74, 91, 158, 214, 219, 290, 306, 350–1, 357, 361–3, 371, 375–7, 379, 388, 591–2
creator 143–7, 149, 161–2, 177–8, 203, 205, 223, 226–7, 229, 273, 277, 282, 287, 323–5, 334, 340–1, 360, 580, 583, 585–6
critical
  current 315–6, 349
  exponent 65, 67, 95–98, 100, 103, 106–7, 110, 113–14, 117, 119, 121, 124–6, 128, 201–2, 225, 233
  fluctuation viii, 92, 188
  phenomena vi, 65, 93, 104, 136, 191, 200, 217, 225, 441, 557
  point 19, 66–8, 87, 92–3, 99, 100, 111, 114, 119, 128, 225, 232–3, 235, 241–2, 263, 286, 295, 327, 355, 357, 361, 363, 369, 391, 398, 457, 459–63, 571, 592
  state 2, 65, 68, 99

temperature 84, 86, 106, 111, 136, 282
crumpling transition 47
Curie temperature 81, 90, 160
Curie–Weiss model 99, 101–2, 106–7, 114
cusp catastrophe 4, 112
Cvitanović, P. 15

d'Alembertian 292
Dalton, J. 351
Davey–Stewartson equation 38, 466, 492, 529–30
Davis, J. H. T. 564, 567–8
Davydov model 222
Dawson's integral 585
DBAR problem 529
de Broglie, L.
 relation 291
 wavelength 211, 295
Debye frequency 181, 307
defect viii, 48, 156, 170–3, 194
Deift, P. 521
Democritus 351
dendritic solidification 63
density
 functional 212
 of states 214, 307
Derrick's theorem 38, 190
Descartes, R. 351
devil's staircase 54
diamagnetism 85
dielectric constant 164
diffusion 20, 441
 limited aggregation (DLA) 63–4
diffusive transition 89
Dikii, L. A. 520
dilation 254, 294
Dirac, P. A. M. 211, 555
 delta 228, 494, 513
 equation 291
disclination 94
discommensuration 94
dislocation 94, 171
disorder 56, 72, 91
dispersion 20, 29, 30, 33–4, 153, 244
 relation 30, 159, 164, 208, 290–2, 294, 316, 331, 493–4
dissipation 13–15, 131, 463
dissipative term 3, 40, 131
DNA 33
Dodd, R. K. 328
domain wall 33, 90, 94–5, 166, 170–1, 252, 264, 329
Domb, C. 98
Dorodnitsyn, V. A. 455
double well potential 8, 78
dromion 38, 530

Dryuma, V. 492
Duffing oscillator 13–15, 17, 20, 41
Dym equation 482–3
dynamical
 instability 136
 matrix 152–3
Dyson–Maleev transformation 344
 operator 335
Dyson's model 111
Dyson–Schwinger equation 216

Eckhaus equation 431
Ehrenfest equation 88–9
Eilbeck, J. C. 328
Einstein, A. 300, 401, 555
 equation 464
electrodeposition 63
electromagnetic
 field 31, 164, 197, 291, 577
 signal 31
 wave 164, 295, 462
electron–phonon interaction 165, 181, 212, 289, 291
electron–proton plasma 301
elementary
 excitation viii, 139, 141, 151, 160, 163, 165, 173, 219, 264, 316
 particle 33, 189, 191, 270
elliptic
 differential equation 16, 460
 focus point 16
 function 15, 246, 263, 266–7, 270, 289, 396, 406–7, 433, 435, 443–5, 459–61, 464, 561–3, 565–7
 integral 262, 311, 564
 umbilic catastrophe 4
 wave 166, 168, 194, 255, 266, 302, 316, 328–9, 343–5, 397, 429
enthalpy 89
entrainment field 44
entropy 72, 89, 109, 113, 242
Epicurus 351
equation of state 3, 6, 20, 88, 93, 99, 100, 102, 104, 106, 114, 116, 130, 346
equilibrium 2, 60, 92, 112–13, 165, 197
 asymptotic 3
 phase viii, 87
 position 151–2, 157, 160, 163
 stable vii, 10
 thermodynamic 63
equipartition theorem 27
Estévez, P. G. 452–3
Euclidean
 metric 260
 signature 246, 253, 263–4, 352
 space 255–6

INDEX

Euler–Lagrange equation 189, 218, 241, 289, 329, 372, 571
evolution equation 20, 44, 207–8, 462, 484–5, 498, 514–20, 522
exchange 322
  integral 374, 377
  interaction 221, 232
exciton viii, 20, 162, 223
external field 3–7, 67, 90, 100, 113, 116, 131, 160, 164, 172–3, 175, 177, 242, 322

Fadeev, L. D. 495
false vacuum 199
Faraday M. 300, 351
Fedyanin 188
Feigenbaum, M. 3, 57, 60, 556
  exponent 60
  map 58
  scenario 57, 61
Fermat's principle 295
Fermi, E. v, 325, 358, 360
  energy 181, 307
  level 214, 320
  sphere 159
  surface 168
Fermi–Dirac 186–8, 218, 226, 228, 283, 287, 582–5
Fermi–Pasta–Ulam (FPU) problem 486
  lattice 33
  system 27
Fermion 85, 146–7, 149, 180, 183–4, 187, 202, 206, 227–8, 234, 276–8, 280–1, 284, 287–9, 300–1, 306, 323–4, 334, 336, 360, 571, 580, 583, 585–6
Fermion–Drone system 335
ferrimagnetism 49, 321
ferrite 49
ferroelastic 5, 78
ferroelectric 5, 33, 78, 91, 93–4, 166, 282
ferromagnet 33, 49, 50, 81–2, 90, 91, 93–4, 97, 101, 102, 107–8, 124, 160, 172, 221, 282, 323
ferromagnetism 49, 73, 101, 104, 111, 166, 321, 327
Feshbach, H. 346
Feynman, R. P. 555
  diagram 1, 216, 239
  integral 239
fibre optics 31, 463, 484
field
  quantization ix, 186
  theory vii, 37, 90, 189, 195–6, 217, 236
  translation 177, 183, 208
Fisher, M. E. 120

Fisher equation 42–3
Fitzhugh–Nagumo equation 446, 452, 454
fixed point 124, 232–3, 466, 471
flame 50–1
Flaschka, H. 491, 524–5
Florjanczyk, M. 436
fluctuation 91, 94, 106, 118–19, 160, 173, 183, 200, 208, 242, 314, 322, 328
  average 2, 3
  critical 47, 92, 188, 322
  dissipation theorem 119
  quantum 200, 331–2
  stochastic 10
  thermal 20, 72, 81, 92, 242
Fock, V. 372
  operator 345, 583
  space 184–5, 234, 265, 572–3, 576
focus point 18–19, 23–4, 363, 390
Fokas, A. S. 483, 491–2, 524–5, 527–9
fold catastrophe 4
Foldy, L. L. 215
Fordy, A. P. 521
form factor 119
Fourier series 222, 325, 588
  component 74, 158, 221
  mode 30
  spectrum 56
  transform 1, 55, 96, 121, 133, 161, 306, 340–1, 490, 492–4, 500, 516
fractal vii, 2, 3, 55, 63–5, 68, 556
  dimension 63–4
Franck–van der Merve Hamiltonian 48
Fredholm 500, 523, 525
free energy 3, 73, 88–9, 104, 106, 108, 112–14, 116–17, 119–21, 123, 131–2, 218, 308, 321
Frenkel exciton 162
Frenkel–Kontorova model 54, 172
Friedberg–Lee model 33
Friedman, M. E. 266, 565–7
Fröhlich, H. 220, 222
  Hamiltonian 221, 291, 306–7
Fuchs, R. 458, 462, 524
fugacity 103

Gagnon, L. 246, 252, 436
Galaktionov, V. A. 454
Galilean invariance 290, 294, 487
  transformation 235, 240, 243–5, 353, 355, 422, 429, 432, 435, 437
Galilei, G. 351
Gambier, B. 461
Gamma function 9

Gardner, C. S. 406, 484, 486–7, 490, 515
Garnier, R. 464, 524
gauge invariance 196, 219
  symmetry 85, 91–2
  transformation 91, 197
Gauss, K. 561
Gaussian 9, 130, 134, 136
  approximation 121–2, 133
  divergence theorem 593
  limit 9
  model 125
  transformation 570
Gel'fand, I. M. 489, 520
Gel'fand–Levitan–Marchenko (GLM)
  equation 471, 489, 499–500, 508, 519, 523, 528
Generalized NLS (GNLS) equation 257, 431, 479, 481–2
Gibbon, J. D. 328
Gibbons, J. 521
Gibbs, J. W. 555
  free energy 88
  phase rule 88
  potential 98
Gilson, C. R. 530
Ginzburg, V. L. 114
  criterion 122
Ginzburg–Landau equation 53
Glauber state 182
Goldstone, J.
  Boson 90, 108, 196–7, 234
  mode 91, 196, 219, 555
  theorem ix, 90–1, 196
Gordoa, P. R. 453
Gor'kov, L. P. 308
Görtler vortex 463
Gradshteyn, I. S. 354
Grammian 530
grand canonical ensemble 103–4
Grassmann, M. 183, 583
Green function 1, 216
Greene, T. M. 406, 484, 487, 490, 515
group velocity 29–30
Gutenberg–Richter law 68

Haberman, R. 492, 521
Haken, H. 3
Haldane, F. D. M. 333, 346, 398
Haldane gap ix, 269, 300, 333, 347, 576
Hamilton, W. R. 555
harmonic oscillator 1, 142, 144, 153, 182, 186, 213, 396
  approximation 154
  damped driven 13
  motion vii

Hartree–Fock (HF) method 214, 216, 218, 323, 352, 371–2, 376–8, 380–3, 386–7, 389, 398, 594
Hartree method 213–18, 352, 372
heat
  equation 441, 515
  flux 156
Heaviside function 195
Heisenberg, W. 322, 326, 331, 344, 554–5, 586
  equation 188–9, 191, 211, 226, 272, 277, 280, 285, 287–8, 291, 326, 340, 360, 390, 571, 588
  Hamiltonian 49, 101, 161
  model 107, 125–6, 169, 221, 332
  picture 225–6
  uncertainty principle 182
helical
  order 49
  phase 110
helium 58, 73, 170, 221, 281
Helmholtz equation 244
Hénon–Heiles equation 24
Hereman, W. 403, 446
Herman, F. 372, 379, 381
Hermitian conjugate 176, 187, 277, 292, 337
Hietarinta, J. 530
Higgs Boson ix, 555
Higgs–Anderson model 33
high-Tc superconductor 84–5, 320, 390, 397
Hilbert space 193
Hill, J. M. 403, 406
Hirota, R. 454, 478, 492, 521, 530
  equation 39
  method 446, 454, 530
  transform 490, 525
Hlavatý, L. 479
hodograph transform 482
Hohenberg, P. C. 108, 215
Holstein–Primakov transformation 161, 221, 334–5
Hopf bifurcation 24–5, 41, 56, 61, 85
Huang, K. 151
Hubbard model ix, 397
  Hamiltonian 301, 322, 390, 397
Hunter D. 98
Huxley T. H. 554
  equation 449
hydrodynamic mode 92
hydrogen 211, 351–2, 362
hyperbolic
  instability point 17
  singularity 12
  umbilic catastrophe 4

hypergeometric function 421–2
  equation 456, 465, 501
hysteresis 2, 6, 88, 94, 118
  bifurcation 21–2
  double 7
  field induced 117–8
  single 22
  thermal 7, 116

Ibragimov, N. H. 403–4
Ichikawa, Y. H. 522
Ince, E. L. 429–30, 460, 474
instability viii, 2, 6, 12–13, 51, 53, 55, 78, 207, 282, 294, 556
instanton 38, 191, 198–9, 264
integrable equation 38, 40, 492
Intermediate Long Wave (ILW) equation 491, 526–7
intermittency 54–5, 59, 60
inverse scattering ix, 185, 246, 428, 469, 482–4, 488–92, 495, 497, 499–500, 515–16, 519–20, 523–5, 527–30
ionization 365, 369, 381, 431
Ising model 1, 78, 104, 107–9, 111, 123, 125–6, 200–1, 331, 462
  Hamiltonian 47, 104, 200
itinerant electron 81, 321
Ito, M. 454

Jackiw, R. 191, 225, 233–4
Jacobi, C. G. J. 555, 561–2
  elliptic modulus 106, 311–12, 344, 563
  function 259, 263, 266, 329, 344, 406, 433, 442, 460, 465, 563, 565–9
  polynomial 269
Jahn–Teller effect 80, 154, 157
Jaulent, M. 530
Joets, A. 46
Jones, G. L. 104
Jones, K. 152
Jordan, P. 335
Josephson, B. D.
  fluxon 173
  junction 33, 58, 167
Joshi, N. 462

Kac–Hubbard–Stratanovich transformation 201
Kadanoff, L. P. 3, 98, 122–4, 232, 556
Kadomtsev–Petviashvilli (KP) equation 38–9, 452, 466, 491, 527–30

Kamran, N. 455
Kato's theorem 388, 390
Kaup, D. J. 490, 516, 520, 523
Kaup–Kuperschmidt equation 520
Kawahara, T. 446
Kawasaki, K. 344
Kelvin, Lord 556
Kerr effect 167, 208
  medium 51, 207
kinetic energy 20, 35, 140, 151, 172, 194–5, 219–21, 271, 274, 307, 357, 359, 361
kink 15–16, 31, 35–6, 38, 130, 166–7, 190, 192, 245, 262, 264, 269, 347, 534–6, 560
Klein–Gordon equation 30
Kodama, Y. 491
Kohn, W. 215
Konno, K. 522
Konopelchenko, B. G. 530
Korteweg–de Vries (KdV) equation 33, 39–40, 406–8, 466, 469–70, 482–90, 492, 494–5, 499–500, 503–5, 507, 509–10, 512, 514–16, 519–20, 523–5, 527–8
  regularized KdV equation 33
Kosterlitz, J. M. 108
Kowalevski, S. 466, 471
Kronecker delta 145, 509
Kruskal, M. D. 27, 33, 406, 451–2, 462, 484, 486–9, 490, 509, 515–16
Kubo, R. 333
Kumei, S. 403, 406
Kuramoto–Shivashinsky equation 50

Ladik, J. F. 491
Lagrange, J. L. 238, 555
  multiplier 373
Lagrangian 34, 36–8, 189, 196, 218, 240, 571
Lamb shift 352
lambda transition 89
Lamé equation 267, 345
  function 267–8
Landau, L. D. vii, 2–4, 56, 72, 74, 93, 99, 112–14, 295, 321, 557
  scenario 56
  theory 111, 117–120, 128
Landau–Ginzburg (LG) ix, 122, 216, 237, 239, 289, 300, 308, 322, 329, 331, 343
  equation 42, 313
  free energy 120, 128, 130
  Hamiltonian 47–8, 125, 128, 129, 133, 201, 236, 305, 557
  model viii, 43, 49, 72, 78, 99, 119, 121, 125, 132, 201, 282, 309
Landau–Lifshitz equation 39

INDEX 603

Landau–Stuart expansion 44
Landé factor 101
Landen's transformation 570
Langmuir oscillations 428
   waves 522
Laplace–Beltrami operator 129, 231
Laplace transform 1, 523
Laplacian 140, 240, 243, 246
laser 51, 56, 58, 93, 132, 176–7, 182, 206–9, 556
   instability 60
latent heat 88–9
Laurent series 468, 476, 478
Lax pair 478, 490, 492, 495, 514–15, 518–19
   equation 514, 516
LC circuit 27, 29
Lebeau, G. 525
Legendre equation 501
   polynomial 502
Leggett, A. J. vi
Lenard–Jones potential 74
Leon, J. J. P. 530
Levanyuk, A. P. 114
Levi, D. 452, 455
Levitan, B. M. 489
Lie, S. 402–3, 410, 424, 426, 438, 440–41, 451, 455
   algebra 422–3, 441, 466
   Bäcklund transformation 404
   bracket 423
   group 403, 409, 411, 431, 438, 454, 476
   point transformation 403
Lieb, E. H. 333
Lifshitz, E. M. 295
limit cycle vii, 15, 22–5, 27
linear
   perturbation 29
   response 1
   wave 28
Liouville equation 473
liquid crystal 45, 73, 86, 91, 94, 132, 172, 282
Lochak, P. 525
logistic map 58–59
long-range order viii, 2, 72–4, 106, 108, 180, 219, 235, 295, 314, 332
Lorentz gauge 197
   invariance 196, 294
Lorentzian approximation 119
   form 120
Lorenz equation 60, 62, 208, 555–6
Lucretius 351
Lukashevich, N. A. 464
Lyapunov stability analysis 24
   characteristic exponent 57

Lyman series 352

Ma, S.-k. 225, 233
Maclaurin series 565
Maeda, S. 456
magnet 49, 89, 96
magnetic
   flux 169
   moment 27, 29, 81, 321
magnetism 22, 139, 276
magnon viii, 162, 222, 282, 344
Makhankov, V. G. 188
Maleev transformation 336
Manakov, S. V. 492, 525, 528
Mandelbrot, B. 3, 63, 556
Manna, M. 530
Mansfield, E. L. 440, 453
March, N. H., 152, 215, 388–90
Marchenko, V. A. 495
Martina, L. 530
Martinez-Alonso, L. 530
Martynov, I. P. 464
Mason, L. J. 463
master equation 130
Mathieu function 270
Mattis, D. C. 333, 335
Maxwell, J. C. 555
   equation 168, 207, 295
Maxwell–Bloch equation 60, 208
McLeod, J. B. 471–3, 478
mean field approximation (MFA) 109–10, 126, 131, 191, 311
   exponent 103, 106
   model 1, 67, 111, 114, 120, 128
Meissner effect 85, 94, 197, 305
Meleshko, S. V. 454
membrane 222
Mendeleev, D. 351
Mermin, N. D. 108
meson 192–4
metal–insulator transition 86, 168
metamagnet ix, 81–2, 300, 321–2, 331, 333–4, 340, 342, 350
metastability 6, 199
Method of Coherent Structures (MCS) ix, x, 139–41, 173, 211, 213, 215, 217–18, 237, 242, 255, 263–4, 270–3, 275, 277, 282–3, 285–6, 288–90, 300–1, 303, 306–8, 321, 323, 326, 332–4, 336, 342, 351, 358–60, 380–1, 383–4, 386, 388, 390–1, 397–8, 428, 431, 437, 557, 571, 591
Minkowski metric 260
   signature 253, 263
   space 255
Miura, R. M. 406, 484, 487, 490, 515, 521

Möbius transformation 460–1
mode–mode coupling ix, 121, 133, 331
modulational instability 313–14, 320, 329
molecular crystal 20
Moloney, J. V. 40
Morikawa, C. K. 486
Morris, H. C. 328
Morse, P. M. 346
Moseley's law 385, 389
Müller, K. A. 84
multifractality 63
multi-kink solution 36, 172
multi-soliton 167, 245, 492
multistability vii, viii, 3, 7
Münster, A. 89

Nagamiya, T. 333
Nakamoto, T. 451
Néel, L. E. F. 332
  temperature 81
nematic 86, 94, 172
Newell, A. C. 40, 490–1, 516, 523–4
Newell–Lanadu–Ginzburg equation 46
Newton I. 300, 351, 554–5
Nicolis, G. 3
Niemeijer, T. 322–3, 332
Nijhoff, F. W. 451
Nimmo, L. 530
node 24, 131
Noether, E. 402
noise 65–6
non-classical method 438–41, 451–4
non-equilibrium phase transition 131
non-Gaussian approximation 133
  distribution 9
  statistics 136
nonlinear
  elastic wave 33
  excitations viii
  field equation 191, 226
  force 3, 20
  Helmhlotz equation 130
  Klein–Gordon equation (NLKG) ix, 36, 39, 129–30, 166, 189, 219, 252, 254–5, 258–9, 263, 294, 310, 343, 478, 474–5
  optics 51, 206, 431
  PDE's 52, 225, 409, 455
  potential 37, 191, 199, 302–3, 346
  Schrödinger equation (NLS) ix, 16, 34–5, 39–40, 51, 130, 168, 189, 202–4, 208, 219, 245–6, 252–3, 264, 289, 294, 302, 308, 328, 342, 428, 431, 436–7, 466, 470, 475, 482, 484, 490–1, 515–16, 518–19, 521, 523
  wave viii, 2, 16, 25, 31–3, 40, 169, 191

normal mode 1, 30
Nucci, M. C. 452
nucleus 140–1, 151–2, 157–9, 162–3, 211, 351, 363, 365, 390
Numerov method 386

occupation inversion 207
Odabasi, H. 374, 378
off-diagonal long-range order 179–80
Olver, P. J. 403, 406, 441, 451, 453–5, 471–3, 478
Ondich, J. 454
Onsager, L.
  equation 131
  relation 47–8
  transition 89
Oppenheimer, R. 154
optical
  fibre 2, 167, 207–8, 428, 436
  instability 50
orbital 214, 369, 371–8, 384, 388, 594
order parameter vii, ix, 2, 3, 6, 8, 10, 20, 22–3, 37, 46–8, 66–7, 74, 80–1, 85–6, 88–98, 112, 119, 121, 124, 129–32, 170, 179, 201–2, 225, 233, 236, 282, 308, 311, 343, 424, 557
Ornstein, L. S. 119, 121, 125
Ovsiannikov, L. V. 403, 454

Painlevé analysis ix, 246, 404, 433–6, 446, 454, 457–8, 460–1, 464–9, 471–4, 479, 482, 525
  conjecture 471
  equation 408, 427–8, 430, 433–4, 438, 461, 463–4, 466, 469–70, 475, 491, 523–5
  test 246, 456, 471–9, 481–3
  transcendent 246, 253, 461, 464
parabolic
  cylinder function 9, 134, 436
  umbilic catastrophe 4, 5
paraelectric 93
paramagnetic 81, 83, 90, 93, 101, 102
partition function 9, 101, 104, 124, 126, 133–4, 202, 329
Paschen series 352
path integral 212, 217
pattern formation vii, viii, 43, 46, 52, 54, 128, 207–8, 290, 294, 556
Pauli, W. 555
  matrix 200
  principle 214, 369
Peierls transition 80–1

pendulum 11
  coupled 27, 29
  damped 19
  damped-driven 16, 18, 41
penetration depth 169, 309
peptide 33
period doubling 15, 17, 40, 54, 57-61
perturbation analysis 40
  expansion 38, 165, 213
  theory 41, 151, 175, 212-13, 239, 367
phase
  diagram 23, 46, 86-8, 160, 200-1, 300, 321-2, 333, 363
  portrait 18, 21, 23-4
  shift 36
  space 2, 10-11, 13, 19-20, 24, 28, 54, 57, 129, 217-18, 289, 303, 363, 390, 392, 396
  phase transition viii, 3, 4, 10, 37, 42, 67, 72-3, 76, 86, 87, 92, 95, 98, 104-5, 107, 111, 119, 124, 128, 165, 179, 191, 200-1, 235, 294-5, 320-1, 556
    first order viii, 5, 7, 46, 54, 88-9, 93, 104-5, 110, 114, 116-18, 128, 321
    second order viii, 5, 6, 66-7, 88-9, 91-3, 104-5, 110-14, 117-18, 303
    weakly first order 47
phasor 94
phi-4 model 38, 180, 190, 192-3, 195
phonon viii, 78, 91, 94, 151, 154-7, 159-160, 166, 176, 180-1, 192, 213, 220-1, 223, 282-3, 288-90, 306-7, 316, 320
photon 164, 290-1, 294
Picard 460
piezoelectric 131, 282
Pimpinelli, A. 332, 530
Pines, D. 175
pitchfork bifurcation 6
Planck, M. 555
plasma 426, 428
  frequency 141, 158
  oscillation 157-8
  physics 31, 363, 407, 462, 486, 528
  wave 306
plasmon 91, 157-60, 282
Poincaré, H. 1, 56, 294, 555
  section 302, 305, 392, 396
Poincaré-Andronov-Hopf equation 24
Poisson distribution 183
polar product system 23-4
polariton 163-5
polarization 78, 93, 153-5, 163-4, 208
polaron 163, 205, 282
polyacetylene 166-7
polymer 46

Pomeau-Manneville scenario 60-1
Potts model 109-110
power spectrum 61-2, 66
pressure 87, 96, 103, 111
Prigogine, I. 3, 23
projection operator 148, 498
prolongation 415-16, 426, 439
protein 428
Prout 351

q-Boson 185
q-deformed algebra 185-6
quantization vi, viii, 41, 191-2, 200, 218, 263-4, 270, 362, 365, 380, 571, 593
quantum
  coherence vii
  excitation x, 128, 139, 179, 194, 334, 342
  field ix-x, 93, 186-8, 191-2, 226, 228, 232-3, 264, 276, 291, 294-5, 326-7, 342, 345, 351, 354, 360-1, 369, 390, 571, 583, 587, 590, 592
  field theory vi, vii, ix, 173, 175, 191, 193, 200, 225, 233-4, 264, 428, 431, 484, 555, 557
  group 185
  lattice soliton ix
quasi-particle 141, 146, 175-7, 179, 188, 200, 211, 281, 285, 290-2, 294-5, 301, 314, 554
Quispel, G. R. W. 451, 455-6

Rajaraman, R. 225, 289, 571
Raman effect 208
Ramani, A. 466, 469, 471, 491, 521
Random Phase Approximation (RPA) 216
Rayleigh, Lord 485
Rayleigh-Bénard convection 43-4, 60, 62
Rayleigh-Ritz method 436
reaction-diffusion 3, 23, 42, 48
Red Spot of Jupiter 33
Reduced Maxwell-Bloch equation 39
reflection coefficient 488, 496, 501, 512-13
regularization 239
regularized long wave (RLW) equation 33
Reichl, L. E. 232
relativity 185, 217, 219, 291, 484
relaxation 2, 20, 191
  constant 20
renormalization group (RG) viii, 3, 99, 122, 124-6, 128, 136, 225, 235, 239
resonance 13-14, 467-9, 477, 481
resonating valence bond (RVB) 332

Reynolds number 45
Ribotta, R. 46
Riccati equation 458–9, 487
Riemann zeta 179
Riemann–Hilbert problem 490, 496–7, 500, 520–1, 524–30
Rogers, C. 403, 533
rolls 43, 46
Röntgen, W. C. 351
Rosenau, P. 441, 451, 454
Rotenberg, M. 353, 592
Ruelle, D. 56
Ruelle–Takens–Newhouse scenario 56, 61
Russell, J. S. 484–6
Rutherford, E. 351
Ryzhik, I. M. 354

S4-model 125–6
Sachdev, P. L. 417
saddle 24
 node bifurcation 8, 21–2, 61
 point 19
Sahadevan, R. 455–6
Santini, P. M. 527, 530
Satsuma, J. 491
saturation 101
Sawada–Kotera equation 520
scaling vii, 3, 67, 72, 95, 98–9, 123–4, 136, 469–70, 472–3, 556
 dynamic 122
 finite-size 67–8
 static 98, 123
scattering 488–90, 495, 498–9, 502, 515, 518–22, 525–9
Schlesinger, L. 524
Schottky anomaly 73
Schrödinger, E. 555
 equation 186–8, 192–3, 218, 231, 265, 269, 290, 345–6, 354, 487–8, 500, 507, 516, 520, 525, 528, 572, 575, 579
 picture 225–6
Schultz, T. 333
Schwarzian 465
Schwarzmeier, J. L. 451
Schwinger, J. 334
Scoufis, G. 464
screw dislocation 48
second quantization 1, 139, 141, 146–7, 150–1, 154, 156, 161, 173, 176, 186–7, 219, 222, 224–6, 233–4, 237
Segur, H. 407, 466, 469, 471, 490–1, 516, 519, 523, 528, 531
self-energy 216
self-focussing 208, 294, 436

self-induced transparency 39
self-organization 3, 65, 556
self-organized criticality (SOC) 63–4, 66–8, 555
self-similarity 55, 59, 63, 65–6, 363, 472, 556
self-trapping 163
semiconductor 20, 56, 73, 140, 221
semi-metal 140
separatrix vii, 11, 16, 18, 25, 303, 396
Shabat, A. B. 490, 515–16
Shadwick, W. F. 533
Sierpinski, W.
 carpet 63
 gasket 63
sigma function 268
similarity transformation 31
Sine–Gordon model 193
 equation (SG) ix, 11, 35–6, 39–40, 48, 167, 199, 466, 470, 473, 477, 490, 516, 519, 531–2, 534–6
 potential 190
single well potential 8, 11
sink 24, 26–7
Skillman, S. 372, 379, 381
Slater, J. C. 379
 determinant 148, 214
slaved mode 3
smectic 85, 94
soft mode viii, 78, 90–2, 94
solitary wave vii, viii, 30–2, 35–6, 38, 130, 191–2, 245, 262, 269, 311, 313, 316, 329, 345, 473, 476, 482, 485–6, 492, 558–60
solitoff 530
soliton vii, viii, 2, 25, 27, 29–31, 33–6, 38, 40–2, 129, 165, 167–70, 190, 192–4, 200, 202–6, 208, 245, 294, 313–14, 328–9, 332, 344, 347, 407, 428, 453–5, 466, 472–3, 476, 482, 484, 486, 490, 502–7, 509–14, 528, 530, 532–3
Sommerfeld, A. 352
source 24, 26–7
specific heat 73, 89, 96, 106, 113, 116, 121, 136, 305
spherical
 harmonic 244, 268, 353
 model 107, 125
spin 47, 81, 86, 94, 101, 106–8, 111, 123, 139, 161, 168, 181, 200–1, 217–18, 221, 224, 232, 270, 275–83, 290, 300, 307–8, 321–3, 326, 331–7, 340, 344, 346–9, 361, 372–4, 376–8, 390, 462, 580, 585, 587
 density wave (SDW) viii, 86, 168–9
 glass 97
 lattice 200
 orbit coupling 140, 352
 spin interaction 161, 322, 352, 462
 wave 160–1, 222, 331, 333

spiral
    domain 49–50, 172
    galaxy 33
squeezed state 184–5, 208
stability 6, 320, 329, 332, 351, 390, 554
    asymptotic 20
    structural 24, 116
Stanley, H. E. 107
steady state 348
Stegun, I. A. 585
Stephani, H. 403, 406
Stephenson, J. 323
stochastic layer 18
Stokes, G. 485
strange attractor 60–2, 208
structural phase transition 77–8, 91, 157, 165, 219
Sturm–Liouville problem 520–1
subharmonics 14
sublimation 111
superconductivity ix, 22, 73, 84–5, 90, 94, 165, 167, 197, 212, 219, 221, 276, 281, 300, 303, 305–9, 328, 556–7
superconductor 33, 85, 91, 93, 132, 140, 169, 197, 200, 224, 281–2, 291, 303, 305, 309, 314, 320, 348
    type I 306, 309
    type II 2, 169, 306, 309
superfluid 33, 73, 85, 91, 93–4, 124, 155, 169–70, 176, 179–80, 219, 281–2
superposition principle 554
susceptibility 9, 67, 89, 96, 102, 116, 119, 134
swallowtail catastrophe 4
Swift–Hohenberg equation 53
symmetry
    breaking vii, 90, 111, 165, 175, 178, 188, 191, 219, 234, 264
    methods 2, 225
    reduction ix, 225, 245–6, 256, 259–60, 263, 282, 401–2, 416–18, 427–8, 431–2, 436, 438, 441, 452–4, 469–76
    variable 190, 245–6, 252–3, 255, 258, 260–1
synergetics 3

Tabor, M. 453, 476, 478–9, 482
tachyonic 189
Tajiri, M. 451
Takeno, S. 344
Takens, F. 56
Talanov lens transformation 437
Taylor, P. L. 139
Taylor–Couette flow 43–4
Taylor expansion 40, 112, 231–2, 235, 571, 592

series 229, 239, 242, 277, 456, 467
thermal
    conductivity 156, 214
    expansion 156–7
thermodynamic
    limit 103–4, 134, 180
    potential 3, 4, 88, 98, 123
Third Law of Thermodynamics 72
Thirring model 491
Thom, R. 3
Thomas, L. H. 358
Thomas–Fermi theory 214–18
Thomson, J. J. 351
Thouless, D. J. 108
three-wave interaction equation 39, 491, 521
time-dependent Landau–Ginzburg equation (TDLG) 45, 48, 131, 441
Toda lattice 28, 39, 455, 491, 525–6
Tomei, C. 521
transistor 58
translation mode 38, 192
transmission amplitude 198
    coefficient 488, 496, 501, 513
tricritical point 87, 114, 116, 118, 294
triple point 87
tunnelling 38, 191, 198–200, 264, 397
turbulence 43–4, 56, 62, 68, 93
turning point 37
twinning 166

Umklapp process 155
unitary transformation 158, 177–8, 220, 223, 234, 271, 277, 291, 307
universality vii
    class 124
    hypothesis 124

van der Waals, J. D. 74, 99–100, 114
van Hove's theorem 104
variational principle 120, 166, 295, 344, 402
vector potential 196, 309
velocity selection 43
vertex function 216, 240
vibronic state 154
Villarroel, J. 526
virial expansion 1
viscosity 85
viscous fingering 63
Volterra integral equation 528
Vorob'ev, E. M. 446
vortex viii, 33–4, 68, 94, 108–9, 169–70, 173
    magnetic 2
    Taylor 44

Wadati, M. 516, 522
Wagner, H. 108
Wannier exciton 162
Ward, R. S. 466, 477
water 58, 407, 426, 428, 431, 462
Watson, G. N. 267, 460
wave 31–2, 37, 42, 162, 345, 361, 389, 407, 419, 426, 428–32, 438, 470, 472–3, 484–6, 514, 528, 530
  deep water 31
  equation 30, 440, 477
  function 80, 85, 93, 124, 132, 139, 179, 181, 189, 200, 213–14, 226, 228, 232, 295, 363
  number 29
  packet 29–30, 295, 494
  shallow water 31
  vector 119, 153, 158–9, 229, 257, 307, 355, 361–2, 586, 589
Weierstrass function 268, 407, 433, 435, 460
Weiss, J. 476, 478–9, 482
Weiss, P. 101, 102
Whittaker, E. T. 267, 460
Widom, B. 98
Wigner, E. 335
Wilson, K. G. vii, 3, 128, 556
Winternitz, P. 246, 252, 452, 455
WKB approximation 198

Woodhouse, N. M. J. 463
Wronskian 530
Wu, T. M. 223

XY model 108–9, 125, 169, 202, 462

Yaffe, L. G. 217
Yanenko, N. N. 454
Yang–Lee theorem 103, 105
Yang–Mills theory 189
  equation 463, 466

Zabusky, N. 27, 33, 486, 509
Zakharov, V. E. 490, 515–16
Zeeman term 101
Zernike, F. 119, 121, 125
zeta function 268
Zhou, X. 525
Ziman, J. 300, 333
zinc 81, 362, 368, 380, 384, 389, 398